Computational Approaches to Materials Design:

Theoretical and Practical Aspects

Shubhabrata Datta
Calcutta Institute of Engineering and Management, India

J. Paulo Davim
University of Aveiro, Portugal

A volume in the Advances in Chemical and Materials Engineering (ACME) Book Series

ENGINEERING SCIENCE REFERENCE
An Imprint of IGI Global

Published in the United States of America by
Engineering Science Reference (an imprint of IGI Global)
701 E. Chocolate Avenue
Hershey PA, USA 17033
Tel: 717-533-8845
Fax: 717-533-8661
E-mail: cust@igi-global.com
Web site: http://www.igi-global.com

Library of Congress Cataloging-in-Publication Data

Names: Datta, Shubhabrata, 1965- editor. | Davim, J. Paulo, editor.
Title: Computational approaches to materials design : theoretical and practical aspects / Shubhabrata Datta and J. Paulo Davim, editors.
Description: Hershey, PA : Engineering Science Reference, [2016] | Includes bibliographical references and index.
Identifiers: LCCN 2016006925| ISBN 9781522502906 (hardcover) | ISBN 9781522502913 (ebook)
Subjects: LCSH: Materials--Mathematical models.
Classification: LCC TA404.8 .C657 2016 | DDC 620.1/10151--dc23 LC record available at http://lccn.loc.gov/2016006925

This book is published in the IGI Global book series Advances in Chemical and Materials Engineering (ACME) (ISSN: 2327-5448; eISSN: 2327-5456)

British Cataloguing in Publication Data
A Cataloguing in Publication record for this book is available from the British Library.

All work contributed to this book is new, previously-unpublished material. The views expressed in this book are those of the authors, but not necessarily of the publisher.

For electronic access to this publication, please contact: eresources@igi-global.com.

Advances in Chemical and Materials Engineering (ACME) Book Series

J. Paulo Davim
University of Aveiro, Portugal

ISSN: 2327-5448
EISSN: 2327-5456

Mission

The cross disciplinary approach of chemical and materials engineering is rapidly growing as it applies to the study of educational, scientific and industrial research activities by solving complex chemical problems using computational techniques and statistical methods.

The **Advances in Chemical and Materials Engineering (ACME) Book Series** provides research on the recent advances throughout computational and statistical methods of analysis and modeling. This series brings together collaboration between chemists, engineers, statisticians, and computer scientists and offers a wealth of knowledge and useful tools to academics, practitioners, and professionals through high quality publications.

Coverage

- Statistical methods
- Coatings and surface treatments
- Composites
- Fracture Mechanics
- Fatigue and Creep
- Artificial Intelligence Methods
- Numerical Techniques
- Metallic Alloys
- Industrial chemistry
- Chemical Engineering

IGI Global is currently accepting manuscripts for publication within this series. To submit a proposal for a volume in this series, please contact our Acquisition Editors at Acquisitions@igi-global.com or visit: http://www.igi-global.com/publish/.

The Advances in Chemical and Materials Engineering (ACME) Book Series (ISSN 2327-5448) is published by IGI Global, 701 E. Chocolate Avenue, Hershey, PA 17033-1240, USA, www.igi-global.com. This series is composed of titles available for purchase individually; each title is edited to be contextually exclusive from any other title within the series. For pricing and ordering information please visit http://www.igi-global.com/book-series/advances-chemical-materials-engineering/73687. Postmaster: Send all address changes to above address.

Titles in this Series

For a list of additional titles in this series, please visit: www.igi-global.com

Position-Sensitive Gaseous Photomultipliers Research and Applications
Tom Francke (Myon, Sweden) and Vladimir Peskov (CERN, Switzerland)
Engineering Science Reference • copyright 2016 • 562pp • H/C (ISBN: 9781522502425) • US $240.00 (our price)

Research Perspectives on Functional Micro- and Nanoscale Coatings
Ana Zuzuarregui (CIC nanoGUNE Consolider, Spain) and Maria Carmen Morant-Miñana (CIC nanoGUNE Consolider, Spain)
Information Science Reference • copyright 2016 • 511pp • H/C (ISBN: 9781522500667) • US $215.00 (our price)

Emerging Research on Bioinspired Materials Engineering
Mohamed Bououdina (University of Bahrain, Bahrain)
Engineering Science Reference • copyright 2016 • 441pp • H/C (ISBN: 9781466698116) • US $215.00 (our price)

Petrochemical Catalyst Materials, Processes, and Emerging Technologies
Hamid Al-Megren (King Abdulaziz City for Science and Technology, Saudi Arabia) and Tiancun Xiao (Oxford University, UK)
Engineering Science Reference • copyright 2016 • 539pp • H/C (ISBN: 9781466699755) • US $225.00 (our price)

Applying Nanotechnology to the Desulfurization Process in Petroleum Engineering
Tawfik A. Saleh (Department of Chemistry, King Fahd University of Petroleum and Minerals, Saudi Arabia)
Engineering Science Reference • copyright 2016 • 555pp • H/C (ISBN: 9781466695450) • US $235.00 (our price)

Quantitative Structure-Activity Relationships in Drug Design, Predictive Toxicology, and Risk Assessment
Kunal Roy (Jadavpur University, India)
Medical Information Science Reference • copyright 2015 • 703pp • H/C (ISBN: 9781466681361) • US $295.00 (our price)

Handbook of Research on Recent Developments in Materials Science and Corrosion Engineering Education
Hwee Ling Lim (The Petroleum Institute, UAE)
Engineering Science Reference • copyright 2015 • 493pp • H/C (ISBN: 9781466681835) • US $325.00 (our price)

Processing Techniques and Tribological Behavior of Composite Materials
Rajnesh Tyagi (Indian Institute of Technology (BHU), Varanasi, India) and J. Paulo Davim (University of Aveiro, Portugal)
Engineering Science Reference • copyright 2015 • 464pp • H/C (ISBN: 9781466675308) • US $225.00 (our price)

IGI GLOBAL
DISSEMINATOR OF KNOWLEDGE
www.igi-global.com

701 E. Chocolate Ave., Hershey, PA 17033
Order online at www.igi-global.com or call 717-533-8845 x100
To place a standing order for titles released in this series, contact: cust@igi-global.com
Mon-Fri 8:00 am - 5:00 pm (est) or fax 24 hours a day 717-533-8661

Table of Contents

Detailed Table of Contents

The chapter primarily deals with brief description of different methods of materials modeling which utilizes the scientific theories in different length scales. It also gives an account of the available tools for situations where data driven models are required. Utilization of imprecise knowledge of a materials system for developing mathematical models is also discussed. A brief account of the use of optimization techniques for designing materials is discussed here.

Recent works have suggested that the defect induced magnetism in Diluted Magnetic Semiconductors (DMSs), Transition Metal Oxides (TMOs) and related materials is facilitated and enhanced by codoping and the synergistic action between the codopants. In the present work we demonstrate that the proposed defect synergy is the result of the interplay among correlated spin-polarization processes which take place in a successive way in neighborhoods centered at the codopants and include their first nearest neighbors. These processes result in a reduction in the superexchange coupling which in turn causes an enhancement in the ferromagnetic coupling (FMC) among the magnetic dopants. The proposed FMC is demonstrated using ab initio calculations of the electronic properties of codoped ZnO, GaN and TiO2.

We briefly present the theoretical framework of a hierarchical multi-scale approach, which is an ab initio-based stochastic method, and its applications to several chemical/physical kinetic processes on metallic surfaces. We first introduce necessary theoretical basis of ab initio and Monte Carlo (MC) methods, and then illustrate different Monte Carlo algorithms for important ensembles, including canonical and grand canonical ensembles. In the following section, we describe two important protocols which are essential to integrate ab initio data and MC models. Two examples are presented in order to elucidate the power of this multi-scale approach. The first example focuses on the combination of kinetic Monte Carlo and transition state theory. We discuss the detailed processes of performing kinetic Monte Carlo simulation on atomic diffusion on alloyed surface, including some technical aspects. In the second example, we presents a different way to account for the local environment-sensitive metal-catalyzed O2 dissociation reactions using combinatory techniques including cluster expansion and grand canonical Monte Carlo methods. This approach provides steady-state rates and rate derivatives that are comparable with experiments. Moreover, the connection between the feasible mechanisms and the observed kinetic behaviors can now be built.

During the last couple of decades, treatment of microstructure in materials science has been shifted from the diagnostic to design paradigm. Design of microstructure is inherently complex problems due to non linear spatial and temporal interaction of composition and parameters leading to the target properties. In most of the cases, different properties are reciprocally correlated i.e., improvement of one lead to the degradation of other. Also, the design of microstructure is a multiscale problem, as the knowledge of phenomena at range of scales from electronic to mesoscale is required for precise composition-microstructure-property determination. In the view of above, present chapter provides the introduction to computationally driven microstructure engineering in the framework of constitutive length scale in microstructure design. The important issues pertaining to design such as phase stability and interfaces has been explained. Additionally, the bird-eye view of various computational techniques in order of length scale has been introduced, with an aim to present the picture of combination of various techniques for solving microstructural design problems under various scenarios.

The major areas of applications of the composite materials today encompass fields as wide as wind energy to marine to construction to aerospace to strategic areas. Apart from such specialized fields the major composite market blooms out of their extensive exploitations in the automotive, sporting goods, pipes, tanks, chemicals, fertilizers and many other industries. As compared to the conventional materials, the composites offer several unique advantages. The list of such advantages may include as a typical example the higher tensile strength, the lighter weight, the greater corrosion resistance, the better surface finish and the easier processability. It is interesting to note that in 2011, the global composite materials market size was \$19.6 billion and the same is estimated now to reach approximately \$34.1 billion in 2018. This amounts to a Compounded Annual Growth Rate (CAGR) of about 10.5% percent. This huge market growth of composites will require a thorough knowledge of the mechanics and mechanical properties of a wide variety of composites under a still wider variety of application dependent constraints. That is why, just to make an humble beginning in this direction, the basics of the composite materials and it's varieties, are briefly described in this Chapter. In addition, the analytical methods of continuum micromechanics have been touched upon. Keeping in view the complex microstructure of most of the composites, and the effect that such complexities can have in the resultant properties, the numerical approaches to continuum micromechanics has been highlighted with emphasis on the fundamentals of the Finite Element Analysis (FEA) based methods. Further, the concepts involved in modelling of the "Failure of composites" have been elucidated with some interesting case studies.

Manufacturing a steel product mix (bar, rod, sheet) involves a series of unit operations - primary steel making, secondary steel making (ladle refining and tundish operation), continuous casting, reheating, rolling and annealing. The properties of the final product depend significantly on how each unit operation is carried out. Each unit operation must be operated to meet the requirements of the subsequent operations. The requirements imposed on a particular unit operation are often conflicting and compromises must be made. Also, there is high degree of uncertainty in the operating parameters of each unit operation, which may lead to considerable deviations from the anticipated performance. To ensure that the final quality specifications of the product is not sacrificed and the customer requirements are met, it is essential to manage the conflict and uncertainty involved in each unit operation of the manufacturing process. In this chapter, we illustrate the use of compromise Decision Support Problem (cDSP) construct and ternary plots to overcome the challenges involved in one of the unit operations, namely, the tundish. The construct can be instantiated for other unit operations to cover the entire manufacturing cycle. Exploring the effects of system variables for each process step through experiments and plant trials is time consuming and very costly. The proposed method allows for faster design exploration of the process and thereby provides a reduced search space to a process designer. The process designer, with reduced experimentation requirements, can explore the narrowed search space to find the operating set points for a tundish. This, in turn, reduces the time and cost involved in production of a steel product mix with a new grade of steel in industry.

This chapter is aimed at readers interested in the topic of informatics-based approaches for accelerated materials discovery, but who are unfamiliar with the nuances of the underlying principles and various types of powerful mathematical tools that are involved in formulating structure–property relationships. In an attempt to simplify the workflow of materials informatics, we decompose the paradigm into several core subtasks: hypothesis generation, database construction, data pre-processing, mathematical modeling, model validation, and finally hypothesis testing. We discuss each task and provide illustrative case studies, which apply these methods to various functional ceramic materials.

Feature selection is considered as an important preprocessing step to data mining and soft computing, whereas regression is a collection of methods to optimally assess the signal from a noisy output. Both seek to arrive at the dependence and relation between different attributes and a target material property. In the present chapter a flock of regression and feature selection techniques are discussed, and the kind of results that can be obtained with each of them has been illustrated with the help of a dataset on steel. The different methods are capable of abstracting data in different forms, thus revealing hidden knowledge from different perspectives. Choosing the most appropriate method depends on the application at hand and the kind of objective that one is looking for.

This chapter highlights the usage of imprecise knowledge of materials systems using fuzzy inference systems. Experts have knowledge of complex materials systems in the imprecise linguistic form. But due to lack of phenomenological relations, material engineers are compelled to depend on empirical models for practical complex systems. This limitation could be overcome to a certain extent through the method of utilizing this imprecise knowledge with the help of fuzzy logic. The case studies presented here have demonstrated that systems with imprecise knowledge but with sparse data could be modeled successfully in this approach.

The recent developments in computational intelligence has enhances the applicability of empirical modelling in different areas particularly in the area of machine learning. These new approaches are based on analysing the data about a system, in particular finding connections between the system state variables (input, internal and output variables) without having precise knowledge about the physical behaviour of the system. These data driven methods explain advances on conventional empirical modelling and include contributions from many overlapping fields like Artificial Intelligence (AI), Computational Intelligence (CI), Soft Computing (SC), Machine Learning (ML), Intelligent Data Analysis (IDA), and Data Mining (DM). The most popular computational intelligence techniques used in process modelling of steel industry

Preface

Materials design is not a new concept. People are doing this consciously or unconsciously from the beginning of the civilization. As the civilization depends on the materials human being learn to use, as described even in the names of the different ages or stages of civilization, designing or inventing new materials was the prime concern for its development. Design of advanced materials is now a purposive effort, and depends on the knowledge of the composition-microstructure-process-property correlation of the system. The evolution of a large number of methods to model this correlation in a material system at the electronic, atomic and microstructural level made computational designing of new materials possible. Materials theory based physical models are being used for designing new materials, though these models are usually constrained by several boundary conditions and assumptions. There are several methods like density function theory, molecular dynamics, phase field theory, finite element methods among others to model any materials system in different length scale. On the other hand empirical materials models try to capture the inherent correlations among the variables on the basis of available database. Several statistical techniques along with the concept of data mining are now capable of informatics based design, leading to the development of novel materials, which are of academic as well as industrial interest. In addition, modeling based on soft computing techniques enabled the design methodologies to cope up with phenomenological perturbations arising due to non-linearity, imprecision, uncertainty, partial truth and approximation inherent in the system knowledge or in the experimental data. Optimization tools also play important role in designing materials with tailor-made properties or developing easily adoptable processes for manufacturing. Evolutionary optimization algorithms are found to be most effective to search the new solutions to the modern day need of better materials.

All these techniques are drawing interest from theoretical as well as practical application point of view. Researchers are using these approaches for discovering new materials, practitioners at industries are using them to improve the performance of their product or reducing the cost of it. This book provides a comprehensive stock of the different approaches of computationally designing new materials and their probable application, with an emphasis on exploring their relevance in practical/industrial conditions. The objective of the book is to provide a comprehensive collection of all possible computational approaches for designing new improved materials. The book will provide a description of different tools and techniques, and their application in the field of materials design, in separate chapters written by experts of the fields. The first chapter provides an introduction to the computational techniques being widely used for designing new materials. From the second chapter onward authors deal with theories and applications of the different concepts and approaches separately. Ab initio methods of materials design are approached in chapters two and three. The fourth chapter discusses different modeling and simulation techniques applied in the area of microstructure. The finite element method and other micromechanical approaches in the area of modeling are described in chapter five. The issue of integrating different concepts of materials modeling in different length scales is dealt in the sixth chapter with the introduction of the concept of integrated computational materials engineering. The concept applying statistical modeling and data mining in materials engineering is a new domain, being called materials informatics. Computational intelligence techniques like neural network and fuzzy logic are also being used for designing new materials. Chapters seven to ten deal with these concepts. The use of optimization tools occupies an important space for materials design, as the authors of chapter eleven describe. Chapter twelve describes two recent developments in data driven materials modeling using technique hybridization. The final chapter deals with a modeling concept applicable for the complex industrial process of steel making.

The editors convey their gratitude to all the authors for their efforts to make this very important compilation on computational materials design. The effort made by the reviewers to improve the quality of the book is also gratefully acknowledged. Both the editors are extremely thankful to their colleagues, friends and family members who have cooperated for the successful completion of the book. The editors are also grateful to their graduate students. The brilliant effort made by the members of IGI Global to shape this compilation in such a nice way is also acknowledged.

Shubhabrata Datta
Calcutta Institute of Engineering and Management, Kolkata India

J. Paulo Davim
University of Aveiro, Portugal

Chapter 1
Computational Materials Design:
Different Concepts and Aspects

Shubhabrata Datta
Calcutta Institute of Engineering and Management, India

Bishnupada Roy
Bankura Unnayani Institute of Engineering, India

J. Paulo Davim
University of Aveiro, Portugal

ABSTRACT

The chapter primarily deals with brief description of different methods of materials modeling which utilizes the scientific theories in different length scales. It also gives an account of the available tools for situations where data driven models are required. Utilization of imprecise knowledge of a materials system for developing mathematical models is also discussed. A brief account of the use of optimization techniques for designing materials is discussed here.

INTRODUCTION

For the last few decades the concept of computational materials design is growing fast due to the advent of high power computational facilities. So many computational concepts, particularly the concepts using the fundamental science of materials systems, could not have been utilized, if strong computer back ups are not provided. As in case of all other computational engineering design procedures, materials design also depends on three aspects, modeling, simulation and optimization. The domain of materials modeling is huge, and it can be divided into three parts. The first and dominant part of materials modeling is the concept of modeling where the fundamental physics and chemistry of the system is utilized in different length scales. The fundamental principles of physics and chemistry, governing the states and properties of condensed matter also paves the way for materials theory for quantitative modeling of structural and

DOI: 10.4018/978-1-5225-0290-6.ch001

functional properties of materials. This consequently provides the path of designing and predicting novel materials and devices with enriched performance. It is an important and fundamental aspect of material science that is becoming more important day by day due to its predictive power at different length and time scale. The length scale varies from atomic to the macroscopic level. This includes electronic structure and properties of materials determined by ab initio and/or semi-empirical methods, atomic level properties of materials, microstructural level phenomena, continuum-level modelling relating to material behavior. It is a scientific approach to technological problems which uses and combines various tools from both fundamental science and materials engineering and bridges length and time scales across many orders of magnitudes by wise selection of quantum mechanical, atomistic, continuum and statistical mechanical approximations. The mathematical and computational tools developed and used are not only accurate, quantitative and predictive as well as robust and reliable. So, modeling is applicable and beneficial to all the established as well as new newly emerging fields of material research which saves huge laboratory resources, and energy by predicting the best possible path to get the best fit product and at the same time gives concrete theoretical background to it (Janssens, Raabe, Kozeschnik, Miodownik, & Nestler, 2007).

When the science behind the behavior of a material system is not fully explored, but several observations are available, then materials scientists and engineers depend on the statistical techniques like regression analysis to develop models correlating the parameters. Several data-driven techniques similar to statistical concepts, like artificial neural network, genetic programming, support vector machines, capable to map highly complex and non-linear relations, are developed by computer scientists. We can think of a domain of modeling in between the physical and empirical modeling concept, for such situation where the knowledge about a system is available, but in imprecise form. In such cases the fuzzy logic can be used to develop models utilizing the system knowledge in such form. This concept is also getting used in materials modeling.

Virtual experimentation using the mathematical models, known as simulation, is advantageous for materials design. Using simulation technique an in-depth idea about the role of different parameters in a materials system could be explored. It can also help to find combination of parameters for a material to achieve its required performance. But the best way to find such solution, in the composition and/or processing parameters for any material, is to go for optimization. The property of a material to be optimized needs an objective function, which is nothing but a correlation between the property and the input parameters like composition and processing parameters. Any model developed through any of the above methods can be utilized as this objective function. Two different approaches of optimization are most common, and they are the classical derivative based optimization approach and the newer non-derivative based evolutionary approach. In this chapter we shall try to introduce some of the common modeling and optimization techniques in brief.

MATERIALS MODELING IN DIFFERENT LENGTH SCALES

As discussed above, the scientific theory behind a material system could be utilized to form mathematical models, which can be utilized for design and development of new materials, as shown in Figure 1. Such modeling could be done using different length scales, starting from atomic or even sub-atomic level to level of microstructure. Depending on length scale there are a number of theories engaged in modelling of materials such as Density Functional Theory (DFT), Molecular Dynamics (MD), Phase Filed Theory and Finite Element method, etc. (Figure 2).

Figure 1. Scientific theory to new materials design

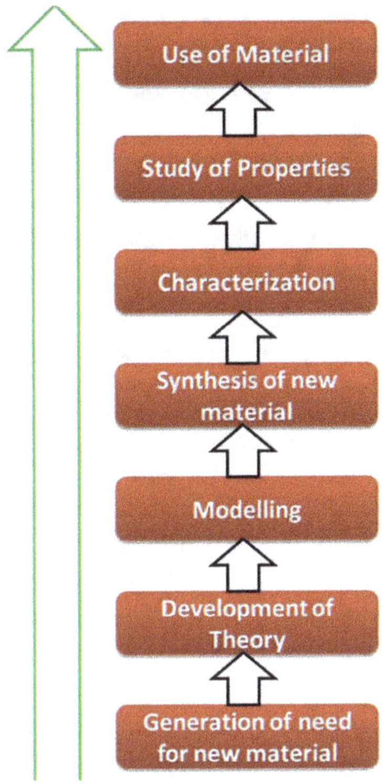

Density Functional Theory (DFT)

Forms the basis of a diversified and very active area of present day research in atomic, molecular, solid state and even nuclear physics (Eschrig, 2003). A computational quantum mechanical modelling method is principally used to investigate the ground state electronic structure of many-body systems. This theory determines the properties of a many-electron system by using functionals i.e. functions of another function, which in this case is the spatially dependent electron density. DFT is among the most popular and versatile methods used in computational chemistry, computational physics and condensed-matter physics. DFT gained popularity for calculations in solid-state physics in the 1970s but was not considered accurate enough for calculations in quantum chemistry until the 1990s, when the approximations used in the theory were greatly refined to better model the exchange and correlation interactions. The results of DFT calculations for solid-state systems match quite satisfactorily with experimental data in most cases. Computational costs are relatively low when compared to traditional methods, such as Hartree–Fock theory and its descendants based on the complex many-electron wave function. This provides a useful balance between accuracy and computational cost (Jones, 2015).

DFT, which has its conceptual roots in the Thomas–Fermi model, was put on a firm theoretical footing by the two Hohenberg–Kohn (H–K) theorems. The original theorems held only for non-degenerate ground states in the absence of a magnetic field but have since been generalized to accumulate these. The first theorem tells us that the ground state properties of a many-electron system are uniquely determined by an electron density that depends on only three spatial coordinates. It gives the foundation of reducing

Figure 2. Some of the common modelling tools in different length scale

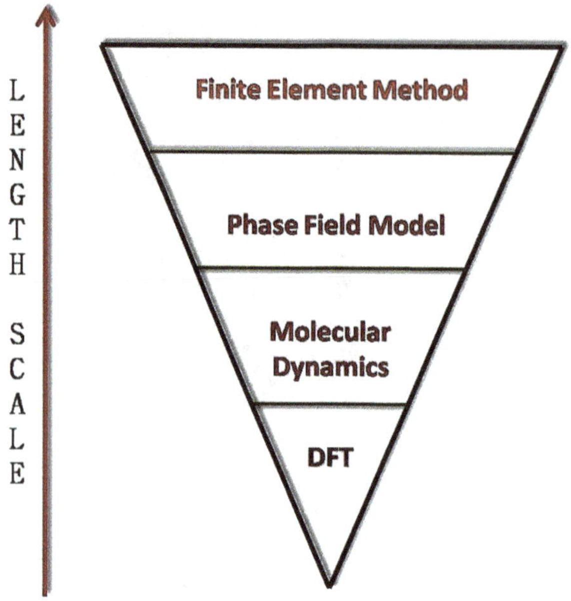

the many-body problem of N electrons with 3N spatial coordinates to three spatial coordinates, through the use of functionals of the electron density. This theorem can also be applied to the time-dependent domain to develop time-dependent density functional theory (TDDFT), which can be used to describe excited states. In the second theorem, an energy functional is defined for the system and it is proved that the correct ground state electron density minimizes this energy functional. The Kohn–Sham DFT (KS DFT) deals with the intractable many-body problem of interacting electrons in a static external potential and reduce it to a tractable problem of non-interacting electrons moving in an effective potential. The effective potential includes the external potential and the effects of the Coulomb interactions between the electrons, e.g., the exchange and correlation interactions. Modeling of the latter two interactions becomes the difficult within KS DFT. The simplest approximation is the local-density approximation (LDA) which is based upon exact exchange energy for a uniform electron gas that can be obtained from the Thomas–Fermi model. Non-interacting systems are relatively easy to solve as the wave function can be represented as a Slater determinant of orbitals and the kinetic energy functional of such a system is known exactly. The exchange-correlation part of the total-energy functional must be approximated as it is still not exactly known. Another approach which is less popular than KS DFT but arguably more closely related to the spirit of the original H-K theorems is orbital-free density functional theory (OFDFT).

The presence of a vector potential i.e. a magnetic field breaks down the DFT formalism described above to various degrees. In such a situation, the one-to-one mapping between the ground-state electron density and wave function is lost. Generalizations to include the effects of magnetic fields have led to two different theories: current density functional theory (CDFT) and magnetic field density functional theory (MFDFT). In both these theories, the functional used for the exchange and correlation is generalized to include more than just the electron density. In CDFT, developed by Vignale and Rasolt, the functionals become dependent on both the electron density and the paramagnetic current density. In MFDFT, developed by Salsbury, Grayce and Harris, the functionals depend on the electron density and the magnetic field. Here the functional form can depend on the form of the magnetic field. In both of these theories it has been difficult to develop functionals beyond their equivalent to local density approximation which are also readily implementable computationally.

So, the application of DFT is broad in the fields of material sciences for the interpretation and prediction of complex system behavior at an atomic scale. DFT computational methods are often applied for the study of systems exhibiting high sensitivity to synthesis and processing parameters. In such systems, experimental studies are often hampered by inconsistent results and non-equilibrium conditions. Examples of contemporary DFT applications include studying the effects of dopants on phase transformation behavior in oxides, magnetic behavior in dilute magnetic semiconductor materials and the study of magnetic and electronic behavior in ferroelectrics and dilute magnetic semiconductors. Difficulties are also there in using density functional theory to properly describe intermolecular interactions, especially Van der Waals forces (dispersion); charge transfer excitations; transition states, global potential energy surfaces, dopant interactions and some other strongly correlated systems and in calculations of the band gap and ferromagnetism in semiconductors. It's incomplete treatment of dispersion can adversely affect the accuracy of DFT (at least when used alone and uncorrected) in the treatment of systems which are dominated by dispersion (e.g. interacting noble gas atoms) or where dispersion competes significantly with other effects (e.g. in biomolecules). The development of new DFT methods to overcome this problem, by alterations to the functional and inclusion of additional terms to account for both core and valence electrons or by the inclusion of additive terms is an area of current research.

Molecular Dynamics (MD)

Molecular dynamics is a technique for computer simulation of complex systems, modeled at the atomic level (Haile, 1997). The equations of motion are solved numerically to follow the time evolution of the system, allowing the derivation of kinetic and thermodynamic properties of interest by means of 'computer experiments'. Biologically important macromolecules and their environments are routinely studied using molecular dynamics simulations. It is a technique that allows one to predict the time evolution of a system of interacting particles (atoms, molecules, granules, etc.). Here the atoms and molecules are allowed to interact for a period of time, giving a view of the motion of the atoms. The basic idea is very simple. First, for a system of interest, one has to specify a set of initial conditions (initial positions & velocities of all particles in the system) and interaction potential for deriving the forces among all the particles. Second, the evolution of the system in time can be followed by solving a set of classical equations of motion for all particles in the system. Within the framework of classical mechanics, the equations that govern the motion of classical particles are the ones that correspond to the second law of classical mechanics formulated by Sir Isaac Newton over 300 years ago:

$$m_i \vec{a}_i = \vec{F}_i$$

Or $m_i \dfrac{d\vec{v}_i}{dt} = \vec{F}_i$

Or $m_i \dfrac{d^2\vec{r}_i}{dt^2} = \vec{F}_i$

As molecular systems consist of a vast number of particles, it is impossible to find the properties of such complex systems analytically. MD simulation avoids this problem by using numerical methods. However, long MD simulations are mathematically ill-conditioned, generating cumulative errors in numerical integration that can be minimized with proper selection of algorithms and parameters, but not eliminated entirely. For systems which obey the ergodic hypothesis, the evolution of a single molecular dynamics simulation may be used to determine macroscopic thermodynamic properties of the system: the time averages of an ergodic system correspond to microcanonical ensemble averages. MD has also been termed "statistical mechanics by numbers" and "Laplace's vision of Newtonian mechanics" of predicting the future by animating nature's forces and allowing insight into molecular motion on an atomic scale.

The method was originally conceived within theoretical physics in the late 1950s but today it is applied mostly in materials science and the modeling of biomolecules. MD is frequently used to refine three-dimensional structures of proteins and other macromolecules based on experimental constraints from X-ray crystallography or NMR spectroscopy. In physics, MD is used to examine the dynamics of atomic-level phenomena that cannot be observed directly, such as thin film growth and ion-subplantation. It is also used to examine the physical properties of nanotechnological devices that have not or cannot yet be created. In biophysics and structural biology, the method is frequently applied for ligand docking, simulations of lipid bilayers, homology modeling and even ab initio prediction of protein structure by simulating folding of the polypeptide chain from random coil. Limitations of the method are related mostly to underlying molecular mechanics force fields. A single run of MD simulation optimizes potential energy, rather than free energy of the protein, meaning that all entropic contributions to thermodynamic stability of protein structure are neglected. The neglected contributions include conformational entropy of polypeptide chain (main factor that destabilizes protein structure) and hydrophobic interactions that are known as main driving force of protein folding. Another important factor is intramolecular hydrogen bonds which are not explicitly included in modern force fields but described as Coulomb interactions of atomic point charges. This is a crude approximation because hydrogen bonds have a partially quantum mechanical nature. Furthermore, electrostatic interactions are usually calculated using the dielectric constant of vacuum, although the surrounding aqueous solution has a much higher dielectric constant. Use of the macroscopic dielectric constant at short interatomic distances is also questionable. Finally, Van der Waals interactions in MD are usually described by Lennard-Jones potentials based on the Fritz London theory that is only applicable in vacuum. However, all types of Van der Waals forces are ultimately of electrostatic origin and therefore depend on dielectric properties of the environment. The direct measurement of attraction forces between different materials (as Hamaker constant) shows that "the interaction between hydrocarbons across water is about 10% of that across vacuum". The environment-dependence of Van der Waals forces is neglected in standard simulations, but can be included by developing polarizable force fields (Baker & Norris, 2015).

Steered molecular dynamics (SMD) simulations, or force probe simulations, apply forces to a protein in order to manipulate its structure by pulling it along desired degrees of freedom. These experiments can be used to reveal structural changes in a protein at the atomic level. SMD is often used to simulate events such as mechanical unfolding or stretching. There are two typical protocols of SMD: one in which pulling velocity is held constant and one in which applied force is constant. Typically, part of the studied system (e.g. an atom in a protein) is restrained by a harmonic potential. Forces are then applied to specific atoms at either a constant velocity or a constant force. Umbrella sampling is used to move the system along the desired reaction coordinate by varying, for example, the forces, distances, and angles manipulated in the simulation. Through umbrella sampling, all of the system's configurations (both high-energy and low-energy) are adequately sampled. Then, each configuration's change in free energy can be calculated as the potential of mean force. A popular method of computing probability mass function (PMF) is through the weighted histogram analysis method (WHAM), which analyzes a series of umbrella sampling simulations.

Design of a molecular dynamics simulation should account for the available computational power. Simulation size (n=number of particles), time step and total time duration must be selected so that the calculation can finish within a reasonable time period. However, the simulations should be long enough to be relevant to the time scales of the natural processes being studied. To make statistically valid conclusions from the simulations, the time span simulated should match the kinetics of the natural process. Otherwise, it becomes analogous to making conclusions about how a human walks when only looking at less than one footstep. Most scientific publications about the dynamics of proteins and DNA use data from simulations spanning nanoseconds (10^{-9} s) to microseconds (10^{-6} s). To obtain these simulations, several CPU-days to CPU-years are needed. Parallel algorithms allow the load to be distributed among CPUs; an example is the spatial or force decomposition algorithm.

Molecular dynamics simulation requires the definition of a potential function or a description of the terms by which the particles in the simulation will interact. In chemistry and biology this is usually referred to as a force field. Potentials may be defined at many levels of physical accuracy; those most commonly used in chemistry are based on molecular mechanics and embody a classical treatment of particle-particle interactions that can reproduce structural and conformational changes but usually cannot reproduce chemical reactions. The reduction from a fully quantum description to a classical potential entails two main approximations. The first one is the Born–Oppenheimer approximation which states that the dynamics of electrons is so fast that they can be considered to react instantaneously to the motion of their nuclei. As a consequence, they may be treated separately. The second one treats the nuclei which are much heavier than electrons, as point particles that follow classical Newtonian dynamics. In classical molecular dynamics the effect of the electrons is approximated as a single potential energy surface, usually representing the ground state. When finer levels of detail are required, potentials based on quantum mechanics are used. Some techniques attempt to create hybrid classical/quantum potentials where the bulk of the system is treated classically but a small region is treated as a quantum system.

Phase Field Model

In recent years, Phase Field Models have become popular to describe a host of free-boundary problems in various areas of research (Provatas & Elder, 2010). The principal characteristic of phase-field models is the diffuseness of the interface between two phases, which are sometimes also called 'Diffuse Interface Models'. The key point of the phase-field approach is that surfaces and interfaces are implicitly described by continuous scalar fields that take constant values in the bulk phases and vary continuously but steeply across a diffuse front. It is a mathematical model for solving interfacial problems. It has mainly been applied to solidification dynamics but it has also been applied to other situations such as viscous fingering, fracture dynamics, vesicle dynamics etc. The method substitutes boundary conditions at the interface by a partial differential equation for the evolution of an auxiliary field (the phase field) that takes the role of an order parameter. This phase field takes two distinct values (for instance +1 and −1) in each of the phases, with a smooth change between both values in the zone around the interface, which is then diffuse with a finite width. A discrete location of the interface may be defined as the collection of all points where the phase field takes a certain value (e.g. 0).

A phase field model correct interfacial dynamics defined in the limit of an infinitesimal interface width (sharp interface limit). This method solves the problem by integrating a set of partial differential equations, avoiding the explicit treatment of the boundary conditions at the interface. The technique was first introduced by Fix and Langer. It has drawn lot of interests from the researchers working in solidification and other areas. Phase field models are used to reproduce a given interfacial dynamics, as in solidification the front dynamics is represented by diffusion equation for either concentration or temperature in the bulk and some boundary conditions at the interface leading to the required sharp interface model. Phase field equations in principle reproduce the interfacial dynamics when the interface width is small compared with the smallest length scale in the problem. Karma and Rappel introduced the thin interface limit for practical quantitative simulations at small dimensions. The increasing power of computers has helped phase field models to become a useful tool for the interfacial problems.

In phase field model free energy functional is computed depending on an order parameter (the phase field) and a diffusive (variational) field. Parameters of the model are then chosen by studying the limit of the model with this width going to zero, in such a way that one can identify this limit with the intended sharp interface model. Other formulations start directly from the phase field equations, without going into any thermodynamical functional. The choice of free energy function, f(Ø), can have a significant effect on the physical behavior of the interface, and should be selected with care. Thus a phase field model can be constructed to purposely reproduce a given interfacial dynamics as represented by a sharp interface model, where the sharp interface limit (the limit when the interface width goes to zero) of the proposed set of phase field equations is usually taken by asymptotic expansions of the fields of the model in powers of the interface width. These expansions are performed both in the interfacial region (inner expansion) and in the bulk (outer expansion) and then are asymptotically matched order by order. Such expansions were in early phase field models performed up to the lower order, but in recent time researchers use higher order asymptotics (thin interface limits) in order to cancel undesired spurious effects or even to include new phenomenon in the model, trying to reach more complexity. In multi-phase field models, microstructure is described by set of order parameters, where each parameter has a specific phase or crystallographic orientation. This model is effective for solid state phase transformations evolving multiple grains, such as recrystallization, grain growth, or first order transformation like austenite to ferrite in steel. In addition to handling multiple grains in a microstructure, multi-phase field models have the capability to describe multiphase microstructure, particularly relevant for high performance steel.

The Finite Element Method (FEM)

Represent a powerful and general class of techniques for finding approximate solutions to boundary value problems for partial differential equations (Suli, 2000). It uses subdivision of a whole problem domain into simpler parts, called finite elements. It also uses variational methods from the calculus of variations to solve the problem by minimizing an associated error function. Analogous to the idea that connecting many tiny straight lines can approximate a larger circle, FEM encompasses methods for connecting many simple element equations over many small sub domains, named finite elements, to approximate a more complex equation over a larger domain. The method thus involves (1) dividing the domain of the problem into a collection of sub domains with each sub domain represented by a set of element equations to the original problem followed by (2) systematically recombining all sets of element equations into a global system of equations for the final calculation. The global system of equations is known as solution techniques and can be calculated from the initial values of the original problem to obtain a numerical answer.

The element equations are locally approximated simplified version of the original complex partial differential equations (PDE). Galerkin method explains the approximation process commonly employed in FEM. The process is to construct an integral of the inner product of the residual and the weight functions, and then set the integral to zero. It is a procedure which minimizes the error of approximation by fitting trial functions into the complex PDE. The residual is the error caused by the trial functions and the weight functions are polynomial approximation functions that project the residual. The process removes all the spatial derivatives from the PDE, thus approximating it locally with a set of algebraic equations for steady state problems and a set of simple differential equations for transient problems. These equation sets are linear if the underlying PDE is linear, and vice versa. Algebraic equation sets developed for the steady state problems are solved using numerical linear algebra methods, and ordinary differential equation sets for the transient problems are solved by standard techniques such as Euler's or the Runge-Kutta methods. Then a global system of equations is generated from the element equations through a transformation of coordinates from the local nodes to the global nodes. The practical application of FEM, known as finite element analysis (FEA), is a computational tool for performing engineering analysis. It includes the use of mesh generation techniques for dividing a complex problem into small elements. In applying FEA, the complex problem is usually a physical system with the underlying physics such as the Euler-Bernoulli beam equation, the heat equation or the Navier-Stokes equations expressed in either PDE or integral equations, while the divided small elements of the complex problem represent different areas in the physical system. FEA works fine when the domain changes (as during a solid state reaction with a moving boundary), when the desired precision varies over the entire domain or when the solution lacks smoothness. A variety of specializations under the umbrella of the mechanical engineering discipline (such as aeronautical, biomechanical, and automotive industries) commonly use integrated FEM in design and development of products. In a structural simulation, FEM helps tremendously in producing stiffness and strength visualizations and also in minimizing weight, materials and costs. FEM allows detailed visualization of where structures bend or twist and indicates the distribution of stresses and displacements. This powerful design tool has significantly improved both the standard of engineering designs and the methodology of the design process in many industrial applications. In the materials domain FEA has made significant inroad in the area of microstructure-property correlation, where complex microstructure is analyzed through FEA and predictive models for different mechanical, thermal and other properties could be developed (Liu & Quek, 2013). In case of complex materials processing such as rolling and forging could also be simulated using this method.

EMPIRICAL MODELING IN THE MATERIALS DOMAIN

In the field of metallurgical engineering empirical models, mostly *statistical regression models*, developed from experimental data, are being used successfully for long time. These models include Hall-Petch equation, predictive models for transformation temperatures (Ac_1, Ac_3, M_S etc), model for carbon equivalent etc. In all these cases the linear regression analysis is employed, which is the most widely used statistical techniques. Here the model is expressed in terms of linear, additive relationships between variables. If Y denotes the "dependent" r or the output, in case of a materials system it may describe any property or any other attribute of the materials system, whose values are to predict from X_1, …,X_n denoting the "independent" variables, then the equation for computing the predicted value of Y is

$$Y = a_0 + a_1X_1 + a_2X_2 + ... + b_nX_n$$

With this type of formula the prediction for Y is a straight-line function of each of the X variables, keeping the others fixed, and the contributions of different X variables to the predictions are additive. The slopes of their individual straight-line relationships with Y are the constants $\mathbf{b}_1$, $\mathbf{b}_2$, …, $\mathbf{b}_k$, and called coefficients of the variables. Such models could be made non-linear or pseudo linear by imparting non-linear functions within such equation, as in case of Hall-Petch equation. As the models are developed from data, the quality of model depends heavily on the quality of data. But such method of modelling are a common solution for materials systems which are highly complex and their attributes cannot be fully explained and mathematically expressed from fundamental.

- **Artificial Neural Network (ANN):** A method of empirical relations developed later by the computer scientists, which is inspired by the way biological nervous systems, such as the brain, process the information. It has a large number of highly interconnected (nonlinear) processing elements, known as neurons (Figure 3). A neuron collects signal from others through the 'dendrites'. Then it sends the signal through an 'axon' to the 'synapse' to pass the signal to the other connected neurons. Thus a signal propagates. In case of ANN neurons are replaced by nodes and the efficiency of the synaptic transmission by weights. As in case of biological systems adjustment to the synaptic connections leads to learning, in case of ANN the weights are adjusted when the relation between the parameters are embedded in the model, also called learning. For the last few decades ANN is becoming increasingly popular for modeling materials systems, particularly for alloys, where the final performance of the materials depends on several parameters related in a complex way (Datta & Chattopadhyay, 2013).

Figure 3. Structure of a multi-layered neural network

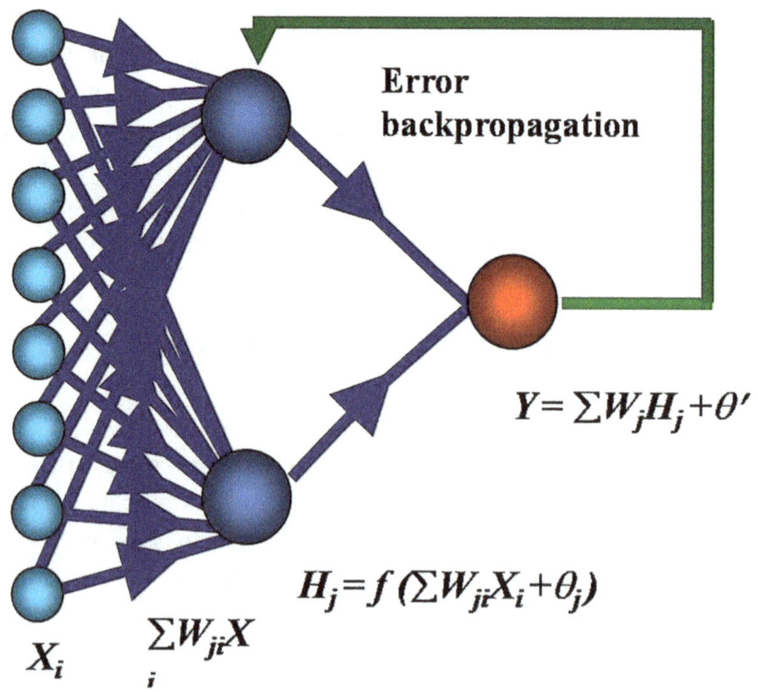

- **Fuzzy Inference System:** Is a method which can exploit imprecise knowledge of a system and being utilized for materials systems successfully. It employs fuzzy logic, which may be expressed as a superset of conventional (Boolean) logic. It was first introduced by Dr. Lotfi Zadeh in the 1960s as a means to model the uncertainty of natural language. Fuzzy logic actually trades off between significance and precision. A fuzzy set is a set without a crisp or clearly defined boundary. The set contains elements with only a degree of membership, as in fuzzy logic the truth of any statement becomes a matter of degree. The membership value (or degree of membership) each point in the input space is mapped to a between 0 and 1 by a membership function (MF), which is a curve whose shape we can define as a function. As in fuzzy inference systems linguistic descriptions in the form of fuzzy IF-THEN rules represent the system to be modeled, the imprecise knowledge of a complex materials system could highly be utilized here for developing mathematical models.

OPTIMIZATION AND MATERIALS DESIGN

Optimization techniques play an important role for designing new materials with improved performance. The proper combination of values of different parameters responsible for the final outcome could be searched using optimization methods. Non-derivative based optimization algorithms, i.e. different evolutionary and swarm algorithms like genetic algorithm, simulated annealing, ant colony optimization, particle swarm optimization, etc are getting special importance in case of designing new materials, due to their flexible approaches for handling the objective functions and the constraints. Sometimes it becomes important to optimize conflicting attributes to develop high performance materials, where multi-objective optimization comes into play and these algorithms are better capable to handle such situations.

- **Genetic Algorithm (GA):** A member of group of optimization algorithms called evolutionary algorithms, is a technique which mimics the principles of natural selection to find the best solution for a problem. GA is actually a stochastic global search method, which operates on a population of feasible solutions and tries to evolve the best solution applying the principle of survival of the fittest. At each generation, a new set of solutions is created by the process of selecting individuals according to their level of fitness and then breeding them. This process leads to the evolution of populations of individuals that are better fit to their environment than their predecessors, as in nature. A single set of parameter string, called a chromosome, is treated as an individual solution and a large population of solutions is created with random parameter values. The fitter solutions are bred with each other, where the probability that an individual solution will pass on some of its parameter values to subsequent children is directly related to the fitness of individual. Many real world optimization problems have multiple conflicting objectives. In case of such multi-objective problems, the genetic search is performed following the concept of Pareto-optimality. In such situation the concept of "optimum" changes, from the unique global optimum, as used in the single objective problems, to a set of solutions providing the best possible compromises between the objectives, known as the Pareto set or front.

WHERE TO GO

The above discussion clearly depicts that there are lots of techniques available for designing materials using scientific theories in different length scales to utilizing experimental observations or even imprecise knowledge expressed in linguistic manner. But the problem lies in the fact that people working in the domain of computational designing of materials tend to work in utilizing any particular modeling tool or particular length scale. The idea of multi-scale modeling has been coined a few decades ago, where the outputs of lower scale of modeling will be utilized as the inputs of higher scale of models. By that way it will be possible to develop mathematical relations between behaviors of materials at the atomic scale to the final microstructure. This will make designing materials with required properties from its atomic level. But such efforts are still limited to some simple systems only. In recent years the concept of integrated computational materials engineering (ICME) is being introduced where the limitation of the previous approach to be overcome through an integrated approach where both physical as well as empirical and semi-empirical methods of modeling to develop practical platforms for solving the issue of designing materials with improved performance.

REFERENCES

Baker, C. H., & Norris, P. M. (2015). Effect of long- and short-range order on SiGe alloy thermal conductivity: Molecular dynamics simulation. *Physical Review B: Condensed Matter and Materials Physics*, *91*(18), 180302. doi:10.1103/PhysRevB.91.180302

Datta, S., & Chattopadhyay, P. P. (2013). Soft Computing Techniques in Advancement of Structural Metals. *International Materials Reviews*, *58*(8), 475–504. doi:10.1179/1743280413Y.0000000021

Eschrig, H. (2003). *The Fundamentals of Density Functional Theory*. Leipzig, Germany: Edition am Gutenbergplatz Leipzig.

Haile, J. M. (1997). *Molecular Dynamics Simulation: Elementary Methods*. New York, NY: Wiley.

Janssens, K. G. F., Raabe, D., Kozeschnik, E., Miodownik, M. A., & Nestler, B. (2007). *Computational Materials Engineering: An Introduction To Microstructure Evolution*. Amsterdam, Netherlands: Elsevier.

Jones, R. O. (2015). Density functional theory: Its origins, rise to prominence, and future. *Reviews of Modern Physics*, *87*(3), 897–924. doi:10.1103/RevModPhys.87.897

Liu, G. R., & Quek, S. S. (2013). *The Finite Element Method: A Practical Course*. Oxford, UK: Butterworth-Heinemann.

Provatas, N., & Elder, K. (2010). *Phase-Field Methods in Materials Science and Engineering*. Weinheim, Germany: Wiley-VCH. doi:10.1002/9783527631520

Suli, E. (2000). *Lecture Notes on Finite Element Methods for Partial Differential Equations*. Oxford, UK: University of Oxford.

Chapter 2
Successive Spin-Correlated Local Processes Underlying the Magnetism in Diluted Magnetic Semiconductors and Related Magnetic Materials

Antonis N. Andriotis
Institute of Electronic Structure and Laser, Greece

Madhu Menon
University of Kentucky, USA

ABSTRACT

Recent works have suggested that the defect induced magnetism in Diluted Magnetic Semiconductors (DMSs), Transition Metal Oxides (TMOs) and related materials is facilitated and enhanced by codoping and the synergistic action between the codopants. In the present work we demonstrate that the proposed defect synergy is the result of the interplay among correlated spin-polarization processes which take place in a successive way in neighborhoods centered at the codopants and include their first nearest neighbors. These processes result in a reduction in the superexchange coupling which in turn causes an enhancement in the ferromagnetic coupling (FMC) among the magnetic dopants. The proposed FMC is demonstrated using ab initio calculations of the electronic properties of codoped ZnO, GaN and TiO2.

1. INTRODUCTION

The origin of magnetism in the technologically important class of the diluted magnetic semiconductors (DMSs) and the transition metal oxides (TMOs) is an unresolved issue and has been the subject of an ongoing debate. While the experimental information in many cases provides contradictory results, the theoretical approaches fail to give a conclusive answer for systems exhibiting magnetic features at dopant concentrations below the percolation threshold.

DOI: 10.4018/978-1-5225-0290-6.ch002

Adding to the problem is the appearance of magnetism in carbon-based materials which lack the presence of d-electrons and/or of dopants of inherent magnetic moments. In these materials, the development of magnetism had to be associated with their inherent properties resulting from structural modifications and imperfections. However, the presence of a single type of structural conformations that could be considered as potential candidates which can provide magnetic moments (as, for example, the presence of carbon vacancies), proved not to be sufficient to explain the appearance of magnetism.

The recent works on the magnetism of Rh-C60 and other carbon-based (Cbased) materials(Andriotis, Menon, Sheetz, & Chernozatonskii, 2003) led to the proposal that the origin of this magnetism is the result of the simultaneous presence of two types of defects. That is, it was proposed that the magnetism in the C-based materials is the result of the synergistic action of defect-induced electronic processes, mostly of local character, which can provide magnetic moments and develop their ferromagnetic coupling (FMC). In particular, it was shown that this synergy is realizable via codoping(Andriotis, Lisenkov, & Menon, 2011; Andriotis & Menon, 2011, 2012) ie., in the simultaneous presence of two kinds of defects which can act in a complimentary way. Specifically, while the defects of one kind are able to provide mainly the magnetic moments, defects of the other kind contribute mainly to the development of the FMC among the magnetic moments. Thus, according to this proposal, the magnetism of the C-based materials is a defect-induced defect-mediated process that is realizable via codoping. (In the following, the term defect will be used to denote any structural and/or topological and/or impurity-atom defects).

Based on our proposed defect-induced defect-mediated approach to the magnetism of the C-based materials, we investigated whether this model approach is applicable in the case of DMSs and TMOs. This idea was further supported by contemporary experimental evidence which demonstrated that the magnetism in DMSs could be enhanced by codoping. In particular, enhanced magnetic features were observed in ZnO codoped with (Fe,Cu)(Han et al., 2002; Park & Min, 2003), (Fe,Co)(Park & Min, 2003), (Co,Cu)(Naeem, Hasanain, Afgan, & Rumaiz, 2008; H. T. Lin et al., 2004), (Co,Li)(Jayakumar, Gopalakrishnan, & Kulshreshtha, 2006; Y. H. Lin, Ying, Li, Wang, & Nan, 2007), (Co,N)(Assadi, Zhang, & Li, 2009), (Co,H)(Shin et al., 2012), (Co,Al)(Chattopadhyay & Natha, 2010), (Co, O-vacancies)(Pemmaraju, Archer, Sanchez-Portal, & Sanvito, 2007), (Mn,Sn)(Norton et al., 2003), (Mn,Co)(Behan et al., 2008; Duan, Rao, Wang, Yu, & Wang, 2008), (Mn,N)(Wang, Sun, Jena, & Kawazoe, 2004), (Mn, Ovacancies)(Mi, Bai, Liu, & Sun, 2007) and (Mn,C)(Yadav, Sanyal, & Mookerjee, 2009). Analogous findings were reported for other TMOs as well as for III-V and II-VI materials as, for example, in GaN codoped with (Mn,Co) and (Cr,Mn)(Santos, Marques, Teles, & Ferreira, 2010); in Cr:TiO2 and Mn:GaAs(Ye & Freeman, 2006); in TiO2(Zhu et al., 2009; Gai, Li, Li, Xia, & Wei, 2009) etc.

Despite the numerous reports on the advantages of codoping in improving the magnetic features of the DMSs and those of the TMOs, our understanding about the origin of their magnetism is far from satisfactory. The majority of the proposed explanations so far is based on the proposed traditional exchange interaction processes including; the double exchange, the superexchange, the s−d exchange or the p − d hopping exchange interaction(Ye & Freeman, 2006), the non-compensated n − p doping(Zhu et al., 2009), the passivation of codopants (Mo,C)(Gai et al., 2009), etc. These model interactions map leading configuration interaction terms onto atomic spin-orbital description terms (see, for example, Ref. (Sato et al., 2010) and references therein). By their nature. these descriptions are focused on processes related to the interactions among the magnetic impurity cations; ie., the central role is attributed to their direct and/or mediated cationic interaction.

In this Chapter, we present our work to uncover the key electronic processes which underlie the development of magnetic features in DMSs and TMOs.

2. COMPUTATIONAL DETAILS

The density functional theory (DFT) in the local spin polarized electron density approximation (LSDA) has been shown to be very useful in predicting many of the properties of interest for a variety of systems. However, it is also well known that DFT tends to underestimate the results for the electronic structure. In particular the energy gap values of semiconductors and TMOs are predicted to be much smaller than the corresponding experimental values. As a result, the location of impurity and defect levels formed in DMSs and TMOs cannot be located correctly relative to the energy gap. This leads to wrong conclusions for the hybridization processes and the underlying physics of the observed magnetic properties.

This picture obtained within the DFT/LSDA does not improve significantly at the level of the generalized gradient approximation (GGA), DFT/GGA (to be denoted as DFT/SGGA). One, therefore, has to employ ab initio calculations that go beyond the DFT/SGGA, while at the same time allowing for efficient computational investigations to be carried out using available computer resources. A compromise that allows us to retain sufficient accuracy while staying within realistic computational limits can be achieved using the DFT/SGGA+U level of approximation where U is a Hubbard-like added potential parameter to the Kohn-Sham DFT Hamiltonian.

The U term in DFT/SGGA+U is a corrective parameter that can be interpreted either as a correction related with the Hartree-Fock (HF) self interaction of the localized (d-) electrons(Solovyev, Dederichs, & Anisimov, 1994) or as a correction of the (unphysical) curvature of the total-energy curve as a function of the occupation numbers(Cococcioni & de Gironcoli, 2005).

As shown by us recently(Andriotis, Mpourmpakis, Lisenkov, Sheetz, & Menon, 2013), the U-parametrization of the DFT/SGGA can be uniquely and with sufficient accuracy can be achieved by fitting the U-parameters (namely those of the d-orbitals of the transition metal atoms (cations) and those of the p-orbitals of the anions to appropriately chosen experimental data, namely the band gap Figure 1: Supercells used in our calculations for codoped ZnO and GaN systems.

Figure 1.

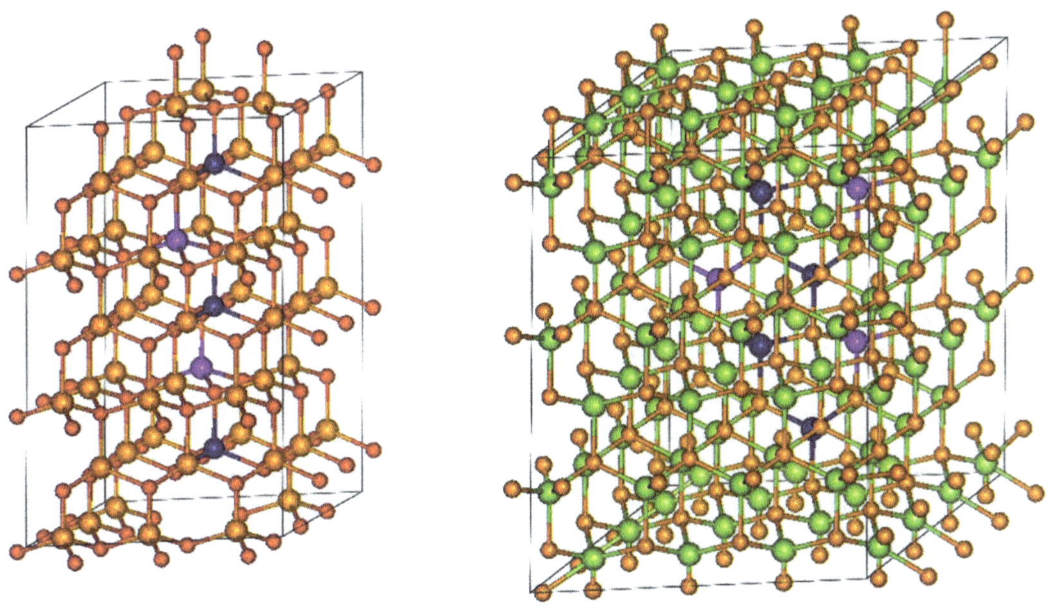

(a) (b)

(a) Relaxed structure of 108 atom Zn(3Co,2Cu)O supercell with a Co-Cu-Co-Cu-Co chain of nearest neighboring codopants over successive Zn planes. The Zn atoms are shown in yellow and O in red. The Co atoms are colored blue, while Cu are in purple. (b) Relaxed structure of 192 atom Ga(4Co,3Cu) N supercell with codopants forming a bipartite lattice. The Ga atoms are shown in green and N in yellow. Coloring of Co and Cu atoms are the same as in (a). value and an additional spectral or structural property. The former can be the location of an energy peak in the electron density of states (DOS) of the doped system while the latter can be the bulk modulus of the system(Andriotis et al., 2013; Khatri & Huda, 2014). In fact, by using simultaneously the positive-U parametrization for the d-orbitals along with a negative-U parametrization for the p-(or the s-) electrons, we can obtain band structures exhibiting energy gap and other spectral features in very good agreement with experiment(Sunkara et al., 2014). This is achieved due to the partial self-interaction correction and the modification that the combined (Ud,Up)-parametrization imposes on the localization (and, respectively, delocalization) of the d-(and, respectively p-) orbitals.

Using the DFT+U formalism we performed a series of calculation on representative DMS and TMOs systems that included ZnO, GaN and TiO2 codoped with various pairs of codopants. We used the DFT in the GGA and the PerdewBurke-Ernzerhof (PBE) functional augmented by including Hubbard-U corrections (GGA+U) based on Dudarev's approach(Dudarev, Botton, Savrasov, Humphreys, & Sutton, 1998) as implemented in the Vienna Ab-initio Simulation Package (VASP)(Kresse & Hafner, 1993; Kresse & Joubert, 1999). Large unit cells were employed for our host materials. Both ZnO and GaN are taken to be in the wurtzite structure using supercells containing 108 and 192 atoms, respectively. TiO2 is taken to be in rutile structure using a super cell of 192 atoms. Co and Cu atoms were used as codopants in ZnO, while for GaN and TiO2, we used the codopant pairs (Co,Cu) and (Mn,Cu). All codopants substituted cationic sites. Up to 5 dopant-atoms were used for ZnO and up to 7 for GaN and TiO2. A2×2×2 Γ-centered pack of k-vectors was used and all structures were fully optimized (including cell optimization) without any symmetry constraints.

We used U-parameters, Ud,X and Up,X for the d and p orbitals, respectively, of the X element as follows (in eV): Ud,Zn =10.5, Ud,Ga =6.5, Ud,T i =9.25, Ud,Cr = Ud,Co = Ud,Ni = Ud,Cu =5.5, Ud,Mn =5.0 Up,O =7.0, Up,N =4.25.

In Figure 1 we show the supercells used in our calculations for codoped ZnO and GaN systems. Magnetic moments of all the atoms in the unit cells were obtained by integrating the difference of (spin-up minus spin-down) electron density of states within the atomic volumes as outlined in the VASP code. More details can be found in Refs.(Andriotis et al., 2003; Andriotis, Sheetz, Richter, & Menon, 2005; Andriotis, Sheetz, & Menon, 2010; Andriotis et al., 2011; Andriotis & Menon, 2011; Lathiotakis, Andriotis, & Menon, 2008a; Lisenkov, Andriotis, Sheetz, & Menon, 2011; Andriotis & Menon, 2012).

3. PROCESSES UNDERLYING THE DEFECT-INDUCED DEFECT-MEDIATED MAGNETISM

It is well understood that the appearance of the (defect-)magnetism is the result of electron-electron correlations which manifest themselves by electronic processes such as charge transfers, (re)hybridizations, induced spin-polarizations, etc. In order to get a better understanding of the processes that underlie the appearance of defect-magnetism, we first examine the electronic properties that take place in monodoped systems (i.e., systems doped with one type of dopants). It is these processes which provide the magnetic moments. We then investigate the features that are introduced by codoping which, according to our findings, lead to the development of the FMC among the magnetic moments.

3.1 Processes Occurring Upon Monodoping

According to our investigations, the key role in the development of magnetism in DMSs and TMOs as well as in C-based materials is played by the electronic processes which are induced by a single defect. These refer to the strength of the charge transfer, the spin-polarization and the rehybridization processes which are induced by a defect mainly on its fist nearest neighbors (1nn). Depending on the strength of these processes, the magnetic defect-defect interaction may be of ferromagnetic (FM) or of antiferromagnetic (AFM) type.

In order to exemplify this, we consider the electronic features that are developed around a magnetic cationic impurity atom which substitutes a cation of the host material. As representative examples, we examine the electronic Figure 2: (a) Parallel spin-polarization induced by a Cu-dopant (which substitutes a Zn atom in a wurtzite ZnO lattice) on its 1nn oxygen atoms. (b) Antiparallel spin-polarization induced by a Mn-dopant (which substitutes a Ga atom in a wurtzite GaN lattice) on its 1nn nitrogen atoms.properties in two systems, namely that of ZnO in which one Zn atom is substituted by a Cu-atom and that of GaN in which a Ga atom is substituted by a Mn-atom.

In Figure 2a, we show the spin-polarization induced by a Cu-dopant (which substitutes a Zn atom in a wurtzite ZnO lattice) on its 1nn oxygen atoms. Similarly, in Figure 2b, we show the spin-polarization induced by a Mn-dopant (substituting a Ga atom in a wurtzite GaN lattice) on its 1nn nitrogen atoms. These figures demonstrate that some dopants (like Cu) induce spin-polarization on their 1nn which are

Figure 2.

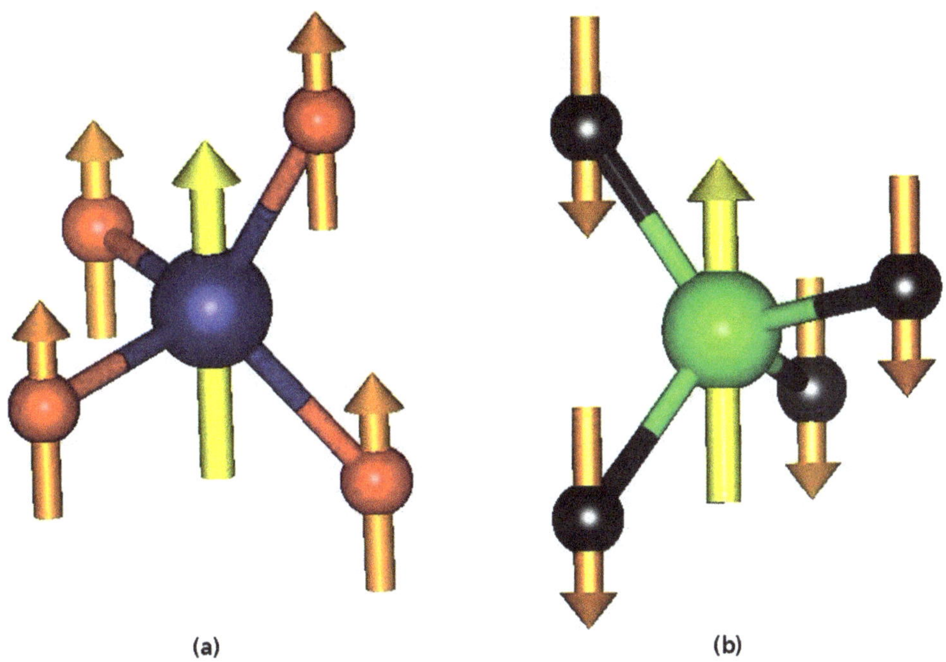

parallel to their own, while others (like Mn) induce antiparallel spin-polarization. This is a local effect that occurs in the neighborhood of each dopant as demonstrated in Ref. (Ye & Freeman, 2006). It depends on the structural relaxation around the dopant as well as on the atomic electronic configurations (namely the orbital filling factors) of the dopant and those of its surrounding 1nn (ligands). The bond strengths of the bonds which are developed between the dopant and its 1nn play a decisive role in the development of the magnetic interactions between the magnetic dopants as will be demonstrated in the following sections (Sec. 3.1.1 and Sec. 3.2).

3.1.1 Interaction Among Magnetic Neighbors -Monodoped Case

In Figure 3, we display a unit cell of the ZnO wurtzite structure in which we substituted two neighboring Zn atoms by a magnetic dopant (say, a Co-atom). Each dopant (Co-atom) with its own 1nn forms a spin-polarized neighborhood shown as a tetrahedron around it in Figure 3. As a result, the spin-polarized tetrahedron acts as an effective magnetic moment which as a unit interacts with Figure 3: Tetrahedral neighborhoods around magnetic dopants within a unit cell of ZnO wurtzite structure in which we substituted two neighboring Zn atoms by Co atoms (shown in turquoise). Zn and Oxygen atoms are shown in green and red color, respectively. The two tetrahedra share a common O-atom.similar effective magnetic moments formed around the other magnetic dopants.

Figure 3.

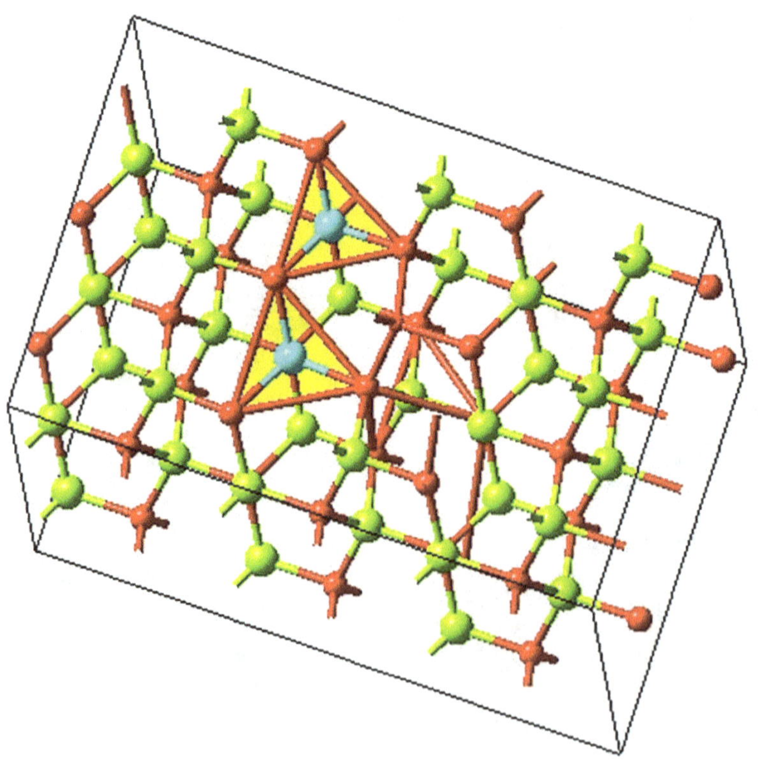

In the monodoped case, two neighboring dopants can develop magnetic interactions which are mediated by the anion which is a common neighbor to both of them. Thus, this interaction can be explained as a result of double exchange or superexchange or a p – d interaction or any other traditional model description which ignores all the ligands of the two dopants except their common one, while it deals solely on processes involving the triad: dopant – common anion – dopant. By contrast, the approach proposed by us generalizes this dopant –common anion –dopant picture by introducing the concept of the interaction among the effective magnetic moments which are developed around the magnetic dopants.

In general, the exchange interaction between two neighboring mono dopants is found to be of AFM type within the ZnO or the GaN hosts. However, with an increase in the mono dopant concentration, the exchange interaction becomes FM for bipartite dopant configurations(Lisenkov et al., 2011). This was found in the case of substitutional doping of ZnO by Mn atoms (see Table 1).

3.2 Processes Occurring Upon Codoping

The key processes which dominate in the development of the FMC among the magnetic moments are system specific; they depend on the symmetry of the host material, its electronic band structure, orbital hybridization, the formation of the impurity states/bands, the type of the codopants, their charge (cationic/anionic) and magnetic state and their concentration. However, our investigations seem to suggest a specific common route for these processes that can be found in most codoped systems. This refers to the development of the interaction between the spin-polarized neighborhoods described in Sec. 3.1. Additional common factors enhancing the interaction between the spin-polarized neighborhoods are also found and are attributed to defect distributions which exhibit bipartite configurations. All these features introduced by codoping are discussed in the following sections (Secs. 3.2.1 and 3.2.2).

3.2.1 Interaction Among Magnetic Neighbors -Codoped Case

In a codoped system, one of the codopants mediates the magnetic interaction between the other type of the codopants. In this case, we are looking at the electronic processes that take place in segments connecting codopants among themselves. For example, if (Co,Cu) are used as codopants in ZnO, we are looking at processes occurring in the segments such as Co-O-Cu-O-Co.

In such a segment, the key role in the Co-Co interaction is sought in the spin-polarization induced on the O-atoms by both the Co atoms on their ligands including the codopant Cu atom. As we discussed in Sec. 3.1, each dopant polarizes its own 1nn anions in a specific way. When the anion is a common neighbor to two dissimilar dopants and experiences the fields of the codopants simultaneously, its resulting polarization depends on the relative strength of the interactions of the two codopants. The stronger the dopant-anion interaction of a dopant is, the more probable it appears to be for the spin-polarization of this dopant to prevail over the other.

Whether the Co-Co magnetic interaction will be of FM or AFM type depends on whether the spin-polarization induced on the common anion by one of the dopants (a Co-atom) is consistent with the spin-polarization the other codopant (the Cu-atom) induces on it.

In the Co-O-Cu-O-Co segment within the ZnO lattice, both Co and Cu induce spin-polarization on their 1nn which is parallel to their own. Thus, the magnetic moment of the Cu atom will be aligned parallel to that of the nearby Co atom as this alignment is consistent with the energetically favorable state of Cu within the ZnO lattice. In turn, the Cu atom will induce parallel polarization to the other nearby Co

Figure 4.

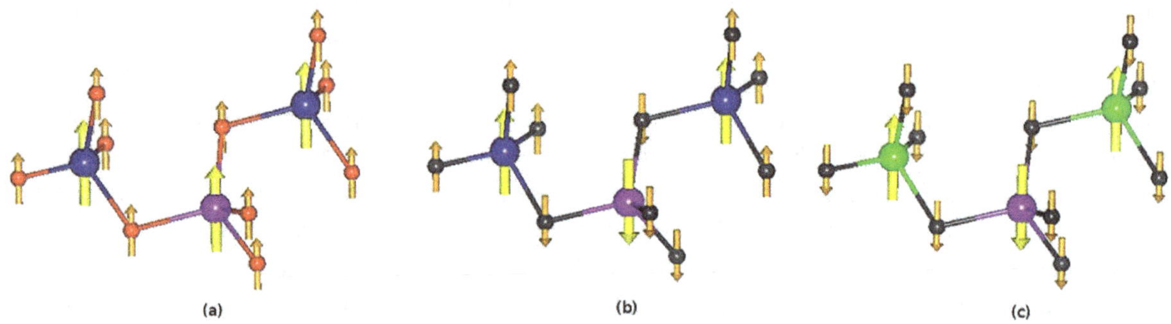

atom leading to an FM type Co-Co interaction as illustrated in Figure 4a. This picture of the successive spin-polarization appears to support, in the same way, the FM coupling among the substitution of Co impurities in graphene(Lisenkov, Andriotis, & Menon, 2012).

It is worth noting that in the AFM configuration Co↑ – O – Cu – O – Co↓ within the ZnO host, the Cu atom appears without any spin polarization. This Figure 4: A close up view of the spin-polarized portions of (a) Zn(Co,Cu)O, (b) Ga(Co,Cu)N and (c) Ga(Mn,Cu)N in their FM configurations. The coloring of Co, Cu and O atoms are the same as in Figure 1. Mn atoms are colored green, while N atoms are shown in black.seems to be quite a general feature as it was also found in the case of Mn-N-CuN-Mn segment in GaN as well as in other systems. On the other hand, in the FM configuration of Co↑ – O – Cu↑ – O – Co↑, not only the Cu atom becomes spin-polarized but it also undergoes a splitting in its d-band(Lathiotakis, Andriotis, & Menon, 2008b).

The type of interaction described above between the codopants is similar in spirit to the generalized superexchange which we had termed as the grand superexchange interaction(Andriotis et al., 2011).

Thus, our findings, while not ruling out the possibility of a carrier mediated coupling (e.g., double exchange, superexchange, etc) among the magnetic cations, nevertheless demonstrate that the FMC accommodates additional significant contributions derived from localized spin-polarizations of the anions induced by the codopants(Andriotis & Menon, 2011). In this view, the non magnetic codopants play a multi-modal role; they become hybridized (with their surrounding ions), while spin polarizing themselves and, in turn, spin polarize their 1st n anions. This facilitates the development of localized spin-polarized molecular orbitals (MOs) over a more extended range. In this picture it is not necessary to have all the 1nn anions to a codopant to be spin-polarized in the same direction.

This is illustrated in Figure 4a where a closeup image of the spin polarized portion of the codoped ZnO system is shown.

3.2.2 Processes Enhancing the Synergy of Codopants

Our extensive investigations have further revealed that bipartite configurations of the codopants can significantly improve the magnetism in these materials(Andriotis et al., 2003, 2005, 2010, 2011; Andriotis & Menon, 2011). Specifically, our findings suggest that the defects activate potential (existing or hidden) bipartite structures and make them amenable to magnetic features as found in a wide range of materials(Lieb, 1989; Ovchinnikov, 1978; Roth, 1966; Kawamoto, 1997; Kawamoto & Suzuki, 1997; Narymbetov et al., 2000; McConnell, 1967, 1963; Breslow, Jaun, Kluttz, & Xia, 1982). Codoping appears as one of the fundamental processes which provides the means to achieve this and can be viewed as a generic approach for the development of magnetism in materials which exhibit potential bipartite structure. This view is entirely different from those views according to which the codoping has been considered as a possible way to tailor the position and the occupancy of the Fermi energy, (EF), of the DMSs(Reed et al., 2005; Kittilstved, Norberg, & Gamelin, 2005; Ozaki et al., 2005; Kane et al., 2006; Ozaki et al., 2006; Kuroda et al., 2007).

More interestingly, bipartite configurations of alternate codopant pathways (of the form... BABABAB. . .) were found to facilitate the localization of the spin polarized MOs and enhance the appearance of the induced magnetism.

4. OTHER EXAMPLE APPLICATIONS

In Table 1, we have included computational results obtained by ab initio methods described in Sec. 2. Our results refer to monodoped and codoped systems and demonstrate our model approach to the defect-induced defect-mediated magnetism.

Three host systems have been chosen, namely the ZnO, GaN and TiO2. As codopants we used transition metal (TM) atoms of the 3d-series. As mono dopants, we used Co in all three hosts. Additionally we performed calculations using Mn and Cu in ZnO.

Our calculations are spin-non-restricted and their outputs converge to the energetically more favorable spin-configuration. A number of spin restricted calculations were used to verify our results and these results are included in Table 1 in terms of the energy difference EFM -EAF M . A negative value for EFM -EAF M indicates FM configuration.

As shown in Table 1, two neighboring mono dopants experience exchange interaction of AFM type in all three hosts. However, as mentioned in Secs. 3.1 and 3.1.1, the AFM exchange interaction could turn to FM type for larger mono dopant concentrations and bipartite dopant configurations as found for Mn in ZnO.

The fact that the AFM coupling prevails in the interaction between two neighboring dopants indicates that the codopant weakens the superexchange interaction in favor of the successive spin-correlation term. Thus, the AFM interaction between two mono dopants can turn to FM type if mediated by a codopant. We demonstrated this with calculations for the following codoped systems:

- In ZnO host with codopants: (Co,Cu) and (Ni,Ag)
- In GaN host with codopants: (Co,Cu), (Cr,Ni), (Ti,Zn), (Ti,Ni), (Co,Ni), (Co,Mn) and (Mn,Cu).
- In TiO2 host with codopants: (Zn,Cu), (Co,Cr), (Co,Cu), (Mn,Cu) and (Cr, Ni).

Table 1. Ab initio computational results. [1]The upper indexes plus/minus denote that the magnetic moment of the corresponding element is very small (positive/negative) in the up/down orientation respectively. [2]Zn atoms are not polarized. [3]The upper index (zero) denotes that the element is not polarized. [4]Another FM configuration was found in which the Cr atoms are not polarized..

All our results are included in Table 1.

Our results for Ga(Co,Cu)N and Ga(Mn,Cu)N are shown in Figures 4b and 4c, respectively. These figures show that in the Cu4 tetrahedron, the Cu-atom and its 1nn N-atoms are spin-polarized ferromagnetically, in accordance with the spin-polarizations that take place within the MnN4 units and thus enabling the FM interaction in Ga(Mn,Cu)N. On the other hand, in the case of AFM coupling of the Mn-atoms of the Mn-N-Cu-N-Mn segment, it is found that all the atoms of the Cu4 unit exhibit no-spin polarization. Similar to the Zn(Co,Cu)O case, we find that the Ga(Co,Cu)N and Ti(Co,Cr)O2 systems also exhibit FMC(Andriotis et al., 2011; Andriotis & Menon, 2011, 2012). The strength of the FMC is found to depend on the way the codopant segments evolve within the host lattices(Andriotis & Menon, 2012).

5. CONCLUSION

Our findings support the spin correlations of localized character contributing to the FMC among the codopants. These result from localized electronic processes that take place in the neighborhood of the codopants. We also note that various recent reports support the contribution to FMC resulting from local processes that take place in the neighborhood of the codopants. For example, Ye et al(Ye & Freeman, 2006) found that, in the case of ZnO doped with Cu, the magnetic moments are localized within the CuO4 tetrahedron with a ferromagnetic coupling between Cu and O. The ferromagnetism emerges from the hybridization of the Cu-3d and O-2p bands. Similarly, in Li doped VO2, the very unusual low-spin VO2 behavior was attributed to the formation of nanosize interacting ferromagnetic spin clusters around intercalated Li ions(Popa et al., 2009). Finally, the localized character is inherent in the recently proposed two-center assisted model of blocked superparamagnetism proposed by Pemmaraju et al.(Pemmaraju, Hanafin, Archer, Braun, & Sanvito, 2008).

In conclusion, our work demonstrates that in DMSs and TMOs the defect-induced defect-mediated magnetism is facilitated and enhanced by codoping, especially in the case in which the codopants form segments exhibiting bipartite configurations(Andriotis et al., 2003, 2005; Lathiotakis et al., 2008a; Andriotis et al., 2010; Lisenkov et al., 2011; Andriotis et al., 2011; Andriotis & Menon, 2011, 2012). The defect induced magnetism appears as a defect synergistic process characterized by defect-induced and defect mediated processes. However, the underlying leading electronic process for the development of the defect magnetism is system specific. In particular, the origin of magnetism in DMSs and TMOs appears to be the result of an interplay between correlated spin-polarization processes that take place in a successive way within neighborhoods centered at the codopants and include their 1nn's. The successive spin-correlated local processes proposed in the present work appear to introduce new contributions underlying the magnetism in diluted magnetic semiconductors and related magnetic materials. The synergistic action of the codopants lead to the reduction of the superexchange interaction, while enhancing the FM coupling between the magnetic dopants.

The proposed codoping mechanism can be considered as a general and generic approach that can be used to justify the defect induced magnetism in a diverse set of non-magnetic systems.

ACKNOWLEDGMENT

MM acknowledges primary funding support from DOE (DE-FG02-07ER46375). The support from NSF SOLAR project (DMS 1125909) was used to develop the computational techniques. A.N.A. acknowledges that this research has been co-financed by the European Union (European Social Fund -ESF) and Greek national funds through the Operational Program 'Education and Lifelong Learning' of the National Strategic Reference Framework (NSRF) -Research Funding Program: THALES. Investing in knowledge society through the European Social Fund.

REFERENCES

Andriotis, A. N., Lisenkov, S., & Menon, M. (2011). Ferromagnetic interactions in hosted bipartite materialsgeneralized-double-exchange and generalized superexchange interactions. *Journal of Physics Condensed Matter*, 23.

Andriotis, A. N., & Menon, M. (2011). Magnetic coupling in dilute magnetic semiconductors: A new perspective.[b]. *Phys. St. Sol.*, *248*, 2032.

Andriotis, A. N., & Menon, M. (2012). The synergistic character of the defect-induced magnetism in diluted magnetic semiconductors and related magnetic materials. *Journal of Physics Condensed Matter*, *24*, 455801. PMID:23086142

Andriotis, A. N., Menon, M., Sheetz, R. M., & Chernozatonskii, L. (2003). Magnetic properties of c60 polymers. *Physical Review Letters*, *90*, 026801. PMID:12570566

Andriotis, A. N., Mpourmpakis, G., Lisenkov, S., Sheetz, R. M., & Menon, M. (2013). U-calculation of the lsda+u functional using the hybrid b3lyp and hse functionals. *Phys. Status Solidi B*, *250*, 356.

Andriotis, A. N., Sheetz, R. M., & Menon, M. (2010). The synergistic character of the defect-induced magnetism in diluted magnetic semiconductors and related magnetic materials. *Journal of Physics Condensed Matter*, *22*, 324210. PMID:23086142

Andriotis, A. N., Sheetz, R. M., Richter, E., & Menon, M. (2005). Defect-originated magnetism in carbon-based and non-traditional inorganic compounds: A new class of magnetic materials. *Europhysics Letters*, *72*, 658.

Assadi, M. H. N., Zhang, Y. B., & Li, S. (2009). N codoping induced ferromagnetism in zno:co (1010) thin films. *Journal of Applied Physics*, *106*, 093911.

Behan, A. J., Mokhtari, A., Blythe, H. J., Score, D., Xu, X. H., Neal, J. R., & Gehring, G. A. et al. (2008). Two magnetic regimes in doped zno corresponding to a dilute magnetic semiconductor and a dilute magnetic insulator. *Physical Review Letters*, *100*, 047206. PMID:18352328

Breslow, R., Jaun, B., Kluttz, R. Q., & Xia, C. Z. (1982). Ground state pi-electron triplet molecules of potential use in the synthesis of organic ferromagnets. *Tetrahedron*, *38*, 863.

Chattopadhyay, S., & Natha, T. K. (2010). Room temperature enhanced positive magnetoresistance in pt and carrier induced zn(fe)o and zn(fe,al)o dilute magnetic semiconductors junction. *Journal of Applied Physics*, *108*, 083904.

Cococcioni, M., & de Gironcoli, S. (2005). Linear response approach to the calculation of the effective interaction parameters in the lda+u method. *Physical Review B: Condensed Matter and Materials Physics*, *71*, 035105.

Duan, L. B., Rao, G. H., Wang, Y. C., Yu, J., & Wang, T. (2008). Magnetization and raman scattering studies of (co,mn) codoped zno nanoparticles. *Journal of Applied Physics*, *104*, 013909.

Dudarev, S. L., Botton, G. A., Savrasov, S. Y., Humphreys, C. J., & Sutton, A. P. (1998). Electron-energy-loss spectra and the structural stability of nickel oxide:an lsda+u study. *Physical Review B: Condensed Matter and Materials Physics*, *57*, 1505.

Gai, Y., Li, J., Li, S. S., Xia, J. B., & Wei, S. H. (2009). Design of narrow-gap tio2: A passivated codoping approach for enhanced photoelectrochemical activity. *Physical Review Letters*, *102*, 036402. PMID:19257373

Han, S. J., Song, J. W., Yang, C. H., Park, S. H., Park, J., Jeong, Y. H., & Rhie, K. W. (2002). A key to room-temperature ferromagnetism in fe-doped zno. *Cu. Appl. Phys. Lett.*, *81*, 4212.

Jayakumar, O. D., Gopalakrishnan, I. K., & Kulshreshtha, S. K. (2006). Surfactant-assisted synthesis of co-and li-doped zno nanocrystalline samples showing room-temperature ferromagnetism. *Advanced Materials*, *18*, 1857.

Kane, M. H., Strassburg, M., Fenwick, W. E., Asghar, A., Payne, A. M., Gupta, S., & Ferguson, I. T. et al. (2006). Correlation of the structural and ferromagnetic properties of ga1-xmnxn grown by metalorganic chemical vapor deposition. *Journal of Crystal Growth*, *287*, 591.

Kawamoto, T. (1997). A theoretical model for ferromagnetism of tdae-c60. *Sol. St. Commun.*, *101*, 231.

Kawamoto, T., & Suzuki, N. (1997). Pressure effects in ferromagnetism of orbital ordering ferromagnet. *Phys. Soc. Japan*, *66*, 2487.

Khatri, P., & Huda, M. N. (2014). Application of attractive potential by dft + u to predict the electronic properties of materials without highly localized bands. *Computational Materials Science*, *81*, 290.

Kittilstved, K. R., Norberg, N. S., & Gamelin, D. R. (2005). Chemical manipulation of high-tc ferromagnetism in zno diluted magnetic semiconductors. *Physical Review Letters*, *94*, 147209. PMID:15904107

Kresse, G., & Hafner, J. (1993). Ab initio molecular dynamics for liquid metals. *Physical Review B: Condensed Matter and Materials Physics*, *47*, 558. PMID:10004490

Kresse, G., & Joubert, D. (1999). From ultrasoft pseudopotentials to the projector augmented-wave method. *Physical Review B: Condensed Matter and Materials Physics*, *59*, 1758.

Kuroda, S., Nishizawa, N., Takita, K., Mitime, M., Bando, Y., Osuch, K., & Dietl, T. (2007). Origin and control of high-temperature ferromagnetism in semiconductors. *Nature Materials*, *6*, 440. PMID:17515918

Lathiotakis, N. N., Andriotis, A. N., & Menon, M. (2008a). Codoping: A possible pathway for inducing ferromagnetism in zno. *Physical Review B: Condensed Matter and Materials Physics*, *78*, 193311.

Lathiotakis, N. N., Andriotis, A. N., & Menon, M. (2008b). Codoping: A possible pathway for inducing ferromagnetism in zno. *Physical Review B: Condensed Matter and Materials Physics*, *78*, 193311.

Lieb, E. H. (1989). Two theorems on the hubbard model. *Physical Review Letters*, *62*, 1201. PMID:10039602

Lin, H. T., Chin, T. S., Shih, J. C., Lin, S. H., Hong, T. M., Huang, R. T., & Kai, J. J. et al. (2004). Enhancement of ferromagnetic properties in zn1-xcoxo by additional cu doping. *Applied Physics Letters*, *85*, 621.

Lin, Y. H., Ying, M., Li, M., Wang, X., & Nan, C. W. (2007). Room-temperature ferromagnetic and ferroelectric behavior in polycrystalline zno-based thin films. *Applied Physics Letters*, *90*, 222110.

Lisenkov, S., Andriotis, A. N., & Menon, M. (2012). Magnetic anisotropy and engineering of magnetic behavior of the edges in co embedded graphene nanoribbons. *Physical Review Letters*, *108*, 187208. PMID:22681116

Lisenkov, S., Andriotis, A. N., Sheetz, R. M., & Menon, M. (2011). Effects of codoping on the ferromagnetic enhancement in zno. *Physical Review B: Condensed Matter and Materials Physics*, *83*, 235203.

McConnell, H. M. (1963). Ferromagnetism in solid free radicals. *The Journal of Chemical Physics*, *39*, 1910.

McConnell, H. M. (1967). Informal remarks. *Proc. R. A. Welch Found. Chem. Res.*, *11*, 144.

Mi, W. B., Bai, H. L., Liu, H., & Sun, C. Q. (2007). Microstructure, magnetic, and optical properties of sputtered mn-doped zno films with high-temperature ferromagnetism. *Journal of Applied Physics*, *101*, 023904.

Naeem, M., Hasanain, S. K., Afgan, S. S., & Rumaiz, A. (2008). Competing effects of cu ionic charge and oxygen vacancies on the ferromagnetism of (zn, co)o nanoparticles. *Journal of Physics Condensed Matter*, *20*, 255223.

Narymbetov, B., Omerzu, A., Kabanov, V. V., Tokumoto, M., Kobayashi, H., & Michailovic, D. (2000). Origin of ferromagnetic exchange interactions in a fullereneorganic compound. *Nature*, *407*, 883. PMID:11057661

Norton, D. P., Pearton, S. J., Hebard, A. F., Theodoropoulou, N., Boatner, L. A., & Wilson, R. G. (2003). Ferromagnetism in mn-implanted zno:sn single crystals. *Applied Physics Letters*, *82*, 239.

Ovchinnikov, A. A. (1978). Multiplicity of the ground state of large alterrnant organic molecules with conjugated bonds. *Theoret. Chim. Acta (Berl.)*, *47*, 297.

Ozaki, N., Nishizawa, N., Marcet, S., Kuroda, S., Eryu, O., & Takita, K. (2006). Significant enhancement of ferromagnetism in zn1xcrxte doped with iodine as an n-type dopant. *Physical Review Letters*, *97*, 037201. PMID:16907538

Ozaki, N., Okabayashi, I., Kumekawa, T., Nishizawa, N., Kuroda, S., & Takita, K. (2005). Suppression of ferromagnetism due to hole doping in zn1xcrxte grown by molecular-beam epitaxy. *Applied Physics Letters*, *87*, 192116.

Park, M. S., & Min, B. I. (2003). Ferromagnetism in zno codoped with transition metals: Zn1-x(feco)xo and zn1-x(fecu)xo. *Physical Review B: Condensed Matter and Materials Physics*, *68*, 224436.

Pemmaraju, C. D., Archer, T., Sanchez-Portal, D., & Sanvito, S. (2007). Atomic-orbital-based approximate self-interaction correction scheme for molecules and solids. *Physical Review B: Condensed Matter and Materials Physics*, *75*, 045101.

Pemmaraju, C. D., Hanafin, R., Archer, T., Braun, H. B., & Sanvito, S. (2008). Impurity-ion pair induced high-temperature ferromagnetism in co-doped zno. *Physical Review B: Condensed Matter and Materials Physics*, *78*, 054428.

Popa, A. I., Vavilova, E., Arango, Y. C., Kataev, V., Taschner, C., Klauss, H. H., & Klingeler, R. et al. (2009). High-temperature ferromagnetism of li-doped vanadium oxide nanotubes. *Europhysics Letters*, *88*, 57002.

Reed, M. J., Arkun, F. E., Berkman, E. A., Elmasry, N. A., Zavada, J., Luen, M. O., & Bedair, S. M. et al. (2005). Effect of doping on the magnetic properties of gamnn: Fermi level engineering. *Applied Physics Letters*, *86*, 102504.

Roth, L. M. (1966). Simple narrow-band model of ferromagnetism due to intraatomic exchange. *Physical Review*, *149*, 306.

Santos, J. P. T., Marques, M., Teles, L. K., & Ferreira, L. G. (2010). Antiferromagnetism with spin polarization of gan-based diluted magnetic semiconductors. *Physical Review B: Condensed Matter and Materials Physics*, *81*, 115209.

Sato, K., Bergqvist, L., Kudrnovsky, J., Dederichs, P. H., Eriksson, O., Turek, I., & Zeller, H. K. R. et al. (2010). First-principles theory of dilute magnetic semiconductors. *Reviews of Modern Physics*, *82*, 1633.

Shin, J. M., Lee, H. S., Cha, S. Y., Lee, S., Kim, J. Y., Park, N., & Jeong, S. et al. (2012). Strong ferromagnetism in pt-coated zncoo: The role of interstitial hydrogen. *Applied Physics Letters*, *100*, 172409.

Solovyev, I. V., Dederichs, P. H., & Anisimov, V. I. (1994). Corrected atomic limit in the localdensity approximation and the electronic structure of d impurities in rb. *Physical Review B: Condensed Matter and Materials Physics*, *50*, 16861. PMID:9976082

Sunkara, S., KalyanVendra, V., Jasinski, J., Deutsch, T., Andriotis, A. N., Menon, M., & Sunkara, M. (2014). New visible light absorbing materials for solar fuels, ga(sbx)n1−x alloy. *Advanced Materials*. doi:10.1002/adma.201305083

Wang, Q., Sun, Q., Jena, P., & Kawazoe, Y. (2004). Carrier-mediated ferromagnetism in n codoped (zn,mn)o (1010) thin films. *Physical Review B: Condensed Matter and Materials Physics*, *70*, 052408.

Yadav, M. K., Sanyal, B., & Mookerjee, A. (2009). Stabilization of ferromagnetism in mn doped zno with c co-doping. *Journal of Magnetism and Magnetic Materials*, *321*, 273.

Ye, L. H., & Freeman, A. J. (2006). Defect compensation, clustering, and magnetism in cr-doped anatase tio2. *Physical Review B: Condensed Matter and Materials Physics*, *73*, 081304.

Zhu, W., Qiu, X., Iancu, V., Chen, X. Q., Pan, H., Wang, W., & Zhang, Z. et al. (2009). Band gap narrowing of titanium oxide semiconductors by noncompensated anion-cation codoping for enhanced visible-light photoactivity. *Physical Review Letters*, *103*, 226401. PMID:20366114

KEY TERMS AND DEFINITIONS

Ab Initio Methods: Computational methods free of any empirical parametrization.

Band Gap: Range of not-allowed electron energies separating valence and conduction bands of a semiconductor.

Band Gap Edges: The lowest and highest energies of the band gap.

Bipartite Lattice: It is a lattice whose atoms can be subdivided into two groups, denoted by (A) and (B), respectively, so that each atom A is surrounded only by B atoms, and vice versa.

Diluted Magnetic Semiconductors (DMS): Materials combining ferromagnetic and semiconducting properties.

Doping: Introduction of impurity or structural defects in a material.

Photoelectrochemical (PEC) Water Splitting: Separation of water elements (oxygen and hydrogen) with the help of sunlight.

Redox Potentials: A measure of the tendency of a material to accept/donate electrons.

Transition Metal Oxides (TMO): Materials consisting of transition metals and oxygen.

Chapter 3
Ab Initio–Based Stochastic Simulations of Kinetic Processes on Surfaces

Zhengzheng Chen
California State University – Northridge, USA

Chao Wu
Xi'an Jiaotong University, China

ABSTRACT

We briefly present the theoretical framework of a hierarchical multi-scale approach, which is an ab initio-based stochastic method, and its applications to several chemical/physical kinetic processes on metallic surfaces. We first introduce necessary theoretical basis of ab initio and Monte Carlo (MC) methods, and then illustrate different Monte Carlo algorithms for important ensembles, including canonical and grand canonical ensembles. In the following section, we describe two important protocols which are essential to integrate ab initio data and MC models. Two examples are presented in order to elucidate the power of this multi-scale approach. The first example focuses on the combination of kinetic Monte Carlo and transition state theory. We discuss the detailed processes of performing kinetic Monte Carlo simulation on atomic diffusion on alloyed surface, including some technical aspects. In the second example, we presents a different way to account for the local environment-sensitive metal-catalyzed O_2 dissociation reactions using combinatory techniques including cluster expansion and grand canonical Monte Carlo methods. This approach provides steady-state rates and rate derivatives that are comparable with experiments. Moreover, the connection between the feasible mechanisms and the observed kinetic behaviors can now be built.

DOI: 10.4018/978-1-5225-0290-6.ch003

INTRODUCTION

Integrated computational simulations have been experiencing booming developments in the last decade. The concept is originated from the requirements of understanding innovative properties of materials with artificial structures and guiding synthesis of high-performance materials at nano-scale. Contemporary computational materials science demands detailed information of electronic structures with high accuracy and the ability to describe dynamical processes of phenomena covering large spatial and temporal scale, such as material fatigue, heterogeneous catalysis, and crystal growth, etc. Although traditional quantum physics/chemistry methods and phenomenological models have been developed very maturely at both ends of the concept of integrated computational simulations, it is only within the recent ten years that researchers start bridging them together and apply the integrated methods to materials engineering (Reuter2003, Rojal2007). In prevailing studies, stochastic methods (mainly Monte Carlo methods) are employed to mimic the complex nature of kinetics, i.e. multiple evolution options from the current state. Key dynamical parameters are then determined by atomistic simulations, e.g., density functional theory (DFT) or molecular dynamics. Piana *et al.* employed molecular dynamics to obtain 40 key kinetic parameters and performed simulations on urea crystal growth in two different solutes. (Piana2005). Honkala *et al.* used *ab initio* calculations to determine reliable parameters even including local chemical environments to calculate the productivity of ammonia synthesis catalyzed by nanoparticle of Ru (Honkala2005). Their calculated turn over frequencies are in good agreement with measured ones. *Ab initio*-based stochastic simulations also have been applied to study mechanical failure of materials (Fu2004), self organization (Mo2005, Negulyaev2008, Ziegler2008), thin film deposition (Huang1998), and electrochemical processes in solid oxide fuel cells (Pornprasertsuk2007, Wang2010), etc. Computational simulations are able to present detailed insight of phenomena or study effects of various factors which are very difficult or even impossible to be obtained by experiments. For example, Zhu *et al.* performed systematic kinetic Monte Carlo simulations to study effects of external stress on the growth of quantum dots on substrates (Zhu2007).

However, due to the diversity of the properties of materials, processes, and purposes of simulations, corresponding models are very different from one to another. General requirements of models are: (1) the model should be able to describe critical steps of kinetics, i.e., the rate determining steps of chemical reactions or atomic piling ups during quantum dot growths, etc., and (2) the model should contain reasonable number of parameters which can be handled by *ab initio* calculations. Different models and/or algorithms have been developed to meet different simulation goals and interested systems. For instance, Voter developed “hyperdynamics” to expedite the system evolution (Voter1997). Henkelman and Jónsson developed on-the-fly dynamical simulation method so that the simulator is able to automatically determine and update the migration list during simulations (Henkelman2001). Parallel algorithms of kinetic Monte Carlo methods have also been developed by several groups (Shim2005, Martínez2011). These developments essentially strengthen the powers of this hierarchical multi-scale approach. In this chapter, we introduce basic theory and several key algorithms of *ab initio*-based stochastic methods. Two examples are presented in order to illustrate the general processes of these stochastic methods, including generating models, incorporating *ab initio* data into models, analyzing outputs, etc. Both examples describe the morphological evolution and chemical/physical kinetic processes on surfaces since surface has become a very important medium in modern science and technology.

THEORETICAL BASIS AND ALGORITHMS

Density Functional Theory and Kohn-Sham Equation

In most cases, computational solid state physics deals with n-body systems with n~10^{23}. By using the well-known Born-Oppenheimer approximation, an n-body system can be described by the following equation:

$$\left\{-\sum_{i=1}^{n}\frac{\nabla_i^2}{2}+\frac{1}{2}\sum_{i,j,i\neq j}^{n}\frac{1}{r_{ij}}-\sum_{I=1}^{N}\sum_{i=1}^{n}\frac{Z_I}{r_{iI}}\right\}\varphi\left(\{\vec{r}_i\}\right)\Big|_{\{\vec{R}_I\}}=\varepsilon\left(\{\vec{R}_I\}\right)\varphi\left(\{\vec{r}_i\}\right)\Big|_{\{\vec{R}_I\}} \tag{1.1}$$

where $\varphi\left(\{\vec{r}_i\}\right)$ is an eigen-wavefunction, ε is the corresponding eigen-energy, and positions of atoms are fixed at $\left\{\vec{R}_I\right\}$. Equation (1.1) is very difficult to solve due to the long-range interaction term $1/r_{ij}$. In modern computational methods, DFT is employed to overcome this complex situation (Hohenberg1964, Kohn1965). DFT rigorously transfers Equation (1.1) to a set of single body equations. In DFT, the ground state energy of a system is determined only by its ground state charge distribution through a universal functional (Hohenberg1964):

$$E^{\mathrm{HK}}\left[\rho(\vec{r});V_{ext}(\vec{r})\right]=T\left[\rho(\vec{r})\right]+E_{ee}\left[\rho(\vec{r})\right]+\int V_{ext}(\vec{r})\rho(\vec{r})d\vec{r}+E_{II} \tag{1.2}$$

Total energy is composed of kinetic energy, electron-electron interaction, electron-ion interaction, and ion-ion interaction. The first three terms depend on charge distribution $\rho(\vec{r})$. E_{II} is a constant for a given atomic configuration. Based on the variational principle, one can get the minimum of E^{HK} by taking the variation of Equation (1.2) with respect to ρ if the analytical expression of this universal functional is known. Unfortunately, by now we do not know it for interacting electron-gas systems.

To bypass this difficulty, Kohn and Sham stepped back a little. They again employed wavefunctions $\phi_j(\vec{r})$ to describe systems. They also have proposed an ansatz: for a real system with charge distribution $\rho(\vec{r})$, there is a virtually non-interacting electron gas system with charge distribution $\rho_0(\vec{r})$ identical to $\rho(\vec{r})$ (Kohn 1965). This $\rho_0(\vec{r})$ can be expresses by $\phi_j(\vec{r})$:

$$\rho_0(\vec{r})=\sum_i\phi_i^*(\vec{r})\phi_i(\vec{r})=\rho(\vec{r}) \tag{1.3}$$

The kinetic energy functional of this virtual system, $T_0\left[\rho(\vec{r})\right]$, can be analytically calculated. Kohn and Sham also calculated Hartree interaction $E_H\left[\rho(\vec{r})\right]$ instead of $E_{ee}\left[\rho(\vec{r})\right]$:

$$E_H\left[\rho(\vec{r})\right]=\int\frac{\rho(\vec{r})\rho(\vec{r}\,')}{|\vec{r}-\vec{r}\,'|}d\vec{r}d\vec{r}\,' \tag{1.4}$$

Apparently, $T_0\left[\rho(\vec{r})\right]+E_H\left[\rho(\vec{r})\right]$ differs from $T\left[\rho(\vec{r})\right]+E_{ee}\left[\rho(\vec{r})\right]$ in Equation (1.2). To compensate it, we need an extra term:

$$E_{\mathrm{xc}}\left[\rho\right]=T\left[\rho\right]+E_{ee}\left[\rho\right]-T_0\left[\rho\right]-E_H\left[\rho\right] \tag{1.5}$$

$E_{\mathrm{xc}}\left[\rho\right]$ is called the exchange-correlation functional. It contains all the many-body effects. Using Equation (1.2) to Equation (1.5), and taking the variation of E^{HK} with respective to $\phi_j(\vec{r})$ under the constraint condition:

$$\sum_j \left\langle \phi_j \middle| \phi_j \right\rangle = N \tag{1.6}$$

We can obtain the equation whose eigen-wavefunction $\phi_j(\vec{r})$ satisfies:

$$\left[-\frac{\nabla^2}{2}+V_H(\vec{r})+V_{ext}(\vec{r})+V_{xc}(\vec{r})\right]\phi_j(\vec{r})=\varepsilon_j\phi_j(\vec{r});\quad V_{xc}(\vec{r})=\frac{\delta E_{xc}\left[\rho(\vec{r})\right]}{\delta\rho(\vec{r})} \tag{1.7}$$

Equation (1.7) is the famous Kohn-Sham (KS) equation. Since $V_H(\vec{r})$ and $V_{xc}(\vec{r})$ require information of $\rho(\vec{r})$, the KS equation has to be solved self-consistently. This is both the power and the limitations of *ab initio* method. The KS equation does not require experienced parameters other than the basic parameters of elements. It is therefore able to deal with materials in a wide range with high accuracy. However, it is very time-consuming and therefore can only treat very small systems. By solving the KS equation, the ground energy $E_0\left[\rho\right]$ is usually expressed as

$$E_0=\sum_j \varepsilon_j-\frac{1}{2}\int\frac{\rho(\vec{r})\rho(\vec{r}\,')}{|\vec{r}-\vec{r}\,'|}d\vec{r}d\vec{r}\,'+E_{xc}\left[\rho(\vec{r})\right]-\int V_{xc}(\vec{r})\rho(\vec{r})d\vec{r} \tag{1.8}$$

$\rho(\vec{r})$ is calculated by Equation (1.3). The first term on the right side of Equation (1.8) is called the band structure energy and the other three terms are called double counting terms.

It is worth emphasizing that $E_{xc}\left[\rho\right]$ is usually expressed as

$$E_{xc}[\rho]=\int \varepsilon_{xc}[\rho(\vec{r})]\rho(\vec{r})d\vec{r} \tag{1.9}$$

which is called the local density approximation (LDA). ε_{xc} divides into two components: $\varepsilon_{xc}=\varepsilon_x+\varepsilon_c$. ε_x can be expressed analytically (Martin 2004). ε_c has been numerically obtained by quantum Monte-Carlo (QMC) method (Ceperley1980). In order to incorporate ε_c into the KS equation, numerical data have been fitted by various analytic functions (Perdew1981, Lee1988, Perdew1992, Perdew1996).

Introduction to Monte Carlo Method and Metropolis Sampling

Monte Carlo method (MC) is a kind of stochastic method which has been widely used in science and engineering. It treats the working system (atomistic systems in this chapter) as one in an ensemble, checks status of every system in the same ensemble following certain rules, and then calculates the ensemble average of specified observables, which usually are the experimental measurable properties. Therefore, if the information of a given system, i.e., atomic interactions, state equations, etc., is available from experiments or *ab initio* calculations, MC is a very powerful tool to bridge the gap between microscopic simulations and macroscopic observations.

Canonical Monte-Carlo Method

NVT-Ensemble

The most common application of MC is to the canonical (*NVT*) ensemble, in which the temperature T, volume V, and number of particles N are kept constant.

For an N-particle *NVT* ensemble, we can easily get its partition function Z_{NVT} (Huang1987):

$$Z_{NVT} = \frac{1}{N!h^{3N}} \int d\mathbf{r}^N d\mathbf{p}^N \exp\left[-\beta\left(\sum_{i=1}^{N}\frac{\mathbf{p}_i^2}{2m} + \mathcal{U}(\mathbf{r}^N)\right)\right] \tag{2.1}$$

in which h is the Plank constant, $U(\mathbf{r}^N)$ is the potential energy of the system, and $\beta=1/k_BT$. We integrate Equation (2.1) with respect to momentum, and get

$$Z_{NVT} = \frac{1}{N!\Lambda^{3N}} \int d\mathbf{r}^N \exp\left[-\beta\mathcal{U}(\mathbf{r}^N)\right] \tag{2.2}$$

where $\Lambda = h/\sqrt{2\pi m k_B T}$ is the de Broglie wavelength. It is easy to have the ratio of probabilities before and after the movement of a particle:

$$\frac{P_{\text{new}}}{P_{\text{old}}} = \frac{\exp\left[-\beta\mathcal{U}(\mathbf{r}_{\text{new}}^N)\right]}{\exp\left[-\beta\mathcal{U}(\mathbf{r}_{\text{old}}^N)\right]} \tag{2.3}$$

We can express the accepting probability of the movement of a particle as

$$A_{\text{Move}} = \min\left\{1, \exp\left(-\beta\Delta\mathcal{U}\right)\right\} \qquad \Delta\mathcal{U} = \mathcal{U}(\mathbf{r}_{\text{new}}^N) - \mathcal{U}(\mathbf{r}_{\text{old}}^N) \tag{2.4}$$

Based on above discussions, we have *Algorithm 1* for *NVT*-MC (Mackay1998):

Algorithm 1

1. Initialize a system: set up initial particle positions $\{\mathbf{r}_i^0\}$, T, V and target sampling number N_{MC}. Set $k = 0$, $\mathbf{r}_{\text{old}}^N = \{\mathbf{r}_i^0\}$;
2. $k = k+1$. Randomly choose a particle i to make a trial movement: $\mathbf{r}_{\text{new}}^N = \mathbf{r}_{\text{old}}^N + \Delta\mathbf{r}_i$. Note that $\Delta\mathbf{r}_i$ can be position, spin on site i, or element type on site i, etc. Calculate the energy change ΔU and A_{Move}. Generate a random number (RN) r_1 distributed uniformly in (0,1), and compare r_1 to A_{Move}. If $r_1 \leq A_{\text{Move}}$, the trial is accepted, set $\mathbf{r}_{new}^N = \mathbf{r}_{trial}^N$, otherwise, the trial is rejected, $\mathbf{r}_{new}^N = \mathbf{r}_{old}^N$;
3. Store $\mathbf{r}_k^N = \mathbf{r}_{new}^N$ and update $\mathbf{r}_{\text{old}}^N = \mathbf{r}_{\text{new}}^N$;
4. If $k<N_{MC}$, go back to step2, otherwise calculate thermo quantities of the ensemble by generated N_{MC} states.

Grand Canonical Monte-Carlo Method

In the previous section, we assume that the total number of particles in the system keeps constant. This assumption, however, is not always valid. For instance, heterogeneous catalysis on surfaces of noble metals and metal oxides, chemical vapor deposition, or nucleation and growth of quantum dots on substrates, etc. To simulate this kind of systems, we have to assume that one system is in contact with a thermo-bath/particle reservoir. Both energy and particles are allowed to exchange between the two parts. As a result, when this extensive system reaches equilibrium, the pressure P, temperature T, and chemical potential of particles μ of the system and the reservoir should be equal. Since we focus on adsorption and diffusion on surfaces in this chapter, it is convenient to choose the volume of the system V as a constant. Therefore, we choose μVT ensemble, which is one kind of grand canonical ensembles (Frenkel2002). We will introduce in this section the MC method for μVT ensemble, which is called the grand canonical Monte-Carlo method (GCMC).

As in a NVT ensemble, we first present the grand partition function Ξ:

$$\Xi(\mu, V, T) = \sum_{N=0}^{\infty} \frac{1}{N!h^{3N}} \int d\mathbf{r}^N d\mathbf{p}^N \exp\left[-\beta\left(\sum_{i=1}^{N} \frac{\mathbf{p}_i^2}{2m} + \mathcal{U}(\mathbf{r}^N) - \mu N\right)\right] \tag{3.1}$$

For convenience, we change coordinates of particles from Cartesian to fractional:

$$\mathbf{s}_i = \mathbf{r}_i / L; \quad L = V^{1/3} \tag{3.2}$$

By integrating Equation (3.1) and expressing the coordinates of particles in fractional form, we get

$$\Xi(\mu,V,T) = \sum_{N=0}^{\infty} \frac{V^N}{N!\Lambda^{3N}} \int d\mathbf{s}^N \exp\left[-\beta\left(\mathcal{U}(\mathbf{s}^N) - \mu N\right)\right] \tag{3.3}$$

Besides the movements of particles, the μVT ensemble allows us to insert one particle into the system or remove one particle from the system. In the case of insertion, assuming that all other N particles keep their positions, the ratio of probabilities before and after the insertion is

$$\frac{P_{\text{new}}(N+1)}{P_{\text{old}}(N)}=\frac{V}{(N+1)\Lambda^3}\exp\left[-\beta\left(\Delta\mathcal{U}_{\text{in}}-\mu\right)\right];\quad \Delta\mathcal{U}_{\text{in}}=\mathcal{U}(\mathbf{s}^N;\mathbf{s}_{\text{in}})-\mathcal{U}(\mathbf{s}^N) \tag{3.4}$$

The accepting probability of inserting a particle is

$$A_{\text{in}}=\min\left\{1,\frac{V}{(N+1)\Lambda^3}\exp\left[-\beta(\Delta\mathcal{U}_{\text{in}}-\mu)\right]\right\} \tag{3.5}$$

Similarly, by keeping other N-1 particle at their positions, the ratio of probabilities before and after removing a particle is

$$\frac{P_{new}(N-1)}{P_{old}(N)}=\frac{N\Lambda^3}{V}\exp\left[-\beta\left(\Delta\mathcal{U}_{out}+\mu\right)\right];\quad \Delta\mathcal{U}_{out}=\mathcal{U}(\mathbf{s}^{N-1})-\mathcal{U}(\mathbf{s}^N) \tag{3.6}$$

The accepting probability of removing a particle is

$$A_{out}=\min\left\{1,\frac{N\Lambda^3}{V}\exp\left[-\beta(\Delta\mathcal{U}_{out}+\mu)\right]\right\} \tag{3.7}$$

Based on Equations(3.4)-(3.7), we have an algorithm of GCMC:

Algorithm 2

1. Set up initial particle number N_0, positions $\left\{\mathbf{s}_i^0\right\}$, chemical potential μ, T, V and target sampling number N_{MC}. Set $k = 0$, $N_{\text{old}} = N_0$, $\mathbf{s}_{old}^N=\left\{\mathbf{s}_i^0\right\}$;
2. $k = k+1$. Generate a RN r_1 distributed uniformly in (0,1), calculate $i=\text{int}\left[r\times(N+1)\right]+1$. If $i\le N$, go to step3. Otherwise, go to step4;
3. Choose particle i to make a trial movement: $\mathbf{s}_{new}^N=\mathbf{s}_{old}^N+\Delta\mathbf{s}_i$. Calculate the energy change ΔU and A_{Move}. Generate a RN r_1 distributed uniformly in (0,1), and compare r_1 to A_{Move}. If $r_1\le A_{\text{Move}}$, the trial is accepted, set $\mathbf{s}_{\text{new}}^N=\mathbf{s}_{\text{trial}}^N$, otherwise, the trial is rejected, $\mathbf{s}_{new}^N=\mathbf{s}_{old}^N$. Go to step7;
4. Change number particles: if $r_1\ge 0.5$,go to step5. Otherwise, go to step6;
5. Inserting particle: Put one particle to an available position, calculate $\Delta\mathcal{U}_{\text{in}}$ and A_{in}. Generate a RN r_2 distributed uniformly in (0,1), and compare r_2 to A_{in}. If $r_2\le A_{\text{in}}$, the trial is accepted, set $\mathbf{s}_{\text{new}}=\left(\mathbf{s}_{\text{old}}^N;\mathbf{s}_{\text{in}}\right)$, $N_{\text{new}}=N_{\text{old}}+1$. Otherwise, go to step7;

6. Deleting particle: Randomly choose a particle i, calculate $\Delta \mathcal{U}_{\text{out}}$ and A_{out}. Generate a RN r_3 distributed uniformly in (0,1), and compare r_3 to A_{out}. If $r_3 \leq A_{\text{out}}$, the trial is accepted, set $\mathbf{s}_{\text{new}} = \left(\mathbf{s}_{\text{old}}^N - \mathbf{s}_i\right)$, $N_{\text{new}} = N_{\text{old}} - 1$. Otherwise, go to step7;
7. Store $\mathbf{s}_k = \mathbf{s}_{new}$, $N_k = N_{\text{new}}$. Update $\mathbf{s}_{\text{old}} = \mathbf{s}_{\text{new}}$, $N_{\text{old}} = N_{\text{new}}$;
8. If $k<N_{\text{MC}}$, go back to step2. Otherwise calculate thermo quantities of the ensemble by generated N_{MC} states.

Kinetic Monte-Carlo Method

Kinetic Monte-Carlo method (KMC) has been attracting intensive attentions since it can simulate the evolution of large systems (~10^7 sites) over a long period (millisecond or second) while presenting microscopic details and mechanisms. We will discuss in this section the theoretical background and technical aspects of KMC.

Figure 1 presents the connection between KMC and molecular dynamics (MD). When a system is stable, it lies in a local minimum (valley) of a 3N-dimensional potential energy surface (PES). At finite temperatures, all atoms vibrate in the vicinity of this valley most of the time. Occasionally, atomic vibration causes the system to overcome some energy barriers and slides into another valley. This is called "migration", which has little probability but is essential to the kinetics of the system. As shown in Figure 1(a), although MD simulations spend lots of time on recording the true atomic trajectories of the evolution, we do not really care about the details of those trajectories. What we are really interested in are: (1) from a given initial state, how many new states can the system migrate to, and (2) how large is the probability of each migration? This is illustrated by Figure 1(b). In this sense, the KMC method can be viewed as a coarse-grained version of MD. Since the period between two migrations is long, it is fair to assume that the two successive migrations are independent from each other. The migration of the system from state i to state j is thus determined only by the migration rate k_{ij}. If we know accurate values of all migration rates, we can perform KMC simulations on the complex atomic evolutions and kinetics.

Figure 1. Coarse-grained process of the KMC method. (a) Illustration of the true atomic trajectories. (b) Topological description of migration: initial state and final state connected to each other by a migration barrier. All details in trajectories are ignored in the KMC method.

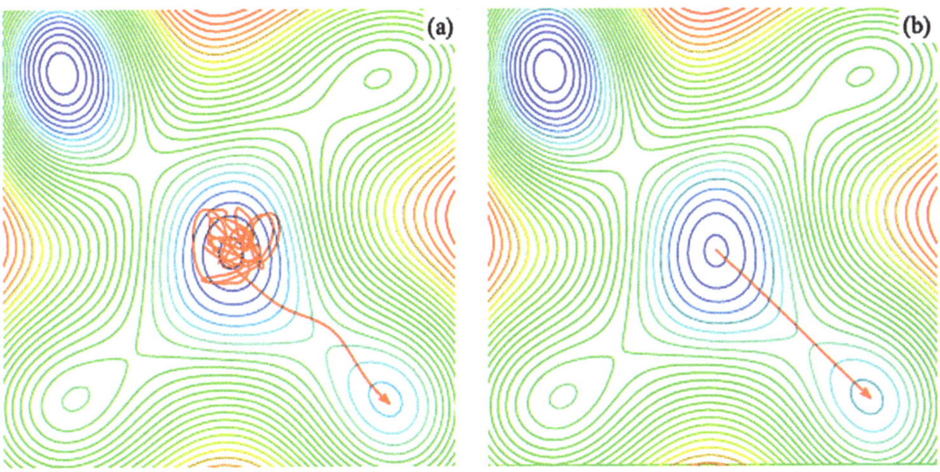

The system is randomly walking on the PES. Assuming total migrating rate is k_{tot}, the probability of the system not migrating within $(t, t+\Delta t)$ is

$$P_{stay}(\Delta t) = 1 - k_{tot}\Delta t + O(\Delta t^2)$$

Similarly, within $(t, t+2\Delta t)$, the probability of the system not migrating is

$$P_{stay}(2\Delta t) = \left(1 - 2k_{tot}\Delta t + O(\Delta t^2)\right)^2$$

Following the same discussions, it is easy to get $\lim_{\tau\to\infty} P_{stay}(\tau)$ with $\tau = K\Delta t$:

$$P_{stay}(\tau) = \lim_{\tau\to\infty}\left(1 - k_{tot}\frac{\tau}{K} + O(K^{-2})\right)^K = \exp\left(-k_{tot}\tau\right) \tag{4.1}$$

From Equation (4.1), we can derive the migration probability of the system in an unit period $p(t)$. The integration of $p(t)$ from t = 0 to t = t should equal $1 - P_{stay}(\tau)$. Therefore $p(t)$ can be expressed as the derivative of $1 - P_{stay}(\tau)$ (Voter1997):

$$p(t) = k_{tot}\exp(-k_{tot}t) \tag{4.2}$$

The above discussions are also valid for a specific migration path k_{ij}. Equation (4.2) shows that $p(t)$ exponentially increases with respect to time. Correspondingly, it is better to choose the time step δt of KMC as an exponential distribution:

$$\delta t = -\frac{1}{k_{tot}}\ln(r) \tag{4.3}$$

where r is a RN uniformly distributed in $(0,1]$.

The most important task in KMC simulations is to determine the set of $\left\{k_{ij}\right\}$. It is usually analytically calculated by harmonic transition state theory (HTST). The basic idea of HTST is that the PES near a minimum is well described by a parabolic function. Therefore the vibrations can be viewed as harmonic oscillators. The vibrating modes which are perpendicular to the reaction path at the saddle point are also harmonic oscillators. This harmonic approximation makes the evaluation of the system entropy easily, which in turn leads to the following simple expression of k_{ij}:

$$k_{ij} = \frac{\prod_{l=1}^{3N}\upsilon_{ij}^{0}}{\prod_{l=1}^{3N-1}\upsilon_{ij}^{sad}}\exp\left(-\frac{\Delta E_{ij}}{k_B T}\right) \tag{4.4}$$

υ_{ij}^{0} and υ_{ij}^{sad} are vibrations with real frequency at state *i* and saddle point of the path *i*→*j*, respectively. $\Delta E_{ij} = E_{ij}^{sad} - E_i$. One difficulty is that the pre-factor in Equation (4.4) needs to be estimated for every migration path. To overcome this difficulty, this pre-factor is usually set as a constant ~ $10^{12\text{-}13}$ Hz. In this chapter, we only deal with systems in which atoms can be mapped to lattice sites. This is called the lattice assumption. One can then view the abstract "migration event" as the change of status of lattice sites (occupation, spin, etc.). Another advantage of the lattice assumption is that we can categorize the migration paths based on *local* configurations instead of *global* ones. It is, however, worth noting that the lattice assumption is not necessary for the KMC method since there will be no lattice at all in some cases (Gillespie2001, Tian2004).

There are several kinds of algorithms of KMC method. We introduce the Bortz-Kalos-Lebowitz (BKL) algorithm (Bortz1975), which has been widely used in research due to its high efficiency and easiness. Figure 2 presents the flowchart of KMC method with the BKL algorithm.

Figure 2. The flowchart of KMC method

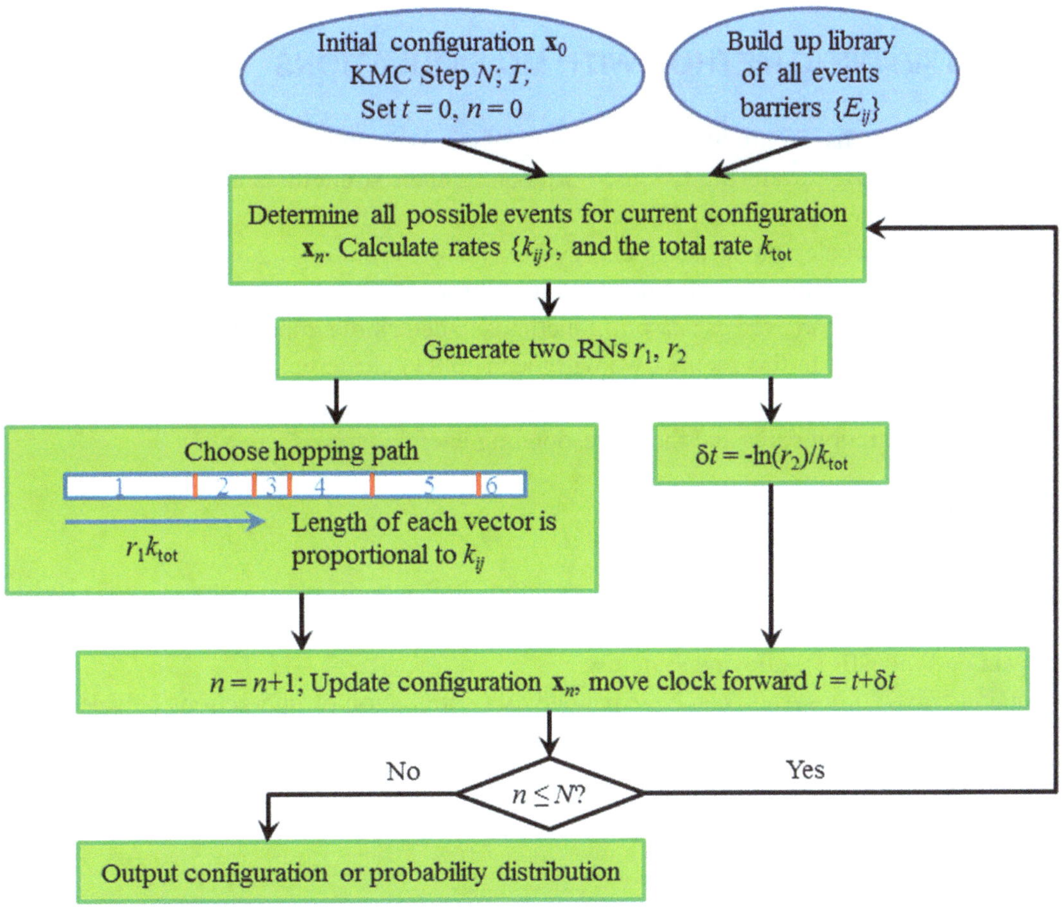

Algorithm 3

1. Set up initial system configuration $\{\mathbf{r}_i^0\}$, temperature T and target simulating steps N_{KMC}. Set k = 0, clock $t = 0$;
2. $k = k+1$. Find out all possible migrating paths from the current state i, and determine corresponding rate k_{ij} based on the event library;
3. Calculate total rate k_{tot} in current state: $k_{\text{tot}} = \sum_j k_{ij}$. As shown in Figure 2, we imagine each path is a vector with length equal to its rate k_{ij}. Then we put all these vectors end to end;
4. Use r_1 to pick up migrating path j_k which satisfies: $\sum_{j=1}^{j_k-1} k_{ij} \le r_1 k_{\text{tot}} < \sum_{j=1}^{j_k} k_{ij}$;
5. Move the clock forward: $t=t+\delta t$, $\delta t = -\ln(r_2)/k_{\text{tot}}$; update the system to the new state j_k. $\mathbf{r}^N(t) = \mathbf{r}_{\text{new}}^N$;
6. If $k<N_{\text{KMC}}$, go back to step2. Otherwise output latest configuration $\{\mathbf{r}_{\text{new}}\}$, serials of snapshots, or probability distribution, etc.

COMBINING *AB INITIO* METHOD WITH MC SIMULATIONS

In previous sections we have introduced basic theory of *ab initio* and MC/KMC methods. As we have discussed, they both have severe limitations. Fortunately, their strengths are complementary to each other. Necessary parameters in MC/KMC methods are usually difficult to obtain, for instance, interactions between the adsorbates on surfaces, energy barriers of migrating paths, even reliable substrate configurations and/or constituents, etc. Information of these aspects is relatively easy to be calculated by *ab initio* methods. We can even use *ab initio* methods to determine important migration paths which might be against intuition. In this section, we will focus on: (1) cluster expansion which is powerful to describe particle interactions, and (2) the nudged elastic method and the dimer method which are suitable to determine migration paths and the corresponding energy barriers.

Cluster Expansion

The cluster expansion method (CE) has been developed very well and becomes a powerful tool to describe interactions of adsorbates or to predict structure phase diagram of binary (multi component see Ref. (van de Walle2013)) alloying systems.

The total energy E of a system can be expanded as a summation of interactions between sites with different orders (one-body, two-body, three-body interactions, etc.):

$$E(\boldsymbol{\sigma}) = c_0 + \sum_i c_i \sigma_i + \sum_{i,j} c_{i,j} \sigma_i \sigma_j + \sum_{i,j,k} c_{i,j,k} \sigma_i \sigma_j \sigma_k + \cdots \tag{5.1}$$

where $\boldsymbol{\sigma}$ is the configuration of lattice sites of the system. It could indicate whether the site is empty or occupied, or the atom on a given site is Element-A or Element-B, etc. c_0, c_i, c_{ij}, c_{ijk} ... are parameters which need to be fitted, and are called effective cluster interactions (ECIs). Assuming that a binary system is composed of N ordered sites, we can in principle always decompose the system in a way containing *m* terms. For instance, these terms include one constant c_0, two one-body ECIs, three two-body ECIs (1st-NN, 2nd-NN and 3rd-NN site-pairs) two three-body ECIs (collinear site-chain and smallest triangle) and one four-body ECI (smallest tetrahedron). Totally, the system has 2^N configurations, each of which corresponds to a specific *E*. We can randomly choose *l* configurations and calculate their energies E_i by *ab initio* methods. Now it is ready to build up an $l \times m$ coefficient matrix **A** satisfying:

$$\mathbf{AC} = \mathbf{E} \tag{5.2}$$

in which **C** is a $m \times 1$ vector consisting of all ECIs, and $\mathbf{E} = \left[E_1, E_2, \cdots, E_l\right]^{\mathrm{T}}$. Using the least square method, we can fit out all ECIs. By using obtained **C** and Equation (5.1), we are able to calculate energies of all these 2^N configurations. Therefore, the phase diagram of binary alloys or stable reconstruction of surface can also be determined if it is necessary.

To fit a set of appropriate CE coefficients, the most convenient method is to employ ATAT toolkit developed by van de Walle and Ceder (van de Walle2002a, van de Walle2002b, van de Walle2002c). It first requires two reference points: $E(\mathbf{0})$ and $E(1)$ corresponding to energies of two "pure" configurations. For example, if the system is a binary alloy with elements *A* and *B*, **0** means the elemental metal *A*, and 1 means the elemental metal *B*. If the system is a surface with adsorbates, **0** means the absence of adsorbates, and 1 means all the surface sites are occupied. All other situations are represented by a parameter θ, i.e. the atomic concentration of element *A* or the coverage of adsorbates. The energy of a given coverage is first transferred by:

$$E(\theta) = E(\boldsymbol{\sigma}) / N_{\text{cell}} - (1 - \theta)E(\mathbf{0}) - \theta E(\mathbf{1}) \tag{5.3}$$

However, at a single coverage there may be multiple configurations ($\boldsymbol{\sigma}$), whose formation energies $E(\boldsymbol{\sigma})$ can be very different due to various molecular level arrangements. When steady state properties are concerned, which is usually the case, only the (near) most stable configurations are selected to form a database pool. Their $E(\boldsymbol{\sigma})$ are used for determining the ECIs. The first step is naturally to choose clusters or figures, i.e., the decomposed interactions. Usually by starting from a minimal set of clusters, we can use Eqs. (5.1) and (5.2) to fit the best evaluation of CE coefficients C. To judge whether C meet requirements, a criterion called "cross validation score" (*CV*) is employed, which is defined as

$$CV = \left(\frac{1}{l}\sum_{i=1}^{l}\left(E_i - \hat{E}_{(i)}\right)^2\right)^{1/2} \tag{5.4}$$

in which E_i is the calculated energy for i^{th} configuration, and $\hat{E}_i$ is the predicted energy for the same configuration using the CE coefficients fitted with all the other (*l*-1) configurations. If the *CV* is smaller than a threshold, we use the fitted C to predict energies of other configurations. Otherwise, we need to either increase number of clusters or change the clusters and repeat the fitting process. If it is necessary, we have to calculate energies of more configurations by *ab initio* methods in order to expand our database.

Figure 3. Iterative determination of optimal Cluster Expansion

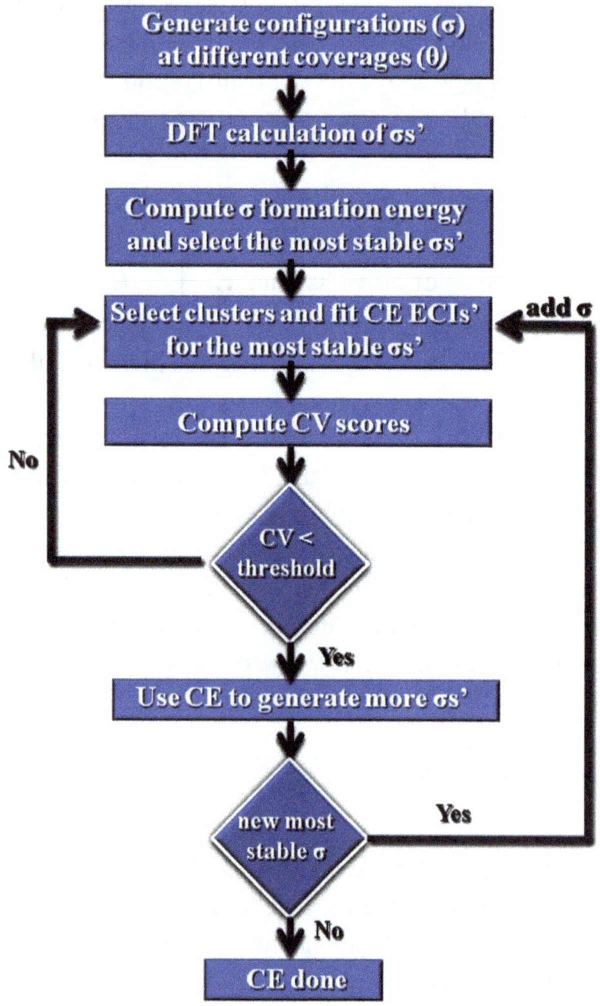

The finalization of a set of optimal clusters and ECIs' is through an iterative way like the one shown in Figure 3. The key to success is to generate a large-enough pool of suitable configurations, which usually are the most stable ones at various compositions (coverages). The *golden* standard to consider the CE has been successfully carried out is that using the set of CE clusters and the ECIs, no new most stable configurations can be predicted.

Nudged Elastic Method and Dimer Method

As we have mentioned in previous sections, a system randomly walks on the PES. If an event of migration happens, the system climbs up from the valley of the initial state (IS) and then slides down to the valley of the final state (FS). The migration path should be the minimum energy path (MEP) whose maximum is a first order saddle point. The energy at this saddle point is maximal along MEP but is minimal along any other directions. Mills and Jónsson *et al.* developed nudged elastic band method (NEB) to get reliable MEPs which are essential to the KMC simulations (Mills1994, Mills1995).

As shown in Figure 4, an instantaneous configuration of the system along the MEP can be viewed as an *Image*. We assume that all these *Images* show along the MEP at the same time and are connected by springs with stiffness *k*. Besides, Images at two ends, IS and FS, are fixed. All other Images are free to relax. Springs prevent images in the middle of MEP from sliding to either IS or FS. But in practice, the distribution of Images images near the saddle point is very low after full relaxation (down-sliding). This makes simulations not able to obtain the MEP accurately. In addition, in the case that the curvature of the MEP is big, those Images will relatively flatly distribute, and thus deviate from the real MEP (corner-cutting). To solve these problems, forces on each Image are projected in the NEB method. To be more specific, one can define a hyper-tangent unit vector $\hat{\tau}$ on every Image. Along $\hat{\tau}$, the Image feels only the projected force from springs; perpendicular to $\hat{\tau}$, the Image feels its projected Hellman-Feynman force. For Image i, $\hat{\tau}_i$ is defined as

$$\tau_i = \frac{\mathbf{R}_i^N - \mathbf{R}_{i-1}^N}{\left|\mathbf{R}_i^N - \mathbf{R}_{i-1}^N\right|} + \frac{\mathbf{R}_{i+1}^N - \mathbf{R}_i^N}{\left|\mathbf{R}_{i+1}^N - \mathbf{R}_i^N\right|}; \quad \hat{\tau}_i = \frac{\tau_i}{\left|\tau_i\right|} \tag{6.1}$$

Correspondingly, the projected Hellman-Feynman force $\mathbf{F}_i^{\perp}$ and spring force $\mathbf{F}_i^{s,\parallel}$ are

Figure 4. Illustration of MEP and NEB method

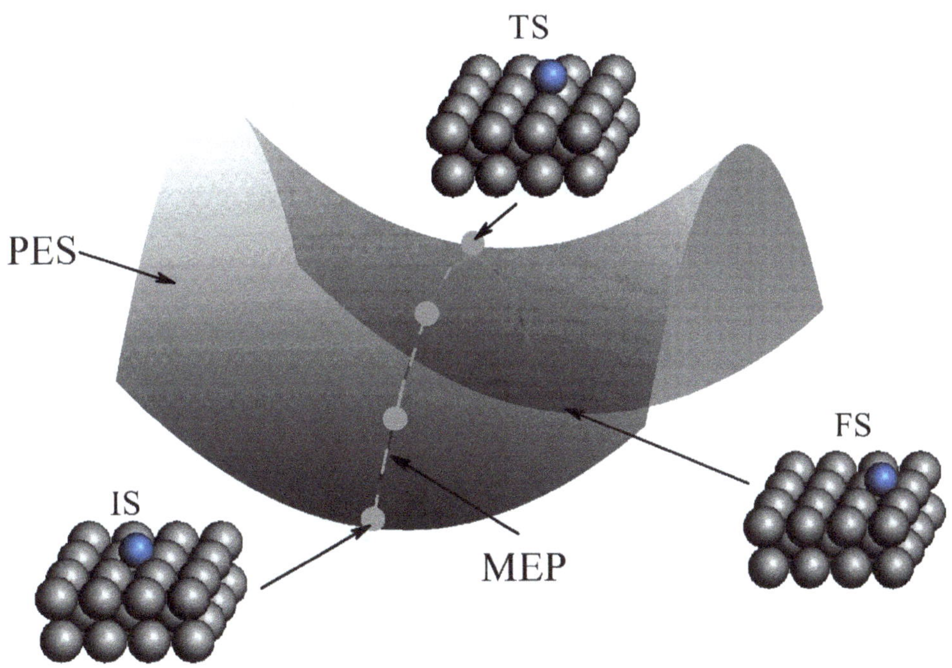

$$\mathbf{F}_i^{\perp} = -\nabla E(\mathbf{R}_i^N) + \nabla E(\mathbf{R}_i^N)\cdot\hat{\tau}_i\hat{\tau}_i;\quad \mathbf{F}_i^{s,\|} = k\left(\left|\mathbf{R}_{i+1}^N - \mathbf{R}_i^N\right| - \left|\mathbf{R}_i^N - \mathbf{R}_{i-1}^N\right|\right)\hat{\tau}_i \tag{6.2}$$

We can initially set configurations of Images by linear interpolation between the IS and FS, and then perform relaxations. Note that in the NEB method, the force on an Image is not consistent with its energy due to the projection and the spring. Therefore, optimizers in the NEB method, i.e. conjugate gradient, quasi-Newton, etc., need to be modified from original algorithms. For detailed discussion, we encourage readers to read References (Henkelman2000, Sheppard2008).

One limitation of the NEB method is that it requires the knowledge of both the IS and FS of a migration path. This requirement, however, means the path has been presumed. There is no guarantee that the lowest barrier can be found by intuition. A rigid way is to determine all migrating paths automatically from a given IS. This is called the Dimer method (Henkelman1999). One dimer consists of two Images, one of which is the IS, the other is generated by applying random displacements on the IS. As shown in Figure 5, energy, force and position of each Image are labeled as E_i, $\mathbf{F}_i^N$ and $\mathbf{R}_i^N$, respectively. Unit vector $\hat{\mathbf{N}}$ points from $\mathbf{R}_2^N$ to $\mathbf{R}_1^N$. The middle point of this dimer is R^N with energy E_0. We can express the curvature C of the PES at $\mathbf{R}^N$:

$$C = \frac{(\mathbf{F}_2^N - \mathbf{F}_1^N)\cdot\hat{\mathbf{N}}}{2\Delta R} = \frac{E - 2E_0}{(\Delta R)^2};\quad E = E_1 + E_2;\quad \Delta R = \frac{1}{2}(\mathbf{R}_2^N - \mathbf{R}_1^N)\cdot\hat{\mathbf{N}} \tag{6.3}$$

Since the total energy of the dimer E is linear to the curvature C, it is easy to get the fact that the dimer rotates till $\hat{\mathbf{N}}$ is parallel to the direction along which C is minimal. This direction is usually the migration path with the lowest barrier. Therefore, every step in the Dimer method contains one rotation and one translation.

Figure 5. (a) The definition of a dimer and key quantities; (b) Rotation step. (Adapted from Figures. 1 and 2 in J. Chem. Phys. (1999) ***111****, 7010.)*

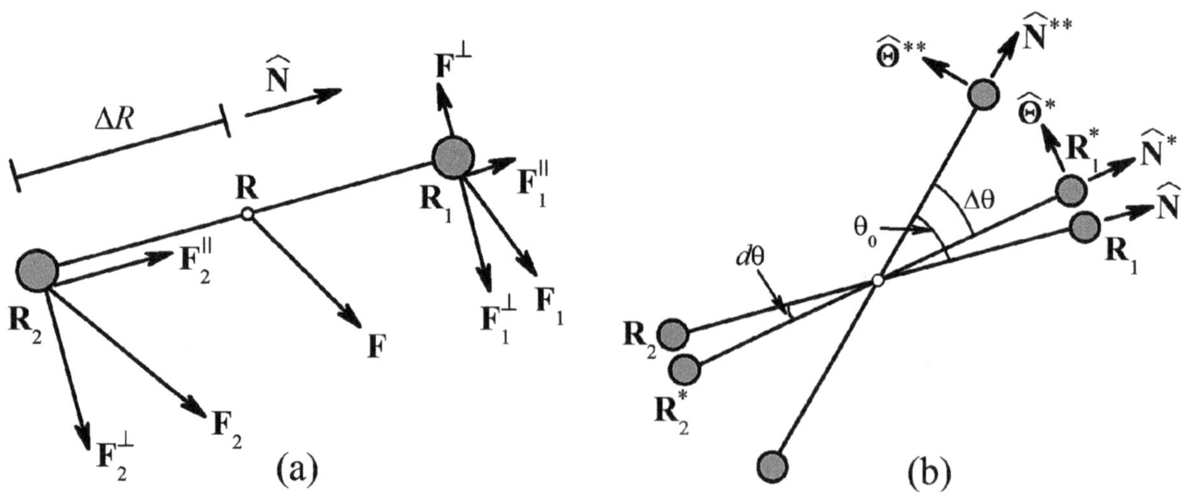

In rotation step, we need to calculate the desirable angle $\Delta\theta$ which leads the dimer parallel to the migrating path. First we calculate the projection of force $\mathbf{F}^{N,\perp}$ perpendicular to $\hat{\mathbf{N}}$:

$$\mathbf{F}^{N,\perp} = \mathbf{F}_1^{N,\perp} - \mathbf{F}_2^{N,\perp} \tag{6.4}$$

The unit vector of $\mathbf{F}^{N,\perp}$ is labeled as $\hat{\mathbf{\Theta}}$. $\hat{\mathbf{\Theta}}$ and $\hat{\mathbf{N}}$ span the plane S for the dimer rotation. Further, we need rotate the dimer in S with a small angle $\delta\theta$. By knowing $\hat{\mathbf{\Theta}}^*$ and $\left(\mathbf{F}^N\right)^*$ at $\theta = \delta\theta$, we can calculate $\Delta\theta$ as (Henkelman1999):

$$\Delta\theta \approx -\frac{\delta\theta}{2}\frac{\mathbf{F}^N \cdot \hat{\mathbf{\Theta}} + \left(\mathbf{F}^N\right)^* \cdot \hat{\mathbf{\Theta}}^*}{\left|\left(\mathbf{F}^N\right)^* \cdot \hat{\mathbf{\Theta}}^* - \mathbf{F}^N \cdot \hat{\mathbf{\Theta}}\right|} \tag{6.5}$$

After rotation, the dimer needs to translate along the current direction till it reaches the saddle point. In the translation step, we use the following force to move the dimer:

$$\left(\mathbf{F}^N\right)^+ = \begin{cases} -\left(\left(\mathbf{F}_1^N + \mathbf{F}_2^N\right)\cdot\hat{\mathbf{N}}\right)\dfrac{\hat{\mathbf{N}}}{2}, & C > 0 \\ \dfrac{\mathbf{F}_1^N + \mathbf{F}_2^N}{2} - \left(\left(\mathbf{F}_1^N + \mathbf{F}_2^N\right)\cdot\hat{\mathbf{N}}\right)\hat{\mathbf{N}}, & C < 0 \end{cases} \tag{6.6}$$

Moving distance Δx can be calculated as following:

$$\Delta x = -\frac{\left(\left.\left(\mathbf{F}^N\right)^+\right|_{\Delta x=\delta x} + \left.\left(\mathbf{F}^N\right)^+\right|_{\Delta x=0}\right)\Big/2}{\left(\left.\left(\mathbf{F}^N\right)^+\right|_{\Delta x=\delta x} - \left.\left(\mathbf{F}^N\right)^+\right|_{\Delta x=0}\right)\Big/\delta x} + \frac{\delta x}{2} \tag{6.7}$$

Therefore, every translation step requires calculating forces twice. There are several works to improve the robust of the Dimer method. Please find discussions in Ref. (Heyden2005).

EXAMPLES OF APPLICATIONS

By now we have systematically introduced the theoretical and technical aspects of *ab initio*-based stochastic methods. We will present in this section two concrete applications. Both examples describe the micro-kinetics on surfaces. We should again emphasize that although the computational power increases fast and allows researchers to build up complicated models, one should try to make the model as simple as possible while retaining the physical reliability. In real research, building models requires tremendous effort and is the most difficult part.

1. Atomic Diffusion on Sn-Decorated Cu(111) Surface

We study the effects of self-diffusion of Sn atoms on the Cu(111) surface. Electromigration (EM) involves the directional diffusion of atoms under the influence of an electric field along the direction of electron flow. The microelectronics industry has recently turned to Cu as the interconnect conductor, due to its superior performance and reliability, particularly in the context of its lower electrical resistivity and its intrinsically higher EM resistance, which is a major mechanism of failure of the integrated circuits that carry large current densities. The EM induced mass transport paths in Cu interconnects at temperatures slower than 100 °C, and occurs preferentially on the (111) surfaces or interfaces. Surface alloying with the heavy *p*-block elements (Sn, Pb, etc.) has been employed as an effective way to improve the EM performance of Cu. A 20 nm thick Cu_3Sn intermetallic compound overlayer on the Cu surface was found to improve an EM lifetime u by one order of magnitude. We employ *ab initio*-based KMC simulations to understand the mechanism of surface Sn alloying on the stability and mobility of a Cu adatom on the Cu-Sn(111) alloy surface (Chen2010).

We first employed *ab initio* calculations to determine the most stable configuration of Sn atoms. Results show that Sn atoms have strong tendency to segregate onto the Cu(111) surface. Sn atoms protrude from the surface by 0.6 Å to accommodate the size mismatch. Sn atoms strongly repel each other. This repulsion decreases quickly with the increase of $d_{\text{Sn-Sn}}$. This trend of the total energy E_{tot} can also be attributed to the oversized Sn atoms which are ~20% larger than Cu atoms. We have chosen the 4th-NN Sn pair as our ground state configuration since its E_{tot} is the lowest.

Due to the presence of Sn atoms, adsorption sites on the Cu(111) surface are not equivalent anymore. We have calculated adsorption energies and migration barriers of a Cu adatom, and plotted the results as the PES of the Sn-decorated Cu(111) surface in Figure 6. The presence of Sn atoms causes deformation of the PES. All atop positions are unstable for adsorption of a Cu adatom, especially at the Sn sites. The more important feature is the absence of local energy minima at the three-fold hollow sites which are nearest to the Sn sites due to the protrusion of Sn atoms from the surface. Consequently, each Sn atom introduces a circular forbidden area (FA) with radius $r_f = \left(\sqrt{2}/2\right)a_0$ (a_0 is the Cu lattice constant), within which a Cu adatom is strictly prohibited to adsorb. Thus, Sn atoms act as adsorption blocking sites which impede the adatom diffusion. Furthermore, Sn atoms weaken the adsorption energy at other three-fold sites in the vicinity. In order to discuss the PES more specifically, we presented in Figure 7 two diffusion paths of an adatom on this alloyed surfaces. The first path (Path-I) passes the vicinity of the only one Sn atom, while the second one (Path-II) gets through the two Sn atoms. Two dark shaded areas centered on each Sn atom denote the FAs. Effects of Sn atoms on migration barriers are local. Beyond $2\sqrt{2}a_0$, a Cu adatom hops just as on a clean Cu(111) surface. When a Cu adatom approaches the Sn sites, it climbs uphill. Not only local minimums, but also migration barriers rise up. This feature is more apparent along the path which goes through the two Sn atoms (Figure 7 (b)): the higher energy barrier occurs when the Cu adatom hopping between two sites in the middle of Sn atoms. Clearly, the mobility of a Cu adatom is substantially reduced close to Sn. These asymmetric migration paths also suggest that an adatom can easily diffuse backward away from Sn atoms. The enhancement of the blocking propensity associated with the Sn pair is due to the concerted formation of low-mobility areas between Sn atoms, as shown by the shaded areas. To explain the PES, Figure 7 (c) presents the local density of states (LDOS) of the surface layer at the Fermi energy $D(E_F)$. $D(E_F)$ is higher between the two Sn atoms because of the contribution from p-orbitals of Sn, and thus strengthens the binding interaction between the surface and the Cu adatom according to the Newns-Anderson model.

We have listed energy barriers E_m of various inequivalent paths in Table 1. By now we are ready to build up kinetic models and perform KMC simulations. As shown in Figure 8, for each adsorption site, we consider three different categories of migration: (1) six paths of 1st-NN FCC↔HCP hopping, (2) three paths of the 2nd-NN FCC↔FCC and HCP↔HCP hopping, and (3) three paths of the 2nd-NN FCC↔HCP hopping, respectively. The migration rate k_{ij} can be calculated by Equation (4.4). The pre-factor is set as a constant $\upsilon^0 \equiv 1.6$ THz for all simulations. k_{ij} for the first category can be calculated directly from Table 1 and Figure 8. For the second category, NEB calculations demonstrate that the migration paths between the 2nd-NN FCC↔FCC and HCP↔HCP sites automatically separates into two consecutive 1st-NN FCC↔HCP hopping steps. The corresponding migration rate k_{ij} thus can be calculated by $k_{ij} = k_{il} \times k_{lj}$ where l is the common 1st-NN adsorption site of both site i and j. For the third category, two different migration mechanisms are considered, as shown in Table 1. The first one is that the Cu adatom goes over the top of the surface atom. The second one is that the adatom exchange with a neighbor surface atom. Clearly, E_m^{atop} and E_m^{ex} are much higher than other barriers. Finally, the rates of migration which involve forbidden areas are set to zero.

As presented in Figure 2, we need first to generate a library of list so that we can use the corrected E_m to inequivalent paths. Therefore, we assign several properties to each path: the names of the IS and FS, and migration barrier E_m. The name of each site is a number. The first digit is called "*type-part*", it denotes FCC (1) or HCP (2). The rest digits are called "*local-part*" which describes the local environment around the site. The length of *local-part* depends on how many Sn atoms the site can see. As we have discussed, effects of Sn on PES is up to $2\sqrt{2}a_0$. Therefore, we plot a circle with a radius $r_{\text{cut}} = 2\sqrt{2}a_0$ around each Sn atom, as shown in Figure 7(a). The ith digit j in *local-part* indicates this site is the j-th

Figure 6. The energy landscape of the system with single Cu adatom on the Cu(111) surface alloyed by two Sn atoms. The brighter (darker) area indicates weak (strong) adsorption sites of the Cu adatom. The lowest energy is set to zero. (©2010, The American Physical Society. Used with permission.)

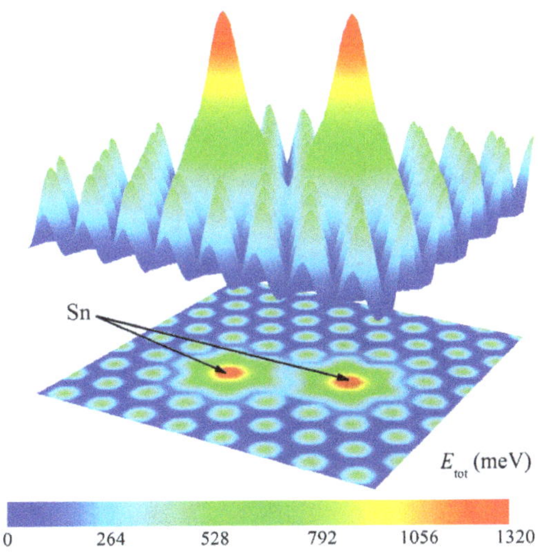

Figure 7. (a) Migration paths I and II for a Cu adatom on the Cu (111) surface with two 4th-NN Sn surface atoms. Yellow (gray) circles denote the Cu (Sn) atoms, while the green (red) circles denote HCP (FCC) sites, denoted by Greek and Latin letters, respectively. (b) Migration energy landscape versus reaction coordinates for pathways I and II, respectively. (c) LDOS contours at the Fermi level, with bright (dark) color indicating high (low) values of LDOS. (d) and (e) Migration path and energy landscape for the Cu_3Sn/Cu interface denoted by dotted line. (©2010, The American Physics Society. Used with permission.)

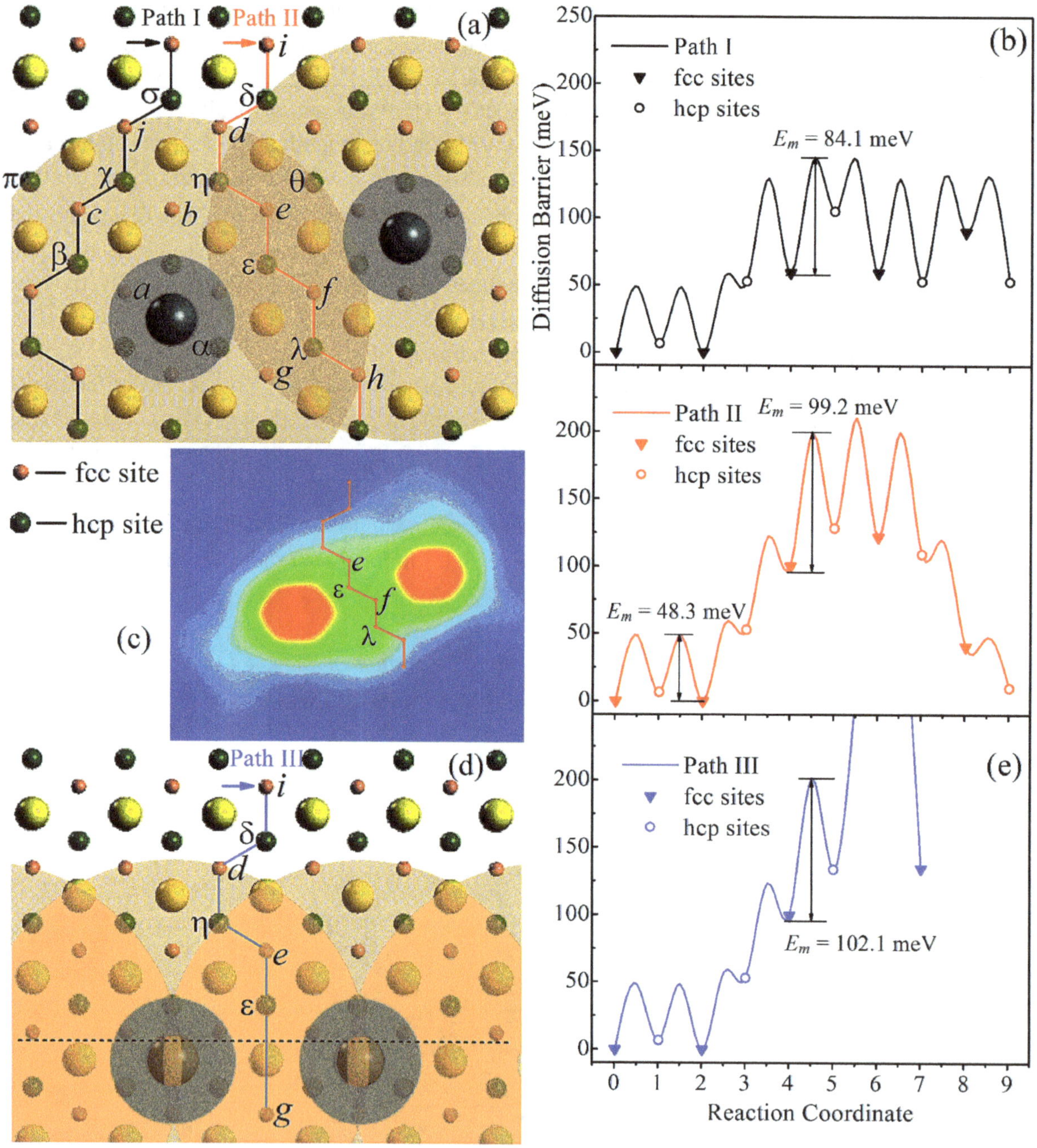

Table 1. Values of migration barrier, E_m, of various inequivalent paths of the Cu adatom on the Cu-Sn surface alloy with two 4th-NN Sn atoms. The labels are shown in Figure 8(a), and the superscripts "atop" and "ex" represent migration paths over the atop site or exchanges with a surface atom in the dilute and ordered alloy cases.

Migration Path	$E_m^{FCC\leftrightarrow HCP}$ (meV)	$E_m^{HCP\leftrightarrow FCC}$ (meV)
FCC↔HCP (pure Cu)	48.3	41.7
$a\leftrightarrow\beta$	∞	∞
$b\leftrightarrow\chi$	42.1	78.9
$c\leftrightarrow\pi$	7.0	36.5
$e\leftrightarrow\theta$	29.7	37.4
$g\leftrightarrow\lambda$	51.4	27.4
$b\leftrightarrow\sigma$ (atop)	427.7	421.1
$b\leftrightarrow\sigma$ (ex)	1289.1	1282.5
$g\leftrightarrow\varepsilon$ (atop) (Cu_3Sn/Cu)	443.1	442.3

Figure 8. Schematic illustration of migration paths in the KMC model. Black solid, red (dark gray) broken and green (light gray) broken arrows represent the first, second and third categories of migration paths.

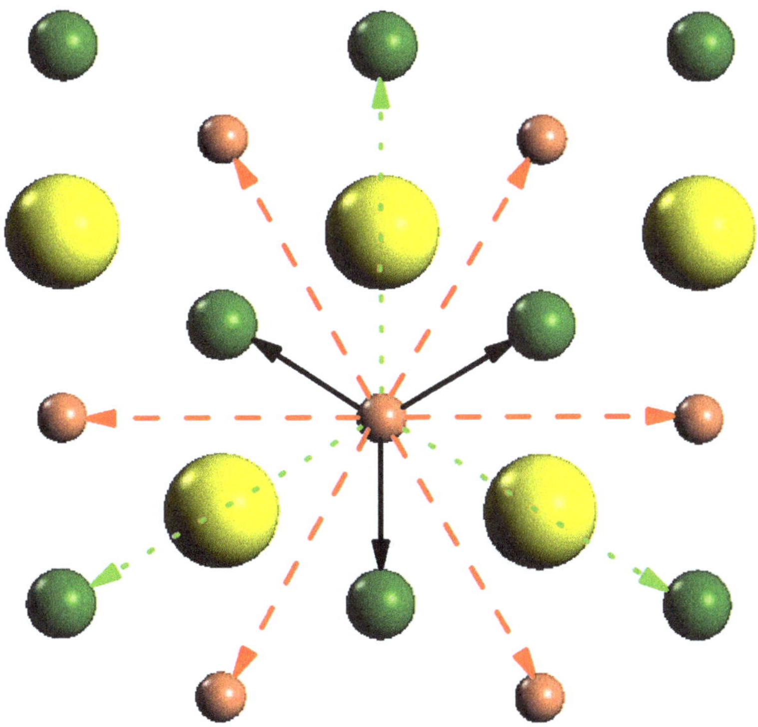

NN from the *i*-th Sn atom that the site can see. Based on the above rules, site ε in Figure 7(a) is labeled as "223", site *f* is labeled as "223" and site *c* is labeled as "13". Further, since the sequence of digits in *local-part* is artificial, one site could be labeled as more than one number, for example, site ε can be either "223" or "232". To avoid this confusion, we permute digits in *local-part* and use the smallest number.

We then employed the BKL algorithm to calculate the distribution probability of a Cu adatom on Cu(111)-2Sn diluted alloying surface (Figure 9(a)-(c)) and the Cu/Cu_3Sn (111) surface (Figure 9(d)-(f)). In the dilute Sn-pair alloy, the diffusion channel for the Cu adatom between the two Sn atoms is closed at 300 K due to the FAs and low-mobility areas between the Sn atoms in Figure 8(a). Thus, the Sn pair hinders the Cu adatom diffusion and redirects it away from or around it. This diffusion path opens with an increasing temperature. Nevertheless, even at 1000 K, the *blocking* effect of Sn remains noticeable.

The *blocking* effect of Sn on adatom diffusion becomes much highly pronounced in the Cu/Cu_3Sn surface. Because of the exclusion of the 1st-NN paths, the Cu adatoms are forced to diffuse solely through atop site and/or exchange migration paths ($b \leftrightarrow \sigma$ or $g \leftrightarrow \varepsilon$ in Figure 8(a) and Table 1), with an energy barrier of about 15 meV higher than that in pure Cu. Thus, Sn surface atoms increase not only the energies of both the FCC and HCP sites but also the energy at the saddle points due to the fact that the adatom is farther away from the surface. The migration rates of these two paths, of ~10^{-7} THz and ~10^{-21} THz at 300 K, respectively, are extremely low compared with that of 0.25 THz on the clean Cu (111) surface. The adatom diffusion on the Cu_3Sn phase is therefore inhibited. Even though more diffusion channels become available at 600 K, the average probability distribution for adatom adsorption remains low. At 1000 K the thermal energy of the adatom is large enough to overcome the migration barrier of the 2nd-NN FCC↔HCP path and the migration rates $\upsilon^{FCC\leftrightarrow HCP}$ of the two paths increase to ~10^{-2} THz and ~10^{-6} THz, respectively, which are still much smaller than that on pure Cu surface. Consequently, at 1000 K the Cu adatom diffuses in the Cu_3Sn phase, but with an extremely low mobility. The average lifetime of

Figure 9. Logarithmic probability distribution for a Cu adatom on the (111) surface at 300 K, 600 K, and 1000 K, respectively, for the dilute Sn-pair alloyed system((a)-(c)), and the surface consisting of an ordered intermetallic Cu $_3$Sn and a clean Cu phase ((d)-(f)). Gray spheres denote Sn atoms. (©2010, The American Physical Society. Used with permission.)

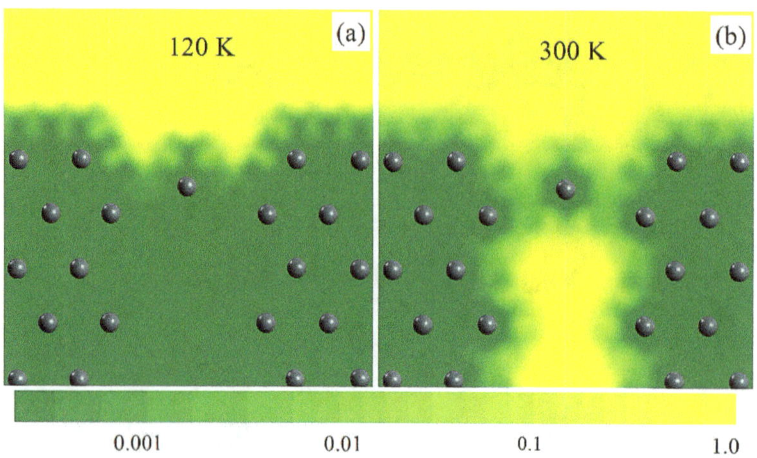

Figure 10. Atomic switch controlled by local temperature

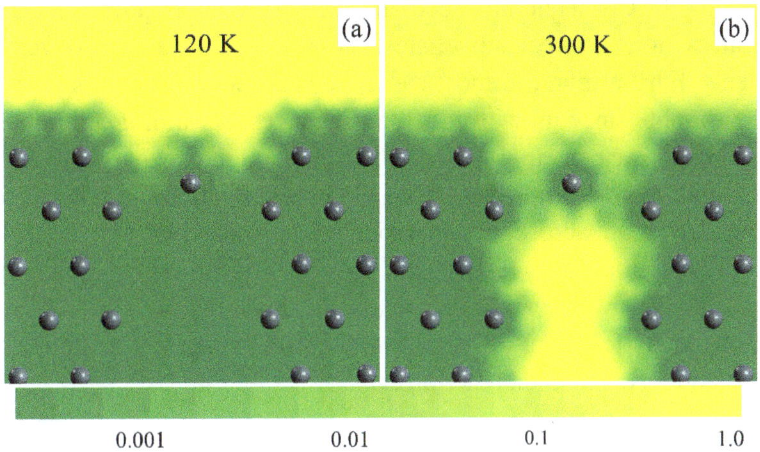

the Cu adatom on the available adsorption sites in Cu_3Sn is ~200 times longer than that in pure Cu, indicating that the ordered Cu_3Sn structure is an excellent electromigration inhibiting candidate.

Based on Figure 9, Sn atoms serve as *blocking* sites for self-diffusion on the Cu(111) surface. The strength of the *blocking* effect depends strongly on the configurations of Sn atoms on the alloyed surface. It is therefore possible to tune the diffusion of adatoms by designing surface alloying pattern. In order to elucidate this possibility, we have designed a Cu-Sn alloying surface as shown in Figure 10. In this system, the clean Cu surface is in contact with two Cu_3Sn ribbons. Further, a Sn atom lies in the middle and is the 4th-NN to both ribbons. By performing KMC simulations at 120 K and 300 K, we obtain the probability distribution of a Cu adatom residency and plot it in Figure 10. It suggests an interesting atomic diffusion switch: at room temperature, the Cu adatom is able to diffuse through the central channel between two ribbons. However, by decreasing the temperature to 120 K, this channel is closed to the adatom due to the *blocking* effect of Sn atoms. These results indicate that a reversible atomic switch can be manipulated by temperature. Although this model needs further experimental demonstration, it clearly shows that maneuverable surface diffusion is possible.

2. Catalytic NO Oxidation on Pt(111) Surface

Quantitatively connecting observed reaction rates, orders, and activation energies to microscopic mechanisms remains one of the great challenges of catalysis science. Such connections are difficult to make even for simple catalytic reactions at surfaces, where interactions between adsorbates can create a myriad of local reaction environments that collectively determine the observed reactivity. These adsorbate interactions influence coverage, adsorbate arrangements, and even the rates of elementary surface reactions. Experiments and density functional theory calculations agree that both the Pt-O* binding energy and rate of O_2 dissociation decrease with increasing coverage, due to destabilizing lateral interactions between O* adsorbates (Getman2009, Smeltz2010,Weaver2005). Various mean-field models, in which the coverage sensitivity is captured in a single coverage variable, θ, have been introduced to build up the connection (Getman2010, Qi2012). While the model is successful at capturing some aspects such as determining kinetic relevant step, predicting reaction coverage window, etc., it fails in predicting kinetic properties. In particular, the model does not reproduce the first-order dependence on of O_2 dissociation on vacant sites (Weiss2009). Consequently, accurate micro-kinetics requires an environment-resolved method to describe interactions between adsorbates. As we have introduced above, the CE method can meet such requirements. Furthermore, in most cases, heterogeneous catalysis should be described by the grand canonical ensemble since it involves atom exchanges. Assuming a reliable set of DFT-based CE parameters are available, we can employ the GCMC method to get equilibrium configurations for a given condition (Tang2004). As a result, the DFT-CE-GCMC methodology keeps the molecular level reaction details, samples over a macroscopic reaction area, and provides kinetic information which are directly comparable to experimental measurements. We will illustrate this general methodology using an example of surface catalysis with a focus on the technical perspective (Wu2012).

We have chosen the catalytic oxidation of NO on the Pt(111) surface as an example elucidate our DFT-CE-GCMC methodology. The total reaction is

$2NO + O_2 \rightarrow 2NO_2$ R1

On the Pt(111) surface, a minimal mechanism includes the following steps

$O_2 + 2^* \rightarrow 2O^*$ R2

$NO + O^* \leftrightarrow NO_2 + {}^*$ R3

where * denotes a surface site. The observed kinetics over both supported Pt catalysts and Pt single crystals suggest that O atoms are the most abundant surface intermediate; and O_2 adsorption and/or dissociation is rate-limiting step (Mulla2005, Mulla2006, Weiss2009, Bhatia2009, Getman2010, Smeltz2010). Based on these observations, we assumed in our model that (a) The only important surface species is O at face centered cubic (FCC) sites; (b) O_2 dissociation requires two neighbor vacant sites simultaneously; and (c) Surface sites do not change during reactions, i.e. O coverage $\theta_O < 0.5$ mono-layer (ML). If $\theta_O > 0.5$, the Pt(111) surface will experience reconstruction (Devarajan2008, Hawkins2009). We do not consider this case in the present example.

We propose a "basis site" approach to modeling surface reactions that permits ready and accurate calculations of macroscopic kinetics (Honkala2005). For given reaction conditions (the oxygen chemical potential μ_O and temperature T), the overall rate of a surface reaction is written as a sum over reaction rates at basis sites $\tilde{r}_i$ weighted by the number density of these sites $\tilde{s}_i$ being observed under given reaction conditions:

$$r(T, P_{O_2}, \mu_O) = \sum_i \tilde{r}_i(T, P_{O_2}) \tilde{s}_i(T, \mu_O) \tag{7.1}$$

in which $\tilde{r}_i$ is a function of temperature T and O_2 flux toward the surface:

$$\tilde{r}_i = F_{O_2} \cdot \kappa \cdot \exp\left(-\frac{E_i^{act}}{k_B T}\right) \tag{7.2}$$

The effective orientational sticking factor κ is empirically taken to be 1/6(Chorkendorff2007, Gokhale2004). The O_2 flux in Equation (6.2) is expressed as:

$$F_{O_2}(T, P_{O_2}) = \frac{P_{O_2}}{L\sqrt{2\pi \cdot m_{O_2} \cdot k_B T}} \tag{7.3}$$

L is the surface total site density (e.g. a DFT estimation is 1.25×10^{19} FCC sites/M²); m_{O_2} is the O_2 molar mass; $E(\sigma) = c_0 + \sum_i c_i \sigma_i + \sum_{i,j} c_{i,j} \sigma_i \sigma_j + \sum_{i,j,k} c_{i,j,k} \sigma_i \sigma_j \sigma_k + \cdots$ is the Boltzmann constant; P_{O_2} is the O_2 pressure. The last parameter needed be determined is the activation energy E_i^{act} which is the dissociation barrier of O_2 within the ith environment. Since NEB calculations are very demanding, it is not practical to directly calculate E_i^{act} of all possible environments. Instead, we employ the Brønsted–Evans–Polanyi (BEP) relationship which indicates the linear correlation between activation barrier and final state binding energy (Nørskov 2002). For the present system, our NEB calculations determine the BEP relationship as

$$E_i^{act} = \max(0, E_{\text{OOBE}}^i + 2.12) \tag{7.4}$$

as shown in Figure 11. E_{OOBE}^i is called O-O pair binding energy (OOBE), indicating the binding energy of two O atoms simultaneously occupying adjacent two FCC sites within the ith environment.

Figure 11. BEP relation between reaction energy and transition energy of O_2 dissociation reaction. (©2011 Elsevier Inc. Used with permission.)

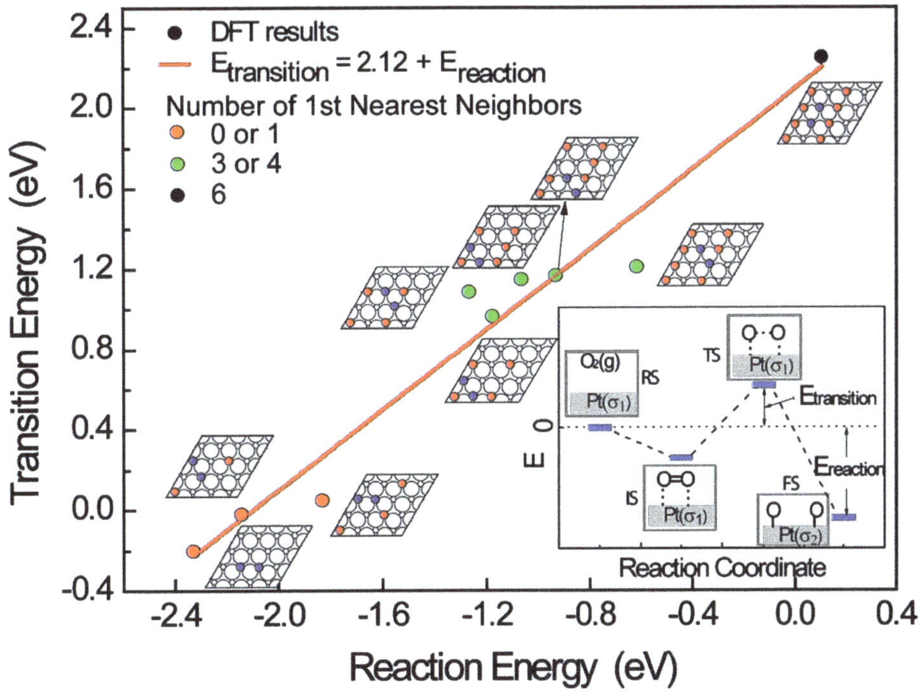

Clearly, it is natural to choose E^{i}_{OOBE} as an indicator to distinguish different environments. With a set of CE parameters, we can perform GCMC simulations to obtain equilibrium O configurations at a given condition, and obtain density of adsorption sites $\tilde{s}_i$ according to E^{i}_{OOBE}. By now, we describe a general procedure of DFT-CE-GCMC methodology. The processes are illustrated in Figure 12.

By fitting energies of more than 50 configurations obtained by DFT, we constructed a high-quality set of CE to describe O-O interactions on Pt(111). There are 12 terms consisting of 8 pairs and 2 triplets, whose *CV* is below 5 meV/FCC site comparing to DFT values. The concrete clusters and ECIs are shown in Figure 13.

GCMC simulations are perform on a 42×42 Pt(111) surface with periodic boundary condition to get equilibrium configurations for different sets of μ_{O} and *T*. For each μ_{O} and *T*, we included ten initial configurations and have them evolve independently, indicated as *threads*. In the present example, only adsorption and desorption of O atoms are included. We did not consider migration of adsorbed O. Each thread is equilibrated until the coverage θ fluctuation is less than 0.15%. The required MC steps per FCC site vary from 7,000 to 30,000 depending on *T*. After equilibration, each thread further flips the configuration 2,000 times per FCC site during production. The configurations were collected for statistics after every 2 flips/FCC site, i.e., 1000 configurations in total for each GCMC thread. For each sampled configuration, all vacant FCC neighbor pairs were recorded first, and E_{OOBE} on each pair was then determined by using CE if two O* were accommodated. An E_{OOBE} distribution (bin size 0.01 eV) was obtained for each configuration. Finally, the statistics was averaged over all the GCMC simulations.

Figure 12. From molecular processes to macro-kinetic observables σ represents configuration. σ-OO is the configuration after O_2 dissociation into neighbor vacancies. E(σ) is the total energy for the configuration. The E_{OOBE} is referenced to O_2 gas phase energy.

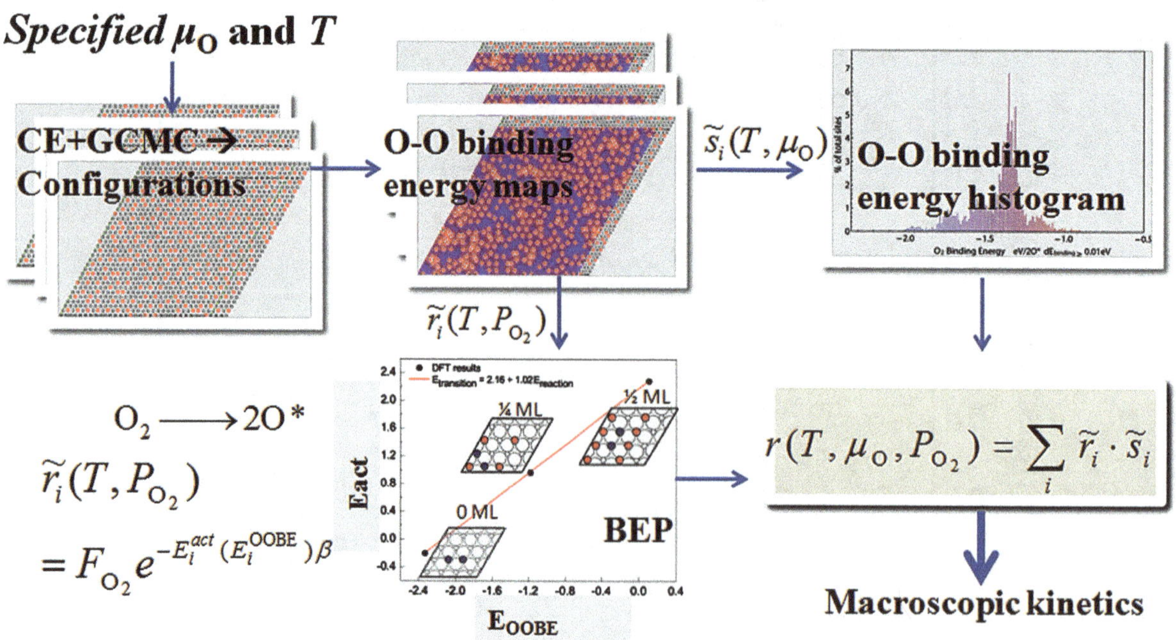

Figure 13. Clusters and their ECIs' used in this work

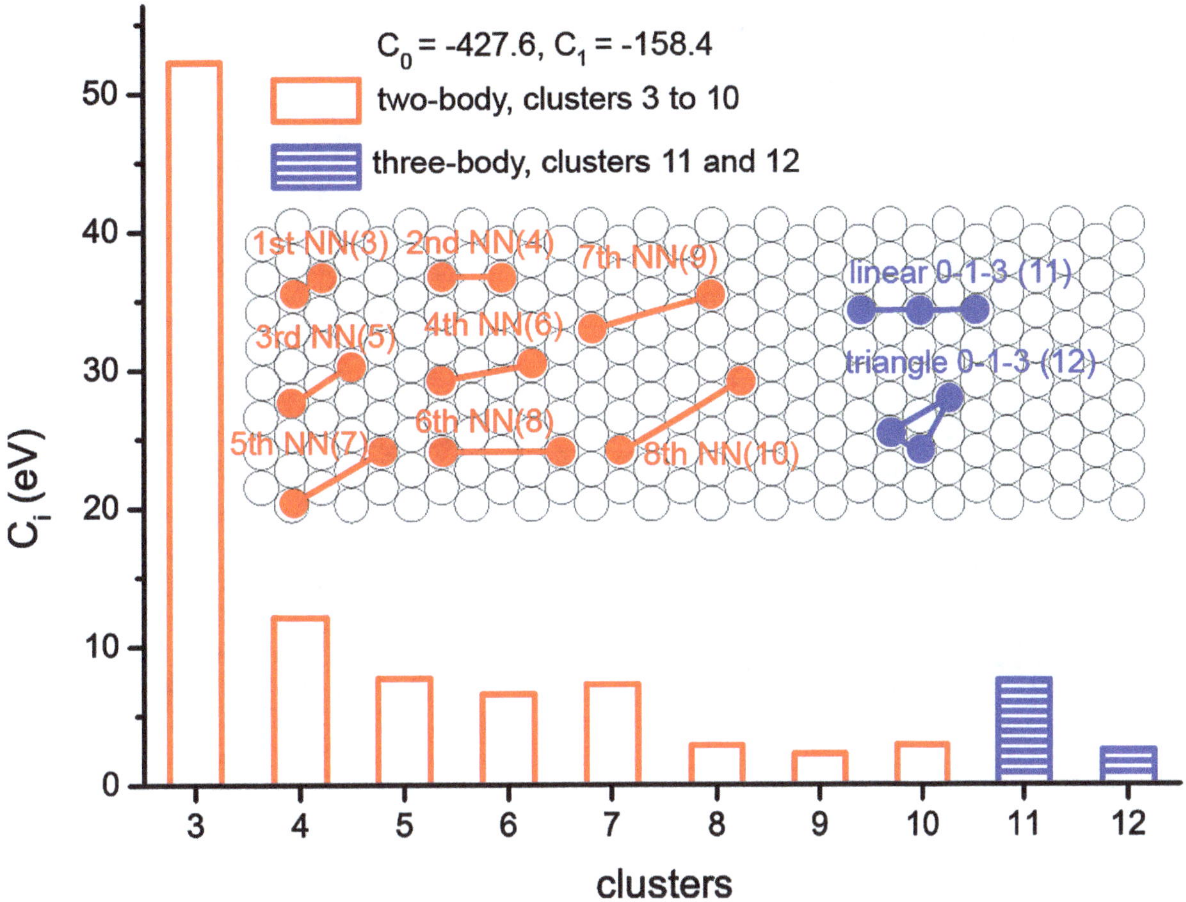

Figure 14 shows the site statistics of two typical equilibrium configurations generated by CE-GCMC at different μ_O. In general, E_{OOBE} are stronger in local O-poor environments, as shown in color map Figure 14(b). This is understandable since O-O interactions on the Pt(111) surface are mainly repulsive, for instance, ECIs shown in Figure 13 are all positive. As μ_O increases, the θ_O increases, and thus the number of available vacant FCC pairs decreases. Further, the E_{OOBE} distribution broadens and moves towards the weaker binding end. Due to the BEP relationship, the weakened E_{OOBE} effectively increases energy barriers of O_2 dissociation. Both factors are detrimental to the catalytic activity of the Pt(111) surface.

Figure 14 has presented enough information of both $\tilde{r}_i$ and $\tilde{s}_i$, which allows us to use Equation (7.1)-(7.3) to calculate macro-properties, i.e. turnover rates (TOR) of oxidation as a function of reaction conditions. μ_O is expressed as the NO_2 to NO pressure ratio:

$$\mu_O(T, P_{NO_2}, P_{NO}) = \mu^0_{NO_2}(T,\ 1\ \text{bar}) - \mu^0_{NO}(T,\ 1\ \text{bar}) + k_B T \ln\left(P_{NO_2}/P_{NO}\right) \quad (7.5)$$

Figure 14. O_2 dissociation site statistics at 600K for two configurations equilibrated with two μ_Os. (a) Configuration snapshots. (b) OOBE color maps. The colors are labeled according to E_{OOBE} values in plot c. (c) Site density versus E_{OOBE}. Surface boundaries are demarcated by the black dashed lines. Numbers under each inset are E_{OOBE} s and the binding sites are indicated by blue bars

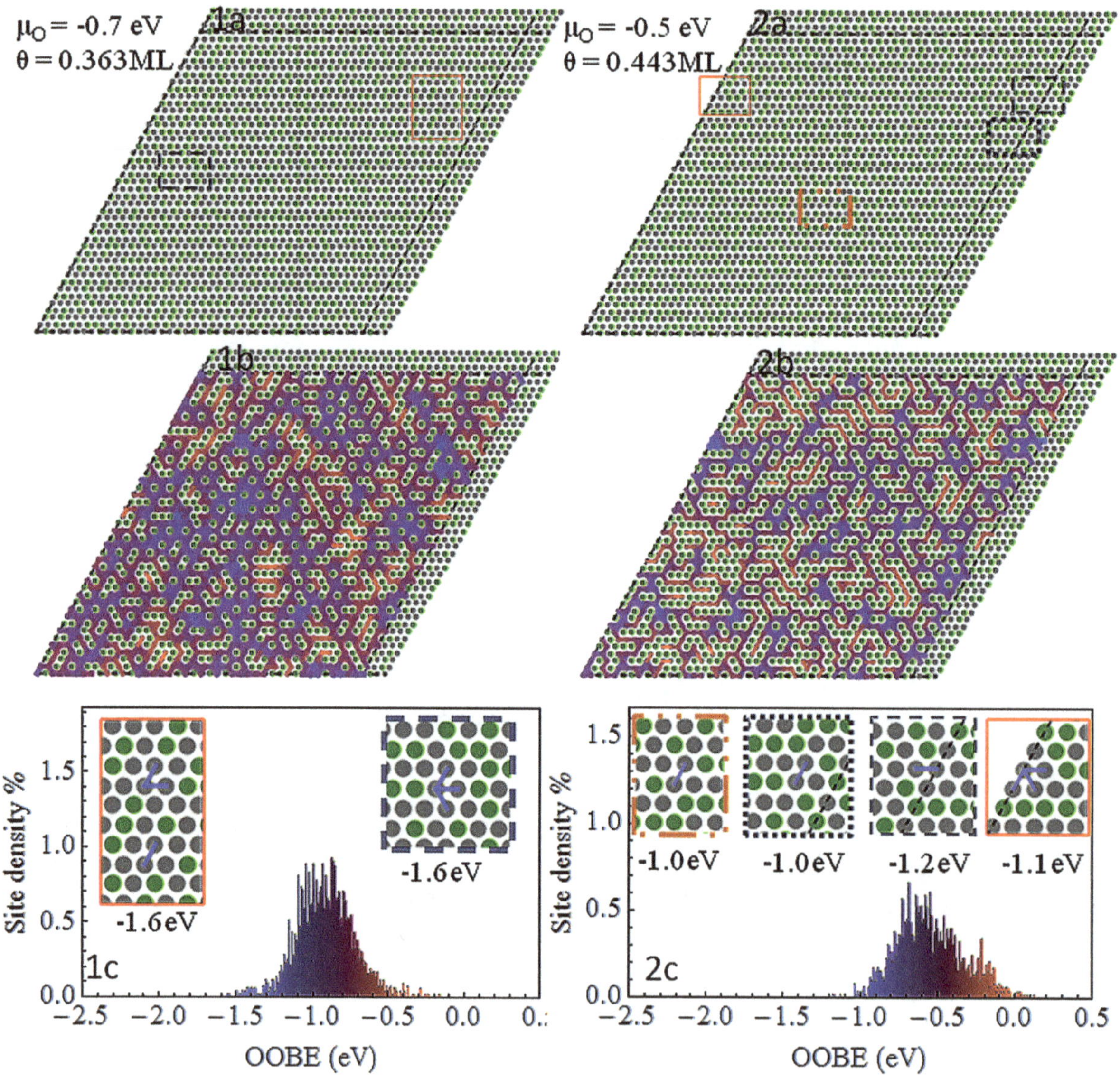

$\mu^0_{NO_2}(T,\ 1\ \text{bar})$ and $\mu^0_{NO}(T,\ 1\ \text{bar})$ can be found from NIST WebBook (Burgess2011). After exploring different pressure ratios and temperatures, we plotted TOR for a series of *T* and for a broad range of pressure ratio in Figure 15. Clearly, under a given *T*, TOR decreases drastically as the NO_2 pressure increases. Macroscopic observation is that the formation of NO_2 inhibits further oxidation by pushing the equilibrium backward to NO and O* (R2). Corresponding microscopic mechanism can be attributed to the fact that O_2 dissociation becomes more difficult on a more crowded surface due to repulsive lat-

Figure 15. Temperature-dependent TOR and apparent activation energy. P_{O_2} = 0.1 bar. (a) Results of TOR(NO_2:NO;T). T ranges from 480K to 680K with an incremental of 40K. Inset NO oxidation TOR at T = 300 °C and P_{O_2} = 0.1 bar versus O coverage. (a) is adapted from Ref. (©2011 Elsevier Inc. Used with permission.)

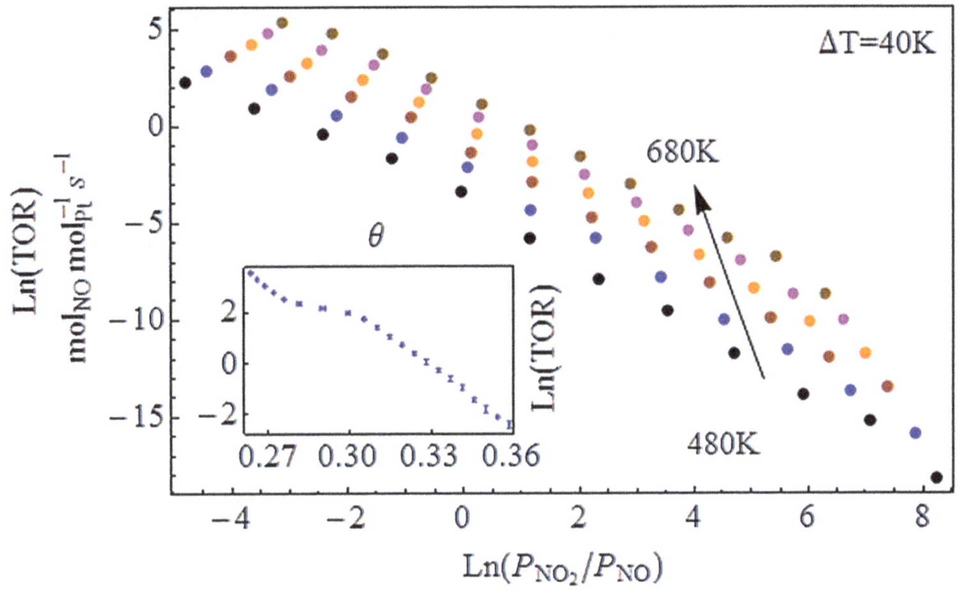

eral interactions as we have discussed before. The apparent activation energy (E_a) can also be derived by Arrhenius plotting of TOR versus T at a given pressure ratio. The slope of the fitting gives the E_a of 77 ±3 KJ/mol, which is consistent with some experimental observations, and is insensitive to pressure ratio (Weiss2009, Smeltz2010, Mulla2006). As shown in Figure 15(Inset), a closer look at the data reveals a slope change: TOR changes slower at θ_O from 0.27 to 0.31, which reflects the transition from ordered $p(2\times2)$-O structure to disordered surface configurations. The reaction order obtained by numerical differentiation is around -1.2 ~ -0.9 at lower pressure ratios (NO_2:NO < 0.3), and increases to -1.6 at about NO_2:NO ≈ 0.6 and stabilizes around that number for higher pressure ratios, which are close to experimental observations.(Weiss2009, Smeltz2010, Mulla2006) In contrast to the apparent activation energy, the reaction order is very sensitive to pressure ratios, *i.e.*, the O configurations.

CONCLUSION

In summary, we have briefly introduced the approaches that integrate *ab initio* methods and common stochastic methods, i.e., KMC and GCMC, in this chapter, with an emphasis on their applications to surface kinetic problems. In order to incorporate the effects of complex surface environments, we also presented brief introductions to the cluster expansion, the nudged elastic band method and the dimer method. These methods are essential to transfer raw data of *ab initio* calculations to parameters that can be directly used by MC models. By discussing in details two simulations on micro-kinetic processes on metallic surfaces, we showed the strength of the specific hierarchical multi-scale model, which combines *ab initio* and MC methods, on complex reactions and/or structural evolution. The first example is a simple *ab initio*-based KMC simulation. The NEB method was employed to directly calculate the PES of alloying Cu(111) surface. Controllable diffusion is predicted by our simulation. The method is powerful to simulate the morphological evolution of crystalline metallic surfaces on which the number of inequivalent migrations paths is small. The second example is more complicated. We introduced DFT-based CE-GCMC method and applied it to catalytic oxidation on the Pt(111) surface. The simulations results of TOR and the apparent activation energy are in good agreement with experiments. This approach described here bridges the gap between microscopic reaction mechanisms and macroscopic kinetic properties. It is general to deal with systems where the PES has a strong dependence on atomic configurations. Although there are still essential limitations for this *ab initio*-based stochastic method, it has been made massive successes on several materials science topics which traditional atomic simulating methods cannot properly handle. Extension and renovation of current models, which aim to deal with larger systems and cover longer time scales, or to decrease artificial assumptions, have been introduced. It is safe to foresee the booming of *ab initio*-based stochastic methods.

REFERENCES

Bhatia, D., McCabe, R. W., Harold, M. P., & Balakotaiah, V. (2009). Experimental and kinetic study of NO oxidation on model Pt catalysts. *Journal of Catalysis*, *266*(1), 106–119. doi:10.1016/j.jcat.2009.05.020

Bortz, A. B., Kalos, M. H., & Lebowitz, J. L. (1975). A new algorithm for Monte Carlo simulation of Ising spin systems. *Journal of Computational Physics*, *17*(1), 10–18. doi:10.1016/0021-9991(75)90060-1

Burgess, D. (2011). Thermochemical Data. In *NIST Chemistry WebBook*. NIST Standard Reference Database Number 69. National Institute of Standards and Technology. Retrieved February 10, 2014, from http://webbook.nist.gov

Ceperley, D. M., & Alder, B. J. (1980). Ground state of the electron gas by a stochastic method. *Physical Review Letters*, *45*(7), 566–569. doi:10.1103/PhysRevLett.45.566

Chen, Z. Z., Kioussis, N., Tu, K., Ghoniem, N., & Yang, J. (2010). Inhibiting adatom diffusion through surface alloying. *Physical Review Letters*, *105*(1), 015703. doi:10.1103/PhysRevLett.105.015703 PMID:20867465

Chorkendorff, I., & Niemantsverdriet, J. W. (2007). *Concepts of Modern Catalysis and Kinetics* (2nd ed.). Wiley-VCH.

Devarajan, S. P., Hinojosa, J. A. Jr, & Weaver, J. F. (2008). STM study of high-coverage structures of atomic oxygen on Pt(1 1 1): *P*(2×1) and Pt oxide chain structures. *Surface Science*, *602*(19), 3116–3124. doi:10.1016/j.susc.2008.08.008

Frenkel, D., & Smit, B. (2002). *Understanding molecular simulation from algorithms to applications*. San Diego, CA: Academic Press.

Fu, C. C., Torre, J. D., Willaime, F., Bocquet, J.-L., & Barbu, A. (2004, December19). Multiscale modelling of defect kinetics in irradiated iron. *Nature Materials*, *4*(1), 68–74. doi:10.1038/nmat1286

Getman, R. B., & Schneider, W. F. (2010). DFT-Based Coverage-Dependent Model of Pt-Catalyzed NO Oxidation. *ChemCatChem*, *2*(11), 1450–1460. doi:10.1002/cctc.201000146

Getman, R.B., Schneider, W.F., Smeltz, A.D., Delgass, W.N., Ribeiro, F.H. (2009). Article. *Physical Review Letters, 78*(102), 076101.

Getman, R. B., Xu, Y., & Schneider, W. F. (2008). Thermodynamics of Environment-Dependent Oxygen Chemisorption on Pt(111). *The Journal of Physical Chemistry C*, *112*(26), 9559–9572. doi:10.1021/jp800905a

Gillespie, D. T. (2001). Approximate accelerated stochastic simulation of chemically reacting systems. *The Journal of Chemical Physics*, *115*(4), 1716–1733. doi:10.1063/1.1378322

Gokhale, A. A., Kandoi, S., Greeley, J. P., Mavrikakis, M., & Dumesic, J. A. (2004). Molecular-level descriptions of surface chemistry in kinetic models using density functional theory. *Chemical Engineering Science*, *59*(22-23), 4679–4691. doi:10.1016/j.ces.2004.09.038

Hawkins, J. M., Weaver, J. F., & Asthagiri, A. (2009). Density functional theory study of the initial oxidation of the Pt(111) surface. *Physical Review B: Condensed Matter and Materials Physics*, *79*(12), 125434. doi:10.1103/PhysRevB.79.125434

Henkelman, G., & Jónsson, H. (1999). A dimer method for finding saddle points on high dimensional potential surfaces using only first derivatives. *The Journal of Chemical Physics*, *111*(15), 7010. doi:10.1063/1.480097

Henkelman, G., & Jónsson, H. (2000). Improved tangent estimate in the nudged elastic band method for finding minimum energy paths and saddle points. *The Journal of Chemical Physics*, *113*(22), 9978. doi:10.1063/1.1323224

Henkelman, G., & Jónsson, H. (2001). Long time scale kinetic Monte Carlo simulations without lattice approximation and predefined event table. *The Journal of Chemical Physics*, *115*(21), 9657. doi:10.1063/1.1415500

Heyden, A., Bell, A. T., & Keil, F. J. (2005). Efficient methods for finding transition states in chemical reactions: Comparison of improved dimer method and partitioned rational function optimization method. *The Journal of Chemical Physics*, *123*(22), 224101. doi:10.1063/1.2104507 PMID:16375464

Hohenberg, P., & Kohn, W. (1964). Inhomogeneous Electron Gas. *Physical Review*, *136*(3B), B864–B871. doi:10.1103/PhysRev.136.B864

Honkala, K., Hellman, A., Remidiakis, I., Logadottir, A., Carlsson, A., Dahl, S., & Nørskov, J. et al. (2005). Ammonia synthesis from First-Principles calculations. *Science*, *307*(5709), 555–558. doi:10.1126/science.1106435 PMID:15681379

Huang, H., Gilmer, G. H., & Díaz de la Rubia, T. (1998). An atomistic simulator for thin film deposition in three dimensions. *Journal of Applied Physics*, *84*(7), 3636–3649. doi:10.1063/1.368539

Huang, K. (1987). *Statistical Mechanics* (2nd ed.). Hoboken, NJ: John Wiley & Sons, Inc.

Kohn, W., & Sham, L. (1965). Self-consistent equation including exchange and correlation effects. *Physical Review*, *140*(4A), A1133–A1138. doi:10.1103/PhysRev.140.A1133

Lee, C., Yang, W. T., & Parr, R. (1988). Development of the Colle-Salvetti correlation-energy formula into a functional of the electron density. *Physical Review B: Condensed Matter and Materials Physics*, *37*(2), 785–789. doi:10.1103/PhysRevB.37.785 PMID:9944570

Mackay, D. (1998). Introduction to Monte Carlo Methods. In M. Jordan (Ed.), Learning in Graphical Models (pp.175-204). Kluwer Academic Press. doi:10.1007/978-94-011-5014-9_7

Martin, R. M. (2004). *Electronic structure: Basic Theory and Practical Methods*. Cambridge, UK: Cambridge University Press. doi:10.1017/CBO9780511805769

Martínez, E., Monasterio, P. R., & Marian, J. (2011). Billion-atom synchronous parallel kinetic Monte Carlo simulations of critical 3D Ising systems. *Journal of Computational Physics*, *230*(4), 1359–1369. doi:10.1016/j.jcp.2010.11.006

McDonald, I. R. (1972). NPT-ensemble Monte Carlo calculations for binary liquid mixtures. *Molecular Physics*, *23*(1), 41–58. doi:10.1080/00268977200100031

Mills, G., & Jónsson, H. (1994). Quantum and Thermal Effects in H_2 Dissociative Adsorption: Evaluation of Free Energy Barriers in Multidimensional quantum systems. *Physical Review Letters*, *72*(7), 1124–1127. doi:10.1103/PhysRevLett.72.1124 PMID:10056623

Mills, G., Jónsson, H., & Schenter, G. K. (1995). Reversible work transition state theory: Application to dissociative adsorption of hydrogen. *Surface Science*, *324*(2-3), 305–337. doi:10.1016/0039-6028(94)00731-4

Mo, Y., Varga, K., Kaxiras, E., & Zhang, Z. (2005). Kinetic Pathway for the Formation of Fe Nanowires on Stepped Cu(111) Surfaces. *Physical Review Letters*, *94*(15), 155503. doi:10.1103/PhysRevLett.94.155503 PMID:15904159

Mulla, S., Chen, N., Cumaranatunge, L., Blau, G., Zemlyanov, D., Delgass, W., & Ribeiro, F. et al. (2006). Reaction of NO and O_2 to NO_2 on Pt: Kinetics and catalyst deactivation. *Journal of Catalysis*, *241*(2), 389–399. doi:10.1016/j.jcat.2006.05.016

Mulla, S., Chen, N., Delgass, W., Epling, W., & Ribeiro, F. (2005). NO_2 inhibits the catalytic reaction of NO and O_2 over Pt. *Catalysis Letters*, *100*(3-4), 267–270. doi:10.1007/s10562-004-3466-1

Negulyaev, N. N., Stepanyuk, V. S., Hergert, W., Bruno, P., & Kirschner, J. (2008). Atomic-scale self-organization of Fe nanostripes on stepped Cu(111) surfaces: Molecular dynamics and kinetic Monte Carlo simulations. *Physical Review B: Condensed Matter and Materials Physics*, *77*(8), 085430. doi:10.1103/PhysRevB.77.085430

Nørskov, J., Bligaard, T., Logadottir, A., Bahn, S., Hansen, L., Bollinger, M., & Jacobsen, C. et al. (2002). Universality in heterogeneous catalysis. *Journal of Catalysis*, *209*(2), 275–278. doi:10.1006/jcat.2002.3615

Perdew, J. P., Burke, K., & Ernzerhof, M. (1996). Generalized Gradient Approximation Made Simple. *Physical Review Letters*, *77*(18), 3865–3868. doi:10.1103/PhysRevLett.77.3865 PMID:10062328

Perdew, J. P., & Wang, Y. (1992). Accurate and simple analytic representation of the electron-gas correlation energy. *Physical Review B: Condensed Matter and Materials Physics*, *45*(23), 13244–13249. doi:10.1103/PhysRevB.45.13244 PMID:10001404

Perdew, J. P., & Zunger, A. (1981). Self-interaction correlation to density-functional approximations for many-electron systems. *Physical Review B: Condensed Matter and Materials Physics*, *23*(10), 5048–5079. doi:10.1103/PhysRevB.23.5048

Piana, S., Reyhani, M., & Gale, J. (2005). Simulating micrometer-scale crystal growth from solution. *Nature*, *438*(7064), 70–73. doi:10.1038/nature04173 PMID:16267550

Pornprasertsuk, R., Cheng, J., Huang, H., & Prinz, F. B. (2007). Electrochemical impedance analysis of solid oxide fuel cell electrolyte using kinetic Monte Carlo technique. *Solid State Ionics*, *178*(3-4), 195–205. doi:10.1016/j.ssi.2006.12.016

Qi, L., & Li, J. (2012). Adsorbate interactions on surface lead to a flattened Sabatier volcano plot in reduction of oxygen. *Journal of Catalysis*, *295*, 59–69. doi:10.1016/j.jcat.2012.07.019

Rak, M., Izdebski, M., & Brozi, A. (2001). Kinetic Monte Carlo study of crystal growth from solution. *Computer Physics Communications*, *138*(3), 250–263. doi:10.1016/S0010-4655(01)00238-7

Reuter, K., & Scheffler, M. (2003). First-Principles atomistic thermodynamics for oxidation catalysis: Surface phase diagrams and catalytically interesting regions. *Physical Review Letters*, *90*(4), 046103. doi:10.1103/PhysRevLett.90.046103 PMID:12570437

Rogal, J., Reuter, K., & Scheffler, M. (2007). First-Principles statistical mechanics study of the stability of a subnanometer thin surface oxide in reactive environments: CO oxidation at Pd(100). *Physical Review Letters*, *98*(4), 046101. doi:10.1103/PhysRevLett.98.046101 PMID:17358787

Schultz, A. J., & Kofke, D. A. (2011). Algorithm for constant-pressure Monte-Carlo simulation of crystalline solids. *Physical Review E: Statistical, Nonlinear, and Soft Matter Physics*, *84*(4), 046712. doi:10.1103/PhysRevE.84.046712 PMID:22181312

Sheppard, D., Terrell, R., & Henkelman, G. (2008). Optimization methods for finding minimum energy paths. *The Journal of Chemical Physics*, *128*(13), 134106. doi:10.1063/1.2841941 PMID:18397052

Shim, Y., & Amar, J. G. (2005). Rigorous synchronous relaxation algorithm for parallel kinetic Monte Carlo simulations of thin film growth. *Physical Review B: Condensed Matter and Materials Physics*, *71*(11), 115436. doi:10.1103/PhysRevB.71.115436

Smeltz, A. D., Delgass, W. N., & Ribeiro, F. H. (2010). Oxidation of NO with O_2 on Pt(111) and Pt(321) Large Single Crystals. *Langmuir*, *26*(21), 16578–16588. doi:10.1021/la101653x PMID:20666498

Tang, H., van der Ven, A. F., & Trout, B. L. (2004, July). Phase diagram of oxygen atoms adsorbed on platinum(111) by first principles investigation. *Physical Review B: Condensed Matter and Materials Physics*, *70*(4), 045420. doi:10.1103/PhysRevB.70.045420

Tian, T., & Burrage, K. (2004). Binomial leap methods for simulating stochastic chemical kinetics. *The Journal of Chemical Physics*, *121*(21), 10356–10364. doi:10.1063/1.1810475 PMID:15549913 van de Walle, A. (2013). Methods for First-Principles alloy thermodynamics. *JOM*, *65*(11), 1523–1532. doi:10.1007/s11837-013-0764-3

van de Walle, A., & Asta, M. (2002c). Self-driven lattice-model Monte Carlo simulations of alloy thermodynamic properties and phase diagrams. *Modelling and Simulation in Materials Science and Engineering*, *10*(5), 521–538. doi:10.1088/0965-0393/10/5/304

van de Walle, A., Asta, M., & Ceder, G. (2002a). The alloy theoretic automated toolkit: A User Guide. *CALPHAD Journal*, *26*(4), 539–553. doi:10.1016/S0364-5916(02)80006-2

van de Walle, A., & Ceder, G. (2002b). Automating First-Principles phase diagram calculations. *Journal of Phase Equilibria*, *23*(4), 348–359. doi:10.1361/105497102770331596

Voter, A. F. (1997). Hyperdynamics: Accelerated molecular dynamics of infrequent events. *Physical Review Letters*, *78*(20), 3908–3911. doi:10.1103/PhysRevLett.78.3908

Wang, X., Lau, K. C., Turner, C. H., & Dunlap, B. I. (2010). Kinetic Monte Carlo simulation of the elementary electrochemistry in a hydrogen-powered solid oxide fuel cell. *Journal of Power Sources*, *195*(13), 4177–4184. doi:10.1016/j.jpowsour.2009.12.130

Weaver, J. F., Chen, J.-J., & Gerrard, A. L. (2005). Oxidation of Pt(1 1 1) by gas-phase oxygen atoms. *Surface Science*, *592*(1-3), 83–103. doi:10.1016/j.susc.2005.07.010

Weiss, B. M., & Iglesia, E. (2009). NO Oxidation Catalysis on Pt Clusters: Elementary Steps, Structural Requirements, and Synergistic Effects of NO_2 Adsorption Sites. *The Journal of Physical Chemistry C*, *113*(30), 13331–13340. doi:10.1021/jp902209f

Wu, C., Schmidt, D., Wolverton, C., & Schneider, W. (2012). Accurate coverage-dependence incorporated into First-Principles kinetic models: Catalytic NO oxidation on Pt(111). *Journal of Catalysis*, *286*, 88–94. doi:10.1016/j.jcat.2011.10.020

Zhu, R., Pan, E., & Chung, W. (2007). Fast multiscale kinetic Monte Carlo simulations of three-dimensional self-assembled quantum dot islands. *Physical Review B: Condensed Matter and Materials Physics*, *75*(20), 205339. doi:10.1103/PhysRevB.75.205339

Ziegler, M., Kröger, J., Berndt, R., Filinov, A., & Bonitz, M. (2008). Scanning tunneling microscopy and kinetic Monte Carlo investigation of cesium superlattices on Ag(111). *Physical Review B: Condensed Matter and Materials Physics*, *78*(24), 245427. doi:10.1103/PhysRevB.78.245427

Chapter 4
Computational Design of Microstructure:
An Overview

G. Anand
The University of Sheffield, UK

P. P. Chattopadhyay
Indian Institute of Engineering Science and Technology Shibpur, India

ABSTRACT

During the last couple of decades, treatment of microstructure in materials science has been shifted from the diagnostic to design paradigm. Design of microstructure is inherently complex problems due to non linear spatial and temporal interaction of composition and parameters leading to the target properties. In most of the cases, different properties are reciprocally correlated i.e., improvement of one lead to the degradation of other. Also, the design of microstructure is a multiscale problem, as the knowledge of phenomena at range of scales from electronic to mesoscale is required for precise composition-microstructure-property determination. In the view of above, present chapter provides the introduction to computationally driven microstructure engineering in the framework of constitutive length scale in microstructure design. The important issues pertaining to design such as phase stability and interfaces has been explained. Additionally, the bird-eye view of various computational techniques in order of length scale has been introduced, with an aim to present the picture of combination of various techniques for solving microstructural design problems under various scenarios.

1. INTRODUCTION

The microstructure can be described as the length scale based ensemble of the representative units, which evolve the solid state material structures through the multi-scale spatiotemporal interaction under the specific energetic pathway. The hierarchical nature of the microstructure originates from the length scale of the constituent phases and defects originated through the path of evolution. At the smallest scale, electronic interaction among the constituent elements decides the nature of coordination among the ele-

DOI: 10.4018/978-1-5225-0290-6.ch004

Figure 1. Framework of microstructure based design (Chattopadhyay, 2014)

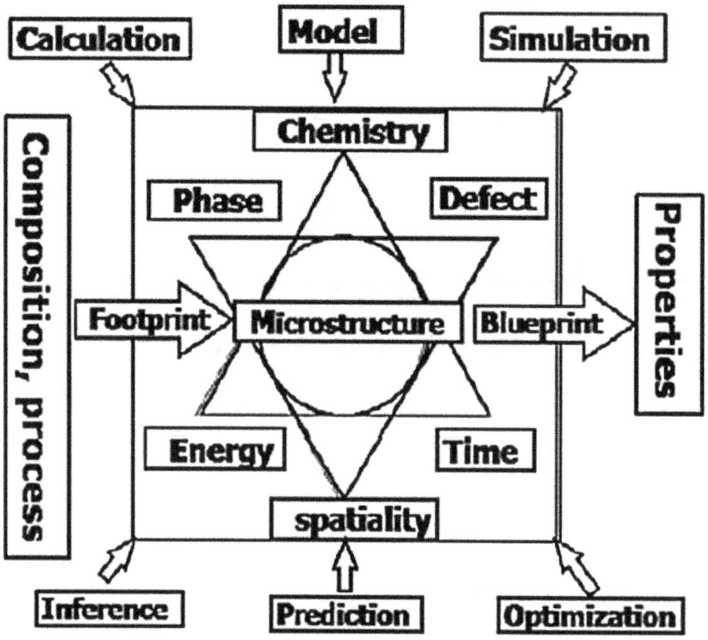

ments. In such length scale, thermodynamically stable one dimensional defect such as vacancy also play the crucial role. In the next level of integration, the coordinating atoms may aggregate themselves in the form of amorphous, quasi-crystalline and non-crystalline or crystalline order. The defect structure in this scale comprises dislocation, stacking faults, grain boundaries etc. The distinct structure on such defects originates through the interactions occurring at lower scale, e.g., stacking fault width.

In the manufacturing framework, microstructures inherit not only the footprints of the process and compositional variables but also the blue-print of the properties. Keeping this in view Figure 1 enumerates the integrated multiscale approach of microstructure based design of materials (Anand, Dey, Kok, Chakraborty, & Chattopadhyay, 2014). Contrary to the conventional imprecise approach of describing the microstructure in terms of the empirical composition-process-property framework, diagnostic framework relies on the deterministic design method, while employing range of experimental and simulation techniques (Anand, Datta, & Chattopadhyay, 2013).

2. COMPUTATIONALLY DRIVEN MICROSTRUCTURE ENGINEERING

The multi-scale nature of the microstructure evolution and the corresponding range of relaxation times result into the dynamic nature of materials structure with the "spatiotemporal hierarchy". This dynamic nature leads to the complexity in the materials design causing the limited predictability (Olson, 1997). In such complex situation, Cohen's "reciprocity principle" is an effective tool to improve the predictability. In this approach, structure-property correlation is considered as the two-way phenomena, as perception of microstructure is also guided by the specific properties to be considered (Cohen, 1976). The structure-property framework can be expanded to include the processing and performance, as shown in Figure 2,

Figure 2. Three link chain model in processing-structure-property-performance framework redrawn from (Olson, 1997)

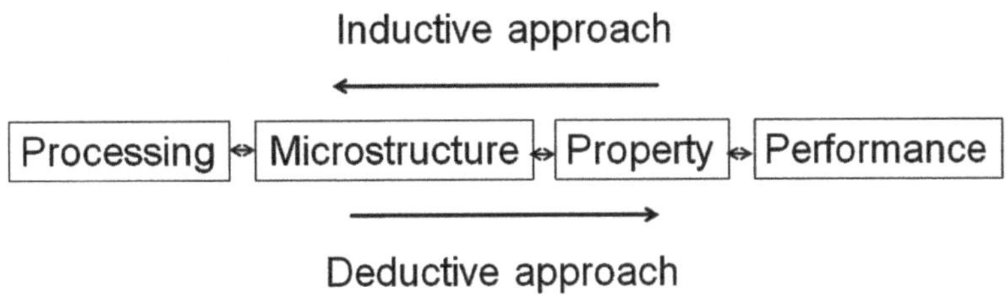

which enumerates the flow of deductive cause and effect approach of science to right, while inductive goal and means relation of engineering have leftwards orientation.

So, the complexity-driven problems of microstructure based design of materials are thought to be solved in the multi-scale framework with the description and integration of the sub-systems (Jenkins, 1972). Figure 3 shows the integrated four-stage multicomponent materials design framework involving increment in length and time scale in downwards direction.

As, shown in Figure 3, the first principle calculations can provide the thermodynamic data of unary, binary or ternary system, along with parameters such interfacial energy and lattice constant, with possibility to study the meta-stable phases, which might be experimentally inaccessible to generate thermodynamic databases. CALculation of PHase Diagram (CALPHAD) takes the data, both from first principle calculations and experiments to predict the phase diagram for the multicomponent systems. In this approach, Gibbs free energy for individual phases can be calculated for thermodynamic database generation. In third stage, phase field technique can be employed to generate the microstructures, even in the case of complex multiphase system. Prediction of microstructure in this technique can be carried out without prior assumptions, with the input data either from CALPHAD (Wu, Chen, Zhang, & Chang, 2009) or first-principle calculations (Vaidyanathan, Wolverton, & Chen, 2002). The last and fourth stage in the above scheme involves continuum or crystal plasticity simulation to predict the macroscopic behaviour.

The following sections aim to provide the brief outline of the techniques employed at different length scale with the proven potential in microstructure based design of materials.

3. CONSTITUTIVE LENGTH SCALES

The bonding of metals can be described on the basis of the primary bonds; namely metallic and covalent bonding. These bonds are responsible for the elastic properties of metals and genesis of the Young's modulus. The energy of the covalent bonds can be well described by the following empirical equation given as Equation (1).

$$E = -\frac{A}{r^m} + \frac{B}{r^n} \ (\mathrm{m} < \mathrm{n}) \tag{1}$$

Figure 3. The different stages of integrated multi-scale modelling approach for design of multicomponent materials [redrawn from (Liu et. al., 2004)

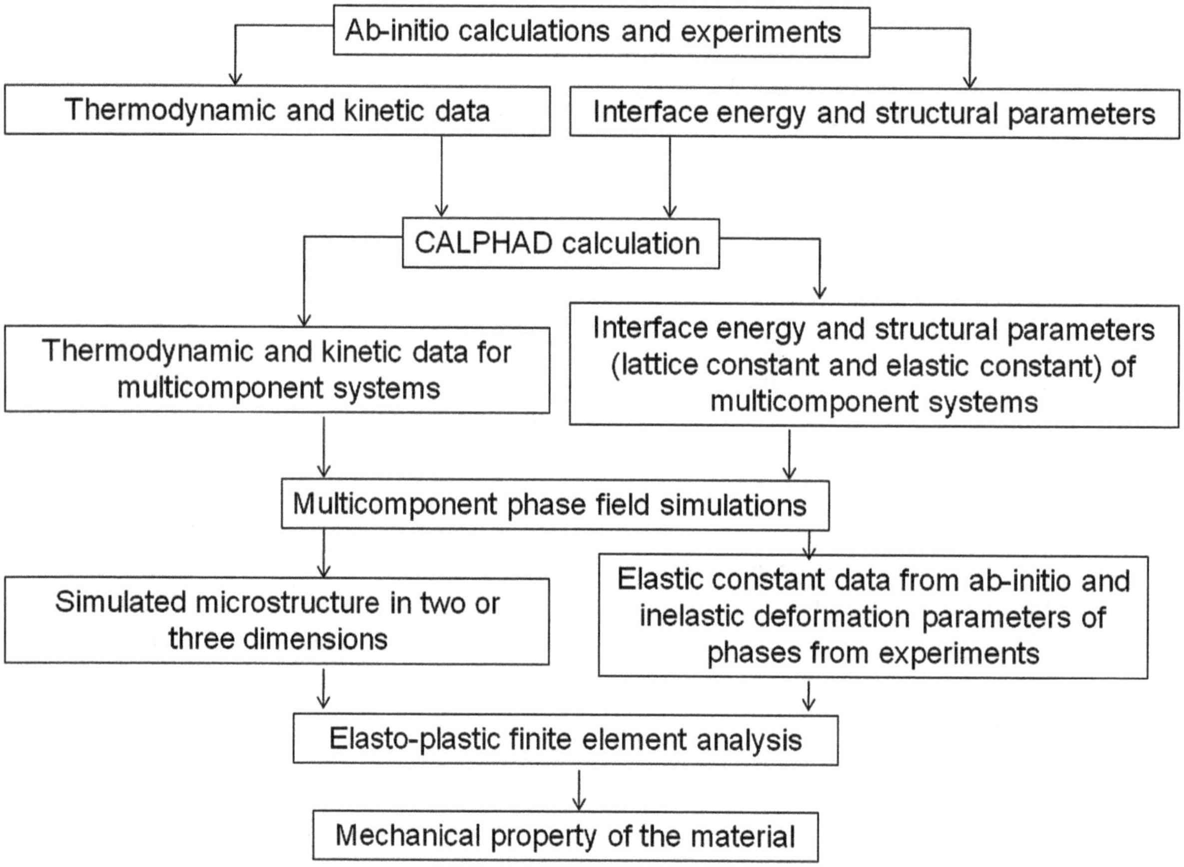

Where, first part of the expression is the attractive part, while the second term represents repulsive part. So, it would lead to the variation, as shown in Figure 4 with minimum energy at the equilibrium lattice spacing.

The metallic bonding is the dominant bonding mechanism in metals and alloys. The metallic bond is described by the 'electron-sea' model, where high energy electron leaves the parent metal atoms to form 'electron-sea'. The characteristic nature of such bonding leads to the high electronic and thermal conductivity in metals. The nature of energy versus lattice spacing can be approximately represented as shown in Figure 4.

Variation in coordination of the constituent atoms leads to the formation of solid solution, intermetallics, amorphous or immiscible systems. The formation of solid solution or intermetallics in simple binary systems is depicted by change in the internal energy during mixing, which is the function of the bond energy. Negative change in the internal energy favours disordered solution, while positive change causes the ordering. The formation of intermetallic compounds is dependent on three factors; namely, atomic size difference, valency and electronegativity difference. The amorphous materials are essentially supercooled liquid, which tend to conserve their high temperature liquid structure at ambient temperature. The amorphous structure can be conceived by considering the random packing of the constituent atoms.

Figure 4. The variation of the potential energy as function of lattice spacing

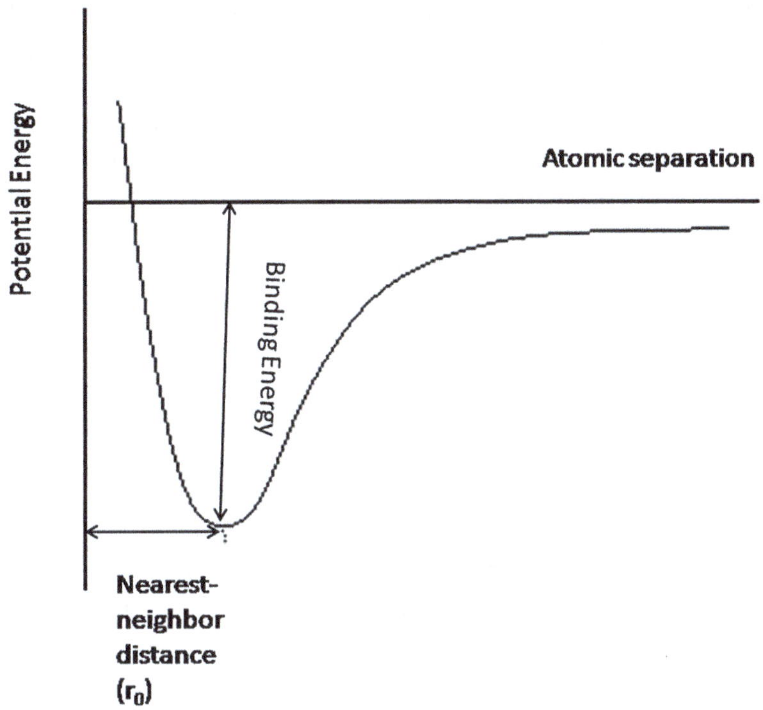

Figure 5. Schematic representation of all possible arrangements of solid that may possibly form from liquid (Stachurski, 2011)

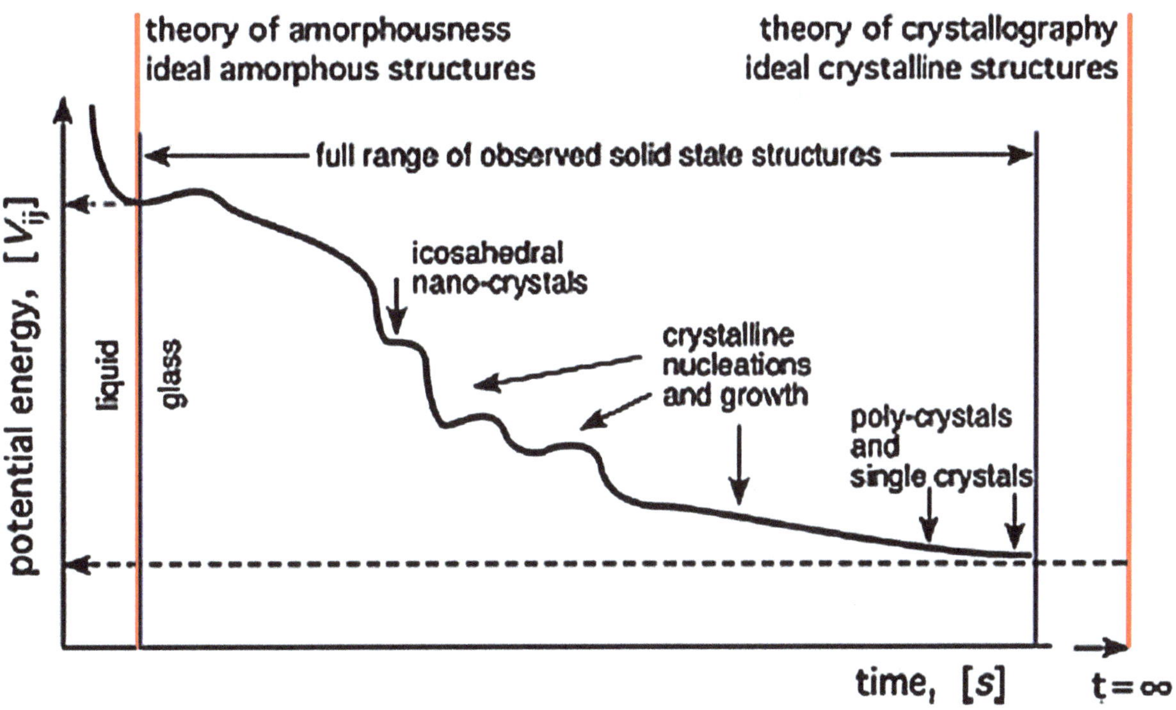

Figure 5 shows the all possible solid configurations, if cooling is continued beyond solidification. It is important to note that, Figure 5 presents the geometric theory of structure of solids, i.e. Crystallographic theory and theory of amorphous structures at two extremes. Crystallographic theory is placed at infinite time and the theory of amorphous structures is placed at time frame of solid-liquid transformation. All solid state atomic arrangements can be explained between these two extremes.

Another approach to understand the structure of solids is the form of structural networks. In conventional Graph theory, network denotes the system of vertices in which any two vertices are connected by the path of constraints (Temperley, 1981). For random dense packing, the constraint can be the minimum distance to be maintained between two atoms, arising from the short range repulsion between atoms. For strongly bonded atoms, such constraint arises due to the covalent bonding of atoms. In the view of above, crystalline structure have long range order with translational periodicity (Figure 6(a)). Similarly, the quasi-crystalline structure is formed with positional order without translational periodicity (Figure 6(b)). It is important to note that in this case, triangles are not equilateral, as in the case of crystalline structure, rather they are isosceles triangle with included angle of 72^0. Such networks are derived from two-dimensional Penrose tiling (Janot, 1992). Non-crystalline networks can be topologically ordered (TO) and topologically disordered (TD), as shown in Figure 6(c) and 6(d) respectively. The TO phase may crystallise readily as topological change (breaking of bonds and reforming) is not required, while TD networks have intrinsic resistance towards crystallisation due to requirement of topological ordering.

Figure 6. (a) crystalline, (b) quasi-crystalline, (c) non-crystalline topologically ordered and (d) topologically disordered structure [a, b, and d redrawn from (Gupta, 1996); (b) drawn from wolfram alpha].

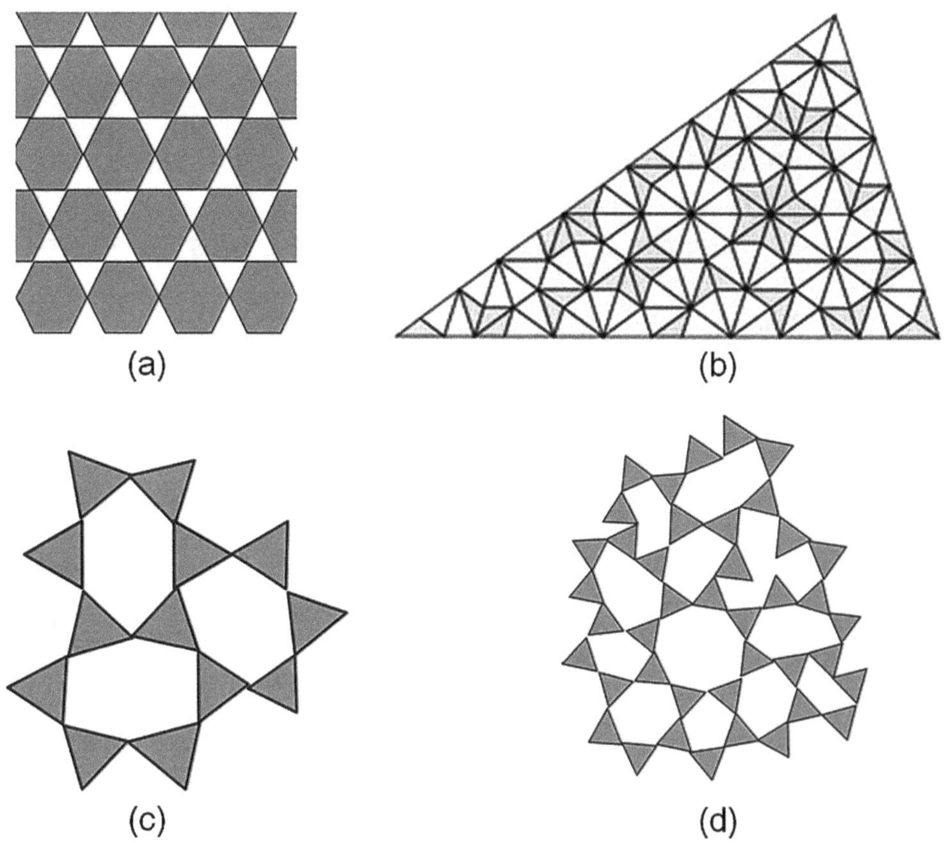

Most of the crystalline metallic alloys are either face-centered cubic (fcc), body-centered cubic (bcc) or hexagonal-closed packed (hcp) structures. The stability of the particular solid solution is dependent upon the electron to atom (e/a) ratio, because the peak of the density of states (DOS) of electrons lies at the contact of Brillouin zone limit and Fermi sphere (Poletti & Battezzati, 2014). So at certain value of e/a ratio, particular crystal structure is stabilized, as it is capable of accommodating more electrons at low energy level than would be possible by another structure. For multi component systems, like high entropy alloys (HEA), which interestingly show simple crystal structure like fcc, bcc (Yeh et. al., 2004) and hcp (Feuerbacher, Heidelmann, & Thomas, 2015) structure, the stability of the particular phase has been proposed to be attributed to valence electron per atom (VEC). Guo et. al. (Guo, Ng, Lu, & Liu, 2011) have attempted to predict the BCC and FCC structured solid solution of HEA in terms of the valence electron concentration (VEC). The VEC is expressed as:

$$VEC = \sum C_i (VEC_i) \quad (2)$$

Where, C_i represents atomic percentage of ith element and VEC_i represents the VEC of the ith element. For bcc structured solid solution, VEC is less than 6.8, while fcc solid solution occurs at VEC greater than 8.

4. CLASSIFICATION

The definition of the microstructure requires the precise description of distribution of shape, size, and spatial distribution of all grains and defects along with quantization of microstructural anisotropy in terms of the orientation distribution of the grains. The variation of the microstructural anisotropy in the different microstructures can be understood in the terms of the aspect ratio of second phase in the case of two-phase microstructures (Hornbogen, 1984).

While describing the microstructure, the distinction between the phase and microstructure need to be understood. The phases are three dimensional zone having continuous and homogeneous physical properties, whereas the microstructure encompasses various phases and defects in it. The phase can be described by the stable or metastable thermodynamics, while in most of the cases elements in the microstructures are non-equilibrium entities (Hornbogen, 1968). The defects in the microstructure are part of the microstructures, unless they are required to minimise the free energy (Presland, 1960).

The microstructure can be fundamentally described in the terms of phase and grain or interface boundaries. On the basis of that, basically three different types of microstructures can be defined in the case of simple two-phase systems; namely dispersion, net and duplex type of microstructures, as shown in Figure 7. The quantitative description of microstructure requires the knowledge of the density of grain or interface boundaries. For ideal dispersion microstructure, having dispersions (β) in matrix (α) (Figure 7(a)), the density of β-β interface is zero, while in the case of 'net-type' microstructure, the dispersion formed at the grain boundaries totally engulf the structure and it has zero density of α-α interface (Figure 7(b)). The third type of microstructure is duplex type microstructure (Figure 7(c)). In this case, the percolation in both α and β takes place. The average grain size in such microstructure is inversely proportional to the sum of the densities of all the boundaries.

Figure 7. Schematic representation of the three typical microstructures, (a) dispersion, (b) net and (c) duplex microstructure

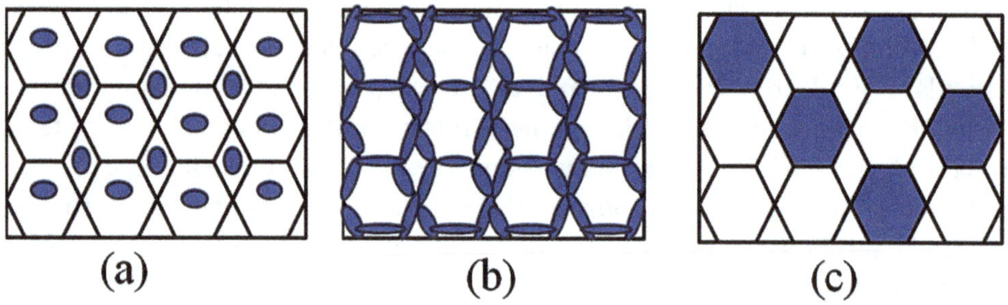

5. INTERFACE STRUCTURE AND PROPERTIES

Most of the engineering materials are polycrystalline in nature with grains being separated by interfaces known as 'grain boundaries'.

The tilt grain boundary (Figure 8(a)) can be described by four degrees of freedom, as two normal vectors, rotation axis and rotation angle (or angle between two planes), while twist grain boundaries (Figure 8(b)) have three degree of freedom, two for normal to its plane and one for twist angle. The mixed grain boundaries can exhibit five degree of freedom. Also, grain boundaries can be symmetrical or asymmetrical. The symmetrical boundaries can have same indices in two grains. $\{hkl\}_A = \{hkl\}_B$, while $\{hkl\}_A \neq \{hkl\}_B$ for asymmetrical boundaries.

In certain cases, under special processing conditions, finite fraction of lattice sites coincides between two lattices, leading to the formation of 'Coincidence Site Lattices (CSL)'. These are also termed as special grain boundaries.

The CSL boundaries are represented in terms of Σ values, where the value of Σ is inversely proportional to the density of coincident points. Hence, lower values imply less grain boundary energy. Σ is expressed as the ratio of area enclosed by unit cell of coincident lattice sites to the area enclosed by unit cell. In the case of cubic structures, whenever even number is obtained while calculating area of

Figure 8. Schematic representation of (a) tilt boundary and (b) twist boundary. A and B are two grains, while ń represents the normal to the interface and θ represents the rotation.

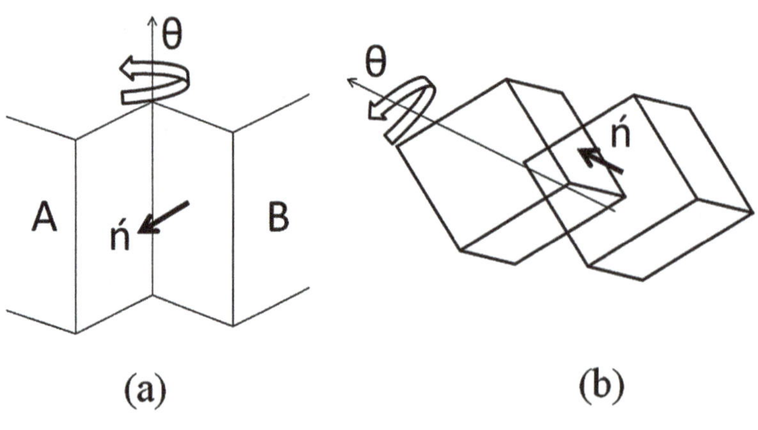

lattice sites, there is CSL lattice sites in the centre of unit cell, which implies that true area is half of the calculated area. Thus, Σ is always odd, i.e., Σ3, Σ5, Σ7, Σ9,.... Σ29.

The Brendon's criteria provide the relationship of proximity of lattice to the CSL structure and its own existence as CSL structure as following relationship (Brandon, 1966):

$$v_m = v_0 \Sigma^{-1/2} \tag{3}$$

Where, v_0 is 15^0, based on low angle to high angle transition. Larger value of Σ implies greater unit cell in the boundary and fewer coincident points, thus larger dislocation densities for same deviation from exact CSL misorientation. So, for larger Σ values, one has to be closer to boundary to be counted in CSL, in comparison for low Σ boundary.

The design approach to achieve greater fraction of CSL boundaries to have better structural, corrosion resistance, and functional properties in materials is termed as "Grain Boundary Engineering". The detailed review and explanation can be found elsewhere (Randle, 2004).

During the last couple of decades grain boundary/interface structures of the nonoscale polycrystalline materials have drawn significant attention owing to the fact that large number of atoms (~50% for 5nm) in such materials are located at the grain boundaries. Figure 9 shows the schematic representation of nanocrystalline materials with dark circles representing atoms in lattice and white showing grain boundary atoms. Such grain boundary structures have earlier been described as 'gas like' with much reduced order (Zhu, Birringer, Herr, & Gleiter, 1987). However, several studies have refuted such claim. For example, quantitative x-ray diffraction analysis of nanocrystalline Pd did not support the 'gas like' structure (Fitzsimmons, Eastman, Müller-Stach, & Wallner, 1991). Subsequently, atomistic simulation of vibrational density of states (Wolf, Wang, Phillpot, & Gleiter, 1995) and high frequency vibrational properties of metallic nanocrystalline boundaries (Derlet, & Van Swygenhoven, 2004) have shown that the nanocrystalline grain boundaries have reduced coordination of atoms with excess free volume at the grain boundary region.

Atomistic simulations have improved the understanding for wide range of phenomena occuring at interfaces such as structural transformations, faceting, dissociation reactions, etc. (Schmidt, Finnis, Ernst, & Vitek, 1998; Sansoz & Molinari, 2005; Tschopp & McDowell, 2007a and Tschopp & McDowell, 2007b). Faceting phenomena of grain boundaries (development of sawtooth profile from flat boundary) have been studied using EAM potentials (Wu, Zhang, & Srolovitz, 2009), and such calculations have predicted that facet migration is function of grain boundary stress. Atomistic modelling has been also applied to extract thermodynamic properties of interfaces. Paticularly in the case of capillary driven processes such as preciptation, grain growth, solute segregation, interface decohesion, etc. (Sutton & Balluffi, 1995). The detailed treatment can be found elsewhere (Rickman & LeSar, 2002; Broughton & Gilmer, 1986; Sutton, Mulheran, & Stoneham, 1992 and Hairie et. al., 1994). For simulating grain growth, thermodynamic driving force is governed by grain boundary stiffness, which is the input for Gibbs-Thomson equation (Herring, 1949). Grain boundary stiffness has been calculated by molecular dynamics calculation from analysis of equilibrium capillary fluctuations above the roughening temperature of boundary (Trautt & Upmanyu, 2005 and Foiles & Hoyt, 2006). The interface segregation is another important issue due to their effect on the structural properties of materials, as well as structural and thermodynamic properties of interfaces (Millet, Selvam, & Saxena, 2007). Segregation in the mutlicomponent systems with heterophase interfaces have been proposed in the case of dilute solutions from

Figure 9. Schematic representation of two-dimensional representation of nanocystalline materials depicting grains (dark) and grain boundaries (light) [redrawn from (Gleiter, 2000)].

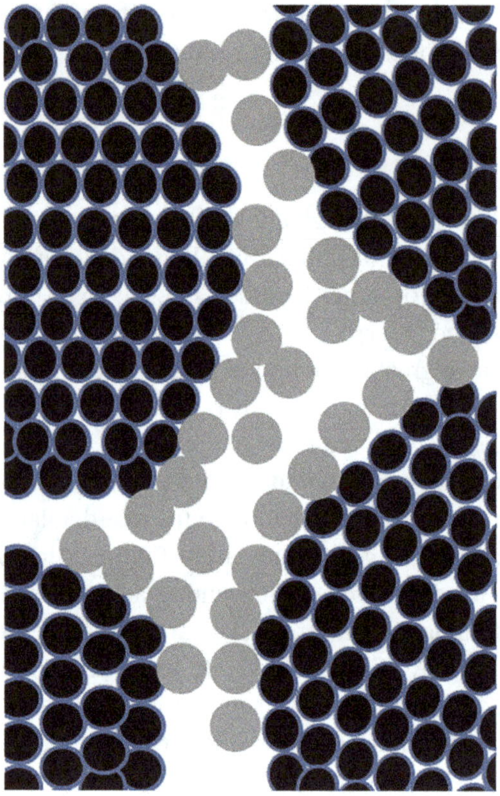

first-principle calculations involving the formation energies of each solute atoms at various locations in the vicinity of the interface. Such formation energies are employed in the standard mean-field statistical thermodynamic models to obtain the interfacial free energy (Marquis, Seidman, Asta, Woodward, & Ozoliņš, 2003; Marquis, Seidman, Asta, Woodward, & Ozoliņš, 2006 and Peng, Soh, Yang, & Hu, 2006).

The interface motion is very critical issue, as it leads to the development of materials microstructure. Depending upon the phenomena of microstructure evolution, properties of particular interface is important. The ultimate development of the morphology of the microstructure is influenced by mobility of the interfaces and non-equilibrium solute partitioning across the interfaces. The grain boundary mobility depends upon the grain boundary crystallography, temperature and impurity content in the material (Mishin, Asta, & Li, 2010). The effect of grain boundary crystallography has been studied with respect to its effect on the mobilities and activation energy of movement, with the results showing maxima or minima in the values at low-Σ misorientations (Upmanyu, Srolovitz, Shvindlerman, & Gottstein, 1999 and Janssens et. al., 2006). The effect of the temperature on the mobility has been studied with molecular dynamics calculation and significant increase in the mobility above roughening temperature (Olmsted, Foiles, & Holm, 2007). it is important to mention here is that the grain boundary mobilities is overestimated, while activation energy of grain boundary migration is underestimated in the simulations in comparison with experimental values. Such discrepancy has been attributed to the impurity (Mendelev & Srolovitz, 2002), as impurity content is high at the interface due to segregation, even when the bulk content is low.

Another important area concerning the grain boundary modelling is the determination of limit of strength of nanocrystalline materials. As Hall-Petch effect (Hall, 1951 and Petch, 1953) predicts the increase in the strength as the grain size is reduced, given by following relationship:

$$\sigma_y = \sigma_0 + \frac{k}{\sqrt{d}} \tag{4}$$

Where, σ_y represents the yield strength of the material, σ_0 is materials constant (stress required for the dislocation to move, when single dislocation is present in the centre of grain or resistance of the lattice for dislocation movement), k is the strengthening coefficient (material specific property) and d is the average grain diameter. But, beyond certain critical grain size (which is specific for particular material) decrease in grain size with the decrease in grain size occurs. This phenomenon is termed as 'Inverse Hall-Petch effect' (Chokshi, Rosen, Karch, & Gleiter, 1989). Molecular dynamics simulations have provided the quantitative understanding of Inverse Hall-Petch relationship, where deformation predominantly occurs through sliding of planes in grain boundaries and by minor dislocation activity, as nanocystalline materials have large fraction of grain boundaries (Schiøtz, Di Tolla, & Jacobsen, 1998).

6. PHASE EQUILIBRIUM

For efficient process design, it is important to know the stability of the phase in the composition-temperature space. The pressure is usually ignored for normal engineering applications. Traditionally, phase diagram have been drawn for unary, binary and ternary systems, through traditional process (Chang et. al., 2004). Since most of the engineering materials are multicomponent systems, development of mutlicomponent phase diagram is the crucial issue. In this direction, CALculation of PHase Diagram (CALPHAD) approach (Kattner, 1997) is so far the most successful for development of such phase diagram using input from single, two or three component phase diagrams.

The CALPHAD method is based upon the minimization of the free energy. For determination of the phase equilibrium, total free energy (G) of the phase need to be defined, which may be written in the terms of contribution from temperature, pressure and magnetic terms.

$$G^{\phi} = G_T^{\phi}(T,x) + G_p^{\phi}(p,T,x) + G_m^{\phi}(T_c,\beta,T,x) \tag{5}$$

Where, $G_T^{\phi}(T,x)$ represents the contribution from temperature, T and composition, x. $G_p^{\phi}(p,T,x)$ is pressure contribution and $G_m^{\phi}(T_c,\beta,T,x)$ is the magnetic contribution, with T_c representing Curie temperature and β is average magnetic moment per atom. For the condensed phases, the pressure contribution can be ignored. Among these terms, the first term is most important for determination of phase stability. The first term of the above expression can be written in terms of T as:

$$G = a + b.T + c.T.In(T) + \sum dn.T^n \tag{6}$$

Where, a, b, c and n are integers. It is important to note that the above expression holds above the Debye temperature (It is the highest temperature that occurs due to oscillation in which all the particles move with same frequency and are in phase).

Also, in the case of systems having more than one element, the compositional contribution to the free energy for disordered solid solution has three components to itself, involving free energy due to mixing of the elements (without interaction, G_m), entropic contribution considering ideal solution behaviour (G_e) and excess free energy term corresponding to the regular solution behaviour (G_{xs}).

$$G = G_m + G_e + G_{xs} \tag{7}$$

For binary solid solution, it may be written as:

$$G = x_A.G_A^0 + x_B.G_B^0 + RT\left(x_A In(x_A) + x_B In(x_B)\right) + x_A x_B \sum_{i=0}^{n} G_i\left(x_A - x_B\right) \tag{8}$$

The last term in above expression represents the regular solution model of the binary solid solution, which can be represented in the form of Redlich-Kister polynomial, used to predict the properties of higher order thermodynamic systems using binary contributions. The sub-lattice description of the ordered solid solution is generalised representation of the free energy, which can be represented as (Kattner, 1997):

$$\begin{aligned}G = {} & x_A.G_A^0 + x_B.G_B^0 + RT\left\{a^1\left(y_A^1 Iny_A^1 + y_B^1 Iny_B^1\right) + a^2\left(y_A^2 Iny_A^2 + y_B^2 Iny_B^2\right)\right\} \\ & + y_A^1 y_A^2 G_{AA}^0 + y_A^1 y_B^2 G_{AB}^0 + y_B^1 y_A^2 G_{BA}^0 + y_B^1 y_B^2 G_{BB}^0 + y_A^1 y_B^1 y_A^2 \sum_{i=0}^{n_{2A}} G_i^{2A}\left(y_A^1 - y_B^2\right)^i + \\ & y_A^1 y_B^1 y_B^2 \sum_{i=0}^{n_{2B}} G_i^{2B}\left(y_A^1 - y_B^1\right)^i + y_A^1 y_A^2 y_B^2 \sum_{i=0}^{n_{1A}} G_i^{1A}\left(y_A^2 - y_B^2\right)^i + y_B^1 y_A^2 y_B^2 \sum_{i=1}^{n_{1B}} G_i^{2A}\left(y_A^2 - y_B^2\right)^i + y_A^1 y_B^1 y_A^2 y_B^2 G^{hp}\end{aligned} \tag{9}$$

The a^1 and a^2 represent site fraction of sub-lattice 1 and 2, given by the number of site in the unit cell. y_A^1, y_B^1, y_A^2 and y_B^2 is the concentration of A and B on sublattices 1 and 2, with the following relations $a^1 y_A^1 + a^2 y_A^2 = x_A$, $a^1 y_B^1 + a^2 y_B^2 = x_B$, $y_A^1 + y_B^1 = 1$ and $y_A^2 + y_B^2 = 1$. The coefficients G_{AA}^0, G_{AB}^0, G_{BA}^0 and G_{BB}^0 is the Gibbs energy of the end member phases.

As it is known that, chemical potential (μ) of the particular element is same in all phases, which may be expressed as:

$$\begin{aligned}\mu_1^{'} &= \mu_1^{''} = \ldots\ldots = \mu_1^{m} \\ \mu_2^{'} &= \mu_2^{''} = \ldots\ldots = \mu_2^{m} \\ &\ldots\ldots\ldots\ldots\ldots\ldots \\ \mu_n^{'} &= \mu_n^{''} = \ldots\ldots = \mu_n^{m}\end{aligned} \tag{10}$$

The above consideration leads to the n number of non-linear equations, which can be solved by the numerical techniques such as Newton-Raphson method to yield chemical potential of each species, which is correlated with the free energy by the following relation:

$$G = \sum_{i=1}^{n} x_i \mu_i \tag{11}$$

Determination of the coefficients (called 'assessment' or 'optimisation') in Gibbs free energy is the crucial issue. The experimental data, such as phase diagram, enthalpy and chemical potential is required to obtain the coefficients. For this purpose, trial and error method or mathematical methods such as Gauss least square or Bayesian estimation method have been found to be useful. However, such methods become computationally expensive as the numbers of elements and /or input data types are increased.

Free energy of the multicomponent systems can be determined by thermodynamic extrapolation of the excess free energy of the constituent systems. Figure 10 shows the CALPHAD methodology for thermodynamic database generation for the multicomponent systems.

Figure 10. CALPHAD methodology for the development of thermodynamic database of multicomponent systems using binary system data [redrawn from (Kattner, 1997)]

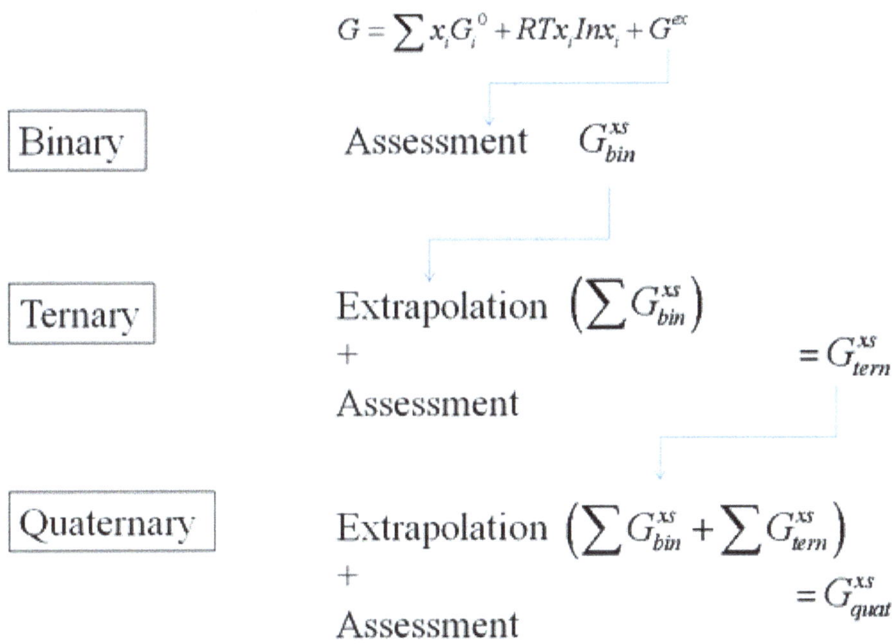

In this description, thermodynamic description of binary solutions is defined. Then, extrapolation schemes are used to extend the thermodynamic function to ternary and higher order systems. These extrapolations are not self consistent in nature, as these need to be validated with experiments and in certain cases, additional parameter known as interaction functions are added for higher order systems.

The parameters to be used for determination of coefficient of the free energy function are determined experimentally. The determination of such parameters is often very tedious. The calorimetric measurements provides the main input for all thermodynamic databases, but there is also possibility of scatter in the range of 0.3 k_B-1 k_B. In the absence of experimental results, *ab initio* methods like Density Functional Theory (DFT) have been found to be extremely useful in estimation of the inputs for calculation of free energy. In view of this it is imperative to enumerate the potential of DFT in the context of the calculation of thermodynamic quantities.

7. DENSITY FUNCTION THEORY

The density functional theory is one of the most appropriate techniques for studying multi-component systems. Such calculation requires atomic number and crystal structure for predicting the electronic and structural properties of solid. DFT allows the calculation of bulk moduli, elastic constants, magnetic moments, spin distributions, etc to name a few. In view of materials design DFT is an effective tool for selection of the composition and/or impurities and studying its effect on the defect structure (DFT) (Paxton, 2014; Tsuchida, Watanabe, Kato, & Seto, 2010; Curtze & Kuokkala, 2010; Grassel, Frommeyer, Derder, & Hofmann, 1997 and Dumay, Chateau, Allain, Migot, & Bouaziz, 2008). The detailed review on this technique can be found elsewhere (Freeman, 1995 and Neugebauer & Hickel, 2013). In this section, the emphasis would be put on the application of DFT for generation of input data for CALPHAD (Liu, 2009).

For CALPHAD modelling, two primary sets of input data are required (Figure 11). Firstly, thermochemical data and secondly phase equilibrium data. The thermochemical data is for single phases, while the phase equilibrium data concerns more than one phase. So, phase equilibrium data is comparatively easier to obtain in comparison with the thermochemical data. Hence, DFT based *ab-initio* methods can be used in conjugation with experimental data for generation of input data for CALPHAD.

The thermodynamic properties of the alloys can be calculated by considering three contributions, which are; firstly the 0 K energy, which is the energy of the system when atoms are fixed, secondly

Figure 11. Schematic showing contribution of thermochemical and phase equilibrium data for CALPHAD modelling [redrawn from (Liu, 2009)]

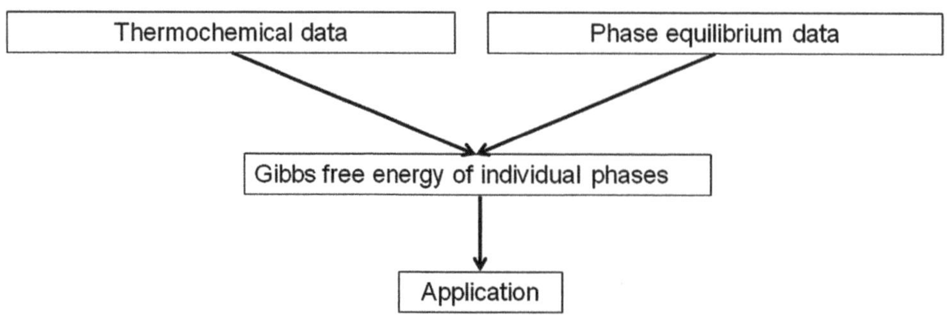

thermodynamic properties of solid at finite temperature, due to lattice vibrations based on the lattice vibration or phonon approach and lastly thermal electronic contribution need to be considered, if the electronic density of states is high at the Fermi level.

In the course of conducting DFT calculations certain approximations are made, so that it is required to solve only a set of one-electron Schrödinger equation for each electron in the system. The first approximation is Born-Oppenheimer or adiabatic approximation (Born & Oppenheimer, 1927), which assumes that electrons are always in instantaneous ground state with nucleus, as nuclei, is much heavy in the comparison with the electrons. So, Schrödinger equation needs to be solved for electron only. The second approximation is termed as 'independent-electron approximation', which implies that the movement of each electron is independent of each other in average effective potential, having contribution from all other electrons. This approximation should be able to describe the many-body electron-electron interaction. Hohenberg-Kohn approach (Hohenberg & Kohn, 1964) considers that energy functional for many-body can be defined on the basis of electron density and potential of interaction electrons is dependent of ground state electron density. The Kohn-Sham approach considers independent electrons with an exchange-correlation functional of electron density and its exchange-correlational energy and potential (Kohn & Sham, 1965). Exchange-correlation energy can be approximated as the local functional of electron density by separating independent electron kinetic energy and long range Coulomb interaction energy. Third approximation involves exchange-correlation energy, according to which exchange-correlation energy density at each point in space is assumed to be same in homogenous electron gas with same electron density. It is termed as Local Spin Density Approximation (LSDA). It implies that exchange-correlation energy have short range (Ceperley & Alder, 1980). Lastly, the fourth approximation is replacement of strong Coulomb potential between core electrons and nucleus by pseudo-potential, which is effective potential acting on valence electrons.

DFT calculation involves solution of the following one-electron Schrödinger's equation:

$$\left[-\frac{h^2}{4\pi me}\nabla^2-\frac{e^2}{4\pi\varepsilon_0}\sum_{i=1}^{N}\frac{Z_1}{|\vec{r}-\vec{R}|}+\frac{e^2}{4\pi\varepsilon_0}\int\frac{\rho(\vec{r}\,')}{|\vec{r}-\vec{r}\,'|}d^3\vec{r}\,'+v_{xc}\left[\rho(\vec{r})\right]\right]\psi_i(\vec{r})=\varepsilon_i\psi_i(\vec{r}) \tag{12}$$

Here, ε_i is ith one-electron energy eigen value with $\psi_i(\vec{r})$ representing the electron wave-function, e is the electron charge, ε_0 is vacuum permeability. The term $-\left(h^2/4\pi me\right)\nabla^2$ is electron kinetic energy with h being Planck's constant and ∇^2 is second order partial derivative along three spatial dimension. The term $\frac{e^2}{4\pi\varepsilon_0}\sum_{i=1}^{N}\frac{Z_1}{|\vec{r}-\vec{R}|}$ represents nuclear attraction potential with $\vec{r}=(x_1,x_2,x_3)$ and $\vec{R}$ is the nuclear coordinate of ith atom having nuclear charge Z_1. The term $\frac{e^2}{4\pi\varepsilon_0}\int\frac{\rho(\vec{r}\,')}{|\vec{r}-\vec{r}\,'|}d^3\vec{r}\,'$ is one-electron Hartree repulsion potential with $\rho(\vec{r})$ being total electronic charge density. $v_{xc}\left[\rho(\vec{r})\right]$ represents the one-electron exchange-correlation potential is given as:

$$v_{xc}\left[\rho(\vec{r})\right]=\frac{\delta}{\delta\rho(\vec{r})}E_{xc}\left[\rho(\vec{r})\right] \tag{13}$$

Figure 12. Schematic representing the typical steps followed in single point DFT simulation

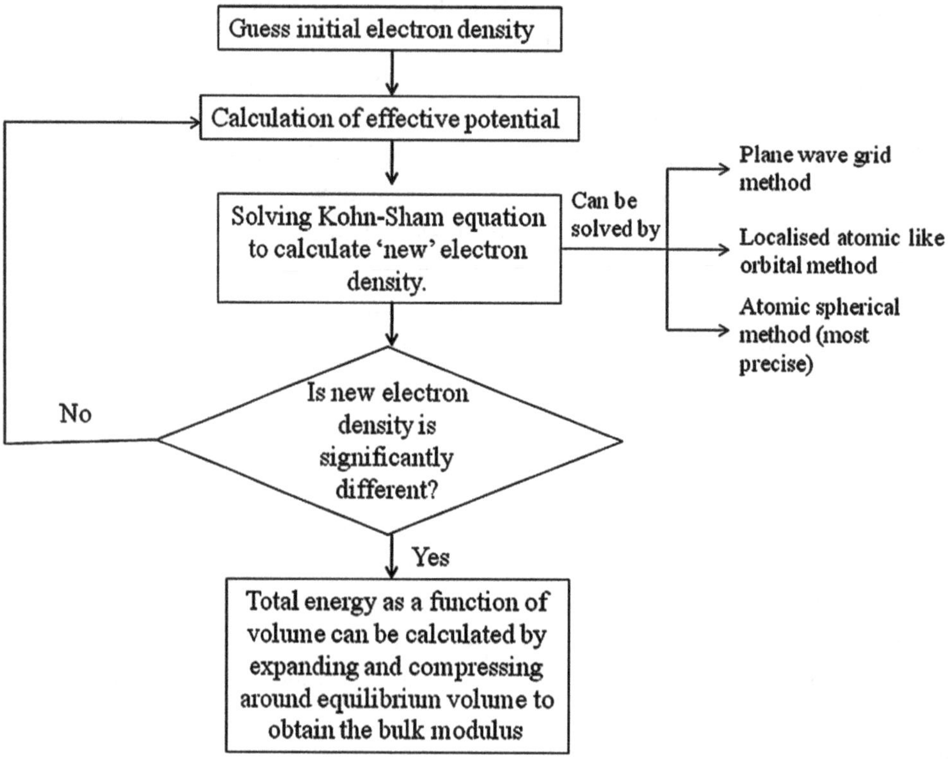

$E_{xc}\left[\rho(\vec{r})\right]$ is exchange correlation energy with the electron density $\rho(\vec{r})$. Figure 12 shows the schematic of the typical single point DFT calculations.

Ideally, numbers of atoms in a cell can be increased to form the supercells for prediction of the thermodynamic properties of random alloys and sample all the possible configurations. However, such approach is computationally expensive. Special periodic quasi-random structures (SQS) method has been put forward to mimic the correlation function of random solution by ignoring the errors due to distant neighbours, while the correlation functions for the first few shells around the given site is only considered (Wei, Ferreira, Bernard, & Zunger, 1990).

For calculation of thermodynamic properties at finite temperature, entropic contributions need to be calculated. Formation enthalpies of pure components and stoichiometric compounds can be calculated from 0K formation energies as the enthalpy of mixing is not strongly dependent upon the temperature (Wang et. al., 2004). As mentioned above, for random phases, SQS method is employed for calculation of formation enthalpies. So, calculation of entropic contribution is the crucial task. The entropy has three components namely; configurational (due to random distribution of chemical species in lattice), vibrational (due to lattice vibrations) and electronic contributions (due to electronic excitations and spin polarisations). It is important to note the length scale in considering the entropy. In terms of the increasing length scales, electronic entropy originates at electron level, while vibrational entropy occurs at interatomic level and configurational entropy plays a role at unit cell level. The configurational entropy $\left(S_{conf}^{mix}\right)$ can be represented as:

$$S_{conf}^{mix} = -R\sum x_i Inx_i \tag{14}$$

It depicts the ideal mixing behaviour with R being Universal gas constant and x_i is the mole fraction. For non-ideal mixing, where short-range ordering occurs, cluster expansion and Monte Carlo methods are used, for which details can be found elsewhere (Sanchez, 1993 and Van de Walle & Asta, 2002).

Vibrational entropy is the manifestation of the elastic waves in the solid, known as 'Phonons', due to lattice vibration at finite temperature. The energy of the phonons is function of the angular frequency of each elastic mode. The phonon density of states (DOS) gives the number of modes of oscillations as the function of frequency. The first principle approach can either be performed with harmonic and quasi-harmonic approximation to obtain the full phonon spectra for given configuration, depending upon the conditions. The restoring force constant tensors $\phi(i,j)$ on atom i due to atom j are calculated from the second derivative of energy change with respect to the displacements. Calculation of the frequencies (υ) of normal mode between two atoms can be obtained from Eigen values of dynamic matrix of $\phi(i,j)/m_i m_j$, where m_i and m_j represents the mass of atom i and j respectively. In harmonic approximation, the vibrational entropy can be calculated by summing up all the microstates in the term of vibrational phonon DOS $(g(v))$ expressed as (Wang, Liu, & Chen, 2004 and Van De Walle & Ceder, 2002):

$$S_{vib} = -k_B \int_0^{\infty} In\left[2\sinh\left(\frac{hv}{2k_B T}\right)\right] g(v)\,dv \tag{15}$$

The force constants can be calculated by either by linear response theory (Gonze, 1997) or 'supercell method', where atoms are slightly displaced from their equilibrium position. The force constant is calculated with their displacement and corresponding reactive forces.

The quasi-harmonic approximation accounts for thermal expansion at higher temperatures and its effect on vibrational entropy. In this case, phonon DOS is made dependent upon volume. As it was stated earlier that the thermally activated electronic entropy is important when electronic DOS is high at Fermi level. In quasi-harmonic approximation, the electronic entropy (Wang, Chen, & Zhang, 2000) can be written as:

$$S_{el} = -k_B \int n(\varepsilon,V)\left[f.\ln(f)+(1-f)\ln(1-f)\right]d\varepsilon \tag{16}$$

Where, $n(\varepsilon, V)$ represents the electronic DOS and f is Fermi distribution (The Fermi function gives the probability at given temperature, whether particular energy state would be occupied).

8. MOLECULAR DYNAMICS (MD) SIMULATION

A number of microstructural phenomena such as phase transformation, plastic deformation, crystallization, etc., involves the large number of atoms and requires the statistical averaging over many large atomic events, which is not possible by quantum mechanical based calculations due to spatiotemporal limitations. Simulations of such phenomena are possible due to the application of semi-empirical inter-atomic potentials. The MD method can be applied to the study the properties of the interfaces, including the structure of grain boundaries (Wu, Zhang, & Srolovitz, 2009 and Hamilton, Siegel, Daruka, & Léonard, 2003), grain boundary segregation and consequent embrittlement. The common approach involves the calculation of binding energies of impurity to the grain boundary and to surface, with expectation that embrittlement impurities should have stronger binding force to the surface than with grain boundaries (Rice & Wang, 1989). Another important application of the study of mechanical behaviour at nano-scale which can be manifestation of interface or surface controlled plasticity (Bulatov, Abraham, Kubin, Devincre, & Yip, 1998). Molecular dynamics has been found to be useful in prediction of deformation mechanisms, which are thought to be highly improbable e.g., in the case of deformation twinning of Al, which is high stacking fault energy material (Yamakov, Wolf, Phillpot, Mukherjee, & Gleiter, 2002).

MD is a simulation technique involving the study of the physical movements of atoms and molecules in the reference of n-body simulation. Usually, the trajectories or projection of atoms and molecules movement are deduced numerically by simply solving Newton's equation of motion for a system of interacting particles, where the interaction or force between the particles is defined by empirical molecular or atomic force fields.

The key issue in MD is the definition of the terms by which particles in the simulation would interact, which is also termed as potential function. Another important aspect of MD fundamental is reduction of fully quantum description to classical potential through two approximations. The first approximation is Born-Oppenheimer approximation, which states the motion of electrons is so fast that they can be considered to react instantaneously to motion of their nuclei. So, electrons and nuclei may be treated separately. The second approximation treats the nuclei as point particles, whose motion follow Newtonian classical dynamics, which is much heavier than electron.

The principle factor involving molecular dynamics simulation involve the precise description of the interatomic potential, which physically describes the nature of the interaction between pair of atoms or the interaction of atom with multiple atoms in condensed state. Figure 4 qualitatively describes the variation of the potential with atomic spacing. It is also important to note that, each atom in crystalline solid experiences same potential due to other atomic fields. The binding energy or cohesive energy of the atom is the depth of the potential well as shown in the Figure 4. The location of potential minima decides the nearest neighbour distance in the lattice. In accordance with the potential, each atom vibrates about the position r_0 with the amplitude depending upon the total energy of the system. It is important to note that, such potential is used only for noble gas elements such as Ne, Xe, Kr, etc. The pair potentials cannot be applied for metals, as pair potentials predict equal values for elastic constants, which is not the case.

The precise description of the potential parameters is very crucial for accurate simulation. During fitting of potentials from experimental input, single physical parameter cannot be used for the generation of whole potential function. So, different kind of properties are required, like equilibrium thermodynamic properties and vibrational properties. Validation of potential can be carried out by calculating the thermodymnamic property, which has not been used in potential fitting and comparison with experimental value can be carried out. This can ensure the transferability and robustness of model. Recently, potentials are fitted using experimental as well as first-principle calculations (Mishin, 2004; Williams, Mishin, & Hamilton, 2006 and Ercolessi & Adams, 1994). The application of first-principle calculations in potential fitting is particularly widespread (Mishin, Asta, & Li, 2010).

In classical MD studies, using embedded atom method (EAM), the energy of the crystal (E) may be given as:

$$E = \frac{1}{2} \sum_{i=j, i \neq j}^{N} \phi_{ij}\left(r_{ij}\right) + \sum_{i} F_i\left(\rho_i\right) \tag{17}$$

Where, ϕ_{ij} denotes the pair energy between atoms i and j, which are separated by distance r_{ij} and F_i represents the embedding energy associated with the embedding an atom i in the local site with electron density ρ_i. The ρ_i may be mathematically represented as:

$$\rho_i = \sum_{i=j, i \neq j}^{N} f_i\left(r_{ij}\right) \tag{18}$$

Where, $f_j(r_{ij})$ represents electron density at the site of atom i arising from atom j at distance r_{ij} away.

For pure element the EAM potential principally comprises three functions namely, pair potential (ϕ_{ij}), electron density (ρ_i) and embedding energy (F). But for an alloy EAM potential should have all three functions for individual elements and also pair energy between different elements. In general, the published EAM potential for atom of single element are defined for different reference states and different cut-off distances, and hence, the functions ϕ, ρ and F fitted or defined for elemental metals cannot be applied or the alloy. But, an innovative method for generalising EAM potential and their cut-off distance have been proposed for alloys (Zhou et. al., 2001). Alloy EAM potential can be constructed from individual elemental EAM potentials, if the normalised potentials are normalised and single cut-off function are employed (Johnson, 1989). The following form of generalised pair potential was deduced to fit the EAM potential set:

$$\phi(r) = \frac{A\exp\left[-\alpha\left(r / r_e - 1\right)\right]}{1 + \left(r / r_e - \kappa\right)^{20}} - \frac{B\exp\left[-\beta\left(r / r_e - 1\right)\right]}{1 + \left(r / r_e - \lambda\right)^{20}} \tag{19}$$

Where, r_e represents the equilibrium spacing between nearest neighbour. A, B, α and β are adjustable parameter. K and λ are additional parameter for cut-off. The form of electron density function is taken with the same as of attractive term of the pair potential with same values of β and λ as:

$$f_r = \frac{f_e \exp\left[-\alpha\left(r / r_e - 1\right)\right]}{1 + \left(r / r_e - \kappa\right)^{20}} \tag{20}$$

The pair potential between different species can be constructed using elemental potential:

$$\phi^{ab}(r) = \frac{1}{2}\left[\frac{f^b(r)}{f^a(r)}\phi^{aa}(r) + \frac{f^a(r)}{f^b(r)}\phi^{bb}(r)\right] \tag{21}$$

9. MONTE-CARLO (MC) SIMULATIONS

The MC simulation is the potential tool for microstructural simulation beside the other applications *viz.*, calculation of properties (such as internal energy, temperature, pressure, volume, etc.), simulating the system of interacting particles, modelling the random processes etc. The most popular example of microstructural modelling using MC is recrystallization and grain growth. In this approach, the mobility of the boundaries can be varied by changing the rate at which sites are sampled for reorientation. Difference in orientation between the two grains in MC simulation can be used to study the abnormal grain growth phenomena (Rollet & Mullins, 1997). MC method have predicted that in static recrystallization, with higher stored energy density continuous nucleation takes place (Srolovitz, Grest, & Anderson, 1986), while low stored energy density may lead to the heterogenous nucleation with dependence on the prior microstructure (Srolovitz, Grest, Anderson, & Rollet, 1988).

In MC simulations, microstructure is represented on discrete two-dimensional lattice. In this method, initialisation is carried out by assigning a random integer (x_i) between 1 and Q, for each lattice site, such that x represents the number of grain orientation in lattice site (or grain orientation) and Q represents the total number grain orientations. In this methodology, the grain is represented by similar orientation number, while the grain boundary is represented by interface between spaces represented by different orientation number. Periodic boundary condition is used to avoid singularity of edges in a finite domain. In this condition, point on the edges of domain is in the symmetry with diagonally opposite edge. Figure

Figure 13. Monte Carlo simulation methodology of recrystallisation [redrawn from (Hore, Das, Banerjee and Mukherjee, 2013)]

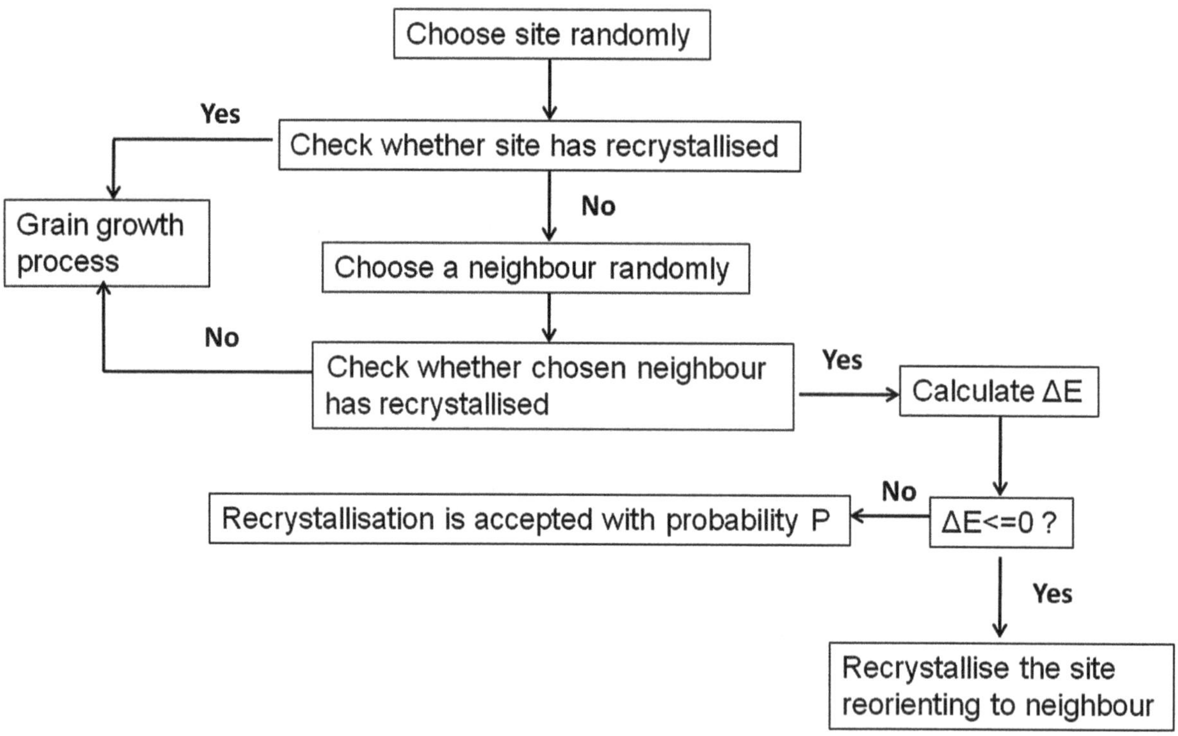

13 shows the MC simulation methodology with reference to recrystallisation simulation. In MC simulation, the energy (E) of lattice for particular orientation numbers is defined as:

$$E = J\sum_{i=1}^{N}\left(1-\delta_{S_iS_j}\right)+H \tag{22}$$

J is the positive constant, which is the function of grain boundary energy. Also, S_i and S_j are orientation number and δ represents Kronecker delta, which assumes the value of one, when $S_i = S_j$ and if S_i and S_j are different, it has value of zero.

The Σ represents the sum is taken all over the nearest neighbour or coordination sites. The number of nearest neighbour depends upon the types of lattice being used. The lattice can be either triangular (Figure 14(a)) or square (Figure 14(b)). In the case of triangular lattice, the number of nearest neighbour is six, while in the case of square or quadratic lattice, the nearest neighbour are eight in number. The type of lattice chosen can have strong effect on the simulation results and might lead to the unphysical results such as grain pinning (Holm & Battaile, 2001). The pinning of grain can be attributed to the MC method itself, where grain growth is simulated as curvature driven process with minimization of the length of boundary leading to the preferable placement of grain boundary along the lattice facets leading to the pinning. Such problem can be either solved by increasing the nearest neighbour points (So, apart from first nearest neighbour, second nearest neighbour in two dimensional lattice and first, second and third nearest neighbour in case of three dimensional lattice) or the simulation can also be performed at higher temperature to roughen up the grain boundary. It is important to note that, if second option is used the real physical temperature of simulation need to be determined by trial and error.

Another important issue in MC simulation is the volumetric energy, which can be chemical free energy of Monte Carlo Unit (MCU) or lattice site (Tong, Li, & Li, 2004) or stored energy of deformation (Radhakrishnan, Sarma, & Zacharia, 1998). As depicted in the flowchart of MC simulation (Figure13), the change in the state (or change in MCU corresponding to random allocation of orientation number) can take place due to decrease in the total energy. The probability of change of state in particular MCU or transitional probability can be expressed as:

Figure 14. Types of lattice typically used in MC simulation (a) triangular lattice and (b) rectangular lattice

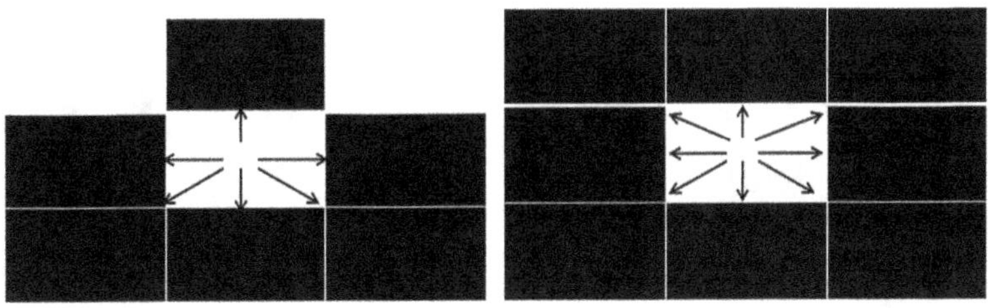

(a) (b)

$$P(\Delta E) = \begin{cases} 1 & \Delta E \leq 0 \\ \exp\left(-\dfrac{\Delta E}{k_B T}\right) & \Delta E > 0 \end{cases} \tag{24}$$

It is important to note that temperature of MC simulation is not real temperature, but it need to be scaled. Also, it has been observed that transitional probability is reduced to 1 for $\Delta E \leq 0$ and 0 for $\Delta E > 0$ (Gao & Thompson, 1996). Also, the process needs to be repeated N times, where N is number of lattice sites (Srolovitz, Grest, Anderson, & Rollet, 1988). In certain modified approach, the assignment of new orientation should be one of the nearest neighbour.

10. CELLULAR AUTOMATA METHOD

Cellular Automata (CA) is the versatile technique for the simulating phase transformation, recrystallization and grain growth phenomena. Compared to other approaches, CA has capability to combine the ability to scale the switching model with physical stringency of boundary dynamics model and computational simplicity (Raabe, 2002). The CA model has been applied to study the topological evolution and competing nuclei during initial stages of recrystallization. CA can improve the understanding of the clustering tendency of nuclei in incipient stage of nucleation.

Cellular Automata is the simulation technique involving spatial and temporal evolution of systems by applying transformation or switching rules. The lattice space defines the regular array of lattice spacing. The state of each lattice point is defined by the dimensionless numbers, dislocation density, crystal orientation, particle densities etc.

CA lattice points are space filling periodic grid of cells, which can be square in two dimensional case and cube in three dimensions. The grid type needs to be defined along with the nearest neighbour distribution. It might be von Neumann type considering nearest neighbour (Hesselbarth & Göbel, 1991) (Figure 15(a)), while Moore type configuration involves nearest neighbour and next nearest neighbour (Figure 15(b)). The selection of particular nearest neighbour configuration affects the shape of grains. The von Neumann configuration lead to the development of diamond shaped grains before pinning (Rappaz & Gandin, 1993), while Moore configuration causes faceted hexagonal shaped grains and in certain cases, modification of Moore configuration to include six nearest neighbour may form octahedral grain shapes before pinning (Davies, 1995). Such problems associated with the morphological development due to lattice neighbouring configuration can be overcome by random grid CA (Janssens, 2003).

Figure 15. The nearest neighbour configurations, (a) von Neumann and (b) Moore configuration

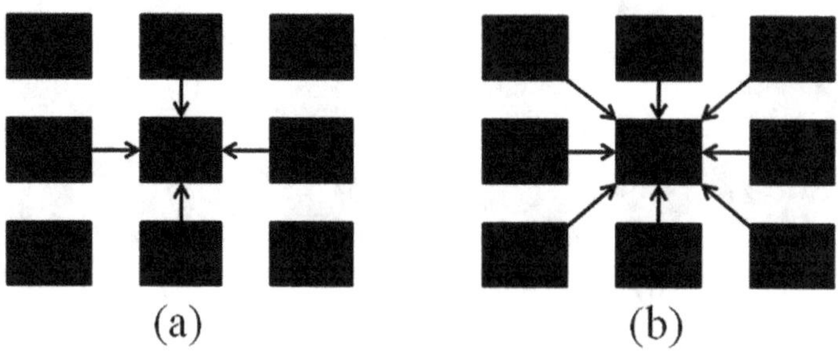

Apart from defining the nearest neighbour configuration, another important issue is defining the transition rule for cells between discrete time steps. These transition rules can be either deterministic or probabilistic which acts on each lattice points. These rules determine the state of particular lattice point as a function of past state, as well as state of neighbouring points. The system is simulated by local rules governing the interaction between cells. The transition rule might be deterministic in nature, but their application is governed by probabilistic rules leading to mimicking of natural phenomena. As in most simulation techniques, the boundary condition needs to be specified. Usually, periodic boundary condition is employed. But, in the case of simulation involving gradients of solute, temperature, or strain at least one boundary pair should be mirror boundary. Application of mirror boundary value to the cases, where CA unit is very small in comparison with the object being modelled, might lead to the significant error.

During the initiation of CA simulation, experimental output (from Electron Back Scattered Diffraction (EBSD)) or theoretical simulation such as crystal plasticity, finite element analysis is mapped on the CA lattice as the state variable of the lattice points. EBSD data can also be scaled for CA simulation. But, inherent decision-making is involved due to the assumption of stored energy distribution, completion of grain boundaries and minimum viable grain size (Raabe, 2001; Zaefferer, 2005 and Raabe, 2007). The crystal plasticity finite element (CPFEM) data can be super-imposed on the CA lattice (Lan, Xiao, Li, & Li, 2005). It is important to note that mapping of data from CPFEM to CA may not be one-to-one. In these cases, averaging of data is required for data transfer. One of the requirements during such process is that the CA grid should be smaller than the node length of CPFEM (Raabe & Becker, 2000). As suggested earlier, certain transformation rule needs to be defined. The number of rules for number for the number of states (k) and the number of neighbours including core cell (n) is $k^{(2n)}$. The physical rules are defined to map the phenomenology considering the physics of the problem. These switching rules define the motion of interface governed by probabilistic form of linearized symmetric rate equation, which physically depicts the diffusion of atoms isotropically through homogenous planer grain boundary segment leading to the negative free energy (Raabe, 2002).

11. PHASE FIELD METHOD

The phase field modelling (PFM) was originally proposed for the modelling of solidification process (Langer, 1986 and Collins & Levine, 1985) to predict the complicated dendritic patterns without tracking the solid-liquid interface (Kobayashi, 1993). The PFM can be used to model diffusional and non-diffusional phase transformations (Chen & Wang, 1996 and Wang & Chen, 1999). The one of the main concerns, while dealing with solid-state phase transformation is dealing with the coherent nuclei in initial stage of the transformation, which leads to the elastic strain energy. It is important to note that phase field models are able to use macroscopic elasticity theory (Khachaturyan, 1983 and Khachaturyan & Shatalov, 1969) to predict the complex microstructure in solid. The solidification and solid-solid phase transformation is guided by the reduction of the free energy, but the driving force for grain growth phenomena is reduction of the interfacial energy of the microstructure. The grain growth phenomenon has been simulated by multi-order parameter PFM with different orders representing different crystallographic orientation (Chen & Yang, 1994).

In the traditional microstructural simulation techniques, the interfaces between the phases are sharp (sharp interface approach) with multi-phase structures defined by the position of interfaces between them (Figure 16(a)). The microstructural simulation is carried out by solving differential equation for

Figure 16. Variation in order parameter in (a) sharp-interface approach and (b) diffused-interface approach (Qin & Bhadeshia, 2010)

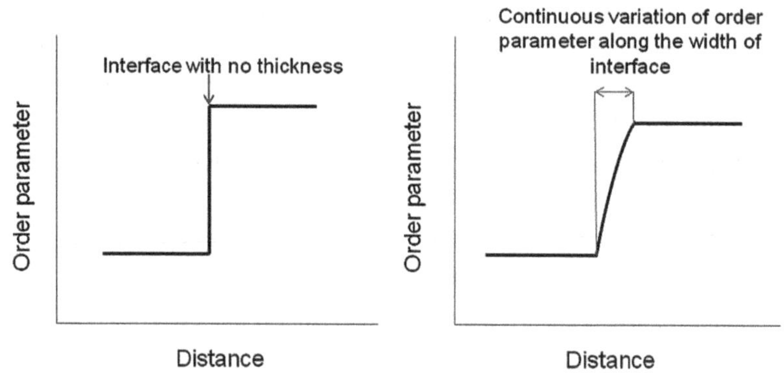

domains with mass balance to incorporate the flux conditions, while constitutive expressions describe the interface with need to track the position of interfaces.

The phase field method (PFM) is essentially the diffused interface approach (Figure 16(b)). In this case, the particular grain has constant order parameter, which is the function of thermodynamic parameters such as composition, orientation etc. The interface between two grains is the region, where order parameter 'gradually' changes its value to order parameter of next grain. So the position of the interface is implicitly determined by the evolution of phase field variables. The temporal evolution of the order parameters involves solution of the concerned partial differential equations. From computational point of view, PFM are created in such a way that in the limit of infinitesimal interface width, the correct interfacial dynamics are recovered. This approach permits to solve the problem by integrating a set of partial differential equation for the whole system, thus avoiding the explicit treatment of boundary condition at the interface.

Another important feature of PFM is the distinction between phase field variables. One set of phase field variables are conserved variable dealing with the conserved quantity such as local composition, while other set of non-conserved variables accounts for non-conserved quantities such as local structure and its orientation. The driving force for the microstructure evolution in PFM is reduction in the free energy (G), which can have various contributions such as bulk energy (G_{bulk}), interfacial energy (G_{int}), elastic strain energy (G_{el}) and energy contributions due to magnetic or electrostatic interactions (G_{me}).

$$G = G_{bulk} + G_{\text{int}} + G_{el} + G_{me} \tag{25}$$

Also, at constant temperature, pressure and composition, the contribution from G_{el} and G_{me} can be ignored. So, the total free energy of the system having inhomogenous microstructure is expressed by considering the set of conserved field variable or composition field (c1, c2,...) and the non-conserved field variable or order parameters (η_1, η_2, ...):

$$G = G_{bulk} + G_{\text{int}} = \int_V \left[g(\phi) + \frac{1}{2}\kappa_\phi (\nabla\phi)^2 \right] dV \tag{26}$$

Where, g is the local free energy density, with ϕ representing field variable which is the function of both conserved (c_i) and non-conserved (η_i) and κ is the gradient energy terms that lead to the diffuse character of the interface and it is dependent upon the interfacial energy and thickness. The gradient energy terms are always positive leading to the field variables to be unstable and giving rise to surface tension.

The type of phase field model depends upon the treatment of various contributions to total free energy and most importantly the local free-energy density function. For solidification modelling, a function having double-form is used, which can be written as (Chen, 2002):

$$g(\phi) = 4\Delta g\left(-\frac{1}{2}\phi^2 + \frac{1}{4}\phi^4\right) \tag{27}$$

Δg is potential difference between two states with low free energy. The free energy function has two minima at -1 and +1. So, in solidification -1 and +1 represents the liquid and solid, respectively. This potential can be used to simulate the isothermal decomposition from parent phase (ϕ=0) to mixture of two phases $(\phi = \pm 1)$. In certain cases, only two order parameters $(\phi = 0,1)$ is required following local free-energy density can be written as:

$$g(\phi) = 4\Delta g\phi^2\left(1-\phi^2\right) \tag{28}$$

For solid-solid phase transformation, the local free-energy density is represented as polynomial of the order parameter as Landau type expansion. Considering the case of precipitation of ordered phase from fcc-binary alloy, it can be written as (Lai, 1990):

$$g(c,\eta_1,\eta_2,\eta_3) = g_d(c,T) + \frac{1}{2}A_2(c,T)\left(\eta_1^2+\eta_2^2+\eta_3^2\right) + \frac{1}{3}A_3(c,T)\eta_1\eta_2\eta_3 + \frac{1}{4}A_{41}(c,T)\left(\eta_1^4+\eta_2^4+\eta_3^4\right) + \frac{1}{4}A_{42}(c,T)\left(\eta_1^2\eta_2^2+\eta_2^2\eta_3^2+\eta_3^2\eta_1^2\right) \tag{29}$$

Where, $g_d(c,T)$ is free energy of the disordered phase and A_2, A_3, A_4, A_{41} and A_{42} are expansion coefficients, which are function of composition and temperature.

In view of the above, there can be basically two pathways in PFM. First pathway involves the transition through discontinuous process of nucleation and growth having inherent activation barrier involved in it. Under the application of external stimulus to generate phase transition, namely temperature difference, certain compositional fluctuation might lead to the increase in free energy in meta-stable region due to increase in nucleation energy. However, sufficiently large and finite compositional fluctuation might lead to decrease in free energy. Second set of phase transition are non-activated processes, where continual decrease in free energy takes place through the course of phase transformation from parent to product phase (Spinodal decomposition). The spinodal decomposition can be mathematically represented by Cahn-Hillard equation, which may be written as:

$$\frac{\partial c}{\partial t} = D\nabla^2\left(c^3 - c - \gamma\nabla^2 c\right) \tag{30}$$

D represents diffusion coefficient, $\sqrt{\gamma}$ is the length of the transition between different zones, and term $\left(c^3 - c - \gamma\nabla^2 c\right)$ is chemical potential and c is the concentration of the parent phase. Above equation describes the phase transition involving the phase separation, in which two components of binary parent phase spontaneously separate and form domains of pure components through diffusion.

For nucleation and growth phenomena, Allen-Cahn model is used to describe the phase transition, which may be expressed as:

$$\frac{\partial \eta}{\partial t} = M_n\left[\varepsilon_n^2\nabla^2\eta - f'(\eta)\right] \tag{31}$$

M_n represents mobility, f is free energy density and η is non-conserved order parameter (usually long range order parameter).

PFM are type of Onsanger type kinetic relation (i.e., expression of the equalities of certain ratios between flows and forces in thermodynamic system out of equilibrium, allowing the notions of local equilibrium) provided with appropriate free energy functional that may depend upon the concentration and order parameter. In these approaches, the state parameters are used as field variables. Such approach allows the study of kinetics of microstructure formation in the terms of variation of field variables in spatial coordinates.

As discussed above, the evolution of the field variables in PFM by solving Cahn-Hillard (Cahn & Hilliard, 1958) and Allan-Cahn (Cahn & Allen, 1977) equation, which is reduced to the finding solutions of following kinetic equations:

$$\begin{aligned} \frac{\partial c_i(r,t)}{\partial t} &= \nabla M_{ij}\nabla\frac{\delta G}{\delta c_j(r,t)} \\ \frac{\partial \eta_p(r,t)}{\partial t} &= -L_{pq}\frac{\delta G}{\delta c_j(r,t)} \end{aligned} \tag{32}$$

Where, M_{ij} and L_{pq} are related to interface mobility. Most of the phase field simulations are carried by using second–order finite difference method on uniform spatial grid and explicit time stepping (Chen, 2002).

Phase field methods need the thermodynamic data such as bulk free energy of the phases, interfacial energies and its anisotropy with lattice mismatch. In this approach, first principles calculations are used to calculate the above stated properties and such information is used to predict the microstructure. This approach can be exemplified by combined atomistic-PFM simulation of precipitation in Al-Cu alloys (Chen, Wolverton, Vaithyanathan, & Liu, 2001) leading to the prediction of precipitate shapes. While in the case of solidification modelling, the liquid-solid interfacial energies and their mobility can be calculated by the atomistic calculations to be used in PFM (Bragard, Karma, Lee, & Plapp, 2002). CALPHAD modelling is also employed to construct free energy function in PFM from thermodynamic and kinetic databases (Grafe, Böttger, Tiaden, & Fries, 2000a and Grafe, Böttger, Tiaden, & Fries, 2000b). Additionally, compositional dependence of the mobilities can be also calculated frim CALPHAD.

12. VORONOI TESSELLATION

The generation of virtual microstructures are often important, in the context of micromechanical simulations. Microstructures obtained from the experimental technique such as electron backscattered diffraction technique (Kim, Yang, Han, Kwak, & Choi, 2012) can be also used as input for micromechanical modelling, but virtual microstructural techniques allows to explore the wide range of morphologies and distribution of phases in microstructures.

Voronoi algorithm is a potential technique to divide the space into polygonal units. Such algorithm has been applied to generate the representative volume element (part of microstructure, adequately large to represent the heterogeneities and small enough to be considered as of infinitesimal volume for application of continuum mechanics framework). The development of such artificial microstructure requires the knowledge of volume fraction of various phases, morphology and two-point correlation functions.

The Voronoi cell can be principally generated by first assigning random points and finding its nearest neighbour by Delaunay triangularization technique. Figure 17 shows the two dimensional case of such procedure.

Periodic structure can be obtained in this approach by choosing point sets from random points and copying it in all neighbours side and ultimate application of triangularization algorithm (Figure 18 (a)). The number of neighbours is six in the case of generation of two-dimensional microstructure, while there are twenty-six neighbours for three dimensional microstructure generation. Figure 18 (b) shows the two and three dimensional case of periodic grain structure.

Another important approach is multi-level Voronoi algorithm, which can generate two-phase or even multiphase structures. Two-phase microstructure, with various spatial and morphological nature of second phase, has been developed with such procedure (Anand, Dey, Kok, Chakraborty, & Chattopadhyay,

Figure 17. Schematic illustration of the Voronoi algorithm. By considering each point and dividing the distance to its nearest neighbours in half, polygons are formed.

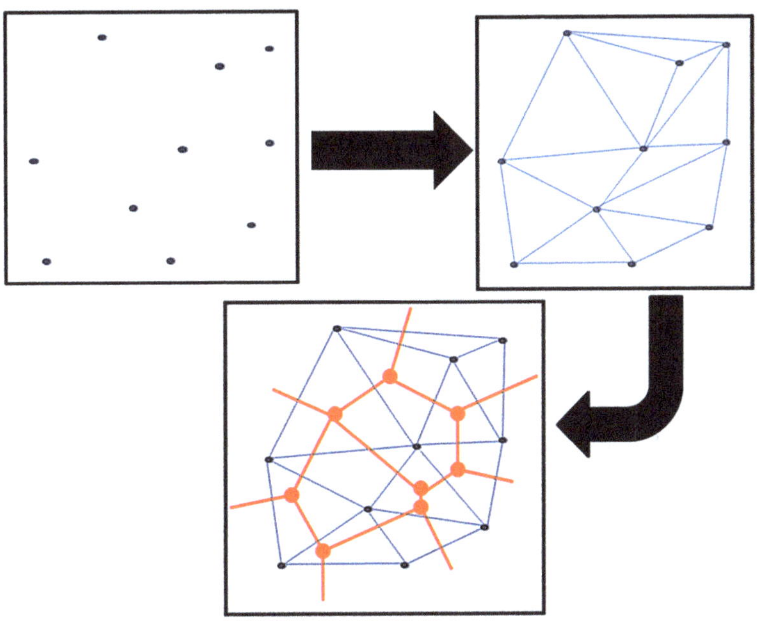

Figure 18. (a) Schematic illustration of the creation of periodic cells. The original point set is copied to all neighbouring cells and (b) two and three dimensional periodic grain structure (Nyagards, 2003).

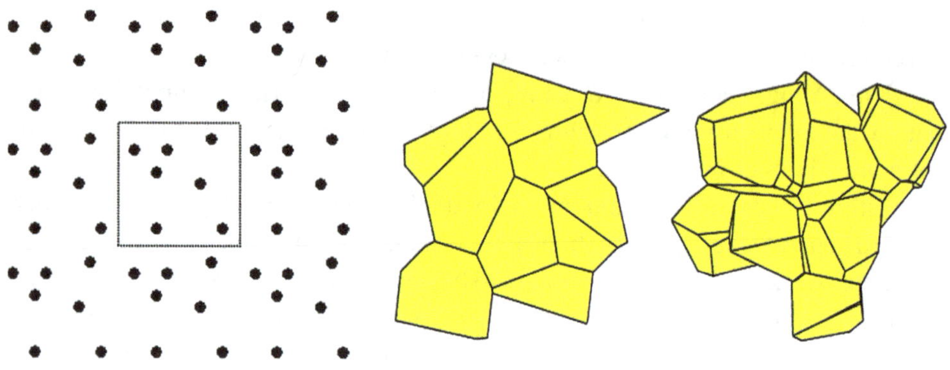

(a) (b)

2014). In first step, a layer of fine tessellation of Voronoi is generated using Poission or hard core point field (say β). In next step, coarse tessellation leading to formation of matrix grains (say α). Ultimately, fine and coarse grains are linked on the basis of desirability for particular type of microstructure, fine tessellation (β) is filtered out to generate particular microstructure.

13. FINITE ELEMENT METHOD

Finite element method has been used for range of mesoscale physical phenomena in solid state, including heat transfer (Reddy & Gartling, 2010), micromagnetic simulations (Fischbacher, Franchin, Bordingnon, & Fangohr, 2007), structural simulations (Zienkiewicz & Taylor, 1977), etc. The present section would emphasize on the continuum finite element analysis (Al-Abbasi & Nemes, 2003). Recently, there has been wide interest in Crystal-plasticity finite element method, details of which can be found elsewhere (Roters, Eisenlohr, Bieler, & Raabe, 2011).

Modelling of deformation behaviour of metals and alloys involves the precise description of the elastic, plastic and failure behaviour. In the study of deformation of metallic systems under the framework of continuum mechanics, the fundamental assumption designates the crystalline material as continuum, i.e., the finite amount of material with properties independent of size and time over which property are measured. The properties of continuum materials can be precisely defined on the basis of relationship between stress and strain. Such relationships are also called constitutive relationship, as it depends upon the property of material. The continuum mechanical models are of two categories. The first one, which can be solved by approximation schemes, involving the assumption, that individual constituent of the structure behaves like the homogenous continuum and plastic strain in each phase is uniform. Another approach is of Finite element analysis, which can provide the fluctuations in stress and strain in inhomogenous continuum structure. The finite element analysis is computational technique to solve boundary value problems to give approximate solution. These boundary value problems can be defined as mathematical problem in which, one or more dependent variables must satisfy a differential equation everywhere in known domain of independent variables and satisfy specific conditions on the boundary of the domain.

These are also called 'field problems', where field is domain of interest, representing physical structure, while field variables are dependent variables of interest governed by differential equation. So, present section provides the basics of methodology to model elasto-plastic deformation of metallic materials.

The constitutive description for elastic deformation is straightforward and can be represented by Equation (33).

$$\sigma_{kl} = C_{ijkl}\varepsilon_{ij} \tag{33}$$

Above can be simply represented as $\sigma = E.\varepsilon^e$, for uniaxial deformation. For elasto-plastic deformation model, the mathematical description of the onset of plasticity may be represented by scaler function termed as 'yield function, F'. The yield function is written in a form, such that, it leads to the condition:

$$F < 0\,(\text{elastic})$$

$$F \geq 0\,(\text{plastic}) \tag{34}$$

For metallic materials, a widely used yield criterion is Von Mises yield criteria, which may be mathematically represented as Equation (35).

$$F = (\sigma_1 - \sigma_2)^2 + (\sigma_2 - \sigma_3)^2 + (\sigma_3 - \sigma_1)^2 - 2\sigma_y^2 \tag{35}$$

σ_1, $\sigma_{2,}$ σ_3 represents the principle stresses and σ_y represents the yield stress of the material.

The post yielding behaviour is defined on the basis of decomposition of total strain in elastic and plastic strain.

$$d\varepsilon = d\varepsilon^e + d\varepsilon^p \tag{36}$$

The plastic behaviour beyond yielding is defined on the basis of 'flow rule'.

The constitutive expression for deformation may be represented as:

$$d\sigma = D_T.d\varepsilon \tag{37}$$

Where, D_T is the tangential materials matrix. The D_T may be derived from known stress and strain tensor.

A consistency condition needs to be fulfilled; involving the stress point must lie on the yield surface, which may be ensured by following expression:

$$dF = \left(\frac{\partial F}{\partial \sigma}\right)^T \partial\sigma + \left(\frac{\partial F}{\partial \varepsilon^p}\right)^T \partial\varepsilon^p = 0 \tag{38}$$

Figure 19. Methodology of continuum elasto-plastic finite element simulation

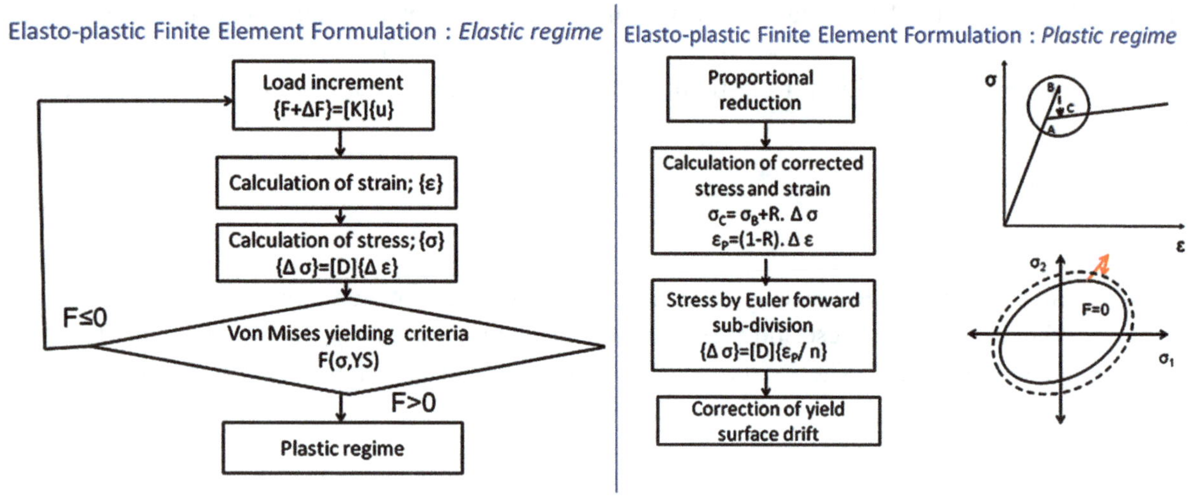

In finite element analysis, we need to work out the finite increments of stress ($\Delta\sigma$) and strain ($\Delta\varepsilon$), for which Equation (4) is approximately holds. Figure 19 shows the flow-chart of elasto-plastic simulation for metallic materials.

One of the crucial parts of plastic part of elasto-plastic simulation is yield surface drift correction (Figure 19), which arise due to discretization. There are various approaches for such corrections such as projecting back along accumulated effective stress direction, projecting back along the plastic flow direction, etc. The detailed treatment of such procedures can be found elsewhere (Roters, Eisenlohr, Bieler, & Raabe, 2011; Simo, 1997 and Gens & Potts, 1988).

F-F Contiguity Parameter (FCP) and M-M Contiguity Parameter (MCP) has been defined by Equation (39) and (40), respectively (where, NFF, NFM and NMM represents the number of F-F, F-M and M-M interfaces intercepted by straight line, as shown in Figure 20) (Anand, Dey, Kok, Chakraborty, & Chattopadhyay, 2014).

$$FCP = \frac{2N_{FF}}{2N_{FF} + N_{FM}} \tag{39}$$

$$MCP = \frac{2N_{MM}}{2N_{MM} + N_{FM}} \tag{40}$$

Contiguity values may range from zero to one. When grains of same phase do not connect at all in a certain direction, the contiguity value is zero in that direction, while the contiguity value of unity means that grains of that phase are connected in that direction without the interruption by the grains of second phase.

Figure 20. Schematic representation of the contiguity parameter. N_{FF}: Number of F-F intersections, N_{FM}: Number of F-M intersections and N_{MM}: Number of M-M intersections (F and M represent two different phases)

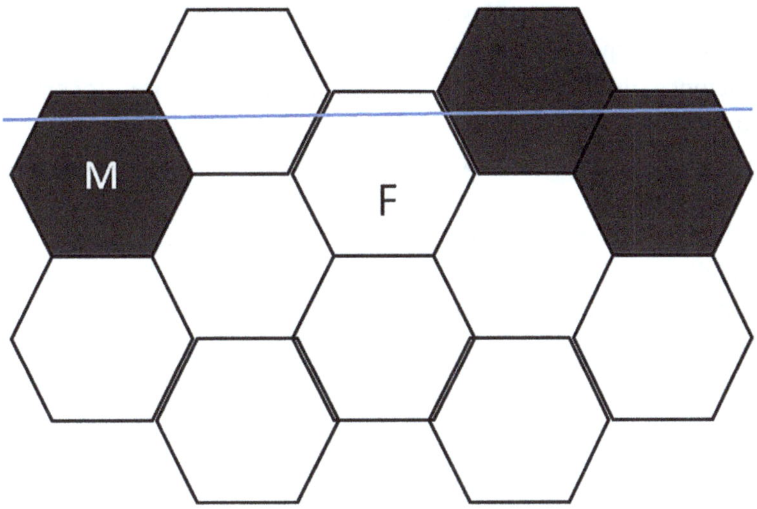

The main motivation to introduce contiguity parameter for the two phase microstructure is to quantitatively correlate the microstructural parameters with the strain partitioning and stress percolation in the course of work hardening.

Based on the result of the micromechanical simulation a mathematical relationship has been developed between the inputs (FCP and MCP) and output (percent area under strain hotspots) using Genetic Programming (Koza, 1992). This is generically called the symbolic regression problem, which assumes no a priori form of the underlying relation. It is a generalization of the linear regression problem, seeking to express the output as just the weighted sum of the inputs. The correlation between the FCP, MCP and fraction area having stress (x) and strain (y) hot spots may be represented as:

$$x = 0.1563 + 0.01766 * \left\{ \frac{(FCP - 0.2724)}{(MCP - 0.2637)} \right\} \tag{41}$$

$$y = 4.1935 + 0.1566 * \left\{ \frac{(MCP - 0.4457)}{(0.5936 - FCP)} \right\} \tag{42}$$

Unanswered Questions in Microstructure Based Design Approach

There has been significant interest in the microstructure driven design of materials. However, there are many fundamental and technological issues remaining to be addressed in the endeavour of deterministic design of microstructure in the multi-scale framework. In a recent attempt of circumventing the reciprocal relation between strength and elongation, development of bimodal microstructure has been found to be the attractive option (Wang, Chen, Sheng, & Ma, 2002). The approach is based on the decoupled contribution of the finer and coarser grains to the strength and elongation, respectively. However, the precise microstructure and the concerned process schedule for achieving the exotic strength-elongation combination is often sporadic than predictive.

In conventional attempt of design, composition-process-microstructure-property correlation is implemented in forward direction. However, in the deterministic framework the reversible predictability is desirable (i.e., from property to microstructure and subsequently to process and composition). Another important issue of concerns is the integration of the contribution of interfaces with the accepted materials simulation techniques. In most of the contemporary approaches, Microstructures are simulated at various scales, while interface properties are calculated separately. There is also emerging need for combined microstructure simulation techniques towards the design of nanostructured and hierarchically architectured microstructures.

REFERENCES

Al-Abbasi, F. M., & Nemes, J. A. (2003). Micromechanical modeling of dual phase steels. *International Journal of Mechanical Sciences*, *45*(9), 1449–1465. doi:10.1016/j.ijmecsci.2003.10.007

Anand, G., Datta, S., & Chattopadhyay, P. P. (2013). Deterministic Approach for Microstructurally Engineered Formable Steels. *International Journal of Metallurgical Engineering*, *2*(1), 69–78.

Anand, G., Dey, P., Kok, P. J. J., Chakraborty, D., & Chattopadhyay, P. P. (2014). Architectured microstructures in steel. *Materials Science and Technology*, *30*(9), 1086–1093. doi:10.1179/1743284714Y.0000000535

Born, M., & Oppenheimer, R. (1927). Zur quantentheorie der molekeln. *Annalen der Physik*, *389*(20), 457–484. doi:10.1002/andp.19273892002

Bragard, J., Karma, A., Lee, Y. H., & Plapp, M. (2002). Linking phase-field and atomistic simulations to model dendritic solidification in highly undercooled melts. *Interface Science*, *10*(2-3), 121–136. doi:10.1023/A:1015815928191

Brandon, D. G. (1966). The structure of high-angle grain boundaries. *Acta Metallurgica*, *14*(11), 1479–1484. doi:10.1016/0001-6160(66)90168-4

Broughton, J. Q., & Gilmer, G. H. (1986). Molecular dynamics investigation of the crystal–fluid interface. VI. Excess surface free energies of crystal–liquid systems. *The Journal of Chemical Physics*, *84*(10), 5759–5768. doi:10.1063/1.449884

Bulatov, V., Abraham, F. F., Kubin, L., Devincre, B., & Yip, S. (1998). Connecting atomistic and mesoscale simulations of crystal plasticity. *Nature*, *391*(6668), 669–672. doi:10.1038/35577

Cahn, J. W., & Allen, S. M. (1977). A microscopic theory for domain wall motion and its experimental verification in Fe-Al alloy domain growth kinetics. *Journal de Physique. Colloque*, *38*(C7), C7–C51. doi:10.1051/jphyscol:1977709

Cahn, J. W., & Hilliard, J. E. (1958). Free energy of a nonuniform system. I. Interfacial free energy. *The Journal of Chemical Physics*, *28*(2), 258–267. doi:10.1063/1.1744102

Ceperley, D. M., & Alder, B. J. (1980). Ground state of the electron gas by a stochastic method. *Physical Review Letters*, *45*(7), 566–569. doi:10.1103/PhysRevLett.45.566

Chang, Y. A., Chen, S., Zhang, F., Yan, X., Xie, F., Schmid-Fetzer, R., & Oates, W. A. (2004). Phase diagram calculation: Past, present and future. *Progress in Materials Science*, *49*(3), 313–345. doi:10.1016/S0079-6425(03)00025-2

Chattopadhyay, P. P. (2014). Microstructure based design of formable steels. *The Banaras Metallurgist*, 43-52.

Chen, L. Q. (2002). Phase-field models for microstructure evolution. *Annual Review of Materials Research*, *32*(1), 113–140. doi:10.1146/annurev.matsci.32.112001.132041

Chen, L. Q., & Wang, Y. (1996). The continuum field approach to modeling microstructural evolution. *JOM*, *48*(12), 13–18. doi:10.1007/BF03223259

Chen, L. Q., Wolverton, C., Vaithyanathan, V., & Liu, Z. K. (2001). Modeling solid-state phase transformations and microstructure evolution. *MRS Bulletin*, *26*(03), 197–202. doi:10.1557/mrs2001.42

Chen, L. Q., & Yang, W. (1994). Computer simulation of the domain dynamics of a quenched system with a large number of nonconserved order parameters: The grain-growth kinetics. *Physical Review B: Condensed Matter and Materials Physics*, *50*(21), 15752–15756. doi:10.1103/PhysRevB.50.15752 PMID:9975940

Chokshi, A. H., Rosen, A., Karch, J., & Gleiter, H. (1989). On the validity of the Hall-Petch relationship in nanocrystalline materials. *Scripta Metallurgica*, *23*(10), 1679–1683. doi:10.1016/0036-9748(89)90342-6

Cohen, M. (1976). Unknowables in the essence of materials science and engineering. *Materials Science and Engineering*, *25*, 3–4. doi:10.1016/0025-5416(76)90043-4

Collins, J. B., & Levine, H. (1985). Diffuse interface model of diffusion-limited crystal growth. *Physical Review B: Condensed Matter and Materials Physics*, *31*(9), 6119–6122. doi:10.1103/PhysRevB.31.6119 PMID:9936623

Curtze, S., & Kuokkala, V. T. (2010). Dependence of tensile deformation behavior of TWIP steels on stacking fault energy, temperature and strain rate. *Acta Materialia*, *58*(15), 5129–5141. doi:10.1016/j.actamat.2010.05.049

Davies, C. H. J. (1995). The effect of neighbourhood on the kinetics of a cellular automaton recrystallisation model. *Scripta Metallurgica et Materialia*, *33*(7), 1139–1143. doi:10.1016/0956-716X(95)00335-S

Derlet, P. M., & Van Swygenhoven, H. (2004). High-frequency vibrational properties of metallic nanocrystalline grain boundaries. *Physical Review Letters*, *92*(3), 035505. doi:10.1103/PhysRevLett.92.035505 PMID:14753887

Dumay, A., Chateau, J. P., Allain, S., Migot, S., & Bouaziz, O. (2008). Influence of addition elements on the stacking-fault energy and mechanical properties of an austenitic Fe–Mn–C steel. *Materials Science and Engineering A*, *483*, 184–187. doi:10.1016/j.msea.2006.12.170

Ercolessi, F., & Adams, J. B. (1994). Interatomic potentials from first-principles calculations: The force-matching method. *EPL*, *26*(8), 583–588. doi:10.1209/0295-5075/26/8/005

Feuerbacher, M., Heidelmann, M., & Thomas, C. (2015). Hexagonal high-entropy alloys. *Materials Research Letters*, *3*(1), 1–6. doi:10.1080/21663831.2014.951493

Fischbacher, T., Franchin, M., Bordignon, G., & Fangohr, H. (2007). A systematic approach to multiphysics extensions of finite-element-based micromagnetic simulations: Nmag. *Magnetics. IEEE Transactions on*, *43*(6), 2896–2898.

Fitzsimmons, M. R., Eastman, J. A., Müller-Stach, M., & Wallner, G. (1991). Structural characterization of nanometer-sized crystalline Pd by x-ray-diffraction techniques. *Physical Review B: Condensed Matter and Materials Physics*, *44*(6), 2452–2460. doi:10.1103/PhysRevB.44.2452 PMID:9999814

Foiles, S. M., & Hoyt, J. J. (2006). Computation of grain boundary stiffness and mobility from boundary fluctuations. *Acta Materialia*, *54*(12), 3351–3357. doi:10.1016/j.actamat.2006.03.037

Freeman, A. J., & Wimmer, E. (1995). Density functional theory as a major tool in computational materials science. *Annual Review of Materials Science*, *25*(1), 7–36. doi:10.1146/annurev.ms.25.080195.000255

Gao, J., & Thompson, R. G. (1996). Real time-temperature models for Monte Carlo simulations of normal grain growth. *Acta Materialia*, *44*(11), 4565–4570. doi:10.1016/1359-6454(96)00079-1

Gens, A., & Potts, D. M. (1988). Critical state models in computational geomechanics. *Engineering Computations*, *5*(3), 178–197. doi:10.1108/eb023736

Gleiter, H. (2000). Nanostructured materials: Basic concepts and microstructure. *Acta Materialia*, *48*(1), 1–29. doi:10.1016/S1359-6454(99)00285-2

Gonze, X. (1997). First-principles responses of solids to atomic displacements and homogeneous electric fields: Implementation of a conjugate-gradient algorithm. *Physical Review B: Condensed Matter and Materials Physics*, *55*(16), 10337–10354. doi:10.1103/PhysRevB.55.10337

Grafe, U., Böttger, B., Tiaden, J., & Fries, S. G. (2000a). Coupling of multicomponent thermodynamic databases to a phase field model: Application to solidification and solid state transformations of superalloys. *Scripta Materialia*, *42*(12), 1179–1186. doi:10.1016/S1359-6462(00)00355-9

Grafe, U., Böttger, B., Tiaden, J., & Fries, S. G. (2000b). Simulations of the initial transient during directional solidification of multicomponent alloys using the phase field method. *Modelling and Simulation in Materials Science and Engineering*, *8*(6), 871–879. doi:10.1088/0965-0393/8/6/308

Grassel, O., Frommeyer, G., Derder, C., & Hofmann, H. (1997). Phase transformations and mechanical properties of Fe-Mn-Si-Al TRIP-steels. *Journal de Physique IV*, *7*(C5), 383-388.

Guo, S., Ng, C., Lu, J., & Liu, C. T. (2011). Effect of valence electron concentration on stability of fcc or bcc phase in high entropy alloys. *Journal of Applied Physics*, *109*(10), 103505. doi:10.1063/1.3587228

Gupta, P. K. (1996). Non-crystalline solids: Glasses and amorphous solids. *Journal of Non-Crystalline Solids*, *195*(1), 158–164. doi:10.1016/0022-3093(95)00502-1

Hairie, A., Hairie, F., Lebouvier, B., Nouet, G., Paumier, E., Ralantoson, N., & Sutton, A. P. (1994). Free energies of interfaces from quasi-harmonic theory: A critical appraisal. *Interface Science*, *2*(1), 17–29. doi:10.1007/BF00188816

Hall, E. O. (1951). The deformation and ageing of mild steel: III discussion of results. *Proceedings of the Physical Society. Section B*, *64*(9), 747–753. doi:10.1088/0370-1301/64/9/303

Hamilton, J. C., Siegel, D. J., Daruka, I., & Léonard, F. (2003). Why do grain boundaries exhibit finite facet lengths? *Physical Review Letters*, *90*(24), 246102. doi:10.1103/PhysRevLett.90.246102 PMID:12857204

Herring, C. (1949). *The physics of powder metallurgy* (R. Gomer & C. S. Smith, Eds.). New York: McGraw-Hill.

Hesselbarth, H. W., & Göbel, I. R. (1991). Simulation of recrystallization by cellular automata. *Acta Metallurgica et Materialia*, *39*(9), 2135–2143. doi:10.1016/0956-7151(91)90183-2

Hohenberg, P., & Kohn, W. (1964). Inhomogeneous electron gas. *Physical Review*, *136*(3B), B864–B871. doi:10.1103/PhysRev.136.B864

Holm, E. A., & Battaile, C. C. (2001). The computer simulation of microstructural evolution. *JOM*, *53*(9), 20–23. doi:10.1007/s11837-001-0063-2

Hore, S., Das, S. K., Banerjee, S., & Mukherjee, S. (2013, February). Monte Carlo simulation of microstructure evolution during thermo-mechanical rolling of steel using grid computing technology. In *Parallel Computing Technologies (PARCOMPTECH), 2013 National Conference on* (pp. 1-7). IEEE. doi:10.1109/ParCompTech.2013.6621406

Hornbogen, E. (1984). Overview no. 37: On the microstructure of alloys. *Acta Metallurgica*, *32*(5), 615–627. doi:10.1016/0001-6160(84)90135-4

Hornbogen, V. I. (1968). Über die Begriffe "Gefiige" und "Phase", Prakt. *Metallographie*, *5*, 51.

Janot, C. (1992). *Quasicrystals*. Oxford, UK: Clarendon Press.

Janssens, K. G. (2003). Random grid, three-dimensional, space-time coupled cellular automata for the simulation of recrystallization and grain growth. *Modelling and Simulation in Materials Science and Engineering*, *11*(2), 157–171. doi:10.1088/0965-0393/11/2/304

Janssens, K. G., Olmsted, D., Holm, E. A., Foiles, S. M., Plimpton, S. J., & Derlet, P. M. (2006). Computing the mobility of grain boundaries. *Nature Materials*, *5*(2), 124–127. doi:10.1038/nmat1559 PMID:16400330

Jenkins, G. A. (1972). 4 The systems approach. *Systems Behaviour*, 56.

Johnson, R. A. (1989). Alloy models with the embedded-atom method. *Physical Review B: Condensed Matter and Materials Physics*, *39*(17), 12554–12559. doi:10.1103/PhysRevB.39.12554 PMID:9948120

Kattner, U. R. (1997). The thermodynamic modeling of multicomponent phase equilibria. *JOM*, *49*(12), 14–19. doi:10.1007/s11837-997-0024-5

Khachaturyan, A. G. (1983). *Theory of Structural Transformations in Solids*. New York: Wiley and Sons.

Khachaturyan, A. G., & Shatalov, G. A. (1969). Elastic-interaction potential of defects in a crystal. *Sov Phys Solid State*, *11*(1), 118–123.

Kim, E. Y., Yang, H. S., Han, S. H., Kwak, J. H., & Choi, S. H. (2012). Effect of initial microstructure on strain-stress partitioning and void formation in DP980 steel during uniaxial tension. *Metals and Materials International*, *18*(4), 573–582. doi:10.1007/s12540-012-4030-y

Kobayashi, R. (1993). Modeling and numerical simulations of dendritic crystal growth. *Physica D. Nonlinear Phenomena*, *63*(3), 410–423. doi:10.1016/0167-2789(93)90120-P

Kohn, W., & Sham, L. J. (1965). Self-consistent equations including exchange and correlation effects. *Physical Review*, *140*(4A), A1133–A1138. doi:10.1103/PhysRev.140.A1133

Koza, J. R. (1992). *Genetic programming: on the programming of computers by means of natural selection* (Vol. 1). MIT press.

Lai, Z. W. (1990). Theory of ordering dynamics for Cu 3 Au. *Physical Review B: Condensed Matter and Materials Physics*, *41*(13), 9239–9256. doi:10.1103/PhysRevB.41.9239 PMID:9993266

Lan, Y. J., Xiao, N. M., Li, D. Z., & Li, Y. Y. (2005). Mesoscale simulation of deformed austenite decomposition into ferrite by coupling a cellular automaton method with a crystal plasticity finite element model. *Acta Materialia*, *53*(4), 991–1003. doi:10.1016/j.actamat.2004.10.045

Langer, J. S. (1986). Models of pattern formation in first-order phase transitions. In Directions in Condensed Matter Physics. World Scientific.

Liu, Z. K. (2009). First-principles calculations and CALPHAD modeling of thermodynamics. *Journal of Phase Equilibria and Diffusion*, *30*(5), 517–534. doi:10.1007/s11669-009-9570-6

Liu, Z. K., Chen, L. Q., Raghavan, P., Du, Q., Sofo, J. O., Langer, S. A., & Wolverton, C. (2004). An integrated framework for multi-scale materials simulation and design. *Journal of Computer-Aided Materials Design*, *11*(2-3), 183–199. doi:10.1007/s10820-005-3173-2

Marquis, E. A., Seidman, D. N., Asta, M., & Woodward, C. (2006). Composition evolution of nanoscale Al 3 Sc precipitates in an Al–Mg–Sc alloy: Experiments and computations. *Acta Materialia*, *54*(1), 119–130. doi:10.1016/j.actamat.2005.08.035

Marquis, E. A., Seidman, D. N., Asta, M., Woodward, C., & Ozoliņš, V. (2003). Mg Segregation at A l/A l 3 S c Heterophase Interfaces on an Atomic Scale: Experiments and Computations. *Physical Review Letters*, *91*(3), 036101. doi:10.1103/PhysRevLett.91.036101 PMID:12906429

Mendelev, M. I., & Srolovitz, D. J. (2002). Impurity effects on grain boundary migration. *Modelling and Simulation in Materials Science and Engineering*, *10*(6), R79–R109. doi:10.1088/0965-0393/10/6/201

Millett, P. C., Selvam, R. P., & Saxena, A. (2007). Stabilizing nanocrystalline materials with dopants. *Acta Materialia*, *55*(7), 2329–2336. doi:10.1016/j.actamat.2006.11.028

Mishin, Y. (2004). Atomistic modeling of the γ and γ'-phases of the Ni–Al system. *Acta Materialia*, *52*(6), 1451–1467. doi:10.1016/j.actamat.2003.11.026

Mishin, Y., Asta, M., & Li, J. (2010). Atomistic modeling of interfaces and their impact on microstructure and properties. *Acta Materialia*, *58*(4), 1117–1151. doi:10.1016/j.actamat.2009.10.049

Neugebauer, J., & Hickel, T. (2013). Density functional theory in materials science. *Wiley Interdisciplinary Reviews: Computational Molecular Science*, *3*(5), 438–448. PMID:24563665

Nyagards, M. (2003). *Microstructural Finite Element Modeling of Metals.* (Unpublished dissertation). Royal Institute of Technology, Sweden.

Olmsted, D. L., Foiles, S. M., & Holm, E. A. (2007). Grain boundary interface roughening transition and its effect on grain boundary mobility for non-faceting boundaries. *Scripta Materialia*, *57*(12), 1161–1164. doi:10.1016/j.scriptamat.2007.07.045

Olson, G. B. (1997). Computational design of hierarchically structured materials. *Science*, *277*(5330), 1237–1242. doi:10.1126/science.277.5330.1237

Paxton, A. T. (2014). From quantum mechanics to physical metallurgy of steels. *Materials Science and Technology*, *30*(9), 1063–1070. doi:10.1179/1743284714Y.0000000521

Peng, P., Soh, A. K., Yang, R., & Hu, Z. Q. (2006). First-principles study of alloying effect of Re on properties of Ni/Ni 3 Al interface. *Computational Materials Science*, *38*(2), 354–361. doi:10.1016/j.commatsci.2005.10.010

Petch, N. J. (1953). The cleavage strength of polycrystals. *Journal of the Iron and Steel Institute*, *174*, 25–28.

Poletti, M. G., & Battezzati, L. (2014). Electronic and thermodynamic criteria for the occurrence of high entropy alloys in metallic systems. *Acta Materialia*, *75*, 297–306. doi:10.1016/j.actamat.2014.04.033

Presland, A. E. B. (1960). *Mechanical Properties of Intermetallic Compounds*. New York: Wiley.

Qin, R. S., & Bhadeshia, H. K. (2010). Phase field method. *Materials Science and Technology*, *26*(7), 803–811. doi:10.1179/174328409X453190

Raabe, D. (2001). Mesoscale simulation of recrystallization textures and microstructures. *Advanced Engineering Materials*, *3*(10), 745. doi:10.1002/1527-2648(200110)3:10<745::AID-ADEM745>3.0.CO;2-C

Raabe, D. (2002). Cellular automata in materials science with particular reference to recrystallization simulation. *Annual Review of Materials Research*, *32*(1), 53–76. doi:10.1146/annurev.matsci.32.090601.152855

Raabe, D. (2007). Multiscale recrystallization models for the prediction of crystallographic textures with respect to process simulation. *The Journal of Strain Analysis for Engineering Design*, *42*(4), 253–268. doi:10.1243/03093247JSA219

Raabe, D., & Becker, R. C. (2000). Coupling of a crystal plasticity finite-element model with a probabilistic cellular automaton for simulating primary static recrystallization in aluminium. *Modelling and Simulation in Materials Science and Engineering*, *8*(4), 445–462. doi:10.1088/0965-0393/8/4/304

Radhakrishnan, B., Sarma, G. B., & Zacharia, T. (1998). Modeling the kinetics and microstructural evolution during static recrystallization—Monte Carlo simulation of recrystallization. *Acta Materialia*, *46*(12), 4415–4433. doi:10.1016/S1359-6454(98)00077-9

Randle, V. (2004). Twinning-related grain boundary engineering. *Acta Materialia*, *52*(14), 4067–4081. doi:10.1016/j.actamat.2004.05.031

Rappaz, M., & Gandin, C. A. (1993). Probabilistic modelling of microstructure formation in solidification processes. *Acta Metallurgica et Materialia*, *41*(2), 345–360. doi:10.1016/0956-7151(93)90065-Z

Reddy, J. N., & Gartling, D. K. (2010). *The finite element method in heat transfer and fluid dynamics*. CRC Press.

Rice, J. R., & Wang, J. S. (1989). Embrittlement of interfaces by solute segregation. *Materials Science and Engineering A*, *107*, 23–40. doi:10.1016/0921-5093(89)90372-9

Rickman, J. M., & LeSar, R. (2002). Free-energy calculations in materials research. *Annual Review of Materials Research*, *32*(1), 195–217. doi:10.1146/annurev.matsci.32.111901.153708

Rollett, A. D., & Mullins, W. W. (1997). On the growth of abnormal grains. *Scripta Materialia*, *36*(9), 975–980. doi:10.1016/S1359-6462(96)00501-5

Roters, F., Eisenlohr, P., Bieler, T. R., & Raabe, D. (2011). *Crystal plasticity finite element methods: in materials science and engineering*. John Wiley & Sons.

Sanchez, J. M. (1993). Cluster expansions and the configurational energy of alloys. *Physical Review B: Condensed Matter and Materials Physics*, *48*(18), 14013–14015. doi:10.1103/PhysRevB.48.14013 PMID:10007810

Sansoz, F., & Molinari, J. F. (2005). Mechanical behavior of Σ tilt grain boundaries in nanoscale Cu and Al: A quasicontinuum study. *Acta Materialia*, *53*(7), 1931–1944. doi:10.1016/j.actamat.2005.01.007

Schiøtz, J., Di Tolla, F. D., & Jacobsen, K. W. (1998). Softening of nanocrystalline metals at very small grain sizes. *Nature*, *391*(6667), 561–563. doi:10.1038/35328

Schmidt, C., Finnis, M. W., Ernst, F., & Vitek, V. (1998). Theoretical and experimental investigations of structures and energies of Σ= 3,[112] tilt grain boundaries in copper. *Philosophical Magazine A*, *77*(5), 1161–1184. doi:10.1080/01418619808214246

Simo, J. C. (1997). TJR Hughes Computational inelasticity, Interdisciplinary Applied Mathematics. *Mechanics of Materials*, 7.

Srolovitz, D. J., Grest, G. S., & Anderson, M. P. (1986). Computer simulation of recrystallization—I. Homogeneous nucleation and growth. *Acta Metallurgica*, *34*(9), 1833–1845. doi:10.1016/0001-6160(86)90128-8

Srolovitz, D. J., Grest, G. S., Anderson, M. P., & Rollett, A. D. (1988). Computer simulation of recrystallization—II. Heterogeneous nucleation and growth. *Acta Metallurgica*, *36*(8), 2115–2128. doi:10.1016/0001-6160(88)90313-6

Stachurski, Z. H. (2011). On structure and properties of amorphous materials. *Materials (Basel)*, *4*(9), 1564–1598. doi:10.3390/ma4091564

Sutton, A. P., & Balluffi, R. W. (1995). *Interfaces in crystalline materials*. Oxford, UK: Clarendon press.

Sutton, A. P., Mulheran, P. A., & Stoneham, A. M. (1992). Direct Free Energy Minimization Methods: Application to Grain Boundaries [and Discussion]. *Philosophical Transactions of the Royal Society of London A: Mathematical, Physical and Engineering Sciences, 341*(1661), 233-245.

Temperley, H. N. V. (1981). *Graph Theory and Applications*. New York: Wiley.

Tong, M., Li, D., & Li, Y. (2004). Modeling the austenite–ferrite diffusive transformation during continuous cooling on a mesoscale using Monte Carlo method. *Acta Materialia*, *52*(5), 1155–1162. doi:10.1016/j.actamat.2003.11.006

Trautt, Z. T., & Upmanyu, M. (2005). Direct two-dimensional calculations of grain boundary stiffness. *Scripta Materialia*, *52*(11), 1175–1179. doi:10.1016/j.scriptamat.2004.12.033

Tschopp, M. A., & McDowell, D. L. (2007a). Structures and energies of Σ3 asymmetric tilt grain boundaries in copper and aluminium. *Philosophical Magazine*, *87*(22), 3147–3173. doi:10.1080/14786430701255895

Tschopp, M. A., & McDowell, D. L. (2007b). Asymmetric tilt grain boundary structure and energy in copper and aluminium. *Philosophical Magazine*, *87*(25), 3871–3892. doi:10.1080/14786430701455321

Tsuchida, Y., Watanabe, T., Kato, T., & Seto, T. (2010). Effect of hydrogen absorption on strain-induced low-cycle fatigue of low carbon steel. *Procedia Engineering*, *2*(1), 555–561. doi:10.1016/j.proeng.2010.03.060

Upmanyu, M., Srolovitz, D. J., Shvindlerman, L. S., & Gottstein, G. (1999). Misorientation dependence of intrinsic grain boundary mobility: Simulation and experiment. *Acta Materialia*, *47*(14), 3901–3914. doi:10.1016/S1359-6454(99)00240-2

Vaithyanathan, V., Wolverton, C., & Chen, L. Q. (2002). Multiscale modeling of precipitate microstructure evolution. *Physical Review Letters*, *88*(12), 125503. doi:10.1103/PhysRevLett.88.125503 PMID:11909473

Van de Walle, A., & Asta, M. (2002). Self-driven lattice-model Monte Carlo simulations of alloy thermodynamic properties and phase diagrams. *Modelling and Simulation in Materials Science and Engineering*, *10*(5), 521–538. doi:10.1088/0965-0393/10/5/304

Van De Walle, A., & Ceder, G. (2002). The effect of lattice vibrations on substitutional alloy thermodynamics. *Reviews of Modern Physics*, *74*(1), 11–45. doi:10.1103/RevModPhys.74.11

Wang, Y., Chen, D., & Zhang, X. (2000). Calculated equation of state of Al, Cu, Ta, Mo, and W to 1000 GPa. *Physical Review Letters*, *84*(15), 3220–3223. doi:10.1103/PhysRevLett.84.3220 PMID:11019055

Wang, Y., Curtarolo, S., Jiang, C., Arroyave, R., Wang, T., Ceder, G., & Liu, Z. K. (2004). Ab initio lattice stability in comparison with CALPHAD lattice stability. *Calphad*, *28*(1), 79–90. doi:10.1016/j.calphad.2004.05.002

Wang, Y., Liu, Z. K., & Chen, L. Q. (2004). Thermodynamic properties of Al, Ni, NiAl, and Ni 3 Al from first-principles calculations. *Acta Materialia*, *52*(9), 2665–2671. doi:10.1016/j.actamat.2004.02.014

Wang, Y. M., Chen, M. W., Sheng, H. W., & Ma, E. (2002). Nanocrystalline grain structures developed in commercial purity Cu by low-temperature cold rolling. *Journal of Materials Research*, *17*(12), 3004–3007. doi:10.1557/JMR.2002.0436

Wang, Y. Z., & Chen, L. Q. (1999). *Methods of Materials Research.* New York: Wiley and Sons.

Wei, S. H., Ferreira, L. G., Bernard, J. E., & Zunger, A. (1990). Electronic properties of random alloys: Special quasirandom structures. *Physical Review B: Condensed Matter and Materials Physics*, *42*(15), 9622–9649. doi:10.1103/PhysRevB.42.9622 PMID:9995203

Williams, P. L., Mishin, Y., & Hamilton, J. C. (2006). An embedded-atom potential for the Cu–Ag system. *Modelling and Simulation in Materials Science and Engineering*, *14*(5), 817–833. doi:10.1088/0965-0393/14/5/002

Wolf, D., Wang, J., Phillpot, S. R., & Gleiter, H. (1995). Phonon-induced anomalous specific heat of a nanocrystalline model material by computer simulation. *Physical Review Letters*, *74*(23), 4686–4689. doi:10.1103/PhysRevLett.74.4686 PMID:10058573

Wu, K., Chen, S., Zhang, F., & Chang, Y. A. (2009). Integrating CALPHAD into phase field simulations for practical applications. *Journal of Phase Equilibria and Diffusion*, *30*(5), 571–576. doi:10.1007/s11669-009-9567-1

Wu, Z. X., Zhang, Y. W., & Srolovitz, D. J. (2009). Grain boundary finite length faceting. *Acta Materialia*, *57*(14), 4278–4287. doi:10.1016/j.actamat.2009.05.026

Yamakov, V., Wolf, D., Phillpot, S. R., Mukherjee, A. K., & Gleiter, H. (2002). Dislocation processes in the deformation of nanocrystalline aluminium by molecular-dynamics simulation. *Nature Materials*, *1*(1), 45–49. doi:10.1038/nmat700 PMID:12618848

Yeh, J. W., Chen, S. K., Lin, S. J., Gan, J. Y., Chin, T. S., Shun, T. T., & Chang, S. Y. (2004). Nanostructured High-Entropy Alloys with Multiple Principal Elements: Novel Alloy Design Concepts and Outcomes. *Advanced Engineering Materials*, *6*(5), 299–303. doi:10.1002/adem.200300567

Zaefferer, S. (2005, July). Application of orientation microscopy in SEM and TEM for the study of texture formation during recrystallisation processes. *Materials Science Forum*, *495*, 3–12. doi:10.4028/www.scientific.net/MSF.495-497.3

Zhou, X. W., Wadley, H. N. G., Johnson, R. A., Larson, D. J., Tabat, N., Cerezo, A., & Kelly, T. F. (2001). Atomic scale structure of sputtered metal multilayers. *Acta Materialia*, *49*(19), 4005–4015. doi:10.1016/S1359-6454(01)00287-7

Zhu, X., Birringer, R., Herr, U., & Gleiter, H. (1987). X-ray diffraction studies of the structure of nanometer-sized crystalline materials. *Physical Review B: Condensed Matter and Materials Physics*, *35*(17), 9085–9090. doi:10.1103/PhysRevB.35.9085 PMID:9941303

Zienkiewicz, O. C., & Taylor, R. L. (1977). *The finite element method* (Vol. 3). London: McGraw-hill.

Chapter 5
Micromechanical and Finite Element Modeling for Composites

Ankur Kumar Gupta
ISRO Satellite Centre, India

Arjun Dey
ISRO Satellite Centre, India

Anoop Kumar Mukhopadhyay
CSIR–Central Glass and Ceramic Research Institute, India

ABSTRACT

The major areas of applications of the composite materials today encompass fields as wide as wind energy to marine to construction to aerospace to strategic areas. Apart from such specialized fields the major composite market blooms out of their extensive exploitations in the automotive, sporting goods, pipes, tanks, chemicals, fertilizers and many other industries. As compared to the conventional materials, the composites offer several unique advantages. The list of such advantages may include as a typical example the higher tensile strength, the lighter weight, the greater corrosion resistance, the better surface finish and the easier processability. It is interesting to note that in 2011, the global composite materials market size was $19.6 billion and the same is estimated now to reach approximately $34.1 billion in 2018. This amounts to a Compounded Annual Growth Rate (CAGR) of about 10.5% percent. This huge market growth of composites will require a thorough knowledge of the mechanics and mechanical properties of a wide variety of composites under a still wider variety of application dependent constraints. That is why, just to make an humble beginning in this direction, the basics of the composite materials and it's varieties, are briefly described in this Chapter. In addition, the analytical methods of continuum micromechanics have been touched upon. Keeping in view the complex microstructure of most of the composites, and the effect that such complexities can have in the resultant properties, the numerical approaches to continuum micromechanics has been highlighted with emphasis on the fundamentals of the Finite Element Analysis (FEA) based methods. Further, the concepts involved in modelling of the "Failure of composites" have been elucidated with some interesting case studies.

DOI: 10.4018/978-1-5225-0290-6.ch005

1. MICROMECHANICS OF COMPOSITES

1.1 Basics of Composite Materials

A composite material is one which is composed of at least two elements working together to produce material properties that are different to the properties of those elements on their own. In practice, most composites consist of a bulk material (the 'matrix'), and a reinforcement of some kind, added primarily to increase the strength and stiffness of the matrix. The most obvious question that comes to the mind in this context is what the typical kinds of man-made composites are. Today, the most common man-made composites can be divided into three main categories:

1. **Metal Matrix Composites (MMC's):** These composites are being increasingly used in the automotive industry. For instance these materials can use a light metal such as aluminium as the matrix, and reinforce it with fibres or particulates such as silicon carbide particles or fibers.
2. **Ceramic Matrix Composites (CMC's):** These materials are meant for use in very high hrash environments e.g., highly corrosive atmosphere, high or very high temperature, high strain rate impact resistant applications, ultra high load, ultra high temperature oxidative environments etc. These materials use a ceramic as the matrix and reinforce it with particles such as aluminium nitride or silicon carbide or zirconia or short/long fibres, or whiskers such as those of alumina, silicon carbide and boron nitride.
3. **Polymer Matrix Composites (PMC's):** These materials represent the most common type of composites. These materials are also known as FRP - Fibre Reinforced Polymers. Contrary to the case of MMCs and CMCs, these materials use a polymer-based resin as the matrix, and a variety of fibres such as glass, carbon, aramid and natural fibres as the reinforcement. The major reason for use of FMC is weight saving. They offer high relative stiffness and strength. For instance, it may be noted that the carbon fibre reinforced polymer composite (CFRP) may be five times as strong as the SS1020 while it has weight one fifth that of the weight of SS1020. Similarly, CFRP has modulus twice as high and strength seven times as high as that of Al6061.

1.2 Basics of Micromechanics

Taken literally, micromechanics refer to mechanics of materials effects at the micrometer scale length. Usually however the term is used more broadly as representing effects at scales somewhat larger than the nano-scale size but much smaller than the macroscopic size. They may center on the micron scale but they can also reach up and down considerably from that size. Micromechanics certainly offer an appealing look at basic effects at intermediate length scales.

Thus, the basic aim of the subject field of "Micromechanics of Composites" is to determine the unknown properties of the composite based on the known properties of the reinforcement phase and the matrix phase. Many of the problems of special interest in the micromechanics domain are related to inclusions that are dispersed into an otherwise continuous phase. Doing this mixing at or near the micron scale affords an enormous variety of materials combinations that can yield a huge array of enhanced properties. When done properly, the combinations can achieve the best attributes of the constituents. When done casually or carelessly it almost always degenerates to the worst from each or all. The properties that are to be improved include stiffness, strength, toughness, processability, stability, durability and a host of other, sometime specialized properties. Of particular interest here are the cases of improving or optimizing stiffness and strength. Stiffness will immediately and briefly be considered and then strength will occupy the major effort in this section. There are many books and countless papers focused upon micromechanics. The vast majority of these deal with the effective stiffness properties. The field of composite materials is the recognized marketplace for such works. For example, the book "Mechanics of Composite Materials", by Christensen (1979) is largely dedicated to the micromechanics analysis of particulate and fibrous reinforcement of a continuous matrix phase. The objective of most books is to determine the effective stiffness properties of the composite material in terms of the corresponding properties of the phases.

There is a very good reason why the effort has been mainly devoted to effective stiffness properties. These stiffness type problems can be and are carefully posed so as to be amenable to complete mechanics type analysis. The micromechanics situation with strength is much more difficult. This requires some explanation. The effective stiffness properties involve volume averages over a representative size or element (VRE). Thus this procedure is an integration type of result with a consequent smoothing effect. In contrast, failure is usually a point specific effect located at a defect or a stress concentration point. Even at a stress concentration point, there still are defects that are operative at much smaller scales at the same point. Macroscopic strength properties are usually due to a distribution of subscale defects. Strength has always been an exceptionally difficult topic for these reasons, but the modulus is not.

All of this relates to micromechanics in the following way. If the micromechanics scale embeds the controlling defect then the micromechanics scale is well suited for the strength studies. To say this way, if the prime defect is best and most clearly characterized at the micron level, then micromechanics is the logical choice for proceeding. Otherwise, the advantages of micromechanics for strength studies are far less clear. To go further with this explanation a particular class of materials is needed to gain a firm grasp of the tradeoffs that must be considered.

Carbon fiber-polymeric matrix composites provide the logical high performance, high value example. Carbon fibers (filaments) are about 5-10 microns in diameter while the separation between them is about an order of magnitude less in dimension. Thus carbon fiber composites may seem like the perfect fit for micromechanics, but it turns out to be more complicated than it first appears.

Now it is also important to estimate the local stress and strain fields in the individual phases which is in particular can provide substantial information about the failure of the composites. As composites are heterogeneous in nature any test would show a long range statistical distribution and those values are strongly dependent on representative volume element, orientation of reinforcement phase etc. In order to simulate the behavior of composite layered structures, one has to choose a modeling scale. There are basically three type of approaches based on scale for modeling of composites: macroscopic, mesoscopic and microscopic (Figure 1).

Figure 1. Scale issues in modeling of composites

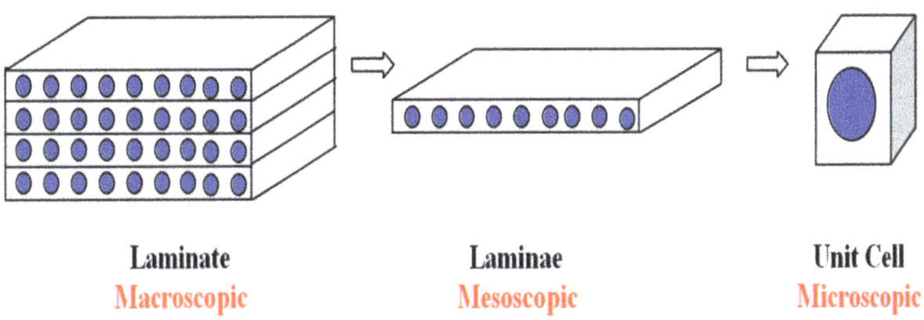

Study by the macroscopic approach doses not allow to take into account the layers position whereas microscopic approach needs larger computation efforts (Germain, Besson & Feyel 2007). In contrast to the macroscopic approach which considers the properties of fibre and matrix separately, the micromechanical approach use the overall properties of composite developed by relating the average stresses and strains. The micromechanical approach may provide much more detail into the true interactions between the fibre and matrix, potentially leading to more accurate model of the composite behavior (Gardner, 1994).

Most composite materials exhibit statistical homogeneity rather than a deterministic arrangement of the constituents, the methods of micromechanics are typically based on the concept of the representative volume element (RVE) (Hashin, 1983). A representative volume element (RVE) is a sample of a heterogeneous material that is entirely typical of the whole mixture on average and contains a sufficient number of inclusions for the apparent properties to be independent of the surface values of traction and displacement (Hill, 1963).

According to Nemat-Nasser and Hori (1993) the RVE for a material point of a continuum is a material volume which is statistically representative for the infinitesimal material neighbourhood of that material point. The term unit cell (UC) is also used extensively in heterogeneous materials and defined as the building block of the heterogeneous material.

Figure 2. Fundamental paradigms of micromechanics analysis utilizing concept of a unit cell (Tang, 2008)

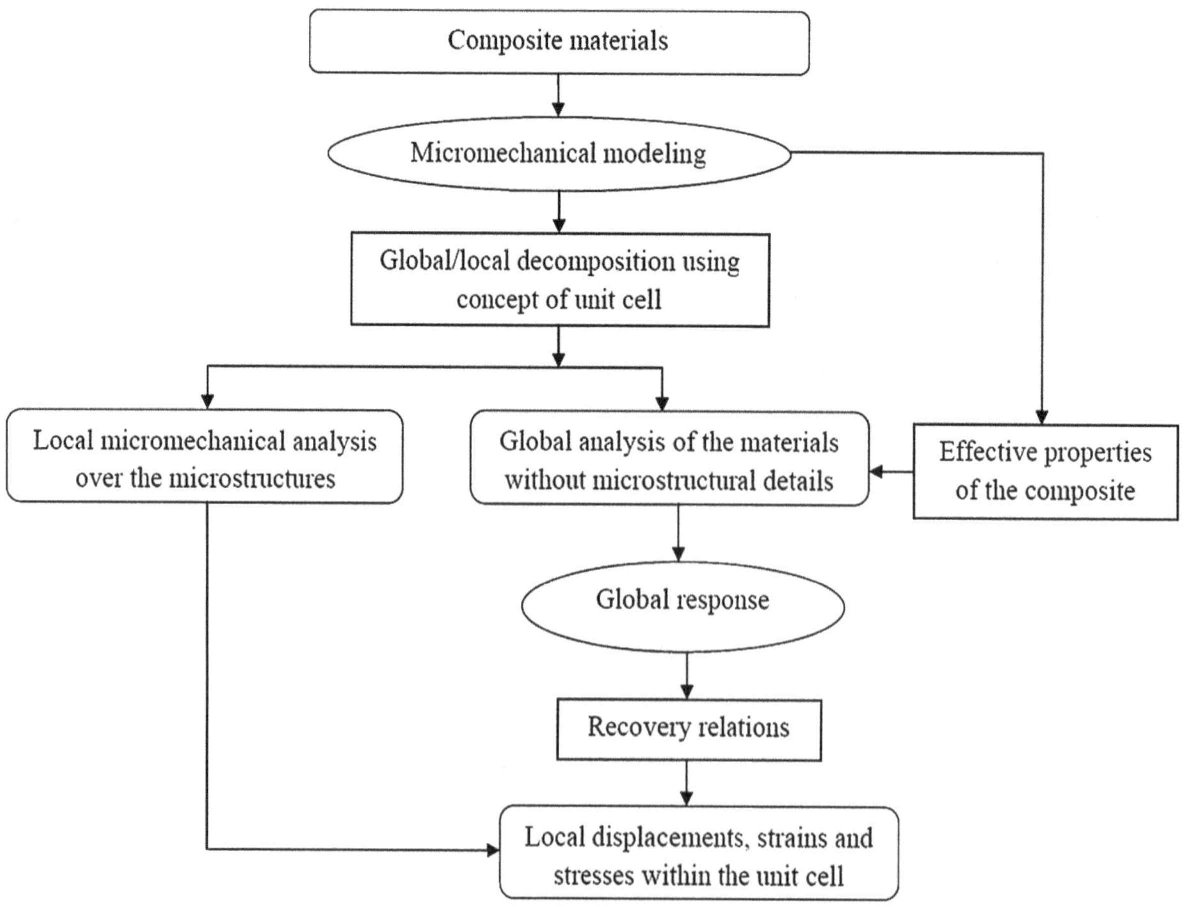

The basic concept of micromechanics analysis of composite material using the concept of unit cell is shown in the Figure 2. First step of the micromechanics models is to decouple the analysis for composites with heterogeneous microstructures into a local micromechanical analysis and second is to provide the effective properties for the global analysis of the material without microstructural details. Final step is to provide local displacement, stress and strain values from global analysis using recovery relations

2. ANALYTICAL METHODS OF CONTINUUM MICROMECHANICS

2.1 Voigt and Reuss Model

Initially, Voigt and Reuss Model (VRM) was introduced to predict upper and lower bound Young's modulus of the composites in particular for fibre reinforced composites (Voigt, 1889; Reuss, 1929). It was assumed that fibre should be in continuous mode and unidirectional in orientation. Later on, it was applied to estimate on hardness, tensile/flexural strength, mass density, thermal and electrical conductivity. Basically, the two models i.e. Voigt model and Reuss Model was based on simple "rule of mixtures" of composites where weighted or volume fractions of the constitute elements was used to predict the various properties of the composite materials. In the case of axial loading, the Voigt model should be applicable where iso-strain condition will be prevailing while Reuss model would be applicable in the case of transverse loading in iso-stress condtion. In other words, Voigt model is appropriate for the axial loading condition where as Reuss model is considered as applicable for transverse loading situation (Figure 3).

Therefore, any property of the composites such as modulus, strength etc. can be predictable if we know the volume fraction of the reinforcement element. The overall property of the composite (P_c) in axial loading condition *i.e.* in the direction parallel to the fibres is represented in the following manner:

Figure 3. Schematic representation of axial loading (Voigt model) and transverse loading (Reuss model) conditions of a continuous fibre reinforced composite

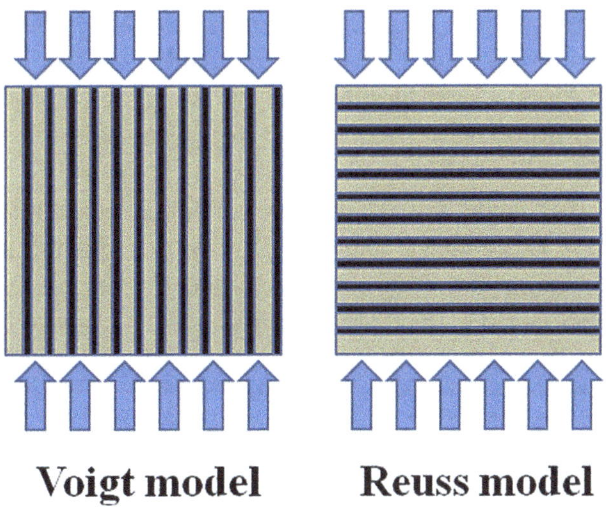

$$P_c = rP_r + (1-r)P_m \tag{1}$$

Thus, Equation (1) represents the mathematical form of Voigt model.

Similarly, P_c in transverse loading condition *i.e.* in the direction perpendicular to the fibres is represented in the following manner:

$$P_c = \left(\frac{r}{P_r} + \frac{1-r}{P_m}\right)^{-1} \tag{2}$$

where r is the volume fraction of reinforcement or second phase. P_r and P_m indicate any property of the corresponding individual phase e.g. reinforcement and matrix, respectively. Equation (1) would predict the upper-bound value while Equation (2) gives us the mathematical form of Reuss Model that predicts the lower-bound value of the overall property of composite, in particular, e.g., for the Young's modulus.

2.2 Hashin-Shtrikman Model

It needs to be appreciated that the Voigt model assumes that all reinforcements are parallel to the direction of loading e.g., uniaxial tension or compression of the composites. In reality though it will be hardly ever possible to guarantee that all the reinforcing elements in a composite microstructure will have their long axis parallel to the direction of loading. What it conveys is that this was too much of an idealization for a real life microstructure of a composite material. The Reuss model also is no exception in this regard. It assumes on the other hand that all reinforcing elements in a composite microstructure are oriented perpendicular to the direction of uniaxial tension or compression. Again it is not very difficult to imagine that it will be really very difficult, if not impossible, to realize a composite with all the reinforcing elements in the microstructure having their long axis perpendicular to the direction of loading. In other word this is again another extreme of an idealization for a real microstructure of a composite material.

What has been discussed already does point out also that the real composite will, in the most likely case, have a microstructure where some of the reinforcing elements will have their long axis parallel while others will have their long axis perpendicular to the direction of loading. Thus, even accepting the original path breaking views of both Voigt (1889) and Reuss (1929), it needs to be appreciated these models had their own limitations as both of these models could predict only wider bounds on Young's moduli. The need for the narrowest or tightest bounds on Young's moduli was thus too crucially and critically important from both historical and technological perspectives. This is exactly what was provided by the Hashin-Shtrikman (H-S) model.

The Hashin-Shtrikman (H-S) model gave the narrowest or tightest possible bounds on Young's moduli (Hashin & Shtrikman, 1963). H-S model is applicable for isotropic composites both for continuously aligned fibre reinforced or random particulate composites. The H-S bounds for the bulk modulus K and shear moduli μ has been represented in the following manner:

$$K_{HS}^{\mp} = K_2 + \frac{f_1}{(K_1 - K_2)^{-1} + f_2\left(K_2 + \frac{4}{3\mu_2}\right)^{-1}} \tag{3}$$

Figure 4. Schematic representation of Hashin-Shtrikman model for isotropic composite

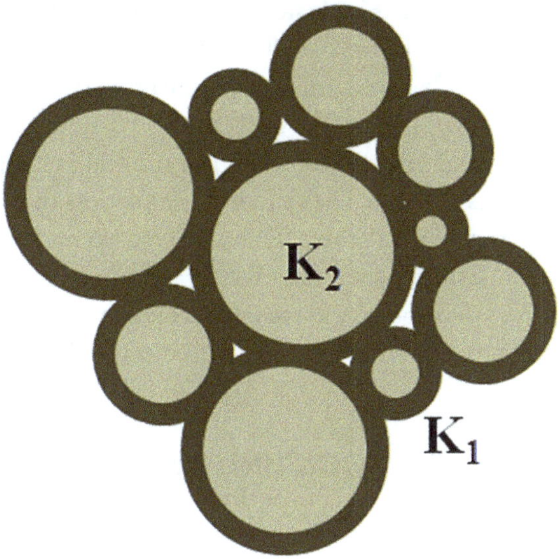

$$\mu_{HS}^{\mp} = \mu_2 + \frac{f_1}{\left(\mu_1 - \mu_2\right)^{-1} + \dfrac{2f_2\left(K_2 + 2\mu_2\right)}{5\mu_2\left(K_2 + 4/3\mu_2\right)}} \quad (4)$$

The schematic representation of H-S model is shown in Figure 4. The upper bound of the modulus values would be predicted while K_2 is greater than K_1. However, the lower bound should be predicted by replacing the indices in the aforesaid equations.

The next question that evolved was to ask naturally what is going to happen to stress and strain sharing between the reinforcement and the matrix in a composite that comprises of more than one type of reinforcing phase e.g., as in a rock microstructure. This very important issue was addressed by Berryman and Berge (1993). It is in fact his seminal work that had introduced the modified H-S model for multiphase composites e.g., a rock microstructure.

2.3 Mori-Tanaka Model

We may also wonder to know what is going to be the internal stress in the matrix phase of a composite that contains a particulate reinforcement phase with elastic properties different from that of the matrix. Due to elastic mismatch the reinforcing particulate phase will be under a characteristic strain called e.g., the eigenstrain. It is interesting to note that the credit of solving this seminal problem of elasticity theory goes to Mori and Tanaka (1973).

They, in fact, have proposed a model (Mori & Tanaka, 1973) by which the average internal stresses in the matrix phase of a composite containing second phase precipitates with eigenstrains can be calculated. Fourteen years later, Benveniste (1987) re-looked into the problem and applied the same for the composite materials with anisotropic and ellipsoidal inclusions. This idea (Benveniste, 1987) was actually based on the concept of "equivalent inclusion". This work of tremendous scientific importance was carried out by Eshelby way back in 1957 (Eshelby, 1957).

All the aforesaid models predict nicely the effective or overall properties of simple, classical composites, with isotropic as well as anisotropic and spherical as well ellipsoidal inclusions as the reinforcing phase(s) in presence/absence of cracks and void growth etc. What is most gratifying to note is this that in spite of the several simplifying assumptions the efficacies of all these models in predicting the experimentally measured data was satisfactory, if not spectacular. This fact also points out that the next level of close match between the prediction and experimental data would require approaches more rigorous than the simple analytical approaches discussed above. This is from where the scope of the numerical approaches to continuum micromechanics emerges. Therefore, we shall discuss the same in the next section.

3. NUMERICAL APPROACHES TO CONTINUUM MICROMECHANICS

3.1 Introduction

There are three basic steps to model any physical phenomena and processes as following:

1. Problem definition
2. Mathematical model
3. Numerical simulation

The first step is to define an idealization of our problem of interest in terms of a set of relevant quantities which we would like to measure. The second step of the idealization of the physical phenomena by a mathematical model in terms of algebraic, differential/integral equations is known as the governing equations of the problem. The final step is to solve these equations under various conditions such as boundary or initial conditions which leads to the understanding of the phenomena.

Exact solutions for these equations are generally difficult to obtain. Numerical methods are adopted to obtain approximate solutions for differential equations. Various numerical solution techniques such as finite difference method (FDM), finite element method (FEM), finite volume method (FVM), boundary element method (BEM), Domain decomposition method (DDM) and differential quadrature method (DQM) are used depending upon the nature of problem.

3.2 Finite Element Method

The finite element method is a more general and powerful numerical method to model the real-world problem that involve complicated physics, geometry, and/or boundary conditions. The method essentially consists of assuming the piecewise continuous function for the solution and obtaining the parameters of the functions in a manner that reduces the error in the solution. In the finite element method a given domain divided into subdomains called finite element and unknown field variables are expressed the in terms of assumed approximating functions (Interpolating functions/Shape functions) within each element. The approximating functions are defined in terms of field variables of specified points called nodes or nodal points.

The finite element analysis method requires the following major steps:

1. Discretization of the entire domain into a finite number of subdomains (elements).
2. Selection of interpolation functions.
3. Derive the element-level governing equations.
4. Assembly of the element matrices for each subdomain to obtain the global matrix for the entire domain (system level governing equations).
5. Imposition of the boundary conditions.
6. Solution of the system of equations involving unknown quantities at the nodes (e.g., displacements).
7. Additional computations (e.g., strains and stresses calculation) at selected elements.

3.2.1 Elements

The entire domain can be discretized by employing line (trusses or beams), area (plates and membranes), or volume (bricks or tetrahedrals) elements depending on the geometry and the physical nature of the problem. Some of the common elements in FEM are shown in Figure 5. Elements may have physical properties such as thickness, coefficient of thermal expansion, density, Young's modulus, shear modulus and Poisson's ratio. Elements can have intrinsic dimensionality of one, two or three space dimensions. There are also special elements with zero dimensionality, such as lumped spring or point masses. Each element, identified by an element number, is defined by a specific sequence of global node numbers. The specific sequence (usually counter clockwise) is based on the node numbering at the element level.

3.2.2 Element Nodes

Each element possesses a set of distinguishing points called nodal point or nodes. A node is a coordinate location in space where the degrees of freedom (DOFs) are defined. Nodes are usually located at the corners or end points of the elements as shown in the Figure 6.

Figure 5. Common elements in FEM

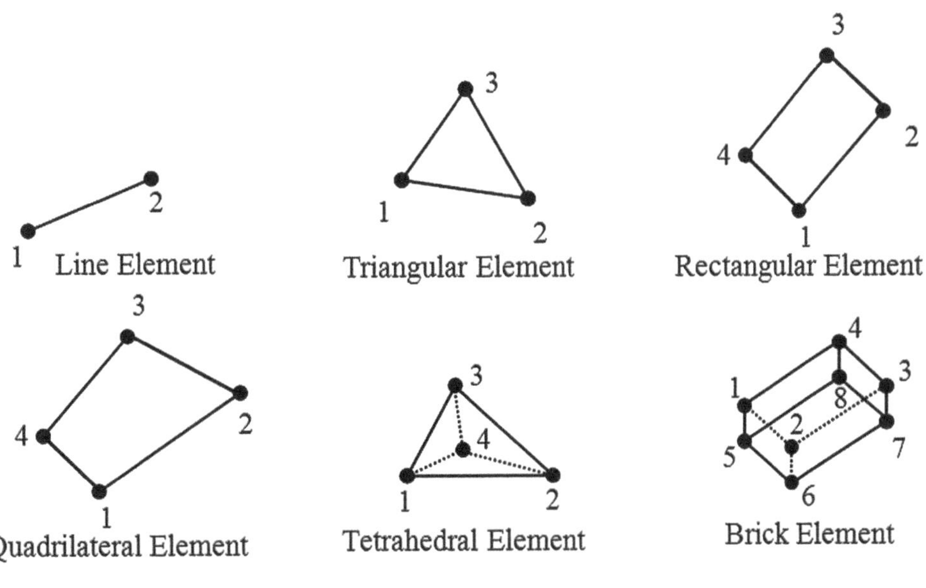

Figure 6. Discretization of domain in to subdomains (i.e., elements)

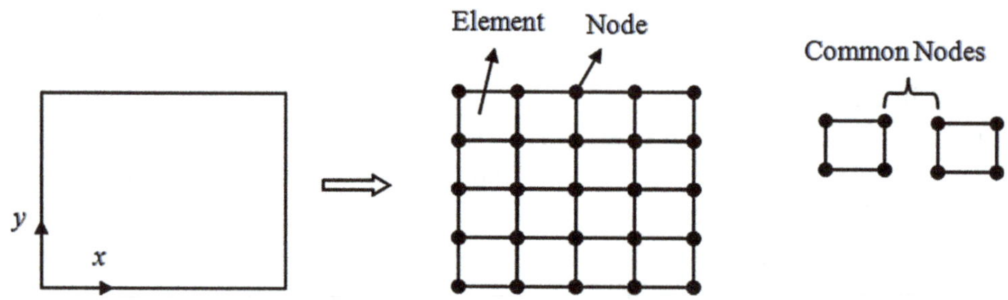

3.3 Approximate Solution Methods

Approximate methods are used to determine solution of the differential and/or integral equations. Consider a problem of structural engineering described in the form of differential equation. The governing differential equation of a bar (Figure 7) having Young's modulus E, cross-section area A, total length L subjected to axial load P is

$$\frac{d}{dx}\left[EA\frac{du}{dx}\right]+q(x)=0 \text{ for } 0<x<L \tag{5}$$

with the boundary conditions $u(0)=0, \left.\frac{du}{dx}\right|_{x=L}=0$, where $u(x)$ and $q(x)$ are axial displacement and distributed axial load per unit length, respectively.

To start with, we assume an approximate/trial solution as:

$$u(x)\cong\hat{u}(x)=\sum_{i=1}^{n}a_i\varphi_i(x) \tag{6}$$

Figure 7. Bar subjected to axial load

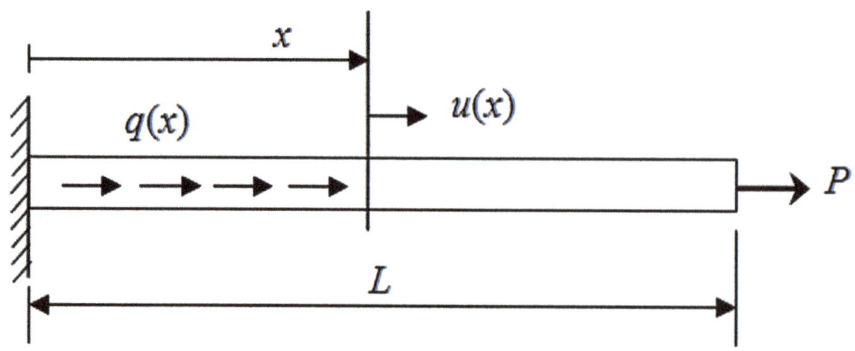

where a_i are coefficients to be determined and $\varphi_i(x)$ is trial function chosen such that specified boundary conditions are satisfied. Now, when approximation of the problem substituted into the differential equation (Equation 5), then an error or residual will evolve:

$$R_d(x, a_i) = \frac{d}{dx}\left[EA\frac{d\hat{u}}{dx}\right] + q(x) \neq 0 \tag{7}$$

The unknown coefficients a_i are determined by forcing the residual (R_d) to vanish in some average sense over the domain and this method is called Method of Weighted Residual (MWR). The Weighted Residual statement can be expressed as:

$$\int_0^L W_k(x)\, R_d(x, a_i)\, dx = 0 \text{ for } k = 1,2,.......n \tag{8}$$

where the number of weight functions $W_k(x)$ are exactly equal to the number of unknown constants a_i . If Equation (8) satisfied for any $W_k(x)$, then residual $R_d(x,a_i)$ will approach zero, and the approximate solution will approach the exact solution. Depending on choice of $W_k(x)$ various methods can be utilized. These are briefly discussed in the next sub-section.

3.3.1 Least Square Error Method

Here the technique is very simple. What is done is this that the integral of square of residual is minimized to derive out the magnitudes of the unknown coefficients (Reddy, 2006).

$$\frac{\partial}{\partial a_k}\int_0^L R_d^2(x, a_i)\, dx = 0 \;\text{ or }\; \int_0^L \frac{\partial R_d}{\partial a_k} R_d\, dx = 0 \tag{9}$$

Comparing Equation (9) with Equation (8), the weight functions can be defined as

$$W_k(x) = \frac{\partial R_d}{\partial a_k}. \tag{10}$$

Equation (10) exhibits that in this method the weight functions are derivatives of the residual with respect to the unknown constants.

3.3.2 Collocation Method

In the collocation method, the residual is forced to be zero at n selected points in the domain as:

$$R_d(x_k, a_i) = 0\, (k = 1,2,.......n) \tag{11}$$

It is well known that a Dirac δ function is a step function. That is to say that it exists only at a single point and vanishes everywhere else. The Dirac δ function is defined as:

$$\delta(x - x_k) = \begin{cases} 1 & x = x_k \\ 0 & \text{otherwise} \end{cases} \tag{12}$$

Therefore, in the collocation method, what is done is simply this that deliberately the weighting functions [e.g., $W_k(x)$] are chosen to be Dirac δ functions in the appropriate domain. Therefore, $W_k(x)$ is expressed as, $W_k(x) = \delta(x - x_k)$.

3.3.3 Sub-Domain Methods

This method can be considered as a modification of the collocation method in which the domain is split into many subdomains Ω_k and it selects the weight functions as indicator functions as:

$$W_k(x) = \begin{cases} 1 & x \in \Omega_k \\ 0 & x \notin \Omega_k \end{cases} \quad k = 1,2,\ldots\ldots n \tag{13}$$

This method minimizes the residual error in each of the chosen subdomains.

3.3.4 Galerkin Method

This method is nothing but an algebraic modification of the Least Squares Method. What is done here is simply this that the trial solution is taken. The derivative of the trial solutions is found out. These derivatives are used in this method where as the derivative of the residual with respect to the unknown a_i is generally used in the conventional least square method. Therefore, in the Galerkin method the weight functions [e.g., $W_k(x)$] can be defined as:

$$W_k(x) = \frac{\partial \hat{u}}{\partial a_k} = \varphi_k(x) \tag{14}$$

These weight functions [e.g., $W_k(x)$] are identical to the original trial functions appearing in Equation (6).

3.4 Weak form of Weighted Residual Statement

The described approximate solution methods require differentiating trial solution as many times as the order of highest derivative in the problem. To reduce the differentiability requirement of the trial solution, a slightly different approach is taken. What is done in this modified approach is this that the integration is done by parts and the general weighted residual statement (Equation (8)) is recast by suitable algebraic manipulation.

Putting the value of residual from Equation (7) in to Equation (8), the weighted residual statement can be expressed:

$$\int_0^L W_k(x)\left[\frac{d}{dx}\left(EA\frac{d\hat{u}}{dx}\right)+q(x)\right]dx=0 \tag{15}$$

$$\Rightarrow \quad \int_0^L W_k(x)\frac{d}{dx}\left(EA\frac{d\hat{u}}{dx}\right)dx+\int_0^L W_k(x)\,q(x)\,dx=0 \tag{16}$$

Integrating first term of Equation (16) by part gives:

$$\left[W_k(x)\left(EA\frac{d\hat{u}}{dx}\right)\right]_0^L-\int_0^L \frac{dW_k}{dx}EA\frac{d\hat{u}}{dx}\,dx+\int_0^L W_k(x)\,q(x)\,dx=0 \tag{17}$$

It can be seen from Equation (17), that the order of differentiability requirement of trial solution $\hat{u}(x)$ has reduced. The original weighted residual statement Equation (15) having term $\frac{d}{dx}\left(EA\frac{d\hat{u}}{dx}\right)$ requires at least quadratic polynomial as trial solution while linear trial function is sufficient for weak form of the statement (Equation (17)).

Now, let us define a primary variable. This is nothing but the dependent unknown $\hat{u}(x)$ when it is cast in the same form as that of the weight function $W_k(x)$ appearing in the boundary expression. The character of this primary variable is such that it gives the essential boundary condition that stems from the bare minimum geometric requirements. Now, let us define a secondary variable. These are nothing but the coefficients of the weight function in the boundary expression. What they do is this that they provide the natural boundary condition due to the presence of a given force. These conditions as discussed above can be briefly expressed as (Reddy, 2006):

Essential boundary condition: $\hat{u}(x=0)=\hat{u}_0$

Natural boundary condition: $EA\left.\frac{d\hat{u}}{dx}\right|_{x=L}=P$

3.4.1 Finite Element Modeling of a Bar Based on Weak Form

Let the bar shown in Figure 7 be dived into N^e elements. The typical element is shown in Figure 8 and it has two nodes, labelled as 1 and 2. These are called the *local node numbers.*

Let each element has a linear axial displacement field and it can be approximated by the following interpolation formula:

Figure 8. Mesh of a bar using typical line element

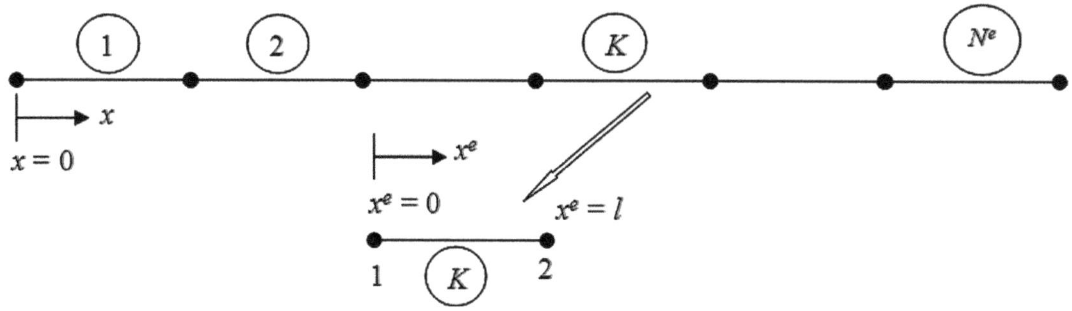

$$\hat{u}(x) = N_1 u_1 + N_2 u_2 = \begin{bmatrix} N_1 & N_2 \end{bmatrix} \begin{Bmatrix} u_1 \\ u_2 \end{Bmatrix} = [H]\{d^e\} \tag{18}$$

where function (N_1, N_2), (u_1, u_2), $[H]$ and $\{d^e\}$ are known as shape function, nodal degrees of freedom, interpolation matrix and vector of elemental degree of freedom.

The shape functions for bar can be written as:

$$N_1 = \left(1 - \frac{x^e}{l}\right) \text{ and } N_2 = \left(\frac{x^e}{l}\right) \tag{19}$$

Utilizing the elemental/local coordinate system, Equation (17) can be recast as:

$$\left[W_k(x^e)\left(EA\frac{d\hat{u}}{dx^e}\right)\right]_0^l - \int_0^l \frac{dW_k}{dx^e} EA\frac{d\hat{u}}{dx^e}\, dx^e + \int_0^l W_k(x^e)\, q(x^e)\, dx^e = 0 \tag{20}$$

Equation (18) is identical to Equation (6), where unknown coefficients a_i are represented by nodal degree of freedom (u_1, u_2) and the trial function $\varphi_i(x)$ is represented by shape functions (N_1, N_2). It can be seen from Equation (14) that the Galerkin method-based finite element method adopts the weighting functions $W_k(x)$ equal to the trial function $\varphi_k(x)$/shape function functions $N_k(x)$. So, for elemental level the weighting function can be written as:

$$i.e.\, W_k(x^e) = \varphi_k(x^e) = N_k(x^e) \tag{21}$$

The weak form of weighted residual statement for an element can be written based on Galerkin finite element method as:

$$\left[N_k(x^e)\left(EA\frac{d\hat{u}}{dx^e}\right)\right]_0^l - \int_0^l \frac{dN_k}{dx^e} EA\frac{d\hat{u}}{dx^e}\, dx^e + \int_0^l N_k(x^e)\, q(x^e)\, dx^e = 0 \tag{22}$$

Equation (22) can be rewritten as:

$$\int_0^l \frac{dN_k}{dx^e} EA \frac{d\hat{u}}{dx^e}\, dx^e = \int_0^l N_k(x^e)\, q(x^e)\, dx^e + \left[N_k(x^e) \left(EA \frac{d\hat{u}}{dx^e} \right) \right]_0^l \tag{23}$$

The term $\left[N_k(x^e) \left(EA \frac{d\hat{u}}{dx^e} \right) \right]_0^l$ represents the external force that acts at the element ends and can be assumed as:

$$\left[N_1(x^e) \left(EA \frac{d\hat{u}}{dx^e} \right) \right]_0^l = \left[\left(1 - \frac{x^e}{l} \right) \left(EA \frac{d\hat{u}}{dx^e} \right) \right]_0^l = -P_0 . \tag{24a}$$

$$\left[N_2(x^e) \left(EA \frac{d\hat{u}}{dx^e} \right) \right]_0^l = \left[\left(\frac{x^e}{l} \right) \left(EA \frac{d\hat{u}}{dx^e} \right) \right]_0^l = P_l \tag{24b}$$

P_0 and P_l are stand for the forces at end of the element.

The term $\frac{d\hat{u}}{dx}$ is known as strain of the bar and can be written as:

$$\frac{d\hat{u}}{dx} = \begin{bmatrix} \frac{\partial N_1}{\partial x} & \frac{\partial N_2}{\partial x} \end{bmatrix} \begin{Bmatrix} u_1 \\ u_2 \end{Bmatrix} = [B]\{d^e\} \tag{25}$$

where $[B]$ is called gradient matrix.

Using Equations (18, 24 and 25), Equation (23) can be written in terms of shape functions and degrees of freedom in the following manner:

$$\left(\int_0^l [B]^T EA [B]\, dx^e \right) \{d^e\} = \int_0^l [H]^T q(x^e)\, dx^e + \begin{Bmatrix} -P_0 \\ P_l \end{Bmatrix} \Rightarrow [K]\{d^e\} = \{F^e\} \tag{26}$$

where stiffness matrix $[K] = \int_0^l [B]^T EA [B]\, dx^e$; the external-internal force vector $\{F^e\} = \int_0^l [H]^T q(x^e)\, dx^e + \begin{Bmatrix} -P_0 \\ P_l \end{Bmatrix}$ and $\{d^e\}$ is the displacement vector, respectively.

3.5 Finite Element Modeling of Laminated Composite Plate

3.5.1 Displacement Field

The geometry and coordinate system of a laminated plate are shown in Figure 9.

Using the first order shear deformation theory (FSDT) based on an assumption that the transverse normals rotate such that they do not remain perpendicular to the mid surface after deformation, the displacements u, v, w at a point (x, y, z) are expressed as:

$$\begin{aligned} u(x,y,z) &= u_0(x,y) + z\theta_x(x,y) \\ v(x,y,z) &= v_0(x,y) + z\theta_y(x,y) \\ w(x,y,z) &= w_0(x,y) \end{aligned} \tag{27}$$

where u_0, v_0, w_0 are middle-surface displacements and θ_x, θ_y are the independent rotations of the normal in the xz and yz planes, respectively.

3.5.2 Stress-Strain Relations

The stress-strain relation of an orthotropic lamina in a state of plane stress can be expressed as:

$$\{\sigma\} = [\bar{Q}]\{\varepsilon\} \tag{28}$$

Figure 9. Coordinate system and geometry of a laminated plate

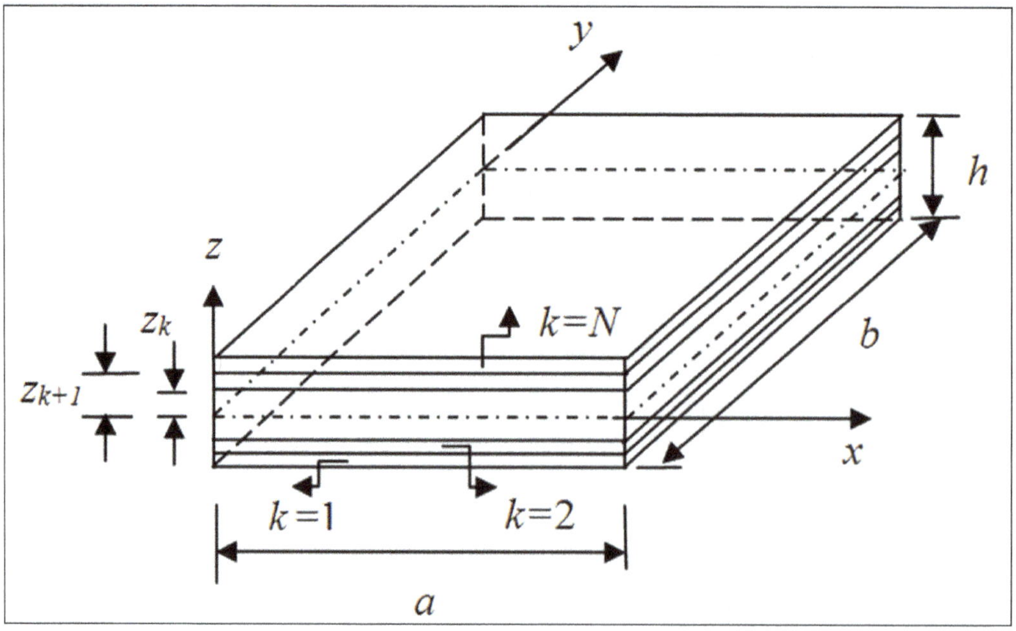

Equation (28) can be expanded in terms of membrane-bending strain $\{\varepsilon^{mb}\}$ and transverse shear strain $\{\varepsilon^{s}\}$ components as (Reddy, 2004):

$$\begin{Bmatrix} \sigma_{xx} \\ \sigma_{yy} \\ \tau_{xy} \end{Bmatrix} = \begin{bmatrix} \bar{Q}_{11} & \bar{Q}_{12} & \bar{Q}_{13} \\ \bar{Q}_{12} & \bar{Q}_{22} & \bar{Q}_{23} \\ \bar{Q}_{13} & \bar{Q}_{23} & \bar{Q}_{33} \end{bmatrix} \begin{Bmatrix} \varepsilon_{xx} \\ \varepsilon_{yy} \\ \gamma_{xy} \end{Bmatrix} \text{ or } \{\sigma^{mb}\} = [\bar{Q}_{mb}]\{\varepsilon^{mb}\}; \tag{29a}$$

$$\begin{Bmatrix} \tau_{xz} \\ \tau_{yz} \end{Bmatrix} = \begin{bmatrix} \bar{Q}_{44} & \bar{Q}_{45} \\ \bar{Q}_{45} & \bar{Q}_{55} \end{bmatrix} \begin{Bmatrix} \gamma_{xz} \\ \gamma_{yz} \end{Bmatrix} \text{ or } \{\sigma^{s}\} = [\bar{Q}_{s}]\{\varepsilon^{s}\} \tag{29b}$$

where

$$\bar{Q}_{11} = Q_{11}\cos^4\theta + 2(Q_{12} + 2Q_{44})\sin^2\theta\cos^2\theta + Q_{22}\sin^4\theta$$

$$\bar{Q}_{12} = (Q_{11} + Q_{22} - 4Q_{44})\cos^2\theta\sin^2\theta + Q_{12}(\sin^4\theta + \cos^4\theta)$$

$$\bar{Q}_{13} = (Q_{11} - Q_{12} - 2Q_{44})\cos^3\theta\sin\theta + (Q_{12} - Q_{22} + 2Q_{44})\cos\theta\sin^3\theta$$

$$\bar{Q}_{22} = Q_{11}\sin^4\theta + 2(Q_{12} + 2Q_{44})\sin^2\theta\cos^2\theta + Q_{22}\cos^4\theta$$

$$\bar{Q}_{23} = (Q_{12} - Q_{22} + 2Q_{44})\cos^3\theta\sin\theta + (Q_{11} - Q_{12} - 2Q_{44})\cos\theta\sin^3\theta$$

$$\bar{Q}_{33} = (Q_{11} - 2Q_{12} + Q_{22} - 2Q_{44})\cos^2\theta\sin^2\theta + Q_{44}(\sin^4\theta + \cos^4\theta)$$

$$\bar{Q}_{44} = Q_{55}\cos^2\theta + Q_{66}\sin^2\theta$$

$$\bar{Q}_{45} = (-Q_{55} + Q_{66})\cos\theta\sin\theta$$

$$\bar{Q}_{55} = Q_{55}\sin^2\theta + Q_{66}\cos^2\theta$$

$$Q_{11} = \frac{E_1}{1 - \nu_{12}\nu_{21}}; Q_{12} = \frac{\nu_{12}E_2}{1 - \nu_{12}\nu_{21}}; Q_{22} = \frac{E_2}{1 - \nu_{12}\nu_{21}}; Q_{44} = G_{12}; Q_{55} = G_{13}; Q_{66} = G_{23}$$

and E, G, ν, θ are Young's modulus, shear modulus, Poisson's ratio and fibre angle, respectively. The subscripts 1, 2, 3 represent the principal material directions.

3.5.3 Strain-Displacement Relations

The strain components can be expressed in terms of the displacements given in Equation (27), in the following manner:

$$\varepsilon_{xx} = \frac{\partial u_0}{\partial x} + z\frac{\partial \theta_x}{\partial x}; \varepsilon_{yy} = \frac{\partial v_0}{\partial y} + z\frac{\partial \theta_y}{\partial y}; \gamma_{xy} = \frac{\partial u_0}{\partial y} + \frac{\partial v_0}{\partial x} + z\left(\frac{\partial \theta_x}{\partial y} + \frac{\partial \theta_y}{\partial x}\right); \gamma_{xz} = \theta_x + \frac{\partial w_0}{\partial x}; \gamma_{yz} = \theta_y + \frac{\partial w_0}{\partial y} \tag{30}$$

Using Equation (30), the membrane-bending $\{\varepsilon^{mb}\}$ and transverse shear $\{\varepsilon^{s}\}$ strains can be written as:

$$\{\varepsilon^{mb}\} = [Z_{mb}]\{\varepsilon_0^{mb}\}; \tag{31a}$$

$$\{\varepsilon^{s}\} = [Z_s]\{\varepsilon_0^{s}\} \tag{31b}$$

where

$$\{\varepsilon_0^{mb}\}^T = \left\{\frac{\partial u_0}{\partial x} \quad \frac{\partial v_0}{\partial y} \quad \frac{\partial u_0}{\partial y} + \frac{\partial v_0}{\partial x} \quad \frac{\partial \theta_x}{\partial x} \quad \frac{\partial \theta_y}{\partial y} \quad \frac{\partial \theta_x}{\partial y} + \frac{\partial \theta_y}{\partial x}\right\}; \{\varepsilon_0^{s}\}^T = \left\{\theta_x + \frac{\partial w_0}{\partial x} \quad \theta_y + \frac{\partial w_0}{\partial y}\right\}$$

$$[Z_{mb}] = \begin{bmatrix} 1 & 0 & 0 & z & 0 & 0 \\ 0 & 1 & 0 & 0 & z & 0 \\ 0 & 0 & 1 & 0 & 0 & z \end{bmatrix}; \quad [Z_s] = \begin{bmatrix} 1 & 0 \\ 0 & 1 \end{bmatrix}$$

The strain energy (S.E.) of composite laminated plates is given as:

$$S.E. = \frac{1}{2}\int_V \{\sigma\}^T \{\varepsilon\}\, dV = \frac{1}{2}\int_V \{\varepsilon\}^T [\bar{Q}]\{\varepsilon\}\, dV \tag{32}$$

Equation (32) can be written in the terms of membrane & bending strain (ε^{mb}) and shear strain components (ε^{s}) as:

$$S.E. = \frac{1}{2}\int_V \{\varepsilon^{mb}\}^T [\bar{Q}_{mb}]\{\varepsilon^{mb}\}\, dV + \frac{1}{2}\int_V \{\varepsilon^{s}\}^T [\bar{Q}_s]\{\varepsilon^{s}\}\, dV \tag{33}$$

$$\begin{aligned} \Rightarrow \quad S.E. = &\frac{1}{2}\int_A \{\varepsilon_0^{mb}\}^T \left(\int_{-h/2}^{h/2} [Z_{mb}]^T [\bar{Q}_{mb}][Z_{mb}]dz\right)\{\varepsilon_0^{mb}\}\, dA \\ &+ \frac{1}{2}\int_A \{\varepsilon_0^{s}\}^T \left(\int_{-h/2}^{h/2} [Z_s]^T [\bar{Q}_s][Z_s]dz\right)\{\varepsilon_0^{s}\}\, dA \end{aligned} \tag{34}$$

$$\Rightarrow \quad S.E. = \frac{1}{2}\iint_A \left\{\varepsilon_0^{mb}\right\}^T \begin{bmatrix} [A] & [B] \\ [B] & [D] \end{bmatrix} \left\{\varepsilon_0^{mb}\right\} dA + \frac{1}{2}\iint_A \left\{\varepsilon_0^{s}\right\}^T [E] \left\{\varepsilon_0^{s}\right\} dA \tag{35}$$

Here, $[A]$, $[B]$, $[D]$ are the coefficient matrices for extensional, bending and bending-extensional coupling, respectively of the composite laminate. Further, the quantity$[E]$ represents the transverse shear stiffness matrix of the composite laminate. These quantities $[A]$, $[B]$, $[D]$ and $[E]$ are defined as:

$$[A] = \int_{-h/2}^{h/2} \left[\bar{Q}_{mb}\right] dz; [B] = \int_{-h/2}^{h/2} z\left[\bar{Q}_{mb}\right] dz; [D] = \int_{-h/2}^{h/2} z^2 \left[\bar{Q}_{mb}\right] dz; [E] = \int_{-h/2}^{h/2} \left[\bar{Q}_{s}\right] dz$$

The total potential energy of composite laminated plate is, let's say, U. It consists of two components, e.g., the strain energy ($S.E.$) stored in the composite laminated plate and the potential energy of the applied load. The expression for the total potential energy U is thus given by:

$$U = \frac{1}{2}\iint_A \left\{\varepsilon_0^{mb}\right\}^T \begin{bmatrix} [A] & [B] \\ [B] & [D] \end{bmatrix} \left\{\varepsilon_0^{mb}\right\} dA + \frac{1}{2}\iint_A \left\{\varepsilon_0^{s}\right\}^T [E] \left\{\varepsilon_0^{s}\right\} dA - \iint_A \begin{Bmatrix} u_0 \\ v_0 \\ w_0 \\ \theta_x \\ \theta_y \end{Bmatrix}^T \begin{Bmatrix} F_x \\ F_y \\ F_z \\ M_x \\ M_y \end{Bmatrix} dA \tag{36}$$

The field variables (u_0, v_0, w_0, θ_x, θ_y) are interpolated using a C^0 continuous, eight-noded serendipity quadrilateral shear flexible plate element with five nodal degrees of freedom as:

$$\begin{Bmatrix} u_0 \\ v_0 \\ w_0 \\ \theta_x \\ \theta_y \end{Bmatrix} = \sum_{i=1}^{8} \begin{bmatrix} N_i^e & 0 & 0 & 0 & 0 \\ 0 & N_i^e & 0 & 0 & 0 \\ 0 & 0 & N_i^e & 0 & 0 \\ 0 & 0 & 0 & N_i^e & 0 \\ 0 & 0 & 0 & 0 & N_i^e \end{bmatrix} \begin{Bmatrix} u_{0i} \\ v_{0i} \\ w_{0i} \\ \theta_{xi} \\ \theta_{yi} \end{Bmatrix} \text{ or } \{d\} = [H]\{d^e\} \tag{37}$$

where, N_i^e's are the shape functions of the eight-noded quadratic serendipity element, u_{0i}, v_{0i}, w_{0i}, θ_{xi}, θ_{yi} are nodal values of field variables and $[H]$ is interpolation matrix.

The strain components given in Equation (31) can be interpolated in terms of nodal values of field variables and shape functions in the following manner:

$$\left\{\varepsilon_0^{mb}\right\}=\begin{Bmatrix}\dfrac{\partial u_0}{\partial x}\\ \dfrac{\partial v_0}{\partial y}\\ \dfrac{\partial u_0}{\partial y}+\dfrac{\partial v_0}{\partial x}\\ \dfrac{\partial \theta_x}{\partial x}\\ \dfrac{\partial \theta_y}{\partial y}\\ \dfrac{\partial \theta_x}{\partial y}+\dfrac{\partial \theta_y}{\partial x}\end{Bmatrix}=\sum_{i=1}^{8}\begin{bmatrix}\dfrac{\partial N_i^e}{\partial x} & 0 & 0 & 0 & 0\\ 0 & \dfrac{\partial N_i^e}{\partial y} & 0 & 0 & 0\\ \dfrac{\partial N_i^e}{\partial y} & \dfrac{\partial N_i^e}{\partial x} & 0 & 0 & 0\\ 0 & 0 & \dfrac{\partial N_i^e}{\partial x} & 0 & 0\\ 0 & 0 & 0 & \dfrac{\partial N_i^e}{\partial y} & 0\\ 0 & 0 & 0 & \dfrac{\partial N_i^e}{\partial y} & \dfrac{\partial N_i^e}{\partial x}\end{bmatrix}\begin{Bmatrix}u_{0i}\\ v_{0i}\\ w_{0i}\\ \theta_{xi}\\ \theta_{yi}\end{Bmatrix} \text{ or } \left\{\varepsilon_0^{mb}\right\}=\left[B_{mb}\right]\left\{d^e\right\} \tag{38}$$

where $[B_{mb}]$ is the curvature gradient matrix.

$$\left\{\varepsilon_0^{s}\right\}=\begin{Bmatrix}\theta_x+\dfrac{\partial w_0}{\partial x}\\ \theta_y+\dfrac{\partial w_0}{\partial y}\end{Bmatrix}=\sum_{i=1}^{8}\begin{bmatrix}0 & 0 & \dfrac{\partial N_i^e}{\partial x} & N_i^e & 0\\ 0 & 0 & 0 & \dfrac{\partial N_i^e}{\partial y} & N_i^e\end{bmatrix}\begin{Bmatrix}u_{0i}\\ v_{0i}\\ w_{0i}\\ \theta_{xi}\\ \theta_{yi}\end{Bmatrix} \text{ or } \left\{\varepsilon_0^{s}\right\}=\left[B_s\right]\left\{d^e\right\} \tag{39}$$

where $[B_s]$ is the shear gradient matrix.

Using Equations (37, 38 and 39), the expression given in Equation (36) can be written in terms of the corresponding shape functions and the respective nodal degrees of freedom in the following manner:

$$U=\frac{1}{2}\int_A\left\{d^e\right\}^T\left[B_{mb}\right]^T\begin{bmatrix}[A] & [B]\\ [B] & [D]\end{bmatrix}\left[B_{mb}\right]\left\{d^e\right\}dA+\frac{1}{2}\int_A\left\{d^e\right\}^T\left[B_s\right]^T[E]\left[B_s\right]\left\{d^e\right\}dA-\int_A\left\{d^e\right\}^T[H]^T\{q\}\,dA \tag{40}$$

where $\{q\}^T=\left\{F_x \quad F_y \quad F_z \quad M_x \quad M_y\right\}$

The minimization of total potential energy with respect to the vector of degrees of freedom leads to the governing equation of the laminated plates as:

$$\left(\left[K_{bm}^e\right]+\left[K_s^e\right]\right)\left\{d^e\right\}=\left\{F^e\right\} \tag{41}$$

where

$$\left[K^e_{bm}\right]=\int_A [B_{mb}]^T \begin{bmatrix} [A] & [B] \\ [B] & [D] \end{bmatrix} [B_{mb}]\, dA; \quad \left[K^e_s\right]=\int_A [B_s]^T [E][B_s]\, dA; \quad \left[F^e\right]=\int_A [H]^T \{q\}\, dA$$

3.6 Advanced Finite Element Methods for Material Modeling

3.6.1 Extended Finite Element Method (XFEM)

The next question that comes is this that how one can tackle situations those involve discontinuities, singularities, localized deformations and complex geometries. For instance there can be need to model propagation of crack, the evolution of dislocations, the modeling of grain boundaries and the evolution of phase boundaries. These situations demand the genesis of a generic FEM which can evolve through the appropriate modifications in the classical finite element method. This is exactly what is done by the Extended Finite Element Method (XFEM). It is nothing but a modified form of what is provided by the classical finite element method. Philosophy adopted to achieve the goal is simple here. In this case, solutions to differential equations with discontinuous functions are used to enrich the solution space (Moes, Dolbow, & Belytschko, 1999). The adaptation of this technique can and really does help to simplify the solutions to many difficult situations encountered when one attempts to do material modelling. Such difficult situation may, for example, include the modelling requirements for evolutions of phase and grain boundaries, the evolution of dislocations, the propagation of a crack, the extents of intergranular and transgranular crack propagations in a microstructure etc.

Let us now examine the differences in the mesh patterns proposed corresponding to the XFEM based model and the FEM based model. It is interesting to note that the mesh for the XFEM based model can be and is completely independent of the location of the grain boundaries (Figure 10a). However, in the conventional FEM based model (Figure 10b) the basic boundary condition is that the element edges must conform to the grain boundaries and the duplicate nodes must be placed (Belytschko, Gracie, & Ventura, 2009). One popular means to enhance the accuracy of approximation is the usage of the enrichment functions that are tailored to suit the solution of a given problem. To achieve the desired accuracy of approximation technique adopted is “the partition of unity”. This technique was proposed about two decades back by Melenk and Babuska (1996). A partition of unity in a domain Ω is a set of functions $\varphi_i(x)$ such that it satisfies the boundary condition (Babuska & Melenk, 1997):

$$\sum_{\forall i} \varphi_i(x) = 1, \ \forall x \in \Omega \tag{42}$$

The property utilized here is very simple. It argues that any function $\psi(x)$ can be reproduced by a product of the partition of unity functions with $\psi(x)$. Thus, in other words additional unknown quantities β_i are introduced to enhance the accuracy of the standard finite element approximation (Belytschko et al., 2009):

Figure 10. Scheme to Discretize a grain boundary using (a) XFEM (b) FEM models

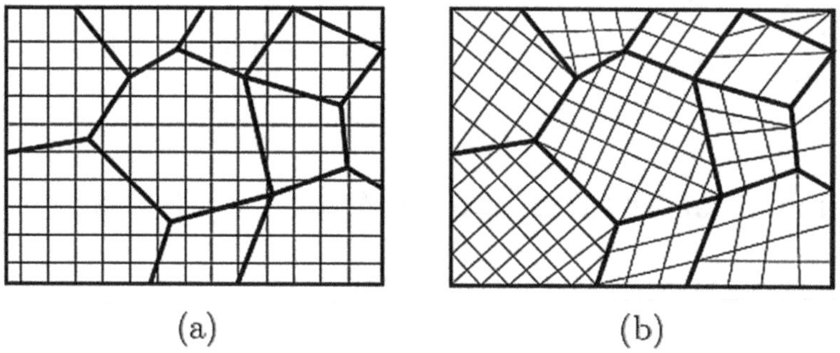

$$u^{\mathrm{XFEM}}(x) = \underbrace{\sum_{\forall i} N_i(x)\, u_i}_{\text{Classical}} + \underbrace{\sum_{\forall i} \varphi_i(x)\, \psi(x)\, \beta_i}_{\text{Enriched}} \tag{43}$$

where N_i's are the standard shape functions of the finite element method and u_i's are the standard degrees of freedom of the problem. The nodal values β_i are additional unknown parameters that adjust for the degree of enrichment.

3.7 Micromechanical Modeling of Composite Materials Using FEM

Though the well-established finite element analysis (FEA) has the ability to analyze a small portion (RVE) of such material most of the FEA based micromechanics models use the homogenization method. The term "homogenization" is the process of determining the effective properties of the homogenous medium and for much of the past half century homogenization was the only task of micromechanics (Daniel & Ishai, 2006). The various finite element methods based homogenization approaches are:

1. Variational asymptotic method for unit cell homogenization
2. Image based computational homogenization
3. Voronoi cell finite element model

3.7.1 Variational Asymptotic Method for Unit Cell Homogenization (VAMUCH)

Yu and Tang (2007a,b) developed a variational asymptotic method for unit cell homogenization based on the variational asymptotic method of Berdichevsky (1979). The technique invokes two essential assumptions within the concept of micromechanics:

1. The exact solutions of the field variables have volume averages over the UC.
2. The effective material properties obtained from the micromechanical analysis of the UC are independent of the geometry, the boundary conditions, and loading conditions of the macroscopic structure, which means that effective material properties are assumed to be the intrinsic properties of the material when viewed macroscopically.

What is done in the VAMUCH method? Well, in this case an asymptotic expansion of the energy functional is done to formulate the variational statements of the unit cell. This variational statement needs to be constructed in a way such that it can reproduce the governing differential equations and periodic boundary conditions of mathematical homogenization theories (MHT). The VAMUCH method has certain unique advantages.

These advantages are not present in the other models. For instance the VAMUCH method requires only two essential assumptions within the concept of micromechanics for heterogeneous material with identifiable unit cells. Further, the VAMUCH method has an inherent variational nature. Therefore, it turns out that one can attain its numerical implementation in a manner that is straight forward and far from complicated. Moreover, it can calculate simultaneously the different material properties in different directions.

There cannot be any doubt that such a technique is bound to be more efficient than that of those approaches which would require multiple runs under different loading conditions. Most importantly, the VAMUCH method calculates the effective properties and the local fields directly with the accuracy that is as good as those of the fluctuation functions. An even more important advantage of the VAMUCH method is that in this case no postprocessing calculations such as stress averaging and strain averaging are needed (Tang, 2008).

3.7.2 Image Based Computational Homogenization

The digital image-based technique is one of method for homogenization to analyze the local stress field in the unit cells of composites. The digital image processing is used to make the microstructural FE model for the homogenization analysis (Terada, Miura, & Kikuchi, 1997). The main procedure of preparation the microstructural FE model can be divided into the following four major stages:

1. Capture and Sampling,
2. Selecting and Thresholding,
3. Exporting, and
4. Stacking.

The first three stages correspond to the pre-processing of FEA, the last process includes both the main part and post-processing of the FEA of the homogenization method (Golanski, Grzes, Kikuchi, & Kozłowski, 2000).

3.7.3 Voronoi Cell Finite Element Model (VCFEM)

Voronoi cell finite element model (VCFEM) evolves the Voronoi tessellation of the microstructure to generate a morphology based network of multi-sided Voronoi cells, each cell containing heterogeneity. Each cell is treated as a FEM element and requires no additional discretization. VCFEM incorporates assumptions from micromechanics theories, as well as adaptive enhancements.

This model has been shown to require significantly reduced degrees of freedom compared to displacement based FEM models and are hence computationally efficient (Moorthy & Ghosh, 1996; 2000). A three-dimensional Voronoi cell finite element model is developed by Ghosh and Moorthy (2004) for analyzing heterogeneous materials containing a dispersion of ellipsoidal inclusions or voids in the matrix. Further, Li and Ghosh (2006) developed a new extended Voronoi cell finite element model (X-VCFEM) that was used for modeling interfacial debonding with arbitrary cohesive cracking in matrix of fiber-reinforced composites.

4. FAILURE OF COMPOSITES

4.1 Introduction

The laminated composites are fastest growing materials, increasingly being used as light weight structural components in modern industry such as aircraft, spacecraft, missile, hydrospace, automobiles, marine and medical prosthetic devices etc., due to capability of attaining higher stiffness/density and strength/density ratios. In order to ensure reliable, safe and failure proof performance, thorough understanding of the prediction of failure initiation, progression and final failure is essential.

The process of failure in such materials, under different loading environment may cause the degradation of material properties and ultimate collapse due to nucleation and growth of defects such as matrix cracks, fibre breakage, fibre matrix debonding and inter-layer delamination. The combination of failure criteria, damage mechanics, and fracture mechanics approaches led to overall failure prediction and is known as progressive failure analysis (PFA). The schematic diagram given in Figure 11 (Zhang & Cai, 2010) depicts the relationship between failure criteria, damage mechanics and fracture mechanics.

4.2 Failure Criteria

Failure criterion refers to set of mathematical equations that predict the initiation of failure i.e. formation of slip planes and dislocations by comparing the current states of stress and strain with strength properties of material (Tsai, & Wu, 1971; Hashin, 1980; Echaabi, Trochu, & Gauvin, 1996; Marat-Mendes & Freitas, 2010). The failure criteria proposed to predict lamina failure could be divided in two main groups:

4.2.1 Failure Mode Dependent Failure Criteria

Failure mode dependent failure criteria are able to predict failure associated with particular failure modes such as matrix cracks, fibre breakage and fibre matrix debonding by considering the non-homogeneous character of composites. Various failure mode dependent criteria are given by: maximum stress criterion, maximum strain criterion and Hashin's failure criterion. These are described below.

4.2.1.1 Maximum Stress Criterion

This criterion predicts failure when any one of the stress components along the principal material axis is equal or greater to the corresponding strength in that direction. In this criterion the interaction between different failure modes is not permitted. Thus, the failure will occur if:

$$\sigma_{11} \geq \begin{cases} X_T & \text{when } \sigma_{11} > 0 \\ -X_C & \text{when } \sigma_{11} < 0 \end{cases}; \quad \sigma_{22} \geq \begin{cases} Y_T & \text{when } \sigma_{22} > 0 \\ -Y_C & \text{when } \sigma_{22} < 0 \end{cases} \quad \text{and} \quad |\tau_{12}| \geq S_{12} \tag{44}$$

Figure 11. Relationship between failure criteria, damage mechanics and fracture mechanics
(Modified from Zhang & Cai, 2010)

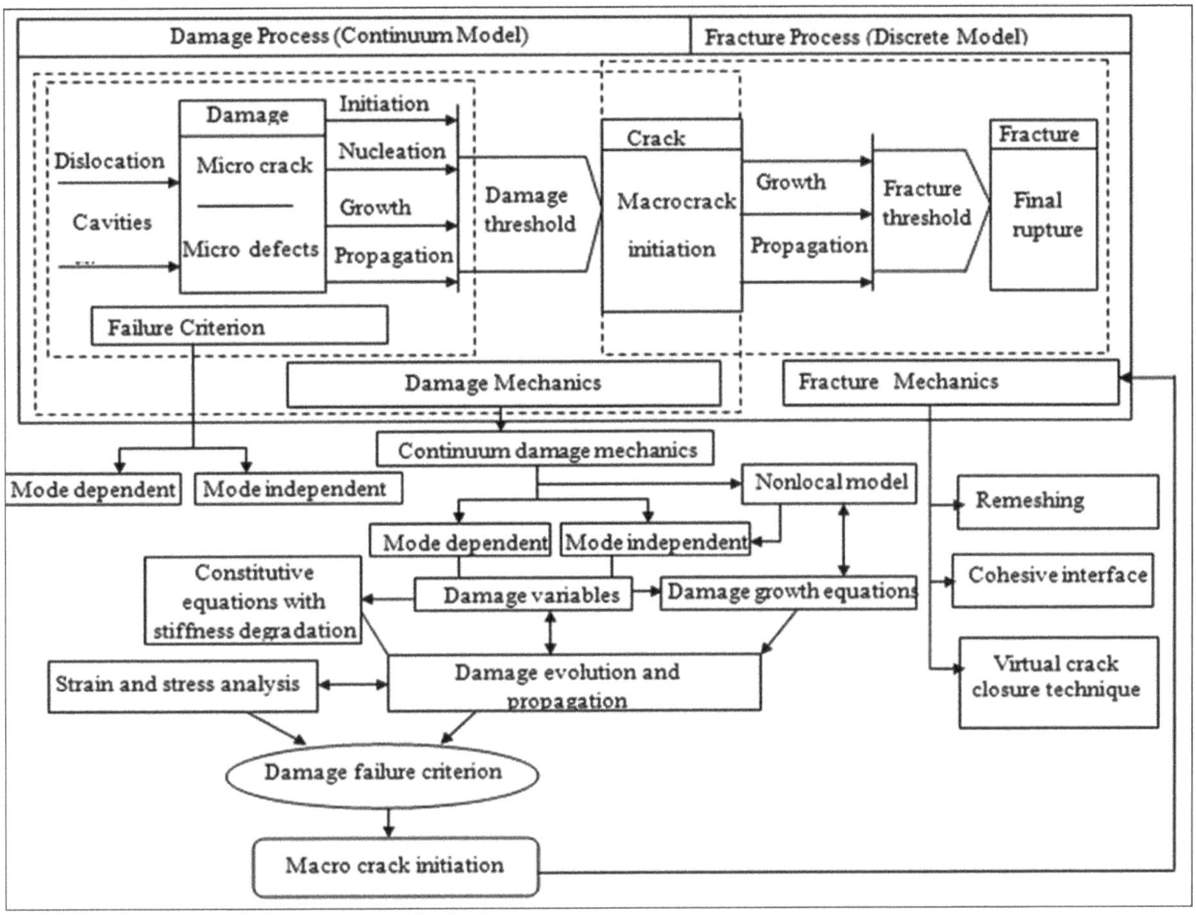

4.2.1.2 Maximum Strain Criterion

This criterion, analogous to maximum stress criterion, predicts failure when the strain components along the principal material axis is equal or greater to the corresponding ultimate strain in that direction. It is expressed as:

$$\varepsilon_{11} \geq \begin{cases} \varepsilon_{1T}^{u} & \text{when } \varepsilon_{11} > 0 \\ \varepsilon_{1C}^{u} & \text{when } \varepsilon_{11} < 0 \end{cases}; \quad \varepsilon_{22} \geq \begin{cases} \varepsilon_{2T}^{u} & \text{when } \varepsilon_{22} > 0 \\ \varepsilon_{2C}^{u} & \text{when } \varepsilon_{22} < 0 \end{cases} \quad \text{and} \quad \left|\gamma_{12}\right| \geq \gamma_{12}^{u} \tag{45}$$

4.2.1.3 Hashin's Failure Criterion

To eliminate the limitation with maximum stress/strain criterion wherein interaction between different failure modes is not permitted, Hashin introduced fiber and matrix failure criteria in which more than one stress components have been used to evaluate the failure modes. Failure indices for Hashin criteria are related to fibre and matrix failures and involve four failure modes. The failure modes included in Hashin's criteria (Hashin, 1980) are given in the following manner:

1. Tensile fibre failure for $\sigma_{11} \geq 0$

$$\left(\frac{\sigma_{11}}{X_T}\right)^2 + \frac{\tau_{12}^2 + \tau_{13}^2}{S_{12}^2} = \begin{cases} \geq 1 & \text{failure} \\ < 1 & \text{no failure} \end{cases} \tag{46a}$$

2. Compressive fibre failure for $\sigma_{11} < 0$

$$\left(\frac{\sigma_{11}}{X_C}\right)^2 = \begin{cases} \geq 1 & \text{failure} \\ < 1 & \text{no failure} \end{cases} \tag{46b}$$

3. Tensile matrix failure for $\sigma_{22} + \sigma_{33} > 0$

$$\frac{(\sigma_{22} + \sigma_{33})^2}{Y_T^2} + \frac{\tau_{23}^2 - \sigma_{22}\sigma_{33}}{S_{23}^2} + \frac{\tau_{12}^2 + \tau_{13}^2}{S_{12}^2} = \begin{cases} \geq 1 & \text{failure} \\ < 1 & \text{no failure} \end{cases} \tag{46c}$$

4. Compressive matrix failure for $\sigma_{22} + \sigma_{33} < 0$

$$\left[\left(\frac{Y_C}{2S_{23}}\right)^2 - 1\right]\left(\frac{\sigma_{22} + \sigma_{33}}{Y_C}\right) + \frac{(\sigma_{22} + \sigma_{33})^2}{4S_{23}^2} + \frac{\tau_{23}^2 - \sigma_{22}\sigma_{33}}{S_{23}^2} + \frac{\tau_{12}^2 + \tau_{13}^2}{S_{12}^2} = \begin{cases} \geq 1 & \text{failure} \\ < 1 & \text{no failure} \end{cases} \tag{46d}$$

where σ_{ii}, τ_{ij}, ε_{ii} and γ_{ij} denote the normal stress, shear stress, normal strain and shear strain components, respectively. The maximum allowable tensile and compressive strengths for lamina are denoted by subscripts T and C, respectively. X_T, Y_T, Z_T denote the allowable tensile strengths in three respective directions in the material. Similarly, X_C, Y_C, Z_C denotes the allowable compressive strengths in three respective directions in the material. Further, S_{12}, S_{13} and S_{23} denote allowable shear strengths in the respective principal directions in the material. Similarly, ε_{1T}^u, ε_{1C}^u are ultimate longitudinal strain in tension and compression, ε_{2T}^u, ε_{2C}^u are ultimate transverse strain in tension and compression and γ_{12}^u is the ultimate in-plane shear strain.

4.2.2 Failure Mode Independent Failure Criteria

It needs to be appreciated that the failure mode independent failure criteria has the limitation that it can predict the failure of the ply in the overall sense but they do not distinguish the failure type or the dominant constituent in the failure process. Generally, these criteria are expressed in polynomial forms and account for the stress interactions. The formulation of the criterion is reduced to a mathematical equation. This is done as a function of different experimental parameters. Such experimental parameters can be e.g., the ply strengths in the principal directions.

4.2.2.1 Tsai-Hill Criterion

The Tsai-Hill theory (Hill, 1948; Azzi & Tsai, 1965) is considered as an extension of the Von Mises failure criterion to anisotropic fibre reinforced composites. According to Tsai-Hill criterion, the failure will occur under a general plane stress state, when the following expression holds:

$$\frac{\sigma_{11}^2}{X^2} + \frac{\sigma_{22}^2}{Y^2} - \frac{\sigma_{11}\sigma_{22}}{X^2} + \frac{\tau_{12}^2}{S_{12}^2} \geq 1 \quad (47)$$

This criterion does not distinguish between tensile or compressive strength, it takes strength values X and Y as X_T or X_C and Y_T or Y_C when the stresses σ_{11} and σ_{22} are tensile or compressive.

4.2.2.2 Tsai-Wu Criterion

Tsai-Wu (1971) proposed the modification of Hill's criterion for lamina which can account for the different tensile and compressive strengths. This failure criterion is a specialization of the general quadratic failure criterion proposed by Gol'denblat and Kopnov (1965) and can be expressed in the form:

$$F_i\sigma_{ii} + F_{ij}\sigma_{ii}\sigma_{jj} = 1 \, i = 1, 2, \ldots\ldots 6 \quad (48a)$$

where failure constant F_i and F_{ij} are the functions of strength parameters. For a state of plane stress (i = 1, 2 and 6), Equation (48a) can be expanded as:

$$F_1\sigma_{11} + F_2\sigma_{22} + F_6\sigma_{66} + F_{11}\sigma_{11}^2 + F_{22}\sigma_{22}^2 + F_{66}\sigma_{66}^2 + 2F_{12}\sigma_{11}\sigma_{22} + 2F_{16}\sigma_{11}\sigma_{66} + 2F_{26}\sigma_{22}\sigma_{66} = 1 \quad (48b)$$

The linear terms in Equation (48b) are used to account for the tension and compression nature. As the strength of lamina loaded in shear along the principal material direction is independent of sign of the shear stress, so, the terms containing the linear shear stress must be vanished, so $F_{16}=F_{26}=F_6=0$. For the sake of consistency of nomenclature σ_{66} is taken as τ_{12}. Considering this condition, Equation (48b) can be rewritten in principal material coordinates as:

$$F_1\sigma_{11} + F_2\sigma_{22} + F_{11}\sigma_{11}^2 + F_{22}\sigma_{22}^2 + F_{66}\tau_{12}^2 + 2F_{12}\sigma_{11}\sigma_{22} = 1 \quad (48c)$$

The terms F_1, F_2, F_{11}, F_{22} and F_{66} can be determined by uniaxial longitudinal/transverse tensile and compressive tests and pure shear test, respectively. These are given by:

$$F_1 = \frac{1}{X_T} - \frac{1}{X_C}; \quad F_2 = \frac{1}{Y_T} - \frac{1}{Y_C}; \quad F_{11} = \frac{1}{X_T X_C}; \quad F_{22} = \frac{1}{Y_T Y_C}; \text{ and } F_{66} = \frac{1}{S_{12}^2} \quad (48d)$$

From Equation (48d) it can be seen that all failure constants except F_{12} can be determined in terms of strength properties using uniaxial tests. The only remaining term to be determined is F_{12} that can be accomplished by biaxial test (Wu, 1972), which is quite complicated to perform. Tsai and Hahn (Tsai & Hahn, 1980) proposed an approximation as $F_{12} = -0.5\sqrt{F_{11}F_{22}}$.

4.2.3 Advanced Failure Criteria

It is interesting to note that according to the results of the Word Wide Failure Exercise (WWFE) the Conventional failure models and criteria are not able to provide a satisfactory description and prediction of failure (Soden, Hinton, & Kaddour, 1998; Hinton, Kaddour, & Soden, 2004). In fact, the failure mechanisms in composite materials are not fully understood yet, especially for compression failure, both for the matrix and fibre-dominated failure modes. It is therefore felt as necessary to revisit the existing failure theories, assess their capabilities and develop new theories.

4.2.3.1 Puck Failure Criterion

Puck's failure criterion (Puck & Schürmann, 1998, 2002; Puck, Kopp, & Knops, 2002) is based on Mohr's hypothesis that fracture is exclusively generated by stresses developed on the fracture plane. This criterion described two basic types of failures as:

1. **Fibre Failure (FF):** The failure condition for the fibre failure for tension and compression is given by:
   a. **Fibre Tension:**

$$\frac{1}{\varepsilon_{1T}}\left(\varepsilon_{11} + \frac{\nu_{f12}}{E_{f1}} m_{\sigma f}\sigma_{22}\right) = 1 \qquad \text{for} \quad \left(\varepsilon_{11} + \frac{\nu_{f12}}{E_{f1}} m_{\sigma f}\sigma_{22}\right) \geq 0 \qquad (49a)$$

   b. **Fibre Compression (kinking):**

$$\frac{1}{\varepsilon_{1C}}\left|\left(\varepsilon_{11} + \frac{\nu_{f12}}{E_{f1}} m_{\sigma f}\sigma_{22}\right)\right| + (10\gamma_{12})^2 = 1 \qquad \text{for} \quad \left(\varepsilon_{11} + \frac{\nu_{f12}}{E_{f1}} m_{\sigma f}\sigma_{22}\right) < 0 \qquad (49b)$$

2. **Inter-Fibre Fracture (IFF):** Based on fracture mechanics and experimental observations, three different criteria (modes A, B and C) are considered for inter-fibre fracture. Mode A, B and C correspond to the combined transverse tension and shear, moderate transverse compression and fairly large transverse compressive loading, respectively. The corresponding mathematical forms are:
   Mode A:

$$\sqrt{\left(\frac{\tau_{12}}{S_{12}}\right)^2 + \left(1 - p_{12}^{(+)}\frac{Y_T}{S_{12}}\right)^2 \left(\frac{\sigma_{22}}{Y_T}\right)^2} + p_{12}^{(+)}\frac{\sigma_{22}}{S_{12}} + \left|\frac{\sigma_{11}}{\sigma_{1D}}\right| = 1 \quad \text{for} \quad \sigma_{22} \geq 0 \qquad (49c)$$

   Mode B:

$$\frac{1}{S_{12}}\left(\sqrt{\tau_{12}^{2}+\left(p_{12}^{(-)}\sigma_{22}\right)^{2}}+p_{12}^{(-)}\frac{\sigma_{22}}{S_{12}}\right)+\left|\frac{\sigma_{11}}{\sigma_{1D}}\right|=1 \quad \text{for} \quad \sigma_{22}<0 \text{ and } 0\leq\left|\frac{\sigma_{22}}{\tau_{12}}\right|\leq\frac{\sigma_{23}^{A}}{\tau_{12}^{c}} \tag{49d}$$

Mode C:

$$\left[\left(\frac{\tau_{12}}{2\left(1+p_{23}^{(-)}\right)S_{12}}\right)^{2}+\left(\frac{\sigma_{22}}{Y_C}\right)^{2}\right]\left(\frac{Y_C}{(-\sigma_{22})}\right)+\left|\frac{\sigma_{11}}{\sigma_{1D}}\right|=1 \quad \text{for } \sigma_{22}<0 \text{ and } 0\leq\left|\frac{\tau_{12}}{\sigma_{22}}\right|\leq\frac{\tau_{12}^{c}}{\sigma_{23}^{A}} \tag{49e}$$

where, v_{f12} and E_{f1} are the Poisson's ratio and fibre modulus of the fibre in fibre direction; ε_{1T} and ε_{1C} are the fracture strains in tension and compression, respectively; $m_{\sigma f}$ accounts for stress magnification effect caused by the different moduli of fibres and matrix and the term σ_{11}/σ_{1D} takes into account the matrix damage that occurs after the fiber breakage. Other terms are expressed as:

$$p_{12}^{(+)}, p_{12}^{(-)} = -\left(\frac{d\tau_{12}}{d\sigma_{22}}\right)_{\sigma_{22}=0}$$

slope of the (σ_{22}, τ_{12}) for $\sigma_{22}\geq 0$, $\sigma_{22}\leq 0$, respectively;

$$p_{23}^{(-)} = p_{12}^{(-)}\frac{\sigma_{23}^{A}}{S_{12}};\ \sigma_{23}^{A} = \frac{S_{12}}{2p_{12}^{(-)}}\left(\sqrt{1+2p_{12}^{(-)}\frac{Y_C}{S_{12}}}-1\right);\ \tau_{12}^{C} = S_{12}\sqrt{1+2p_{23}^{(-)}}$$

4.2.3.2 Langley Research Centre (LaRC) Criteria

The use of existing failure theories failed to predict accurately the initiation and progression of failure in composites. This is why the NASA Langley Research Center (LaRC) has undertaken initiative to develop new theories. These theories (LaRC02, LaRC03 and LaRC04) frame a series of criteria and are based on fracture mechanics, experimental observations and micromecanical modeling. Some of them are briefly described below:

1. **LaRC02:** LaRC02 (Davila, Jaunky, & Goswami, 2003) is a physically based criterion, derived from Hashin's criteria for unidirectional laminates and Puck's action plane concept. This criterion is able to predict matrix and fibre failure accurately without requiring curve-fitting parameters.
2. **LaRC03:** LaRC03 criterion uses the six different equations to predict the matrix and fibre failure, without the curve-fitting parameters. The angle of the fracture plane is found out using Mohr-Coulomb hypothesis for matrix failure under transverse compression. A criterion for matrix failure in tension is developed from fracture mechanics models of matrix cracks and the in situ strengths of the associated crack matrix is predicted by this effort (Davila & Camanho, 2003; Davila, Camanho, & Rose, 2005).
3. **LaRC04:** LaRC04 criterion is capable of predicting the general three-dimensional (3D) loading and in-plane shear non-linearity. It entails a modified version of the LaRC03 plane stress criterion

(Pinho, Davila, Camanho, Iannucci, & Robinson, 2005). Further, it investigates the matrix compressive failure through the utilization of the 3D version of the Puck criterion matrix compression failure (Puck & Schurmann, 1998). This criterion used a failure model proposed by Dvorak and Laws (1987) for predicting matrix cracking in a ply subjected to in-plane shear and transverse tension. This effort was based on the fracture mechanics analysis of cracked plies.

4.3 Continuum Damage Mechanics Approach

Continuum damage mechanics (CDM) is a phenomenological approach that represents the macroscopic effects of microscopic cracks/cavities in the material by introducing internal state variables (damage variables). The damage model is based on a generalized macroscopic continuum theory within the framework of irreversible thermodynamics and enables to predict the progressive damage and failure load. Kachanov (1999) first introduced the concept of continuum factor in his study of fracture of material under creep. To formulate the constitutive relation of damage material, stress equivalence principle (Rabotnov, 1969) and strain equivalence principle (Lemaitre, 1971) were introduced. To resolve the issue of unsymmetric constitutive matrix involving anisotropic damage based on the strain equivalence principle, Sidoroff (1981) proposed the hypothesis of strain energy equivalence.

Using continuum damage mechanics model, distributed microscopic damage can be quantified by the use of damage variables. The damage variables along the principal material directions are defined as:

$$D_i = \frac{A_i - \overline{A_i}}{A_i} \tag{40}$$

where A_i and $\overline{A_i}$ are cross-sectional areas of undamaged and damaged material, respectively, along the principal planes of material symmetry. The values of damage variables are in the range $0 < D_i < 1$ where $D_i = 0$ corresponds to no damage normal to i^{th} principal material direction, while $D_i = 1$ corresponds to a complete separation of the material.

4.4 Fracture Mechanics Approach

In the fracture studies in composites, the basic interest is in the initiation of fracture propagation from the 'localized' imperfections which are known (or assumed) to exist in the material (Erdogan, 1972). The discrete methods which follow the fracture mechanics approach basically consider a composite with a certain amount of damage (macrocrack initiation) and try to predict the conditions for which the amount of damage will increase up to final rapture (Dharan, 1978).

The propagation of a crack leads to displacement discontinuity *i.e.* separation between boundaries of elements. Linear fracture mechanics has been the classical tool to model displacement discontinuities. In addition, the necessity of dealing with the process of formation and propagation of discontinuities led the development of the non-linear (decohesive) fracture mechanics techniques essentially based on the introduction of adhoc discrete constitutive equations (Bazant & Planas, 1998). Oliver (2000) and Oliver, Huespe, Pulido, & Chaves (2002) introduced an analysis of strong discontinuities approach that analyzes the development of displacement discontinuities in media governed by standard continuum constitutive equations (stress vs. strain) by linking the decohesive fracture mechanics with isotropic continuum damage models. Further they extended their continuum strong discontinuity approach (CSDA) to investigate the fracture process in fibre reinforced composite material (reinforced concrete) (Linero Segrera, Oliver, & Huespe, 2010). Several methods have been proposed in literature to model this discontinuity to perform an automatic re-meshing (Rashid, 1998; Bouchard, Bay, Chastel, & Tovena, 2000; Shahani & Amini Fasakhodi, 2009).

5. CASE STUDIES

5.1 Case Study: FEM Based Continuum Damage Modeling of Fibre Reinforced Composites

The implementation of continuum damage mechanics model into a finite element model based on the first order and layerwise shear deformation theories for fibre-reinforced composite laminated plate is presented by Robbins, Reddy, & Rostam-Abadi (2005a, b). Ghosh and Sinha (2004) and Ghosh (2008) have studied impact response of laminated composite plates and spherical panels using damage variable evolution equation for uniaxial loading (Matzenmiller, Lubliner, & Taylor, 1995) to evaluate damage variables in the principal material directions.

In reported literature (Robbins et al., 2005a, b; Ghosh & Sinha, 2004; Ghosh, 2008) geometric nonlinearity was not considered. For the thin structural elements undergoing large deformation, inclusion of geometric nonlinearity is important and needs to be incorporated in the failure and damage analysis of composites. The nonlinear dynamic response of laminated plates with piezoelectric layers is studied by Tian, Fu, & Wang (2009a, b), considering elasto-plastic deformation, damage evolution law similar to plastic flow rule and classical plate theory. The elasto-plastic postbuckling of orthotropic plates is studied by Tian and Fu (2008) using classical plate theory and continuum damage based on damage flow rule. Tian et al. (2009a, b) and Tian and Fu (2008) have used stress tensor as the conjugate force to damage which is thermodynamically inconsistent (Hayakawa, Murakami, & Liu, 1998). Murari and Upadhyay (2012, 2013) proposed a ply-level continuum damage model representing various micro-level damage mechanisms such as fiber break, fiber-matrix debonding and matrix cracking and developed a functional continuum representation of stiffness degradation based on a three dimensional micromechanical analysis of a single cell representative volume element considering different fiber volume fraction and damage levels.

Gupta, Patel, & Nath (2012, 2013a,b) developed a finite element formulation incorporating thermodynamically consistent continuum damage model and geometric nonlinearity for progressive damage analysis of composite laminated plates and cylindrical/conical panels under transverse loading. They also investigated the effect of progressive failure on postbuckling behavior of laminated plates considering geometric nonlinearity and evolving material damage under uniaxial, biaxial compressive and in-plane shear loadings (Gupta, Patel, & Nath, 2014). Further, Patel and Gupta (2014) proposed a nonlocal continuum damage model for progressive failure analysis of thick/moderately thick laminated composite plates/panels using a global higher-order shear deformation theory including zig-zag function.

The overall formulation consists of continuum damage formulation and finite element formulation:

5.1.1 Continuum Damage Modeling

This section deals with the formulation of a continuum damage model for damage initiation and damage evolution.

5.1.1.1 Constitutive Equations

The relation between stress $\{\boldsymbol{\sigma}\}$ in the damaged continuum and effective stress $\{\bar{\sigma}\}$ in the equivalent undamaged continuum can be expressed as (Lee, Peng, & Wang,1985):

$$\{\bar{\sigma}\} = [\mathbf{M}]\{\sigma\} \tag{51}$$

where [**M**] is the effective damage tensor. The non-zero elements of the effective damage tensor (Chow & Wang, 1987) are given by:

$$M_{11}=1/(1-D_1),\ M_{22}=1/(1-D_2),\ M_{33}=1/(1-D_3),$$
$$M_{44} = \sqrt{1/(1-D_1)(1-D_2)},\ M_{55} = \sqrt{1/(1-D_1)(1-D_3)},\ M_{66} = \sqrt{1/(1-D_2)(1-D_3)} \tag{52}$$

where D_1, D_2, D_3 are the damage variables along the principal material directions. The constitutive relations for damaged and equivalent undamaged configurations are given as:

$$\{\sigma\} = [\bar{\mathbf{C}}]\{\varepsilon\} \text{ and } \{\bar{\sigma}\} = [\mathbf{C}]\{\bar{\varepsilon}\} \tag{53}$$

where $[\bar{\mathbf{C}}]$ and [**C**] are constitutive matrices of damaged and undamaged continua, respectively, $\{\varepsilon\}$ and $\{\bar{\varepsilon}\}$ are the respective strains. The principle of equivalence of strain energy in the damaged and undamaged configuration is given by (Sidoroff, 1981):

$$\frac{1}{2}\{\sigma\}^T[\bar{\mathbf{C}}]^{-1}\{\sigma\} = \frac{1}{2}\{\bar{\sigma}\}^T[\mathbf{C}]^{-1}\{\bar{\sigma}\} \tag{54}$$

Substituting Equation (51) in Equation (54), the constitutive matrix of damaged state $[\bar{\mathbf{C}}]$ can be expressed in terms of the principal damage variables and constitutive matrix of undamaged state [**C**] as:

$$[\bar{\mathbf{C}}] = [\mathbf{M}]^{-1}[\mathbf{C}][\mathbf{M}]^{-T} \tag{55}$$

5.1.2.2 Thermodynamic Formulation of Damage

The damage evolution equations are derived by thermodynamics principles and constitutive equations. The Helmholtz free energy density Ψ is considered here as the thermodynamic state potential: a continuous scalar function, concave with temperature, and function of natural state variables given by (Lemaitre, 1996).

$$\Psi = \Psi(\varepsilon, T, \mathbf{D}, \beta) \tag{56}$$

where **ε** is characterized by strain tensor, *T* is temperature, **D** is damage tensor, and β is non dimensional hardening parameter that control the process of damage hardening. The Clausius–Duhem inequality can be stated as (Lemaitre, 1996):

$$\{\sigma\}^T \{\dot{\varepsilon}\} - \rho(\dot{\Psi} - s\dot{T}) \geq 0 \tag{57}$$

where σ is the Cauchy stress, ρ the material density and s is known as the specific entropy per unit mass. Considering isothermal elastic evolutions, Equation (57) can be written as:

$$\left(\{\sigma\} - \rho\frac{\partial \Psi}{\partial\{\varepsilon\}}\right)^T \{\dot{\varepsilon}\} - \left(\rho\frac{\partial \Psi}{\partial\{D\}}\right)^T \{\dot{\mathbf{D}}\} - \rho\frac{\partial \Psi}{\partial \beta}\dot{\beta} \geq 0 \tag{58}$$

where ρ, ψ and β are material density, Helmholtz free energy and overall damage parameter; $\{\dot{\varepsilon}\}$, $\{\dot{\mathbf{D}}\}$ and $\dot{\beta}$ are rate of change of strain, damage and overall damage parameter, respectively. The damage driving force **Y** and damage hardening variable γ are defined as:

$$\{\mathbf{Y}\} = -\rho\frac{\partial \Psi}{\partial\{\mathbf{D}\}} = -\frac{\partial\left(\frac{1}{2}\{\varepsilon\}^T[\bar{\mathbf{C}}]\{\varepsilon\}\right)}{\partial\{\mathbf{D}\}} = -\frac{1}{2}\{\varepsilon\}^T\frac{\partial[\bar{\mathbf{C}}]}{\partial\{\mathbf{D}\}}\{\varepsilon\} \tag{59a}$$

$$\gamma = \gamma(\beta) = \rho\frac{\partial \Psi}{\partial \beta} \tag{59b}$$

5.1.2.3 Damage Evolution Equations

In order to predict the evolution of damage $\{\dot{\mathbf{D}}\}$, a convex damage surface is considered (Barbero & de Vivo, 2001):

$$g(\mathbf{Y}, \gamma) = \sqrt{J_{11}Y_1^2 + J_{22}Y_2^2 + J_{33}Y_3^2} - (\gamma_0 + \gamma(\beta)) \tag{60}$$

where γ_0 is a material constant representing the initial damage threshold of undamaged material, $\gamma(\beta)$ provides the increase in damage hardening with evolution of damage variables, J_{11}, J_{22} and J_{33} are damage tolerance parameters of the material.

The damage state variables D_i and β are determined such that the dissipation function defined by $\phi(\mathbf{Y}, \gamma) = Y_i\dot{D}_i - \gamma\dot{\beta} \geq 0$ is maximum subject to the constraint $g(\mathbf{Y}, \gamma) = 0$. Using Lagrange multiplier approach, evolution equations for $\dot{D}_i$ and $\dot{\beta}$ are obtained as:

$$\dot{D}_i = \lambda \frac{\partial g}{\partial Y_i}, \text{ and } \dot{\beta} = -\lambda \frac{\partial g}{\partial \gamma} \tag{61}$$

where $\lambda = \left(\frac{\partial g}{\partial Y_i} \Big/ \frac{\partial \gamma}{\partial \beta} \right) \dot{Y}_i$ and $\dot{Y}_i$ is the rate of change of damage driving force Y_i.

Rate-type equations of damage evolution are expressed in the form of incremental damage driving force **Y** and damage hardening variable γ, which are functions of incremental strain vectors and overall damage parameter, respectively. The incremental strain vectors, directly evaluated from displacement increments, are used to evaluate damage variables at every Gauss point using Newton-Raphson iterative technique. A linear relationship $\gamma = k_h \beta$ is used to determine the damage hardening variable γ representing the damage hardening characteristics of composite laminates where k_h is a constant that defines the rate of damage hardening.

5.1.3 Finite Element Formulation

A laminated composite cylindrical panel is considered with the coordinates *s*, *θ* and *z* along the meridional, circumferential and radial/ thickness directions, respectively, as shown in Figure 12. The total thickness, length, radii of the panel are *h*, *a*, *r*, respectively, and *ϕ* is sector angle. The displacements *u*, *v* and *w* at a point (*s*, *θ*, *z*) are expressed as:

$$\begin{aligned} u(s,\theta,z) &= u_0(s,\theta) + z\psi_s(s,\theta) \\ v(s,\theta,z) &= v_0(s,\theta) + z\psi_\theta(s,\theta) \\ w(s,\theta,z) &= w_0(s,\theta) \end{aligned} \tag{62}$$

Figure 12. Coordinate system and geometry of laminated panel

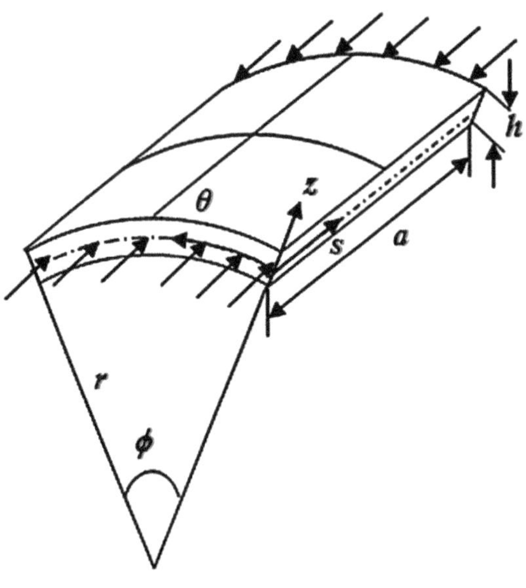

where u_0, v_0, w_0 are middle-surface displacements and ψ_s, ψ_θ are the independent rotations of the normal to the reference surface about the θ and s axes, respectively.

The strains are written in terms of the middle surface displacement and rotations as:

$$\{\varepsilon\} = \begin{Bmatrix} \varepsilon_{\mathbf{p}}^{\mathbf{L}} \\ 0 \end{Bmatrix} + \begin{Bmatrix} \mathbf{z}\varepsilon_{\mathbf{b}} \\ \varepsilon_{\mathbf{s}} \end{Bmatrix} + \begin{Bmatrix} \varepsilon_{\mathbf{p}}^{\mathbf{NL}} \\ 0 \end{Bmatrix} \tag{63}$$

where $\{\varepsilon_{\mathbf{p}}^{\mathbf{L}}\}$, $\{\varepsilon_{\mathbf{b}}\}$, $\{\varepsilon_{\mathbf{s}}\}$, and $\{\varepsilon_{\mathbf{p}}^{\mathbf{NL}}\}$ are linear membrane, bending, transverse shear, and nonlinear membrane strain vectors, respectively, and are given by:

$$\{\varepsilon_p^L\} = \begin{Bmatrix} \dfrac{\partial u_0}{\partial s} \\ \dfrac{\partial v_0}{r\partial\theta} + \dfrac{w_0}{r} \\ \dfrac{\partial u_0}{r\partial\theta} + \dfrac{\partial v_0}{\partial s} \end{Bmatrix}; \; \{\varepsilon_b\} = \begin{Bmatrix} \dfrac{\partial\psi_s}{\partial s} \\ \dfrac{\partial\psi_\theta}{r\partial\theta} \\ \dfrac{\partial\psi_s}{r\partial\theta} + \dfrac{\partial\psi_\theta}{\partial s} \end{Bmatrix}; \; \{\varepsilon_s\} = \begin{Bmatrix} \psi_s + \dfrac{\partial w_0}{\partial s} \\ \psi_\theta + \dfrac{\partial w_0}{r\partial\theta} - \dfrac{v_0}{r} \end{Bmatrix}; \; \{\varepsilon_p^{NL}\} = \begin{Bmatrix} \dfrac{1}{2}\left(\dfrac{\partial w_0}{\partial s}\right)^2 \\ \dfrac{1}{2}\left(\dfrac{\partial w_0}{r\partial\theta} - \dfrac{v_0}{r}\right)^2 \\ \dfrac{\partial w_0}{\partial s}\left(\dfrac{\partial w_0}{r\partial\theta} - \dfrac{v_0}{r}\right) \end{Bmatrix} \tag{64}$$

where r is the principal radius of curvature in the circumferential direction.

The stress resultant vector $\{\tilde{\mathbf{N}}\} = \{N_{ss}N_{\theta\theta}N_{s\theta}\}^T$, the moment resultant vector $\{\tilde{\mathbf{M}}\} = \{M_{ss}M_{\theta\theta}M_{s\theta}\}^{\mathrm{T}}$ and the transverse shear stress resultant vector, $\{\tilde{\mathbf{Q}}\} = \{Q_{sz}Q_{\theta z}\}^T$, can be expressed in terms of the membrane $\{\varepsilon_{\mathbf{p}}\} = \{\varepsilon_{\mathbf{p}}^{\mathbf{L}}\} + \{\varepsilon_{\mathbf{p}}^{\mathbf{NL}}\}$, bending $\{\boldsymbol{\varepsilon}_{\mathbf{b}}\}$ and transverse shear $\{\boldsymbol{\varepsilon}_{\mathbf{s}}\}$ strain vectors through the constitutive relation as:

$$\begin{Bmatrix} \{\tilde{\mathbf{N}}\} \\ \{\tilde{\mathbf{M}}\} \\ \{\tilde{\mathbf{Q}}\} \end{Bmatrix} = \begin{bmatrix} [\tilde{\mathbf{A}}] & [\tilde{\mathbf{B}}] & [0] \\ [\tilde{\mathbf{B}}] & [\tilde{\mathbf{D}}] & [0] \\ [0] & [0] & [\tilde{\mathbf{E}}] \end{bmatrix} \begin{Bmatrix} \{\varepsilon_{\mathbf{p}}\} \\ \{\varepsilon_{\mathbf{b}}\} \\ \{\varepsilon_{\mathbf{s}}\} \end{Bmatrix} \tag{65}$$

where $[\tilde{\mathbf{A}}]$, $[\tilde{\mathbf{B}}]$, $[\tilde{\mathbf{D}}]$, and $[\tilde{\mathbf{E}}]$, , and are extensional, bending, bending-extensional coupling and transverse shear stiffness coefficients matrices of the composite laminate, respectively and are calculated using stiffness coefficients (Reddy, 2004) for a composite laminated panels of total thickness h consisting of N layers with stacking angles θ_k ($k = 1,\dots.\ N$) and layer thicknesses h_k ($k = 1, \dots, N$).

The field variables (u_0, v_0, w_0, ψ_s and ψ_θ) are interpolated using a C^0 continuous, eight-noded serendipity quadrilateral shear flexible shell element with five nodal degrees of freedom as:

$$(u_0, v_0, w_0, \psi_s, \psi_\theta) = \sum_{i=1}^{8} N_i^0 \left(u_{0i}, v_{0i}, w_{0i}, \psi_{si}, \psi_{\theta i}\right) \tag{66}$$

where N_i^0 are the shape functions of the eight-noded quadratic serendipity element and u_{0i}, v_{0i}, w_{0i}, ψ_s, ψ_θ are nodal values of field variables.

The total potential energy U, consisting of strain energy and potential energy of external loads, is given by

$$U = \frac{1}{2}\iint_A \Big[\{\varepsilon_p\}^T [\tilde{\mathbf{A}}]\{\varepsilon_p\} + \{\varepsilon_p\}^T [\tilde{\mathbf{B}}]\{\varepsilon_b\} + \{\varepsilon_b\}^T [\tilde{\mathbf{B}}]\{\varepsilon_p\} + \{\varepsilon_b\}^T [\tilde{\mathbf{D}}]\{\varepsilon_b\}$$
$$+ \{\varepsilon_s\}^T [\tilde{\mathbf{E}}]\{\varepsilon_s\}\Big] dA - \int_A q w_0 dA + \int_0^{r\phi} P u_0(s,\theta)\Big|_0^a \, r d\theta \tag{67}$$

where q is the distributed transverse load, P is meridional compressive force per unit length of curved edges of the panel.

Following the procedure of Rajasekaran and Murray (1973), the total potential energy U can be expressed as:

$$U = \{\delta\}^T \left[\frac{1}{2}[\mathbf{K}] + \frac{1}{6}[\mathbf{K}_1] + \frac{1}{12}[\mathbf{K}_2]\right]\{\delta\} - \{\delta\}^T \{\mathbf{F}\} \tag{68}$$

where the stiffness matrices $[\mathbf{K}]$, $[\mathbf{K}_1]$ and $[\mathbf{K}_2]$ are dependent on displacement vector due to damaged state constitutive matrix and geometric nonlinearity; $\{\mathbf{F}\}$ is mechanical load vector. The minimization of total potential energy with respect to the vector of degrees of freedom $\boldsymbol{\delta}$ leads to the governing equation of the panel as:

$$\left[[\mathbf{K}] + \frac{1}{2}[\mathbf{K}_1] + \frac{1}{3}[\mathbf{K}_2]\right]\{\delta\} = \{\mathbf{F}\} - \{\tilde{\mathbf{F}}\} \tag{69}$$

where $\{\tilde{\mathbf{F}}\} = \int_A \sum_{k=1}^{N} \int_{z_k}^{z_{k+1}} \left[\frac{\partial D_K}{\partial \delta_i}\right]\left\{\frac{\partial U}{\partial D_K}\right\} dz dA$

The solution of Equation (69) is obtained using the Newton–Raphson iteration technique and adaptive displacement control method to efficiently trace the equilibrium path (Patel, Shukla, & Nath, 2005). The solution starts with the load increments; and whenever during marching, the tangent stiffness matrix becomes semi-positive or negative definite, the solution approach is switched to the adaptive displacement control. The degree of freedom having the highest rate of change in the previous step is selected as the control parameter and is updated in each step. The step size is decided based on the increment in the previous step and the number of equilibrium iterations with suitable error norms. Using incremental displacements, incremental damage variable values are calculated and damage variables are updated.

This iterative procedure is repeated till the second norm of the incremental displacement vector between two consecutive iterations is less than the specified convergence tolerance (0.001%). The loading increments are continued till the maximum value of the damage variables reach unity or convergence failure.

5.1.4 Results and Discussion

The postbuckling response and failure load obtained from the present formulation with geometric nonlinearity and evolving damage is validated with the results of Singh and Kumar (1999) for sixteen layered cross-ply $(0^0/90^0)_{4S}$ and angle ply $(-45^0/45^0)_{4S}$ laminated plates under uniaxial compression (Figure 13) with material properties: E_1 = 132.58 GPa, $E_2 = E_3$ = 10.8 GPa, $G_{12} = G_{13}$ = 5.7 GPa, G_{23} = 3.4 GPa, $\nu_{12} = \nu_{13} = 0.24$, $\nu_{23} = 0.49$; F_{1t} = 1515 MPa, F_{1c} = 1697 MPa, $F_{2t} = F_{2c}$ = 43.8 MPa, F_6 = 86.9 MPa; $J_{11} = 0.2752567 \times 10^{-15}$, $J_{22} = J_{33} = 0.42365 \times 10^{-13}$, $k_h = 2.75 \times 10^{-7}$and geometric parameters: a = 279 mm, b = 279 mm, h = 2.16 mm. For evaluating the damage parameters J_{22} and k_h, in-plane shear strain at failure point and shear stress versus strain curve are required (Barbero & de Vivo, 2001) in addition to strength properties given by Singh and Kumar (1999). These parameters are assumed appropriately such that the predicted failure load is close to the available results of Singh and Kumar (1999) for one case. The results are compared in Figure 14. It can be seen from Figure 14 that the load deflection curve predicted using the present formulation is in good agreement with the available results of Singh and Kumar (1999). They performed progressive failure analysis based on Tsai Hill criterion to predict progressive failure of lamina and the maximum stress criterion to predict the onset of delamination at the interface of two adjacent layers.

The effect of radius-to-span ratio (r/a) and evolving damage on the static response characteristics is studied considering eight-layered angle ply $(45^0/-45^0)_{2S}$ SSSS laminated cylindrical panels (a = 2 m, a/h = 100, ϕ = 0.4 rad.) under uniaxial compression and the results are shown in Figure 14. The material properties considered are (Robbins et al., 2005a): E_1 =167 GPa, $E_2 = E_3$ = 8.13 GPa, $G_{12} = G_{23} = G_{13}$ = 8.8252 GPa, $\nu_{12} = \nu_{13} = \nu_{23} = 0.27$; $J_{11} = 0.9524\text{x}10^{-15}$, $J_{22} = J_{33} = 0.4381\text{x}10^{-12}$, $k_h = 7.595\text{x}10^{-7}$. The distribution of normalized displacements (w/h) over the surface of the panel is also shown in Figure 14 corresponding to the respective failure loads

Figure 13. Laminated plates under uniaxial loading

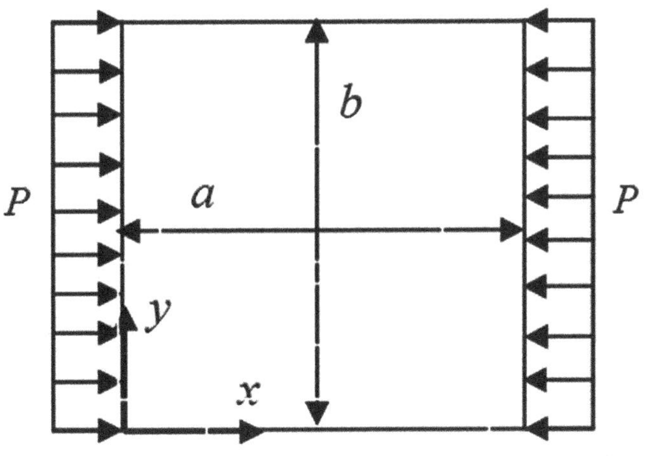

Figure 14. Comparison of postbuckling response for different lay-ups under uniaxial

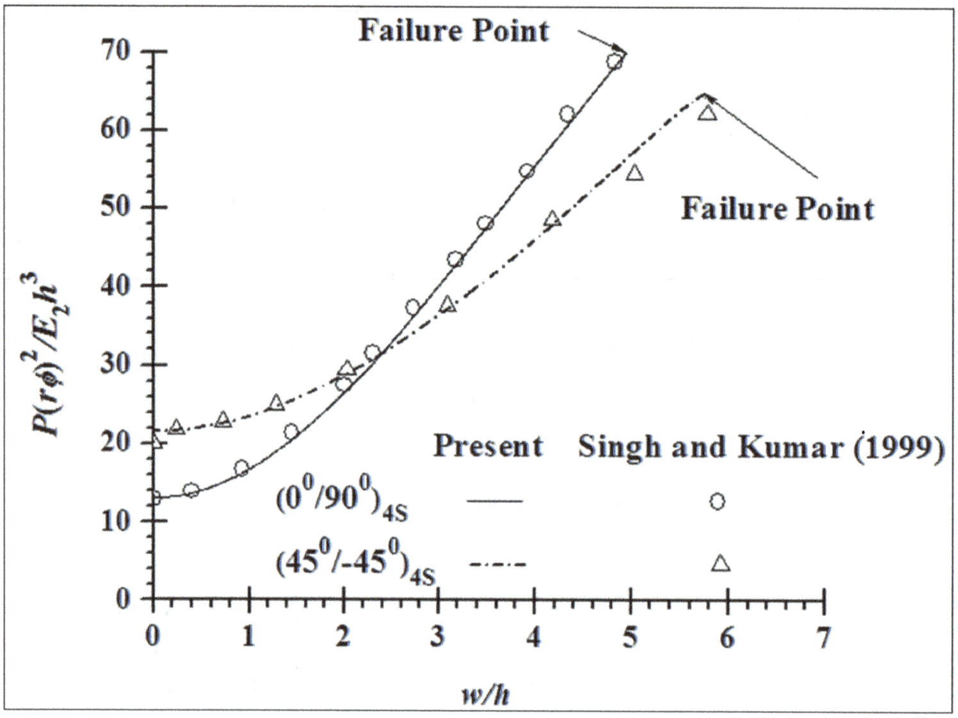

It can be observed from Figure 15 that the deflection without the damage evolution is found to be smaller than that with the damage due to continuous degradation of material properties. It can be inferred from Figure 15 that the failure load decreases with increases radius-to-span ratio (*r/a*). This is due to the curvature induced bending-stretching coupling leading to decrease in stiffness of panels with increasing radius of curvature. It can also be observed from these figures that for curved panels (r/a = 1.5, 3.0) transverse central displacement is non-zero at the onset of loading due to curvature induced bending-stretching coupling whereas for flat panel ($r/a = \infty$) the transverse central displacement is non-zero only after load reaches buckling load.

5.2 Case Study in Analytical Modeling of CMCs

5.2.1 Introduction

Ceramics are brittle by nature. Particulate and Fiber reinforced ceramic matrix composites (CMC) were developed to solve this problem. However, the improvement in toughness in particulate CMCs as well as in fiber reinforced composites demand stringent processing control, difficult to be accommodated in an industrial production process. So, multilayer ceramic composites (MLC) were developed. In such laminar structures the aim is to tailor the microstructure through appropriate control of the interface such that; the typical brittle fracture nature can be at least partially avoided and a pseudo-ductility introduced (Figure 16) into the ceramic system.

Figure 15. Effect of radius-to-span ratio (r/a) and damage on the response of SSSS angle-ply laminated $(45^0/-45^0)_{2S}$ cylindrical panels (a = 2 m, a/h = 100, ϕ = 0.4 rad.)

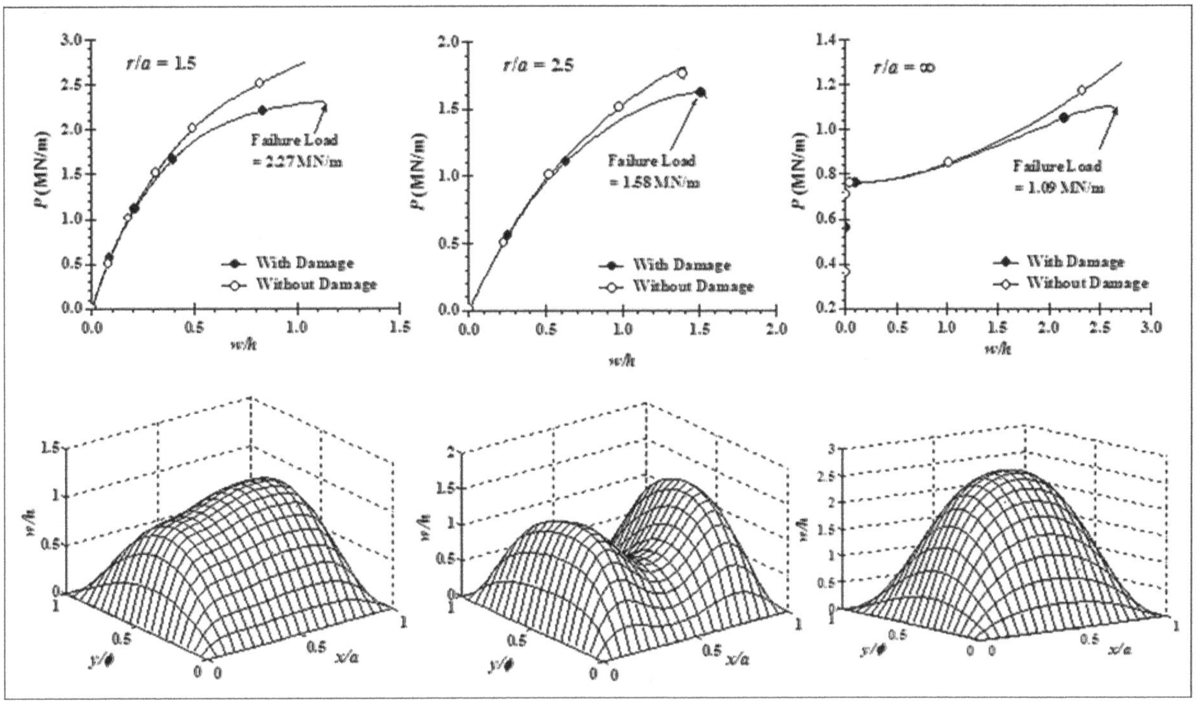

The concept of pseudo-ductility is explained schematically in Figure 16. Multilayered, Laminar, Ceramic Composites (MLC) are capable of sustaining enhanced strain to failure. In such composites, the sequential failure of each layer is, usually, accompanied by a load drop i.e. stress decrement. To initiate failure in the next layer, subsequent reloading is necessary. As such, the event of load drop is associated with the failure of a single layer and how the presence of the interface affects the mode of crack propagation. The crack propagation can take place along the interface, or within the interface or across the interface depending on certain critical parameters such as the ratio of thickness of the layers, the interlaminar fracture energy, presence of residual compressive stress at the interface etc. However, after each decrement in load, the stress in the next, intact layer is generally lower than its fracture strength. This is why, the remaining portion of the MLC system needs to be reloaded to initiate again the failure event thus, affecting the overall enhancement in failure strain. This process results in a stepped stress-strain response, termed as the pseudo-ductile behaviour. The origination of the term "pseudo-ductile" is linked to the significant enhancement in failure strain of the MLC system, in comparison to that of the brittle, monolithic material (Figure 16).

The multilayer ceramic composites provide the unique advantage that here, the system can be pre-designed layer by layer to suit a given end application (Clegg, Kendall, Alford, Button, & Birchall, 1990; Bermejo, Baudín, Moreno, Llanes, & Sánchez-Herencia, 2007; Chen, Pascual, Fischer, Kolednik, & Danzer, 2007; Cutler, Zok, & Lange, 1996; Folsom, Zok, Lange, & Marshall, 1992; Hirvonen et al., 1998; Jiménez-Piqué, Ceseracciu, Chalvet, Anglada, & de Portu, 2005; Kim & Kriven, 2003; Liu & Hsu, 1996; Park, Roh, Han, Kim, & Park, 2002; Park, Yang, & Stevens, 2003; Requena, Moreno, & Moya, 1989; Shigegaki, Brito, Hirao, Toriyama, & Kanzaki, 1996; Shobu, Watanabe, & Tsuji, 1989;

Figure 16. Typical pseudo-ductile stress-strain behaviour of a multilayered, laminar, ceramic composites. For the purpose of comparison only, the typical stress-strain behaviour during brittle failure of a monolithic ceramic is also included in the same figure.

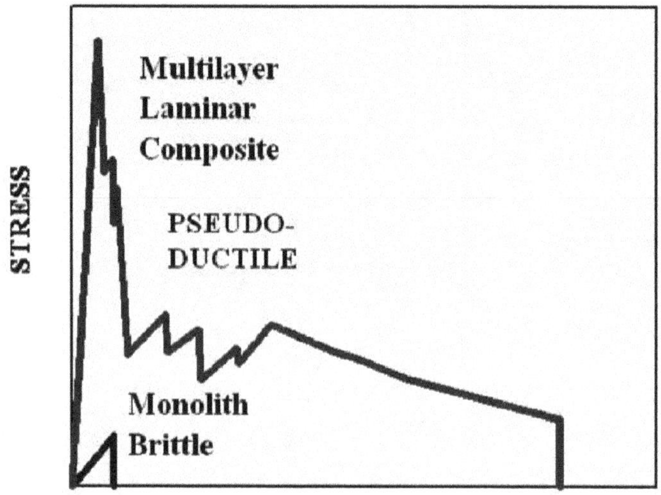

Tomaszewski, Weglarz, Boniecki, & Rećko, 2000; Tomaszewski, Węglarz, Wajler, Boniecki, & Kalinski, 2007; Tuffé & Wilkinson, 1995; Vandeperre, Kristofferson, Carlström, & Clegg, 2001; Zhang, Yue, & Watanabe, 1999; Dey et al., 2007). In particular, the MLCs with weak interfaces can sustain stresses even after the onset of failure process. So, they possess better flaw tolerance. That makes them more reliable for structural applications. A host of tape cast and / or slip cast MLC system have been studied during the last two decades. These include SiC / Graphite, (Clegg et al., 1990) Si_3N_4/ BN (Liu & Hsu, 1996), Si_3N_4 (dense) / Si_3N_4 (porous) (Shigegaki et al., 1996), SiC / (glass + fiber) (Cutler et al., 1996), Al_2O_3/ (Al_2O_3 + 4 vol. % ZrO_2) (Requena et al., 1989), Al_2O_3/ fiber-reinforced epoxy composite (Folsom et al., 1992), $MoSi_2$/ ($MoSi_2$ + $Al_2O_{3(pl)}$) (Tuffé & Wilkinson, 1995), $MoSi_2$ / SiC (Hirvonen et al., 1998), Al_2O_3 / $MoSi_2$ + 20 wt.% of Mo_2B_5 (Shobu, Watanabe, & Tsuji, 1989), Al_2O_3 / TiC / $MoSi_2$ + Mo_2B_5 (Zhang et al., 1999), 3.4Y-PSZ / Al_2O_3, 12Ce-PSZ / Al_2O_3 (Tomaszewski et al., 2000), Al_2O_3/ Al_2O_3-37 wt.% m-ZrO_2 (Vandeperre et al., 2001), Si_3N_4 (seeded) / Si_3N_4 (non-seeded) (Park et al., 2002), Y-TZP/ Ce-TZP (Park et al., 2003), Mullite / $AlPO_4$ (Kim & Kriven, 2003) etc. systems. Recent work include MLC systems like Al_2O_3 / ZrO_2 (Jiménez-Piqué et al., 2005), Al_2O_3 / 60 vol. % Al_2O_3-40 vol. % 3Y-TZP (Chen et al., 2007), Al_2O_3-YTZP / Al_2O_3 - Pure ZrO_2 (Bermejo et al., 2007), Al_2O_3 / $LaPO_4$, (Tomaszewski et al., 2007; Dey et al., 2007) etc.

The main design feature of these MLCs is that there is always an interface between the matrix layers. This interface could be weak with a low strength, low toughness material such as graphite (Clegg et al., 1990), boron nitride (Liu & Hsu, 1996), porous materials (Shigegaki et al., 1996), or a low strength but more damage tolerant plastic material such as GFRP (Cutler et al, 1996), GFRP and / or CFRP (Folsom et al., 1992). The interface can also contain a soft, easily deformable phase like $MoSi_2$ (Tuffé & Wilkinson, 1995; Hirvonen et al., 1998; Shobu et al., 1989; Zhang et al., 1999). Other type of interface in MLCs could be designed as both strong and tough like Al_2O_3 (Tomaszewski et al., 2000), Al_2O_3-37 wt. % m-ZrO_2 (Vandeperre et al., 2001), long grained silicon nitride (Park et al., 2002), Ce-TZP (Park et al., 2003), Al_2O_3 / ZrO_2 composite (Jiménez-Piqué et al., 2005), Al_2O_3 / Al_2O_3-40 vol. % 3Y-TZP (Chen et al., 2007), and Al_2O_3 - Pure ZrO_2 (Bermejo et al., 2007). In these later designs, however, the management and tuning of the residual stress between the matrix and interface layers, to enhance toughening, has emerged as a critical control parameter.

An equally important approach in MLC design issue is to use a weak interface with low interfacial toughness, like $AlPO_4$ (Kim & Kriven, 2003), $LaPO_4$ (Tomaszewski et al., 2007; Dey et al., 2007), etc. In fact earlier work done in our laboratory has already demonstrated that a porous alumina (Ghosh, Guha, Mukhopadhyay, & Maiti, 2005) or a porous nano-zirconia interface layer (Ghosh, Guha, Krishna, Mukhopadhyay, & Maiti, 2006) can also be used as a weak interface to enhance toughening in MLC.

However, there has not been much work done on modeling the mechanical behaviour of MLCs. Phillipps, Clegg, and Clyne (1993a) developed a model to establish the optimum lamina thickness in a given MLC material with weak interface as a function of interfacial toughness, lamina strength and Young's modulus. The predictions of the model were successfully verified from experimental data on a SiC / C / SiC MLC system (Phillipps, Clegg, & Clyne, 1993b), but these authors did not take into account the role of interfacial layer thickness. On the other hand, the work of Folsom, Zok, and Lange (1994), that assumed, the interfacial toughness to be zero, made the major prediction that single-phase MLC beams with a weak interface can withstand flexural load beyond the initial cracking event under displacement-controlled conditions and that they would withstand higher failure strains in flexure than in tension, Later, experimental data from Al_2O_3/ GFRP and / or CFRP based MLC system (Folsom et al., 1992) were used to verify predictions of the model.

In the case of strongly bonded MLC systems, the model of Sbaizero and Lucchini (1996) predicted that for a alumina/zirconia MLC system under the action of residual stress, crack bifurcation would occur after the crack propagating from the inner side to the outer side of a MLC, has propagated about 5% of the outer layer thickness, where the typical inner layer composition was zirconia + 15% alumina and the typical outer layer composition was alumina + 50% zirconia. Cai, Green, and Messing (1998) developed a model for maximum surface stress in an Alumina/ Ce-PSZ-Alumina MLC system assuming that all layers have the same thickness and that the surface layer has a Young's modulus higher than that of the adjacent interface layer. However, in most of the published work by other researchers e.g. references (Clegg et al., 1990; Bermejo, Baudín et al., 2007; Chen et al., 2007; Cutler et al., 1996; Folsom et al., 1992; Hirvonen et al., 1998; Jiménez-Piqué et al., 2005; Kim & Kriven, 2003; Liu & Hsu, 1996; Park et al., 2002; Park et al., 2003; Requena et al., 1989; Shigegaki et al., 1996; Shobu et al., 1989; Tomaszewski et al., 2000; Tomaszewski et al., 2007; Tuffé & Wilkinson, 1995; Vandeperre et al., 2001; Zhang et al., 1999; Dey et al., 2007), the thicknesses of the matrix and the interface layers were unequal.

In a recent work, Sglavo and Bertoldi (2006) assumed the presence of a discontinuous stepwise stress profile in a beam of strongly bonded lamina, with each lamina having same elastic modulus and a constant intrinsic toughness value, to predict the toughness curve of the MLC beam as a function of crack length and residual stress such that; the surface defects are forced to grow in a stable way before becoming critical; thus; finally leading to achievement of MLC materials from alumina/ mullite and alumina/ zirconia systems; with a high and reliable strength of 700 ± 28 MPa. On the other hand, Bermejo et al., (2007) used the standard beam theory (Timoshenko & Gere, 1972) to predict the stress distribution under four point bending of a prismatic MLC beam formed by layers having different elastic moduli and thermal expansion coefficients.

Chen et al. (2007) developed a numerical model to yield the fracture toughness of MLC as a function of the crack length, assuming the presence of spatially varying material forces which can induce an additional crack driving force term. However, the application of this model necessarily required precise, independently measured experimental data on intrinsic fracture toughness, Young's modulus, Poisson's ratio, and the coefficient of thermal expansion of all the constituent layers as well as those of the singular phases involved in the layer composition.

The presence of weak interfaces transverse to a crack growing in a brittle material causes the crack to be deflected with a consequent increase in the resistance to crack growth (Cook et al., 1964; Gupta, Yuan, & Martinez, 1993; Kendall, 1975; Lee, Howard, & Clegg, 1996; Mammoli, Graham, Reimanis, & Tullock, 1995; Ming-Yuan & Hutchinson, 1989; Ming Yuan, Evans, & Hutchinson, 1994; Phillipps, Clegg, & Clyne, 1993a; Pompidou & Lamon, 2007; Tullock, Reimanis, Graham, & Petrovic, 1994). Two main approaches to crack deflection at weak interface have been proposed.

The first approach is that of that of Cook et al. (1964). These authors consider that a crack is nucleated at the interface ahead of the matrix crack and deflection of the main crack results from coalescence with the interfacial crack.

The second approach is that of Ming-Yuan and Hutchinson (1989). These authors consider that the matrix crack is supposed to reach the interface and to remain stationary there and any further crack extension is the result of a competition between two phenomena: propagation along interface between the matrix and the reinforcement, and propagation through the reinforcement phase. This approach leads to a relationship where the strength or toughness of interface and reinforcing phase are compared to a ratio of stresses or strain energy release rates.

Cook et al. (1964) predicted that an interface with a theoretical tensile strength of less than 20% of that of the matrix will de-bond ahead of the elliptical main crack allowing crack deflection to occur. In the case of atomically sharp cracks, however, Gupta et al. (1993) proposed that assuming absence of elastic mismatch across the interface, the requirement for crack deflection to occur is that, the interface strength must be less than 35% of the strength of the bulk material.

On the other hand, the energy based analysis of Kendal (1975) predicted that, crack deflection would occur if the fracture energy of the interface is less than about 10-20% of that of the bulk, the exact value depending on the relative thickness of the two layers, which was supported by model experiments. The validity of the Cook and Gordon concept have been extended by Pompidou and Lamon (2007) by a finite element analysis based model, that was supported by experimental data. The model took into account the presence of an interphase and several interfaces in CMC's and multilayers. Finally, the model obtained a master curve relating interface and un-cracked constituent strengths to Young's moduli of the constituent phases.

In the case of the second approach, work of Ming-Yuan and Hutchinson (1989) and that of Martinez and Gupta (1994), for the case where a primary crack terminates at an interface between two semi-infinite planes, predicted that crack deflection would occur when, the ratio of fracture energies of the interface, to the matrix, was less than or equal to the ratio of the strain energy release rates of the crack deflected at the interface to that of the crack propagating straight-through and that the critical condition would be achieved when the strain energy release rate of the deflected crack was about 25-26% of that of the crack propagating straight-through.

The role of residual stress in this scenario, was later included in the problem (Ming Yuan et al., 1994) which got further support from numerical analyses (Tullock et al., 1994). Mammoli et al. (1995) found that the presence of defects in the interface can increase the critical ratio of fracture energies of the interface to that of the matrix for crack deflection. However, the excellent work of Lee et al. (1996) demonstrated that initial crack deflection is possible when the interfacial fracture energy is less than ~

64% of that of the bulk material. It also identified that the growth of interface defects was not sufficient to ensure toughening through crack deflection, but that a further condition, that any growing interface defects must not kink out of the interface, was needed to be also satisfied. One must keep in mind that it is not unusual for a brittle MLC material, to contain defects especially at the interface. Despite all these works, thus, it is still not clear, how a crack is deflected at the interface in a ceramic MLC beam and how other factors such as global loading states and layer thicknesses influence the crack deflection criteria.

So, the existing works do not adequately identify the design variables and the related material properties important for development of MLCs with high failure energy as well as fracture toughness. Therefore, it was conceived that, in the present work, a model will be developed to highlight the key issues and parameters involved in the design of high toughness MLCs. The model shall be based on compliance and energy considerations of the cracked MLC system. Finally, the model shall evaluate the amount of energy consumed per unit volume, $E_{con\text{MLC}}$ during controlled propagation of interfacial cracks. Ultimately the predictions of the model shall be verified from literature data as well as from experimental data obtained during the course of this research work.

5.2.2 Development of the Model

In the present work a new theoretical model was proposed and developed to optimize the fracture behavior of multilayered ceramic matrix composites (MLCs). For all the MLC systems considered in the present work, the matrix layer thickness was planned to be kept at a fixed thickness of 150 μm and the plan was to take the interface layer thickness fixed at 7 μm. The architecture considered was of a typical MLC structure with 20 matrix layer separated by 19 very thin (e.g. 7μm) interfacial layers. The outer layer on both sides of the MLC system was the matrix layer. The predictions of the model were made for a MLC system having brittle matrix / weak interface / brittle matrix architecture e.g., thick 150 μm alumina matrix layers separated by a very thin (~7 μm), weak, monazite interface. Finally, the predictions were verified with some experimental data from literature as well as our own work.

This micromechanical parameters based model was presented to prevent sudden catastrophic failure and also controlled extension of interfacial crack and through-thickness matrix crack. The strategy of utilizing crack deflection and controlled crack-path enhancement directly increased the fracture toughness of MLCs.

A multilayer composite beam was taken. The thickness of the beam comprised of *N* number of matrix layers of thickness t_m each and (*N-1*) very thin interfacial layer of thickness t_i ($<< t_m$) each. The matrix layers can be brittle, strong or strong and tough or only tough. It is loaded, for the sake of simplicity, in three point bend test configuration (Figure 17).

The interface layers are primarily considered as weak such that symmetrically propagated cracks of length C_k can be present in a given k^{th} interface layer. This beam of width w is loaded (Figure 18) under a centrally applied load P with a loading span of length of $2l$. The composite beam had width w and as mentioned earlier thickness, t_c. We further assume that the 1st matrix layer was pre-notched (Figure 19) to produce initial crack generation.

We mentioned earlier that the multilayer composite (MLC) beam was subjected to three-point bending with rollers arrangement. The load, P was given in the middle point of the loading span with the crosshead displacement of d.

Figure 17. Schematic of a multilayer composite beam with thick matrix / thin interfacial layers architecture before bending during three point loading (Dey, A., Dey, P., Datta, & Mukhopadhyay, 2008; Dey et al., 2008).

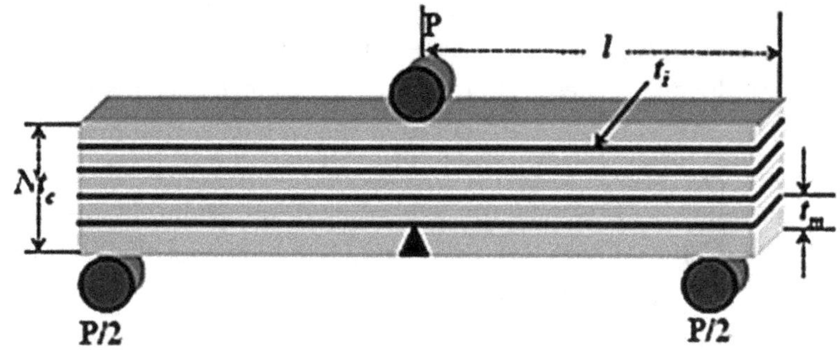

Figure 18. Schematic of a multilayer composite beam with thick matrix / thin interfacial layers architecture after bending during three point loading.

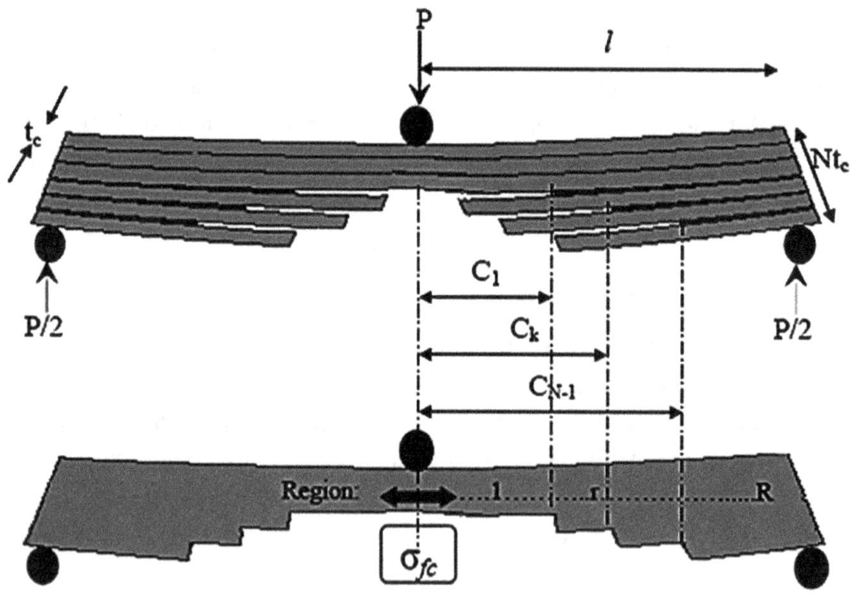

It was assumed that doubly deflected interfacial cracks were produced by bending and extended through very thin interfacial zone and reached till end of the loading span without a complete failure of beam i.e. the beam is, in principle, locally delaminated in a controlled manner but the failed parts are not separated fully. We visualized the failure process as a sequence of the following events.

It was assumed that, starting from the starter notch, there would be a growth of crack through the interface when the local strain energy release rate of the interface, G_i, was greater than the critical strain energy release rate of the interface G_{ic} (Figure 19). Next, there would be a through thickness crack through the matrix layer to reach the adjacent interface layer. It was assumed, to develop the model that the same sequence of events would continue till all the layers in the MLC had failed. The matrix crack

Figure 19. Schematic of crack growth both interfacial and through-thickness of a MLC beam during failure subjected to a three point bending.

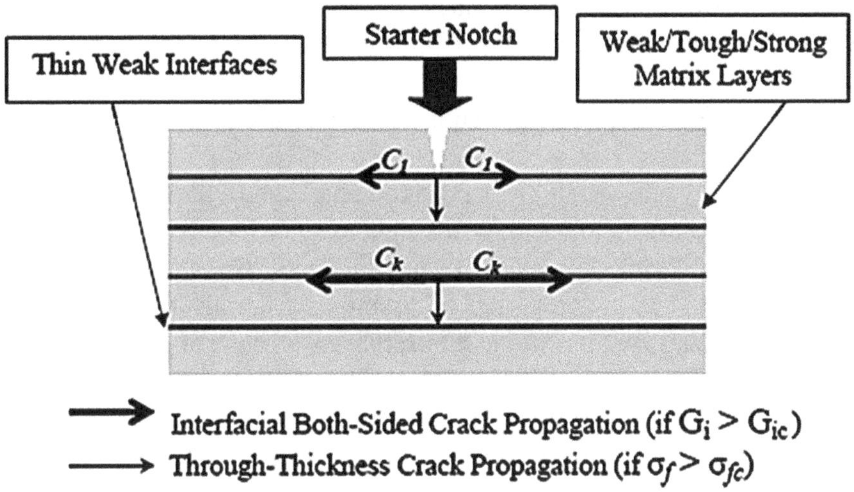

growth was assumed to occur when the load on the intact portion of the beam gives rise to a fracture stress, σ_f equal the critical failure stress, σ_{fc}, of an un-notched bar of same dimension, composition and same loading condition. The matrix cracks and interface cracks were assumed to occur and propagate symmetrically from the center of the beam. It was also assumed that two interface cracks of similar dimension and nature would propagate in a symmetric manner from near the tip of the starter notch in the first matrix layer.

The model developed in the present work estimates the effective failure energy, E_{conMLC}, per unit volume of the MLC beam. The basic assumptions of the model are (Dey et al., 2007; Phillipps et al., 1993a; Phillipps et al., 1993b; Dey et al., 2008a; Dey et al., 2008b):

1. A multilayer composite beam comprised of N number of matrix layers of thickness t_m each and $(N-1)$ very thin interfacial layer of thickness t_i ($<< t_m$) each, is considered. The effective total thickness of the multi layer composite beam is $T = Nt_m + (N-1)\, t_i \approx Nt_m + Nt_i$ (assuming that for very large no of N, $N-1 \approx N$) $= N(t_m + t_i) = Nt_c$ where, $t_c (= t_m + t_i)$ is the effective thickness of a matrix plus interface pair in MLC.
2. The multilayer composite beam of length L, thickness T and width w is loaded, for the sake of simplicity, in a three point bend configuration with a loading span of length $2l$ along with roller arrangements under a centrally applied load. The load gradually increases from a value of zero up to a critical value P at which, the critical fracture process gets initiated in the beam.
3. Here, the 1st matrix layer is pre-notched through the entire thickness up to the beginning of the first interface layer. This is done to produce initial crack generation. The interface layers are primarily considered as weak such that symmetric interfacial cracks of length C_k can be propagated through the thin interface under an appropriate applied stress in a given k^{th} interface layer. It is also assumed that the interface layers are generally smooth and free of any extraneous processing defect.
4. Starting from the starter notch, there will be growth of the very first crack through the first interface layer when the local strain energy release rate of the interface, G_i, will be greater than the critical

strain energy release rate of the interface G_{ic}. Next, there will be a through thickness crack through the second matrix layer to reach the adjacent second interface layer and so on.

5. The symmetrical interfacial cracks are produced by bending. They are extended through very thin interfacial layer. They can also reach till end of the loading span without a complete failure of the MLC beam.
6. In principle, the MLC beam is, locally delaminated in a controlled manner but the failed parts are not fully separated.
7. The critical strain energy release rate G_{ic} depends on phase angle, Ψ. The tangent of the quantity Ψ equals the ratio of the shearing to opening stress intensity factors. It can vary between pure mode I (crack opening mode, $\Psi = 0^0$) and mode II (shearing mode, $\Psi = 90^0$) and can also depend upon the loading geometry and elastic properties of dissimilar materials. However, following the work of Phillipps et al. (1993a, 1993b), it was assumed that any change of phase angle would have negligible effect on G_{ic}. So, G_{ic} was taken as constant and uniform for all interfacial layers throughout the MLC beam.
8. Propagation of the interfacial crack will be governed by the elastic strain energy released due to the crack propagation. During this period no through-thickness cracking will occur.
9. The matrix crack growth will occur when the applied load on the intact portion of the beam gives rise to a local failure stress, σ_f equal to or greater than the critical fracture stress, σ_{fc}, of an un-notched matrix bar of same dimension and composition and under the same loading condition. For the sake of simplicity, it is also assumed that σ_{fc} will remain unaffected by the statistical distribution of the characteristic flaw size distribution in the matrix layer.
10. As a result of the step (i) above, the crack can start propagating through the thickness of a given layer in the MLC beam when and only when the stress generated on that particular layer equals the failure stress of that layer.
11. G_{bc}, the strain energy release rate for bulk matrix is uniform for all bulk layers throughout the specimen.
12. The through-thickness crack is allowed to propagate only in the center of the beam and the interfacial cracks are allowed to symmetrically propagate along both side of the center point.

The energy spent per unit volume by the loading system, E_{sp}, is stored in the MLC beam. Thus, E_{sp} is given by the expression, $E_{sp} = \frac{\sum_{n}^{N-1} P_n \cdot \partial d_n}{2Nt_c wl}$, where all the terms have been already defined, and P_n, ∂d_n stands respectively, for the load and incremental change in crosshead displacement of any n^{th} layer.

13. The effective total failure energy, E_{conMLC}, consumed per unit volume during failure of the given MLC beam is the sum of the contributions from the interfacial and through-thickness cracking processes.
14. Based on considerations of frictional and other irreversible processes of energy dissipation, and keeping in mind the fact that the energy consumed per unit volume during failure of the MLC beam (i.e. E_{conMLC}) will have to be provided by the amount of energy, E_{sp}, spent per unit volume by the loading system, we have $E_{conMLC} \leq E_{sp}$.

Here, the 1st matrix layer is pre-notched through the entire thickness, as mentioned above, to produce initial crack generation. Now, the total failure energy, E_{conMLC}, consumed per unit volume of the MLC beam can be written as,

$$E_{conMLC} = \frac{1}{\alpha}G_{ic} + \frac{1}{\beta}G_{bc} \tag{70}$$

where G_{ic}/α and G_{bc}/β are the failure energy per unit volume of the interface and failure energy per unit volume of the bulk matrix material, respectively. Here, α and β are appropriate scaling parameters, since G_{ic} and G_{bc} are critical strain energy release rate terms, which are usually expressed as per unit area. As G_{bc} is the inherent property of the bulk ceramic matrix material, and can not be changed for a given ceramic matrix material, the design aspect of MLC depends on increasing G_{ic} in order to improve the overall failure energy of the composite. However, the geometric factors of the interface e.g., thickness of the interface etc. are also important parameters and shall be discussed later in this paper. Thus, we have,

$$E_{conMLC} = \frac{1}{\alpha}\frac{K_{ic}^{\ 2}}{Y_i}\left(1-\upsilon_i^{\ 2}\right) + \frac{1}{\beta}\frac{K_{bc}^{\ 2}}{Y_b}\left(1-\upsilon_b^{\ 2}\right) \tag{71}$$

because G is related to the toughness K as,

$$G = \frac{K^2}{Y}\left(1-\upsilon^2\right) \tag{72}$$

Here, in Equation (71), K_{ic}, Y_i, ν_i are the fracture toughness, Young's modulus, and Poisson's ratio of interface material, respectively. Similarly, K_{bc}, Y_b, ν_b are the fracture toughness, Young's modulus and the Poisson's ratio of bulk matrix material, respectively. So, our main objective here is to increase the interfacial fracture toughness K_{ic}, so that E_{conMLC} can be maximized. In order to evaluate the compliance (χ) of the beam, at any stage during bending, it is assumed that zero loads would be carried by the portion of a layer between the through-thickness crack and the tip of an interfacial crack.

Therefore, the failure energy consumed per unit volume of the MLC beam can be written from Equation (71) above, as:

$$E_{conMLC} = \frac{1}{\alpha}\frac{3}{8}\frac{t_c^{\ 3}Y_i\left(N-1\right)^2 d'^2}{l^4} + \frac{1}{\beta}\frac{K_{bc}^{\ 2}}{Y_b}\left(1-\upsilon_b^{\ 2}\right) \tag{73}$$

From simple geometric consideration it can be shown easily that, both α and β have dimension of length. In particular β is given by $2l$, where as, the term "α" can have any non zero value $\leq 2l$.

5.2.3 The Crack Propagation Algorithm

Following the above model, an algorithm was developed in the present work to compute the failure energy ($E_{con\ MLC}$) consumed per unit volume by the MLC beam and the total amount of energy (E_{sp}) spent per unit volume by the applied loading system. To develop the algorithm, the sequential steps of operation were in the following manner:

The crosshead displacement d was progressively increased by Δd amount in a quasi-static manner such that the new crosshead displacement was $d+\Delta d$. The compliance χ was computed from the loading span, beam stiffness and a given crosshead displacement d. Using the change of compliance with interfacial crack length, $\left(\frac{d\chi}{dC_k}\right)$, for any load P; G_i can be calculated. Interfacial crack growth was then simulated, subject to satisfaction of the criterion, $G_i \geq G_{ic}$. If this criterion was satisfied, interfacial crack growth would occur. This would induce further interfacial crack growth C_I *to* $C_I+\Delta C_I$, say, for the first interface. This may happen, in general, in any given region R. When the region R is exhausted, the process should repeat for the next layer in the sequence i.e. R proceeds to become then $R+1$. After this event occurred, the next sequence was that of a through thickness crack growth. This was assumed to occur, when the local stress, $\sigma_f = -\frac{Plt_c}{4M_r}(N-k)$, would be $\geq \sigma_{fc}$ for that layer. The same sequence of event was repeated in the MLC beam to make the total beam failure to occur, when all of the total number of layers would be failing in the process.

The algorithm used for the progressive generation and propagation of crack through the interface and the bulk alternately through all the layers, are given below:

Step 1: enter the parameters: w, t_m, t_l, l, N, Y_l, Y_b, G_{ic}, G_{bc}, σ_{fc}

Step 2: calculate χ, P, $\frac{d\chi}{dc}$, G_i, σ_f

Step 3: if $G_i > G_{ic}$

then $C_1 = C_1 + \Delta C_1$ (interfacial crack occurs), repeat step 2 onward

else

Step 4: if $\sigma_f > \sigma_{fc}$

then $R = R + 1$, $K = R - 1$, $E_{sp} = E_{sp} + \Delta E_{sp}$, $E_{conMLC} = E_{conMLC} + \Delta E_{conMLC}$ (through thickness crack occurs), go to step 5

else $d = d + \Delta d$, repeat step 2 onward

Step 5: if $K = N$

then calculate output E_{conMLC} and E_{sp}

end

else repeat step 2 onward

5.2.4 Validation of the Model

5.2.4.1 Alumina / Lanthanum Phosphate / Alumina MLCs

Let us consider an alumina / lanthanum phosphate / alumina MLC beam with 20 alumina matrix layers and 19 thin interfacial lanthanum phosphate layers in between the alumina matrix layers. The matrix layer thickness was assumed ~ 150 μm. The interfacial layer thickness was assumed as ~ 7 μm. The justification for such choice has been elaborated later in this work. Also, other researchers working on SiC / Graphite / SiC MLC system have reported similar data for matrix layer thickness, interface layer thickness and similar number of layers of the matrix and interface (Phillipps et al. 1993a; 1993b). Here alumina is chosen as the matrix layer as it is the most studied material of all the common ceramic materials. The basis for choice of lanthanum phosphate (LP) as a weak interface component of the MLC system is that, it has strain energy release rate G_{ic} ($LaPO_4$) of only ~ 10 Jm^{-2} (Mawdsley, Kovar, & Halloran, 2000), i.e. sufficiently low to allow a crack deflection during failure propagation through LP to alumina; which has a much higher strain energy release rate e.g. G_{bc} (Al_2O_3) ~ 20 Jm^{-2} (Mawdsley et al., 2000). The pertinent mechanical properties used for making predictions according to our model are given in Table 1.

Figures 20-23 show the data on variation of load *(P)* as a function of the crosshead displacement which is a main testing variable as well as those of normalized interfacial crack length (C/*l*), the amount of energy (E_{sp}) spent per unit volume by the applied loading system and total energy (E_{conMLC}) consumed during failure of the alumina / lanthanum phosphate (LP) / alumina MLC system as a function of the crosshead displacement and the number (*N*) of failed matrix alumina layer, which is a main processing variable. The data presented in Figure 20 confirmed that the alumina / LP / alumina MLCs exhibit a characteristic load-unload-reload behaviour as the crosshead displacement increased. As the number of failed matrix alumina layer increased, the drop in magnitude of load to cause failure was evident from Figure 21. For instance, the failure of the 1st alumina matrix layer occurred at a load of about 171 N, while those of the 5th and 10th alumina matrix layers occurred at about 106.6 and 47.4 N loads. It was interesting to note that initially the normalized interfacial crack length (C/*l*) was nearly equal to 1, until the 7th matrix alumina layer had failed in the MLC beam, Figure 22. This value of 1 for the normalized interfacial crack length (C/*l*) would mean, that with increase in load, during the initial "crack burst-out process", there would be interfacial cracks propagating in both directions from the notch tip up to the point where the two supporting rollers were present (Phillipps et al. 1993a; 1993b). Physically, such a process would greatly aid in energy spent by the applied loading system. Subsequently, this energy would be consumed during the evolution of the entire failure process of the various layers in the MLC beam. Following the failure of the 7th matrix layer, gradually the extent of interfacial crack propagation

Table 1. Mechanical Properties of Al_2O_3 and $LaPO_4$ Components

Alumina	Lanthanum Phosphate
Young's Modulus, E = 370 GPa Matrix Layer Failure Stress, σ_{fc} = 400 MPa Critical Strain Energy Release Rate, G_{bc} = 20 J/m^2, Thermal Expansion Coefficient, $\alpha = 9 \times 10^{-6}/ {}^0C$ Poisson's Ratio, $\nu = 0.22$	Young's Modulus, E = 155 GPa Critical Interfacial Strain Energy Release Rate, G_{ic} = 10 J/m^2, Thermal Expansion Coefficient, $\alpha = 10 \times 10^{-6}/ {}^0C$ Poisson's Ratio, $\nu = 0.275$

Figure 20. Load vs. Crosshead displacement of alumina / LP /alumina MLC
Dey et al. (2008a)

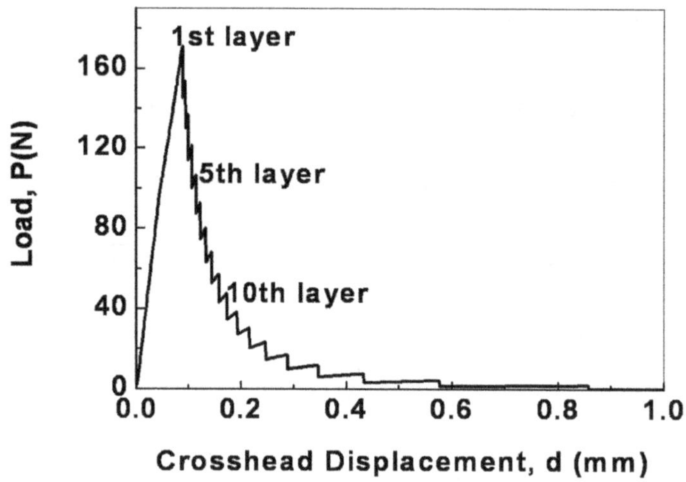

Figure 21. Variation of failure load as a function of the number of matrix alumina layers
Dey et al. (2008a)

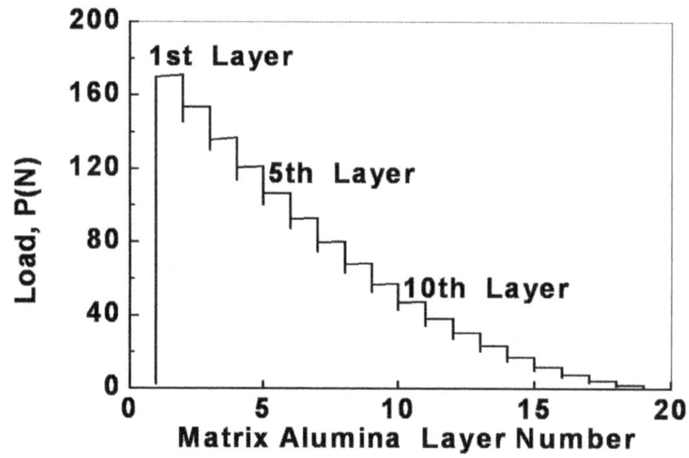

decreased in the MLC system, so far so to be about 60% of the half-loading span *(l)*, at the instant when the 19th matrix alumina layer failed, Figure 22. This particular sequence of crack propagation in MLC system is commonly referred to as the "crack creeping in" process (Phillipps et al. 1993a; 1993b).

The main advantage of such a controlled delamination crack growth was that the amount of energy consumed (E_{conMLC}) by the MLC also increased substantially with the number of failed alumina matrix layers, Figure 23. Initially, during the crack burst out process, the amount of failure energy consumed (E_{conMLC}) was much lesser compared to the total amount of energy, E_{sp}, spent per unit volume by the applied loading system to cause sequential crack growth in the MLC. However, as the number of failed matrix alumina layer numbers increased to e.g., > 10, the gap between the E_{sp} and E_{conMLC} terms narrowed considerably. Finally, when the 19th matrix alumina layer had failed the total amount of failure energy

Figure 22. Variation of normalized interfacial crack length (C/l) as a function of the number of matrix alumina layers
Dey et al. (2008a)

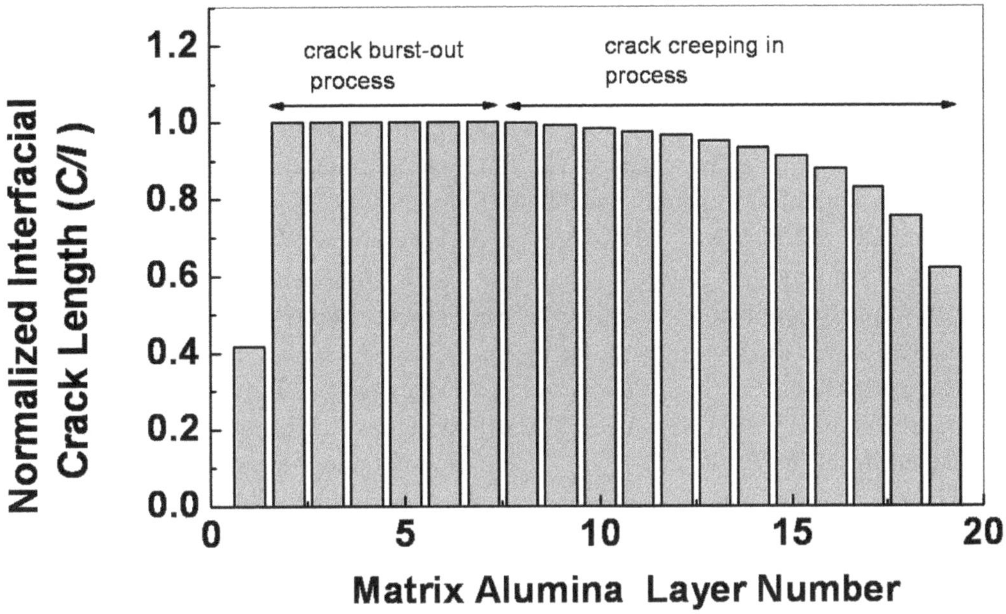

Figure 23. Variation of work done and failure energy dissipated per unit volume as a function of the number of matrix alumina layers
Dey et al. (2008a)

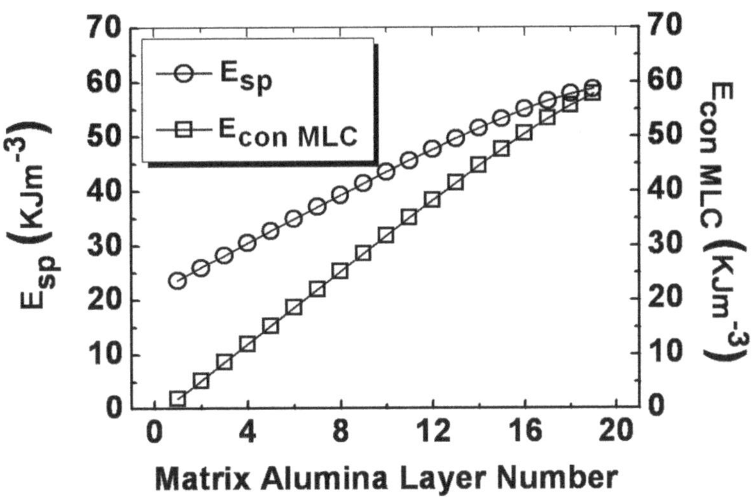

consumed (E_{conMLC}) was estimated as 57.9 KJm^{-3} that was nearly 99% of the total amount of energy, E_{sp}, (58.8 KJm^{-3}) spent by the applied loading system, Figure 23. Thus, the effective failure energy of the present alumina / LP based MLC beam considered, was predicted as 57.9 KJm^{-3}.

5.2.5 Experimental Validation of the Model

The failure energy of the matrix alumina layers was 6.67 KJm^{-3}, Figure 24. Therefore, in the present alumina / LP / alumina MLC beam, the predicted failure energy (57.9 KJm^{-3}), improved by a factor of almost ~ 9 times (8.68 times), over that of the matrix alumina layers, Figure 24. The experimentally measured value of $E_{con,MLC}$ of such an MLC prepared in our laboratory was ~ 55.6 KJm^{-3} (Dey et al., 2008a; Dey et al., 2008b), which matched well with the predicted value of 57.9 KJm^{-3}.

Similarly, the literature data (Mawdsley et al., 2002) showed that the experimental value of E_{conMLC} of tape cast 45 layer alumina / lanthanum phosphate / alumina MLC beam (t_m =150 μm and t_i = 15 μm) was ~ 57.88 KJm^{-3} . In a similar way, for hot pressed 55 layer alumina/ lanthanum phosphate/ alumina MLC beam (t_m =125 μm and t_i = 15 μm) E_{conMLC} was ~77.77 KJm^{-3} (Mawdsley et al., 2002). These experimental data were also comparable to our own experimental data (Phillipps et al. 1993a; 1993b) as well as to that predicted in the present work for $E_{con\ MLC}$.

Further, these experimental $E_{con\ MLC}$ values were ~ 9 times and ~ 12 times higher respectively than that of bulk matrix alumina (Mawdsley et al., 2002) Both of these experimental values of E_{conMLC} were, however, obtained by 4 point bend test. E_{conMLC} value may vary depending upon the processing route of composite layers, purity of material and type of loading configurations during mechanical testing of MLC (Phillipps et al. 1993a; 1993c).

6. CONCLUSION

This chapter has dealt mainly with the micromechanics of composites and the relevance of the same for technological applications. At the beginning the basics of composite materials have been discussed with a special emphasis on the Metal Matrix Composites, the Ceramic Matrix Composites and the Polymer

Figure 24. Energy consumed per unit volume ($E_{con\ MLC}$) during failure of tape cast alumina matrix layers and the alumina / LP / alumina MLC beam
Dey et al. (2008a)

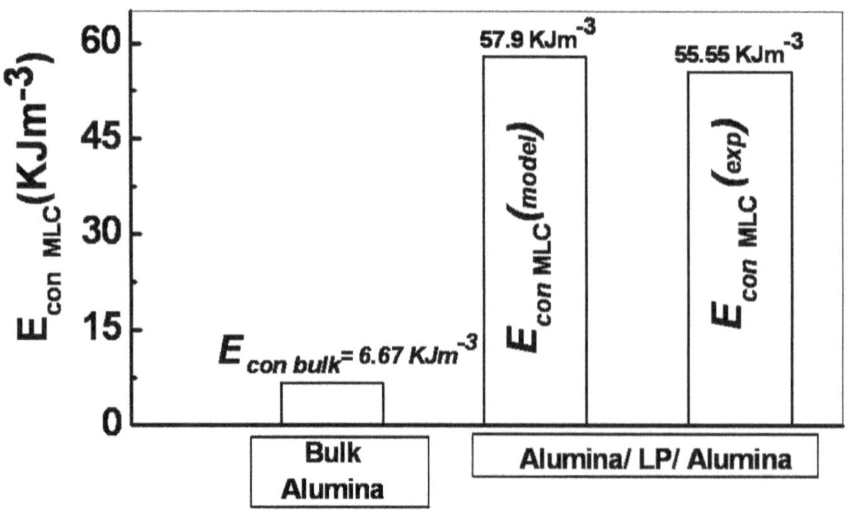

Matrix Composites. Next the basic of micromechanics has been discussed with special reference to macroscopic, mesoscopic and microscopic sales of modelling. Subsequently, the analytical methods of continuum micromechanics for composites e.g., the voigt, the Reuss, the Hashin-Shtrikman and the famous Mori-Tanaka models were discussed. Further, it was explained that depending up-on the nature of problem various numerical solution techniques such as finite difference method (FDM), finite element method (FEM), finite volume method (FVM), boundary element method (BEM), domain decomposition method (DDM) and differential quadrature method (DQM) can be used. The basic concepts of the finite element method, how the elements, nodes, the methods e.g., the least square error method, collocation, sub-domain and Galerkin methods etc., of finding the approximate solutions have been briefly discussed. Next, the concept of weak form of weighted residual statement was applied to the case of mesh for a bar for the finite element modelling based on the weak form. The next example was chosen as the case of the finite element modelling of a laminated composite plate. Subsequently, the advanced finite element methods e.g., the extended finite element method, the micromechanical modelling of composite materials using FEM for material modelling etc. have been very briefly touched upon. These discussions were followed up by the discussion on failure criteria, damage mechanics and fracture mechanics based criterion based modelling for failure of composites.

Subsequently, the case study on FEM based continuum damage modelling of a fibre reinforced composite e.g., SSSS angle-ply laminated $(45^0/-45^0)_{2S}$ cylindrical panels; was presented as a typical, demonstrative example. The final case study has been deliberately chosen from the analytical modelling of a multilayer ceramic composite (MLC) e.g., an alumina / lanthanum phosphate / alumina MLC beam. The MLC comprised of 20 alumina matrix layers and 19 thin interfacial lanthanum phosphate layers in between the alumina matrix layers. The matrix layer thickness was assumed ~ 150 μm while the interfacial layer thickness was assumed as ~ 7 μm. The effective failure energy per unit volume of the present alumina / LP based MLC beam considered, was predicted as about 57.9 KJm^{-3}. This value matched closely with the experimentally measured value of about 55.6 KJm^{-3}.

REFERENCES

Azzi, V. D., & Tsai, S. W. (1965). Anisotropic strength of composites. *Experimental Mechanics*, *5*(9), 283–288. doi:10.1007/BF02326292

Babuska, I., & Melenk, J. M. (1997). The partition of unity method. *International Journal for Numerical Methods in Engineering*, *40*(4), 727–758. doi:10.1002/(SICI)1097-0207(19970228)40:4<727::AID-NME86>3.0.CO;2-N

Barbero, E. J., & de Vivo, L. (2001). A constitutive model for elastic damage in fiber-reinforced PMC laminae. *International Journal of Damage Mechanics*, *10*(1), 73–93. doi:10.1106/6PQ6-31JW-F69K-74LU

Bazant, Z. P., & Planas, J. (1998). *Fracture and size effect in concrete and other quasibrittle materials*. Boca Raton: CRC Press.

Belytschko, T., Gracie, R., & Ventura, G. (2009). A review of extended/generalized finite element methods for material modeling. Modelling and Simulation in Materials Science and Engineering, 17(4).

Benveniste, Y. (1987). A new approach to the application of Mori-Tanaka's theory in composite materials. *Mechanics of Materials*, *6*(2), 147–157. doi:10.1016/0167-6636(87)90005-6

Berdichevskii, V. (1979). Variational-asymptotic method of constructing a theory of shells. *Journal of Applied Mathematics and Mechanics*, *43*(4), 711–736. doi:10.1016/0021-8928(79)90157-6

Bermejo, R., Baudín, C., Moreno, R., Llanes, L., & Sánchez-Herencia, A. J. (2007). Processing optimisation and fracture behaviour of layered ceramic composites with highly compressive layers. *Composites Science and Technology*, *67*(9), 1930–1938. doi:10.1016/j.compscitech.2006.10.010

Berryman, J. G., & Berge, P. A. (1993). Rock elastic properties: Dependence on microstructure. In C. S. Chang & J. W. Ju (Eds.), Homogenization and constitutive modeling for heterogeneous materials. AMD.

Bouchard, P. O., Bay, F., Chastel, Y., & Tovena, I. (2000). Crack propagation modelling using an advanced remeshing technique. *Computer Methods in Applied Mechanics and Engineering*, *189*(3), 723–742. doi:10.1016/S0045-7825(99)00324-2

Cai, P. Z., Green, D. J., & Messing, G. L. (1998). Mechanical characterization of Al2O3/ZrO2 hybrid laminates. *Journal of the European Ceramic Society*, *18*(14), 2025–2034. doi:10.1016/S0955-2219(98)00151-4

Chen, C. R., Pascual, J., Fischer, F. D., Kolednik, O., & Danzer, R. (2007). Prediction of the fracture toughness of a ceramic multilayer composite – Modeling and experiments. *Acta Materialia*, *55*(2), 409–421. doi:10.1016/j.actamat.2006.07.046

Chow, C. L., & Wang, J. (1987). An anisotropic theory of elasticity for continuum damage mechanics. *International Journal of Fracture*, *33*(1), 3–16. doi:10.1007/BF00034895

Christensen, R. M. (1979). *Mechanics of composite materials*. New York: Wiley.

Clegg, W. J., Kendall, K., Alford, N. M., Button, T. W., & Birchall, J. D. (1990). A simple way to make tough ceramics. *Nature*, *347*(6292), 455–457. doi:10.1038/347455a0

Cook, J., Gordon, J. E., Evans, C. C., & Marsh, D. M. (1964). A mechanism for the control of crack propagation in all-brittle systems. *Proceedings of the Royal Society of London. Series A, Mathematical and Physical Sciences*, *282*(1391), 508–520. doi:10.1098/rspa.1964.0248

Cutler, W. A., Zok, F. W., & Lange, F. F. (1996). Mechanical behavior of several hybrid ceramic-matrix-composite laminates. *Journal of the American Ceramic Society*, *79*(7), 1825–1833. doi:10.1111/j.1151-2916.1996.tb08002.x

Daniel, I. M., & Ishai, O. (2006). *Engineering mechanics of composite materials* (2nd ed.). New York: Oxford University Press.

Davila, C., Jaunky, N., & Goswami, S. (2003). *Failure criteria for FRP laminates in plane stress*. Norfolk, VA: Structural Dynamics, and Materials Conference.

Davila, C. G., & Camanho, P. P. (2003). *Failure criteria for FRP laminates in plane stress*. Technical Report NASA/TM-2003-212663. NASA Langley Research Center.

Davila, C. G., Camanho, P. P., & Rose, C. A. (2005). Failure criteria for FRP laminates. *Journal of Composite Materials*, *39*(4), 323–345. doi:10.1177/0021998305046452

Dey, A., Dey, P., Datta, S., Bandyopadhyay, N. R., Mukhopadhyay, A. K., & Maiti, H. S. (2007). A new theoretical model for development of damage tolerant composite. In *International Seminar on Ceramics, "CeraTec 2007"*. Indian Ceramic Society.

Dey, A., Dey, P., Datta, S., Bandyopadhyay, N. R., Mukhopadhyay, A. K., & Maiti, H. S. (2008b). A new theoretical model for development of damage tolerant composites. *Transaction of the Indian Ceramic Society*, *67*(2), 63–74. doi:10.1080/0371750X.2008.11078643

Dey, A., Dey, P., Datta, S., & Mukhopadhyay, A. K. (2008a). A new model for multilayer ceramic composites. *Materials and Manufacturing Processes*, *23*(5), 513–527. doi:10.1080/10426910802104278

Dharan, C. K. H. (1978). Fracture mechanics of composite materials. *Journal of Engineering Materials and Technology*, *100*(3), 233–247. doi:10.1115/1.3443485

Dvorak, G. J., & Laws, N. (1987). Analysis of progressive matrix cracking in composite laminates II. first ply failure. *Journal of Composite Materials*, *21*(4), 309–329. doi:10.1177/002199838702100402

Echaabi, J., Trochu, F., & Gauvin, R. (1996). Review of failure criteria of fibrous composite materials. *Polymer Composites*, *17*(6), 786–798. doi:10.1002/pc.10671

Erdogan, F. (1972). Fracture problems in composite materials. *Engineering Fracture Mechanics*, *4*(4), 811–840. doi:10.1016/0013-7944(72)90018-5

Eshelby, J. D. (1957). The determination of the elastic field of an ellipsoidal inclusion, and related problems. *Proceedings of the Royal Society of London. Series A, Mathematical and Physical Sciences*, *241*(1226), 376–396. doi:10.1098/rspa.1957.0133

Folsom, C. A., Zok, F. W., & Lange, F. F. (1994). Flexural properties of brittle multilayer materials: I, Modeling. *Journal of the American Ceramic Society*, *77*(3), 689–696. doi:10.1111/j.1151-2916.1994.tb05350.x

Folsom, C. A., Zok, F. W., Lange, F. F., & Marshall, D. B. (1992). Mechanical behavior of a laminar ceramic/fiber-reinforced epoxy composite. *Journal of the American Ceramic Society*, *75*(11), 2969–2975. doi:10.1111/j.1151-2916.1992.tb04373.x

Gardner, J. P. (1994). *Micromechanical modeling of composite materials in finite element analysis using an embedded cell approach* (Masters thesis, MIT, USA). Retrieved from http://dspace.mit.edu/bitstream/handle/1721.1/43275/31369665.pdf?...1

Germain, N., Besson, J., & Feyel, F. (2007). Composite layered materials: Anisotropic nonlocal damage models. *Computer Methods in Applied Mechanics and Engineering*, *196*(41–44), 4272–4282. doi:10.1016/j.cma.2007.04.009

Ghosh, A. (2008). Hygrothermal effects on the initiation and propagation of damage in composite shells. *Aircraft Engineering and Aerospace Technology*, *80*(4), 386–399. doi:10.1108/00022660810882755

Ghosh, A., & Sinha, P. K. (2004). Dynamic and impact response of damaged laminated composite plates. *Aircraft Engineering and Aerospace Technology*, *76*(1), 29–37. doi:10.1108/00022660410514982

Ghosh, S., Guha, A., Krishna, K. M., Mukhopadhyay, A. K., & Maiti, H. S. (2006). Tape cast multilayer composite of nano zirconia with high toughness. *Materials and Manufacturing Processes*, *21*(7), 662–668. doi:10.1080/10426910600611680

Ghosh, S., Guha, A., Mukhopadhyay, A. K., & Maiti, H. S. (2005). Tape cast porous nano alumina multilayer composites. *Transactions of the Indian Ceramic Society*, *64*(2), 101–108. doi:10.1080/0371750X.2005.11012195

Ghosh, S., & Moorthy, S. (2004). Three dimensional Voronoi cell finite element model for microstructures with ellipsoidal heterogeneties. *Computational Mechanics*, *34*(6), 510–531. doi:10.1007/s00466-004-0598-5

Gol'denblat, I. I., & Kopnov, V. A. (1965). Strength of glass-reinforced plastics in the complex stress state. *Polymer Mechanics*, *1*(2), 54–59. doi:10.1007/BF00860685

Golanski, D., Grzes, J., Kikuchi, N., & Kozłowski, M. (2000). Modeling of residual stresses in metal matrix composites using digital image based technique and homogenization method.*Proceedings of the International Conference on Mathematical Modeling and Simulation of Metal Technologies.*

Gupta, A. K., Patel, B. P., & Nath, Y. (2012). Continuum damage mechanics approach to composite laminated shallow cylindrical/conical panels under static loading. *Composite Structures*, *94*(5), 1703–1713. doi:10.1016/j.compstruct.2011.12.002

Gupta, A. K., Patel, B. P., & Nath, Y. (2013a). Nonlinear static analysis of composite laminated plates with evolving damage. *Acta Mechanica*, *224*(6), 1285–1298. doi:10.1007/s00707-013-0875-5

Gupta, A. K., Patel, B. P., & Nath, Y. (2013b). Continuum damage modeling of composite laminated plates using higher order theory. *Composite Structures*, *99*, 141–151. doi:10.1016/j.compstruct.2012.11.038

Gupta, A. K., Patel, B. P., & Nath, Y. (2014). Postbuckling response of composite laminated plates with evolving damage. *International Journal of Damage Mechanics*, *23*(2), 222–244. doi:10.1177/1056789513489322

Gupta, V., Yuan, J., & Martinez, D. (1993). Calculation, measurement, and control of interface strength in composites. *Journal of the American Ceramic Society*, *76*(2), 305–315. doi:10.1111/j.1151-2916.1993.tb03784.x

Hashin, Z. (1980). Failure criteria for unidirectional fiber composites. *Journal of Applied Mechanics*, *47*(2), 329–334. doi:10.1115/1.3153664

Hashin, Z. (1983). Analysis of Composite Materials-A Survey. *Journal of Applied Mechanics*, *50*(3), 481–505. doi:10.1115/1.3167081

Hashin, Z., & Shtrikman, S. (1963). A variational approach to the theory of the elastic behaviour of multiphase materials. *Journal of the Mechanics and Physics of Solids*, *11*(2), 127–140. doi:10.1016/0022-5096(63)90060-7

Hayakawa, K., Murakami, S., & Liu, Y. (1998). An irreversible thermodynamics theory for elastic-plastic-damage materials. *European Journal of Mechanics. A, Solids*, *17*(1), 13–32. doi:10.1016/S0997-7538(98)80061-2

Hill, R. (1948). A Theory of the yielding and plastic flow of anisotropic metals. *Proceedings - Royal Society. Mathematical, Physical and Engineering Sciences*, *193*(1033), 281–297. doi:10.1098/rspa.1948.0045

Hill, R. (1963). Elastic properties of reinforced solids: Some theoretical principles. *Journal of the Mechanics and Physics of Solids*, *11*(5), 357–372. doi:10.1016/0022-5096(63)90036-X

Hinton, M. J., Kaddour, A. S., & Soden, P. D. (2004). A further assessment of the predictive capabilities of current failure theories for composite laminates: Comparison with experimental evidence. *Composites Science and Technology*, *64*(3–4), 549–588. doi:10.1016/S0266-3538(03)00227-6

Hirvonen, J. P., Torri, P., Lappalainen, R., Likonen, J., Kung, H., Jervis, J. R., & Nastasi, M. (1998). Oxidation of MoSi2/SiC nanolayered composite. *Journal of Materials Research*, *13*(04), 965–973. doi:10.1557/JMR.1998.0135

Jiménez-Piqué, E., Ceseracciu, L., Chalvet, F., Anglada, M., & de Portu, G. (2005). Hertzian contact fatigue on alumina/alumina-zirconia laminated composites. *Journal of the European Ceramic Society*, *25*(15), 3393–3401. doi:10.1016/j.jeurceramsoc.2004.09.011

Kachanov, L. (1999). Rupture time under creep conditions. *International Journal of Fracture*, *97*(1-4), 11-18.

Kendall, K. (1975). Transition between cohesive and interfacial failure in a laminate. *Proceedings of the Royal Society of London. Series A, Mathematical and Physical Sciences*, *344*(1637), 287–302. doi:10.1098/rspa.1975.0102

Kim, D. K., & Kriven, W. M. (2003). Mullite–Aluminum phosphate laminated composite fabricated by tape casting. *Journal of the American Ceramic Society*, *86*(11), 1962–1964. doi:10.1111/j.1151-2916.2003.tb03590.x

Lee, H., Peng, K., & Wang, J. (1985). An anisotropic damage criterion for deformation instability and its application to forming limit analysis of metal plates. *Engineering Fracture Mechanics*, *21*(5), 1031–1054. doi:10.1016/0013-7944(85)90008-6

Lee, W., Howard, S. J., & Clegg, W. J. (1996). Growth of interface defects and its effect on crack deflection and toughening criteria. *Acta Materialia*, *44*(10), 3905–3922. doi:10.1016/S1359-6454(96)00068-7

Lemaitre, J. (1971). Evaluation of dissipation and damage in metals submitted to dynamic loading. *Proceedings of the ICM-1*.

Lemaitre, J. (1996). A course on damage mechanics (2nd ed.). Berlin: Springer.

Li, S., & Ghosh, S. (2006). Extended Voronoi cell finite element model for multiple cohesive crack propagation in brittle materials. *International Journal for Numerical Methods in Engineering*, *65*(7), 1028–1067. doi:10.1002/nme.1472

Linero Segrera, D. L., Oliver, J., & Huespe, A. E. (2010). Numerical modeling of the fracture process in reinforced concrete by means of the continuum strong discontinuity approach. Part I: Formulation. *Ingeniería e Investigación*, *30*(2), 5–15.

Liu, H., & Hsu, S. M. (1996). Fracture behavior of multilayer silicon nitride/boron nitride ceramics. *Journal of the American Ceramic Society*, *79*(9), 2452–2457. doi:10.1111/j.1151-2916.1996.tb08996.x

Mammoli, A. A., Graham, A. L., Reimanis, I. E., & Tullock, D. L. (1995). The effect of flaws on the propagation of cracks at bi-materials inferfaces. *Acta Metallurgica et Materialia*, *43*(3), 1149–1156. doi:10.1016/0956-7151(94)00308-5

Marat-Mendes, R. M., & Freitas, M. M. (2010). Failure criteria for mixed mode delamination in glass fibre epoxy composites. *Composite Structures*, *92*(9), 2292–2298. doi:10.1016/j.compstruct.2009.07.017

Martínez, D., & Gupta, V. (1994). Energy criterion for crack deflection at an interface between two orthotropic media. *Journal of the Mechanics and Physics of Solids*, *42*(8), 1247–1271. doi:10.1016/0022-5096(94)90034-5

Matzenmiller, A., Lubliner, J., & Taylor, R. L. (1995). A constitutive model for anisotropic damage in fiber-composites. *Mechanics of Materials*, *20*(2), 125–152. doi:10.1016/0167-6636(94)00053-0

Mawdsley, J. R., Kovar, D., & Halloran, J. W. (2000). Fracture behavior of alumina/monazite multilayer laminates. *Journal of the American Ceramic Society*, *83*(4), 802–808. doi:10.1111/j.1151-2916.2000.tb01278.x

Melenk, J. M., & Babuska, I. (1996). The partition of unity finite element method: Basic theory and applications. *Computer Methods in Applied Mechanics and Engineering*, *139*(1–4), 289–314. doi:10.1016/S0045-7825(96)01087-0

Ming Yuan, H., Evans, A. G., & Hutchinson, J. W. (1994). Crack deflection at an interface between dissimilar elastic materials: Role of residual stresses. *International Journal of Solids and Structures*, *31*(24), 3443–3455. doi:10.1016/0020-7683(94)90025-6

Ming-Yuan, H., & Hutchinson, J. W. (1989). Crack deflection at an interface between dissimilar elastic materials. *International Journal of Solids and Structures*, *25*(9), 1053–1067. doi:10.1016/0020-7683(89)90021-8

Moës, N., Dolbow, J., & Belytschko, T. (1999). A finite element method for crack growth without remeshing. *International Journal for Numerical Methods in Engineering*, *46*(1), 131–150. doi:10.1002/(SICI)1097-0207(19990910)46:1<131::AID-NME726>3.0.CO;2-J

Moorthy, S., & Ghosh, S. (1996). A model for analysis of arbitrary composite and porous microstructures with voronoi cell finite elements. *International Journal for Numerical Methods in Engineering*, *39*(14), 2363–2398. doi:10.1002/(SICI)1097-0207(19960730)39:14<2363::AID-NME958>3.0.CO;2-D

Moorthy, S., & Ghosh, S. (2000). Adaptivity and convergence in the Voronoi cell finite element model for analyzing heterogeneous materials. *Computer Methods in Applied Mechanics and Engineering*, *185*(1), 37–74. doi:10.1016/S0045-7825(99)00349-7

Mori, T., & Tanaka, K. (1973). Average stress in matrix and average elastic energy of materials with misfitting inclusions. *Acta Metallurgica*, *21*(5), 571–574. doi:10.1016/0001-6160(73)90064-3

Murari, V., & Upadhyay, C. S. (2012). Micromechanics based ply level material degradation model for unidirectional composites. *Composite Structures*, *94*(2), 671–680. doi:10.1016/j.compstruct.2011.09.001

Murari, V., & Upadhyay, C. S. (2013). Micromechanics based diffuse damage model for unidirectional composites. *Composite Structures*, *96*, 419–432. doi:10.1016/j.compstruct.2012.09.008

Nasser, S., & Hori, M. (1993). *Micromechanics: overall properties of heterogeneous materials*. Amsterdam: North-Holland.

Oliver, J. (2000). On the discrete constitutive models induced by strong discontinuity kinematics and continuum constitutive equations. *International Journal of Solids and Structures*, *37*(48–50), 7207–7229. doi:10.1016/S0020-7683(00)00196-7

Oliver, J., Huespe, A. E., Pulido, M. D. G., & Chaves, E. (2002). From continuum mechanics to fracture mechanics: The strong discontinuity approach. *Engineering Fracture Mechanics*, *69*(2), 113–136. doi:10.1016/S0013-7944(01)00060-1

Park, D. S., Roh, T. W., Han, B. D., Kim, H. D., & Park, C. (2002). Microstructure and mechanical properties of silicon nitride with tri-laminate structures. *Journal of the European Ceramic Society*, *22*(4), 535–543. doi:10.1016/S0955-2219(01)00310-7

Park, H. C., Yang, T. Y., & Stevens, R. (2003). Fracture strength and damage resistance of slip cast Y-TZP/Ce-TZP layered composites. *Journal of Materials Science*, *38*(3), 561–564. doi:10.1023/A:1021849909830

Patel, B. P., & Gupta, A. K. (2014). An investigation on nonlocal continuum damage models for composite laminated panels. *Composites. Part B, Engineering*, *60*, 485–494. doi:10.1016/j.compositesb.2013.12.073

Patel, B. P., Shukla, K. K., & Nath, Y. (2005). Thermal postbuckling characteristics of laminated conical shells with temperature-dependent material properties. *AIAA Journal*, *43*(6), 1380–1388. doi:10.2514/1.13259

Phillipps, A. J., Clegg, W. J., & Clyne, T. W. (1993a). Fracture behaviour of ceramic laminates in bending—I. Modelling of crack propagation. *Acta Metallurgica et Materialia*, *41*(3), 805–817. doi:10.1016/0956-7151(93)90014-J

Phillipps, A. J., Clegg, W. J., & Clyne, T. W. (1993b). Fracture behaviour of ceramic laminates in bending—II. Comparison of model predictions with experimental data. *Acta Metallurgica et Materialia*, *41*(3), 819–827. doi:10.1016/0956-7151(93)90015-K

Phillipps, A. J., Clegg, W. J., & Clyne, T. W. (1993c). The correlation of interfacial and macroscopic toughness in SiC laminates. *Composites*, *24*(2), 166–176. doi:10.1016/0010-4361(93)90013-X

Pinho, S. T., Davila, C. G., Camanho, P. P., Iannucci, L., & Robinson, P. (2005). *Failure models and criteria for FRP under in-plane or three dimensional stress states including shear nonlinearity*. Technical Report NASA/TM-2005-213530. NASA Langley Research Center.

Pompidou, S., & Lamon, J. (2007). Analysis of crack deviation in ceramic matrix composites and multilayers based on the Cook and Gordon mechanism. *Composites Science and Technology*, *67*(10), 2052–2060. doi:10.1016/j.compscitech.2006.11.013

Puck, A., Kopp, J., & Knops, M. (2002). Guidelines for the determination of the parameters in Puck's action plane strength criterion. *Composites Science and Technology*, *62*(3), 371–378. doi:10.1016/S0266-3538(01)00202-0

Puck, A., & Schürmann, H. (1998). Failure analysis of FRP laminates by means of physically based phenomenological models. *Composites Science and Technology*, *58*(7), 1045–1067. doi:10.1016/S0266-3538(96)00140-6

Puck, A., & Schürmann, H. (2002). Failure analysis of FRP laminates by means of physically based phenomenological models. *Composites Science and Technology*, *62*(12–13), 1633–1662. doi:10.1016/S0266-3538(01)00208-1

Rabotnov, Y. N. (1969). *Creep problems in structural members*. Amsterdam: North-Holland Press.

Rajasekaran, S., & Murray, D. W. (1973). Incremental finite element matrices. *Journal of the Structural Division*, *99*(12), 2423–2438.

Rashid, M. M. (1998). The arbitrary local mesh replacement method: An alternative to remeshing for crack propagation analysis. *Computer Methods in Applied Mechanics and Engineering*, *154*(1–2), 133–150. doi:10.1016/S0045-7825(97)00068-6

Reddy, J. N. (2004). *Mechanics of laminated composite plates and shells: theory and analysis* (2nd ed.). Boca Raton, FL: CRC Press.

Reddy, J. N. (2006). *An introduction to the finite element method* (3rd ed.). New York: McGraw-Hill.

Requena, J., Moreno, R., & Moya, J. S. (1989). Alumina and alumina/zirconia multilayer composites obtained by slip casting. *Journal of the American Ceramic Society*, *72*(8), 1511–1513. doi:10.1111/j.1151-2916.1989.tb07692.x

Reuss, A. (1929). Berechnung der Fließgrenze von Mischkristallen auf Grund der Plastizitätsbedingung für Einkristalle. *ZAMM - Journal of Applied Mathematics and Mechanics. Zeitschrift für Angewandte Mathematik und Mechanik*, *9*(1), 49–58. doi:10.1002/zamm.19290090104

Robbins, D. H. Jr, Reddy, J. N., & Rostam-Abadi, F. (2005a). An efficient continuum damage model and its application to shear deformable laminated plates. *Mechanics of Advanced Materials and Structures*, *12*(6), 391–412. doi:10.1080/15376490500259202

Robbins, D. H. Jr, Reddy, J. N., & Rostam-Abadi, F. (2005b). Layerwise modeling of progressive damage in fiber-reinforced composite laminates. *International Journal of Mechanics and Materials in Design*, *2*(3-4), 165–182. doi:10.1007/s10999-006-9001-3

Sbaizero, O., & Lucchini, E. (1996). Influence of residual stresses on the mechanical properties of a layered ceramic composite. *Journal of the European Ceramic Society*, *16*(8), 813–818. doi:10.1016/0955-2219(96)00005-2

Sglavo, V. M., & Bertoldi, M. (2006). Design and production of ceramic laminates with high mechanical resistance and reliability. *Acta Materialia*, *54*(18), 4929–4937. doi:10.1016/j.actamat.2006.06.019

Shahani, A. R., & Amini Fasakhodi, M. R. (2009). Finite element analysis of dynamic crack propagation using remeshing technique. *Materials & Design*, *30*(4), 1032–1041. doi:10.1016/j.matdes.2008.06.049

Shigegaki, Y., Brito, M. E., Hirao, K., Toriyama, M., & Kanzaki, S. (1996). Processing of a novel multilayered silicon nitride. *Journal of the American Ceramic Society*, *79*(8), 2197–2200. doi:10.1111/j.1151-2916.1996.tb08959.x

Shobu, K., Watanabe, T., & Tsuji, K. (1989). Effects of Mo2B5 addition to MoSi2 ceramics. *Journal of the Ceramic Society of Japan*, *97*(10), 1311–1314. doi:10.2109/jcersj.97.1311

Sidoroff, F. (1981). Description of anisotropic damage application to elasticity.*Proceedings of the IUTAM Colloquium on Physical Nonlinearities in Structural Analysis*. doi:10.1007/978-3-642-81582-9_35

Singh, S. B., & Kumar, A. (1999). Postbuckling response and strength of laminates under combined in-plane loads. *Composites Science and Technology*, *59*(5), 727–736. doi:10.1016/S0266-3538(98)00125-0

Soden, P., Hinton, M., & Kaddour, A. (1998). A comparison of the predictive capabilities of current failure theories for composite laminates. *Composites Science and Technology*, *58*(7), 1225–1254. doi:10.1016/S0266-3538(98)00077-3

Tang, T. (2008). *Variational asymptotic micromechanics modeling of composite materials* (Ph.D. thesis, Utah State University, Logan, Utah). Retrieved from http://digitalcommons.usu.edu/cgi/viewcontent.cgi?article=1071&context=etd

Terada, K., Miura, T., & Kikuchi, N. (1997). Digital image-based modeling applied to the homogenization analysis of composite materials. *Computational Mechanics*, *20*(4), 331–346. doi:10.1007/s004660050255

Tian, Y. P., & Fu, Y. M. (2008). Elasto-plastic postbuckling of damaged orthotropic plates. *Applied Mathematics and Mechanics*, *29*(7), 841–853. doi:10.1007/s10483-008-0702-y

Tian, Y. P., Fu, Y. M., & Wang, Y. (2009a). Nonlinear dynamic response of piezoelectric elasto-plastic laminated plates with damage. *Composite Structures*, *88*(2), 169–178. doi:10.1016/j.compstruct.2008.02.019

Tian, Y. P., Fu, Y. M., & Wang, Y. (2009b). Nonlinear dynamic response and vibration active control of piezoelectric elasto-plastic laminated plates with damage. *Journal of Vibration and Control*, *15*(10), 1463–1492. doi:10.1177/1077546309103265

Timoshenko, S., & Gere, J. M. (1972). *Mechanics of materials*. New York: Van Nostrand Reinhold Co.

Tomaszewski, H., Weglarz, H., Boniecki, M., & Rećko, W. M. (2000). Effect of barrier layer thickness and composition on fracture toughness of layered zirconia/alumina composites. *Journal of Materials Science*, *35*(16), 4165–4176. doi:10.1023/A:1004823429336

Tomaszewski, H., Węglarz, H., Wajler, A., Boniecki, M., & Kalinski, D. (2007). Multilayer ceramic composites with high failure resistance. *Journal of the European Ceramic Society*, *27*(2–3), 1373–1377. doi:10.1016/j.jeurceramsoc.2006.04.030

Tsai, S. W., & Hahn, H. T. (1980). *Introduction to composite materials*. Westport, CT: Technomic Publishing.

Tsai, S. W., & Wu, E. M. (1971). A general theory of strength for anisotropic materials. *Journal of Composite Materials*, *5*(1), 58–80. doi:10.1177/002199837100500106

Tuffé, S., & Wilkinson, D. S. (1995). MoSi2-based sandwich composite made by tape casting. *Journal of the American Ceramic Society*, *78*(11), 2967–2972. doi:10.1111/j.1151-2916.1995.tb09071.x

Tullock, D. L., Reimanis, I. E., Graham, A. L., & Petrovic, J. J. (1994). Deflection and penetration of cracks at an interface between two dissimilar materials. *Acta Metallurgica et Materialia*, *42*(9), 3245–3252. doi:10.1016/0956-7151(94)90422-7

Vandeperre, L. J., Kristofferson, A., Carlström, E., & Clegg, W. J. (2001). Thermal shock of layered ceramic structures with crack-deflecting interfaces. *Journal of the American Ceramic Society*, *84*(1), 104–110. doi:10.1111/j.1151-2916.2001.tb00615.x

Voigt, W. (1889). Uber die Beziehung zwischen den beiden Elastizitätskonstanten isotroper Körper. *Wiedmanns Annalen Physik and Chemie*, *38*(12), 573–587. doi:10.1002/andp.18892741206

Wu, E. M. (1972). Optimal experimental measurements of anisotropic failure tensors. *Journal of Composite Materials*, *6*(4), 472–489. doi:10.1177/002199837200600404

Yu, W., & Tang, T. (2007a). A variational asymptotic micromechanics model for predicting thermoelastic properties of heterogeneous materials. *International Journal of Solids and Structures*, *44*(22–23), 7510–7525. doi:10.1016/j.ijsolstr.2007.04.026

Yu, W., & Tang, T. (2007b). Variational asymptotic method for unit cell homogenization of periodically heterogeneous materials. *International Journal of Solids and Structures*, *44*(11–12), 3738–3755. doi:10.1016/j.ijsolstr.2006.10.020

Zhang, G. J., Yue, X. M., & Watanabe, T. (1999). Al2O3/TiC/(MoSi2+Mo2B5) multilayer composites prepared by tape casting. *Journal of the European Ceramic Society*, *19*(12), 2111–2116. doi:10.1016/S0955-2219(99)00031-X

Zhang, W., & Cai, Y. (2010). *Continuum damage mechanics and numerical applications*. Berlin: Springer. doi:10.1007/978-3-642-04708-4

Chapter 6
Integrated Computational Materials Engineering for Determining the Set Points of Unit Operations for Production of a Steel Product Mix

Rishabh Shukla
TCS Research, India

Ravikiran Anapagaddi
TCS Research, India

Amarendra K. Singh
TCS Research, India

Janet K. Allen
University of Oklahoma, USA

Jitesh H. Panchal
Purdue University, USA

Farrokh Mistree
University of Oklahoma, USA

ABSTRACT

Manufacturing a steel product mix (bar, rod, sheet) involves a series of unit operations - primary steel making, secondary steel making (ladle refining and tundish operation), continuous casting, reheating, rolling and annealing. The properties of the final product depend significantly on how each unit operation is carried out. Each unit operation must be operated to meet the requirements of the subsequent operations. The requirements imposed on a particular unit operation are often conflicting and compromises must be made. Also, there is high degree of uncertainty in the operating parameters of each unit operation, which may lead to considerable deviations from the anticipated performance. To ensure that the final quality specifications of the product is not sacrificed and the customer requirements are met, it is essential to manage the conflict and uncertainty involved in each unit operation of the manufacturing process. In this chapter, we illustrate the use of compromise Decision Support Problem (cDSP) construct and ternary plots to overcome the challenges involved in one of the unit operations, namely, the tundish. The construct can be instantiated for other unit operations to cover the entire manufacturing cycle. Exploring the effects of system variables for each process step through experiments and plant trials is time consuming and very costly. The proposed method allows for faster design exploration of the process and thereby provides a reduced search space to a process designer. The process designer, with

DOI: 10.4018/978-1-5225-0290-6.ch006

reduced experimentation requirements, can explore the narrowed search space to find the operating set points for a tundish. This, in turn, reduces the time and cost involved in production of a steel product mix with a new grade of steel in industry.

1. FRAME OF REFERENCE

Consider the following scenario: A steel mill day-in-day-out is involved in production of steel product mix using a particular grade of steel. Due to a change in customer needs, the steel mill has to produce the same product mix with a different grade of steel. To be able to switch the production, the personnel at the steel mill will need to spend a lot of time and money in carrying out experiments and plant trials, to determine the operating set points for each unit operation of manufacturing process chain. In this chapter we propose a method for simulating the behaviour of the unit operation, explore the solution space and determine the operating set points of a unit operation. The method is demonstrated for the tundish.

Manufacturing a steel product mix involves series of unit operations - ladle refining, tundish operation, continuous casting, rolling and annealing. The sequence and flow of information from one unit operation to another is shown in Figure 1. Industry continues to look for ways to reduce development costs and time required when a different grade of steel is to be used for manufacturing a new product mix. To meet this need, we suggest that integrated modelling of the entire manufacturing chain is required (Singh, Pardeshi, & Goyal, 2011). Integration is critical as the processes are linked and the outcome of one unit operation influences the way in which other operation needs to be carried out and there may be alternative paths which yield the same end-product. However before integration, it is necessary to develop models to account for uncertainty and essential trade-offs for each of the constituent processes.

Each unit performs a specified task and meets the requirements of subsequent unit operation. Typically, there is some conflict in the requirements for each unit that constitutes the manufacturing process chain. Further, all the processes are transient and parameters are dynamically adjusted throughout the operation giving rise to significant levels of uncertainty. Typically, in an industrial setting it is difficult to obtain information about batch sizes, exact temperature, etc. As the properties of the end-product are significantly affected by the way each unit operation is carried out, it becomes essential to manage the uncertainty and conflict involved in each unit operation of the manufacturing chain.

The horizontal integration of series of unit operations comes under the umbrella of Integrated Computational Materials Engineering (ICME) (Allison, Li, Wolverton, & Su, 2006). The ICME construct bridges the gap between design and manufacturing by facilitating the horizontal integration of unit operations and the vertical integration of models at different length scales. Several groups have used ICME to enable faster development of materials and manufacturing routes for them. Luo (2013) used computational thermodynamics and CALPHAD modeling in the selection and development of new magnesium alloys. Smith et al. (2013) used the ICME approach for predicting pitting corrosion. Their integrated approach uses thermodynamic modelling to identify the chemistry of the corrosive environment, atomistic modelling to understand the kinetics of corrosion reactions, and experimentation to validate the model predictions. Maiwald-Immer et al. (2013) demonstrated the use of ICME to enable the development of rapid manufacturing technologies by using simulation techniques to understand the effects of process parameters on the properties and quality of the final product. Agarwal and Shivpuri (2013) developed a hybrid computational framework for the integrated process design of high performance parts. They integrated materials and process design to predict the process design parameters for maximizing

Figure 1. Steel product manufacturing chain

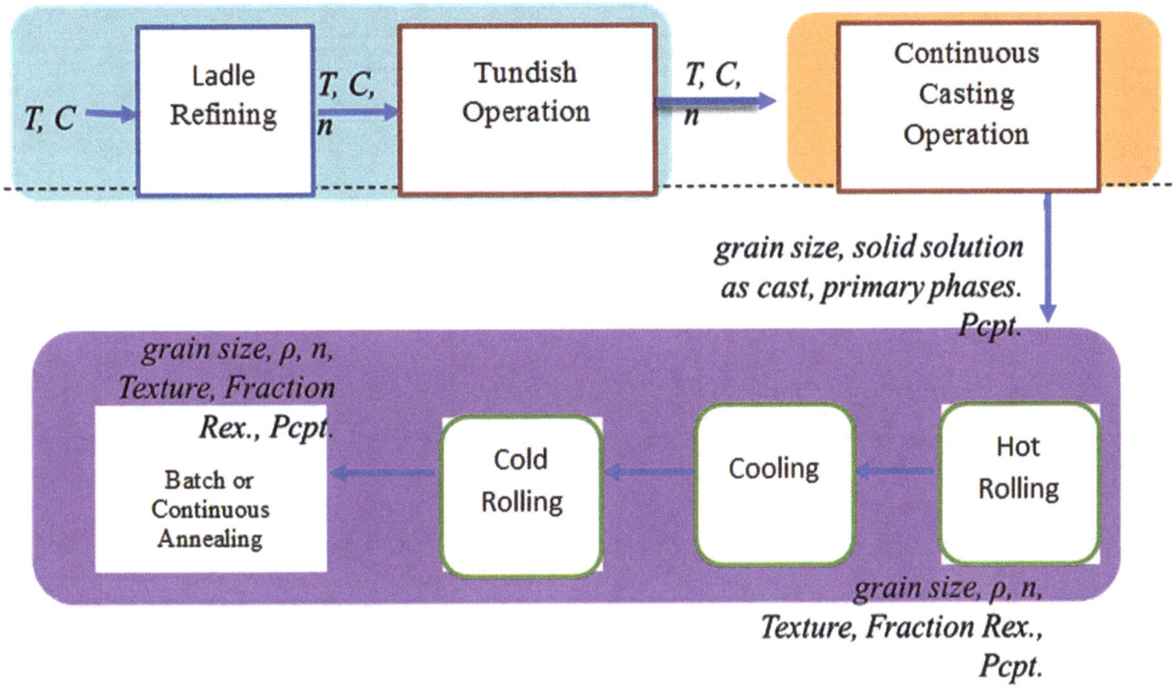

inclusion sensitive life of an aeroengine disk. An industrial application of ICME with computer aided manufacturing of Virtual Aluminum Castings, bridging the gap between manufacturing and design steps and enabling faster development of cast aluminum components have been demonstrated (Allison et al., 2006). A majority of the papers that embody the ICME construct focus on the design of material and *not* the integrated realization of the material and product. In comparison there is relatively little published on ICME for the integrated realization of the material and the product (Sinha, 2013; McDowell, 2010).

In this chapter, we propose a method to enable rapid and cost effective development of steel products for a new grade of steel with reduced experimentation. Our method facilitates the vertical integration of models at different length scales and horizontal integration of different unit operations to pass information from one unit operation to the subsequent one. For example, coupling of fluid flow with inclusion evolution model to find the inclusion distribution in the steel melt within tundish is an instance of vertical integration, used in the proposed method. Flow of this information to the subsequent unit operation (which is continuous casting) for prediction of spatial inclusion distribution of inclusions in a cast slab, is an example of horizontal integration. The horizontal integration of unit operations and vertical integration of models at different length scales is carried out using a platform, namely, Platform for Realization of Engineered Materials and Products (PREMAP) (Allen, 2013; Gautham, 2013).

Our focus in this chapter is to demonstrate our method for managing the uncertainty and conflicts involved in one unit operation, namely, the tundish and then instantiate the same for other unit operations. A tundish is a vessel which is inserted between the ladle and caster. The tundish serves as a bridge between the batch process of ladle refining and the casting operation. Schematic of a tundish is shown in Figure 2; key parameters of a typical tundish are shown in the figure.

Figure 2. Schematic of Tundish (a) bottom (b) side view

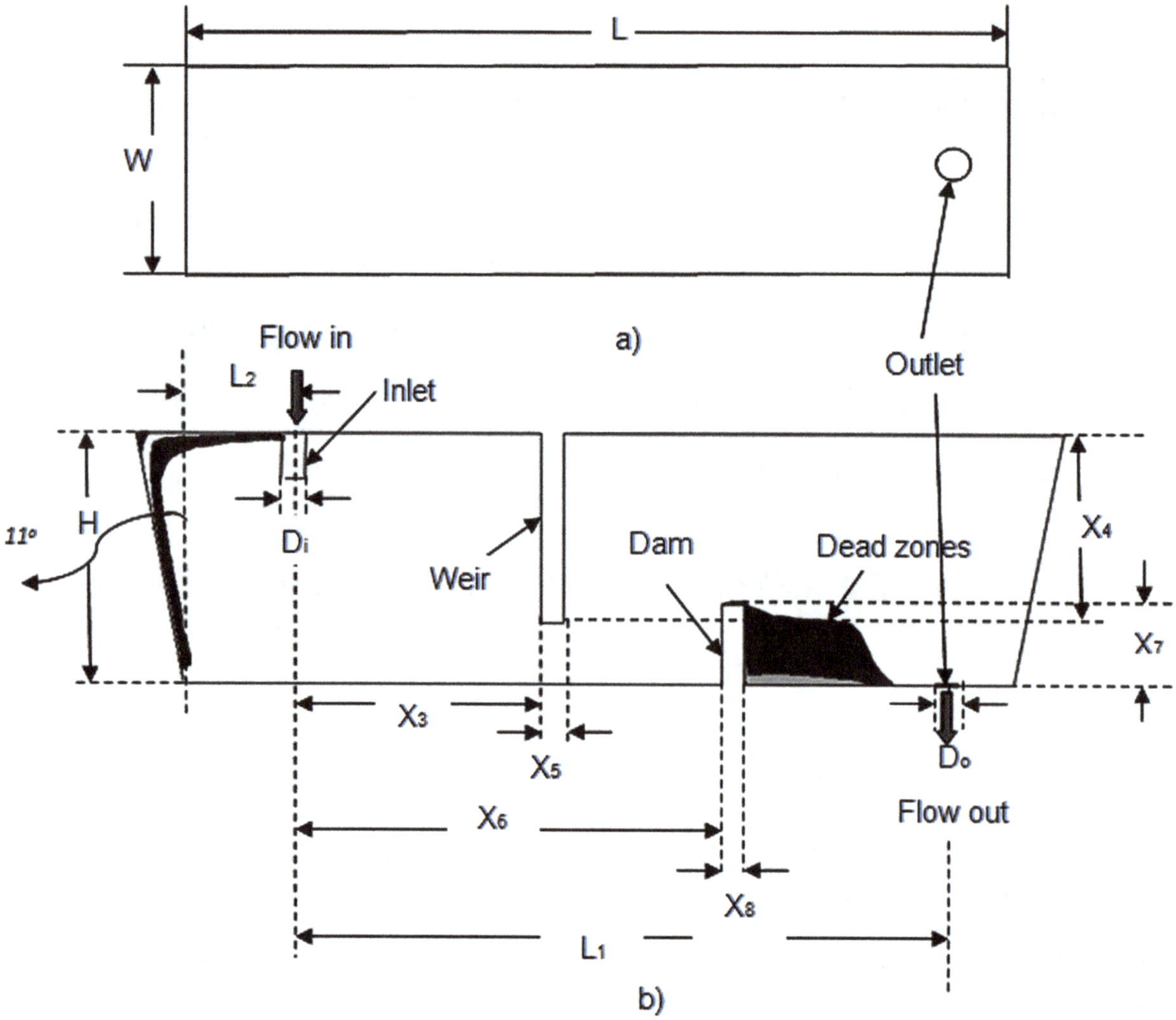

Steel is tapped from ladle furnace to the tundish via the inlet; it spends sometime within the tundish, and subsequently flows out into the mold through the outlet. The flow profile of the steel within tundish governs the efficiency in performing the intended functions. Flow modifiers such as dam and weir (shown in Figure 2) are used to modify and get suitable flow characteristics of the steel melt. The area shown in black represents dead zone volume region in the tundish.

Over the years, with increasing emphasis on the cleanliness and quality of the steel product, the functions to be performed by the tundish have increased. The purpose of the tundish is to maintain chemical and thermal homogeneity of steel and facilitate inclusion removal. To maintain homogeneity, the dead zone volume within tundish must be as low as possible. But, low dead zone volume results in a low residence time of steel within tundish, thereby resulting in lower inclusion removal efficiency. This implies that the two requirements imposed on the tundish are conflicting in nature. Also, temperature, flow rate and inclusion distribution of the incoming steel from ladle to tundish varies with respect to time. The process therefore needs to be dynamically adjusted and the operating parameters are changed throughout the process. Managing the uncertainty in the operating parameters is critical so as to ensure that the deviation from the achievement of output goal is small under all circumstances. In this chapter, we build on what we present in our previous work by providing additional information for the development of the model and the use of ternary plots to determine the operating set points (which includes design parameters of flow modifiers and temperature and flow rate of incoming steel) for the unit operation (Anapagaddi et al., 2014).

2. PROPOSED METHOD

Our aim is to explore the design space of tundish operation to account for conflicting requirements and variability involved in the process. In Figure 3 we illustrate our method.

Detailed Computational Fluid Dynamics (CFD) based mathematical models are used to simulate various phenomena involved in the tundish operation. The detailed models are exercised to generate data for development of Response Surface Models (RSM). The RSMs are used to instantiate the compromise Decision Support Problem (cDSP) described in Section 2.6. The cDSP is exercised to define the design space (feasible range of operating set points) for the tundish operation. Ternary plots that embody ranges of feasible weights for each goal provide a means to determine the operating set points that are most appropriate.

2.1 Mathematical Models for Calculating Inclusion Removal Efficiency and Dead Zone Volume

In this section we provide a brief explanation of the mathematical models used for predicting mixed, plug and dead zone volumes and inclusion distributions within the tundish. The incompressible mass conservation equation, Navier-Stokes equation for fluid flow and the Fourier equation for heat transfer are solved using the k-ε model of turbulence to obtain the steady state three-dimensional velocity fields and the temperature profile within the tundish (Llegbusi, 1989; Zhang, 2006). The governing equations are taken from references (Mazumdar, 1999; Zha, 2001) and are given below:

Continuity equation: $\frac{\partial u_i}{\partial x_i} = 0$ (1)

Momentum equation:

$$\rho u_j \frac{\partial u_i}{\partial x_j} = -\frac{\partial p}{\partial x_i} + \frac{\partial}{\partial x_j}\left[\mu_{eff}\left\{\frac{\partial u_i}{\partial x_j} + \frac{\partial u_j}{\partial x_i}\right\}\right] + g\left(1 - \beta\left(T_o - T\right)\right) \quad (2)$$

Turbulent viscosity:

$$\mu_t = \rho C_\mu \frac{k^2}{\varepsilon} \quad (3)$$

Transport equations for k and ε :

$$\frac{\partial\left(\rho u_j k\right)}{\partial x_j} = \frac{\partial}{\partial x_j}\left[\frac{\mu_{eff}}{\sigma_k} \times \frac{\partial k}{\partial x_j}\right] + G - \rho k \quad (4)$$

$$\frac{\partial\left(\rho u_j \varepsilon\right)}{\partial x_j} = \frac{\partial}{\partial x_j}\left[\frac{\mu_{eff}}{\sigma_\varepsilon} \times \frac{\partial \varepsilon}{\partial x_j}\right] + \frac{\left(C_1 G\varepsilon - C_2 \rho \varepsilon^2\right)}{k} \quad (5)$$

where,

$$G = \mu_t \frac{\partial \mu_j}{\partial x_i}\left[\frac{\partial \mu_i}{\partial x_j} + \frac{\partial \mu_j}{\partial x_i}\right] \quad (6)$$

and,

$$C_1 = 1.44, C_2 = 1.92, \sigma_k = 1, \sigma_\varepsilon = 1.3$$

The heat equation:

$$\frac{\partial\left(\rho u_i h\right)}{\partial x_i} = -\frac{\partial}{\partial x_i}\left[k_{eff} \times \frac{\partial T}{\partial x_i}\right] \quad (7)$$

Effective thermal conductivity:

$$k_{eff} = k_o + \frac{C_p \mu_t}{\Pr_t} \quad (8)$$

The flow fields and temperature profile, obtained solving equations 1 through 8 using a commercial software package Ansys, are then used to carry out RTD analysis and predict inclusion evolution. Brief descriptions of the models for RTD analysis and inclusion evolution are given in Section 2.2 and 2.3 respectively.

2.2 Residence Time Distribution (RTD) Analysis

The transient transport equation for a tracer (scalar), given in equation 9, is solved to predict the time vs. concentration profile at the outlet of tundish. This information is then used to compute the concentration curve, which in turn is used to calculate the dead, mixed and plug zone volume inside the tundish. The calculations of dead, mixed and plug zone volumes are important for quantification of flow characteristics of steel melt within the tundish. Additional details of the method used for predicting the flow characteristics within the tundish are provided in the literature (Kumar, 2008; Damle, 1995).

$$\frac{\partial\left(\rho C\right)}{\partial t} + \frac{\partial}{\partial x_i}\left(\rho \mu_i C\right) = \frac{\partial}{\partial x_i}\left[\rho D_{eff} \frac{\partial C}{\partial x_i}\right] \quad (9)$$

where,

$$D_{eff} = D + \frac{\mu_{eff}}{\rho Sc} \tag{10}$$

2.3 Inclusion Evolution Model

An inclusion evolution model is used to predict changes in inclusion size distribution in a uniform volume of steel with respect to time. The behavior of inclusions within tundish and their subsequent evolution have been evaluated using the Population Balance Method (PBM) method (Cwudzinski, 2012; Miki, 1999). The behavior of inclusion within the tundish has been studied considering different mechanisms of collisions such as Brownian, turbulent, Stokes collision, floatation and adhesion. The mathematical equations representing inclusion number density changes by different mechanisms of inclusion removal is taken from above cited literature and is summarized:

Floatation:

$$\frac{dn(r)}{dt} = -\frac{n(r)w_i}{H} \tag{11}$$

Stokes velocity:

$$w_i = \frac{2}{9} g r_i^2 \frac{\Delta\rho}{\mu} \tag{12}$$

Adhesion:

$$\frac{dn(r)}{dt} = -Mr^2 N(r) \tag{13}$$

Where,

$$M = \frac{0.62 \times 10^{-2} \varepsilon^{1.5}}{\upsilon^{1.25}} \frac{S}{V} \tag{14}$$

The rate constants for Brownian, turbulent and Stokes collision are provided in equation 15-17 respectively.

$$\beta_1(r_i, r_j) = \frac{2kT}{3\mu}\left\{\frac{1}{r_i} + \frac{1}{r_j}\right\}(r_i + r_j) \tag{15}$$

$$\beta_2(r_i, r_j) = 1.3(r_i + r_j)^3 \sqrt{\frac{\varepsilon}{\upsilon}} \tag{16}$$

$$\beta_3\left(r_i, r_j\right) = \frac{2g\Delta\rho}{9\mu}\left(\left|r_i - r_j\right|\right)\Pi\left(r_i + r_j\right)^3 \tag{17}$$

The number of collisions between two different size inclusions in a unit time and volume is expressed as:

$$N_{ij} = \beta\left(i, j\right)n\left(r_i\right)n\left(r_j\right) \tag{18}$$

where,

$$\beta\left(r_i, r_j\right) = \beta_1\left(r_i, r_j\right) + \beta_2\left(r_i, r_j\right) + \beta_3\left(r_i, r_j\right) \tag{19}$$

To calculate change in inclusion distribution, apart from the number of collisions between inclusions, the velocity and trajectory of inclusions traversing the liquid steel is required. This trajectory is obtained using a Lagrangian particle tracking method, where the transport equation for each inclusion during its traverse through the liquid steel, has been solved (Miki & Thomas, 1999). The velocity of inclusions traversing through the liquid steel is derived from a force balance, which includes drag (F_D) and buoyancy forces (F_g) relative to the steel:

$$F = m_p a_p = m_p \frac{du_p}{dt} = F_D - F_g = \frac{\Pi}{8} d_p^2 \rho C_D \left(u_p - u\right)^2 - \frac{\Pi}{6} d_p^3 g\left(\rho - \rho_p\right) \tag{20}$$

And, the velocity of inclusion traversing the liquid steel:

$$\frac{du_{pi}}{dt} = \frac{3}{4}\frac{1}{d_p}\frac{\rho}{\rho_p}\left(u_{pi} - u_i\right)^2 - \frac{\rho - \rho_p}{\rho_p} g_i \tag{21}$$

The overall change in inclusion distribution is represented by Equation 22, which is based on the conservation of mass for inclusions of each size range (5-75 micron).

$$\frac{dn\left(r\right)}{dt} = \frac{1}{2}\int_0^r n(r_i)\alpha\beta\left(r_i, r_j\right)n\left(r_j\right)\left(\frac{r}{r_j}\right)^2 dr_i - n\left(r\right)\int_0^{r_{\max}} n\left(r_j\right)\alpha\beta\left(r_i, r_j\right)dr_j - \frac{n(r)w_i}{H} - Mr^2 n(r) \tag{22}$$

where,

$$r^3 = r_i^3 + r_j^3$$

2.4 Model Validation

The coupled CFD-RTD and CFD-PBM model described above for predicting flow characteristics (Section 2.2) and inclusion distribution (Section 2.3) inside the tundish have been validated by comparison with the published literature. In Figure 4 we observe that the results predicted by the model are in close agreement with the reported findings in the published literature (Zhang, 2000; Ding, 2011).

Figure 4. Model validation (a) CFD-PBM Model and, (b) CFD-RTD Model

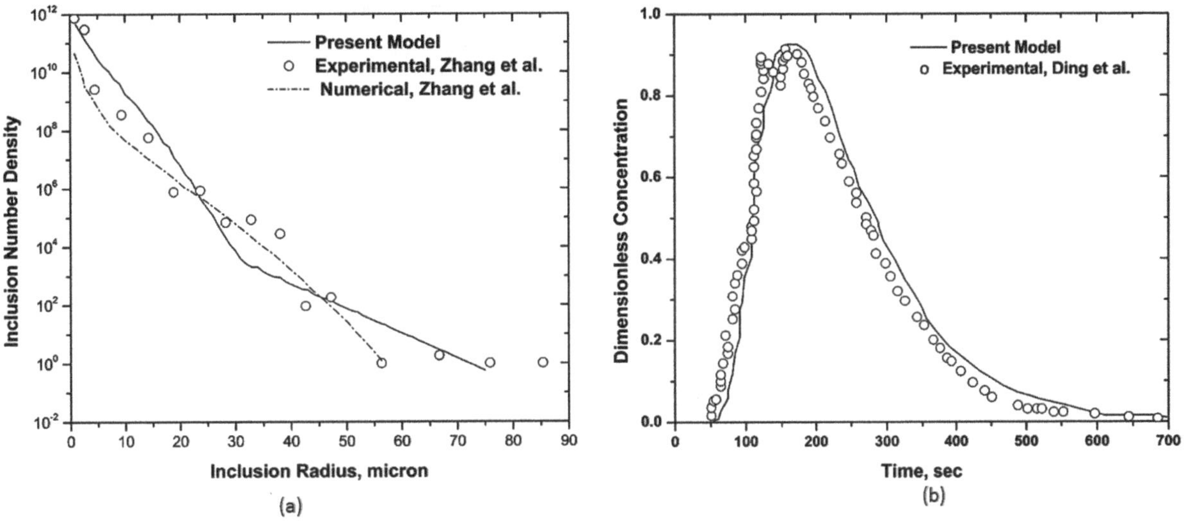

The detailed models described in the preceding sections are then used to develop the Response Surface Models (RSM), which is explained in the next section.

2.5 Response Surface Models (RSM)

Ideally, comprehensive mathematical models should be used for design exploration but these results in high computational time and costs. To reduce the time and costs, simplified models with known levels of accuracy are used to instantiate cDSP thereby facilitating design space exploration.

In Figure 3 the steps involved in developing a Response Surface Model are shown. Orthogonal arrays are used to determine a set of system variables to capture the effect of explanatory variables (system variables) on the responses. For each set of system variables in the orthogonal array, numerical simulations have been carried out using detailed models explained in Section 2.1-2.3. Data generated as a result of these numerical simulations are used to develop RSM using non-linear regression techniques (Myers, Montgomery, & Anderson-Cook, 2009).

The responses represented as a function of input system variables are outlet temperature of steel, characteristics of inclusion distribution and flow characteristics within the tundish. The corresponding equations are given in Table 1. The developed equations (in Table 1) are valid for a given range of system variables, which is shown in Table 2.

Although all RSMs have reasonable R^2 values, to determine the accuracy of the RSM we compare the results from the RSM with those using CFD-RTD and CFD-PBM model (explained in Sections 2.2 and 2.3). A comparison is shown in Figure 5. We note that the deviation in the prediction of output parameters by the two models (detailed model and RSM) is less than 5%, implying the accuracy of the RSM is greater than 95%, which means the developed RSM can be used for design exploration.

Figure 3. Proposed Method

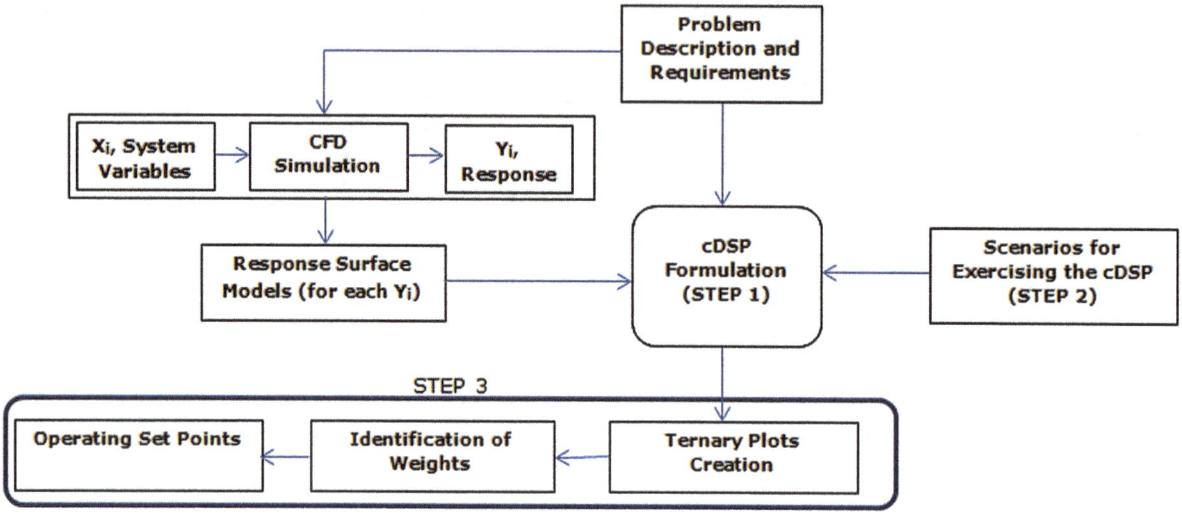

Table 1. Response surface models

		Response Surface Models
Superheat	ΔT ($R^2 = 0.98$)	$= 14.89 - 4619X_2 + 0.024X_3 + 2.53X_4 + 16.24X_5 + 0.25X_6 + 4.6X_7 + 8.3X_8 + 336988X_2^2 - 17.6X_4X_5 - 32.83X_7X_8 - 0.0015\frac{X_6 - X_3}{X_2}$
Characteristics of outlet inclusion number density	$\log(a_{out})$ ($R^2 = 0.994$)	$= -8.3 - 695.6X_2 + 0.59X_3 + 42.43X_4 - 0.15X_5 - 0.71X_6 - 4.7X_7 - 0.027X_8 + 2.04X_9 + 15.8X_{10} + 103000X_2^2 - 0.032X_9^2 - 6.85X_{10}^2 - \frac{0.72}{X_3} + \frac{31}{X_4} + \frac{1.9}{X_6} - \frac{0.36}{X_7} - 36.83\frac{X_2}{X_6 - X_3} + 0.0033\frac{X_6 - X_3}{X_2}$
Characteristics of outlet inclusion number density	b_{out} ($R^2 = 0.996$)	$= -0.19 - 59.63X_2 - 0.04X_3 + 0.21X_4 - 0.008X_5 - 0.0017X_6 + 0.23X_7 - 0.002X_8 + 0.0013X_9 + 1.2X_{10} + 4555X_2^2 - 2.4\times10^{-5}X_9^2 - 0.3X_{10}^2 - \frac{0.009}{X_3} + \frac{0.12}{X_4} - \frac{0.06}{X_6} + \frac{0.0145}{X_7} + 1.3\frac{X_2}{X_6 - X_3} - 6\times10^{-5}\frac{X_6 - X_3}{X_2}$
Dead Zone Volume (%)	V_d ($R^2 = 0.97$)	$= 54.05 - 1028.5X_2 - 13.13X_3 + 265.4X_5 + 13.13X_6 + 77.8X_8 - 21523X_2^2 - 332X_4X_5 - 247.35X_7X_8 - \frac{32.93}{X_4} + \frac{18.26}{X_6} - \frac{2.72}{X_7} - 0.017\frac{X_6 - X_3}{X_2}$
Mixed Zone Volume (%)	V_m ($R^2 = 0.99$)	$= 280.19 - 37547.5X_2 - 151.9X_3 - 96.85X_4 + 2.62X_5 + 89.7X_6 + 386X_7 - 13.1X_8 + 2157683X_2^2 - \frac{39.9}{X_3} - \frac{115.8}{X_4} - \frac{44.12}{X_6} + \frac{23.4}{X_7} + 4171.03\frac{X_2}{X_6 - X_3} - 0.235\frac{X_6 - X_3}{X_2}$
Plug Zone Volume (%)	V_p ($R^2 = 0.98$)	$= 431.96 + 46414.2X_2 + 186.7X_3 + 238.47X_4 + 9.08X_5 - 111.48X_6 - 506.6X_7 + 9.84X_8 - 2583719X_2^2 + \frac{41.73}{X_3} + \frac{229.8}{X_4} + \frac{60.61}{X_6} - \frac{30.85}{X_7} - 5104.9\frac{X_2}{X_6 - X_3} + 0.302\frac{X_6 - X_3}{X_2}$

Table 2. Range of system variables

Sr. No.	System Variables	Symbol	Range	
			Min	Max
1	Inlet Temperature, K	X_1	1843	1873
2	Inlet steel flow rate, m^3/s	X_2	2.83×10^{-3}	7.54×10^{-3}
3	Weir Position, m	X_3	0.65	1.44
4	Weir depth, m	X_4	0.8	0.9
5	Weir Thickness, m	X_5	0.1	0.2
6	Dam position, m	X_6	1.05	1.85
7	Dam height, m	X_7	0.2	0.35
8	Dam thickness, m	X_8	0.1	0.2
9	Logarithm of a_{in}	X_9	13.613	27.9645
10	b_{in}	X_{10}	0.18	0.65

Figure 5. RSM validation (a) inclusion removal efficiency, and (b) flow characteristics prediction

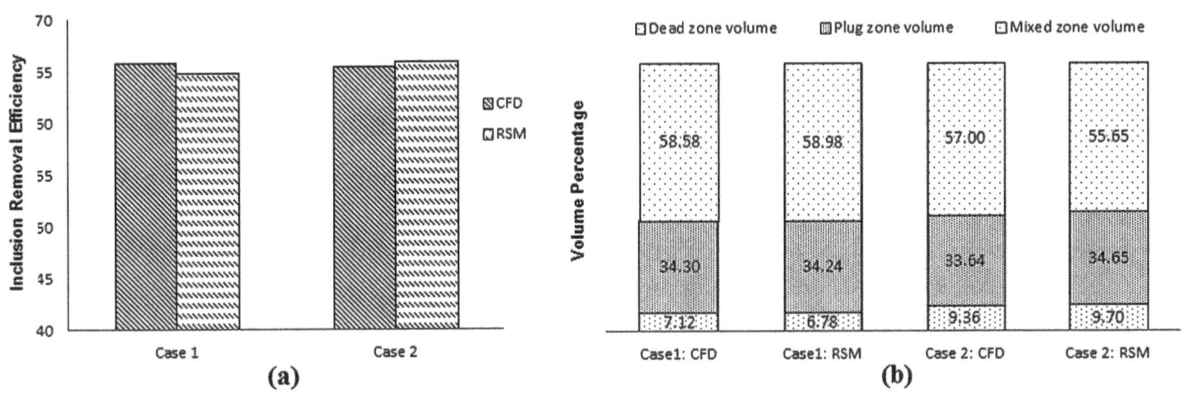

2.6 Compromise Decision Support Problem (cDSP) Formulation (Table 3, Step 1)

The cDSP is a hybrid formulation involving concepts from both traditional mathematical programming and goal programming. The cDSP construct has been used for a variety of engineering applications (Hernandez, 2001; Smith, 2014).

In Table 3 we show the values for the fixed parameters of tundish (refer to Figure 2). The other system variables (design parameters of the tundish flow modifiers and conditions of incoming steel) can vary within a specified range as given in Table 2. Our aim is to design a tundish that will minimize dead

Table 3. Fixed parameters of tundish

Parameter	Value
Tundish length (L)	3.7 m
Tundish width (W)	0.8 m
Tundish height (H)	1.1 m
Angle of tundish walls $(^{\circ})$	11
Inlet and outlet diameter (D_i, D_o)	0.1 m
Distance of inlet from wall (L_2)	0.5 m
Distance of inlet and outlet (L_1)	2.7 m

zone volume and maximize the inclusion removal efficiency. The cDSP for determining the set points for the operation of the tundish follows:

Given

- Tundish geometryis fixed and is given in Table 3.
- Liquidus temperature depends on the grade of steel, i.e., composition of the constituent elements and for this problem it is, $T_{liquidus} = 1805\ K$.
- $\eta_{Target} = 100$, $V_{d,\max} = 35$ *and* $\sigma^2_{V_{d,\max}} = 1 \times 10^4$.
- The ***RSMs***; see Section 2.5. When we exercise the cDSP, we do not use the detailed mathematical models presented in Sections 2.1-2.3, for calculation of the output parameters of the tundish operation. Reason being time and cost involved in computation is significantly high. Instead, RSMs (presented in Section 2.5) are used to compute the output parameters, i.e., goals of the formulation, for a particular set of system variables. The results predicted by RSMs are in close agreement with that of the detailed models (see Section 2.5)

Find

The system variables, which includes flow rate and temperature of incoming steel from ladle furnace and design parameters of flow modifiers, i.e., dam and weir (as shown in Figure 2).

- Position of dam (X_6) and weir (X_3) from inlet, m
- Thickness of dam (X_8) and weir (X_5), m
- Height of dam (X_7) and depth of weir (X_4), m
- Inlet flow rate of molten steel (X_2), m^3/sec
- Inlet temperature of molten steel (X_1), K

- The value of deviation variables which indicate the degree which a proposed solution varies from the target performance values: $d_i^+, d_i^-; i = 1, \ldots, 3$

where, d_i^+, d_i^-, are the deviation variables; d_i^+ is a measure of the over achievement and d_i^- is a measure of the under achievement in a particular goal. The aim is to get as small a value of deviation variable as possible, as means less deviation of the achieved value from the target value of the specified goal.

Satisfy

Process Design Constraints: Listed down are the constraints that need to be satisfied while carrying out the tundish operation. The constraints are taken based on industrial experiences or operating practices. The constraints on maximum and minimum superheat, dead zone, plug zone and mixed zone volume is required to ensure smooth casting operation.

Minimum Superheat Requirement (K): $\left(X_1 - \Delta T\right) - T_{liquidus} - 10 \geq 0$ (23)

Maximum Superheat Requirement (K): $40 + T_{liquidus} - \left(X_1 - \Delta T\right) \geq 0$ (24)

Minimum Plug Zone Volume (%): $V_p(X_i) - 5 \geq 0$ (25)

Maximum Plug Zone Volume (%): $35 - V_p(X_i) \geq 0$ (26)

Minimum Mixed Zone Volume (%): $V_m(X_i) - 35 \geq 0$ (27)

Maximum Mixed Zone Volume (%): $75 - V_m(X_i) \geq 0$ (28)

Difference between Mixed and Plug Zone Volume (%): $V_m(X_i) - V_p(X_i) \geq 0$ (29)

Difference between Mixed and Dead Zone Volume (%): $V_m(X_i) - V_d(X_i) \geq 0$ (30)

Difference between Plug and Dead Zone Volume (%): $V_p(X_i) - V_d(X_i) \geq 0$ (31)

(X_i), denotes that the parameter is a function of system variables, $i = 1, \ldots, 10$ (listed in Table 2).

Goals

Goal 1 $(G1)$: Maximize Inclusion Removal Efficiency

$$\frac{\eta(X_{i)}}{\eta_{Target}} + d_1^- - d_1^+ = 1; (here, d_1^+ = 0) \tag{32}$$

where $\eta(X_i)$ is calculated using RSMs developed for characteristics of inclusion number density, i.e., $\log(a_{out})$ and b_{out} (see Table 1).

Goal 2 $(G2)$: Minimize Dead Zone Volume

$$\frac{V_d(X_i)}{V_{d,\max}} + d_2^- - d_2^+ = 0; (here, d_1^- = 0, V_{d,Target} = 0) \tag{33}$$

Goal 3 $(G3)$: Minimize Variability in Dead Zone Volume

$$\frac{\sigma^2_{V_d}(X_i)}{\sigma^2_{V_{d,\max}}} + d_3^- - d_3^+ = 1; (here, d_3^- = 0, \sigma^2_{V_{d,Target}} = 0) \tag{34}$$

Where, $V_d(X_i)\, and\, \sigma^2_{V_d}(X_i)$ are calculated using RSM developed for dead zone volume (see Table 1).

Bounds

- *Bounds on Process System variables:*

  $0 \leq X_i \leq 1 : i = 1,...,10$ (X_i is the normalized value of system variables listed in Table 2)

- *Bounds on Deviation Variables:*

$$d_i^- . d_i^+ = 0; d_i^-, d_i^+ \geq 0 : i = 1,...,3$$

Minimize the Deviation Function

In the compromise DSP, the objective is to minimize the difference between the value which is desired and the value which is obtained. The Archimedean form of the deviation function follows:

$$Z = \sum_{i=1}^{m} W_i.(d_i^+ + d_i^-); \sum_{i=1}^{3} W_i = 1, W_i \geq 0 : i = 1,...,3 \tag{35}$$

Figure 6. Scenarios exercised

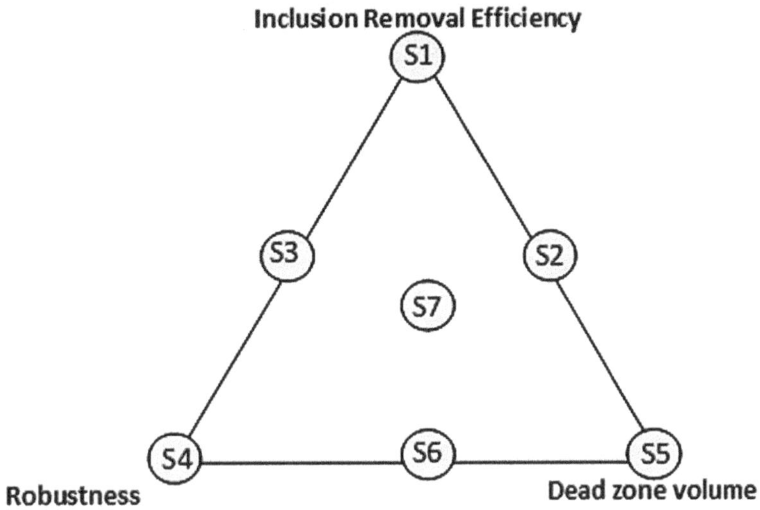

3. RESULTS AND DISCUSSION

The DSIDES software is used to solve the cDSP presented in Section 2.6 for many scenarios; the results are collated and used to create ternary plots (Reddy, 1992; Mistree, 1992).

3.1 Scenarios for Exercising the cDSP (Figure 6, Step 2)

The feasible solution space is determined by exercising the cDSP for various scenarios; see Figure 6. Scenarios shown in Figure 6 represent different cases where the emphasis on the different goals is varied. Different weights are assigned to goals in each scenario in order to cater to the requirements; see Table 4.

Scenario 1: Steel cleanliness is most critical and so the entire weight is assigned to the goal of inclusion removal efficiency.

Scenario 2: Represents a trade-off between on the goals of dead zone volume and inclusion removal efficiency.

Scenario 3: Represents a trade-off between on the goals of inclusion removal efficiency and robustness goal (minimization of variability in dead zone volume).

Scenario 4: The unit operation is carried out in the presence of variability in the input parameters. So, managing the variability of the operation is most critical, as a result of which, the entire weight is assigned to the robustness goal.

Scenario 5: Represents a situation where maintaining the thermal and chemical homogeneity of the steel melt is critical. The entire weight is assigned to goal on dead zone volume.

Scenario 6: Represents a trade-off between dead zone volume and robustness.

Scenario 7: Represents a case where equal importance has been given to all the three goals, namely, inclusion removal efficiency, dead zone volume and robustness (variability in dead zone volume) goal.

Table 4. Scenarios considered for process design exploration

Scenario	Weights Assigned to Goal 1 (G1) (Inclusion Removal)	Weights Assigned to Goal 2 (G2) (Dead Zone Volume)	Weights Assigned to Goal 3 (G3) (Variability in Dead Zone Volume)
S1	1	0	0
S2	0.5	0.5	0
S3	0.5	0	0.5
S4	0	0	1
S5	0	1	0
S6	0	0.5	0.5
S7	0.33	0.33	0.33

In Figure 7, we show the variation of deviation variables (Goal G2 – minimize the dead zone volume) with iterations for two scenarios, namely, Scenario 1 and 5.

In Figure 7 we observe that the deviation variable (d_2^+) for Scenario 1 fluctuates and value at the last iteration increases when compared with the first iteration value whereas the deviation variable (d_2^+) decreases for Scenario 5. We therefore conclude that dead zone volume is closer to the target value in Scenario 5than in Scenario 1. This behaviour is explained in the context of weights assigned to a spe-

Figure 7. Deviation variable of dead volume for Scenarios 1 and 5

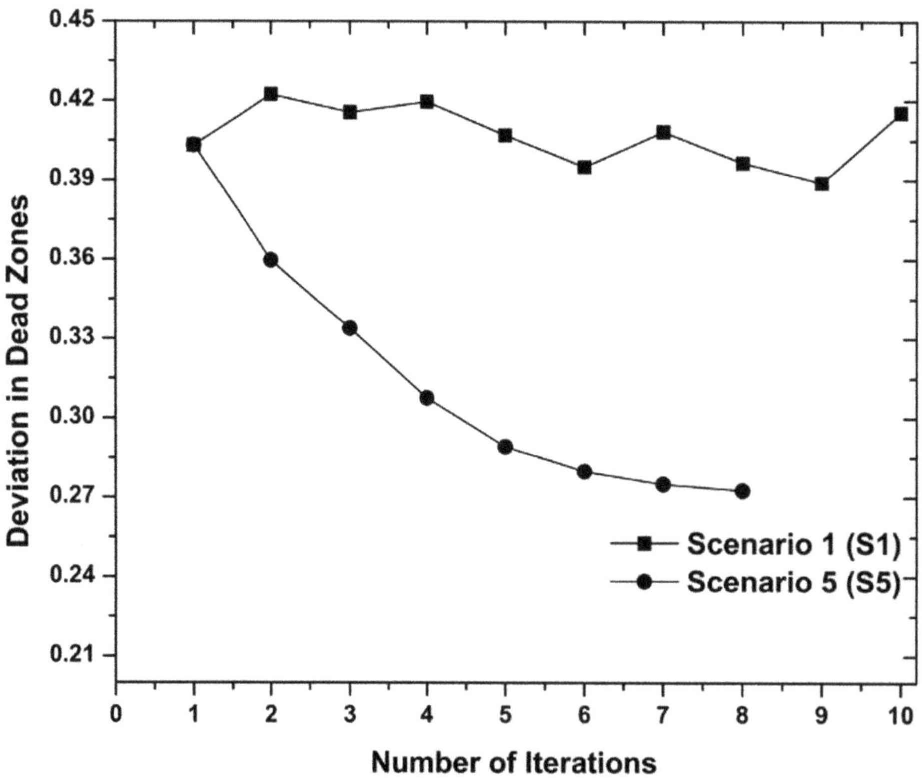

cific goal. A weight has been assigned to dead zone (Goal 2) in Scenario 5 while no weight has been assigned to it in Scenario 1; see Table 4. The effect of assigning weights to a goal for achievement of a target specification on a goal is qualitatively explained with the help of deviation variables variation in Figure 7. To show the effects of varying weights on the degree of achievement of goals quantitatively, the achieved value of goals for different weights assigned to goals are plotted. In Figure 8 we show the results for goals G2 (minimum dead zone volume) and G1 (maximum inclusion removal efficiency) for different scenarios. This variation in outcomes for different scenarios is a consequence of choosing different weights in the deviation function (Equation 35).

In summary, the outcomes shown in Figures 7 and 8 are the result of different weights being used in the deviation function. Hence, the question: *What weights should be used to determine the operating set points?* In the next section we demonstrate an approach, using ternary plots to determine quickly whether it is possible to select the operating set points from the available window of system variables to ensure minimum deviation from the target for each goal of the unit operation or not. If yes, to find out the operating set points that must be used for the unit operation.

3.2 Ternary Plots to Determine Operating Set Points Quickly (Table 5, Step 3)

Typically, exercising the cDSP for a limited number of scenarios is insufficient for completely understanding the interactions among goals or determining the weights to be assigned. The exploration of design space with the consideration of many scenarios, thus, becomes important. Our aim is to determine the operating set points that will lead to the achievement of all the goals as closely to the target value

Figure 8. Achieved values of goals for different scenarios

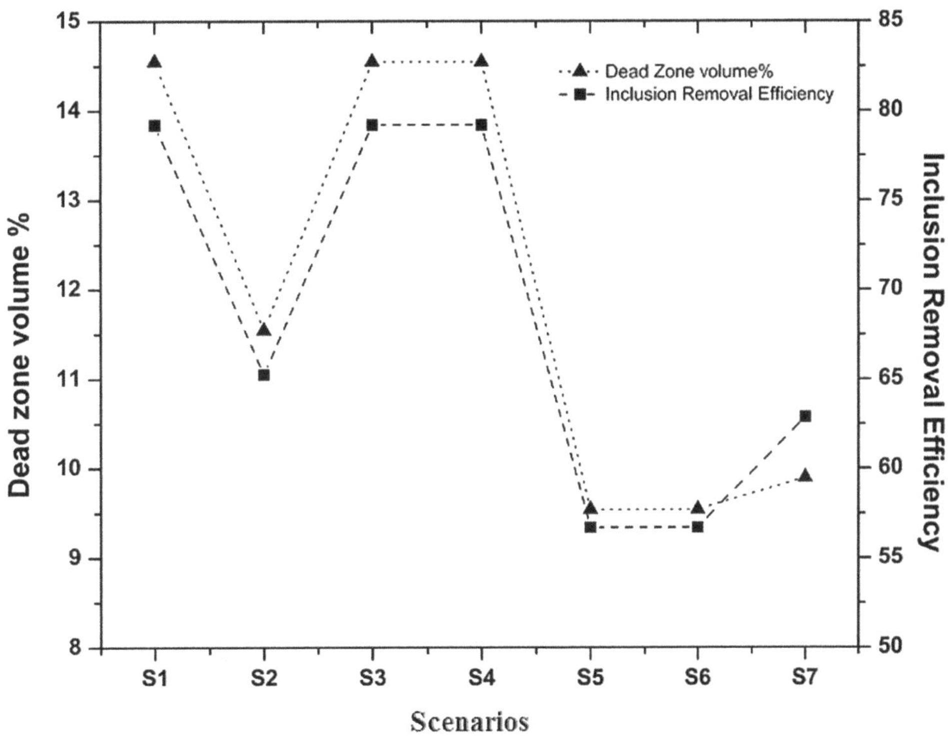

Table 5. Maximum and minimum achieved values of the goals

Goal	Units	Minimum Achieved Value	Maximum Achieved Value
$G1$	(%)	56.71	79.21
$G2$	(%)	9.54	14.55
$G3$	(%)	0.015	0.04734

(specified in Section 2.6) as possible. Meeting the exact target specifications is usually difficult, if not impossible. The cDSP is exercised for different scenarios for available window of system variables (see Table 2) to calculate maximum and minimum possible achieved values for each goal. Normalization of the achieved values over the design search space is done and is then used to plot the ternary plots. Ternary plots allow designers to visualize different possible process scenarios thereby facilitating quick decisions. The axes of the triangular ternary plot represent weights assigned to each goal and the colour inside the triangular region indicates the achieved value of the goal which is represented in that particular plot.

Ternary plots are constructed for the tundish operation. Three goals, namely inclusion removal efficiency (G1), dead zone volume (G2) and variability in dead zone volume or robustness (G3) have been considered. A number of scenarios, varying the weightages of each goal, have been solved using DSIDES. The maximum and minimum achieved values of each goal over these scenarios are reported in Table 5.

The values in Table 5 are used for normalization, and the normalized value of each goal has been used for creating the ternary plots. Ternary plots for inclusion removal efficiency, dead zone volume and robustness (variability in dead zone volume) are shown in Figure 9 (a), (b), and (c) respectively.

We have identified maximum and minimum achievable values of each goal over the design search space. Let us consider a situation where a process designer wants 80% of the maximum achievable value for inclusion removal efficiency and 20% of the maximum achievable value for dead zone volume and variability in dead zone volume is desired. In the latter case, lower value is desired because aim is to minimize these two goals during the tundish operation.

1. **Inclusion Removal Efficiency (G1):** The aim is to maximize the value of inclusion removal efficiency. The requirement put above by the process designer is to achieve a value either equal to more than 80 of the maximum achievable value. The region depicting the set of weights for attainment of the specified requirement on inclusion removal efficiency is shown in Figure 10 (region bounded by axis of plot and curve in the direction of specified arrow). If the designer assigns weights to the goals as per this region, an operator will end up meeting the imposed requirement on efficiency.
2. **Dead Zone Volume (G2):** As the aim of tundish is to minimize the value of dead volume, so we want a lower achieved value of dead volume. The requirement put by the process designer (explained above) is to achieve a value either equal to or lower than 20% of the maximum achievable value. The region depicting set of weights for attaining the specified requirement on dead zone volume is shown in Figure 11 (region bounded by axis of plot and curve in the direction of specified arrow). If the designer assigns weights to the goals in this region, an operator will end up meeting the imposed requirement on dead zone volume.

Figure 9. Ternary plots (a) inclusion removal efficiency (b) dead zone volume (c) variability in dead zone volume (robustness)

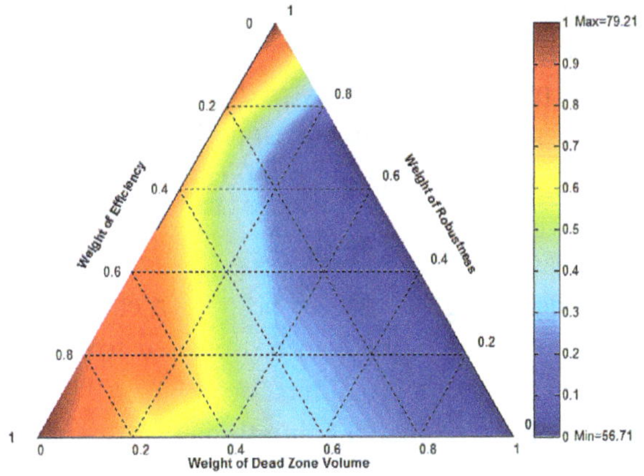

(a)

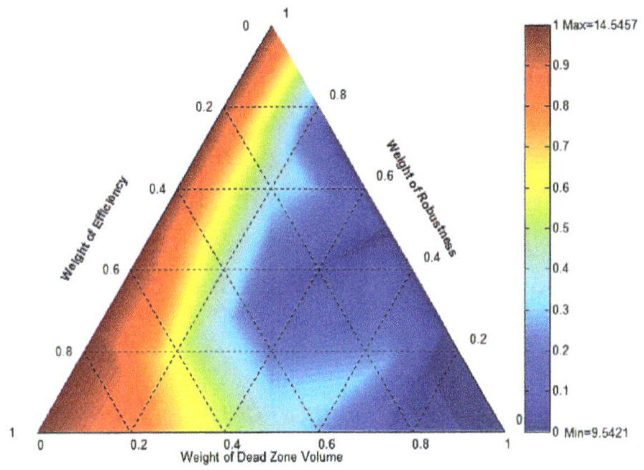

(b)

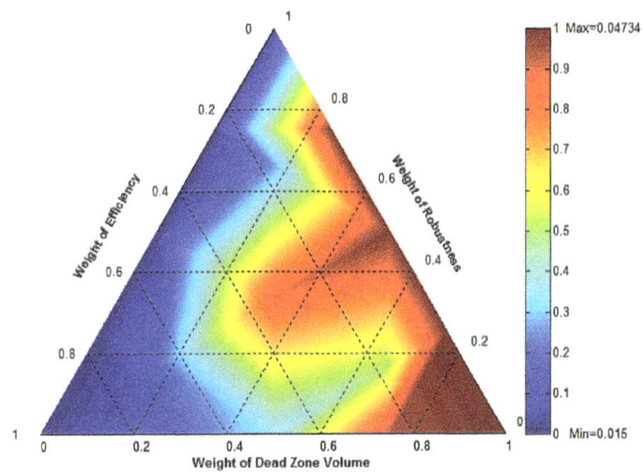

(c)

3. **For Robustness (G3):** The aim is to maximize the robustness of the process, i.e., minimize the variability in the dead zone volume with any change in the input parameters, so lower value of variability is desired. The process designer puts a requirement of achieving a value either equal to or lower than 20%of the maximum achievable value. The region depicting set of weights for attaining the specified requirement on variability in dead zone volume is shown in Figure 12 (region bounded by axis of plot and curve in the direction of specified arrow). If the designer assigns weights to the goals in this region, an operator will end up with meeting the imposed requirement on robustness.

We have identified the region depicting weights to be put on each goal for achieving a particular specified value of either of the goal. The weights can be selected from the identified region, run the cDSP construct and get the operating set points that must be used for the required purpose (an example is shown in the last). Now, let us look for a practical situation encountered in an industry where all the goals of a process are important and have to be met simultaneously. This implies, now a process designer is interested to find the design set points that must be used to meet all the three requirements discussed above. To analyse such a situation, the identified regions (for weights to be put on the goals, as discussed above) have been superimposed in a single ternary plot as shown in Figure 13. The observations about the simultaneous achievement of requirements on each goal are as follows:

1. **G1 and G2:** From Figure 13, we see that there is no overlap between the *region depicting set of weights to be assigned to goal*G1 and G2. This implies that there is no scenario where we can run the unit operation to satisfy both the requirements, i.e., getting 80% of maximum achievable value for efficiency and 20% of maximum achievable value for dead zone volume. .

Figure 10. Region depicting set of weights for meeting requirement on inclusion removal efficiency

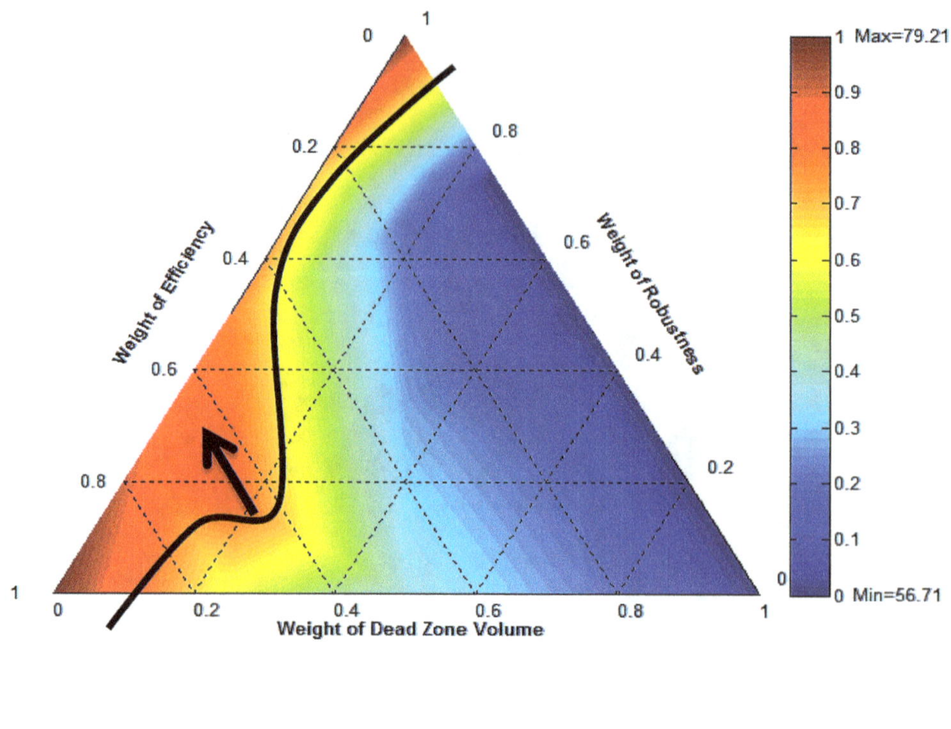

Figure 11. Region Depicting Set of Weights for meeting Requirement Dead Zone Volume

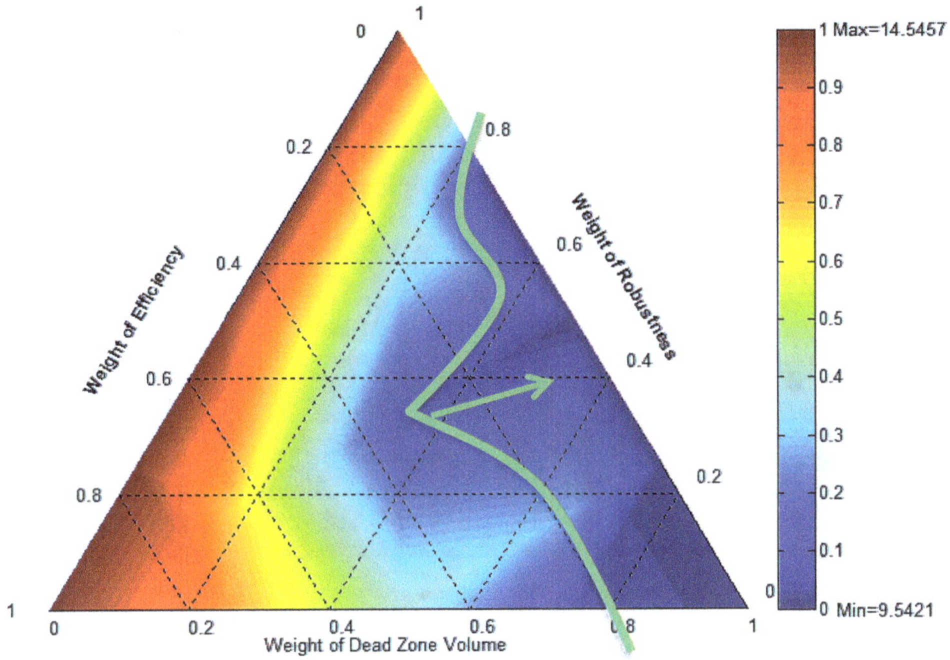

Figure 12. Region Depicting Set of Weights for meeting Requirement on Robustness

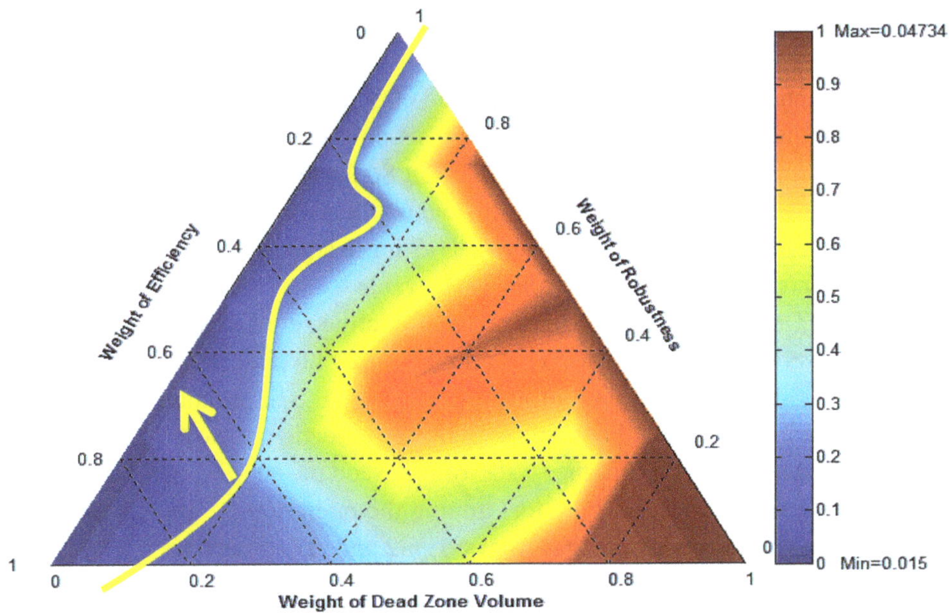

Figure 13. Superimposition of feasible spaces

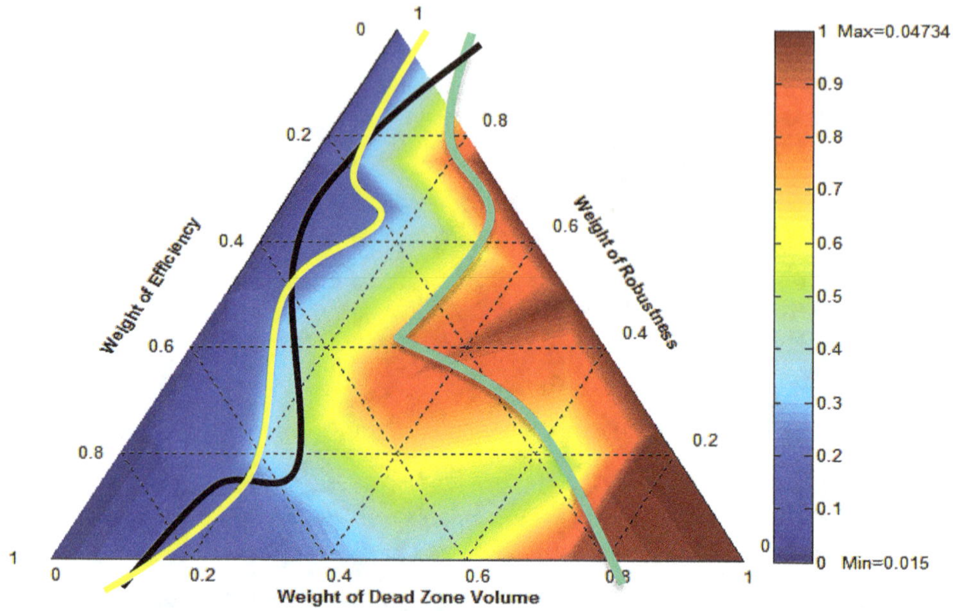

2. **G2 and G3:** From the Figure 13, we see that there is no overlap between the *region depicting set of weights to be assigned to goal*G2 and G3. This implies that there is no such scenario where we can run the unit operation to satisfy both the requirements.
3. **G1 and G3:** From the Figure 13, we see that region depicting set of weights to be assigned to goal G1 and G3 do overlap. This region is the area bounded by the weight of efficiency axis of the ternary plot, yellow and the black curves. If designer assigns the weight to the goals within this area, the requirement on both G1 and G3 goals will occur. Weights for each goal have been selected from this region and are used to run the formulated cDSP construct using DSIDES software. The system variables, thus predicted, are the operating set points at which the process should be operated to have inclusion removal efficiency more than 80% of the maximum achievable value and variability in dead volume less than 20% of the maximum achievable value. The weights selected are 7/12 for efficiency, 4/12 for robustness and 1/12 for dead zone volume. The system variables (operating set points), thus predicted, are shown in Table 6.
4. **G1, G2 and G3:** There is no region where region depicting set of weights to be assigned to goal G1, G2 and G3 overlap simultaneously. This implies, under no circumstances with the given window of system variables, the unit operation can be operated to have inclusion removal efficiency more than 80% and dead zone volume and variability in dead zone volume less than 20% of the maximum achievable value. This implies trade-off between the achievements of goals needs to be done. For example, a designer can look for running the operation to attain a value of 80%, 40% and 20% of the maximum achievable value for G1, G2 and G3 respectively.

The above discussion highlights how ternary plots can be used by a designer to determine whether a simultaneous achievement of a preferred value of goals is possible or not. And, if possible, how to assign the weights to goals, that will be used for running the cDSP construct and predicting the operating set points or system variables.

Table 6. Set points (system variables) predicted

Sr. No.	System Variables	Symbol	Predicted Value
1	Inlet Temperature, K	X_1	1826
2	Inlet steel flow rate, m^3/s	X_2	5.805E-03
3	Weir Position, m	X_3	0.91
4	Weir depth, m	X_4	0.85
5	Weir Thickness, m	X_5	0.15
6	Dam position, m	X_6	1.53
7	Dam height, m	X_7	0.32
8	Dam thickness, m	X_8	0.15
9	Logarithm of a_{in}	X_9	20.78
10	b_{in}	X_{10}	0.415

4. CONCLUSION

In this chapter, a method to determine the set points for a unit operation in a manufacturing process chain is proposed. The use of cDSP construct and tools such as ternary plots are demonstrated to perform the design exploration, thereby determining system variables for achievement of specified targets of different goals. The goals can be on functions that the process is intended to perform or requirements related to uncertainty or variability of the process (as discussed for the case of tundish). The proposed method also helps to find the feasible space for managing the conflict and variability involved. The weights for each goal can be determined from the identified feasible solution space and used in the cDSP construct for obtaining the design set points of the process that, if implemented, will result in attaining these requirements. Although the method is illustrated for one unit operation, namely, the tundish it can be used to determine the set points for unit operations as well.

The weights for the goals are dependent on the preference of the decision maker. If the decision maker has a clear idea of his/her preference structure, he/she can pre-assign these weights without going through the procedure described in Section 3.2, and use the results from a single execution of the cDSP. However, if the decision maker is not completely clear about how to assign weights, then the approach presented in this section can help in identifying what is achievable, and how the different weights would affect the achievement of goals. Therefore, from a decision making standpoint, the primary advantage of the proposed approach is that it allows the exploration of the design space, while at the same time, facilitates the decision makers in clarifying their preferences for trade-offs among different goals.

The scope of the problem being discussed is huge and will add value to the steel industry. Suppose an industry is set to produce steel product mix from a new grade of steel. This will require a lot of experimentation and plant trials to determine the operating set points for each unit operation involved in the entire manufacturing chain. The proposed method of faster and approximate design exploration for determining the operating set points can be instantiated for all the unit operations involved. The individual cDSP based models can be then linked together using an ICME based platform, namely, PREMAP to carry out the integrated process modelling. This, in turn, will help in identifying the operating set points for each process step that will result in production of steel product with a new grade of steel, with reduced time and cost involved. A part of the problem is solved and demonstrated in the current work. The results are motivating and further work needs to be done for completion of the stated integrated process modelling of steel product manufacturing process. The scope for future research work is briefly listed below:

1. Using the cDSP construct to manage the conflict and uncertainty involved in other unit operations, such as ladle refining, continuous casting, rolling, annealing.
2. Using Inductive Design Exploration Method (IDEM) [27] to facilitate horizontal integration of unit operations and thereby assist a process designer in integrated design exploration of an entire steel product manufacturing chain (Choi, McDowell, Allen, Rosen, & Mistree, 2008).
3. As property of steel significantly depends on the microstructure, based on the property specifications of the final product, the desired microstructure can be predicted. This can be used to put requirement on the downstream unit operations of steel product manufacturing; the operations should be carried out in a way to ensure the evolution of desired microstructure. The information on requirements will further flow from downstream to the upstream processes. In this way, each unit operation can be designed so as to meet the property requirements of the final product.

ACKNOWLEDGMENT

The authors acknowledge the support and encouragement given by the management of Tata Consultancy Services Ltd., India. F. Mistree and J.K. Allen gratefully acknowledge support from the NSF Grant CMMI 1258439 and the L.A. Comp and John and Mary Moore Chairs, respectively. J.H. Panchal acknowledges support from NSF CMMI Grants 1265622 and 1261860.

REFERENCES

Agarwal, K., & Shivpuri, R. (2013). A Computational Framework for Integrated Process Design of High Performance Parts. 2nd *World Congress on Integrated Computational Materials Engineering*.

Allen, J., Mistree, F., Panchal, J., Gautham, B., Singh, A., Reddy, S., . . . Kumar, P. (2013). Integrated Realization of Engineered Materials and Products: A Foundational Problem. 2nd *World Congress on Integrated Computational Materials Engineering*.

Allison, J., Li, M., Wolverton, C., & Su, X. M. (2006). Virtual Aluminium Castings: An Industrial Application of ICME. JOM, 28-35.

Anapagaddi, R., Shukla, R., Goyal, S., Singh, A. K., Allen, J. K., Panchal, J. H., & Mistree, F. (2014). Exploring the Design Space to Determine the Operating Set Points for a Tundish. *International Design Engineering Technical Conferences & Computers and Information in Engineering Conference IDETC/CIE 2014*.

Choi, H., McDowell, D. L., Allen, J. K., Rosen, D., & Mistree, F. (2008). An Inductive Design Exploration Method for Robust Multiscale Materials Design. *Journal of Mechanical Design*, *130*(031402), 1–13.

Cwudzinski, A., & Jowsa, J. (2012). Mathematical Modeling of Behaviour a Liquid Steel Flow and a Non-metallic Inclusions in the One Strand Wedge Type Tundish with a Subflux Turbulence Controller and a Low Dam. *Materials Science Forum*, *706-709*, 1379–1384. doi:10.4028/www.scientific.net/MSF.706-709.1379

Damle, C., & Sahai, Y. (1995). The Effect of Tracer Density on Melt Flow Characterization in Continuous Casting Tundishes – A Modeling Study. *ISIJ International*, *35*(2), 163–169. doi:10.2355/isijinternational.35.163

Ding, N., Bao, Y., Sun, Q., & Wang, L. (2011). Optimization of Flow Control Devices in a Single-Strand Slab Continuous Casting Tundish. *International Journal of Minerals, Metals and Materials*, *18*(3), 292–297. doi:10.1007/s12613-011-0436-7

Gautham, B., Kulkarni, N., Khan, D., Zagade, P., Reddy, S., & Uppaluri, R. (2013). Knowledge Assisted Integrated Design of a Component and Its Manufacturing Process. 2nd *World Congress on Integrated Computational Materials Engineering*.

Hernandez, G., Allen, J. K., & Mistree, F. (2001). The Compromise Decision Support Problem: Modelling the Deviation Function as in Physical Programming. *Engineering Optimization*, *33*(4), 445–471. doi:10.1080/03052150108940928

Jha, P. K., Dash, S. K., & Kumar, S. (2001). Fluid Flow and Mixing in a Six Strand Billet Caster Tundish: A Parametric Study. *ISIJ International*, *41*(12), 1437–1446. doi:10.2355/isijinternational.41.1437

Kumar, A., Mazumdar, D., & Koria, S. C. (2008). Modeling of Fluid Flow and Residence Time Distribution in a Four-Strand Tundish for Enhancing Inclusion Removal. *ISIJ International*, *48*(1), 38–47. doi:10.2355/isijinternational.48.38

Llegbusi, O. J., & Szekely, J. (1989). Effect of Magnetic Field on Flow, Temperature and Inclusion Removal in Shallow Tundishes. *ISIJ International*, *29*(12), 1031–1039. doi:10.2355/isijinternational.29.1031

Luo, A. (2013). Application of Computational Thermodynamics and CALPHAD in Magnesium Alloy Development. 2nd *World Congress on Integrated Computational Materials Engineering*.

Maiwald-Immer, T., Gohler, T., Fischersworring-Bunk, A., Korner, C., Osmanlic, F., & Bauerei, A. (2013). Application of ICME Methods for the Development of Rapid Manufacturing Technologies. 2nd *World Congress on Integrated Computational Materials Engineering*.

Mazumdar, D., & Guthrie, R. I. L. (1999). The Physical and Mathematical Modelling of Continuous Casting Tundish System. *ISIJ International*, *39*(6), 524–547. doi:10.2355/isijinternational.39.524

McDowell, D. L., Panchal, J. H., Choi, H. J., Seepersad, C. C., Allen, J. K., & Mistree, F. (2010). *Integrated Design of Multiscale Materials and Products*. New York: Elsevier.

Miki, Y., & Thomas, B. G. (1999). Modeling of Inclusion Removal in a Tundish. *Metallurgical Transaction B*, *30B*(4), 639-654.

Mistree, F., Hughes, O. F., & Bras, B. (1992). Compromise Decision Support Problem and the Adaptive Linear Programming Algorithm. In M. P. Kamat (Ed.), *Structural Optimization: Status and Promise* (pp. 251–290). Washington, DC: AIAA.

Myers, R. H., Montgomery, D. C., & Anderson-Cook, C. M. (2009). *Response Surface Methodology: Process and Product Optimization using Designed Experiments*. New York: John Wiley & Sons.

Reddy, R., Smith, W. F., Mistree, F., Bras, B. A., Chen, W., Malhotra, A., & Patel, P. et al. (1992). *DSIDES. User Manual Systems*. Design Laboratory, Department of Mechanical Engineering, University of Houston.

Singh, A. K., Paradeshi, R., & Goyal, S. (2011). Integrated Modeling of Tundish and Continuous Caster to meet Quality Requirement of Cast Steels.*1st World Congress on Integrated Computational Materials Engineering*. doi:10.1002/9781118147726.ch11

Sinha, A., Allen, J., Panchal, J., & Mistree, F. (2013). Microstructure Mediated Design of Material and Product.*2nd World Congress on Integrated Computational Materials Engineering*. doi:10.1002/9781118767061.ch7

Smith, K., Chen, L., Darling, R., Garosshen, T., Jaworowski, M., Opalka, S., . . . Zafiris, G. (2013). ICME Approach to Corrosion Pit Growth Prediction. 2nd *World Congress on Integrated Computational Materials Engineering*.

Smith, W. F., Milisavljevic, J., Sabeghi, M., Allen, J. K., & Mistree, F. (2014). Accounting for Uncertainty and Complexity in the Realization of Engineered Systems.*International Conference on Complex Systems Design & Management*.

Zhang, L. (2006). Fluid Flow, Heat Transfer and Inclusion Motion in Molten Steel Continuous Casting Tundishes. *Fifth International Conference on CFD in the Process Industries*.

Zhang, L., Taniguchi, S., & Cai, K. (2000). Fluid Flow and Inclusion Removal in Continuous Casting Tundish. *Metallurgical and Material Transaction B*, *31B*, 253-266.

ADDITIONAL READING

Allen, J. K., Seepersad, C. C., & Mistree, F. (2006). A Survey of Robust Design with Applications to Multidisciplinary and Multiscale Systems. *Journal of Mechanical Design*, *128*(4), 832–843. doi:10.1115/1.2202880

Bambach, M., Scmitz, G., & Prahl, U. (2013). ICME- A Mere Coupling of Models or a Discipline of Its Own? 2nd*World Congress on Integrated Computational Materials Engineering, Vol. 1. (pp. 285-290).* Salt Lake Marriott Downtown at City Creek, Salt Lake City, Utah.

Bensouici, M., Bellaouar, A., & Talbi, K. (2009). Numerical Investigation of the Fluid Flow in Continuous Casting Tundish using Analysis of RTD Curves. *Journal of Iron and Steel Research*, *16*(2), 22–29. doi:10.1016/S1006-706X(09)60022-4

Bulko, B., & Kijac, J. (2010). Optimization of Tundish Equipment. *ActaMetallurgicaSlovaca*, *16*(2), 76–83.

Chen, W., Allen, J. K., Tsui, K.-L., & Mistree, F. (1996). A Procedure for Robust Design: Minimizing Variations Caused by Noise Factors and Control Factors. *ASME Journal of Mechanical Design*, *118*(4), 478–485. doi:10.1115/1.2826915

Choi, H-J., McDowell, D. L., Rosen, D., Allen, J. K., & Mistree, F. (2008). An Inductive Design Exploration Method for Robust Multiscale Materials Design. *Journal Mechanical Design*, 130(3), 031402-1/13.

Liang-cai, Z., Ming-an, W., Bo-yu, C., Can-rong, W., & Ying-xiong, Z. (2010). Flow Control in Six-Strand Billet Continuous Casting Tundish with Different Configurations. *Journal of Iron and Steel Research, International*, 17(7), 07-12.

Mistree, F., Smith, W. F., & Bras, B. A. (1993). A Decision-Based Approach to Concurrent Engineering. In H. R. Paresai & W. Sullivan (Eds.), *Handbook of Concurrent Engineering* (pp. 127–158). New York: Chapman & Hall. doi:10.1007/978-1-4615-3062-6_8

Panchal, J. H., Kalidindi, S. R., & McDowell, D. L. (2013). Key Computational Modeling Issues in Integrated Computational Materials Engineering. *Journal Computer-Aided Design*, *45*(1), 4–25. doi:10.1016/j.cad.2012.06.006

Pardeshi, R., Basak, S., Singh, A. K., Basu, B., Mahashabde, V., Roy, S. K., & Kumar, S. (2004). Mathematical Modeling of the Tundish of a Single – Strand Slab Caster. *ISIJ International*, *44*(9), 1534–1540. doi:10.2355/isijinternational.44.1534

KEY TERMS AND DEFINITIONS

cDSP: Compromise Decision Support Problem is a hybrid formulationinvolving concepts of both traditional mathematical programming and goal programming. The construct aims at providing system variables to manage the conflict and variability involved in a particular process. The solution, thus obtained, may be sub-optimal but are robust in nature.

Response Surface Models: RSM's are simplified models, developed based on the data generated by experiments or numerical models, to empirically represent the effect of explanatory variables on the responses of the process. These are not physics based models.

Robust: A process is said to be robust if the variations in the input operating parameters does not significantly influence the output parameters. This implies process is less affected by the variations and uncertainties of the operations.

APPENDIX

Table 7. Nomenclature

a, b	Characteristics of inclusion number density	V	Volume of tundish, m³
a_i, a_{ij}, b_{ij}	Coefficients of response surface equations	V_d	Percentage of dead zone volume
a_{in}, b_{in}	Characteristics of inlet inclusion number density	V_m	Percentage of mixed zone volume
a_{out}, b_{out}	Characteristics of outlet inclusion number density	V_p	Percentage of plug zone volume
a_p	Acceleration of the particle, m/s²	W_i	Weight of ith deviation variable
C	Concentration of tracer	x_i	Special direction, m
C_D	Drag coefficient	X_i	System variable
C_p	Specific heat	X_1	Inlet temperature, K
D_{eff}	Effective diffusivity	X_2	Inlet steel flow rate, m³/s
d_i^-, d_i^+	Deviation variables	X_3	Weir position, m
d_p	Diameter of the particle, micron	X_4	Weir depth, m
F_D	Drag force, N	X_5	Weir thickness, m
F_g	Gravitational force, N	X_6	Dam position, m
G	Gravitational constant, m/s²	X_7	Dam height, m
H	Enthalpy, J	X_8	Dam thickness, m
K	Turbulence kinetic energy, m²/s²	X_9	Logarithm of a_{in}
K_{eff}	Effective thermal conductivity	X_{10}	b_{in}
m_p	Mass of the particle, kg	w_i	Stokes velocity, m/s
$n(r_i)$	Number density of inclusion of size r_i, number of inclusion/ m³ of steel	Z	Deviation function

$n_{in}(r_i)$	Inlet number density of inclusion of size r_i, number of inclusion/ m^3 of steel	β	Coefficient of Thermal expansion, 1/K
$n_{out}(r_i)$	Outlet number density of inclusion of size r_i, number of inclusion/ m^3 of steel	$\beta_1(r_i, r_j)$	Coefficient of Brownian collision
P	Pressure, atm	$\beta_2(r_i, r_j)$	Coefficient of Turbulent collision
Pr	Prandtl number	$\beta_3(r_i, r_j)$	Coefficient of Stokes collision
Q	Volumetric flow rate of steel, m^3/s	μ_{eff}	Effective viscosity, pa-sec
r_i	Radius of inclusion, micron	μ_t	turbulent viscosity, pa-sec
S	Surface area, m^2	ρ	Density of steel, kg/m^3
Sc	Sherwood Number	ε	Turbulence eddy dissipation, m^2/s^3
t	Time, sec		
$T_{liquidus}$	Liquidus temperature, K	ΔT_i	Temperature difference between inlet and outlet
u	Averaged velocity of steel, m/s	η	Inclusion removal efficiency
u_i	Velocity in special direction x, y, z, m/s	$\sigma_k, \sigma_\varepsilon, C_1, C_2, C_\mu$	Constants
u_p	Velocity of particle, m/s	υ	Kinematic viscosity, m^2/s

Chapter 7
Informatics-Based Approaches for Accelerated Discovery of Functional Materials

P. V. Balachandran
Los Alamos National Laboratory, USA

J. M. Rondinelli
Northwestern University, USA

ABSTRACT

This chapter is aimed at readers interested in the topic of informatics-based approaches for accelerated materials discovery, but who are unfamiliar with the nuances of the underlying principles and various types of powerful mathematical tools that are involved in formulating structure–property relationships. In an attempt to simplify the workflow of materials informatics, we decompose the paradigm into several core subtasks: hypothesis generation, database construction, data pre-processing, mathematical modeling, model validation, and finally hypothesis testing. We discuss each task and provide illustrative case studies, which apply these methods to various functional ceramic materials.

INTRODUCTION

Advancements in materials synthesis, processing, characterization, theory, and computation have significantly changed the way in which we generate, share, and analyze data in materials science. Modern day combinatorial experiments enable routine screening of large areas of composition space in a high-throughput fashion generating large volumes of data in an accelerated manner (Koinuma & Takeuchi, 2004; Potyrailo *et al.* 2011). In the arena of computational materials research, the last two decades have also seen tremendous progress in high-performance and high-throughput *ab initio* quantum mechanical-based calculations, enabled by ever increasing computational power at lower costs and improvements in parallel and optimization algorithms. The culmination of these efforts is a new virtual platform for the discovery, design, and development of functional materials and devices (Curtarolo *et al.* 2012; Jain *et al.* 2013; Saal *et al.* 2013). They are used in more than 10,000 publications annually, providing a bal-

DOI: 10.4018/978-1-5225-0290-6.ch007

ance between accuracy and efficiency, solving the N-electron Schrodinger equations primarily *via* the density-functional theory (DFT) formalism (Hohenberg & Kohn, 1964; Kohn & Sham, 1965). They have an outstanding track record of enabling theorists to predict properties of materials that are subsequently validated experimentally (Parr & Wang, 1995; Ceder, 1998; Rondinelli *et al.* 2008 & 2012; Mannhart & Scholm, 2010; Setyawan & Curtorolo, 2012; Olson, 2013).

Aided by the advancements in various levels of materials theory, including methods based on Green's functions (GW approach, Hedin, 1965) and dynamical mean field theory (DMFT, Georges *et al.* 1996; Kotliar *et al.* 2006) for excited state and correlated electron materials problems, improved exchange-correlation functionals, user-friendly computational codes [such as VASP (Kresse & Furthmuller, 1996; Kresse & Joubert, 1999) and Quantum ESPRESSO (Gianozzi *et al.* 2009) to name a few], and the growing infrastructure of massive supercomputing user-facilities and cloud computing, high-throughput and accurate *ab initio* calculations are generating volumes of data at an unprecedented level. For instance, digital databases such as "The Materials Project" (Jain *et al.* 2013), "Aflowlib.org" (Curtorolo *et al.* 2012) and "OQMD.org" (Saal *et al.* 2013) store *ab initio* calculations performed on more than 38,151, 17,200 and 284,000 entries, respectively, and counting.

Despite the encouraging progress, we have only scratched the surface when it comes to navigating the vast chemical space of inorganic compounds. There is a massive amount of compounds remaining to be explored beyond those reported in experimental structure databases. Multicomponent inorganic systems, which are the focus of this chapter, form the crux of modern materials science – inorganic compounds with more than two cationic sites appear to be the minimal requirement (Cheetham *et al.* 1994) for crystalline materials with functional electronic, magnetic, and optical properties. The flexibility of atomic structures allow fine tuning of these properties by altering, (i) interatomic distances (by using cations with carefully selected ionic radii), (ii) degree of covalency (by judicious choice of elements) and (iii) active frontier orbitals (by varying the oxidation states of the cations and anions contributing states to the low energy electronic structure in a controlled manner).

The attractive structural and functional features, which arise upon going from a binary to ternary to higher-order multi-element compound, however, introduces an enormous challenge in the rational search and discovery paradigm. To understand the complexity of the problem, consider the 60 elements highlighted in the periodic table (see Figure 1). Consider the case of a three-component (ternary) ABC materials system, where A, B and C are one of the 60 highlighted elements in Figure 1. Then, there are $C_3^{60} = 34,220$ possible combinations. In addition, Muller and Roy have documented more than 20 different major structure-types in their compendium on *The Major Ternary Structural Families* (Muller & Roy, 1974). This increases the number of possible candidates to 0.68×10^6 combinations. Furthermore, three-dimensional crystallography indicates there are 230 space groups in which the atoms can be periodically and uniquely arranged. Even if we intelligently refine our choice to (on an average) 5 space groups per candidate composition, then the number of calculations increases to 3.4×10^6 structures!

Now, consider the possibility of solid solutions ($BaTiO_3$-$SrTiO_3$, where the Sr and Ba atoms are distributed randomly in the unit cell lattice), defects ($La_2NiO_{4+\delta}$, where δ is the excess interstitial oxygen atoms in the lattice), and ordered cationic sublattices—the chemical space explodes. Even with the most powerful supercomputers, many compositions remain beyond the scope of *ab initio* calculations in part due to the large number of atoms in the unit cell. It is well known that the complexity of DFT calculations scale on the order of N^3, where N is the number of electrons (Parr & Wang, 1995). Informatics-based approaches have the potential to aid both high-throughput experimental and computational studies, by

Figure 1. The vast combinatorial chemical search space afforded by the periodic table in systematically designing new materials chemistries. We have highlighted 60 useful elements to illustrate to the enormous space requiring exploration (see text).

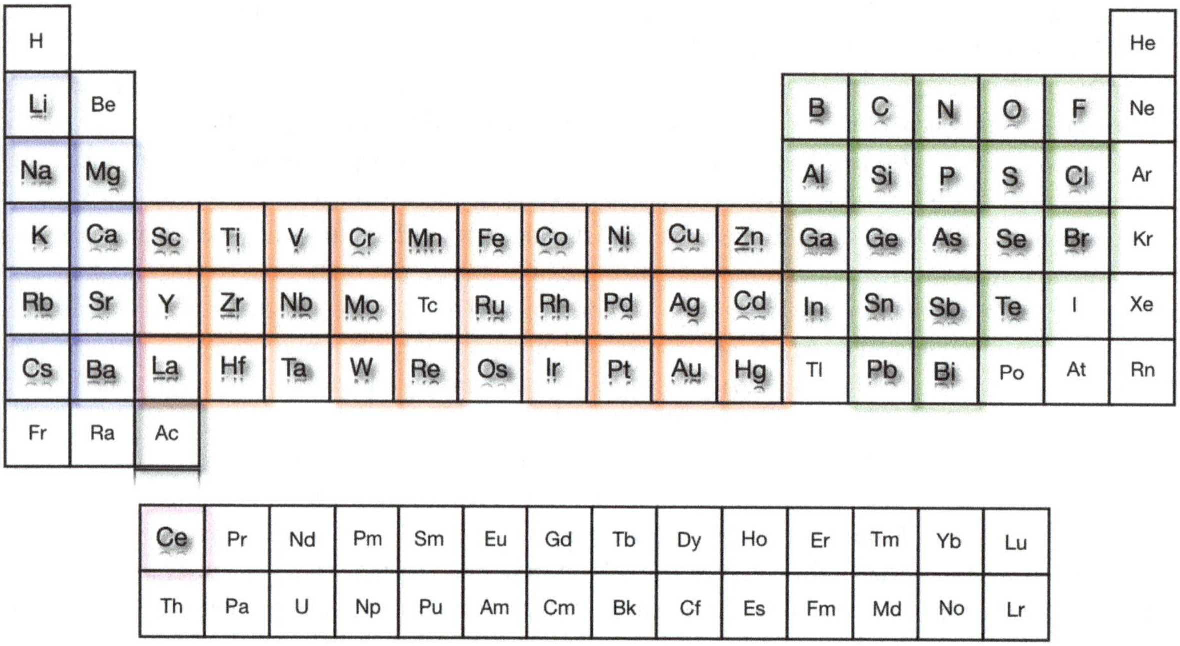

reducing the complexity of the problem into a manageable proportion through rapidly screening the vast chemical search space and identifying (down-selecting to) fewer candidate compositions for targeted experimental and computational explorations.

In this chapter, we describe the informatics strategy from a holistic viewpoint, starting with discussions involving how to construct a database, where to find attributes (we also list a wide variety of attributes that are commonly used in materials informatics study) and the steps involved in data mining. We then describe how to relate data mining methods with specific materials science and engineering questions, and finally show how to validate the predictions or outcomes using computational and/or theoretical methods. We describe four case studies that encompass key components concerning a materials informatics study discussed in this chapter. We conclude by offering insights on challenges that ought to be overcome to establish materials informatics firmly as a computational materials design tool in the materials science and engineering arena.

MATERIALS INFORMATICS: BACKGROUND

Materials informatics is an emerging computational approach in the field of materials science, which is built on the foundations of data science, information science, databases, statistics, optimization, machine learning and data mining. One of the key objectives is to transform raw data of known materials systems obtained from diverse experimental measurements and/or computer simulations into testable hypotheses. As noted by Rajan (2005), an analogy for understanding the importance of developing materials informatics can be found in systems biology, where the combination of systems-level experimentation and computational tools have made it possible to integrate, visualize and analyze genomic and protein expression data in a systematic manner, thereby uncovering critical signaling networks at the molecular-level that could be used to address important issues in the field of life sciences (Rajan, 2005). In a similar fashion, these approaches can be applied to materials science problems, to understand how chemistry, atomic structure, and microstructure interact to yield functional properties. Because most material classes can be understood within a multilevel interaction framework, as their functionality originates from structural and spatial evolution concomitant with assorted mechanisms and interlinked processes, the hierarchical structure and coupled interactions are situated to be understood with informatics-based methods.

In Figure 2(a), we show the statistics indicating steady increase in popularity of informatics-based methods in materials science, quantified using the number of scientific publications (based on ISI web of science) as a metric, on various topics including materials informatics, machine learning, statistical learning, and data mining. This increase over the last decade, and recently since 2010 can be in part attributed to the national research programs aimed at accelerating materials discovery and deployment into new technologies; for example, the Materials Genome Initiate (MGI) in the United States (Council, 2011).

In Figure 2(b), we reproduce a pie-chart from Le *et al.* (2012) that clearly distinguishes how different class of materials, such as polymers, ceramics, catalysts, etc., have been studied based on the topic of quantitative structure-property relationships. We find that soft materials (biomaterials and polymers), catalysts, and nanomaterials together form a major fraction of the scientific investigations. However inorganic ceramic materials (which are the focus of this chapter) represent only 2%, relative to that of the other fields. One of the key reasons has been attributed to the fact that earlier efforts (Rajan, 2010) had been focused predominantly on ways to acquire, organize and manage that data in large databases, such as the International Crystal Structure Database (ICSD, Belsky *et al.* 2002). While extremely useful, these large databases cannot be used "as-is" when one hopes to glean patterns of behavior that would offer insight into the complex chemical and geometrical relationships governing the crystal structure stability or measured property of a given compound.

Even now, constructing an "informatics-ready" database for performing informatics-based analysis in the field of ceramics is a challenging task. In this chapter, we attempt to address how one may practically overcome that obstacle, simplifying the workflow of materials informatics by decomposing the paradigm into several core subtasks: database construction, data pre-processing, data mining, model validation and finally, hypothesis testing. We discuss each subtask at sufficient length and then provide case studies, which have been successful in representative functional oxides.

INFORMATICS-BASED MATERIALS DISCOVERY PARADIGM

Materials Data Structure

The first step involved in any materials informatics study is the "hypothesis generation." What is the materials science question and how we intend to address it? For example, in a materials informatics study involving prediction of band gaps, some of the important data to be considered besides the size of the gap are: thermodynamic stability of the compound, crystal symmetry, temperature at which the band gap is measured or calculated, gap nature (direct or indirect), bond geometry, and crystal chemistry. These data put together would form the "materials data structure." One of the critical aspects in

Figure 2. (a) Statistic showing the growing popularity within a span of 10 years, quantified in terms of number of scientific publications (obtained from an ISI Web of Science search), in informatics-based methods and approaches in the field of materials science. (b) Pie-chart reproduced from the work of Le et al. (2012), where they summarize the relative number of studies performed on the topic, quantitative structure-property relationships. Notice that ceramics represent only 2% of the total, indicating existence of potential challenges.

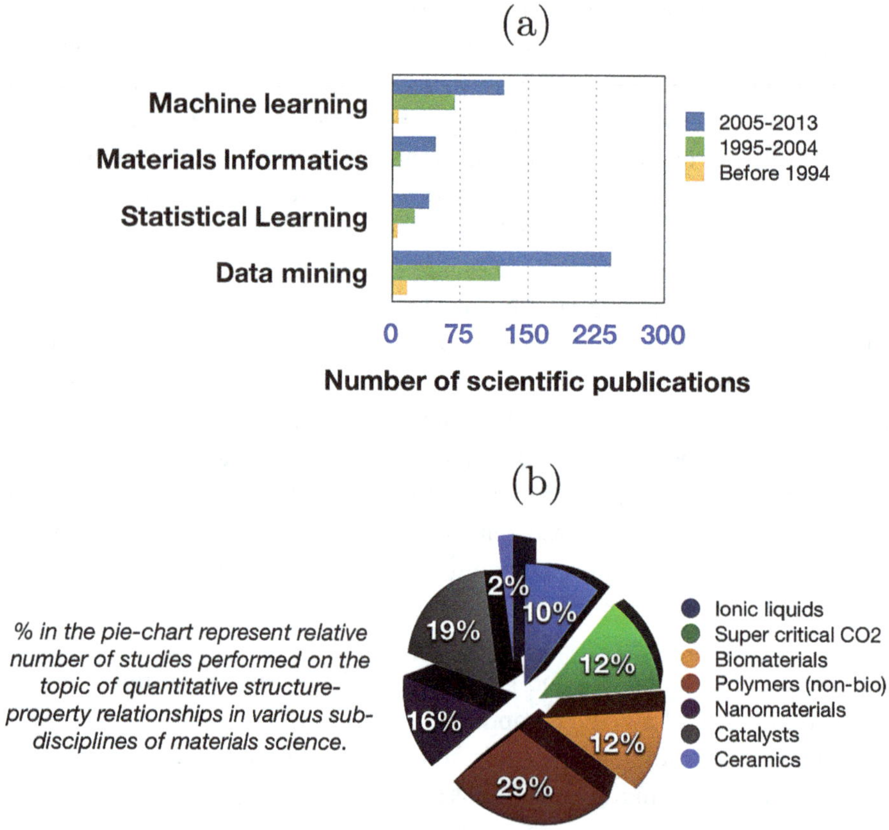

hypothesis generation step is the understanding of constraints imposed by the dataset. In any informatics study, besides data provenance (*i.e.* documenting the sources) it is important to highlight the boundary conditions. For example, in the aforementioned band gap prediction problem, if the crystal symmetries of the materials were not explicitly included in the dataset, then highlighting it would allow for reliable interpretation of the model and subsequent theoretical or experimental validation.

Once the problem statements and the boundary conditions are identified, then we embark on the process of database construction detailed next. The database construction is an informed process, where one includes evidences or features believed to be important in supporting different mechanisms governing the underlying materials properties. The mathematical model evaluates the evidences and succinctly represents the complex materials system. The models are then validated with the help of known (often experimental) outcomes to ensure its generality and reveal any shortcomings. Once models are validated, it is then used to extract new insights, translating data into useful information and ultimately understanding (or knowledge). From the model, new materials chemistries could be predicted and subsequently tested using physics-based computational or experimental approaches. In the next section, we discuss how to construct materials data structure for performing a materials informatics study.

Database Construction

We discuss the database construction for materials informatics from data mining perspective. We consider a tuple $X = (x_1, x_2, ..., x_m)$ depicting m independent or unique chemical compositions. Here, each composition x_i is represented by an n-dimensional attribute vector, $A = (A_1, A_2, ..., A_n)$. Each attribute in A would represent a feature of X. For instance, in most ceramic materials, the crystal structure serves an important role in governing the compounds electronic properties. Therefore, one of the primary tasks requires finding factors that determine stable crystal structures and atomic geometries. It has been postulated that there are five inter-related factors responsible for the formation of a particular crystal structure (Pettifor, 1995; Villars *et al.* 1998): size, atomic number, electrochemical factor, valence electron count and cohesion energy. In the materials informatics paradigm, descriptors (also referred to as variables or property proxies) representing these factors would become the attributes, A_i, in the database. In this section of the chapter, we elaborate on how to build the A such that it can account for the trends on which various aspects of chemical bonding and structural characteristics are studied.

In Table 1, we have enumerated some of the most commonly used descriptors that capture the relative orbital energies of atoms belonging to crystal chemistries at various levels of approximations. For readers interested in gaining an understanding of the physical meaning behind each descriptor, we have provided the reference alongside each descriptor. Note that Table 1 is by no means exhaustive. Descriptor development is an active area of research. In addition, one could also employ more specialized descriptors such as bond lengths, bond angles and irreducible representations of lattice modes.

Example Database Construction

Given these descriptors, we will now discuss a simplified example on how to build X and A for a materials informatics study. Consider a binary compound made of two elements, which we denote as *P* and *Q*. For simplicity, let us assume that the stoichiometry is *PQ*. If we consider the 60 main elements in the periodic table (see Figure 1), then in principle we can enumerate $C_2^{60} = 1,770$ compositions belonging to *PQ* stoichiometry. The C_2^{60} compositions would represent the X, which could be regarded as a column vector. For each composition x_i in X there is an associated set of attributes A_i; A could then be constructed using the attributes given in *Computational Approaches to Materials Design: Theoretical and Practical Aspects.* For argument sake, if we assume that all 22 descriptors are applicable for the hypothetical binary compound, then it is possible to define a string of length 44 (22 descriptors for both elements *P* and *Q*) representing each x_i in X.

Notice the high-dimensionality of each composition x_i in our database. One of the hypotheses is that the number of A_i needed to predict even simple structures (such as binary compounds discussed) can be large. Since there are no universal scales that quantify the relative and absolute electronic energies of all chemical elements in the periodic table in different atomic environments, the suggested approach, although exploratory, seems appropriate if one has to capture both geometrical and bonding characteristics of the crystal chemistries spanning the entire periodic table (Rajan, 2010). Indeed there

Table 1. List of discrete scalar descriptors available for materials informatics along with their references.

Descriptors	Reference
Shannon's Ionic Radius	Shannon, 1976
Zunger's Pseudopotential Core Radii Sum	Zunger, 1980
Orbital radii	Weber & Cromer, 1965
Nominal bond distance	Brese & O'Keeffe, 1991
Electron affinity	Hotop & Lineberger, 1980
Ionization energy	Mulliken, 1934
Effective Nuclear Charge – Slater Scale	Slater, 1930
Effective Nuclear Charge – Clementi Scale	Clementi & Raimondi, 1963
Effective Nuclear Charge – Froese-Fischer Scale	Froese-Fischer, 1972
Wigner-Seitz cell radius	Skriver, 2004
Electronegativity – Pauling's Scale	Pauling, 1932
Electronegativity – Mulliken's Scale	Mulliken, 1934
Electronegativity – Gordy's Scale	Gordy, 1946
Electronegativity – Allred-Rochow's Scale	Allred & Rochow, 1958
Electronegativity – Sanderson's Scale	Sanderson, 1952
Electronegativity – Martynov-Batsanov's Scale	Rabe *et al.* 1992
Electronegativity – Atomic Solid State Energy Scale	Pelatt *et al.* 2011
Chemical hardness	Pearson, 1986
Enthalpy of Formation	Saxena, 1993
Polarizing Power	Ahren, 1952
Valence Electron Number	Rabe *et al.* 1992
Pettifor scale	Pettifor, 1995
Mendeleev number	Villars *et al.* 2001

are several reports in the literature, where this hypothesis has been tested for complex families of crystal structures (Mooser & Pearson, 1959; Rabe *et al.* 1992; Pettifor, 1995; Villars *et al.* 1998 & 2001; Rajan, 2005; Suh & Rajan, 2009; Rajan, 2010; Balachandran *et al.* 2011; Dey *et al.* 2014).

Alternatively, one could also use variables for A that are *categorical* in nature (instead of being numerical). For instance, one of the simplest examples is to use the elemental symbol, such as "Ca" for calcium, "Fe" for iron, and so forth. In principle, the elements in A could take any form, but the key thing is to ensure that a database of X and A is available that "sufficiently" describes the materials science question of interest so that the data mining/machine learning methods will be able to extract patterns capturing new insights for accelerating the discovery of new materials.

Pre-Processing

After database construction and prior to data mining, pre-processing is an intermediary step that serves to prepare the data for mining or model building (Witten & Frank, 2000). One of the common reasons for pre-processing arises from the fact that the materials databases contain attributes of different units and scales. For example, consider bond geometrical attributes, such as bond lengths and bond angles that are expressed in Angstrom units and degrees, respectively. From a numerical standpoint, the bond lengths and angles differ by at least one or two orders of magnitude, which would introduce unintentional bias in data mining methods. The role of data normalization is to remove the unintended bias, so that all attributes are treated on equal footing (Sola & Sevilla, 1997). Also note that the applicability of data normalization methods is data mining algorithm dependent. For instance, data pre-processing is recommended in analysis involving regression, whereas in some classification algorithms such as decision trees, data pre-processing is not a mandatory requisite.

Data Mining

So far, we have discussed some general principles concerning building the database and data pre-processing for the materials informatics study. The next step is data mining. There are two types of learning schemes, namely supervised and unsupervised materials informatics learning. The names for these two schemes are borrowed from the conventional nomenclature used in the data mining literature (Witten & Frank, 2000; Tan *et al.* 2005; Han & Kamber, 2006).

In *supervised learning*, each tuple, X, is assumed to belong to a predefined variable as determined by another attribute, which is called the target variable, Y. The objective of supervised learning is to map a function, $Y = F(A)$ that can predict the associated Y of a given tuple in X. An example for supervised learning includes prediction of the melting point (Y) as a function of factors governing the cohesive energies (A). Typically mapping, F, in supervised learning is represented in the form of classification rules, hierarchical tree structures, or mathematical formulae.

The key difference between supervised learning and *unsupervised learning* is that one has no *a priori* information on the predefined target variable of each tuple X in the unsupervised learning scheme. The objective of unsupervised learning is to assign a class to each tuple in X without having a known target variable. For example, in a database containing materials that are ferroelectrics (although not explicitly identified) and that are unknown (meaning, it is not determined from prior experimental or computational studies whether it is a ferroelectric system or not), "assigning a class" refers to labelling those unknown compositions as ferroelectrics or not based on an unsupervised learning algorithm. This approach is used to find correlations and similarities in data sets, and the outcome is typically represented in the form of data clustering.

Table 2 summarizes various supervised and unsupervised learning methods in the context of materials science and engineering. When a database containing both structure and properties of materials may be constructed, then supervised learning methods are useful to establish the quantitative relationship between them. When the property of interest is a numerical quantity, such as yield strength or melting point, then prediction methods are well suited. On the other hand, when the property of interest is crystal structure-type or crystal symmetry, where it is a categorical quantity, then classification methods are better suited. There is a plethora of prediction and classification data mining methods available in the computer science and statistics literature. However the challenge is to identify appropriate data mining tools for the constructed materials database. In Table 2, we relate data mining methods with some of the common research objectives in materials science and engineering problems.

Table 2. A summary of learning schemes, materials science and engineering research objectives, and data mining methods for materials informatics.

Learning Scheme	Materials Science Problems	Data Mining/ Statistical Learning Methods
Classification (supervised)	• Classification of crystal structure-type of compounds • Crystal symmetry • Band gap (direct or indirect) • Compound formability • Metal or insulator • Solid solution or phase separation	• Artificial neural networks • Support vector machine • Decision trees • Random forests • Rough sets • *k*-nearest neighbors • Discriminant analysis
Prediction (supervised)	• Melting point • Strength (tensile or yield) • Lattice constants • Band gap value • Transition temperatures (ferroelectric/ferromagnetic/superconducting) • Piezoelectric constants	• Artificial neural networks • Support vector machine • Regression tree • Linear regression • Partial least squares • Genetic programming
Data dimensionality reduction and clustering (unsupervised)	• High throughput analysis of spectral data • Problems where gathering property data is nontrivial • Problems involving searching for similarity between materials compositions in the absence of property data	• Principal component analysis • IsoMap • *k*-means clustering • Hierarchical clustering
Ranking of attributes (supervised and unsupervised)	• Coordinates of a structure map • Discovery of key attributes impacting a materials property	• Principal component analysis • Information or Shannon entropy • Genetic algorithm

Figure 3 depicts various data mining algorithms available for performing unsupervised and supervised learning. Formal discussions of data mining algorithms are beyond the scope of this chapter and we recommend the standard texts (Witten & Frank, 2000; Tan *et al.* 2005; Han & Kamber, 2006) for an in-depth coverage. In this chapter, we describe three methods that are used in the case studies: principal component analysis (PCA), *k*-means clustering and Bayes' rules.

Principal Component Analysis

Principal component analysis (PCA) is one of the well-known unsupervised data-dimensionality reduction methods (Ringnér, 2008). PCA assumes the dataset consists of a large number of inter-correlated descriptors that lie on a linear manifold. The purpose of this method is to reduce the dimensionality of a data set, while retaining maximum variability in the data. Data-dimensionality reduction is achieved by transforming the *original* set of variables to a new set of *derived* variables, called the principal components (PC's). Each PC is a linear combination of the weighted contribution of all attributes and the magnitude of the weight determines the relative impact of each descriptor in affecting the PC. The derived variables are then ordered such that that the first few parameters retain most of the variation present in all of the original variables. This process is akin to the eigenvalue problem, where we have a square correlation (or covariance) matrix that is decomposed into eigenvalues and eigenvectors. Thus, the first PC accounts for the maximum variance (highest eigenvalue) in the dataset. The second PC is orthogonal to the first and accounts for most of the remaining variance. Similarly, the mth PC is orthogonal to all others and has the mth largest variance in the set of PCs. In Figure 4, the pseudo-algorithm is given.

Figure 3. Some common data mining algorithms for materials informatics broadly classified into two types: unsupervised and supervised learning, based on the nature of the learning scheme (details discussed in main text). PCA stands for principal component analysis, IsoMap or isometric mapping (Tenanbaum et al. 2000), LLE stands for local linear embedding (Roweis & Saul, 2000), and H in H-LLE stands for Hessian (Donoho & Grimes, 2003). A vast majority of these algorithms (and many more) are implemented in commercial codes such as Matlab® (Matlab, 2012), and open-source platforms such as R (R, 2012).

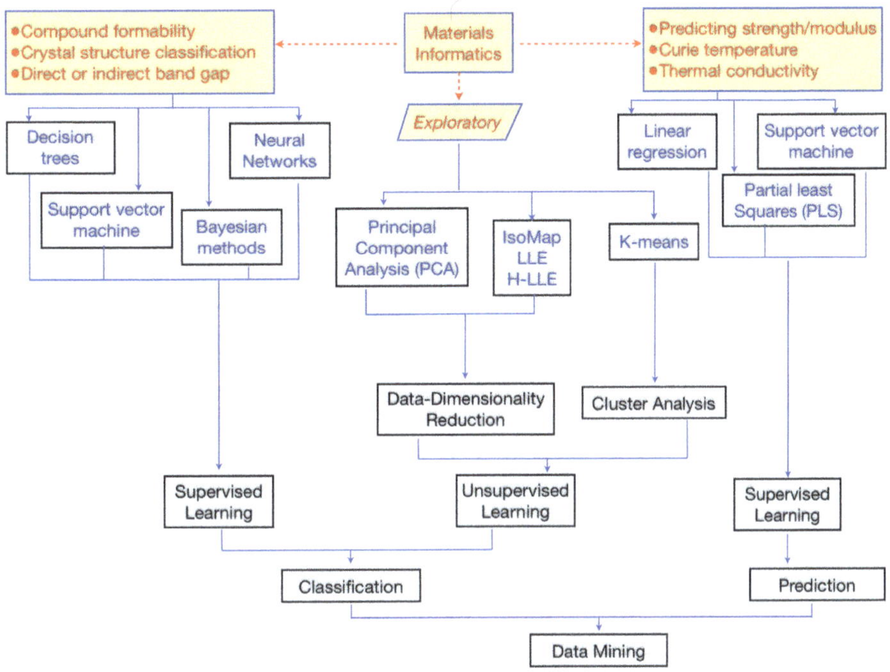

Once all the PC's have been calculated, only those with eigenvalues above a critical level are retained. A standard data-mining rule of thumb is to retain only those PC's whose eigenvalue is greater than or equal to 1. From the knowledge of the calculated PC's, one can determine the relative importance of each descriptor and the correlation between any two descriptors by examining the relative magnitude of the weights associated with each attribute and the cosine of the angle between two attributes, respectively. Information pertaining to the relative importance of descriptors will be helpful in identifying the dominant descriptors; the correlation information will aid in screening the dominant descriptors to avoid choosing redundant ones. Thus from the evaluation of the variance-covariance information (without including any information about the predefined class label), it is possible to screen a large library of descriptors and identify from that set the most important contributions to a physical problem.

K-Means Clustering

k-means clustering is a another well-known methodology used in pattern recognition studies for discovering natural grouping(s) in complex data sets (Jain, 2010). It is useful in materials science problems because it provides a way to rapidly gain insight into the behavior of unknown materials based on its clustering with known materials whose properties are well studied.

Figure 4. Pseudo-algorithm for performing principal component analysis (taken from Balachandran, 2011)

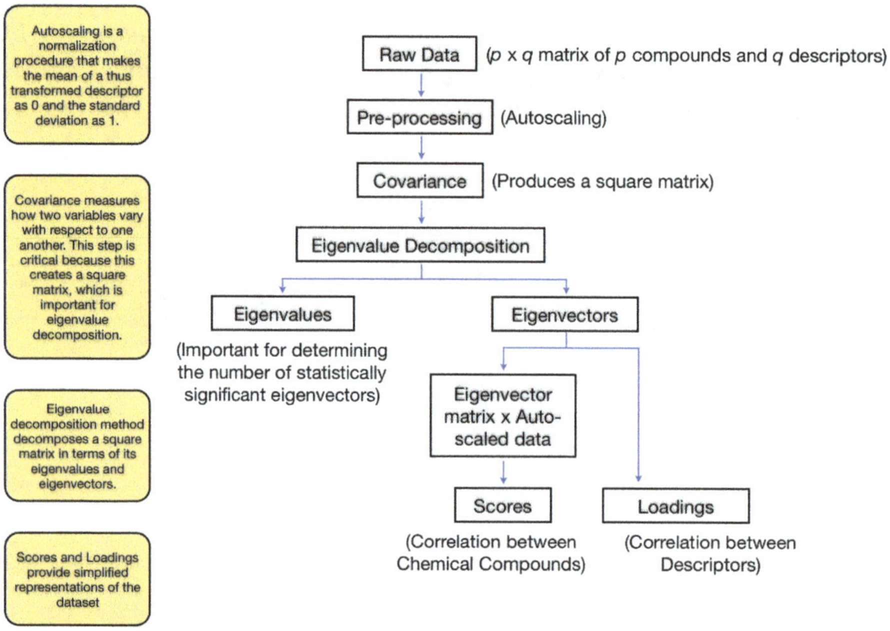

Given a representation of n objects or data points, the k-means clustering attempts to find k groups (or clusters) based on a "measure of similarity," such that the similarities between data points in the same cluster are high while the similarities between data points in different clusters are low. Particularly when the size of the dataset is small for supervised learning, k-means clustering serves as an exploratory approach to rapidly access the property of an unknown material. The exploratory nature of the clustering approach creates a degree of uncertainty in model interpretation, and therefore, the recommended procedure is to explore various "measures of similarities" and then interpret the results carefully with some physical basis to determine the best measure for a given dataset.

The pseudo-algorithm is given in Figure 5. The value of k must be provided at the beginning of the analysis and the algorithm randomly assigns its center. It is good practice in materials science problems to follow the criterion suggested by Tibshirani *et al.* (2001) to determine the optimal number of clusters, where the sum of squared error distance is plotted as a function of k and the "elbow" in the plot provides a reasonable metric for convergence. An object (or data point) is assigned to the cluster to which it is most similar, based on the "distance" between the object and the cluster mean. There are different similarity metrics, namely the Euclidean distance (L_2-norm), Cityblock distance (L_1-norm), and Cosine metric, to name a few that can be employed for pattern recognition. After a particular distance metric is chosen, the algorithm computes a new mean for each cluster. This process iterates until the squared-error criterion function converges,

$$E = \sum_{i=1}^{k} \sum_{p \in C_i} \left| p - m_i \right|^2$$

Figure 5. Pseudo-algorithm for performing k-means clustering analysis (taken from Balachandran, 2011).

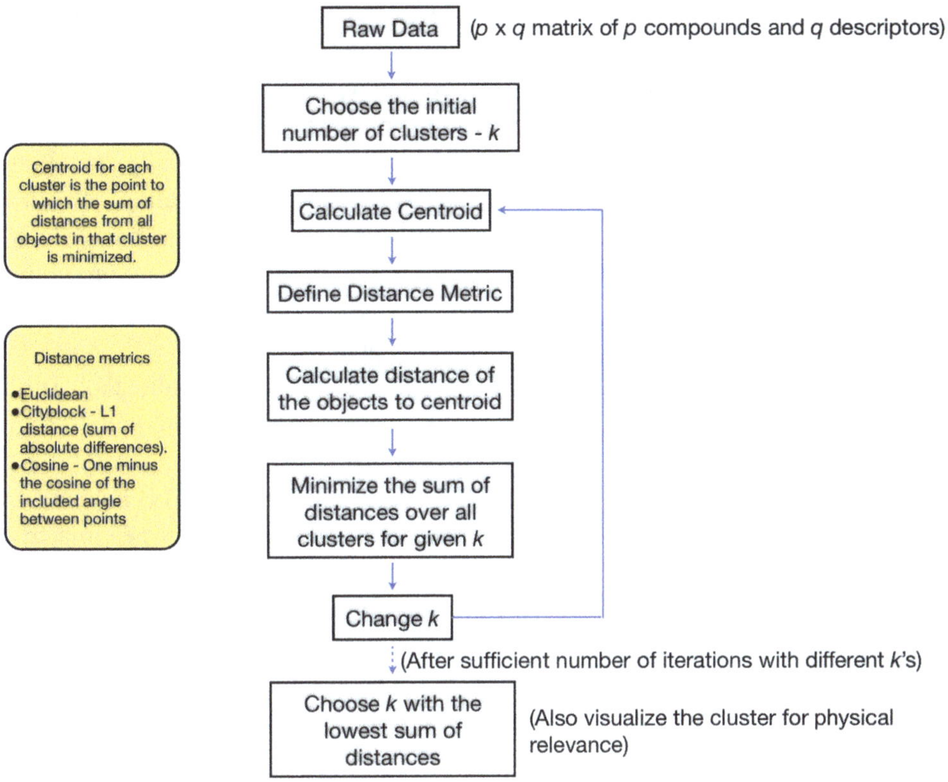

where E is the sum of the square error for all objects in the data set; p is the point in the space representing a given object (e.g., chemical composition); and m_i is the mean of cluster C_i.

Bayes' Rules

Bayes' rule, also known as Bayes' theorem, is one of the well-known data-mining methods that combine prior experience, $p(\theta)$, with observed data or current evidence, $p(D)$, to make inference [*posterior*, $p(\theta|D)$] about the data (Krushke, 2011; Efron, 2013). Mathematically, the theorem is expressed as,

$$p(\theta \mid D) = \frac{p(D \mid \theta)\, p(\theta)}{p(D)}$$

The prior, $p(\theta)$, is the strength of the belief in θ *without* the data D about chemical elements. The *posterior*, $p(\theta|D)$, is the strength in the belief θ when data D about chemical elements *are given*. The term $p(D|\theta)$ is usually referred to as the *likelihood*. The objective is to estimate the *posterior*, $p(\theta|D)$, to rapidly screen for chemical compositions and identify new materials.

Model Validation

Model validation is an important step during model building, particularly in the supervised learning scheme. It is required to test the performance of the model and avoid the condition called *overfitting* (Hawkins, 2004). Overfitting refers to a scenario, where the data mining model performs exceedingly well on the training set that was used to build it, but when applied to a new and unseen data set that was not originally present in the training set, its performance deteriorates dramatically due to poor generalization. The ideal goal is to develop a generic model that both describes the training set with sufficiently high accuracy and also provides reliable estimates on unseen datasets. This is particularly important when trying to uncover new mechanisms for functional electronic behaviour, as it provides a means to discover new material compositions that would *a priori* not be considered as potential useful materials despite being able to support a given property based on the predictive model.

Broadly speaking, there are two ways in which data mining models are validated: (i) n-fold cross-validation and (ii) independent test sets. The model performance is based on how well the training model is able to represent the data in the test set. In n-fold cross-validation approach, one takes the training set and divides it randomly into n-folds (normally, $n=10$ is acceptable). Each fold is held out and the data mining scheme is then performed on the remaining nine-tenths. The error rate is estimated on the held out fold. The final model performance is quantified by averaging the error rate n-times.

On the other hand, in the case of independent test set, the original training dataset is divided into two sets: (i) a smaller training set and a (ii) test set. The numbers of compounds to be used for training and test sets depend on the size of the dataset and there are no strict requirements. Generally, if a large dataset is available (> 1,000 compositions), then it would be best to use 50% of the data for training and the remaining 50% for testing. With smaller datasets, 75% training – 25% testing or 66% training – 34% testing divisions are common. Advanced methods based on bootstrapping, sampling with replacement, (Efron & Gong, 1983) and Bayesian approaches could also be utilized for building optimal models (Dalton & Dougherty, 2011).

NEW MATERIALS DISCOVERY

After having built a dataset for materials informatics that permits a predictive study, data mined the set to obtain an optimized model, and evaluated the model performance using cross-validation or independent test sets, the next important step is use the developed models for making new predictions on compounds that are yet-to-be synthesized or formulate theories for materials behavior. These tasks are by no means simple, because it is important to recognize the boundary conditions of the datasets that were used during model building. For instance, if a dataset was built for predicting properties of (say) perovskite oxides, then the analysis might yield erroneous results when oxide compositions stabilizing in structure types other than perovskites are used. We recommend the readers to be wary of this key limitation arising from databases, when involved in an informatics-based study.

Although cross-validation or independent test sets are accepted norms to validate a data-driven model, we recommend testing the data-driven predictions using *ab initio* or related computational methods, where appropriate. The main reason being that one must remember that statistically derived correlations do not imply causation; they rather provide guidance into understanding and where to look for new understanding. In the case studies discussed next, we describe how the data mining predictions are complemented with density functional theory calculations to directly assess the underlying cause of the statistically discovered correlations and informatics models. We also show how this could be done for both unsupervised and supervised learning problems.

Case Studies

We select four case studies that serve as good examples to illustrate different materials data structures that have been integrated for both unsupervised and supervised learning. Although, our case studies include different structural families of ceramic materials, we recommend the readers to focus on the diversity of the constructed data sets and the data mining methods used for extracting new materials chemistry and physics understanding. Thus the examples should motivate the readers to be creative when constructing a database for materials informatics. More often than not, the available data resources dictate the hypothesis generation and database construction for materials informatics. Therefore, the ability to formulate the problem in different ways in search of previously unexplored insights serves as a powerful repertoire in a materials informatics study.

The first study is on a class of materials known as apatites, where we demonstrate unsupervised learning (Balachadran & Rajan, 2012; Balachandran *et al.* 2014). Materials with the apatite crystal structure are actively studied for a wide range of applications such as lasers, biomaterials, radiation detection materials, hosts for immobilizing toxic elements, and fast oxide-ion conductors (White & ZhiLi, 2003). Earlier works have shown that in apatites, crystal symmetry plays a dominant role in determining their properties (White & ZhiLi, 2003; White *et al.* 2005). The objective of the study was to explore the relationship between chemical elements that occupy the apatite lattice and crystal symmetry. Principal component analysis (PCA) and clustering analysis is used to detect patterns underlying a complex high-dimensional data structure with more than 20 descriptors. We also show how to interpret the analysis and rationalize the findings using density functional theory calculations.

The second study focuses on oxide perovskites (Balachandran & Rondinelli, 2013). These materials have attracted significant interests because their properties are exceedingly sensitive to details of local chemistry and bonding, whereby subtle changes in composition enable them to exhibit a range of physical phenomena – superconductivity, multiferroicity, and magnetoresistance – making them ideal materials for fundamental research (Bhalla *et al.* 2010; Grinberg *et al.* 2013). Their societal impact is tantamount to applications in energy (battery electrodes, solid oxide fuel cells), information technology (data storage), and health (ultrasound imaging). Perovskites are of interests not only in bulk form, but also as thin films that are epitaxially grown on substrates for device integration and as a practically means to control bonding through mechanical constraints. We describe a new descriptor set that captures the complexity of structure in these materials, based on irreducible representations (irreps) of the crystallographic space groups that describe the symmetry properties of the lattice modes and crystal structure (Stokes *et al.* 2013). A unified framework for treating both thin film and bulk materials on equal footing with the informatics methods is then possible, as both a bulk and thin film material can have the same composition but vastly difference crystal structure because they are subjected to different boundary conditions.

Unlike the first two studies, where all attributes in the data structure were numerical quantities, in the third case study we describe how to employ *categorical* variables as attributes and use that information, in turn, to identify design rules linking site-chemistry with crystal symmetry in layered ceramics known as Ruddlesden-Popper oxides (Balachandran *et al.* 2014). These materials have recently garnered renewed interest as low-loss tunable dielectrics (Lee *et al.* 2013), and oxygen evolution/reduction reaction catalysis (Takeguchi *et al.* 2013).

The final study focuses on a regression problem studied by Bucholz *et al.* (2012), where they utilized regression trees to estimate the friction coefficients of candidate ceramic materials that are promising high temperature solid-state lubricants. In lieu of this chapter, this case study presents an interesting example, because it integrates different data mining methods to accomplish the objective of building a predictive regression model.

Unsupervised Learning: Structure Map for Apatites

The hypothesis or materials science question associated with this study was to uncover the relationship between chemical elements that occupy the apatite lattice and the equilibrium crystal symmetry (or space group). A predictive understanding of the relationship between chemistry and structure would permit us to discover new apatite chemistries with tailored properties. The raw data for this study were taken from the work Mercier *et al.* (2005), where they quantify the bond distortions in the $P6_3/m$ apatite compounds using fifteen algebraically independent bond lengths and bond angles. The objective is to screen the bond distortion descriptors, rank them, and identify a subset of dominant descriptors that capture a significant portion of variance in the dataset. Since the geometrical parameters carry important information pertaining to the crystal structure, understanding their trends allows for tracking of structural modifications brought about by chemical substitutions (Mercier *et al.*, 2005 & 2007).

In addition to these structural parameters, the ionic radii and electronegativity were included using values from the work of Shannon (1976) and Pauling (1932), respectively. Table 3 summarizes all twenty-nine attributes (elements of A) used in the unsupervised learning scheme carried out by Balachandran and Rajan (2012).

A data set of twenty-five compounds (elements of X) with the stoichiometry $A^{I}_4A^{II}_6(BO_4)_6X_2$ was developed. Note that, there is no Y (a target variable) in this data structure, which qualifies the current study as an unsupervised learning materials informatics.

The informatics analysis involved three steps:

1. Pre-processing step, where an autoscaling method was used to normalize the data. All column vectors were mean-centered and standardized.
2. Principal component analysis then reduces the dimensionality of the dataset from 29 to 4, *i.e.*, four derived descriptors are able to describe 90% of variance in the data set. The result was the discovery of two key bond distortion angles that explains the majority of variance in the dataset.
3. *k*-means clustering analysis for detecting clusters of compositions that behave similarly.

The results of these tasks are summarized in Figure 6. In the "structure map" [Figure 6(a)], clusters of compositions are highlighted. For example, clusters 1 and 2 (*k*=1 and *k*=2 respectively) correspond to F-apatites [red circles in Figure 6(a)]. They are well localized in the structure map and are characterized by relatively low values of the structural parameters α_{AII} and $\psi^{AIz=0}{}_{AI\text{-}O1}$ [see Figure 6(b) for a schematic of the bond distortion angles].

Interestingly, the cluster analysis reveals that the two fluoro-apatites do not belong to the clusters *k*=1 and *k*=2 are $Hg_{10}(PO_4)_6F_2$ (in *k*=4) and $Zn_{10}(PO_4)_6F_2$ (in *k*=7). The relative position of $Zn_{10}(PO_4)_6F_2$

Table 3. List of twenty-nine descriptors (A) for unsupervised learning of apatite materials.

Descriptor	Brief Description
a (Å)	Lattice constant of the hexagonal unit cell
c (Å)	Lattice constant of the hexagonal unit cell
c/a	Variable axial ratio (no unit)
r_{AI}(Å)	Shannon's ionic radii of A^{I}-site ion (9-coordination)
r_{B}(Å)	Shannon's ionic radii of B-site ion
r_{AII}(Å)	Shannon's ionic radii of A^{II}-site ion (7-coordination for F^{-} and 8-coordination for Cl^{-} and Br^{-} - Dordević *et al.*, 2008)
r_{X}(Å)	Shannon's ionic radii of X-site ion
Av CR (Å)	Average crystal radius = $[(r_{AI}x4)+(r_{AII}x6)+(r_{B}x6)+(r_{O}x24)+r_{X}x2)]/42$
A_{EN}-O_{EN}	Electronegativity difference A-atom and O-atom
B_{EN}-O_{EN}	Electronegativity difference B-atom and O-atom
A_{EN}-X_{EN}	Electronegativity difference A-atom at $A^{[II]}$-site and X-atom
A_{EN}-B_{EN}	Electronegativity difference A-atom at $A^{[I]}$-site and B-atom
A^{I}-O1 (Å)	Distance between A^{I} and O1 atom
A^{I}-$O1^{AIz=0}$ (Å)	Distance between A^{I} and O1 atom with the constraint $z=0$ *at* A^{I}
Δ_{AI-O} (Å)	Difference in the lengths A^{I}-O1 and A^{I}-O2
$\Delta_{AI-O}^{AIz=0}$ (Å)	Difference in the lengths A^{I}-O1 and A^{I}-O2 with the constraint $z=0$ *at* A^{I}
ψ_{AI-O} (°)	The angle that A^{I}-O1 bond makes with respect to c
$\psi_{AI-O}^{AIz=0}$ (°)	The angle that A^{I}-O1 bond makes with respect to c with the constraint $z=0$ *at* A^{I}
δ_{AI}(°)	Counter-rotation angle of $A^{I}O_{6}$ structural unit
φ_{AI} (°)	Metaprism twist angle ($\pi/3-2\delta_{AI}$)
α_{AI} (°)	Orientation of $A^{I}O_{6}$ unit with respect to a
$<B\text{-}O>$ (Å)	Average B-O bond length
$<\tau_{O-B-O}>$ (°)	Average O-B-O bond-bending angle
ρ_{AII} (Å)	A^{II}-A^{II} triangular side length
A^{II}-X (Å)	Distance between A^{II} and X atom
α_{AII} (°)	Orientation of A^{II}-A^{II}-A^{II} triangles with respect to a
A^{II}-O3 (Å)	Distance between A^{II} and O3 atom
$\Phi_{O3-AII-O3}$ (°)	O3-A^{II}-O3 angle
E_{total} (eV)	Total energy calculated from *ab initio* calculations

compound could be rationalized on experimental grounds, because all fluoro-apatites in clusters k=1 and 2 have been experimentally synthesized, whereas Grisafe and Hummel (1970) showed that it is difficult to synthesize fully stoichiometric $Zn_{10}(PO_4)_6F_2$ in the apatite phase. The puzzling outcome from clustering analysis was the relative position of $Hg_{10}(PO_4)_6F_2$ in cluster k=4. Note that unlike $Zn_{10}(PO_4)_6F_2$ compound, $Hg_{10}(PO_4)_6F_2$ has not been synthesized before. Furthermore, the ionic sizes of Hg^{2+} and Ca^{2+} cations are similar, yet they are in two different clusters suggesting more suitable differences in the chemistry of Hg and Ca.

Figure 6. Case study 1-Unsupervised learning to create structure maps for apatites. (a) Structure map for apatites in $P6_3/m$ crystal symmetry, whose coordinates were identified using principal component analysis (PCA); data clustering analysis was performed using k-means clustering. Seven clusters (k) emerge in the structure map. Compositions within the same cluster are anticipated to behave similarly. Notice cluster k=4, where $Hg_{10}(PO_4)_6F_2$ is located. It appears as an anomaly, because unlike other fluoro-apatites (red circles in k=1 and 2) its relative position in the structure map is in cluster k=4, where there are no other fluoro-apatites. (b) A schematic showing the two bond distortion angles used as coordinates in the structure map. In (c) low-energy phonon frequencies from the lattice dynamical density functional calculations for the high symmetry $P6_3/m$ apatite phases for $Ca_{10}(PO_4)_6F_2$ and isovalent and isostructural $Hg_{10}(PO_4)_6F_2$. All low-energy phonons at the zone-center (Γ-point) in $Ca_{10}(PO_4)_6F_2$ have positive frequencies, indicating that the high symmetry structure is dynamically stable. On the other hand, phonons with negative frequencies, indicating imaginary wavenumbers, are found for $Hg_{10}(PO_4)_6F_2$, suggesting that the high symmetry $P6_3/m$ structure is dynamically unstable. Indeed, it distorts to a lower symmetry triclinic $P\overline{1}$ structure. Reprinted with permission from Balachandran, P.V. and Rajan, K. Acta Crystallogrpahica B 68 24-33 (2012) and Balachandran, P.V. et al. Acta Crystallographica B 70, 612-615 (2014). Copyright 2014 International Union of Crystallography.

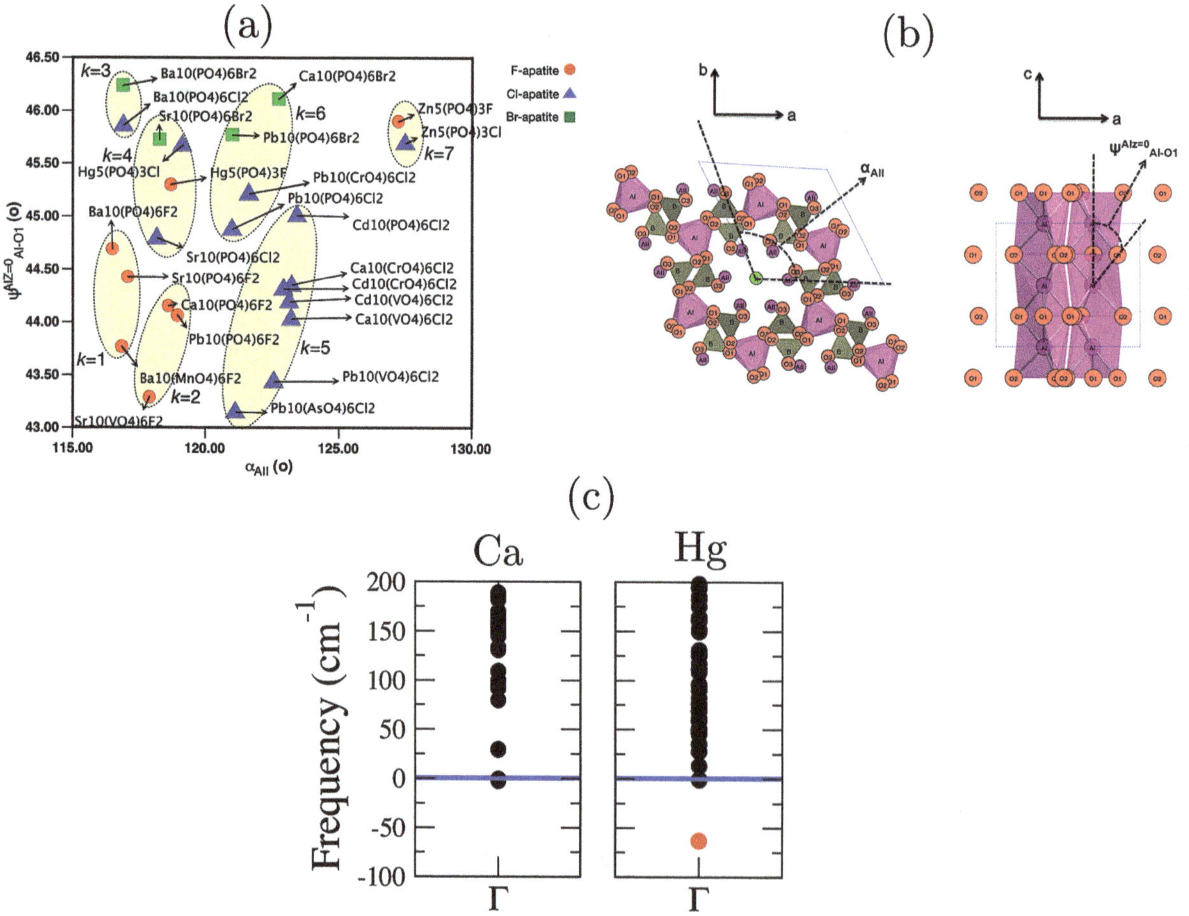

To understand this behaviour, Balachandran *et al.* (2014) performed lattice dynamical calculations within density functional theory (DFT) for $Hg_{10}(PO_4)_6F_2$ and $Ca_{10}(PO_4)_6F_2$ in the $P6_3/m$ crystal symmetry – the same one used in the informatics study. It was found that the $Hg_{10}(PO_4)_6F_2$ in $P6_3/m$ structure was dynamically unstable, indicated by an imaginary phonon at the zone-center. In contrast, $Ca_{10}(PO_4)_6F_2$ was found to be stable, all vibrational modes have positive frequencies [Figure 6(**c**)]. The origin of the imaginary frequency was attributed to the electronic configuration of the Hg^{2+} cations, which have fully occupied $5d$-orbitals and an empty $6s^0$ manifold in the valence configuration. In contrast, the valence configuration of Ca^{2+} is $3p^6$ orbitals and $4s^0$ manifold. The band gaps of $Hg_{10}(PO_4)_6F_2$ and $Ca_{10}(PO_4)_6F_2$were calculated to be 1.12 and 5.67 eV, respectively. The reduced band gap of $Hg_{10}(PO_4)_6F_2$ allows for cross-gap hybridization between the $5d^{10}$ and $6s^0$ electronic states facilitated by a symmetry reduction from the hexagonal $P6_3/m$ to $P\bar{1}$ triclinic structure. The lattice dynamical studies and DFT calculations provide insights into the electronic structure origins that made the $Hg_{10}(PO_4)_6F_2$ in $P6_3/m$ symmetry an outlier in the structure map. Thus by combining informatics with DFT calculations, it was possible to rapidly uncover previously unknown structure–chemistry relationships in the complex fluoro-apatite materials system.

Descriptors for Bulk Materials and Thin Films: Perovskite Nickelates

In this example, we focus on perovskite nickelates ($RNiO_3$, where R is a rare earth element). The structure of these ceramic oxides are characterized by a three-dimensional network of corner-connected metal (often d-block elements) oxygen octahedra, with alkali, alkaline-earth or lanthanide elements filling holes in the body centers of the octahedral network. These nickelate oxides exhibit non-trivial changes in structure and physical properties, including sharp first-order temperature-driven metal to insulator transitions, unusual antiferromagnetic order in the ground state, and site- or bond-centered charge disproportionation owing to the valence and spin-state flexibility of the Ni^{3+} cation (Catalan, 2008). Furthermore, it has been shown that both the electronic and magnetic transition temperatures can be modified by applying epitaxial strain when these materials are grown as thin films (Fong *et* al. 2004; Chakhalian *et al.* 2011; Rondinelli *et al.* 2012).

In Figure 7, we show the rare earth cation-temperature phase diagram of bulk $RNiO_3$ nickelate perovskites. We focus on two important characteristics in the phase diagram: the (i) metal to insulator transition temperature (T_{MI}) and the (ii) paramagnetic to antiferromagnetic phase transition temperatures (T_N). Note that when R=Nd and Pr, $T_{MI} = T_N$. The question posed is: what aspects of the "structure" govern the temperatures of the two phase transitions? The need for this study came from the observation that none of the known crystal chemistry attributes were able to fully describe the variance in the T_N as a function of rare earth cation.

In the literature, tolerance factor is a well-known attribute, which could be considered as an aggregate descriptor that encodes the mismatch in the bond lengths of R–O and Ni–O bonds in the perovskite lattice, which has been used to rationalize the T_{MI} trend of $RNiO_3$ compounds. However, tolerance factor does not explain why $T_{MI} = T_N$ in $NdNiO_3$ and $PrNiO_3$, indicating that the encoded bond lengths does not contain all of the necessary physics in describing the magnetic phase boundaries of $PrNiO_3$ and $NdNiO_3$ compounds—no known structural attributes could fully rationalize the T_N trend.

Distortion-mode decomposition analysis (Howard & Stokes, 1998; Campbell *et al.* 2006; Stokes *et al.* 2007; Orogengoa *et al.* 2009; Perez-Mato *et al.* 2010) provides an alternative basis set for studying

Figure 7. Case study 2-Tolerance factor–Temperature phase diagram of $RNiO_3$ nickelate perovskites after Catalan (2008). T_{MI} stands for metal-to-insulator transition and T_N stands for the antiferromagnetic Néel temperature. Pnma and $P2_1/c$ indicate space groups belonging to orthorhombic and monoclinic crystal systems, respectively.

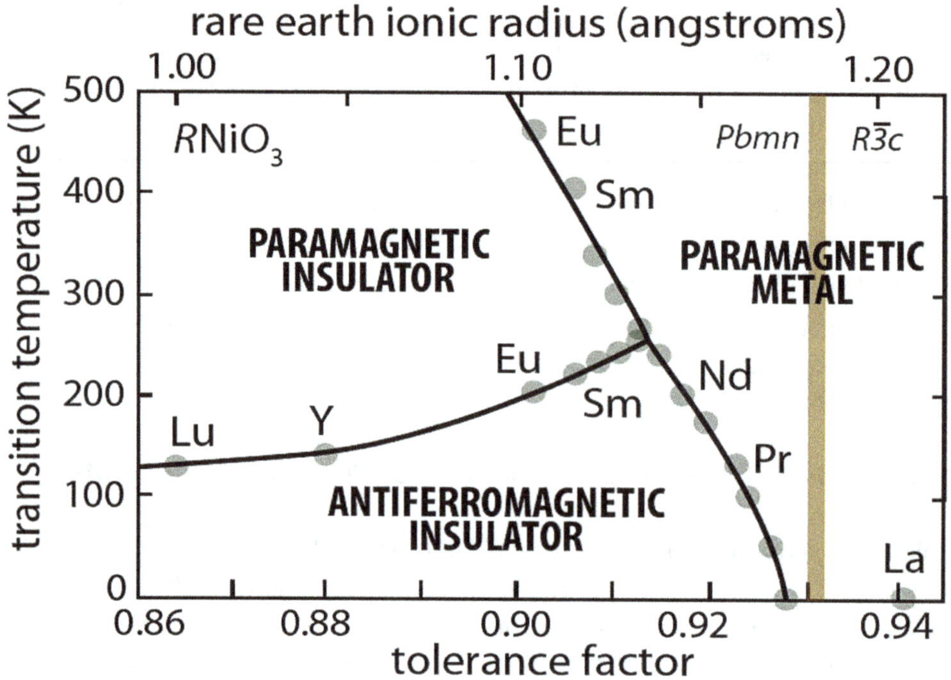

crystal structures. It involves describing a distorted (low-symmetry) structure as arising from a (high-symmetry) parent structure with one or more static symmetry-breaking structural distortions [Figure 8(a)-(i)]. In the undistorted parent structure, symmetry-breaking distortion-modes have zero amplitude. The low-symmetry phase, however, will have finite amplitudes for each mode described by an irreducible representation (irrep) of the high-symmetry structure compatible with the symmetry breaking. Additional details regarding distortion-mode decomposition analysis may be found in the literature (Howard & Stokes, 1998; Campbell *et al.* 2006; Stokes *et al.* 2007; Orogengoa *et al.* 2009; Perez-Mato *et al.* 2010; Balachandran & Rondinelli, 2013).

The distortion-mode analysis is powerful, because it provides a complete and systematic way to isolate multiple and complex distortions in crystals. By comparison of the amplitude of various modes, it is possible to directly assess each modes contribution to the mechanism underlying the structural and electronic phase transition. Furthermore, the distortion-mode analysis relies solely on crystal structural data, which enables both bulk and thin film stabilized structures with identical compositions to be evaluated on equal footing. Note that such direct comparison is not possible through aggregate parameters widely followed in the literature such as the tolerance factor, ionic radius, or electronegativity, *i.e.,* when the composition is fixed.

In terms of this data structure, the X contains the $RNiO_3$ nickelates, A contains the distortion modes, and Y contains the T_{MI} and T_N data. Since we have Y, this is a supervised learning materials informatics study. In Figure 8(j), we show a heat map capturing the statistical correlation between different

Figure 8. Case study 2-(a)-(i) A schematic of the modes capturing different structural distortions in the ground state of perovskite nickelates sorted by irreps of cubic ($Pm\bar{3}m$). The undistorted structure is shown in (a). In (j), the heat map showing the statistical correlation between the irreps and two target variables, T_{MI} and T_N for bulk $RNiO_3$ compounds. Heat map captures important insights linking the bond distortions shown in (a)-(i) and macroscopic transition temperatures (T_{MI} and T_N). Reprinted with permission from Balachandran, P.V. and Rondinelli, J.M. Physical Review B 88 054101 (2013). Copyright 2013 American Physical Society.

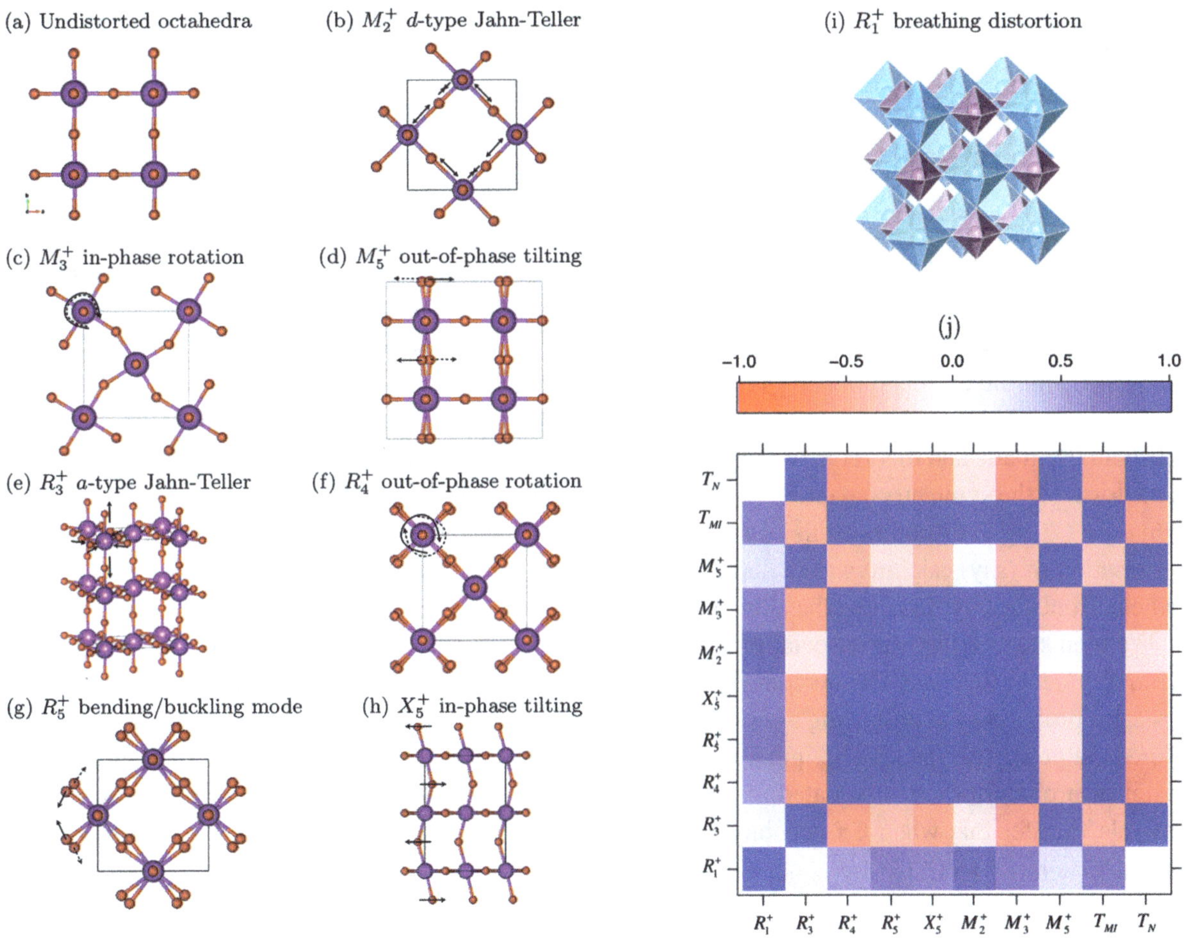

distortions in the structure and the phase transition temperatures (T_{MI} and T_N). A strong positive correlation is found between irreps that describe distortions to the NiO_6 octahedra: M_2^+, M_3^+, X_5^+, R_4^+, and $R_5^+$and T_{MI}. These five irreps fully describe the *Pnma* crystal structure relative to the cubic phase found in the metallic nickelates at high temperature, reinforcing the concept that the orthorhombic distortions are largely responsible for the bandwidth-controlled transport behavior in nickelates (Zhou *et al.* 2005).

Our analyses also reveal the existence of a strong linear relationship between T_N and two irreps, R_3^+ and M_5^+. The linear relationship is valid for both $T_{MI} = T_N$ and $T_{MI} > T_N$ nickelates, indicating that R_3^+*and* M_5^+ contain additional information not captured by either the conventional ⟨Ni–O–Ni⟩ angle or tolerance factor descriptors. Using correlation analysis, we uncovered the existence of a statistical correlation between irreps R_3^+ and M_5^+ and the antiferromagnetic transition temperature, T_N. The M_5^+ irrep describes out-of-phase tilting (see Figure 8d), whereas the R_3^+ irrep (see Figure 8e) refers to a set of cooperative atomic displacements, where the four equatorial oxygen atoms are brought closer to the central Ni-atom, whereas the two apical oxygen atoms move apart (and vice-versa). The R_3^+ distortions are commonly referred to as a-type Jahn-Teller distortions in the perovskite literature. One of the important findings is that R_3^+ and M_5^+ irreps capture the T_N trends that were previously unknown and the implications are that these distortions are not encoded in the widely recognized tolerance factor metric. This work brings out the critical importance of the choice of data structures in an informatics study, when describing complex materials systems.

Supervised Learning in Materials Informatics: Composition Predictions for Noncentrosymmetric Oxides

In this example, unlike the previous two cases, we demonstrate how categorical variables may serve as attributes for matrix A. Here, the material of interest has the stoichiometry A_2BO_4 and is commonly referred to as Ruddlesden-Popper (214-RP) compounds (Ruddlesden& Popper, 1957). The objective of this study is to discover the relationship between cations that occupy A- and B-sites in A_2BO_4 (where O is oxygen) and the crystal symmetry (Balachandran *et al.* 2014).

For this study, the X contains the compositions of various A_2BO_4 chemistries that have been experimentally made and theoretically predicted; Y contains the crystal symmetry information, encoded in the form of the symmetry-adapted modes (irreps) described above [see the original paper of Balachandran *et al.* (2014) for further details]. However, A was constructed differently; for example, the element information of (say) calcium was included as "Ca" in A, instead of using attributes such as ionic radii and electronegativity that describe element Ca.

We then searched for the associations capturing how frequently will an element and specific crystal symmetry occur together. The outcome was quantified in terms of a probability using Bayes' rules. The posterior probability of observing a crystal symmetry given a specific element in the lattice is estimated to create a predictive model. In Figure 9(a) and (b), we show the outcome from Bayes' rules where the posterior probabilities are estimated for each A- and B-site element in the dataset. From Figure 9, it is possible to infer that when Ca occupies the A-site, two types of distortions should be present in the crystal structure, described by notations $X_2^+ \oplus X_3^+$ and P_4. Similarly, when Ir occupies the B-site it prefers P_4 distortion. When Ca and Ir are combined together to form Ca_2IrO_4 in 214-RP crystal structure, it was predicted that in the ground state crystal structure one of the three following distortions derived from the tendencies of the individual cations to distort would co-exist: $X_2^+ \oplus X_3^+$, P_4 and $X_2^+ \oplus X_3^+ \oplus P_4$.

Our hypothesis from Bayesian analysis was that Ca_2IrO_4 could be a potential ferroelectric material. Our study was further motivated by the fact that ferroelectricity in 214-RP phases is rare (Balachandran *et al.* 2014). We then employed density functional theory to determine the ground state structure of Ca_2IrO_4 for *ten* different crystal symmetries. Our calculated total energy results are shown in Figure 9(c). We find that there is a strong competition between the centrosymmetric $Pbca$ symmetry and the polar (or ferroelectric) $Pca2_1$ symmetry. Note that we predicted both $Pbca$ and $Pca2_1$ phases using our Bayes' rules, which are in excellent agreement with our DFT calculations.

Figure 9. Case study 3-Posterior probabilities that a particular mode distortion, described by irreps of $I4/mmm$ in 214 RP oxides, will be observed given a specific (a) A-site or (b) B-site element. ϕ indicate no tendency to octahedral distortions. Balachandran et al. (2014) used the insights gained from (a) and (b) to discover a new chemical composition, Ca_2IrO_4, and predict its crystal symmetry. In (c) we show the total energy calculated from density functional theory for various crystal symmetries to validate the predictions from Bayes' rules. Structures with lowest energies are considered to be more stable relative to those that show high energies. The ground state of Ca_2IrO_4 could be $I4_1/acd$, $Pbca$ or $Pca2_1$ and DFT calculations indicate that the ground state is $Pbca$ with a strong energetic competition between $Pbca$ and $Pca2_1$ crystal symmetries, suggesting excellent agreement between our informatics-based Bayes' rules and DFT calculations. Reprinted with permission from Balachandran, P.V. et al. Inorganic Chemistry 53 pp. 336-348 (2014). Copyright 2014 American Chemical Society.

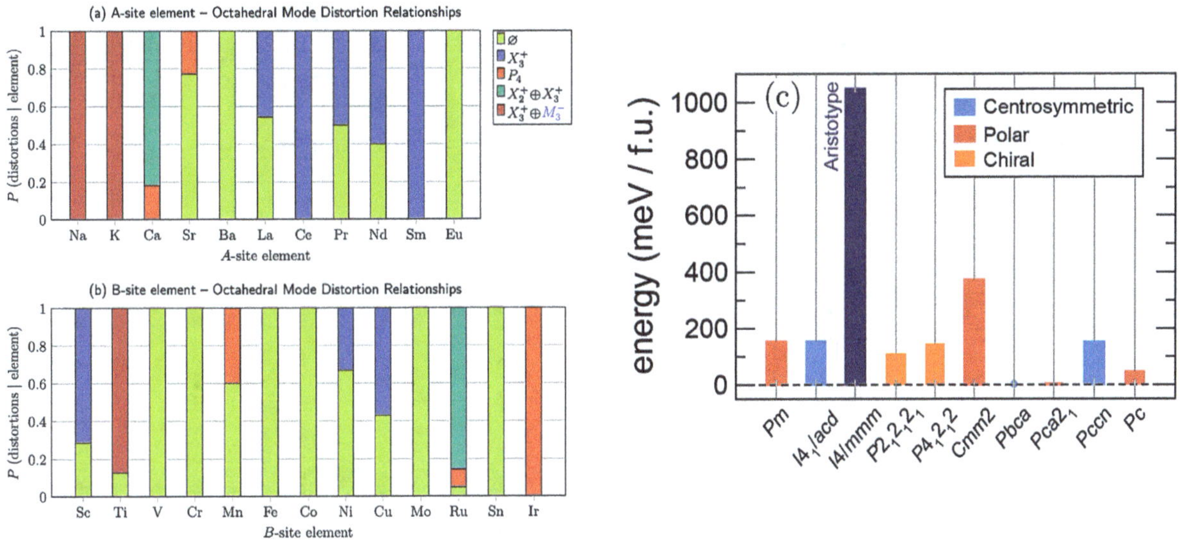

Regression Trees for Predicting Friction Coefficients

In this study Bucholz *et al.* (2012) focused on 38 different binary ceramic materials (X) and described each ceramic using 16 attributes (A). The objective of the work was to develop a data mining model that could predict the friction coefficients (Y , target variable) of the ceramic materials, which is a numerical quantity. Therefore, this is a supervised learning and prediction study. The authors performed pin-on-disk experiments on 24 out of 38 ceramics to measure the friction coefficients and the data for the remaining 14 materials were taken from the literature. Motivated by earlier research on solid-state lubricant materials, where friction coefficients have been linked to attributes such as polarizing power of a cation and electronegativity, the authors constructed a database containing 16 attributes. A materials informatics study was formulated based on the hypothesis that statistical correlations exist between some of the attributes and measured friction coefficients of the binary ceramic materials.

Instead of predicting friction coefficients directly from 16 attributes, the authors first performed a "feature selection" study, where the aim was to screen all 16 attributes and identify a reduced set of dominant attributes that capture a majority of variance in the dataset. In fact, an unsupervised principal component analysis (PCA) indicated that some of the attributes were strongly correlated to one another.

Figure 10. Case study 4-(a) Informatics-based strategy adapted by Bucholz et al. (2012) for modeling friction coefficient, (b) Predictive regression tree model and (c) performance of the regression tree model. In (a), abbreviations PCA, VIP, and SS refer to principal component analysis, variable importance in projection, and sum of squares, respectively. Reprinted with permission from Bucholz et al. Tribology Letters 47(2), 211-221 (2012). Copyright 2012 Springer.

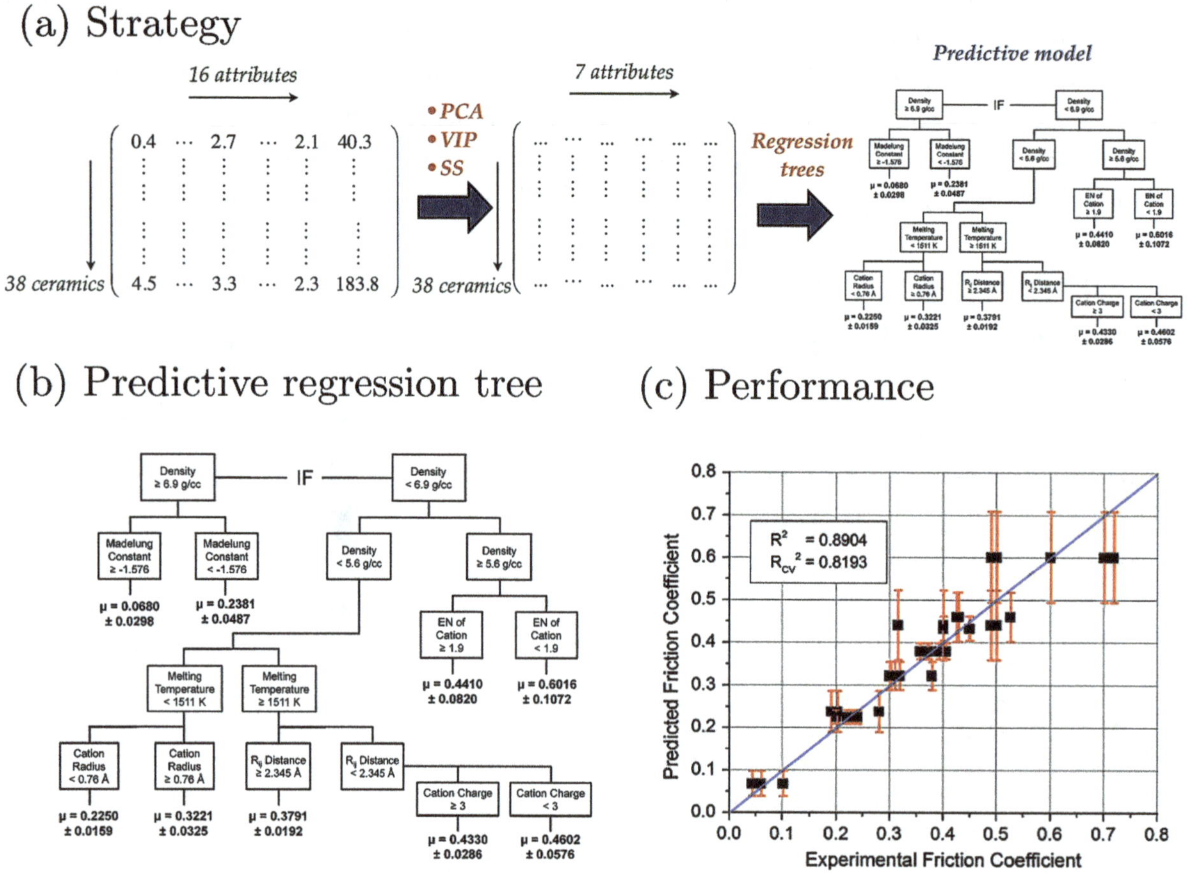

The authors then employed two metrics, namely variable importance in projection (VIP) and sum of squares measure to quantify the relative importance of each attribute [details can be inferred from Bucholz *et al.* (2012)]. From this analysis, the number of attributes was down-selected from 16 to 7. Regression trees were then created to formulate a data-driven model for predicting friction coefficient with the 7 down-selected attributes. The overall strategy of their approach is schematically shown in Figure 10(a).

In Figure 10(b) and (c), the final regression tree model and the experimental vs. predicted friction coefficients (reproduced from the original work) are also shown, respectively. The outcome of this work has two important implications in the design of new high friction coefficient solid-state materials. First, the authors were able to rank the relative importance of key descriptors that impact the friction coefficient – a macroscopic property that is difficult to construct from equilibrium and microscopic quantities. They find that the density of the ceramic, its melting point, and the cation radius have a larger impact relative to other descriptors. Second, the authors uncover a design rule that could be potentially utilized as a screening metric for high-throughput calculations and the identification of new solid-state inorganic lubricants.

FUTURE DIRECTIONS

Although there has been steady interest in the field of materials informatics in the last decade or so (see Figure 2a), the research is still in its infancy. Unlike the fields of genomics and proteomics, where there are established protocols, supported by advancements in instrumentations, in materials science we haven't yet converged on a common materials data structure. There are a number of schools of thought, when it comes to the definition of a "materials genome." We propose that a focus on *function-by-design*, by collecting the necessary data, which is wide ranging in type, number and complexity, is necessary to close the data gap between basic research and the development of new engineered materials through creation of new physical models that guide construction of descriptors. We have been able to achieve remarkable progress in informatics-based studies involving materials belonging to a particular structure-type. However, the challenging question is to develop a generalized scheme, where both chemical and structural diversity could be included in the same dataset. Informatics-based data analysis alone is likely insufficient to create this knowledge, and the methods described should be guided by physical models to make materials design tractable. This knowledge will ultimately lead to a broader understanding of materials properties, enabling the discovery of compounds well beyond the set of materials from which the informatics-based understanding was conceived.

From the algorithm standpoint, we need to develop robust methods that could handle small datasets and quantify uncertainty. Unlike protein expression and microarray data in life sciences wherein each sample is defined by thousands of measurements, synthesis and characterization of materials is still an expensive endeavor. Therefore, from a practical standpoint, we have to be content with relatively smaller sample sizes. Nevertheless, this opens up new opportunities in the field of algorithm development for materials informatics, which remains an unchartered territory.

CONCLUSION

To summarize, we have introduced how to apply materials informatics methods for understanding material structure and function. We discussed it from a practical standpoint for the benefit of interested practitioners and researchers. The entire informatics workflow was outlined into subtasks, and explained some of the intricacies involved in each step. A comprehensive account of how to construct the data structure for materials informatics learning schemes was presented, and more importantly, provided a list of key references for those interested to build robust databases. The critical role of data mining was addressed through some illustrative case studies, and it was showed how different algorithms make it possible to answer difficult questions facing materials scientists and engineers.

ACKNOWLEDGMENT

The authors acknowledge fruitful discussions with Dr. Turab Lookman. This research was supported by the Defense Advanced Research Projects Agency (DARPA) under Grant No. N66001-12-4224 and the National Science Foundation under Grant No. DMR-1454688.

REFERENCES

Ahrens, L. H. (1952). Anion Affinity and Polarizing Power of Cations. *Nature*, *169*(4298), 463–463. doi:10.1038/169463a0

Allred, A. L., & Rochow, E. G. (1958). A scale of electronegativity based on electrostatic force. *Journal of Inorganic and Nuclear Chemistry*, *5*(4), 264–268. doi:10.1016/0022-1902(58)80003-2

Balachandran, P. V. (2011). *Statistical learning for chemical crystallography.* (Doctoral dissertation). Iowa State University, Ames, IA.

Balachandran, P. V., Broderick, S. R., & Rajan, K. (2011). Identifying the "inorganic gene" for high-temperature piezoelectric perovskites through statistical learning. *Proceedings of the Royal Society A: Mathematical, Physical, and Engineering Science, 465*, 2155-2175.

Balachandran, P. V., Puggioni, D., & Rondinelli, J. M. (2014). Crystal-chemistry guidelines for noncentrosymmetric A_2BO_4 Ruddlesden-Popper oxides. *Inorganic Chemistry*, *53*(1), 336–348. doi:10.1021/ic402283c PMID:24320755

Balachandran, P. V., & Rajan, K. (2012). Structure maps for $A^{I}_{6}A^{II}_{4}(BO_4)_6X_2$ apatites via data mining. *Acta Crystallographica. Section B, Structural Science*, *68*(1), 24–33. doi:10.1107/S0108768111054061 PMID:22267555

Balachandran, P. V., Rajan, K., & Rondinelli, J. M. (2014). Electronically driven structural transitions in $A_{10}(PO_4)_6F_2$apatites (A=Ca, Sr, Pb, Cd, and Hg). *Acta Crystallographica. Section B, Structural Science*, *70*(3), 612–615. doi:10.1107/S2052520614003710 PMID:24892609

Balachandran, P. V., & Rondinelli, J. M. (2013). Interplay of octahedral rotations and breathing distortions in charge ordering perovskite oxides. *Physical Review B: Condensed Matter and Materials Physics*, *88*(5), 054101. doi:10.1103/PhysRevB.88.054101

Belsky, A., Hellenbrandt, M., Karen, V. L., & Luksch, P. (2002). New developments in the inorganic crystal structure database (ICSD): Accessibility in support of materials research and design. *Acta Crystallographica. Section B, Structural Science*, *58*(3), 364–369. doi:10.1107/S0108768102006948 PMID:12037357

Bhalla, A. S., Guo, R., & Roy, R. (2000). The perovskite structure - a review of its role in ceramic science and technology. *Materials Research Innovations*, *4*(1), 3–26. doi:10.1007/s100190000062

Brese, N. E., & O'Keeffe, M. (1991). Bond-valence parameters for solids. *Acta Crystallographica. Section B, Structural Crystallography and Crystal Chemistry*, *47*(2), 192–197. doi:10.1107/S0108768190011041

Bucholz, E. W., Kong, C. S., Marchman, K. R., Sawyer, W. G., Phillpot, S. R., Sinnott, S. B., & Rajan, K. (2012). Data-driven model for estimation of friction coefficient via informatics methods. *Tribology Letters*, *47*(2), 211–221. doi:10.1007/s11249-012-9975-y

Campbell, B. J., Stokes, H. T., Tanner, D. E., & Hatch, D. M. (2006). ISODISPLACE: A web-based tool for exploring structural distortions. *Journal of Applied Crystallography*, *39*(4), 607–614. doi:10.1107/S0021889806014075

Catalan, G. (2008). Progress in perovskite nickelate research. *Phase Transitions*, *81*(7-8), 729–749. doi:10.1080/01411590801992463

Ceder, G. (1998). Computational materials science: Predicting properties from scratch. *Science*, *277*, 1237–1242.

Chakhalian, J., Rondinelli, J. M., Liu, J., Gray, B. A., Kareev, M., Moon, E. J., & Freeland, J. W. et al. (2011). Asymmetric orbital-lattice interactions in ultrathin correlated oxide films. *Physical Review Letters*, *107*(11), 116805. doi:10.1103/PhysRevLett.107.116805 PMID:22026694

Cheetham, A. K. (1994). Advanced inorganic materials: An open horizon. *Science*, *264*(5160), 794–795. doi:10.1126/science.264.5160.794 PMID:17794719

Clarke, R., Ressom, H. W., Wang, A., Xuan, J., Liu, M. C., Gehan, E. A., & Wang, Y. (2008). The properties of high-dimensional data spaces: Implications for exploring gene and protein expression data. *Nature Reviews. Cancer*, *8*(1), 37–49. doi:10.1038/nrc2294 PMID:18097463

Clementi, E., & Raimondi, D. L. (1963). Atomic Screening Constants from SCF Functions. *The Journal of Chemical Physics*, *38*(11), 2686. doi:10.1063/1.1733573

Council, N. S. T. (2011). *Materials Genome Initiative for global competitiveness*. Washington, D.: Office of Science and Technology Policy.

Curtarolo, S., Setyawan, W., Wang, S., Xue, J., Yang, K., Taylor, R. H., & Levy, O. et al. (2012). AFLOWLIB.ORG: A distributed materials properties repository from high-throughput *ab initio* calculations. *Computational Materials Science*, *58*, 227–235. doi:10.1016/j.commatsci.2012.02.002

Dalton, L. A., & Dougherty, E. R. (2011). Bayesian minimum mean-square error estimation for classification error – Part 1: Definition and the Bayesian MMSE error estimator for discrete classification. *IEEE Transactions on Signal Processing*, *59*(1), 115–129. doi:10.1109/TSP.2010.2084572

Dey, P., Bible, J., Datta, S., Broderick, S., Jasinski, J., Sunkara, M., & Rajan, K. et al. (2014). Informatics-aided bandgap engineering for solar materials. *Computational Materials Science*, *83*, 185–195. doi:10.1016/j.commatsci.2013.10.016

Donoho, D. L., & Grimes, C. (2003). Hessian eigenmaps: Locally linear embedding techniques for high-dimensional data. *Proceedings of the National Academy of Sciences of the United States of America*, *100*(10), 5591–5596. doi:10.1073/pnas.1031596100 PMID:16576753

Efron, B. (2013). Bayes' theorem in the 21st Century. *Science*, *340*(6137), 1177–1178. doi:10.1126/science.1236536 PMID:23744934

Efron, B., & Gong, G. (1983). A leisurely look at the Bootstrap, the Jackknife, and Cross-Validation. *The American Statistician*, *37*(1), 36–48.

Fong, D. D., Stephenson, G. B., Streiffer, S. K., Eastman, J. A., Auciello, O., Fuoss, P. H., & Thompson, C. (2004). Ferroelectricity in ultrathin perovskite films. *Science*, *304*(5677), 1650–1653. doi:10.1126/science.1098252 PMID:15192223

Froese, F. C. (1972). Average-energy-of-configuration Hartree-Fock results for the atoms helium to radon. *Atomic Data and Nuclear Data Tables*, *4*, 301–399. doi:10.1016/S0092-640X(72)80008-1

Georges, A., Kotliar, G., Krauth, W., & Rozenberg, M. J. (1996). Dynamical mean-field theory of strongly correlated fermion systems and the limit of infinite dimensions. *Reviews of Modern Physics*, *68*(1), 13–125. doi:10.1103/RevModPhys.68.13

Giannozzi, P. (2009). Quantum espresso: A modular and open-source software project for quantum simulation of materials. *Journal of Physics Condensed Matter*, *21*, 395502. PMID:21832390

Gordy, W. (1946). A New Method of Determining Electronegativity from Other Atomic Properties. *Physical Review*, *69*(11-12), 604–607. doi:10.1103/PhysRev.69.604

Grinberg, I., West, D. V., Torres, M., Gou, G., Stein, D. M., Wu, L., & Rappe, A. M. et al. (2013). Perovskite oxides for visible-light-absorbing ferroelectric and photovoltaic materials. *Nature*, *503*(7477), 509–512. doi:10.1038/nature12622 PMID:24213630

Han, J., & Kamber, M. (2006). *Data Mining: Concepts and Techniques. The Morgan Kaufmann Series in Data Management Systems* (2nd ed.). San Francisco: Morgan Kaufmann.

Hawkins, D. M. (2004). The problem of overfitting. *Journal of Chemical Information and Computer Sciences*, *44*(1), 1–12. doi:10.1021/ci0342472 PMID:14741005

Hedin, L. (1965). New method for calculating the one-particle Green's function with application to the electron-gas problem. *Physical Review*, *139*(3A), A796–A823. doi:10.1103/PhysRev.139.A796

Hohenberg, P., & Kohn, W. (1964). Inhomogeneous electron gas. *Physical Review*, *136*(3B), B864–B871. doi:10.1103/PhysRev.136.B864

Hotop, H., & Lineberger, W. C. (1985). Binding Energies in Atomic Negative Ions: II. *Journal of Physical and Chemical Reference Data*, *14*(3), 731. doi:10.1063/1.555735

Howard, C. J., & Stokes, H. T. (1998). Group-theoretical analysis of octahedral tilting in perovskites. *Acta Crystallographica. Section B, Structural Crystallography and Crystal Chemistry*, *54*(6), 782–789. doi:10.1107/S0108768198004200

Jain, A., Ong, S. P., Hautier, G., Chen, W., Richards, W. D., Dacek, S., & Persson, K. A. et al. (2013). Commentary: The Materials Project: A materials genome approach to accelerating materials innovation. *APL Materials*, *1*(1), 011002. doi:10.1063/1.4812323

Jain, A. K. (2010). Data clustering: 50 years beyond K-means. *Pattern Recognition Letters*, *31*(8), 651–666. doi:10.1016/j.patrec.2009.09.011

Kohn, W., & Sham, L. J. (1965). Self-consistent equations including exchange and correlation effects. *Physical Review*, *140*(4A), A1133–A1138. doi:10.1103/PhysRev.140.A1133

Koinuma, H., & Takeuchi, I. (2004). Combinatorial solid-state chemistry of inorganic materials. *Nature Materials*, *3*(7), 429–438. doi:10.1038/nmat1157 PMID:15229491

Kotliar, G., Savrasov, S. Y., Haule, K., Oudovenko, V. S., Parcollet, O., & Marianetti, C. (2006). Electronic structure calculations with dynamical mean-field theory. *Reviews of Modern Physics*, *78*(3), 865–951. doi:10.1103/RevModPhys.78.865

Kreidler, E. R., & Hummel, F. A. (1970). The crystal chemistry of apatite: Structure fields of fluor and chlorapatite. *The American Mineralogist*, *55*, 170–184.

Kresse, G., & Furthmuller, J. (1996). Efficient iterative schemes for *ab initio* total-energy calculations using a plane-wave basis set. *Physical Review B: Condensed Matter and Materials Physics*, *54*(16), 11169–11186. doi:10.1103/PhysRevB.54.11169 PMID:9984901

Kresse, G., & Joubert, D. (1999). From ultrasoft pseudopotentials to the projector augmented-wave method. *Physical Review B: Condensed Matter and Materials Physics*, *59*(3), 1758–1775. doi:10.1103/PhysRevB.59.1758

Krushke, J. K. (2011). *Doing Bayesian Data Analysis*. Oxford, UK: Academic Press, Elsevier.

Le, T., Epa, C., Burden, F. R., & Winkler, D. A. (2012). Quantitative structure-property relationship modeling of diverse materials properties. *Chemical Reviews*, *112*(5), 2889–2919. doi:10.1021/cr200066h PMID:22251444

Lee, C.-H., Orloff, N. D., Birol, T., Zhu, Y., Goian, V., Rocas, E., & Schlom, D. G. et al. (2013). Exploiting dimensionality and defect mitigation to create tunable microwave dielectrics. *Nature*, *502*(7472), 532–536. doi:10.1038/nature12582 PMID:24132232

Mannhart, J., & Schlom, D. (2010). Oxide interfaces – an opportunity for electronics. *Science*, *327*(5973), 1607–1611. doi:10.1126/science.1181862 PMID:20339065

MATLAB and Statistics Toolbox Release. (2012b). TheMathWorks, Inc.

Mercier, P. H. J., Dong, Z., Baikie, T., Le Page, Y., White, T. J., Whitfield, P. S., & Mitchel, L. D. (2007). *Ab initio* constrained crystal-chemical Rietveld refinement of $Ca_{10}(V_xP_{1-x}O_4)_6F_2$ apatites. *Acta Crystallographica. Section B, Structural Crystallography and Crystal Chemistry*, *63*(1), 37–48. doi:10.1107/S0108768106045538 PMID:17235192

Mercier, P. H. J., Le Page, Y., Whitfield, P. S., Mitchell, L. D., Davidson, I. J., & White, T. J. (2005). Geometrical parameterization of the crystal chemistry of $P6_3/m$ apatites: Comparison with experimental data and *ab initio* results. *Acta Crystallographica. Section B, Structural Crystallography and Crystal Chemistry*, *61*(6), 635–655. doi:10.1107/S0108768105031125 PMID:16306671

Mooser, E., & Pearson, W. B. (1959). On the crystal chemistry of normal valence compounds. *Acta Crystallographica*, *12*(12), 1015–1022. doi:10.1107/S0365110X59002857

Muller, O., & Roy, R. (1974). *The major ternary structural families.Crystal chemistry of non- metallic materials*. New York: Springer-Verlag. doi:10.1007/978-3-642-65706-1

Mulliken, R. S. (1934). A New Electroaffinity Scale: Together with Data on Valence States and on Valence Ionization Potentials and Electron Affinities. *The Journal of Chemical Physics*, *2*(11), 782. doi:10.1063/1.1749394

Olson, G. B. (2013). Genomic materials design: The ferrous frontier. *Acta Materialia*, *61*(3), 771–781. doi:10.1016/j.actamat.2012.10.045

Orogengoa, D., Capillas, C., Aroyo, M. I., & Perez-Mato, J. M. (2009). AMPLIMODES: Symmetry-mode analysis on the Bilbao Crystallographic Server. *Journal of Applied Crystallography*, *42*(5), 820–833. doi:10.1107/S0021889809028064

Parr, R. G., & Yang, W. (1995). Density-functional theory of the electronic structure of molecules. *Annual Review of Physical Chemistry*, *46*(1), 701–728. doi:10.1146/annurev.pc.46.100195.003413 PMID:24341393

Pauling, L. (1932). The nature of the chemical bond. IV: The energy of single bonds and the relative electronegativity of atoms. *Journal of the American Chemical Society*, *54*(9), 3570–3582. doi:10.1021/ja01348a011

Pearson, R. G. (1986). Absolute electronegativity and hardness correlated with molecular orbital theory. *Proceedings of the National Academy of Sciences of the United States of America*, *83*(22), 8440–8441. doi:10.1073/pnas.83.22.8440 PMID:16578791

Pelatt, B. D., Ravichandran, R., Wager, J. F., & Keszler, D. A. (2011). Atomic Solid State Energy Scale. *Journal of the American Chemical Society*, *133*(42), 16852–16860. doi:10.1021/ja204670s PMID:21861503

Perez-Mato, J. M., Orobengoa, D., & Aroyo, M. I. (2010). Mode crystallography of distorted structures. *Acta Crystallographica. Section A, Foundations of Crystallography*, *66*(5), 558–590. doi:10.1107/S0108767310016247 PMID:20720321

Pettifor, D. G. (1995). *Bonding and Structure of Molecules and Solids*. Oxford, UK: Oxford University Press.

Phillpot, S. R., & Sinnott, S. B. (2009). Simulating multifunctional structures. *Science*, *325*(5948), 1634–1635. doi:10.1126/science.1177794 PMID:19779181

Potyrailo, R., Rajan, K., Stoewe, K., Takeuchi, I., Chislom, B., & Lam, H. (2011). Combinatorial and high-throughput screening of materials libraries: Review of state of the art. *ACS Combinatorial Science*, *13*(6), 579–633. doi:10.1021/co200007w PMID:21644562

R Development Core Team. (2012). *R: A language and environment for statistical computing*. Vienna, Austria: R Foundation for Statistical Computing.

Rabe, K. M., Phillips, J. C., Villars, P., & Brown, I. D. (1992). Global multinary structural chemistry of stable quasicrystals, high-Tc ferroelectrics, and high-Tc superconductors. *Physical Review B: Condensed Matter and Materials Physics*, *45*(14), 7650–7676. doi:10.1103/PhysRevB.45.7650 PMID:10000573

Rajan, K. (2005). Materials Informatics. *Materials Today*, *8*(10), 38–45. doi:10.1016/S1369-7021(05)71123-8

Rajan, K. (2010). Data Mining and Inorganic Crystallography. In D. W. M. Hoffmann & L. N. Kuleshova (Eds.), *Data Mining in Crystallography134, Structure & Bonding. Springer Berlin/Heidelberg*.

Ringner, M. (2008). What is principal component analysis? *Nature Biotechnology*, *26*(3), 303–304. doi:10.1038/nbt0308-303 PMID:18327243

Rondinelli, J. M., May, S. J., & Freeland, J. W. (2012). Control of octahedral connectivity in perovskite oxide heterostructures: An emerging route to multifunctional materials discovery. *MRS Bulletin*, *37*(03), 261–270. doi:10.1557/mrs.2012.49

Rondinelli, J. M., Stengel, M., & Spaldin, N. (2008). Carrier-mediated magnetoelectricity in complex oxide heterostructures. *Nature Nanotechnology*, *3*(1), 46–50. doi:10.1038/nnano.2007.412 PMID:18654450

Roweis, S. T. & Saul, L.K. (2000). Nonliner dimensionality reduction by local linear embedding, *Science, 22*, 2323-2326.

Ruddlesden, S. N., & Popper, P. (1957). New compounds of the K_2NiF_4 type. *Acta Crystallographica*, *10*(8), 538–539. doi:10.1107/S0365110X57001929

Saal, J. E., Kirklin, S., Aykol, M., Meredig, B., & Wolverton, C. (2013). Materials Design and Discovery with High-Throughput Density Functional Theory: The Open Quantum Materials Database (OQMD). *Journal of Metals*, *65*, 1501–1509.

Sanderson, R. T. (1952). Electronegativities in inorganic chemistry. *Journal of Chemical Education*, *29*(11), 539. doi:10.1021/ed029p539

Saxena, S. K. (1993). *Thermodynamic data on oxides and silicates: an assessed dataset based on thermochemistry and high pressure phase equilibrium*. Berlin: Springer-Verlag. doi:10.1007/978-3-642-78332-6

Setyawan, W., & Curtarolo, S. (2010). High-throughput electronic structure calculations: Challenges and tools. *Computational Materials Science*, *49*(2), 299–312. doi:10.1016/j.commatsci.2010.05.010

Shannon, R. (1976). Revised effective ionic radii and systematic studies of interatomic distances in halides and chalcogenides. *Acta Crystallographica. Section A, Foundations of Crystallography*, *32*(5), 751–767. doi:10.1107/S0567739476001551

Skriver, H. L. (2004). *Materials Science Databases, CAMP DTU*. Retrieved from http://guglen.dk/databases/hlsPT/ptable.php

Slater, J. C. (1930). Atomic Shielding Constants. *Physical Review*, *36*(1), 57–64. doi:10.1103/PhysRev.36.57

Sola, J., & Sevilla, J. (1997). Importance of input data normalization for the application of neural networks to complex industrial problems. *IEEE Transactions on Nuclear Science*, *44*(3), 1464–1468. doi:10.1109/23.589532

Stokes, H. T., Campbell, B. J., & Cordes, R. (2013). Tabulation of irreducible representations of the crystallographic space groups and their superspace extensions. *ActaCrystallographica Section A*, *69*, 388–395. PMID:23778095

Stokes, H. T., Hatch, D. M., & Campbell, B. J. (2007). *Isotropy*. Retrieved from stokes.byu.edu/isotropy.html

Suh, C., & Rajan, K. (2009). Informatics for chemical crystallography. *JOM Journal of the Minerals. Metals and Materials Society*, *61*(1), 48–53. doi:10.1007/s11837-009-0009-7

Takeguchi, T., Yamanaka, T., Takahashi, H., Watanabe, H., Kuroki, T., Nakanishi, H., & Ueda, W. et al. (2013). Layered perovskite oxide: A reversible air electrode for oxygen evolution/reduction in rechargeable metal-air batteries. *Journal of the American Chemical Society*, *135*(30), 11125–11130. doi:10.1021/ja403476v PMID:23802735

Tan, P.-N., Steinback, M., & Kumar, V. (2005). *Introduction to Data Mining*. Addison-Wesley.

Tenenbaum, J. B., Silva, V., & Langford, J. C. (2000). A Global Geometric Framework for Nonlinear Dimensionality Reduction. *Science*, *290*(5500), 2319–2323. doi:10.1126/science.290.5500.2319 PMID:11125149

Tibshirani, R., Walther, G., & Hastie, T. (2001). Estimating the number of clusters in a data set via the gap statistic. *Journal of the Royal Statistical Society. Series B, Statistical Methodology*, *63*(2), 411–423. doi:10.1111/1467-9868.00293

Villars, P., Brandenburg, K., Berndt, M., LeClair, S., Jackson, A., Pao, Y.-H., & Iwata, S. et al. (2001). Binary, ternary, and quaternary compound former/nonformer prediction via Mendeleev number. *Journal of Alloys and Compounds*, *317-318*, 26–38. doi:10.1016/S0925-8388(00)01410-9

Villars, P., Onodera, N., & Iwata, S. (1998). The Linus Pauling file (LPF) and its application to materials design. *Journal of Alloys and Compounds*, *279*(1), 1–7. doi:10.1016/S0925-8388(98)00605-7

Waber, J. T., & Cromer, D. T. (1965). Orbital radii of atoms and ions. *The Journal of Chemical Physics*, *42*(12), 4116. doi:10.1063/1.1695904

White, T.J. & ZhiLi, D. (2003). Structural derivation and crystal chemistry of apatites. *ActaCrystallographica Section B*, *59*(1), 1–16. PMID:12554967

White, T. J., Ferraris, C., Kim, J., & Madhavi, S. (2005). Apatite - An Adaptive Framework Structure. *Reviews in Mineralogy and Geochemistry*, *57*(1), 307–401. doi:10.2138/rmg.2005.57.10

Witten, I. H., & Frank, E. (2000). *Data Mining: Practical Machine Learning Tools and Techniques with Java Implementation. The Morgan Kaufmann Series in Data Management Systems*. San Francisco: Morgan Kaufmann.

Zhou, J.-S., Goodenough, J. B., & Dabrowski, B. (2005). Exchange interaction in the insulating phase of $RNiO_3$. *Physical Review Letters*, *95*(12), 127204. doi:10.1103/PhysRevLett.95.127204 PMID:16197104

Zunger, A. (1980). Systematization of the stable crystal structure of all AB-type binary compounds: A pseudopotential orbital-radii approach. *Physical Review B: Condensed Matter and Materials Physics*, *22*(12), 5839–5872. doi:10.1103/PhysRevB.22.5839

ADDITIONAL READING

Bhadeshia, H. K. D. H. (2010). Neutral networks in materials science. *ISIJ International*, *39*(10), 966–979. doi:10.2355/isijinternational.39.966

Ceder, G., Morgan, D., Fischer, C., Tibbetts, K., & Curtarolo, S. (2006). Data-mining-driven quantum mechanics for the prediction of structure. *MRS Bulletin*, *31*(12), 981–985. doi:10.1557/mrs2006.224

Geurts, P., Irrthum, A., & Wehenkel, L. (2009). Supervised learning with decision tree-based methods in computational and systems biology. *Molecular BioSystems*, *5*(12), 1593–1605. doi:10.1039/b907946g PMID:20023720

Hautier, G., Fischer, C., Ehrlacher, V., Jain, A., & Ceder, G. (2011). Data mined ionic substitutions for the discovery of new compounds. *Inorganic Chemistry*, *50*(2), 656–663. doi:10.1021/ic102031h PMID:21142147

Jaqaman, K., & Danuser, G. (2006). Linking data to models: Data regression. *Nature Reviews. Molecular Cell Biology*, *7*(11), 813–819. doi:10.1038/nrm2030 PMID:17006434

Mackay, A. L. (1999). From "The Dialects of Nature" to the inorganic gene. *Foundations of Chemistry*, *1*(1), 43–56. doi:10.1023/A:1009924107380

Noble, W. S. (2006). What is a support vector machine? *Nature Biotechnology*, *24*(12), 1565–1567. doi:10.1038/nbt1206-1565 PMID:17160063

Rajan, K. (Ed.). (2013). *Informatics for Materials Science and Engineering*. Waltham, MA: Butterworth-Heinemann.

Reshef, D. N., Reshef, Y. A., Finucane, H. K., Grossman, S. R., McVean, G., Turnbaugh, P. J., & Sabeti, P. C. et al. (2011). Detecting novel associations in large data sets. *Science*, *334*(6062), 1518–1524. doi:10.1126/science.1205438 PMID:22174245

Rodgers, J. R., & Cebon, D. (2006). Materials Informatics. *MRS Bulletin*, *31*(12), 975–980. doi:10.1557/mrs2006.223

Service, R. F. (2012). Materials scientists look for a data-intensive future. *Science*, *335*(6075), 1434–1435. doi:10.1126/science.335.6075.1434 PMID:22442457

Srinivasan, S., & Rajan, K. (2013). "Property phase diagrams" from compound semiconductors through data mining. *Materials (Basel)*, *6*(1), 279–290. doi:10.3390/ma6010279

Tropsha, A. (2010). Best practices for QSAR model development, validation, and exploitation. *Molecular Informatics*, *29*(6-7), 476–488. doi:10.1002/minf.201000061

Chapter 8
Applications of Feature Selection and Regression Techniques in Materials Design: A Tutorial

Partha Dey
Academy of Technology, Adisaptagram, Hooghly, India

Joe Bible
Eunice Kennedy Shriver National Institute of Child Health and Human Development, USA

Swati Dey
Indian Institute of Engineering, Science and Technology Shibpur, India

Somnath Datta
University of Florida, USA

ABSTRACT

Feature selection is considered as an important preprocessing step to data mining and soft computing, whereas regression is a collection of methods to optimally assess the signal from a noisy output. Both seek to arrive at the dependence and relation between different attributes and a target material property. In the present chapter a flock of regression and feature selection techniques are discussed, and the kind of results that can be obtained with each of them has been illustrated with the help of a dataset on steel. The different methods are capable of abstracting data in different forms, thus revealing hidden knowledge from different perspectives. Choosing the most appropriate method depends on the application at hand and the kind of objective that one is looking for.

DOI: 10.4018/978-1-5225-0290-6.ch008

INTRODUCTION

Design of new and superior materials involves data with a large number of variables. It is very often advisable to assess the importance of these input variables (or predictors) and draw a quantitative interrelation between them and the output (or target) variable. This substantially improves any prediction or design model from different ways. Firstly it gives an overview of the role of different factors in the desired properties of the material and makes design and control of experiments easier by chopping off redundant variables. Secondly, in the case of predicting a material property from its composition and processing variables, selecting the most relevant variables renders the prediction model simpler, more efficient, and often even more accurate (by impeding over-fitting). Thirdly, where it is desirable to obtain a closed-form relationship between the input variables and the final property, a simple expression relating them makes the system and its underlying process explainable in physical terms.

Regression and feature selection have served as important tools to statistical learning and data mining. There are various feature selection and regression techniques for assessing the importance and interrelation of variables. In this chapter we discuss a number of techniques which will be useful for a practitioner or researcher in materials design and engineering, though they are equally applicable to a data set from any field of science, engineering as well as humanities. Finally, we illustrate the working and kind of results that these techniques generate, through a numerical example, focusing on the kind of inference that may be drawn from each method, along with a comment on choosing the most appropriate method.

REGRESSION TECHNIQUES BASED ON STATISTICAL LEARNING

The most classical regression technique is the ordinary least squares (OLS), which works well when the number of predictors is small, the correlation between the predictors is weak and their contributions to the conditional mean response are linear and additive. Departure from these assumptions would require considering more sophisticated methods. As for example, when the number of predictors is greater than the number of observations, OLS is not applicable. The aforementioned task of addressing model creation in the presence of a large number of descriptors can be accomplished via the implementation of one or several applicable regression techniques; Principle Component Regression (PCR), Partial Least Squares (PLS), Sparse Partial Least Squares (SPLS), Least Absolute Shrinkage and Selection Operator (Lasso), Ridge Regression and Elastic Net are examples of such techniques. The first three belong to a class of techniques referred to as latent factor techniques, the last three a class referred to as regularized regression techniques. Both are capable of handling a large number of descriptors with the distinction being that latent factor regression accomplishes this end by performing the analysis using a set of derived components (latent factors) whereas regularization does so by allowing descriptors to contribute disproportionately to the analysis.

Principle Component Analysis constructs a set of vectors (principle components), each of which describes the projection of each descriptor onto a p-dimensional space. PCR performs regression analysis using this set of principle components. PLS unlike PCR takes into account the response (target variable/property) as well as the response variables in constructing the latent factors used in analysis. PLS derives the latent factors in a way that maximizes the covariance between the latent factors and the target vector. SPLS introduces a sparsity parameter into PLS analysis which adds another dimension to the investigation.

Lasso can be characterized in terms of a penalized least squares analysis, through the introduction of a penalty on the absolute magnitude of the coefficients. Placing a penalty on the absolute magnitude of the coefficients has a desirable side effect in that it allows descriptors to discretely enter the model as well as contribute disproportionately. Ridge regression accomplished a similar end by introducing a penalty on the size of the squared magnitude of the coefficients, the quadratic penalty allows disproportionate contribution but not discrete selection. Elastic Net places both an absolute and quadratic penalty on the size of the coefficients.

Each technique we have discussed has its own redeeming qualities. For example, PCR is popular when there is no need to interpret the resulting model in terms of the original descriptors. Lasso and Elastic net are capable performing descriptor selection while maintaining interpretability in terms of the original descriptors. While each of these techniques is applicable in the $p>n$ scenario (where p is the number of descriptors and n is the number of records belonging to the data),they are also attractive candidates for the ease with which they are capable of performing model selection regardless of the number of available descriptors. When an analyst is first presented with the data it is seldom evident which technique would provide the most desirable model. It is for this reason that multiple techniques are often employed in an effort find the best models from a number of techniques and compare. In the following sections we will discuss each of the proposed methods in detail.

Let, $X \in \mathbb{R}^p$ and $Y \in \mathbb{R}^q$ denote a set of descriptor and response variables. In regression analysis, we seek to estimate a function, say $f(X, \beta)$ such that $f(X, \beta) = E(Y \mid X)$. In this parameterization, X and Y are observed for a training set of size n and we seek to estimate $\beta \in \mathbb{R}^p$ given the assumed relationship f and the observed information X and Y. In the sequel, we will slightly abuse the notation to denote the observed data by $X_{n\times p}$ and $Y_{n\times 1}$, *i.e.* we will assume univariate Y unless otherwise indicated.

For designing new materials with target characteristics expressed in terms of a response Y, the use of a regression model could be useful. Suppose we are looking for a material with a target value of Y = c. Suppose, we have fitted a regression equation of the form $\hat{Y} = f(X, \hat{\beta})$, where X denotes certain (elemental) descriptors of the constituent elements. Then the design problem translates to solving a mathematical equation, $f(X, \hat{\beta}) = c$, in the search space of appropriate X values. The details of the numerical techniques that can be used to find such solutions will depend on the problem at hand and is beyond the scope of this overview article.

$$\hat{Y} = f(X, \hat{\beta})$$
$$f(X, \hat{\beta}) = c$$

Ridge Regression

The first three methods we will discuss can be viewed as solutions of penalized least squares regression problems. Ridge regression (Hastie, Tibshirani & Friedman, 2001) accomplishes shrinkage through the imposition of a penalty on sum of the squared components of β. Estimates are obtained as the solution to an optimization problem:

$$\hat{\beta}^{Ridge} = \arg\min \{(X\beta - Y)^T (X\beta - Y) + \lambda \sum_{j=1}^{p} \beta_j^{\,2}\}. \quad (1)$$

The quadratic penalty on β admits a closed form solution for fixed λ. However, the nature of the quadratic penalty does not allow ridge regression to perform variable selection. Model selection is carried out by varying λ. Another important characteristic of ridge regression can be made apparent from investigating the characterization of the closed form solution, $\hat{\beta}^{Ridge} = (X^T X + \lambda I)^{-1} X^T Y$. The addition of the λ term to the diagonal of $X^T X$ alleviates the singularity issue encountered in the estimation of β in the $p > n$ scenario which plagues OLS estimation.

Least Angle Shrinkage and Selection Operator

In the case of Least Angle Shrinkage and Selection Operator (Lasso) the penalization is a constraint placed on the absolute size of β (Tibshirani 1996). Lasso has a very desirable property in that the penalty imposed on the absolute size of β allows a less influential variable to not contribute to the model (variable selection) as well as force less influential variables to contribute less to the final β estimate (this lesser contribution is what is referred to as shrinkage). β^{Lasso} estimates are obtained as the solution to

$$\hat{\beta}^{Lasso} = \arg\min \{(X\beta - Y)^T (X\beta - Y) + \lambda \sum_{j=1}^{p} |\beta_j|\}, \tag{2}$$

where λ is a shrinkage parameter used to control the amount of shrinkage imposed on β. Model selection is carried out by varying λ. Lasso solution paths are obtained via implementation of the LARS algorithm and can be interpreted in terms of λ or what is referred to as a Lasso step. The LARS algorithm (Efron, Hastie, Johnstone & Tibshirani, 2004) seeks to find a path of variable inclusion whereby a single variable is entered into the model at a time and its contribution to the model is increased until there exists another variable more correlated with the response. The point at which we enter another variable into the model is what is referred to as a Lasso step. It is with respect to Lasso steps which we will interpret our example.

Elastic Net

We have seen two penalized regression methods one with a penalty placed on the absolute magnitude of β and one with a penalty on the sum of squares of the components of β. Elastic Net (Zou & Hastie, 2005) is a penalized regression which utilizes both to perform both shrinkage and selection. Estimates are obtained as the solution to

$$\hat{\beta}^{ElasticNet} = \arg\min (X\beta - Y)^T (X\beta - Y) + \lambda_1 \sum_{j=1}^{p} |\beta_j| + \lambda_2 \sum_{j=1}^{p} \beta_j^2, \tag{3}$$

where λ_1 and λ_2 are the absolute and quadratic penalty parameters respectively. It has been suggested that through the introduction of the quadratic penalty, Elastic Net is a stabilization of the Lasso. Model selection is carried out by varying λ_1 and λ_2. In practice we interpret λ_1 as in the Lasso, that is in terms of Lasso steps, whereas λ_2 the quadratic penalty will be tuned similarly to the ridge regression over a fixed set of values determined a priori.

Principle Component Regression

Principle Component Regression (PCR) relies on an analysis technique known as Principle Component Analysis (PCA). It is often the case that many variables within a set will be correlated. Heuristically, this means that two or more variables will convey overlapping information. PCA seeks to address this redundancy by transforming a set of random variables into a set of derived independent variables. Independence is used in this sense to convey that the new derived variables are not correlated. This is accomplished by projecting the set of variables onto a new space such that each coordinate direction in the new space (principle component) is orthogonal to the others such that the first principle component captures the greatest possible amount of variation within the original set of variables (post transformation), the second captures the second greatest amount and so on. In practice it is often possible to capture 90% of the variation in a large set of variables using only a few principle components. It should be noted that PCA only projects X onto this new space. Formally, the mth principle component directions are constructed such that given X and $\Sigma_X$the covariance matrix of X we obtain eigenvalues λ as the solutions to $det(\Sigma_X - aI) = 0$. The corresponding eigenvectors e are then obtained such that $\Sigma_X e_i = \lambda_1 e_i$, subject to the constraint $|e_i| = 1$. The eigenvectors are then reordered according to their corresponding eigenvalues and the principle components may be obtained as $S_i = Xe_i$.

Let, X_m^{pc} denote the set containing the first m principle components. Then the estimate of β_{PCR} can be obtained as the solution to

$$\hat{\beta}_{PCR} = \arg\min_{\beta}(X_m^{pc}\beta - Y)^T(X_m^{pc}\beta - Y). \tag{4}$$

While PCR addresses colinearity issues within the descriptor set it has a more apparent utility as a dimension reduction technique. These two properties make PCR an ideal candidate for high dimensional analyses. However, while not immediately apparent a downside to the use of PCR is that the derived components are not readily interpretable in terms of the original descriptor set. As such further analysis is necessary in order to convey any physical relationship between the original descriptive and response variable. Model selection is facilitated by varying the number of principle components retained and evaluating the performance of each model, usually through cross validation.

PARTIAL LEAST SQUARES

Partial Least Squares Regression (PLS) is another latent variable regression technique. Unlike PCR which seeks to derive a new set of descriptors disregarding the response variables, PLS includes the response in the derivation of the new descriptor set. Whereas PCR performs regression using a set of descriptors chosen to capture the maximum amount of variability within the descriptor set, PLS uses descriptors chosen to capture both the maximum variability within the descriptor set while also maximizing the covariance between the descriptors and the response variables (Hastie Tibshirani and Friedman). The details of PLS component construction are a bit more involved than for principle component construction. However we will briefly consider them here from a latent decomposition standpoint. Let, $Y = TQ^T + F$ and $X = TP^T + E$ where Q and P are score and loading matrices respectively and E and F are vectors of random errors. Finding T such that $T=XW$ with the elements of W (matrix of direction vectors)being obtained iteratively as

$$w_k = \arg\max_{w} w^T X^T Y Y^T X w, \tag{5}$$

such that $w^T w = 1$ and $w\mathbf{\Sigma}_{\mathbf{X}} w_j = 0$, for $j=1,\dots,k$-1 (Chun & Keles, 2010) . Estimation of Q and P are obtained via ordinary least squares and finally we obtain estimates of β as $\hat{\beta}_{PLS} = WQ^T$. Model selection is facilitated by varying the number of PLS components retained and making a comparison of the performance of each. Similar to PCR the PLS components are not easily interpreted in terms of the original descriptor variables.

Sparse Partial Least Squares

While PLS functions as a dimension reduction technique overcoming the singularity issues encountered when implementing OLS it fails at being able to address the $p >> n$ case. It can be shown that in the case that p is much greater than n the PLS estimator is inadequate. Like in the PCR problem all components of X contribute at least somewhat to each of the latent components used for PLS modeling. In addition to creating issues associated with the interpretability of the resultant model this uncensored contribution of noise variables exacerbates the inadequacies of the PLS formulation for the large p small n problem. One solution to this problem is to enforce a sparsity constraint into the formulation of the PLS direction vectors.

The sparsity of SPLS is obtained by imposing a penalty (λ_1) on the absolute size of surrogate direction vectors (c), and singularity issues are resolved by enforcing a quadratic penalty (λ_2) on the same and introducing a third parameter (κ) to address concavity issues. SPLS components are acquired as the solution to

$$\min_{\alpha, c} \left\{ -\kappa \alpha^T M \alpha + (1-\kappa)(c-\alpha)^T M (c-\alpha) + \lambda_1 \sum_{j=1}^{p} |c_j| + \lambda_2 \sum_{j=1}^{p} c_j^2 \right\} \tag{6}$$

s.t. $\alpha^T \alpha = 1$ (Chun & Keleş, 2010). Obtaining direction vectors α and surrogate direction vectors a appears to be dependent; from the formulation of (6), on four tuning parameters namely, κ, λ_1, λ_2 and K. However, in the case of univariate Y the solution is not dependent on κ, similarly in practice λ_2 is set to ∞. Thus, the relevant tuning parameters in the univariate Y case are reduced to λ_1 and K the *i.e.* sparsity parameter and the number of retained components. In order to properly evaluate candidate SPLS models one will need to fix λ_1 and obtain performance measures according to the number of components retained, vary the value of λ_1 and repeat. In the univariate case this is accomplished in practice by varying a sparsity parameter η relating to λ_1 and *K*. SPLS components are difficult to interpret in terms of the original descriptors.

It is not possible to know *a priori* which regression technique will be the best for predicting a given property of a material system. It is therefore suggested that one tries all these techniques and uses the cross-validated mean squared error to select the best model. We illustrate the use of these techniques in an illustrative example later in this chapter. Another recent example can be found in Dey et al. (2014).

FEATURE SELECTION TECHNIQUES IN DATA MINING AND SOFT COMPUTING

In order to use data mining tools effectively, data preprocessing is essential in order to gain successful knowledge from data. Feature selection is one of the important and frequently used preprocessing modules in data mining. It reduces the number of features, removes irrelevant or redundant data, reduces noise in the data, and brings immediate effects like speeding up a learning algorithm, improving mining performance, increasing prediction accuracy and brushing up result comprehensibility. Feature selection has been an active field of research for more than four decades in statistical pattern recognition, machine learning and data mining (Liu & Yu, 2005).

Feature selection algorithms have seen applications in many areas of machine learning such as classification, clustering or association rules. While *classification* deals with data having class labels, and has been the subject of early research on feature selection, later efforts have adopted feature selection procedures for *clustering* as well, where data comes with no *a priori* class labels (Figure 1).

As the number of features increase, finding an optimal feature subset becomes intractable and the task of feature selection turns NP-hard (Ramaswami & Bhaskaran, 2009). Consequently some heuristic approach is taken to find the optimal subset of features that would entail on the task at hand. Classically, a feature selection process consists of four basic steps, namely subset generation, subset evaluation, stopping criterion and validation (Dash & Liu, 1997). Subset *generation* is a search process that produces candidate feature subsets based on a certain search strategy. It may start with the empty set, the set of all features, or some random subset, to which features are iteratively added and/or removed, or randomly regenerated. Each candidate subset is evaluated and compared with the previous best one according to a certain *evaluation* criterion. If the new subset turns to be better, it replaces the existing best. As the optimality of a feature subset is greatly dependent on the evaluation function, there is little unanimity as to which criterion is the best. As a result, different evaluation functions are employed in data mining that rely on perspectives from diverse fields of mathematics, statistics, computation, or even thermodynam-

Figure 1. Steps to the classical Feature Selection problem

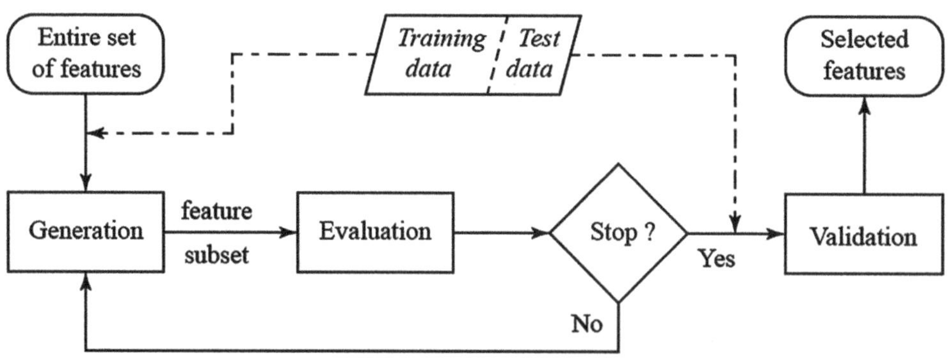

ics. Dash and Liu (1997) has broadly classified evaluation criteria into five categories, namely distance measure, dependence measure, information measure, consistence measure and classification error rate measure. While the first three are computational effective, the last achieves greater accuracy, whereas consistence measure may be regarded as a compromise between the two. This process is repeated until a given *stopping* condition is satisfied. A successful stopping criterion ensures that the search for better subsets does not continue exhaustively or indefinitely. The algorithm can be set to stop, depending on the subset generation procedure, on reaching a given number of features in the subset, or completing a certain number of iterations. Alternatively, it can be set to stop, depending on the evaluation criteria, if the value of the evaluation function reaches a threshold, or fails to improve after successive iterations. Finally, the subset selected needs some kind of *validation* to ratify its effectiveness or accuracy through different tests or comparisons. The simplest form of validation divides the dataset into a training and a test set, and compares the error induced from either sets. Overfitting is said to set in if the testing error is considerably (or consistently) greater than the training error. Sometimes such a measure of overfitting is also regarded as a stopping criteria, which ensures that the testing error doesn't far surpass the training error. Particularly when the dataset is too small or highly biased or imbalanced, *m*-fold cross validation is employed—where the dataset is divided into *m* equal subsets, each one in turn serving as the test set, and the others merged to form the training set. The average of these *m* errors give the cross validation test error. A special case of *m*-fold is leave-one-out, where *m* is equal to the number of objects in the data.

Feature selection algorithms can also be classified, according to the evaluation criteria adopted, as: *filter* models and *wrapper* models. While the former relies on the general characteristics of the data to select feature subsets, a wrapper model uses the performance of some data mining algorithm as the evaluation criteria. Performance of the latter is usually better with respect to the mining algorithm, but at the same time computationally more expensive. However, *hybrid* models take advantage of both models by exploiting the advantages of each in different stages of the search(Liu and Yu, 2005).

The following methods based on different data mining/ soft computing paradigms, can be regarded as wonderful techniques for feature selection:

1. Decision Tree
2. Artificial Neural Network
3. Rough Set Theory
4. Genetic Algorithm

The Relativity of Redundancy

A feature is said to be *irrelevant* if it 'does not affect the target concept in any way,' and *redundant* if it 'does not add anything new to the target concept' (John et al., 1994). The simplest feature selection methods select few best individual features one by one. A feature evaluation function is used to rank individual features, then the highest ranked *m* features are selected. Although these methods can exclude irrelevant features, the selected feature set may often include redundant features. "The *m* best features are not the best *m* features" (Peng et al., 2005). One way of avoiding this is using principal components (as in PCR), where each constructed feature is orthogonal to (and hence independent of) all other features. But this happens at the cost of losing transparency and physical significance of original variables. A genetic algorithm search may sometimes lead to an optimal feature subset which is in fact the 'best *m* features', but even there, getting an optimized result is not always ensured.

Decision Tree

Decision Tree is a well-known method for classifying data based on a series of questions of the form $x_i \overset{?}{\leq} v_k^i$ at each node of the tree, which leads to the left if the value of the input variable x_i is less than or equal to a certain value v_k^i, and to the right if greater than v_k^i. It is possible to ask questions of a different nature, sometimes leading to more than two branches, particularly in case of nominal or categoric variables. Each question however partitions the dataset into recursively smaller and *purer* subsets, until every branch ends up in a class label, which is the class assigned to the data (or record) in question.

The number of decision trees for a certain set of variables is exponentially large. So, finding the optimal decision tree for a given dataset is usually NP-hard. Most often a 'greedy' algorithm is used to select the best variable for partitioning the objects. There are several measures for selecting the variable and a corresponding value to split the data on its basis. The most popular of these are:

$$\begin{aligned} Entropy &= -\sum_{i=1}^{n_c-1} p(i\,|\,t)\log_2\left[p(i\,|\,t)\right] \\ Error &= 1-\max_i\left[p(i\,|\,t)\right] \\ Gini &= 1-\sum_{i=1}^{n_c-1}\left[p(i\,|\,t)\right]^2 \end{aligned} \tag{7}$$

where $p(i\,|\,t)$ is the probability of finding an object of class i at some node t. Though these different impurity measures will have different values for the same variable in a dataset, yet using either of these would yield the same result in most cases.

The point at which to stop further branching a decision tree is itself a difficult decision. The classical criteria is to go on branching till the node contains a completely *pure* subset of the data, *i.e.* where all objects under that node belong to only one class. But this criteria often fails to work, as it may lead to very few data in support of every terminal node (too stringent rules), which subsequently may not predict the class of an unknown data object with sufficient precision (overfitting), or the tree may grow unnecessarily large, involving redundant parameters or asking too many questions (overgrown tree). So researchers have often sought to early stopping, sometimes based on a threshold measure of impurity, which signals the tree to stop branching as soon as the value of impurity at a node falls below this threshold. Some researchers prefer post-pruning a tree, where a part of a tree consisting of several branches is replaced by a smaller sub-tree or a single terminal node.

Artificial Neural Network

Artificial Neural Network (ANN) has been known as an effective tool for machine learning, pattern recognition, and particularly prediction of outputs from multiple inputs. ANN mimics the complex network of neurons in the human brain, where sensory inputs modify the connection weights of hidden neurons inside our brain, subsequently leading to some output in the form of reflexes, actions or decisions. A generic ANN model has one input layer (containing a number of input variables in the model), one or more hidden layers, and usually one node in the final or output layer. Each node in the hidden and output

Figure 2. A typical Artificial Neural Network showing the different layers and their connecting weights.

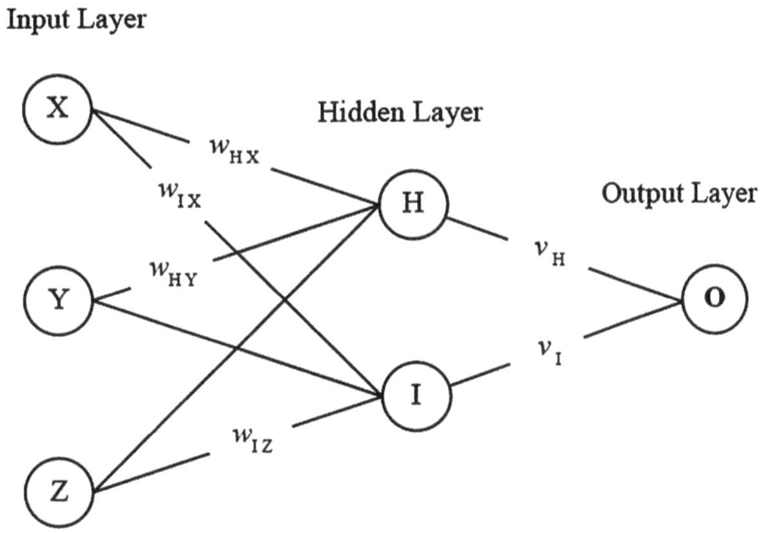

layer possesses a certain numerical value which is a function of the nodal values in the previous layer and the *weights* connecting those nodes. These weights are *trained* via the records in the dataset, so that the output, a functional value of the weights and previous layer nodes, matches the desired output for each data point. Figure 2 shows a typical network of nodes in different layers connected through a primary set of weights w_{ij} connecting a hidden node *i* from an input node *j*, and a subsequent set of weights v_i connecting the output node O from the hidden node*i* (i = H, I; j = X, Y, Z).

The Connection Weight Method (Olden 2004) illustrates a way to use the connection weights to calculate the *significance* of the input variables. For the sake of convenience, the weights connecting the hidden layer from the input layer may be expressed as a matrix **W**, and the weights connecting the output node from the hidden layer may be expressed as a vector v as follows:–

$$\mathbf{W} = \begin{bmatrix} w_{\mathrm{HX}} & w_{\mathrm{IX}} \\ w_{\mathrm{HY}} & w_{\mathrm{IY}} \\ w_{\mathrm{HZ}} & w_{\mathrm{IZ}} \end{bmatrix}, \qquad \mathbf{v} = \begin{Bmatrix} v_{\mathrm{H}} \\ v_{\mathrm{I}} \end{Bmatrix} \tag{8}$$

which, when multiplied, directly gives the significance of the input variables:–

$$S = \mathbf{W}\mathbf{v} \tag{9}$$

It may be convenient to express the significances values on a percent scale S^* by multiplying the significance vector S by a suitable factor:–

$$S^* = \psi\, S \tag{10}$$

where the scaling factor $\psi = 100/\sum|S_i|$, S_i being the significance of the ith variable, i= X, Y, Z.

Rough Set Theory

Rough Set Theory is a new mathematical tool dealing with information systems. After going through three decades of research along theoretical and various inter-disciplinary fields, Rough Set Theory has attained a certain degree of maturity. In recent years a rapid growth of interest in rough set theory and its application has been observed worldwide (Swiniarski & Skowron, 2003).

An information system $\mathcal{I}$ essentially consists of a set of objects U. Each object has definite values for a set of conditional attributes A, depending on which the value of a decision attribute d is ascertained. Formally it is the ordered pair $\mathcal{I} = (U, A \cup \{d\})$.We can select a subset of features $S \subseteq A$, and proceed to examine whether the features in it are necessary and sufficient to classify all objects in U.

Let $[u]_S$ be the set of objects indiscernible from any object $u \in U$ using the information in S. That is, every object which has the same values for every feature in S will constitute $[u]_S$.

Now, let U be partitioned on d, to generate the class of objects $\{C_1, C_2, \ldots, C_l\}$. The *S-lower approximation* of C_i is defined as

$$\underline{S}(C_i) = \{ u \mid [u]_S \subseteq C_i \} \tag{11}$$

Intuitively, $\underline{S}(C_i)$ contains those objects whose values for every features in S belongs to the ith class. The *positive region* of S is the union of all such lower approximations, while its *classification accuracy* is defined by the fraction of objects in the positive region.

$$\gamma(S) = \frac{|POS_S(d)|}{|U|} = \frac{|\bigcup_i \underline{S}(C_i)|}{|U|} \tag{12}$$

Rough Set Theory performs feature selection through a *reduct*, which is the minimal feature subset that has a classification accuracy of 1.Thus, $R \subseteq A$ is a reduct if and only if both these conditions are satisfied:

$$\begin{aligned} &\gamma(R) = 1, \qquad \text{and} \\ &\gamma(S) < 1 \qquad \forall\, S \subset R \end{aligned} \tag{13}$$

The *significance* of an attribute a is measured by the relative decrease in γon removing a from R.

$$\begin{aligned} \sigma(a) &= \frac{\gamma(R) - \gamma(R - \{a\})}{\gamma(R)} \\ &= 1 - \gamma(R - \{a\}) \qquad \text{if } \gamma(R) = 1 \end{aligned} \tag{14}$$

Genetic Algorithm

Genetic Algorithm (GA)is a heuristic search algorithm inspired by Darwin's theory of natural selection. It offers efficient solutions to NP-hard problems by exploring large multidimensional search spaces through simultaneous search in many regions. A population of solutions is randomly initialized and subsequently evaluated using some *fitness function*. The individuals' fitnesses are then used to select individuals that produce offsprings constituting the next generation. This continues for several generations till an acceptable solution is found. In feature selection, GA may be used as an appropriate *subset generation* tool, whose fitness function would normally call an external classifier to evaluate the population's individuals (feature subsets).

A solution in GA (here a feature subset) is usually *encoded* in binary as a string of 0s and 1s. For example, if there are *n* features in a dataset, a string with *n* bits would be sufficient to describe all possible feature subsets. For a particular feature (say the ith) the corresponding (i^{th}) bit would be zero if the feature is not included in the subset, and one if it is. The GA cycle starts by randomly *initializing* a population of (say *N*) feature subsets where every bit in each of the individuals is randomly given a value 0 or 1.

The *fitness* of each individual solution is then calculated, which needs to be sourced out to some external function or routine. This function would evaluate the efficiency of the particular subset of variables in correctly predicting the pertinent value or class of each data point. The degree to which the particular feature subset is able to correctly classify or predict the data may be assigned as the fitness of that particular solution. Of course, one way of calculating the classification or prediction accuracy is by using Rough Set or ANN.

Based on the fitnesses, individual solutions are *selected* on a simple proportional basis. If F_i is the fitness of the ith solution, a random number *r* is generated between 0 and 1, and the kth solution is selected if and only if:

$$S_{k-1} \leq r \cdot S_n < S_k, \qquad S_0 = 0 \quad \text{and} \quad S_l = \sum_{i=1}^{l} F_i, \ l = 1,2,\ldots,n$$

Figure 3. A generic flowchart and the crossover and mutation operators in Genetic Algorithm

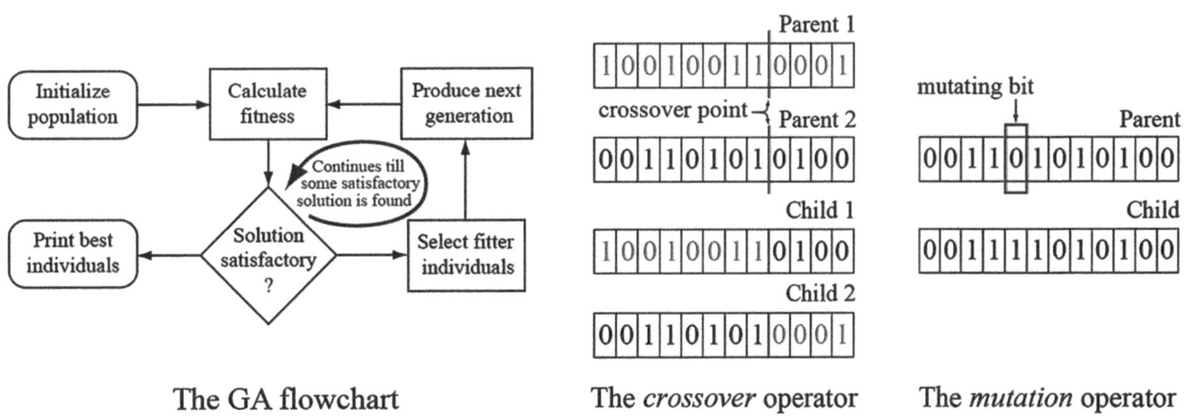

Solutions thus selected (inter)change their 'genetic' information through different operators like *crossover* or *mutation*. While crossover randomly selects a point beyond which genetic material between the two parents are interchanged, crossover chooses a single bit whose binary value is toggled. Each solution is given a certain predefined probability of undergoing one or more of these operations, which ultimately give rise to a new generation of offsprings. While the probability of crossover is usually kept high in the range of 0.6 to 0.8, the probability of mutation is usually set to a very low value of around 0.01 to 0.05. Figure 3 depicts the compact flowchart of a typical Genetic Algorithm, together with the principles of crossover and mutation.

ILLUSTRATIVE EXAMPLE

We will discuss each of the above mentioned methods, and illustrate the kind of results that can be obtained from each kind of analysis. For our analysis we use a dataset on steel that relates the composition and process parameters to the ultimate tensile strength (UTS) of the steel. Transformation induced plasticity (TRIP) steels, first reported by Zackay et al. (1967), exhibit a superior combination of strength and ductility. This characteristic has made TRIP steel a potential material for various automobile parts requiring high strength with adequate formability. As the phenomenon of TRIP is quite complicated and differentially dependent on the composition and processing parameters, there is sufficient scope for exploring the role of different variables on the strength of the steel, using data mining, soft computing and statistical learning techniques.

The TRIP steel data set contains 90 objects, most of which was collected from published literature of several researchers (Chen et al., 1989, Sakuma et al., 1991, Meyer et al, 1999, Papaefthymiou et al., 2004, Bandopadhyay and Datta, 2004, Mukherjee et al., 2006). For each of the TRIP steel specimens, three compositions (Carbon, Manganese and Silicon) and five processing parameters (cold deformation, Intercritical annealing temperature and time, and Bainitic transformation temperature and time) have been compiled in the dataset, whose symbols and range of values are given below in Table 1.

Table 1. List of variables in the dataset with their range of values

Sl. No.	Variable Name (Unit)	Symbol	Minimum	Maximum
	Input Variables			
1.	Carbon (wt%)	C	0.12	0.291
2.	Manganese (wt%)	Mn	1	2.39
3.	Silicon (wt%)	Si	0.48	2
4.	Cold deformation (%)	d	56.25	77.14
5.	Intercritical annealing Temp. (°C)	T_a	750	860
6.	Intercritical annealing time (sec)	t_a	51	1200
7.	Bainitic transformation Temp. (°C)	T_a	350	500
8.	Bainitic transformation time (sec)	t_a	30	1200
	Output (Target) Variable			
1.	Ultimate tensile strength (MPa)	UTS	580.72	887.44

EXPERIMENTAL RESULTS

The performance of a regression model can be considered from one of two perspectives. The ability to estimate known values i.e. the accuracy with respect to the data used to create the model and the ability to predict unknown qualities. A common measure to quantify a model's performance with respect to known values is mean square error (MSE). MSE is a standard metric which can be used to compare competing models. However, in many cases an investigator will be developing a model to estimate new (unknown) values. In the case that a model is to be used to predict unknown quantities MSE is inadequate as it will tend to underestimate the expected error of the model with respect to new predictions. One method for adjusting for MSE's underestimation of expected error is cross validation (CV).

Briefly, MSE is defined as the expected squared error given the data that is,

$$MSE = E_{X,Y}[(f - Y)^2] = \frac{\sum (f - Y)^2}{n}.$$

As mentioned before this tends to underestimate the error associated with making new predictions. Through the implementation of CV we can calculate more accurate estimates of the predictive error. CV involves partitioning the data into two sets, a training set and a test set. The process proceeds as follows:

1. Partition the data into two sets, a training set $(X_{Tr}, Y_{Tr}) \subset (X, Y)$ and a test set $(X_{Te}, Y_{Te}) \subset (X, Y)$ such that $P((X_{Tr}, Y_{Tr}) \cap (X_{Te}, Y_{Te})) = 0$.
2. Estimate β given (X_{Tr}, Y_{Tr}) i.e. $f_{X_{Tr},Y_{Tr}}(X, \beta)$. That is use the training set to estimate the desired model.
3. Estimate the test set values using the model obtained via the training data i.e. $f_{X_{Tr},Y_{Tr}}(X_{Te}, \beta)$.
4. Obtain error estimates based on the fitted test values for example, $e = f_{X_{Tr},Y_{Tr}}(X_{Te}, \beta) - Y_{Te}$.
5. Calculate the desired metric (for example MSE $= \frac{\sum e^2}{n}$).

There are several cross validation schemes, K-fold, 2-fold and Leave One Out (LOO) are common examples. In the case of K-fold CV the data are partitioned into K approximately equal sized subsets steps one through five would be performed K times using each of the K subsets as the test data and the remaining K-1 subsets as the training sets. The resulting error estimate is then taken as the average of the K error estimates. 2-fold is a variation in K-fold where the data set is split in half instead of K partitions. LOO is a derivation on this theme whereby the training set is obtained by removing one element from the data. The trained model is then used to calculate the error of estimation for the test element, this process is repeated for every element in the data and the results are averaged in order to obtain the CV error. LOO is desirable in cases where there is little data available and is less desirable in cases where there is a large amount of data as calculating LOO CV error estimates in this case will become computationally intensive.

In this example we use CV error estimates to compare the performance of the six regression methods discussed earlier. The data we present is described in the previous section. The results are presented using the Mean Square Cross Validated Error (MSCVE) as the metric of comparison. The method of

CV employed was LOO, for clarification let $e_i = f_{X_{-i},Y_{-ir}}(X_i, \beta) - Y_i$ denote the error estimate obtained on the ith element from the model trained on the data with the ith element removed then MSCVE is obtained as

$$MSCVE = \frac{\sum_{i=1}^{n} e_i^2}{n}.$$

Regression Results

We have normalized the data so that each descriptor in the sample has zero mean and unit variance. This is a customarily step in data processing since most regression algorithms work on normalized variables for greater numerical stability. Thus, we report the results in the transformed scale. In the case of PCR, PLS and Lasso the tuning parameters are predetermined that is to say that PCR and PLS tuning is carried out by retaining any of the 8 components and Lasso tuning being performed by varying the number of steps. Ridge, SPLS and Elastic Net, however, require specifying user defined tuning parameters. This can be problematic as poor selection of these tuning parameters can result in poor model performance. For this example the quadratic Ridge penalty was varied over a logarithmic scale such that $\lambda = \{10^z : z = -2, -1.8, -1.6, \ldots, 5\}$. The SPLS penalty η was evaluated over the set $\{0.01, 0.11, \ldots, 0.91\}$ and the Elastic Net penalty λ_2 was evaluated over the set of values $\{0.1, 0.2, \ldots, 0.9\}$.

Using these parameterizations we find that the 4 component PCR performs the best with MSCVE = 2504.6 followed by the SPLS model with $\eta = 0.81$ and $k = 8$. The best performing model results for each method are presented in Table 2. All regression analysis was conducted using R(R Core Team, 2013). Lasso and Elastic Net model estimates were obtained using the <enet()> function, available in the R package<elasticnet> (Zou & Hastie, 2012). PLS and SPLS models were estimated using the <pls> and <spls> packages (Mevik et al., 2013 and Chung et al., 2013) (Table 2).

Similarly, Figures 4, 5, 6, 7, 8 and 9 graphically demonstrate MSCVE for each method according to the various models investigated.

So far as an individual descriptor's contribution to the respective models is concerned it should be noted that direct comparison is difficult. Methods such as Elastic Net and SPLS which seek sparse solutions will select a subset of the descriptors and return coefficients corresponding to the 'relevant' subset. In contrast methods such as Ridge Regression and PCR produce coefficients for the entire set of descriptors and one is left to interpret the importance of each descriptor relative to the others. In the case of PCR, the best performing regression technique (see Table 2), the first four principle components load heavily towards C, Si and Tb. In the case of the ridge model the four descriptors with the largest absolute contribution are C, Mn, d and Si, in that order. It should be noted that PLS and Lasso are special cases of SPLS and Elastic Net, respectively and as such we will consider the Elastic Net and SPLS models as they outperformed the Lasso and PLS models. The best performing SPLS model consisted of three variables C, Mn and d. In the case of elastic net the model selects all of the descriptors but C, Mn, Si and d have the largest standardized coefficients and thus contribute disproportionately in relation to the other descriptors.

Table 2. Tabular summary of cross validated regression example using PCR, SPLS, Ridge, PLS, Elastic Net and Lasso.

Method	Parameterization	MSCVE
PCR	4 Components	2504.6
SPLS	$\eta = 0.81, k = 2$	2544.6
Ridge	$\lambda_2 = 15.85$	2610.2
PLS	3 Components	2615.6
Elastic Net	$\lambda_2 = 0.3$, Steps = 8	2649.4
Lasso	Steps = 8	2745.0

Feature Selection Results

Rough Set Theory

The decision attribute was discretized into 3, 5, and 10 classes respectively, assigning approximately equal number of objects to each class. Subsequently, we sought to find an attribute subset (or reduct) which would exactly classify every object in the dataset, with at least one cut on each of those attributes.

Figure 4. PCR results, plotting MSCVE versus the number of PC's retained. The four component model yields the best PCR candidate model (and consequently the best performing overall model) with MSCVE=2504.6.

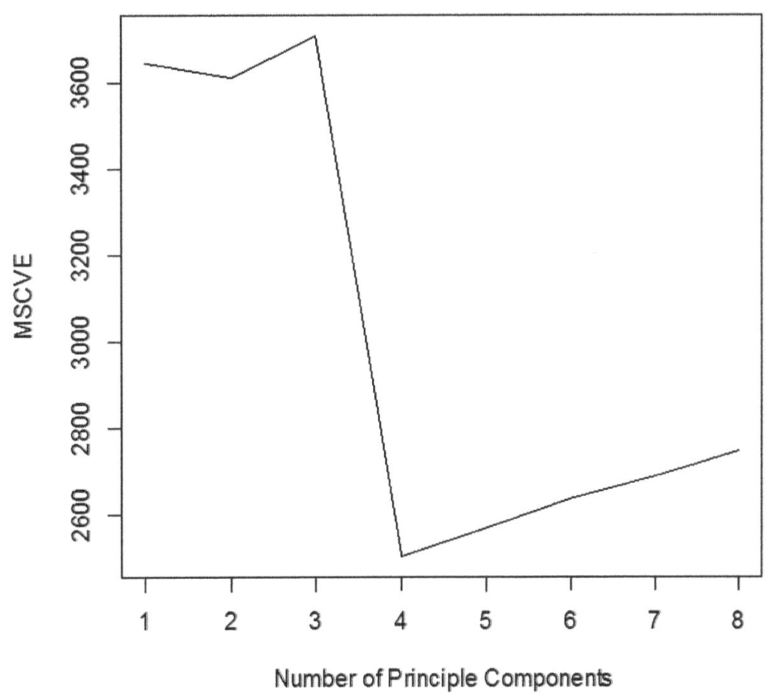

Figure 5. PLS results, plotting MSCVE vs. the number of retained PLS components. The three component PLS model yields the best candidate PLS model with MSCVE=2615.6.

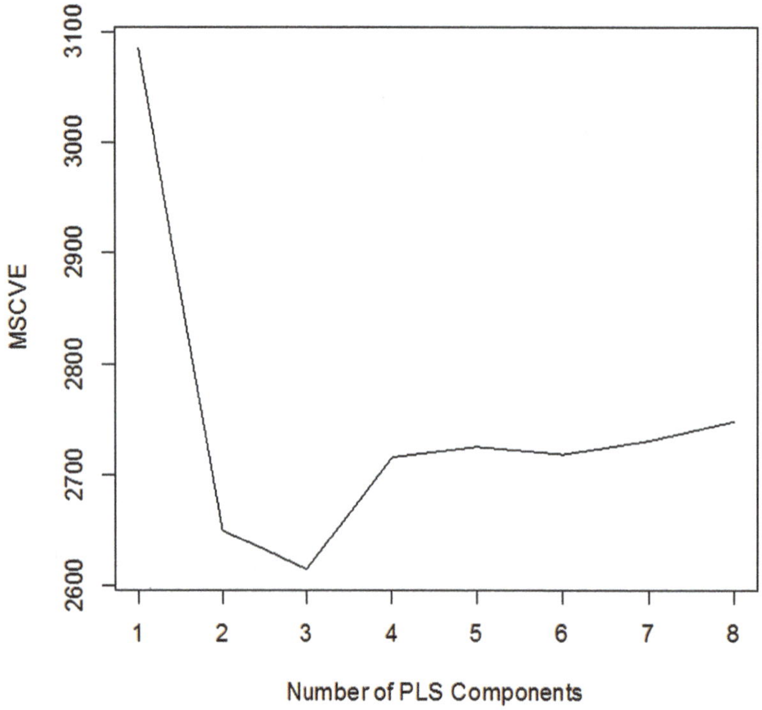

Figure 6. SPLS results, plotting MSCVE against η and the number of retained components. The 2 component model with η = .81 is the best candidate SPLS model with MSCVE=2544.6.

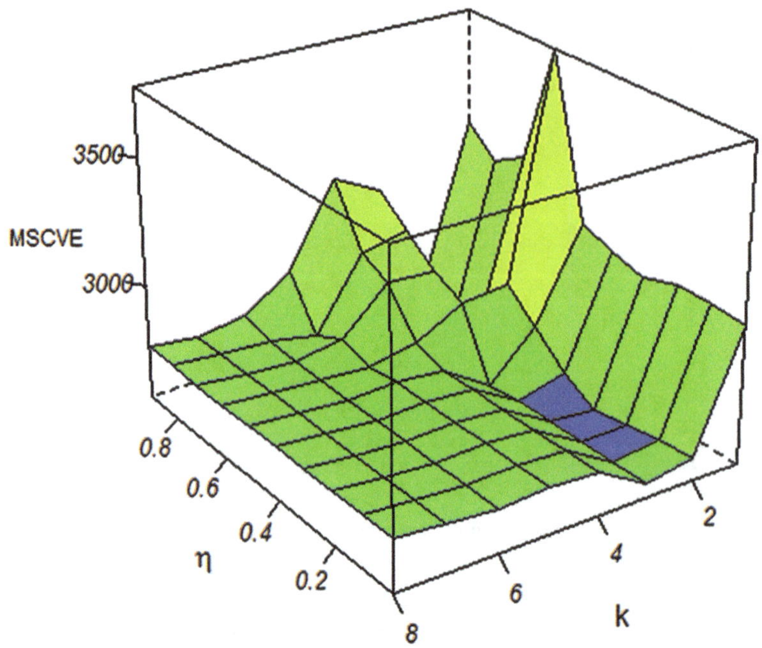

Figure 7. Lasso Results, MSCVE plotted against Lasso steps. The 8 step lasso yeilds the best candidate Lasso model with MSCVE=2745.0.

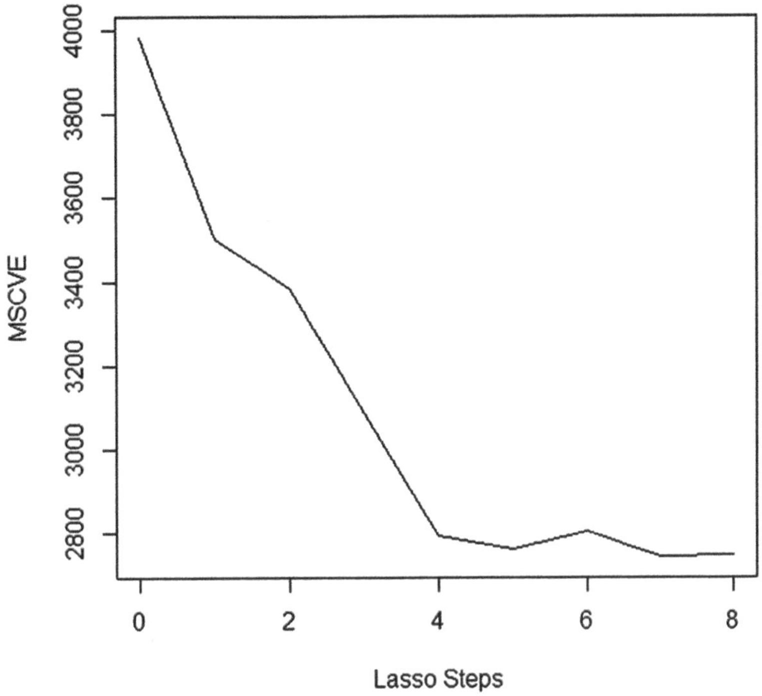

Figure 8. Ridge results, MSCVE plotted against λ.λ=15.85 produces the best candidate Ridge model with MSCVE = 2610.2.

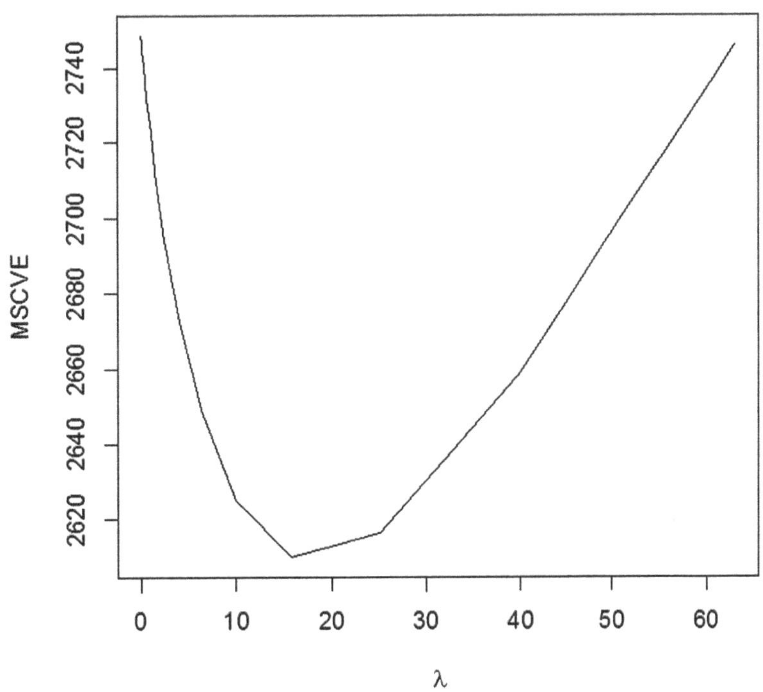

Figure 9. Elastic Net results, MSCVE plotted against λ and the number of steps. The best candidate Elastic net model is the 8 step model with λ= 0.3 and MSCVE=2649.4.

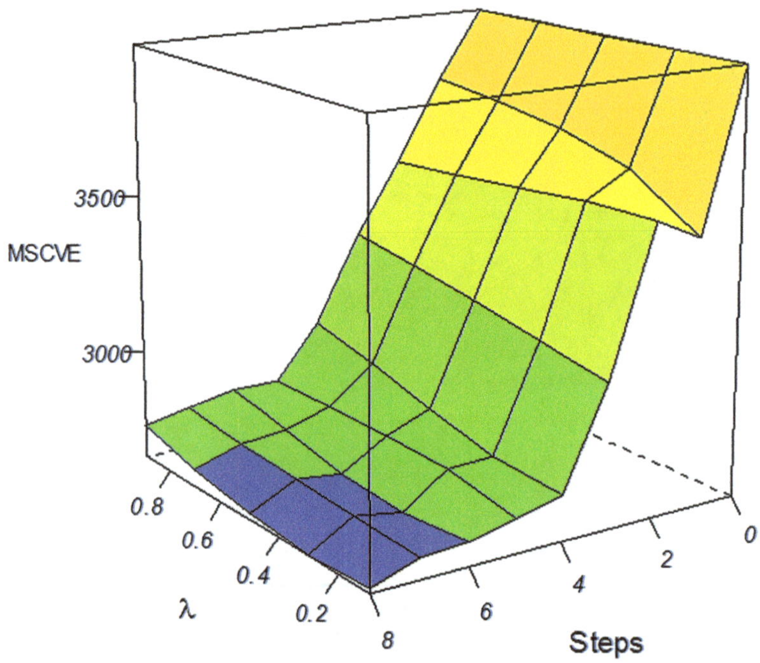

Where the number of variables is small (and the reduct is expected to be even smaller), it is possible to search for reducts by exhaustively taking all attribute subsets. The search starts with the singletons, then all two-attribute subsets are evaluated, and so on, stopping at the point where one reduct is found that exactly classifies all objects in the dataset. This gives a minimal reduct (with lowest cardinality). For the given dataset, the minimal reduct found is:

$$R = \left\{ \text{Si}, T_a, T_b, t_b \right\}$$

Figure 10. Significance of attributes in the minimal reduct, found through Rough Set by categorizing the decision attribute(Strength)into 3, 5 or 10 classes. The significance is calculated using equation (14) and expressed in per-cent by multiplying σby 100.

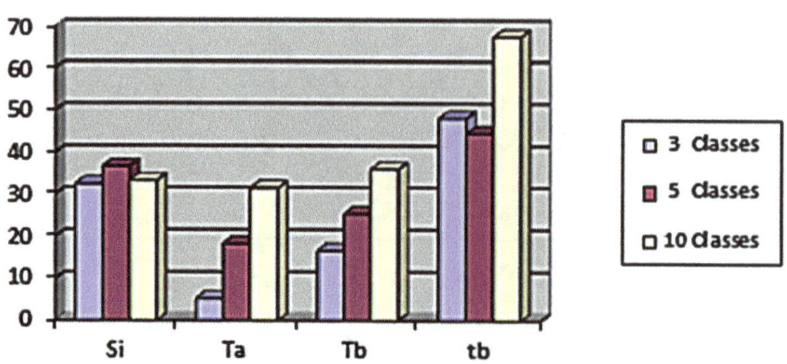

The other four attributes, *viz.* C, Mn, *d*, and t_a, are regarded as *redundant*, as it is possible to achieve 100% classification accuracy without these attributes.The significance of the attributes in the reduct have been found from equation (14) and are plotted in Figure 10.

When the number of variables is too large, or there appears to be a greater number of variables in the reduct, the exhaustive search for a reduct may be too computationally intensive. In that case it is advisable to use a special module for feature subset generation, which suggests some subset of features either randomly or using the knowledge and experience of how some of the subsets fare as a reduct. In a later subsection we discuss one such feature subset generation method, *viz.* Genetic Algorithm, which relies on the classification accuracy of initially guessed reducts to arrive at an optimum sized reduct with minimal cardinality and maximal classification accuracy.

Decision Tree

The target variable, Ultimate Tensile Strength (UTS), was primarily discretized into three classes, and an ordered pair (x_i, v_k^i) consisting of an input variable and its corresponding value (cut) is identified which best partitions the dataset into the *purest* subsets. Complete purity requires objects from just a single class to be present at a node. There are various measures of impurity, the most commonly used being Gini coefficient (see eqn. (7)).

A decision tree has been formed by partitioning the dataset recursively into smaller and smaller subsets through two-way branching, each time asking a question of the form $x_i \leq v_k^i$, and proceeding to the left or right branch of the tree accordingly as the answer is yes or no. Here v_k^i is a value in the range of the variable x_i, based on which the data (sub)set is partitioned. The possible set of values of v_k^i consists of the set of all midpoints of two consecutive distinct values of x_i, sorted in ascending order. The best variable-value pair (x_i, v_k^i) is chosen as the one which gives the *purest* partition of the relevant objects at that particular node, determined by the criteria of minimum Gini coefficient (Figure 11).

Each input variable that takes part in classifying objects from different classes are deemed important. However, the nodes (variables) which produce leaves with more objects in its support are more important than others. On this basis, the importance of each variable is calculated by adding the number of objects that fall hierarchically under the respective nodes. The results are presented in Figure 12.

Results clearly indicate the overwhelming role of silicon in determining the strength of TRIP steel. Bainitic transformation time and temperature shows up to be the next important variables influencing strength. The role of carbon, manganese and intercritical annealing time should also be considered with care in the design of this genre of steel.

In addition to leaving a note on the importance of variables, the decision tree informs the designer of TRIP steel to control the important variables so that they are kept within certain bounds. For example, looking at the extreme left branch of the tree (Figure 11), one can conclude that a silicon level of approximately less than 1.2 wt%, a Bainitic transformation temperature of less than 460°C and an intercritical annealing time of less than 4 min would give a high strength TRIP steel. Of course there are other constraints too, imposed by the range and general nature of the dataset on which the decision tree is based. These can be more specifically found in the experimental conditions adopted in the 20 examples from which the decision tree derive the rule, and these conditions need to be followed to get the desired steel.

Figure 11. Decision Tree formed for the TRIP dataset. Each elliptical node represents an input variable and a corresponding value which partitions the dataset successively, finally reaching a rectangular leaf, denoting the Strength class (High, Low, or Medium). The figures below the leaves represent the number of objects supporting the particular decision, and the figures in parenthesis denote the misclassified objects, present in just a few cases.

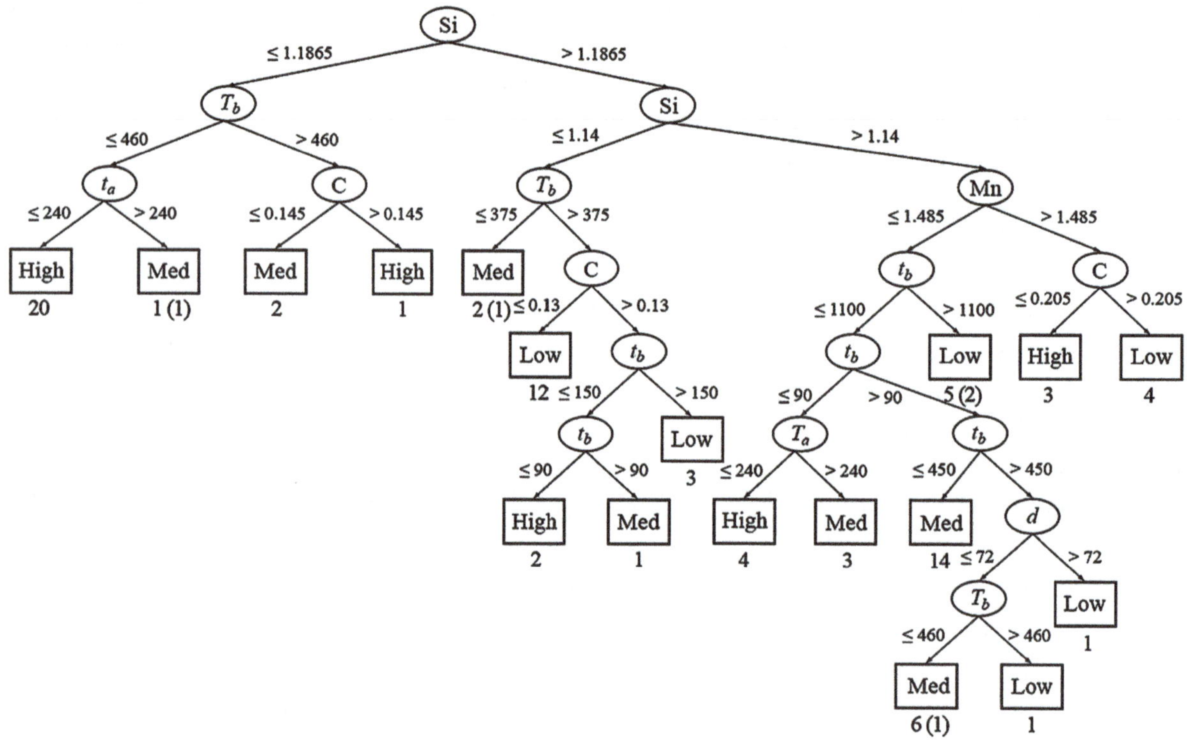

Figure 12. Importance of variables calculated from the Decision Tree by adding up the objects hierarchically falling under the respective node pertaining to that variable. The figures are normalized and expressed as per-cent by multiplying the fraction shared by each variable with 100.

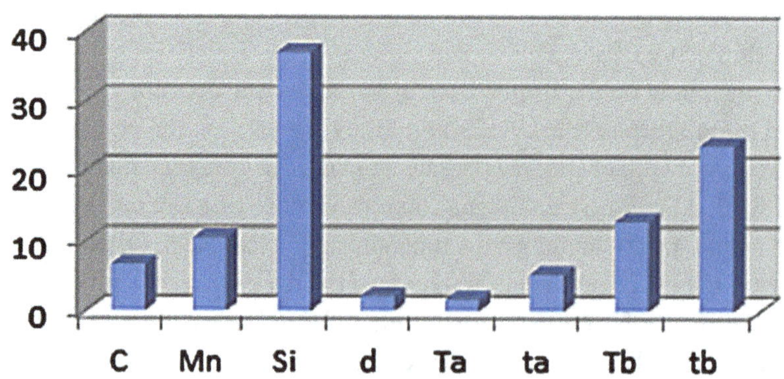

Table 3. Predictions from decision tree with leave-one-out (LOO) cross validation.

Number of Correct Classifications	Number of Dubious Classifications	Number of Wrong Classifications	Overall *Prediction Accuracy*
69	5	16	79.4%

Further, leave-one-out (LOO) cross-validation was used to test the classification accuracy of the decision tree. In this analysis, one datum (or record) was removed from the data set, and the remaining data were used to form a decision tree. The ensuing decision tree was utilised to classify the left-out datum. This was repeated for each of the 90 data points, leading to the following results in table 3:

The results in Table 3 show that there were 69 cases where the decision tree correctly predicted the class of the left-out (test) data. In 16 cases, the predictions were wrong, whereas in 5 cases the decision tree could not make unique predictions, as the final node contained objects from two different classes—one of which was the actual class of the test data. Considering the dubious classifications as half-correct, the classification accuracy of the decision tree model was found to be (69 + 5/2)/90 or 79.4%.

Artificial Neural Network

The ANN used in this work is a feed forward multi-layered perceptron type architecture trained with scale conjugate gradient backpropagation algorithm. A model has been developed for the dataset having one hidden layer with 5 nodes, and training for the whole dataset. At the end of the training, the model

Figure 13. The Artificial Neural Network model and Significance of variables calculated by the Connection Weight method. (a) The prediction accuracy of the Neural Network model in terms of the scatter plot of the actual UTS values versus the predicted UTS values, showing a correlation coefficient (R) of 93.9%, and (b) the significance of the individual variables expressed in percent (i.e. scaled up to get a total absolute value of 100— see Eqn. (10)).

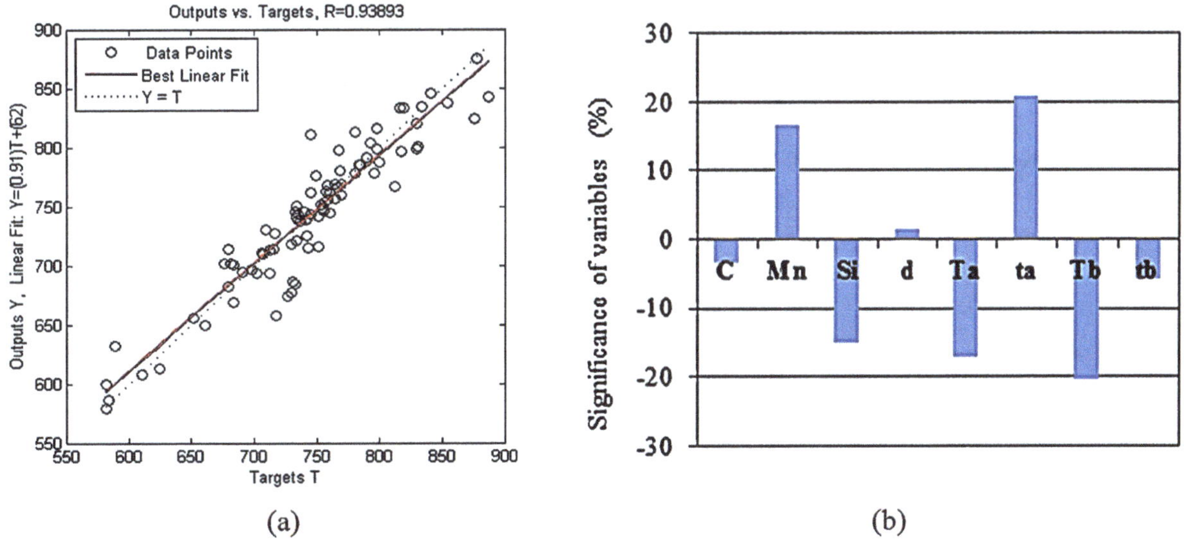

(a) (b)

that resulted gave an excellent prediction of the output (UTS) values with $R = 0.9389$ (See Figure 13(a)). Using Equations (8), (9) and (10) the significances of the 8 different input variables were found (see Figure 13(b)).

DISCUSSION

It might be interesting to note that the different regression techniques and feature selection methods originate from diverse fields of mathematics, statistics, computation or information theory. However still, one common objective of all these different methods is to arrive at a certain abstraction of the data. While regression attempts to do this by striking a kind of summary in terms of a function or a compact formula, feature selection tries to reduce the dimensionality of the data by selecting certain representative variables which speak for themselves as well as the others. To give a very loose analogy, regression is like drafting a constitution for a society (read data) to perform optimally, whereas feature selection is more like forming a senate or representative body who would carry out certain prerogatives for the wellbeing all its members (variables).

Moreover, it is never complete to go with only one of these techniques, since each has its own way of looking at the data and its inherent information. The no-free-lunch theorem suggests that whichever way (or method) one may adopt, there may be something very important that one might gain, but that always comes at the cost of something else. In other words, it is very difficult to say, without actually testing each method, which technique reveals the most needed information or knowledge from a dataset. So, it is advisable not to rely on just one rigid technique for extracting the required knowledge from a dataset, but try out several of them and compare between their results to find out the one that best suits the application and objective at hand. In the following subsections we discuss the advantages as well as their limitations, and cautions necessary for implementing the different regression and feature selection techniques.

Regression Techniques

It was noted at the beginning of this chapter that data preprocessing is of considerable importance when implementing any particular data mining technique. Similarly, a brief cautionary note was made regarding the fact that many regression techniques are not invariant to transformations of the data as such standardization of data prior to implementation was recommended. However, in the context of regression methods standardization is generally the last step of data preprocessing. Often the data, at least the response, will be measured in units which are not ideal for regression analysis. Situations where the data could require transformations other than standardization arise from either an a priori knowledge of the relationship between descriptors and the response or to satisfy various assumptions associated with proper implementation of a specific regression technique.

Of the six regression methods discussed in this chapter, each is what is referred to as a linear method. That is to say that it is assumed that the true relationship between the response Y and the descriptors X is that Y is a linear combination of the components of X. It is often the case that this assumption is not tenable and careful consideration should be given to choosing appropriate transformations when applying such techniques. In a similar vein, it should be noted that linear models and nonlinear models can be compared in the same analysis (regardless of the transformation) provided the objective measures are calculated in the same unit of measure. In the case of latent factor techniques it was noted that model interpretation in terms of components of the original data is often difficult. While it is true that these methods make interpretation difficult they do not make it impossible. The field of factor analysis is devoted specifically to problems of this type. In the case of PCR the interpretation of the resultant components is standard practice. While construction of PLS and SPLS components is more involved than that of PCR components interpretation of PLS and SPLS components can be made following similar reasoning to PCR components with respect to the importance of the contributing variables.

To summarize a few of the requisite steps to proper implementation of regression methods as data mining tools one should consider the following necessary though not necessarily sufficient components of the analysis:

1. Appropriate preprocessing: At this point the analyst should ask themselves "are the data appropriately scaled; given a priori expectations, to facilitate the desired modeling?" Things to consider at this stage of the analysis include but are not limited to transformations of the data to accommodate modeling of the desired relationship between the response and descriptors and transformations to address assumptions associated with the various regression techniques employed.
2. Tuning and Comparison: Given that various regression techniques may produce models with varying structures with responses in different scales i.e. $Y = f(X)$ whereas another may model $\log(Y) = f(X)$, ensure that the target values (error measures) are calculated in the same units.
3. Interpretation: In the case of linear methods one can through varying degrees of difficulty identify the relevant components of a regression method. Depending on the target audience or intended application of the resultant model not all regression techniques will provide adequate or interpretable information as to the relationship between the response and the descriptors. In theory the model which performs best with respect to the target measure is the desirable one. In practice it is often the case that a reasonable compromise between interpretability and performance is made.

Feature Selection Methods

While Decision Tree and Rough Set are methods to look at the data in a discrete or stepped fashion, Artificial Neural Network tries to form a functional relationship between the input and target variables. DT and RST, by discretizing the data, tends to lose some (contingent) information through clubbing different values of a variable into a single category, but presents a sharp and clear overview of the path leading to a particular decision or class of the target (property). On the other hand, ANN deals every single value as distinct from the others, and models the network on a continuous function of the inputs. But the cost one has to pay for the infinite variation of the variables is the clarity of the model, where the functional relationship is obscured, or at least the dependence of the target property on the input variables is able to render very little physical explanation. The methods may be considered as different points of balance on the tradeoff line between accuracy and simplicity.

Decision tree is a simple yet powerful mode for arriving at a decision (often the class of a specimen or data point) by asking a minimal set of questions about certain of its variables. The example illustrated above considers all the variables, and ranks the different variables based on how many questions are asked about it. It is however possible to choose a certain number of variables (a feature subset) and form a decision tree based only on those few variables. Comparing the classification accuracy of each of these different DTs with different feature subsets would be another way of testing which feature subset renders the highest accuracy. This detailed analysis was beyond the scope of the present chapter.

It is similarly possible to develop a sophisticated algorithm for feature selection using Rough Set, by taking samples from the data, and finding the reduct for each sample. Finally, the features that appear in most of the samples may be collected to form, what is called a *dynamic reduct*. This enhanced technique would serve to filter out the noise or reduce the effect of isolated outliers.

A dedicated tutorial on Rough Set by Komorowski, Pawlak, Polkowski, & Skowron (2002) will provide the interested reader with further details on this, while an article by Dey, Dey, Datta, & Sil (2011) depicts several examples (including one on the TRIP steel data) using an advanced method for finding the optimal set of variables that filter out noise more effectively without losing essential information.

The Connection Weight method discussed in the Artificial Neural Network subsection is only one of the many methods used to assess the importance of an ANN model. There are certain other effective methods, like Garson's algorithm, method of partial derivatives, input perturbation or sensitivity analysis, which might work better in certain conditions. The interested reader is referred to Olden (2004) for further explanation of these methods.

Finally, it may be worth noting that Genetic Algorithm has not found place in any of the illustrative examples presented in this chapter. The reason behind this was that the example used for illustrating the different techniques contained very few variables, where it was possible and very much feasible to deal with the separate variables and their possible subsets individually. When the number of variables is large enough (say in the order of thousands, or even in the range of hundred), as is the case in many real world materials engineering applications, this would become too cumbersome. It is particularly in that situation that GA would be computationally beneficial, owing to the numerous subsets that would then need to be considered to select the most suitable feature subset. For instance, just 30 variables would amount to over a billion feature subsets, where it would be impossible to conduct an extensive search for reducts through RST alone. GA and RST acting in unison, would then be a perfect match, with GA proposing a subset, and RST evaluating its fitness— which would be able to find the optimal feature subset (reduct) in just a few iterations. Srikrishna et al. (2013) provide one instance of an application of GA to feature selection.

Explanation of the Physical Phenomena from Computational Models

Transformation Induced Plasticity (or TRIP) is a complicated phenomena in terms of physical metallurgy. There are different mechanisms acting in different stages of the production process which take effect at different scales of the steel microstructure, that are subsequently responsible for the wonderful combination of strength, ductility and other mechanical properties— thus making it an unique and highly demanded steel in the industry. One way of decoding or understanding these phenomena is to search bottom up, either from existing phenomenological models, or through ab initio calculations. Unfortunately, not much success has been met with in the past few decades in explaining the TRIP phenomena satisfactorily (Bandyopadhyay & Datta, 2004). On the other hand data mining, soft computing and statistical learning have in their arsenal a lot of techniques which can bring out some insight into the interrelationships between the composition (and/ or processing) variables and the target mechanical property. It might be a rhetoric to claim that the latter computational techniques bring out the physical phenomena hidden in scores of experimental results, but when intelligently applied, often taking help of existing expert knowledge, they can definitely prompt the physicist, metallurgist, or materials designer where to look for to get to the underlying theory.

So far as the feature selection and regression techniques are concerned, the present chapter illustrates how some of the features evolve to be more important in the course of analysis, and how the strength of the steel can be predicted based on derived or selected variables. There is a general agreement on the features selected through different methods adopted here with respect to the TRIP steel example. Most of the regression models seem to select the three compositions C, Mn and Si as important, in addition to the two processing variables cold deformation (d) and bainitic transformation temperature (Tb). The feature selection methods mostly seem to favor Mn and Si, together with intercritical annealing temperature (Ta) and bainitic transformation temperature and time (Tb and tb) as the important composition and processing variables. From the metallurgical point of view, C, Mn and Si are all deemed to be very important compositional parameters in deciding the strength of any steel, particularly in the range of values pertaining to the dataset. C and Mn are potential austenite stabilizers and play a significant role in the hardenability of the retained austenite, whereas the amount of Si in the steel determines the formation of carbides in the microstructure which consequently gives high strength to the steel. The bainitic transformation temperature and time are jointly responsible for the formation of bainite in the microstructure, which significantly contributes to the strength of the TRIP steel.

CONCLUSION

Feature selection and regression are very important components of data mining and statistical learning. As selecting the relevant variables (and removing the redundant ones) can improve classification or prediction accuracy, thereby enhancing model accuracy and facilitating design of experiments, choosing the correct tuning parameters and number of components in various regression techniques can help to reveal the nature and degree of dependence of certain material properties by hinting at the underlying physical phenomena in the science of materials. Some advanced regression techniques have been discussed in the present chapter, and their implementation shown through an illustrative example. A few recent techniques for feature selection have also been presented along with their demonstrative working using the same example. Comparison among the methods show that the different methods agree to a great extent in their results, though each method has its own way of looking at and interpreting the same dataset.

ACKNOWLEDGMENT

We gratefully acknowledge support from National Science Foundation (Grant # DMS-11-25909 – SOLAR: New Materials Search for Solar Energy Conversion to Fuels).

REFERENCES

Bandyopadhyay, N. R., & Datta, S. (2004). Effect of manganese partitioning on transformation induced plasticity characteristics in microalloyed dual phase steels. *ISIJ International*, *44*(5), 927–934. doi:10.2355/isijinternational.44.927

Chen, H. C., Era, H., & Shimizu, M. (1989). Effect of phosphorus on the formation of retained austenite and mechanical properties in Si low-carbon steel sheet. *Metallurgical Transactions. A, Physical Metallurgy and Materials Science*, *20*(3), 437–445. doi:10.1007/BF02653923

Chun, H., & Keleş, S. (2010). Sparse partial least squares regression for simultaneous dimension reduction and variable selection. *Journal of the Royal Statistical Society. Series B, Statistical Methodology*, *72*, 3–25. doi:10.1111/j.1467-9868.2009.00723.x PMID:20107611

Chung, D., Chun, H., & Keleş, S. (2013). *spls: Sparse Partial Least Squares (SPLS) Regression and Classification*. R package version 2.2-1. Retrieved from http://CRAN.R-project.org/package=spls

Dash, M., & Liu, H. (1997). Feature Selection for Classification. *Intelligent Data Analysis*, *1*(1-4), 131–156. doi:10.1016/S1088-467X(97)00008-5

Dey, P., Bible, J., Datta, S., Broderick, S., Jasinski, J., Sunkara, M., & Rajan, K. et al. (2014). Informatics-aided band gap engineering for solar materials. *Computational Materials Science*, *83*, 185–195. doi:10.1016/j.commatsci.2013.10.016

Dey, P., Dey, S., Datta, S., & Sil, J. (2011). Dynamic discreduction using Rough Sets. *Applied Soft Computing*, *11*, 3887–3897. doi:10.1016/j.asoc.2011.01.015

Efron, B., Hastie, T., Johnstone, I., & Tibshirani, R. (2004). Least angle regression. *Annals of Statistics*, *32*, 407. doi:10.1214/009053604000000067

Hastie, T., Tibshirani, R., & Friedman, J. (2001). *The Elements of Statistical Learning*. New York, NY, USA: Springer New York Inc.

John, G. H., Kohavi, R., & Pfleger, K. (1994). Irrelevant features and the subset selection problem. In *Proceedings of the Eleventh International Conference on Machine Learning*. doi:10.1016/B978-1-55860-335-6.50023-4

Komorowski, J., Pawlak, Z., Polkowski, L., & Skowron, A. (2002). *Rough Sets: A Tutorial*. Retrieved from http://alfa.mimuw.edu.pl/prace/1999/D5/Tutor06_09.ps

Liu, H., & Yu, L. (2005). Towards integrating feature selection algorithms for classification and clustering. *IEEE Transactions on Knowledge and Data Engineering*, *17*(4), 491–502. doi:10.1109/TKDE.2005.66

Mevik, Wehrens, & Liland (2013). *PLS: Partial Least Squares and Principal Component regression*. R package version 2.4-3. Retrieved from http://CRAN.R-project.org/package=pls

Meyer, M. D., Vanderschueren, D., & Cooman, B. D. (1999). The influence of the substitutionof Si by Al on the properties of cold rolled C-Mn-Si TRIP steels. *ISIJ International*, *39*(8), 813–822. doi:10.2355/isijinternational.39.813

Mukherjee, M., Singh, S. B., & Mohanty, O. N. (2006). Neural network analysis of straininduced transformation behaviour of retained austenite in TRIP-aided steels. *Materials Science and Engineering A*, *434*(1-2), 237–245. doi:10.1016/j.msea.2006.06.076

Olden, J. D., Joy, M. K., & Death, R. G. (2004). An accurate comparison of methods for quantifying variableimportance in artificial neural networks using simulated data. *Ecological Modelling*, *178*(3-4), 389–397. doi:10.1016/j.ecolmodel.2004.03.013

Papaefthymiou, S., Bleck, W., Kruijver, S., Sietsma, J., Zhao, L., & van der Zwaag, S. (2004). Influence of intercritical deformation on microstructure of TRIP steels containingAl. *Materials Science and Technology*, *20*, 201–206.

Peng, H., Long, F., & Ding, C. (2005). Feature selection based on mutual information: Criteria of max-dependency, max-relevance, and min-redundancy. *IEEE Transactions on Pattern Analysis and Machine Intelligence*, *27*, 1226–1238. doi:10.1109/TPAMI.2005.159

R Core Team. (2013). *R: A language and environment for statistical computing*. R Foundation for Statistical Computing. Retrieved from http://www.R-project.org/

Ramaswami & Bhaskaran. (2009). A Study on Feature Selection Techniques in Educational Data Mining. *Journal of Computing*, *1*, 7–11.

Safavian, S. R., & Landgrebe, D. (1991). A survey of decision tree classifier methodology. *IEEE Transactions on Systems, Man, and Cybernetics*, *21*, 660–674. doi:10.1109/21.97458

Sakuma, Y., Matsumura, O., & Akisue, O. (1991). Influence of C content and annealing temperatureon microstructure and mechanical properties of 400°C transformed steel containing retained austenite. *ISIJ International*, *31*(11), 1348–1353. doi:10.2355/isijinternational.31.1348

Srikrishna, A., Reddy, B. E., & Srinivas, V. S. (2013). Automatic Feature subset selection using Genetic Algorithm for clustering. *International Journal on Recent Trends in Engineering and Technology*, *9*, 85–89.

Swiniarski, R. W., & Skowron, A. (2003). Rough set methods in feature selection and recognition. *Pattern Recognition Letters*, *24*, 833–849. doi:10.1016/S0167-8655(02)00196-4

Tibshirani, R., Johnstone, I., Hastie, T., & Efron, B. (1996). Regression shrinkage and selection via the lasso. *Journal of the Royal Statistical Society. Series A, (Statistics in Society)*, *58*, 267.

Zackay, V. F., Parker, E. R., Fahr, D., & Bush, R. (1967). The enhancement of ductility in high strength steels. *Transactions of the American Society for Metals*, *60*, 252–259.

Zou, H., & Hastie, T. (2005). Regularization and variableselection via the elastic net. *Journal of the Royal Statistical Society. Series A, (Statistics in Society)*, *67*(2), 301–320. doi:10.1111/j.1467-9868.2005.00503.x

Zou, H., & Hastie, T. (2012). *elasticnet: Elastic-Net for Sparse Estimation and Sparse PCA*. R package version 1.1. Retrieved from http://CRAN.R-project.org/package=elasticnet

Chapter 9
Imprecise Knowledge and Fuzzy Modeling in Materials Domain

Subhas Ganguly
National Institute Technology Raipur, India

Shubhabrata Datta
Calcutta Institute of Engineering and Management, India

ABSTRACT

This chapter highlights the usage of imprecise knowledge of materials systems using fuzzy inference systems. Experts have knowledge of complex materials systems in the imprecise linguistic form. But due to lack of phenomenological relations, material engineers are compelled to depend on empirical models for practical complex systems. This limitation could be overcome to a certain extent through the method of utilizing this imprecise knowledge with the help of fuzzy logic. The case studies presented here have demonstrated that systems with imprecise knowledge but with sparse data could be modeled successfully in this approach.

1. INTRODUCTION

A critical issue in design and development of new materials, for advance and complex systems, is that in most cases we do not have complete understanding of the fundamental basis of co-relationships of compositional and processing space, microstructure space and materials response space. The attributes of these variable spaces are too difficult for direct measurements and too difficult to describe their interaction when a particular physical phenomenon is occurring. They exhibit extreme nonlinear behavior across the different length and time scale. As a consequence imprecision and uncertainties are inherent characteristics of many materials phenomenon. Therefore the plausible methods for solving problems, where knowledge or information is imprecise, approximate and incomplete, are developing as an area of interest of the materials designers. Different modeling techniques are there to model the structure and properties of materials either from first principles or from other physical or chemical theories. But in most of the cases practical materials systems are complex enough to be modeled through such theories. Absence of precise scientific knowledge to describe such multivariate materials systems makes it

DOI: 10.4018/978-1-5225-0290-6.ch009

difficult to develop physical models. In some cases databases developed experimentally or from data warehouses of industries could be used to develop empirical relations among the variables. Such data driven models developed through statistical or computational intelligence based techniques viz. artificial neural network could also be used to extract useful knowledge from the data. In recent time, significant success has been achieved in implementation of informatics and machine learning in the effort to describe the features of different complex materials systems. Employment of computational intelligence or soft computing techniques for solving critical materials development issues lead to highly intelligent and more realistic results. Hybrid intelligent systems deal with the integration of two or more of the techniques. The combined use of techniques has been found to result in effective problem solving in comparison to the stand alone applications of such techniques. Hence stochastic models, though difficult to design and manipulate, can provide insights into real-world processes and offer new computational opportunities.

Thus there exist two distinct approaches, knowledge driven physical models and data driven empirical models. In between these two domains remains the sphere of modeling using imprecise knowledge. In case of materials systems there are many such complex cases where the knowledge can be expressed linguistically, but not through mathematical expressions. Fuzzy inference system based on fuzzy set theory is capable of converting such imprecise linguistic expressions to predictive models (Zadeh, 1965). Generally speaking model predictions are never certain. Understanding uncertainty in a model also is important to infer its solutions. So far many techniques are described for solving certain materials problems. The focus of the current chapter is the implementation of a computational technique to solve materials design problem where knowledge information is imprecise, approximate and incomplete by invoke of the tractable information processing and computational learning approaches. Fuzzy set theory, rough set theory, neural network, neuro-fuzzy hybridization, rough-fuzzy hybridization etc are such techniques to name a few. Here the fuzzy inference system and its hybridized form with neural network will be covered with an emphasis on their current abilities and limitations taking few real application example problems. The most important aspect of these techniques is that in addition to their capability to develop relationship between the variables, they can extract knowledge from data through developing comprehensible linguistic if-then rules. Both the aspects are described separately with case studies on their applications in the materials system.

2. IMPRECISE KNOWLEDGE IN MATERIALS SYSTEM

The expert knowledge of many complex materials systems are expressed through imprecise statements. It may be in the form of 'steel with high volume fraction of martensite have high strength' or 'high glass forming ability alloys are multi component alloy'. In these statements high volume fraction or high glass forming ability words are used to denote some imprecise description of states. In the present context these kinds of imprecise knowledge, commonly used in explaining materials system knowledge, are used to process the information and compute to solve some problem. It should be noted that we are not concerned here with words that are used to denote inherently impreciseness of the system rather the techniques to be used to solve the problem managing the concept of imprecise entities in the information processing and computing framework. Unlike the other engineering design task, materials design problems distinctly encompass the systematic interpretation and optimization of complex and nonlinear systems. These complexities mainly arise from the relationship among composition and process with the materials response or performance. The model based strategy for solving complex materials design

Figure 1. Elements of prediction of complex material phenomena

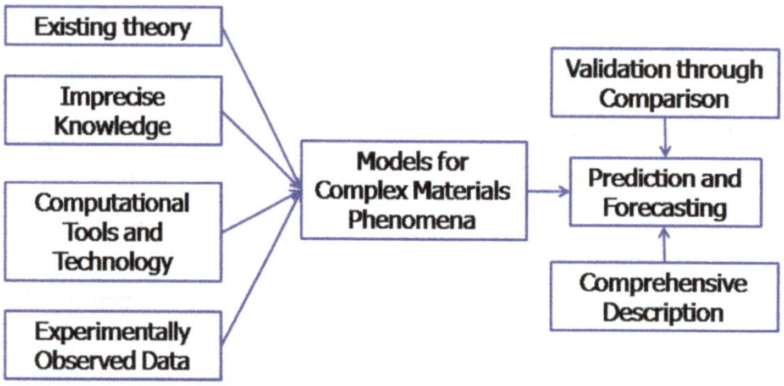

problems became a matured approach in materials research. The methods begin with simplification of the problem followed by formulation of mathematical theory for simplified problem. All the components of the problem is modeled and integrated without losing the complexity of the problem.

The elements of the modeling a complex material phenomenon associated with imprecise domain knowledge are explained in Figure 1. In the figure "existing theory" means the fundamental mechanism that we use to describe the concept. "Imprecise knowledge" means the full description of the imprecise information that inputs into any correlation in a materials system. "Computational tools and technology" includes both fundamental and practical algorithm issues, as well as accomplishment of mathematical basis of the modeling activity. "Prediction" includes constraints, such as associated consequences of the prediction, timeliness and accuracy requirements, as well as the formal predictive content of the modeling itself. "Experimental Data" denotes already performed experimental information that can be used in both the simulation of the phenomenon, as well as in assessments of the accuracy of the prediction, "validation through comparison." Finally, "Comprehensive description" implies that the goal for performing a prediction to begin with is not isolated. Rather, we seek to use prediction to achieve better understanding of the phenomenon, and then iterate to improve our next exercise of advancement.

Imprecision in materials phenomena are mainly arises from the insufficient knowledge of the physical phenomena at different length scales, the lack of understanding of the way information propagates from one length scale to another and the presence of inherent uncertainties involve with materials behavior in macroscopic scale. Consequently the uncertainties associated with various materials attributes lead to material response that cannot be accurately predicted using deterministic models. As an example microstructure can be viewed as the frame of reference for quantification and measurement of high dimensional space, characterized by chemistry, structure and defects attributes of the materials of interest. Figure 2 describes the fundamental basis for microstructural engineering. The direct measurement of the microstructural features and correlating them with materials response is a practical problem. However, microstructure is actually a qualitative representation of the effect of these attributes. Interpretation of this microstructure is the most important step to describe the appropriate structure property correlation, and thus judicious design of materials chemistry and the synthesis and processing steps. It is obvious that indirect method of measuring the structure, chemistry and defect spaces encounter associated imprecision and uncertainty inherently. It should be noted that microstructure engineering is an efficient method of dealing such uncertainty and imprecise knowledge, however, developing the computational framework and technologies managing this kinds of uncertainty and imprecise knowledge information is practical challenge for exploring the better understanding of the subjects and advancement.

Figure 2. Microstructure engineering as the framework of reference for measuring the properties in terms of its structure chemistry and defects information.

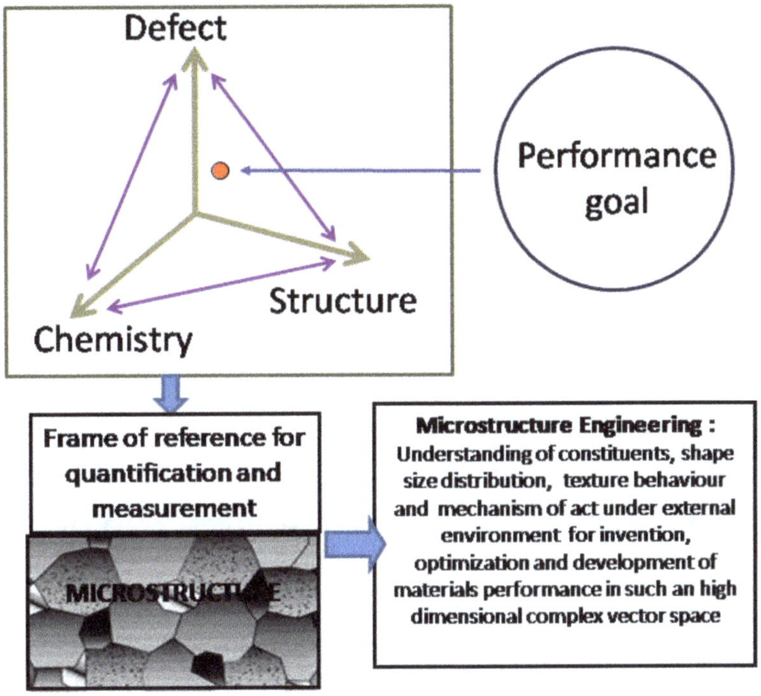

3. FUZZY LOGIC IN DEALING IMPRECISION

Fuzzy Logics have been proposed in the literature as a way to represent and reason with vague and imprecise knowledge. Their decidability, the empirically tractable and efficient reasoning algorithms, that carry over to fuzzy description logics, have attracted the attention of many research communities and domains that deal with imprecise knowledge and information. The fuzzy inference system is a process that enables mapping from a given set of inputs to an output space by the use of fuzzy logic. The FIS comprises (a) membership function (b) fuzzy logic operator and (c) if-then rule. A membership function (MF) is a curve that defines mapping of each point in the input space to a membership value between 0 and 1, called the degree of membership (μ). There are several types of membership functions, viz. triangular, trapezoidal, Gaussian, sigmoidal, asymmetrical polynomial etc, of which the sigmoidal function is the most commonly used function. There are several types of fuzzy logic operator, of which the Mamdani-type and Sugeno-type are commonly used (Jang, Sun, & Mizutani, 1997). Figure 3 describes the two types of FIS using the two different operators. FIS also uses if-then rule statements, which are used to formulate the conditional statements between the inputs and the outputs. The if-then rule assumes the form:

If x is A then y is B

where A and B are linguistic values defined by the fuzzy sets on the specific ranges. In a case of plain carbon steel, we can say if carbon is low then strength is low. The above concept of FIS is used for developing several systems, where the input and output knowledge remain imprecise.

Figure 3. Fuzzy inference systems using (a) Mamdani-type and (b) Sugeno-type fuzzy logic operators.

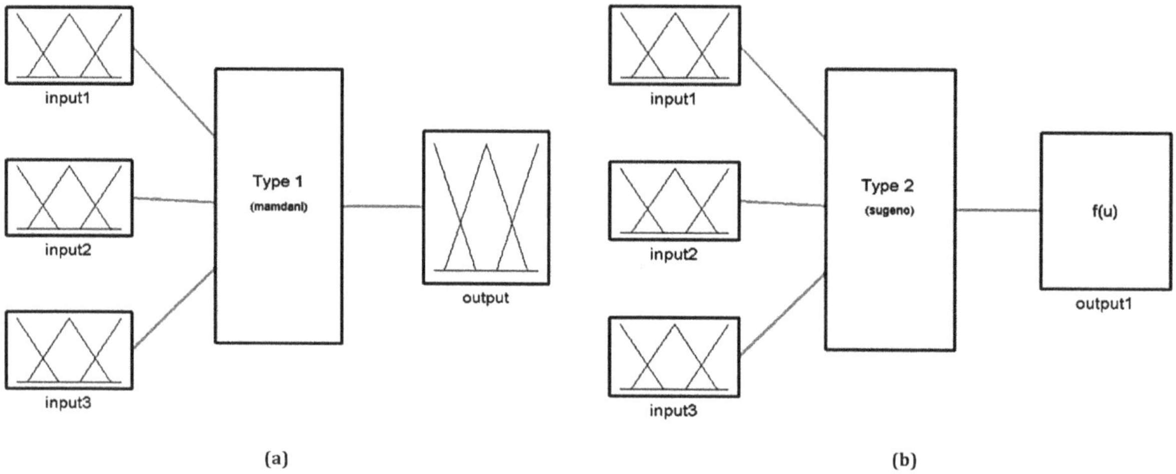

4. DESIGN OF MATERIALS MANAGING IMPRECISION AND UNCERTAINTY

As discussed earlier, in complex systems, knowledge may be expressed in terms of imprecise linguistic statements. The final property of any steel depends on some inherent phenomena or microstructural features, e.g. different phase fractions and distribution, grain size, different strengthening mechanisms, which come out as the complex effect of the composition and processing parameters. If we try to incorporate the imprecise knowledge of experts on these aspects within the purview of the model, the model structure becomes complex in nature. Let us take the case of transformation induced plasticity (TRIP) assisted steel. The unique beauty of TRIP steels is the presence of retained austenite (RA) in the microstructure that, upon straining, transforms to martensite. This transformation absorbs energy and improves the work hardening of the steel, allowing the delay of localized necking and thus giving improved elongation (Zackay, Parker, Fahr, & Busch, 1967). Typical TRIP steel is a low carbon steel with a multiphase microstructure of ferrite with bainite, retained austenite and with or without martensite. The alloying constituents of typical TRIP steels are mainly C, Mn, Si and/or Al. The guiding principles in designing the alloy chemistry for TRIP steel are avoiding pearlite formation and by the process enriching the retained austenite phase with carbon reasonably. Carbon is the most significant element required to stabilize the austenite; it also plays a role in the transformations required to produce TRIP steels. Carbon slows the transformation of austenite to ferrite or bainite at temperature, and delays the transformation to martensite on cooling by lowering the martensite start (Ms) temperature. An increase in the carbon content of the steel increases the amount of retained austenite to be present at room temperature. Manganese is an austenite stabilizing element as it increases the solubility of carbon in austenite. Mn delays the formation of pearlite, as it lowers the start temperature of cementite precipitation. According to Takahashi, manganese addition may lower Ar_3 and T_0 temperatures, so that the carbon concentration in austenite at T_0 temperature reduces relatively although the volume fraction of austenite phase increases. On the other hand, an excess adding of manganese can lower so much the carbon concentration in austenite at the T_0 temperature that the bainite transformation starts preferentially prior to formation of carbon-enriched austenite. Therefore, the volume fraction of bainite increases considerably resulting in the decrease in volume fraction and carbon concentration of retained austenite. Silicon is a well known carbide inhibiter in steels. The low solubility of silicon in carbide results in rejection of silicon at the carbide interface during the transformation, and thus the silicon concentration increases during an early stage of growth. It also increases the local activity of carbon and thus the carbon flux is reduced to inhibit development of the carbide embryos. In order to design the final microstructure with retained austenite of required amount and composition, precise processing is essential. There are number of ways to produce TRIP steels, but in all of them the key is to enrich austenite with carbon during the processing so that it becomes stable at room temperature. The simplest way is a two-stage heat treatment. First heat to the intercritical range and hold, then cool to a temperature in the bainite transformation range and hold, and finally quench to room temperature. In the first stage, a two-phase mixture of austenite and ferrite will form with the ratio of phase volumes determined by the temperature. The carbon will partition to the austenite, and be evenly distributed in that phase. During the isothermal holding in bainitic region, carbide-free bainite will form from the austenite phase and, again, the carbon will partition to the remaining austenite. When the steel is cooled to room temperature the carbon enrichment of the austenite will allow it to be retained. If the enrichment is not sufficient some amount of martensite may also form. The retained austenite can be present in a range of volume fractions, as well as in various sizes, shapes and locations in the microstructure. All these factors of retained austenite in the microstructure control its potential for transformation under stress or strain.

The above discussion may help the authors appreciate the complexity of the system. An attempt to make a model of the strength of TRIP-assisted steels using the imprecise knowledge described above using fuzzy inference system (FIS) (Dey, Datta, Chattopadhyay, & Sil, 2008). Here the problem domain had been divided into several sub-problems and the schematic block diagram of the approach is presented in Figure 4. In the three layered model the independent variables, viz. the composition of the steel and the processing variables, are related to the microstructural phenomena, which are finally related to the properties. The outputs of a layer act as the input of the next layer. In this way the knowledge, though in imprecise linguistic form, could be utilized describing the system intrinsically.

The first layer (extreme right of Figure 4) represents the composition and the process variables of the first stage of heat treatment. This forms certain intermediate microstructural features, which along with the variables of the second stage of heat treatment makes the final microstructure. Finally the microstructure is correlated to the properties. The correlations between the variables in different layers are made using the domain knowledge expressed in the form of if-then rules. A small portion of such correlations is described below (Dey, Datta, Chattopadhyay, & Sil, 2008):

1. If (the intercritical time is high) and (the intercritical temperature is low) and the (C content is high) and (the Mn content is high) then (the composition of intercritical austenite is high).
2. If (the intercritical time is low) and (the intercritical temperature is high) and (the C content is low) and (the Mn content is low) then (the composition of intercritical austenite is low).
3. If (the intercritical time is high) and (the intercritical temperature is medium) and (the C content is high) and (the Mn content is low) then (the composition of intercritical austenite is medium).
4. If (the intercritical time is high) and (the intercritical temperature is medium) and the (C content is low) and (the Mn content is medium) then (the composition of intercritical austenite is low).
5. If (the intercritical time is medium) and (the intercritical temperature is low) and (the C content is medium) and (the Mn content is high) then (the composition of intercritical austenite is medium).

Figure 4. The structure of FIS for prediction model of strength of TRIP aided steel (Dey, Datta, Chattopadhyay, & Sil, 2008)

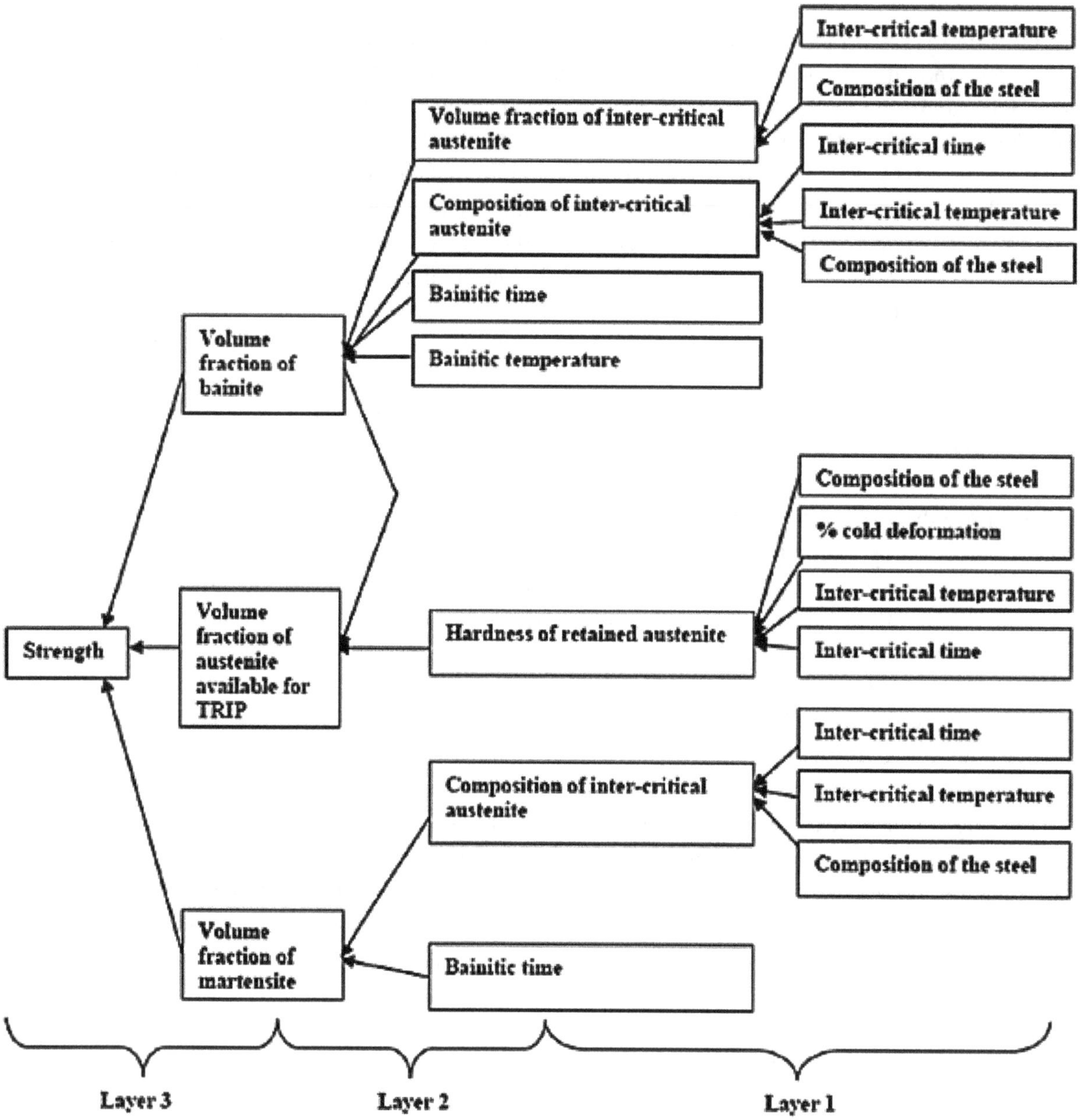

Such correlations using the knowledge in the form of linguistic rules are used to develop the model predicting the mechanical properties of TRI steel. The model is validated by data collected from the published literature as shown in Figure 5 and found to be sound enough, considering the limitations of a rule-based model.

Similar effort is made to model complex high strength low alloy (HSLA) steel system, which is thermomechanically controlled processed (Datta, & Banerjee, 2005a). The mechanical properties of

Figure 5. Comparison of prediction strength with published literature data (Dey, Datta, Chattopadhyay, & Sil, 2008)

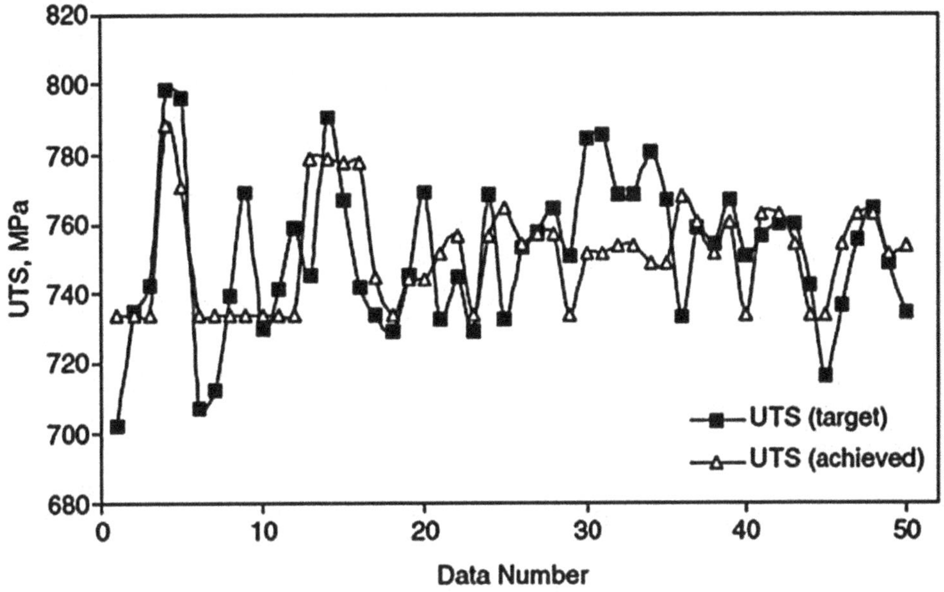

such steel depend on the chemical composition including minor additions of carbide formers and the thermomechanical processing parameters, which include slab reheating temperature, deformation at different temperature regime, finish rolling temperature and cooling rate. The alloying additions and this particular processing lead to different types of strengthening mechanisms in the steel, in highly complex and non-linear manner. Variations in these parameters change the property in a very subtle way due to the complex effect of these parameters individually as well as in combination. No physical model could describe such a complex system which could make a correlation between the independent inputs as stated above with the mechanical properties of the steel. In such cases the normal practice is to go for statistical or other kind of data driven black box models. But the roles of the variables are not unknown to the expert. They are known imprecisely and can be transformed to if-then rules, e.g. 'if Cu is high and Nb is high and B is high then yield strength of the steel is high'. Such rules could be transformed to form a fuzzy inference system to correlate between the inputs and outputs based on the imprecise knowledge based if-then rules. But here the main problem seemed to be handling so many variables (16 in total) in the format of such rules. To solve this issue here also, similar to the TRIP steel modeling, the inputs are divided into several groups. Unlike the previous model where the groups were formed to relate the independent variables with the microstructural features, here groups are made as per their role in the final outcome. The common elements, who mostly contribute in the solid solution strengthening are placed in one group, the carbide formers are placed in another, and the processing variables are placed in a separate group. This makes the rule formation also easier than handling all the variables together.

5. MODELING USING IMPRECISE KNOWLEDGE AND EXPERIMENTAL DATA TOGETHER

The models based on rules will have a moderate predictability compared to any data driven model. Thus the fuzzy models discussed so far also had moderate predictability. But the model accuracy may be improved to a certain extent by adjusting the model parameters using available data. The parameters of the fuzzy models can be modified through a kind of training as in case of neural network, where the prediction errors are back-propagated to adjust the parameters associated to each membership function of Sugeno type FISs. This type of modeling is called adaptive neuro-fuzzy inference system (ANFIS), which is basically a hybridization of neural network and fuzzy inference system. In cases where system knowledge in imprecise and experimental data both are available, the advantage of both could be utilized using neuro-fuzzy modeling scheme to develop a sort of semi-empirical model. Any data driven modeling technique requires good amount of a data for developing good models. So in case of modeling a materials system where the available data is not large enough, but the imprecise system knowledge could be utilized, the strategy of neuro-fuzzy modeling could successfully be implemented. In this system the inherent knowledge of the data could also be incorporated in the model, and thus the prior knowledge might be refined or enriched. In a complex materials system often this becomes a necessity. Neuro-fuzzy inference system could also be generated without prior knowledge. Here the if-then rules are developed from the data using fuzzy clustering techniques. The inputs and output data are clustered in to several fuzzy sets through data clustering, and the correlation between the input and output clusters are expressed in term of if-then rules. This is a highly effective method of for extracting useful knowledge from data, popularly known as data mining. The knowledge embedded in data can be extracted in the form of easily comprehensible linguistic rules. Sometime the user can even modify the extracted rules with prior knowledge to avoid possible bias in the data. This may treated as the reverse process of refining the rules using data, as stated earlier. The above approaches definitely show good performance from the prediction point of view. It is seen that the increase in number of cluster and thus the number of rules improves the accuracy and predictability of the system.

The previous problem of modeling the mechanical properties of HSLA steels is dealt by neuro-fuzzy systems and correlation between the effect of composition and process parameters with the properties are developed using both prior knowledge, in the form of if-then rules, and the experimental data (Datta, & Banerjee, 2005b). As discussed earlier, it is difficult and in some way impractical to develop a single neuro-fuzzy system relating all these variables. The number of rules that will be required to define such a system will be enormously high to make the ANFIS predict reasonable results. So, all the inputs and their relations with the strength were divided into four subsystems, the outputs of those subsystems were then compounded into a final ANFIS to give the final strength of the steel (Figure 6). Similar variables are clubbed to make several subsystems. Experimental data is then used judiciously to develop the rules. The outputs may be considered as the contribution of elements (in the first three ANFISs) and process parameters (in the fourth ANFIS) towards the overall yield strength of the steel.

It is seen that the outputs of these four systems have individually described the contributions of carbon-manganese-silicon-nickel-chromium (C *etc*), copper-molybdenum (Cu-Mo), niobium-titanium-boron (Nb-Ti-B) and the contribution of process parameters (proc.param.) towards the final yield strength value. These outputs (i.e., the output of subsystems) are then integrated in to find out the final yield strength of the experimental steels. Separate sub-systems of FISs have separate rule bases either describing the prior knowledge or developed using fuzzy clustering technique. Let us take, for example, the FIS describing

Figure 6. The structure of the ANFIS (Datta, & Banerjee, 2005b)

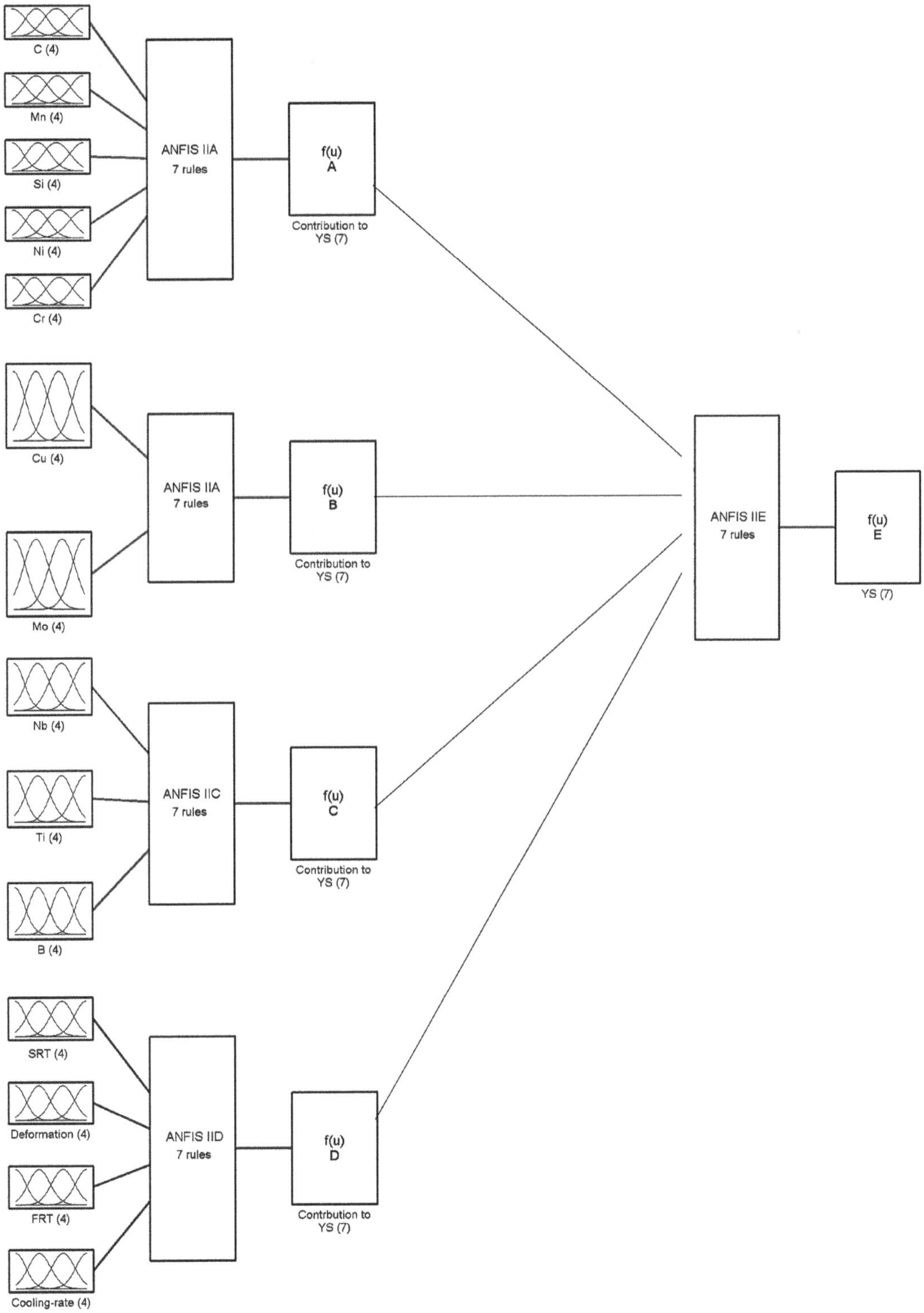

the role of the minor additions, i.e. niobium, titanium and boron. Nb, Ti and B are the most important micro-alloying elements in HSLA steel due to their ability to strengthen steel through grain refinement as well as precipitation hardening. Similar to the earlier case, the inputs are divided into four membership functions and the normalised output is divided into seven members. The seven rules designed to explain the system is stated below (Datta, & Banerjee, 2005b):

1. If Nb is low and Ti is low and B is low then YS is very low,
2. If Nb is low and Ti is low-medium and B is low-medium then YS is low,
3. If Nb is low-medium and Ti is low-medium and B is low-medium then YS is low-medium,
4. If Nb is low-medium and Ti is high-medium and B is high-medium then YS is medium-medium,
5. If Nb is high-medium and Ti is high-medium and B is high-medium then YS is high-medium,
6. If Nb is high and Ti is high-medium and B is high then YS is high,
7. If Nb is high and Ti is high and B is high then YS is very high.

The model developed using the above if-then rules is found to have captured the synergistic effect of these elements quite successfully as predicted by the model (Figure 7). The outputs of the sub-systems constitute the inputs of final, which assigns four memberships to each input. However, the output is divided into seven members. The rules formulated are as follows (Datta, & Banerjee, 2005b):

Figure 7. Surface view of the relations between yield strength and (a) Nb-Ti, (b) Nb-B and (c) Ti-B (Datta, & Banerjee, 2005b)

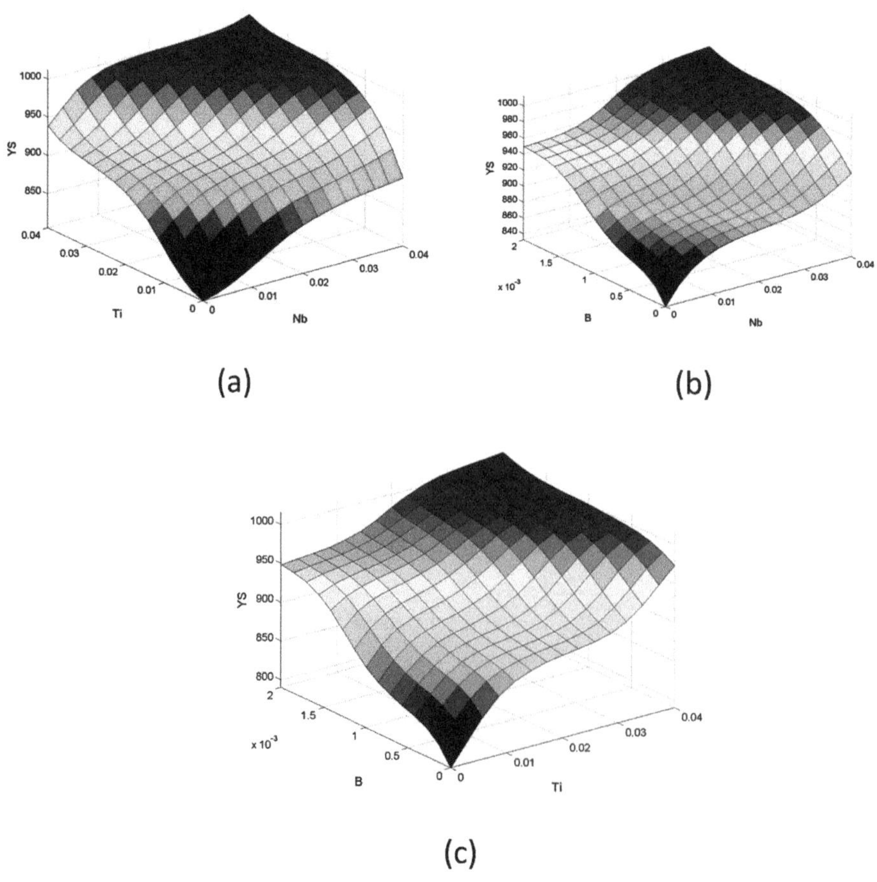

1. If C *etc.* is low and Cu-Mo is low and Nb-Ti-B is low and process parameters is low then YS is very low,
2. If C *etc.* is low-medium and Cu-Mo is low-medium and Nb-Ti-B is low and proc.param. is low then YS is low,
3. If C *etc.* is low-medium and Cu-Mo is low-medium and Nb-Ti-B is low-medium and proc.param. is low-medium then YS is low-medium,
4. If C *etc.* is high-medium and Cu-Mo is high-medium and Nb-Ti-B is low-medium and proc.param. is low-medium then YS is medium-medium,
5. If C *etc.* is high-medium and Cu-Mo is high-medium and Nb-Ti-B is high-medium and proc.param. is high-medium then YS is high-medium,
6. If C *etc.* is high and Cu-Mo is high and Nb-Ti-B is high-medium and proc.param. is high-medium then YS is high,
7. If C *etc.* is high and Cu-Mo is high and Nb-Ti-B is high and proc.param. is high then YS is very high.

The trained ANFIS prediction is found to have average error of 12.5 MPa (Figure 8), which is quite acceptable for all practical purposes.

There are certain practical issues need to be kept in mind during application of the present tool for modeling any such complex materials systems. The increase in the number of rules leads to an improvement in the predictability of the system. This envisages the need to gather a thorough understanding of the relations between the inputs and outputs in order to successfully design a neuro-fuzzy system. This is in total contradiction to the conventional concept of artificial neural network, where the network draws the relations between the inputs and the output without having a consideration of the physical significance

Figure 8. Actual vs. predicted yield strength by ANFIS after training (Datta, & Banerjee, 2005b)

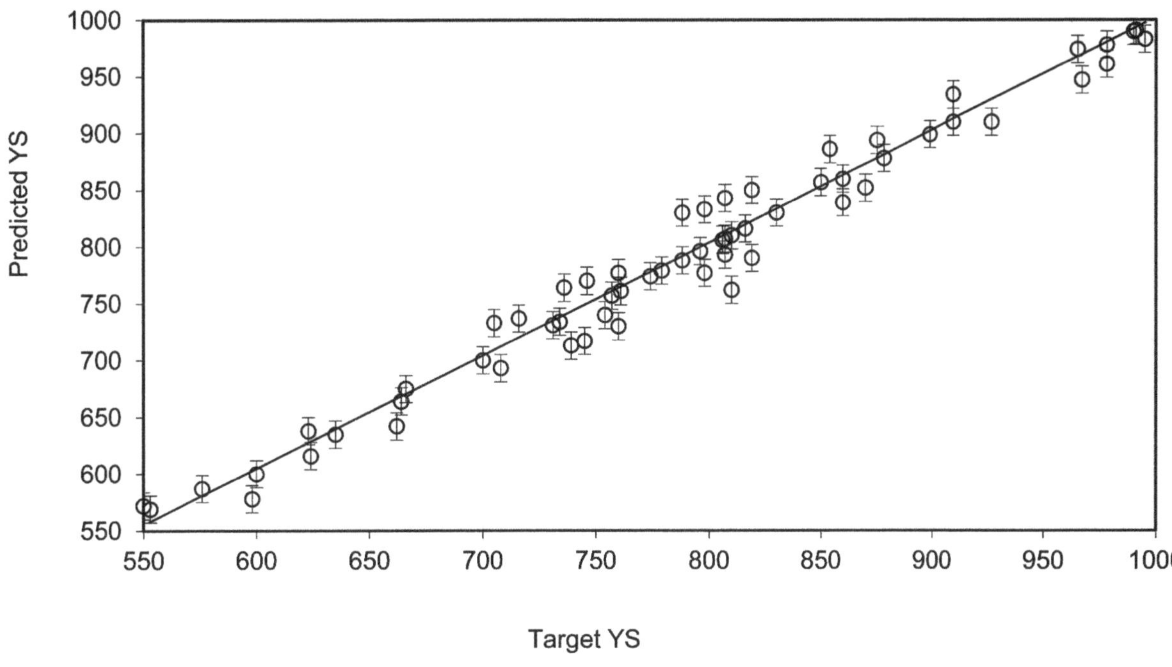

of the system. Hybridization of neural network and fuzzy system has however, acquired the ability to overcome the so-called limitation of neural network in modeling a materials process with due emphasis on the existing knowledge. On the other hand, the dependence of a neuro-fuzzy system on the if-then rules, i.e. the prior understanding of the relations, pose a definite limitation for modeling a neuro-fuzzy system with large number of input variables in the field of materials science. As in the case of the present exercise, the complex relationships between the composition and process variables with the mechanical properties of HSLA steel is difficult to be defined through a few if-then rules. It is necessary to formulate a large number of rules to design an effective neuro-fuzzy system. Unfortunately, this becomes a rigorous exercise and may not be considered suitable for practical use. With a view to overcome this limitation, a process of developing sub-systems has been introduced here. All the input variables are divided into four sub-systems according to their resemblance in their manner of strengthening the steel.

In another interesting application of fuzzy rule based modelling where prior imprecise knowledge and published experimental data are used in tandem, the mechanical properties of titanium alloys are modelled with the purpose of designing improved materials for orthopaedic implants (Datta, Mahfouf, Zhang, Chattopadhyay, & Sultana, 2016). It is known that Ti alloys are considered the most suitable for replacing failed bones, due to their excellent specific strength and the best biocompatibility. Though pure titanium and Ti–6Al–4V occupy the market of titanium biomaterials, but due to lower elastic modulus, β type titanium alloys are found to be useful. The implant material should posses low elastic modulus, as it reduces stress shielding of the neighbouring bone. But pure the common Ti alloys have elastic modulus of around 110 GPa, while the human cortical bone has a stiffness of around 20 GPa.

Figure 9. Schematic of the fuzzy models (composition-process-microstructure-property correlation) (Datta, Mahfouf, Zhang, Chattopadhyay, & Sultana, 2016)

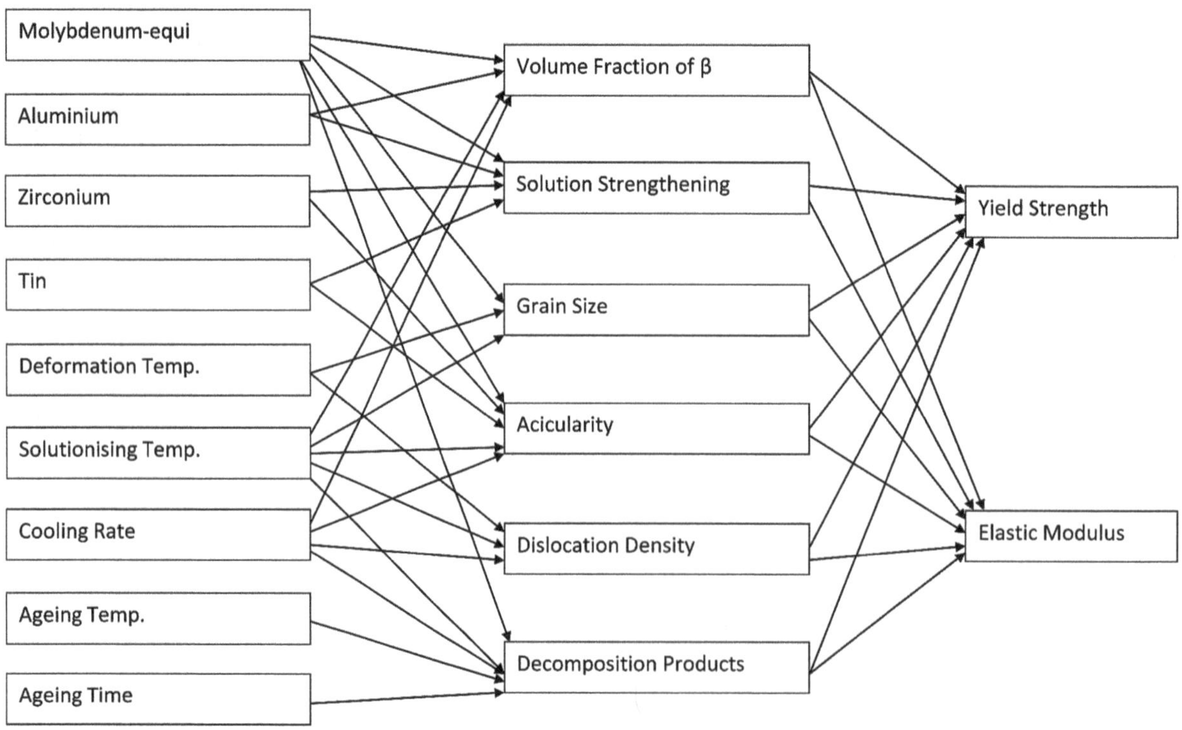

Thus selection of composition and shifting toward β alloys could be an effective proposition for further improvement of the properties of the Ti alloys appropriate for utilization as bone replacement. Thus the prior knowledge of the system along with the previous experimental findings are utilized to develop models for yield strength and elastic modulus of Ti alloys using a two-layered FIS, where the first layer correlates the composition and process variables with the microstructural features, and is developed using prior-knowledge (Figure 9). The second layer correlating the microstructure with properties is developed using published experimental data.

8. HYBRID MODEL USING PRECISE AND IMPRECISE CONCEPTS

Till now we have been dealing with imprecise knowledge and data. We discussed several options for utilizing both for developing a good predictive materials model. Now the issue of dealing both precise and imprecise knowledge together will be dealt. This situation may arise when the knowledge on one aspect of the materials system is precise, for which physical model may be developed, and knowledge on the other aspect of the system is imprecise, which might be handled through fuzzy modeling. Let us recall the complex system of TRIP steel discussed earlier. In a previous work, the tensile behavior of the steel was simulated using constitutive models describing the TRIP behavior (Bouquerel, Verbeken, & DeCooman, 2006). But the model related with the microstructural features viz. the constituent phases, grain size, dislocation density etc. But designing steel needs correlation of the property of the steel with the controllable independent variables, i.e. its composition and process variable. As it is not possible to correlate the independent variables with the microstructural parameters mentioned above through precise mathematical modeling, a fuzzy layer could be introduced before the constitutive model. Thus constitutive model inputs are linked with the controllable parameter by developing distributive FIS agent for each of the input, which are flexible enough to tolerate the uncertainty exist in the primary input layer (Ganguly, Datta, Chattopadhyay, & Chakraborti, 2009). Necessary assumptions are made using domain knowledge to generalize the tensile model for the considered realm of the problem.

CONCLUSION

The examples furnished herein essentially depict that systems with limited or imprecise knowledge and adequate or to some extent availability of information could be effectively modeled using fuzzy inference systems. As it is a common situation for many complex materials systems, knowledge both in precise and imprecise form along with experimental data could be utilized in tandem to develop a comprehensive model. Fuzzy logic has the flexibility to be hybridized with different system and can handle imprecision and uncertainty in the domain of materials.

REFERENCES

Bouquerel, J., Verbeken, K., & DeCooman, B. C. (2006). Microstructurebased model for the static mechanical behavior of multiphase steels. *Acta Materialia*, *54*(6), 1443–1456. doi:10.1016/j.actamat.2005.10.059

Datta, S., & Banerjee, M. K. (2005a). Fuzzy modeling of strength–composition–process parameters relationships of HSLA steels. *Materials and Manufacturing Processes*, *20*(5), 761–776. doi:10.1081/AMP-200055136

Datta, S., & Banerjee, M. K. (2005b). Neuro-fuzzy modeling of the strength of thermomechanically processed HSLA steels. *Indian Journal of Physics*, *79*, 473–483.

Datta, S., Mahfouf, M., Zhang, Q., Chattopadhyay, P. P., & Sultana, N. (2016). Imprecise knowledge based design and development of titanium alloys for prosthetic applications. *Journal of the Mechanical Behavior of Biomedical Materials*, *53*, 350–365. doi:10.1016/j.jmbbm.2015.08.039

Dey, S., Datta, S., Chattopadhyay, P. P., & Sil, J. (2008). Modeling the properties of TRIP steel using AFIS: A distributed approach. *Computational Materials Science*, *43*(3), 501–511. doi:10.1016/j.commatsci.2007.12.009

Ganguly, S., Datta, S., Chattopadhyay, P. P., & Chakraborti, N. (2009). Designing the Multiphase Microstructure of Steel for Optimal TRIP Effect: A Multiobjective Genetic Algorithm Based Approach. *Materials and Manufacturing Processes*, *24*(1), 31–37. doi:10.1080/10426910802540398

Jang, J.-S. R., Sun, C.-T., & Mizutani, E. (1997). *Neuro-Fuzzy and Soft Computing*. Englewood Cliffs, NJ: Prentice Hall.

Zackay, V., Parker, E., Fahr, D., & Busch, R. (1967). The enhancement of ductility in high-strength steels. *Transactions of the ASM*, *60*, 252–259.

Zadeh, L. A. (1965). Fuzzy Sets. *Information and Control*, *8*(3), 338–353. doi:10.1016/S0019-9958(65)90241-X

Chapter 10
Artificial Neural Network and Its Application in Steel Industry

Itishree Mohanty
Tata Steel Limited, India

Dabashish Bhattacherjee
Tata Steel Limited, The Netherlands

ABSTRACT

The recent developments in computational intelligence has enhances the applicability of empirical modelling in different areas particularly in the area of machine learning. These new approaches are based on analysing the data about a system, in particular finding connections between the system state variables (input, internal and output variables) without having precise knowledge about the physical behaviour of the system. These data driven methods explain advances on conventional empirical modelling and include contributions from many overlapping fields like Artificial Intelligence (AI), Computational Intelligence (CI), Soft Computing (SC), Machine Learning (ML), Intelligent Data Analysis (IDA), and Data Mining (DM). The most popular computational intelligence techniques used in process modelling of steel industry includes neural networks, fuzzy rule-based systems, genetic algorithms as well as approaches to model integration. This chapter describes mainly the application of Artificial Neural Network (ANN) in steel industry. ANN has extensively used in improving and controlling different processes of steel industry like steel making, casting and rolling which lead to indirect energy savings through reduced product rejects, improved productivity and reduced down time. The efficiency of artificial neural network tool in handling steel plant processes has been discussed in detail. ANN based models are found to be very potential to handle very complex, dynamic and non-linear problems.

1. INTRODUCTION

The combination of properties in various steel grades ensures its applicability in different areas. Strength, toughness, ductility, formability, weldability, corrosion resistance and surface quality are all important properties, depending on the application. The performance of steel can be defined as the balance between these properties while remaining cost-effective. So as per the requirements or applications, the

DOI: 10.4018/978-1-5225-0290-6.ch010

demanding features of steel sheets in the market are high formability, adequate strength, excellent dent resistance, adequate workability, good welding properties and excellent surface finish. Many researchers have attempted to find an optimum balance of properties (requirements) using ANN and other soft computing tools (S. Datta (2006), S. Datta (2007), S. Datta (2004), S. Dey (2009), A. Sinha (2013), S K Ghosh (2005), S K Ghosh (2009), P Das (2009), S. Ganguly (2007), S. Ganguly (2009), A. Mukhopadhyay (2005), I Mohanty (2008)).

The recent trend in automotive industry is use of multi-phase, high strength innovative material with excellent surface quality. In order to compete with today's market steel companies are putting their effort in developing new products and improving the process, yield and product quality with optimum energy and cost. This is only possible by significantly optimizing and controlling the process. Computational methods based models will be able to help the production process to achieve its market demands and product quality posed by the customers. In modern manufacturing plants the data that characterize the plant process is collected automatically through sensors, PLCs, etc and stored in large databases or data warehouses. Data mining tools can be utilized to extract interesting and useful information in the manufacturing process. Neural network based models are extremely useful in such circumstances, not only in the study of processes but wherever the complexity of the problem is massive and dynamic in nature and at the same time it is very difficult to develop a first principle based model (Bhadeshia, 1999; H. K. D. H. Bhadeshia, 2008). Several attempts have already been made in this direction (Z. Guo (2004), S. M. K. Hosseini(2004), P. Korczak (1998), J. Kusiak (2002), J. Larkiola(1998), H. Monajati (2010), S. K. Ghosh (2008),Gladman T (1975),L. A. Dobrza´nski (2005), P. Das (2007), C. Dumortier (1998), V. Narayan (1999), S. Datta (2006), A. Mukhopadhyay (2005), S.B. Singh (1998), I Mohanty, 2011, 2014; G Majumdar, 2010, Sbarbaro-Hofer, D. (2002), L Rokach (2008)). It is not surprising that numerous attempts have been made to use neural network modelling to deal with complex properties such as fatigue (H. Fujii, 1996; T. Goswami, 1996; J. M. Schooling, 1999; T. Svensson, 2004), toughness (H. K. D. H. Bhadeshia, 1995 ;T. Cool, 1997; S. H. Lalam; D, 2000. Dunne, 2004), corrosion resistance (J. Cai, 1999) and creep rupture life (M. Evans, 1999; F. Brun, 1999; D. Cole, 2000; Y. S. Yoo, 2002).

Artificial neural network (ANN) has been established as a potential tool in steel industries in prediction of properties and design of alloys. It has been used successfully to control the mill process for achieving target properties with available steel composition and mill constraints by capturing the complex interactions among them (Martinetz, 1995; Döll, 1999, M.J. Willis, 1992; Andreas Draeger, 1995). Various models have been developed for different parts of the rolling mill to control like furnace model, deformation model, roll-force model, mill set up calculation model, run-out-table cooling model, mechanical property prediction model and deformation and joining process model. Different kinds of optimization models are available like draft scheduling, roll force ratio distribution, etc. A few physical models have been developed to predict phase transformation and mechanical property. Each model has been developed with a particular aim which is localized to that part of the mill.

This chapter gives an idea regarding the application of ANN in steel industry in key areas related to design of structural steels and alloys including composition–process–property correlation, control and optimization. First part of this chapter gives a detail overview of ANN. Second part describes the applications in the area of mechanical property prediction, modelling strength of TMCP HSLA steel, optimisation of tensile properties to facilitate new product development and accretion of yield strength accuracy in skin pass mill. These sections aim to establish the potential of ANN in capturing the complex correlation among the materials and process variables to determine the productivity and quality of the products. It also explains the benefits achieved by implementations of this technology in different processes.

2. SOFT-COMPUTING TECHNIQUES

Soft-computing techniques are techniques which use computing methods capable of representing imprecise, uncertain and vague concepts (Odetunji A, 2009). It is economical and capable of using human expert knowledge unlike hard computing. The field which encompasses these new approaches is called data-driven modelling (DDM). DDM is very efficient in analysing the data of a system by extracting information, finding correlations particularly when the knowledge about the physical behaviour of the system is very poor and the system is very complex and dynamic in nature. These methods represent large advances on conventional empirical modelling

The domain of soft computing encompasses a number of techniques having certain specialities in performing a specific function. The different techniques in this soft computing basket are ANN, GA, fuzzy logic, PCA, rough set and SVM... etc. Sometimes to improve the performance these techniques are used in combination of the tools mentioned above. Examples of these hybridized tools are ANN-fuzzy, ANN-GA, rough-fuzzy, rough-GA etc. However among these techniques those with high potential for application in steel industries are ANN, fuzzy rule-based systems, genetic algorithms.

3. ARTIFICIAL NEURAL NETWORK

An Artificial Neural Network (ANN) is an information processing tool designed to model a particular task based on the philosophy of human biological nervous systems. A neural network is a complex, non-linear and massively parallel information processing system that has a natural tendency of storing and making use of experiential knowledge. It is composed of a large number of highly interconnected processing elements (neurons) working in union to solve specific problems. Each part of the neuron plays a role in communicating the information in the system. The interconnections between these nodes are called synapses or weights (Haykin, 1994). This is a function approximation tool which receives an input, processes the data and provides an output. It has the capacity to learn the relationship between the dependent and independent variables. Like brain it acquires knowledge through learning examples and interneuron connection strengths are used to store knowledge. The resemblance of ANN with human nervous system is shown in Figure 1. Like biological systems, learning in ANN is also the result of adjustments of the synaptic connections exist between neurons. The process of learning can be supervised or unsupervised depending on the information provided. Most artificial neural networks bear some resemblance to their biological counterparts but are very efficient at their intended tasks.

Neural networks can use a variety of topologies and learning algorithms. While creating a functional model of the biological neuron, the synapses of the neuron are modeled as weights and the strength of the connection between an input and a neuron is noted by the value of the weight. All the inputs, modified by their respective weights which can be positive (excitatory connections) or negative (inhibitory connections), are summed up. This activity is a linear combination and an activation function controls the amplitude of the output of the neuron. An output generally lies either between 0 and 1 or -1 and 1. Mathematically ANN based models are represented as

Figure 1. Resemblance of ANN with human nervous system

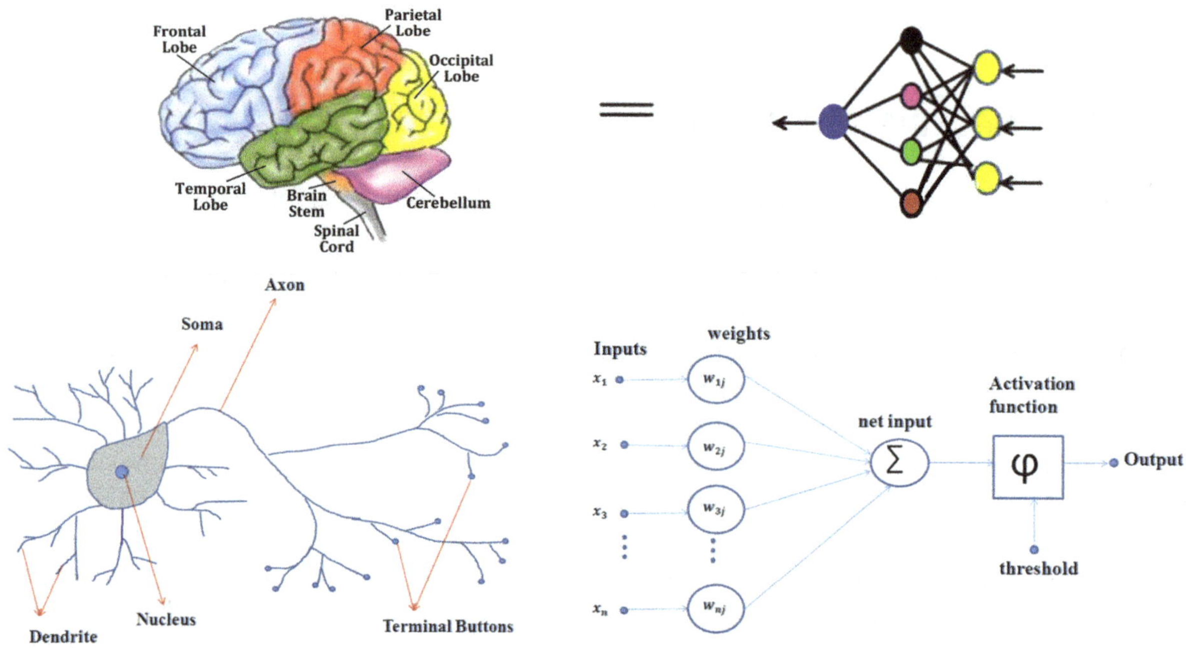

$$V_k = \sum_{j=1}^{n} W_{kj} X_j \tag{1}$$

$$y_k = \varphi(V_k) \tag{2}$$

Where y_k Output Neuron.

V_k Weighted sum of all synaptic inputs and bias
W_{kj} Synaptic weight connecting the output of neuron j to the input of neuron k.
X_j Input matrix.
φ Activation function.

3.1 Feed Forward Network

According to McCulloch and Pitts (1943) a neuron is a logical threshold element with two possible states. Such a threshold element has some input channels and one output channel. An input channel can be at one state at any point of time which is either active (1) or inactive (0). The activity states of all input channels thus encode the input information as a binary sequence of a number of bits as the number of input channels. After that all input signals are linearly summed up and the summation value is compared against a threshold value to give the state of the threshold element. If the sum exceeds the threshold value, neuron will be excited otherwise in inactive state. The excited state refers to firing and the quiet state refers to not firing of an action potential of biological neurons. These states are represented in the model by the binary values 1 and 0 for the activity of the output channel (Bose and Liang, 1996). Excitatory and inhibitory input signals are modulated by synaptic strengths. This model of McCulloch and Pitts for the first time gave an idea about the method to carry out logical operations and has found a way for numerous models in the later stage. But this methodology:

- Failed to explain the necessary interconnections between neurons that might occur through learning.
- Depend on error-free functioning of all their components and did not display the error tolerance of biological neural networks.

Hebbian Learning

To overcome the limitations of McCulloch and Pitts theory, Hebb (1949) suggested that the connection between two neurons is plastic and changes in proportion to the activity correlation between the presynaptic and the postsynaptic cell. If two neurons on either side of a connection are activated simultaneously then the strength of that particular connection is increased and if activated asynchronously then that particular connection is weakened or eliminated. The main aim of this theory is the adjustment of weights between learning nodes for better representation of the relationship between the nodes (Kohonen, 1989). One of its simplest mathematical formulations is

$$\Delta W_{jk} = \gamma y_j y_k \tag{3}$$

Where $\gamma \rightarrow$ learning rate $(0 < \gamma \leq 1)$,

$y_j \rightarrow$The input of the neuron j,
$y_k \rightarrow$The output (postsynaptic response) and
$W_{jk} \rightarrow$The weight that connects the neuron j and neuron k.

$$y_k = \sum_j w_j y_j \tag{4}$$

Where $y_j \rightarrow j^{th}$ input

Another common rule uses not the actual activation of unit *k* but the difference between the actual and desired activation for adjusting the weights:

$$\Delta W_{jk} = \gamma y_j (d_k - y_k) \tag{5}$$

in which d_k is the desired activation provided and y_k is the actual activation. This is often called the Widrow-Hoff rule or the delta rule. With the help of the computer, it has become possible to simulate in more detail the learning capacity of networks made of neurons subject to rules of the above kind and to demonstrate practical applications of such systems.

Back Propagation

This type of model is used as function approximation tool which receives an input, processes the data and provides an output. During this process it learns the relationship between the dependent and independent variables. In order to predict the output a corresponding desired or target response is set against each input in this process (Hajek, 2005). An error is deduced from the difference of the desired response and the real system output (predicted output).This calculated error value is fed back to the system to do the necessary adjustments in the system in a systematic fashion with an aim to reduce this error value (commonly known as the learning rule). This process is repeated until the targeted output is achieved. This learning process is called supervised learning. In unsupervised learning the goal is to construct feature variables from which the observed variables (both input and target) can be predicted. It is used to develop its own representation of input activity.

Back propagation algorithm is based on the generalization of Widrow-Hoff learning rule in multiple layer networks and nonlinear differentiable transfer functions. This is the most commonly used supervised training algorithm for the update of weights of a multi-layer perceptron model which is based on error correction rule.This algorithm aims to achieve the minimum value of the error function in weight space using the method of gradient descent. The combination of weights which minimizes the error function is considered to be a solution of the learning problem (Rumelhart, 1986). Since this method requires computation of the gradient of the error function at each iteration step gives guarantee the continuity and differentiability of the error function. It explains the change in behavior of the system by changing the weights and biases. Suppose a multilayer network is represented as

- $\vec{x}$ input vector for unit j (x_{ji} = i^{th} input to the jth unit)
- $\vec{w_j}$ weight vector for unit j (w_{ji} = weight on x_{ji})
- $z_j = \vec{w_j} \cdot \vec{x_j}$ the weighted sum of inputs for unit j
- o_j = output of unit j ($o_j = \sigma(z_j)$)
- t_j = target for unit j
    - *Downstream(j)* = set of units whose immediate inputs include the output of j
    - *Outputs* = set of output units in the final layer

The goal of back-propagation is to compute the partial derivatives $\partial E / \partial w$ and $\partial E / \partial b$ of the cost function E with respect to any weight w or bias b in the network.

If the training set consists of exactly one example, the error can simply be denoted by E. We need to calculate $\partial E / \partial w_{ji}$ for each input weight w_{ji} for each output unit j. Since z_j is a function of w_{ji} regardless of where in the network unit j is located,

$$\frac{\partial E}{\partial w_{ij}} = \frac{\partial E}{\partial z_j} \cdot \frac{\partial z_j}{\partial w_{ji}} \tag{6}$$

$$= \frac{\partial E}{\partial z_j} x_{ji} \tag{7}$$

Furthermore, $\frac{\partial E}{\partial z_j}$ is the same regardless of which input weight of unit j. To update this quantity by δ_j. Consider the case when $j \in Outputs$, sum squared error overall output signals is given by

$$E = \frac{1}{2} \sum_{k \in Outputs} \left(t_k - \sigma(z_k)\right)^2 \tag{8}$$

3.2 Radial Basis Network

Broomhead and Lowe (1988), and Moody and Darken (1989) were the first to introduce Radial Basis Function Network (RBFN) in the design of neural networks. The strategy used in RBFNs is approximating an unknown function with a linear combination of non-linear functions, called basis functions. The basic functions are radial functions, i.e. they have radial symmetry with respect to a centre (Khaled, 2005). RBF emerged as a variant of artificial neural network in the late 80's.

RBFs are embedded in two layer network where each hidden unit implements a radial activated function. RBFN is another member of the feed-forward neural networks and has both unsupervised and supervised phases. In the unsupervised phase radial basis functions of the inputs are calculated with the help of centre and standard deviation of the clusters after sending the input data clusters and the cluster details to the hidden neurons. Unlike back propagation in RBFN hidden nodes use a Gaussian or some kernel basis function instead of sigmoidal or S-shaped activation function. These hidden units act as a processor to fetch a score against the match between input vector and its connection weights or centres. The linear combination of the hidden nodes based on the output weights (weights connecting the basis units to the outputs) are used to get the final classification.

The structure of an RBF networks in its most basic form involves three entirely different layers. The radial basis function of the hth neuron is given by (Ramuhalli et al. 2002)

$$f_h(x_i) = \exp\left(-\frac{x_i - c_h{}^2}{\sigma_h{}^2}\right) \tag{9}$$

Where σ_h and c_h are the width and the centre of the basis function, respectively, and $\|x_i - c_h\|$ is the Euclidean distance between the ith input vector and the centre of the basis function. RBF networks include three layers. The input layer only distributes the input variables to all hidden layer neurons. The output of the hidden layer neurons is obtained by (9). These outputs are weighted and fed to the output layer neurons where the weighted values are summed to obtain the output of the network. The network output is thus given by

$$y = \sum_{h=1}^{H} w_h f(x_i) \tag{10}$$

Above, H is the number of neurons and w_h is the weighting coefficients.

Identification of appropriate number of neurons and the centres and widths of the radial basis functions is very much necessary to build a RBF network. This may be a complex and computationally expensive task. Clustering and dynamic stepwise selection methods are widely accepted for identification of these parameters (Wang and Xiang, 2007). As the parameters highly cross-correlated, finding the appropriate values by hit-and-trial method is not recommended and thus need to be defined simultaneously. The width or degree of sensitivity of all the hidden nodes towards inputs is same in the RBF network. However, hidden units with a wide area of reception are very much useful in portions of the input space where there are few patterns. But in contrary very highly tuned units with narrow reception fields are efficient in case of crowded portions of the input space. Computing these individual widths increases the performance of the RBF network at the expense of a more complicated training process.

In a back propagation network, adjustment of weights in all layers is done at the same time. In radial basis function networks, however, first the weights into the hidden layer basis units are usually set which is followed by the second layer of weights. As the input moves away from the connection weights, the activation value falls off. This behaviour forces to the use of the term "centre" for the first-layer weights. centre weights are calculated by Kohonen feature maps, statistical methods such as K-Means clustering techniques. In any case, they are then used to set the areas of sensitivity for the RBF network's hidden units, which then remain fixed. This process typically uses the standard steepest descent algorithm. The training problem becomes quadratic once if c_i's (radial basis function centres) are known.

3.3 Recurrent Network

A recurrent neural network (RNN) is a class of artificial neural network where connections between units form a directed cycle (feedback arc set) by creating an internal state of network to exhibit dynamic temporal behaviour. In contrast to feed-forward network, this directed cycle is a network with feedback elements that enable signals from one layer to be fed back to a previous layer. Unlike feed-forward neural networks, RNNs are very much applicable to tasks such as unsegmented connected handwriting recognition because of their capability to process arbitrary sequence of inputs. A simple recurrent network is one with three layers, an input, an output and a hidden layer. A set of additional context units are added to the input layer that receive input from the hidden layer neurons. The feedback paths from the hidden layer to the context units have a fixed weight of unity. A recurrent neural network (RNN) is useful in technological applications. The main feature of a recurrent neural network (RNN) is that the network with at least one feed-back connection for the activation to flow round in a loop. This enables the networks to do temporal processing and learn sequences (e.g., perform sequence recognition/reproduction or temporal association/prediction). RNN is more powerful and acceptable tool than other approaches. The architectures of recurrent neural networks can take many different forms, but they all share two important common features:

1. Incorporation of a multi-layer perceptron as a sub-system.
2. Exploitation of the powerful non-linear mapping capabilities of the multi-layer perceptron with some form of memory.

Figure 2 is a schematic representation of RNN. A fully recurrent network is one where every neuron receives input from all other neurons in the system. Such networks cannot be easily arranged into lay-

Figure 2. A schematic of RNN (S. Haykin, 1994)

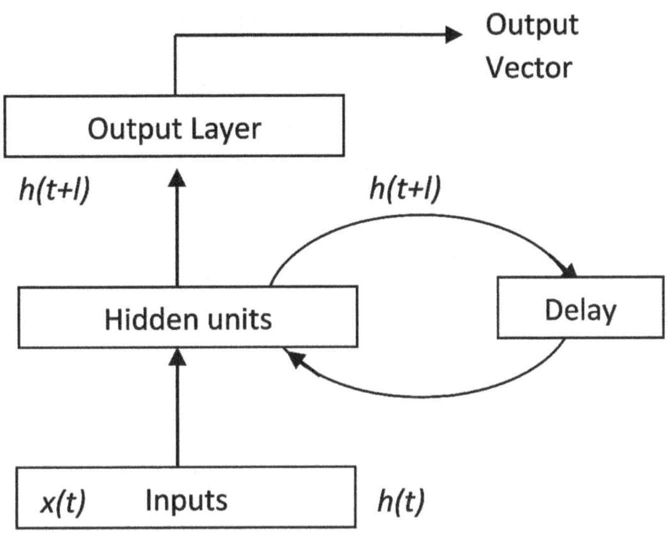

ers. A small subset of neurons is responsible for receiving external input and another small subset is responsible for producing system output. It is necessary to discretize the time t and update the activations with one time step at a time. This might correspond to the time scale at which real neurons operate or for artificial systems it can be any time step size appropriate for the given problem. A delay unit is introduced which simply delays the signal/activation until the next time step.

Recurrent networks, in terms of their properties, act as dynamic systems. The main features of this network are stability, controllability and observability (Robinson, 1994). Stability is the measure of the behaviour of the network output to small changes occur either in network inputs or weights. It also gives bounds over time on the network output in terms of maximal amplification of the network's input. Controllability is concerned whether one can control the dynamic behaviour of the recurrent network or not. A recurrent neural network is said to be controllable if an initial state is steerable to any desired state within a finite number of time steps.

Observability refers to the results of the control applied to the recurrent network. A recurrent network is said to be observable if the state of the network can be determined from a finite set of input/output measurements (S. Haykin, 1994).

3.4 Self Organising Map (SOM)

An important aspect of an ANN model is its learning mechanism. Some cases this learning needs guidance and in some cases it does not. Based on this learning philosophy, all artificial neural networks can be divided into two categories - supervised and unsupervised. Supervised training technique is a learning process in which a target output exists for each input pattern and the network learns to produce the desired outputs. In case of unsupervised training, the networks learn to form their own classifications of the training data without any external guidance. In this case class membership is broadly defined by the input patterns having common features and that the network will be able to identify those features across the range of input patterns. One interesting class of unsupervised system is based on competitive learning, in which competition is taken place among the output neurons for the right to respond to a subset of the input data. In this mechanism only one neuron per group (output neuron) is active at any particular point of time. The winning neuron gets activated (Kohonen, 1989). Such competition can be induced by having lateral inhibition connections (negative feedback paths) between the neurons by forcing them to organise themselves, for which such a network is called a Self Organizing Map (SOM).

In supervised learning, a desired output is required for each input vector during the training of network. An ANN of the supervised learning type, such as the multi-layer perceptron, uses the target result as a guide for the formation of the neural parameters. It helps the neural network to learn the behaviour of the process under study. In unsupervised learning, the training of the network is entirely data-driven and target results for the input data vectors are not provided. An ANN of the unsupervised learning type, such as the self-organizing map, can be used for clustering the input data and extracting features inherent to the problem. The self-organizing map was developed by professor Kohonen. The SOM has been proven useful in many applications because no human intervention is needed during the learning and those little needs to be known about the characteristics of the input data. In addition to that SOM is used in data clustering without the knowledge of the class memberships of the input data.

Self-organization often involves both competition and correlative learning. Competitions among neurons help in getting possession or ownership of this input when presented with a stimulus. The winners strengthen their weights or their relationships with this input (Datta, 2004). Hebbian learning is the most common rule for unsupervised or self-organized learning.

3.5 Sensitivity Analysis

In neural network sensitivity is reaction of outputs with variation in inputs. It is difficult to identify influences of the input variables on the output of a system, particularly in case of ANN models, due to its complex hidden relationships. Sensitivity analysis provides way to automatically identify all relevant parameters from a set of potential parameters. The sensitivity analysis is applied on the trained neural network model (Hunter, 2000).

In a three layer feed forward neural network, $\vec{z} = (z_1, z_2, z_3, \ldots, z_t)$, $\vec{y} = (y_1, y_2, y_3, \ldots, y_j)$ and $\vec{o} = (o_1, o_2, o_3, \ldots, o_k)$ represents the input, hidden and output layers respectively. Defining a training pair P as tuple $P = (\vec{z}(p), \vec{t}(p))$ where $\vec{t} = (t_1, t_2, t_3, \ldots, t_k)$ represents the target vector. Given any training pair p, the sensitivity $S_{ki}^{(p)}$ of a trained output o_k with respect to an input z_i can be defined as

$$S_{ki}^{(p)} = \frac{\partial o_k}{\partial z_i} \tag{11}$$

$$= o_k' \sum_{j=1}^{J} w_{kj} \frac{\partial y_j}{\partial z_i} \tag{12}$$

$$= o_k' \sum_{j=1}^{J} w_{kj} y_j' v_{ji} \tag{13}$$

where y_j denotes the output of jth hidden neuron of the hidden layer $\vec{y}$, $o_k^{'}$ is the value of the derivative of the output layer activation function

$$O_k = f\left(\sum_{j=1}^{J} w_{kj} y_j\right) \quad (14)$$

and $y_j^{'}$ is the value of the derivative of the hidden layer activation function

$$Y_j = f\left(\sum_{i=1}^{I} v_{ji} z_i\right) \quad (15)$$

where w_{kj} denotes the weight value between hidden neuron y_j and output o_k, v_{ji} is the weight value between input z_i and hidden neuron y_j. Equation (11) only defines the sensitivity of one output o_k with respect to one input z_i for pattern p. For training pattern p define the sensitivity matrix $S^{(p)}$ which consists of entries $S_{ki}^{(p)}$ as

$$S^{(p)} = O^{'} W Y^{'} V \quad (16)$$

Where $W^{(k\times j)}$ and $V^{(J\times I)}$ are respectively the output and hidden layer weight matrices. Eq (13) defines the sensitivity matrix for a specific training pattern p. however each training pair p produces a different sensitivity matrix $S^{(p)}$. The sensitivity matrix, $S^{(p)}$, must be evaluated over the entire training set. The mean square average sensitivity matrix is defined as S_{avg} is defined as

$$S_{ki,avg} = \sqrt{\frac{\sum_{p=1}^{P} \left(S_{ki}^{(p)}\right)^2}{p}}$$

3.6 Performance of ANN Model

It is important to observe that the performance of the ANN based systems hinges heavily on the data. That's why this data should be pre-processed before considering it as meaningful information. It is developed with the help of a systematic step-by-step procedure which optimizes certain criteria by following certain learning rules. The basic input/output training data is used to find the most optimal operating point. During designing phase of ANN models, parameters like network topology, performance function, learning algorithm and stopping criteria for training phase are decided by the developer. So it is not so simple to determine the size and parameters of the network as there is no theorem or formula to do it. The optimum network is designed by playing with these parameters. Main difficulty with this method is to refine the solution when the system does not work properly. Despite these issues neural networks based solutions are very efficient in terms of development, time and resources. ANN provides real solutions that are difficult to match with other technologies (H.K.D.H. Bhadeshia, 2008). In addition to learning process the selection of an appropriate learning algorithm is also very important to get a better solution. Learning implies that a processing unit is capable of changing its input/output behaviour as a result of changes in the environment. Since the activation rule is usually fixed when the network is constructed and since the input/output vector cannot be changed that's why to change the input/output behaviour the weights corresponding to that input vector need to be adjusted (S.B. Singh, 1998; I. Mohanty, 2008). A method is thus needed by which, at least during a training stage, weights can be modified in response to the input/output process. A number of such learning rules are available for neural network models (Haykin, 1994).

4. PROS AND CONS OF ARTIFICIAL NEURAL NETWORK

Neural networks, with their remarkable ability to derive meaning from complicated or imprecise data, can be used to extract patterns and detect trends that are too complex to be noticed by either humans or other computer techniques. A trained neural network can be thought of as an "expert" in the category of information it has been given to analyse. Other advantages include:

1. Adaptive learning: An ability to learn how to do tasks based on the data given for training or initial experience.
2. Self-Organisation: An ANN can create its own organisation or representation of the information it receives during learning time.
3. Real Time Operation: ANN computations may be carried out in parallel, and special hardware devices are being designed and manufactured which take advantage of this capability.
4. Fault Tolerance via Redundant Information Coding: Partial destruction of a network leads to the corresponding degradation of performance. However, some network capabilities may be retained even with major network damage.

Neural networks are able to detect similarities in inputs, even though a particular input may never have been seen previously. This property allows for excellent interpolation capabilities, especially when the input data is noisy (not exact). When a data stream is analysed using a neural network, it is possible to detect important predictive patterns that were not previously apparent to a non-expert. Thus the neural network can act as an expert.

Despite the excellent performance of neural networks, it is perceived as a "black box" because of its limitation to provide a comprehensible explanation of the process through which a given output has been obtained from a neural network (Andrews, 1995). Network training helps to derive the knowledge and store implicitly in numerical forms as synaptic weights in the network. Unlike the parameters of regression functions, these weights have no obvious meaning in most cases. It is difficult, if not impossible, to interpret these weights due to their complex nature (Lein, 1997). Therefore, a neural network is often accused of being a black box, which hides the relation between inputs and outputs in the weights of the neurons of its 'hidden' layers (Benitez et al. 1997). As a result, we cannot gain any understanding of the problem at hand due to the lack of an explanatory capability to provide insight into the characteristics of the dataset.

It is not very easy and straight forward to develop and use neural network based models in the field. The ability of a neural network model can be a disadvantage because of its "black box" nature, greater computational burden, proneness to over fitting and the empirical nature of model development. The proneness to over fitting can be handled by limiting the training using cross-validation.

5. APPLICATION IN STEEL INDUSTRY

The development and processing of materials is very complex. Although many research works on materials have been done to understand the underlying phenomena, there remain many problems where quantitative treatments are dismally lacking. ANN based tools have been applied in prediction, modelling, control, identification, design and optimisation areas of materials engineering (Datta, 2004;Datta, 2006 ;Datta, 2007 ;S. Dey, 2008 ; A. Sinha, 2013;S. K. Ghosh, 2005, 2009;P. Das, 2009; S. Ganguly, 2007, 2009; I Mohanty, 2008, 2011, 2014) . The majority of the applications have been in the area of materials properties prediction. In addition to this other application areas are optimization of properties, alloy design, process control, etc. The practicability of such models in materials engineering application in terms of economy of tool is difficult to justify as it is well known that ANN with one layer will produce a good approximation. In most of the work it is seen that, ANN models have been shown to generate better predictions than the classical linear regression or first principle based models. One of the reasons frequently cited for the use of ANN in materials engineering is its ability to recognise and learn the underlying non-linear relations between input and output without the need to construct an explicit mathematical model. This is why the ANN technique is considered to be model free and useful in the modelling of complex input-output relationships. Also the ANN is tolerant to data containing noise and measurement errors. Conventional ANNs have a number of limitations which must be borne in mind when applying them in materials engineering. First, there is the need to manage a large number of parameters used for controlling variable in ANN model. This process is not systematic but intuitive. Inability to appropriately manage ANN model parameters accounts for the difficulty in obtaining stable solutions and danger of over- fitting resulting in the lack of generalization capability.

Neural networks have broad applicability to real world business problems. In fact, they have already been successfully applied in many industries. Since neural networks are best at identifying patterns or trends in data, they are well suited for prediction or forecasting needs like sales forecasting, industrial process control, customer research, data validation, risk management and target marketing. But to give you some more specific examples; ANN are also used in the following specific paradigms: recognition of speakers in communications; diagnosis of hepatitis; recovery of telecommunications from faulty software; interpretation of Chinese words; undersea mine detection; texture analysis; three-dimensional object recognition; hand-written word recognition; and facial recognition. As the aim here is to concentrate on application of ANN in steel industry, details of few case studies related steel industry is explained as follows.

Case 1: On-Line Mechanical Property Prediction System for Hot Strip Mill by Composition-Process-Property Correlation

Rolling is a metal forming process in which metal stock is passed through one or more pairs of rolls to reduce the thickness and to make the thickness uniform. Rolling is classified according to the temperature of the metal rolled. If the temperature of the metal is above its recrystallization temperature, then the process is known as hot rolling. The initial breakdown of slabs (210 mm thickness) into plates or sheets (30-40 mm approx.) is done by hot rolling process. The objective is to decrease the thickness with increase in length and little increase in width. This process is carried out in different stages like reheating, rolling and cooling. In hot rolling process slabs are heated up initially to 1100- 1300 °C before feeding to the roughing rolling mill. The temperature in last finishing stand in finishing mill varies from 800-955 °C, but should be above the austenite to ferrite start temperature (Ae3) to produce uniform equiaxed ferrite grains. One of the key determinants of the quality of the hot rolled steels is its mechanical properties which must lie within stretched tolerance range. Once rolling is done, samples are collected for testing which is costly and time consuming. The speed of this process needs to be improved to handle the production pressure. In that case if the test result is made ready as soon as the rolling is done, it would save time and money and enhance the despatch process. For high-end applications precise control of qualities is done by controlling composition and process parameters. A slight variation in these parameters will result in variation in mechanical properties. It is very difficult to predict the mechanical properties accurately with the first-principles based mathematical model. So to have knowledge of the strip quality immediately after rolling, a neural network based model is developed. In this case study the way a neural network based online system is developed which helps in predicting mechanical properties of IF steel strip is described in detail. It also explains the way this model helps in understanding various other metallurgical phenomena during rolling.

It has been found that, mechanical property of the hot rolled steel is influenced by process parameters like reheating furnace parameters, roughing mill process parameters, FRT, CT, speed of rolling in addition to chemistry and coil dimension. Tracking and simulation of the complicated metallurgical phenomena like the effect of slab drop-out temperature, coiling temperature and also the impact of micro-alloying elements in hot rolled steel has been done by ANN model. The geometrical, compositional, processing parameters and the mechanical property data covering all grades of IF steel have been collected for one year. As small sized network cannot learn a problem properly and on the other hand an oversized neural network will lead to an over fitting of the training data set and subsequent poor generalization performance. To take care of these issues data is divided into three sets such as training set, validation set and test set (I. Mohanty, 2008). A neural network model with one hidden layer is developed in which weights of the neurons are updated using the gradient descent algorithm. Several ANN models have been generated separately for prediction of tensile strength (UTS), yield strength (YS) and percentage elongation (%el) of hot rolled steel with number of hidden units varying from 8 to 40 to find the appropriate network to capture the complex relationship between parameters and process dynamism of IF grade steel. Among the alternate models the best was chosen for each output variable. A multilayer supervised learning based artificial neural network based model has been developed. This model predicts mechanical properties for various grades of IF steel containing various combinations of Ti, Nb and B. The predicted results show reasonable good agreement with measured values.

These models were used to study the effects of some of the variables on YS, UTS and % elongation (%el). These simulation studies are carried out by varying two variables at the same time and keeping other variables at constant at an average value. It is important for processing of steels as in many occasions two or more parameters in combination affect the final properties. For practical constraints and costly experimental techniques involved, this study is best done using NN techniques. For example, as we know the effective utilization of Nb for strengthening steel depends on slab drop out temperature (SDOT). Suitable SDOT helps in keeping the Nb in solution which during rolling can effectively retard the recrystallization of austenite. However, wrong SDOT might lead to partial dissolution of Nb which will lessen its effectiveness (I. Mohanty, 2014). A detailed analysis using NN could be helpful in preparing a processing window with optimum combination of these two input parameters. As a second example the case of FRT and CT can be presented. These two are most important rolling parameters that can be controlled easily. Both of them have pronounced effect on the mechanical property. However they are also dependent on each other. It is understood that the difference between FRT and CT indicates the cooling rate on the run-out-table (ROT). It is well known that the cooling rate has high impact on the mechanical property of steel. It is due to this we need to examine the effect of FRT and CT together. It is almost impossible to develop a first principle model to study this, it is also extremely difficult to experimentally generate data in plant as rolling with different combination of FRT and CT would be very costly. However using NN this can be done very easily and effectively. Sometime NN analysis also throws light on areas which is almost impossible to find using subject knowledge. Ti and Nb are the two most important micro alloying agents added to steel. The effect of these two parameters on YS, UTS, % elongation and YS/UTS ratio is shown in Figure 3. The dissolution temperature of both Ti and Nb could depend on the amount of N and C in steel. Quite often these two micro alloying elements strengthen the steel by same mechanism i.e. grain size refinement and precipitation hardening. Both these alloying elements also effectively fix C and N and improve the properties of IF steel. However as these are costly elements, it is important to use them in right combination. It is for this reason we have studied the combined effect of Ti and Nb on the mechanical properties of steel. Some interesting results are displayed here.

The biggest benefit of NN analysis of the process data is its quantitative assessment of the process which can be tried in practice for improving the productivity. On the other hand other modeling techniques often give a quantitative analysis which cannot be used directly for improving process capabilities. The two main challenges of this property prediction system are prediction accuracy and on-line availability

Figure 3. Effect of Titanium and niobium on (a) YS, (b) UTS, (c) % elongation and (d) YS/UTS ratio (I. Mohanty, 2014)

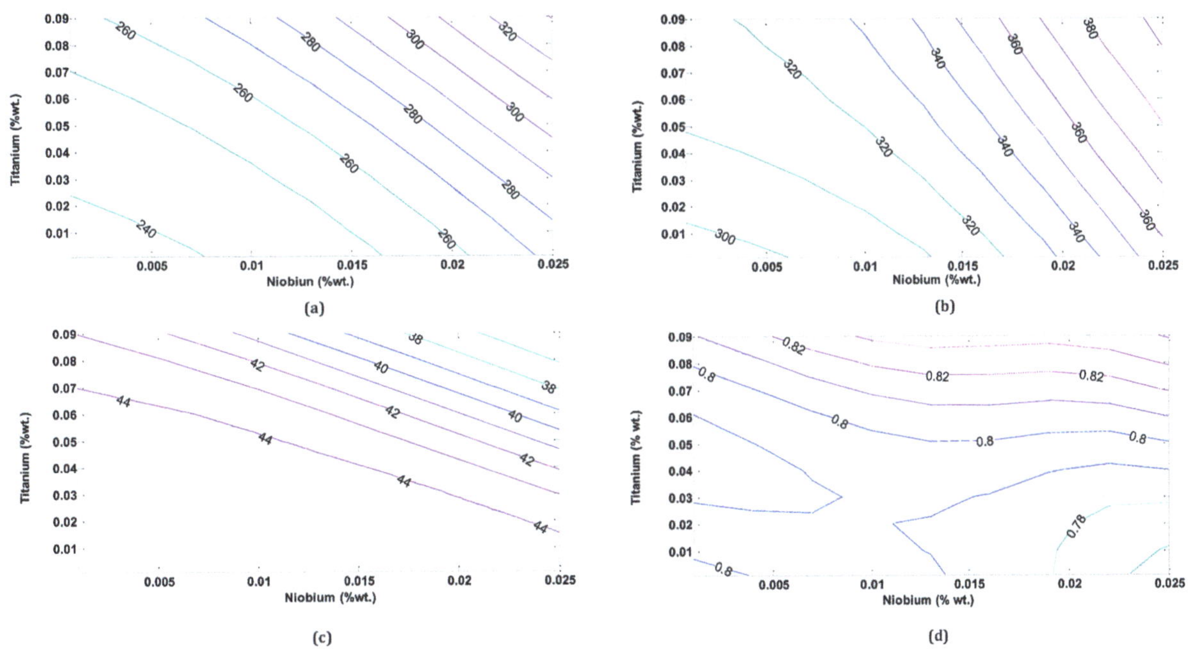

of the system. As the HSM automation system has many sub systems which serves different stages of hot rolling, it is necessary to receive all the necessary input parameters (slab information, Chemistry, roughing mill parameters and finishing mill parameters) from the information system as soon as the rolling is over, so that materials can be dispatched to the customer on time. After receiving the input data the system calculates the mechanical properties and sends it to a particular system which is used by the technical services for dispatching any coil to the market. The online property prediction system helps in reducing mechanical testing time. Indirectly it helps in reducing product cycle time and better utilization of manpower. It is an un-manned system where very little or no maintenance is required. It has the capacity to store data for one month which can help in further analysis. This online system is based on an off line model. MATLAB 7.0.1 is considered as basic programming software for the design of neural network architecture and development of algorithms. The results of this work indicate that the off-line ANN model can be a good tool for studying the role of various processing and composition variables on the final properties of IF steel. Findings from this study can be successfully utilized to design the composition and processing of steel with desirable properties. This model can be used as a guideline for further experimentation.

Case 2: Kohonen Network Modelling for the Strength of Thermo-Mechanically Processed HSLA Steel

In high strength low alloy (HSLA) steels thermo mechanical controlled processing (TMCP) is a widely used process. Apart from chemistry, the microstructure and mechanical properties of thermo-mechanically processed HSLA steel depend on various other parameters such as reheating temperature, amount of deformation in different stages of rolling, finish rolling temperature and cooling rate. The final properties of the steel are governed by the basic strengthening mechanism which is the combination of the parameters described above.

In this case study various groups of the steel data have been made according to its alloying contents. This is done by developing a model through the Self Organising Map (SOM) or Kohonen network which uses unsupervised learning of the neural network for clustering. Learning Vector Quantisation (LVQ) has been applied to separate the datasets of the steel in accordance with the dominant strengthening mechanisms operative in the steel due to its particularity in the stoichiometry and processing variables. This method is employed for training competitive layers in somewhat supervised manner. Generally, no tool is more intelligent than human brain; In addition to that it is not possible to have an error and bias free decision with certainty. Reasoning does not work properly under such circumstances as it yields inaccurate measure in the inferences. To overcome this uncertainty in prediction unsupervised learning technique can be used in some of such cases to handle.

The SOM is very much effective in situations where it is important to analyse large number of cases and identify groups with similar features. This methodology makes the process simpler by projecting high dimensional data sets to a low dimensional space. The basic idea behind methodology is that whenever there are two cells with similar activations, the weights that connect them should be increased so that the connected neurons reinforce each other to support learning. The main aim of this learning method is the adjustment of weights between learning nodes to better represent the relationship between the nodes.

Self Organizing Maps helps to learn and classify input vectors according to the way they are assembled in the input space. Neighbouring neurons in the SOM learn to recognize neighbouring sections of the input space and learn both the distribution and topology of the input vectors they are trained on (T. Kohonen, 1989).

In the present case a SOM network was developed with the aim to have efficient design of composition and process parameters of the steel. As it has been described earlier that the inputs for the model are composition (C, Mn, Si, Ni, B, Cr, Mo, Cu, Ni, Ti), slab reheating temperature, deformation in three different zones, finish rolling temperature and cooling rate. As there are six strengthening mechanism, the model is designed such that number of neurons are six and each neuron gets a definite bias. LVQ method has been used to impart a certain amount of supervision in the learning process and to handle the poor performance of the model towards unknown data. The input vectors are initially divided into 16 sub classes which are linearly designed to six target classes. Three networks were developed with 3, 6 and 8 neurons. But after analysis, it is found that this model is capable of clustering the steel data into five groups according to its alloying additions. The network with six neurons is proved to be very good with respect to the number of data clusters it has map onto. The drawback of four-neuron network is that it is not capable of mapping the data in five clusters. In case of eight-neuron network the number of unused neuron is 3 which is 1 in case of six-neuron network. The number of unused neuron increases the chances of randomly identifying themselves with the five cluster specific neuron increases by creating more errors in mapping. The input weights assigned to the three inputs by the six neurons are plotted in Figure 4. The figure shows that the first two neurons, which map the clusters having single addition.

When a large set of data with all the input parameters (sixteen) are fed to map based on their their dominant strengthening mechanism, the process of self organising mapping is found less effective. Neurons are required to connect to correct destinations of data clusters by training them with supervisory mode. That's why the network is trained with linear vector quantisation process. Strengthening mechanism by precipitation hardening is caused by the presence of elements like carbon, manganese, nickel, copper, niobium, titanium and boron in steel composition; as regard to the process parameters, deformation in the non-recrystallisation and the two-phase regions as well as the finish rolling temperature have such effects also. Elements like manganese, copper, niobium and titanium and process parameters like low

Figure 4. The input weights connected to the neurons of the SOM of niobium, titanium and boron with six neurons (S Datta, 2004)

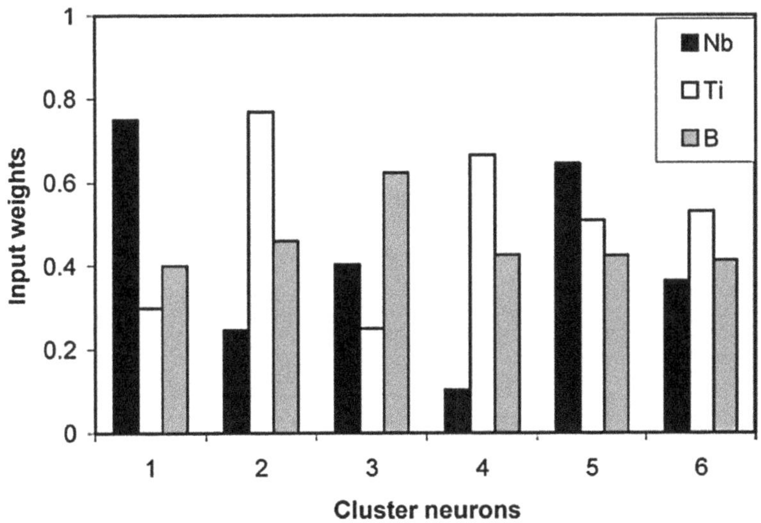

slab reheating temperature, high deformation in various stages and low finish rolling temperature are the enablers for the refinement of grains which helps in increasing the strength of steel. In addition to chemistry strength of HSLA steel also depends on the deformation in the non-recrystallisation region and also the cooling rate. On the other hand, carbon, manganese, silicon, nickel, copper, molybdenum and chromium influence the solid solution hardening. Strengthening due to dislocation hardening is achieved by high amount of deformation in the non-recrystallisation and the two-phase region and by subsequent cooling at a higher rate. Combined effects of molybdenum, nickel and titanium help to enhance it. The texture hardening is the result of high deformation in the two-phase region.

Considering the above facts, the datasets of input variables are chosen in such a way that only one of the strengthening mechanisms will be active. The neurons are trained to map these data clusters and are found to be very effective.

Prediction of the dominant strengthening mechanism acting on a thermo mechanically processed HSLA steel effectively found out through the network training process by mapping the neurons to clusters which is given in Table 1. Table 1 demonstrates that grain refinement is the dominant strengthening mechanism for alloy 1 where all the process parameters are highly favourable for achieving smaller grain size. It is known that low slab reheating and finish rolling temperatures along with high amount of deformation percentage in the three stages followed by a high cooling rate lead to considerable grain refinement in a steel, which contains high microalloying elements *viz.* niobium and titanium. Incidentally the major solid solution strengthening elements like molybdenum and chromium are absent in this alloy. The present network is therefore found to corroborate with the existing knowledge of physical metallurgy.

In this work it has been observed that solid solution strengthening mechanism takes place when an alloy contains higher amount of molybdenum, chromium and silicon but contains neither niobium or titanium, provided that the slab reheating temperature is high enough to dissolve all the substitutional elements and the deformation in each stage is small enough to prohibit strain induced precipitation. This metallurgical belief is amply supported by the network prediction that solid solution hardening is the dominant strengthening mechanism. From this study it is clear that the network prediction with unknown data is compatible with the theoretical predictions, which follow from physical metallurgy principles.

Table 1. Some of the prediction results for some steels with their process parameters in respect of the dominant strengthening mechanism

Alloys	Composition(Normalised)										Process Parameters(Normalised)						Prediction For Dominating Mechanism
	C	Mn	Si	Ni	Cr	Cu	Mo	Nb	Ti	B	SRT	D1	D2	D3	FRT	CR	
1.	0.15	0.63	0.28	0.18	0.00	0.82	0.00	0.75	0.65	0.58	0.22	0.71	0.89	0.84	0.21	0.90	Grain Refinement
2.	0.82	0.78	0.69	0.64	0.77	0.76	0.88	0.00	0.00	0.00	0.52	0.21	0.32	0.15	0.58	0.42	Solid solution hardening
3.	0.74	0.65	0.18	0.59	0.14	0.82	0.00	0.74	0.68	0.81	0.53	0.18	0.88	0.74	0.24	0.35	Precipitation hardening
4.	0.32	0.24	0.26	0.31	0.16	0.00	3.72	0.52	0.67	0.00	0.62	0.21	0.81	0.85	0.22	0.92	Dislocation Hardening
5.	0.78	0.66	0.81	0.00	0.71	0.00	0.64	0.74	0.68	0.00	0.75	0.21	0.69	0.15	0.65	0.32	Micro structural change

Case 3: Model for Optimizing Tensile Properties and Facilitating Design of New Alloys

Cold rolling of steel sheets is done with an aim of having thin sheets with better mechanical properties and smooth surface finish which is not possible by hot rolling. Due to the emerging applications of steels, the production of cold rolled sheets has been improved over the period of time. However the contribution of these cold rolled steels in automobile production is decreasing because of requirements for lower fuel consumption and higher body rigidity. We need to reduce the weight of the body for improving fuel economy. Materials with lower specific weights than steel are the enablers for weight reductions. It can also be achieved by lowering the thickness of the steel sheet. The reduction in thickness must be compensated for, by using steel of higher yield strength. So the automobile industry is using high-strength steel sheet to reduce the body weight. Because of high end applications, cold rolled micro-alloyed, phosphorus-alloyed, interstitial free and high strength low alloy steel sheets require very rigid control of the mechanical properties.

The cold rolled steels are made to have high ductility which is a necessity for their applications. Increase in strength of such steel definitely helps the purpose, but it could not be done in the cost of the ductility. So the objective here is to improve the strength of cold rolled sheets without compromising the ductility. As strength and ductility are two conflicting requirements, this case can be treated as a multi-objective optimization problem. The relationship between various compositional and processing parameters and the mechanical properties is not easy to establish from the mathematical models developed on the basis of rolling technology. Various interaction between the numerous input and output parameters in a rolling mill makes this types of models often unusable in practical sense. Improving and controlling the properties depends on the control of compositional and process parameters. To arrive at a solution for this one should have knowledge regarding the relationship between various parameters and the mechanical properties. ANN based models (Bhadeshia, 1999) perform better in handling the problems mentioned above due to its capability in solving non-linear problems. An effectively designed ANN model can be successfully used to control the mill process for achieving target properties with available steel composition and mill constraints. In this case ANN models are developed for predicting properties by correlating the composition and process parameters. These developed ANN models are used to extract the role of the variables on the final properties of the cold rolled steels, which will increase the domain knowledge and help to achieve a set of optimized properties for further improvement of the process as well as the product. .

To handle the optimization issues in engineering problems (Deb, 2001), genetic algorithm (GA) (Goldberg, 2002) is being efficiently used for, which includes several studies involving materials property and processing. In the present work attempts are made to find an optimum balance of YS, UTS, %elongation and r-bar in cold rolled IF steel sheets using ANN and GA. The cost functions for the optimisation is formulated using the ANN models developed (I. Mohanty, 2011). Both ANN and GA model parameters are given in Tables 2 and 3, respectively.

The ANN used here is supervised multilayered feed forward networks trained with scaled conjugate gradient back propagation algorithms. The input variables comprising of ten compositional and seven process variables (the finish rolling temperature (FRT) and coiling temperature (CT) of hot strip mill, percentage cold deformation, batch annealing parameters and skin pass mill parameters) are defined as input nodes and the four property variables (YS, UTS, %el, r-bar) are taken as output nodes. The inputs and outputs are connected through hidden units. The inputs are multiplied by weights for a hidden node input layer weights; summation of all those is then added to a bias value and finally operated by a suitable transfer function. Similar operations are repeated for varying number of hidden units (5 to 40) in order to find out suitable network architecture. The transfer function used in this work is tanh.

Separate ANN models for YS, UTS, %elongation and r-bar are developed and used as the objective functions for developing GA based optimization. Following equations describing the trained networks are used to maximize four objectives (YS, UTS, %elongation and r-bar), expressed in the general form:

$$\text{Max} \sum_{i}^{j} w_j \left\{ \tanh\left(\sum w_{ji} x_i + b_j \right) \right\} + b$$

$$\text{Subject to } x_i^{LB} \le x_i \le x_i^{UB}$$

Where w_{ji} is the input to the hidden layer weight matrix, w_j is the hidden layer weight vector, b_j and b denotes the input to hidden layer and hidden layer to output bias value for the trained network. x_i represent the parameters for which the maximization search is attempted, whereas x_i^{LB} and x_i^{UB} are the variable bounds defining the search space for the optimization Having several objective functions which are to be minimized/maximized, the idea of "optimum" changes, because in multi-objective problems, we are really searching for good compromises (or "trade-offs") between the objectives following the concept of pareto-optimality rather than a single solution as in global optimization.

Table 2. ANN Model Parameters

No Of Hiddel Layer	1
No Of Input	17
No Of Output	4
No Of Nodes in hidden layer	5-40
Transfer Function	tanh

Table 3. GA Model Parameters

Population size	100
Chromosome length	111
Number of generations	500
Number of functions	2
Number of constraints	0
No. of binary-coded variables	17
Selection strategy	Tournament selection
Crossover type on binary strings	Uniform cross-over
Crossover probability	0.9
Mutation probability for binary strings	0.007

According to pareto-optimality, no other solution could exist in the feasible range that is at least as good as some member of the pareto set, in terms of all the objectives, and strictly betters in terms of at least one. The binary coding scheme is used to represent the variables, where one string (chromosome) contained all the compositional variables as well as the other processing variables. A substring of size 5–10 bits, depending on the precision requirement, is used to represent each decision variable.

The non-dominated pareto solutions, developed through the multi-objective optimization give a range of combination of strength and ductility. The optimal solutions are also analysed to extract further knowledge on the role of various parameters, which could be used for designing the chemistry of the alloy as well as process parameters for producing cold rolled strips with tailor made property.

Figure 5 shows the set of non-dominated solutions called pareto fronts resulted from multi-objective optimization of the strength and formability parameters of the ultra low carbon cold rolled steel sheets. The pareto fronts deliver several solutions for achieving cold rolled steel having improved mechanical properties. The pareto solutions in Figure 5 (a) and 5 (b) describes that the variation in YS is in the range of 250–300 MPa. This implies that without much compromise to the formability considerable improvement of YS is still possible through proper selection of composition and process parameters.

Pareto solutions also helped in explaining the knowhow regarding the variation of the composition and process parameters. It is seen that the variation of optimum solutions in most of the parameters are quite less, even fixed for some cases. It has also been observed that some parameters are there where the variation is absent or negligible. Except Ti and Nb all other compositional variables do not have much variation in the pareto solutions. This shows that most critical factor is choosing the Ti, Nb and V content to design the alloy steel with right balance of YS, UTS and %elongation. Among the process variables, the finish rolling temperature (FRT) and coiling temperature (CT) have some variation. All the pareto solutions are numbered as sorted in order of increasing strength and decreasing formability for all cases. In Figures 6-7, the pareto solutions are plotted for variation of the composition and process parameters with the solutions. This gives a better understanding of the role of variables in the optimum solutions, where one objective is increasing and the other is decreasing. When the multi-objective optimization is done for the UTS, YS with %elongation and r-bar, it is seen that Nb has a important role, as the UTS value increases with increasing Nb content (Figures 6 and 7). It shows that the role of Nb is significantly different from Ti in case of IF steel (Tamehiro, 1987). Though it is not that significant in case of YS, but does have a role in work hardening of the steel. As in case of YS, it is seen that higher FRT is preferred for higher tensile strength. The above exercise generates some additional information

Figure 5. Pareto solution (a) YS and %elongation (b) YS and r-bar (I. Mohanty, 2011)

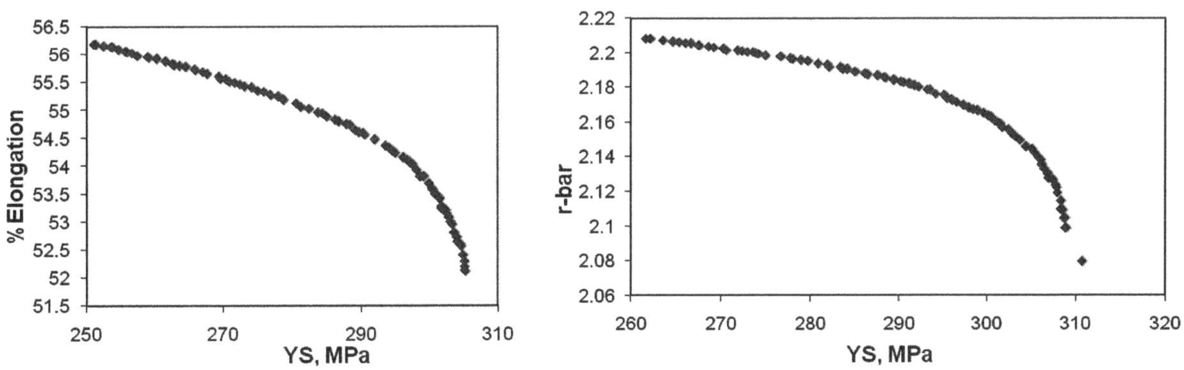

Figure 6. Variation in (a) Nb and (b) FRT with increase in UTS and decrease in %elongation as found in Pareto solutions (I. Mohanty, 2011)

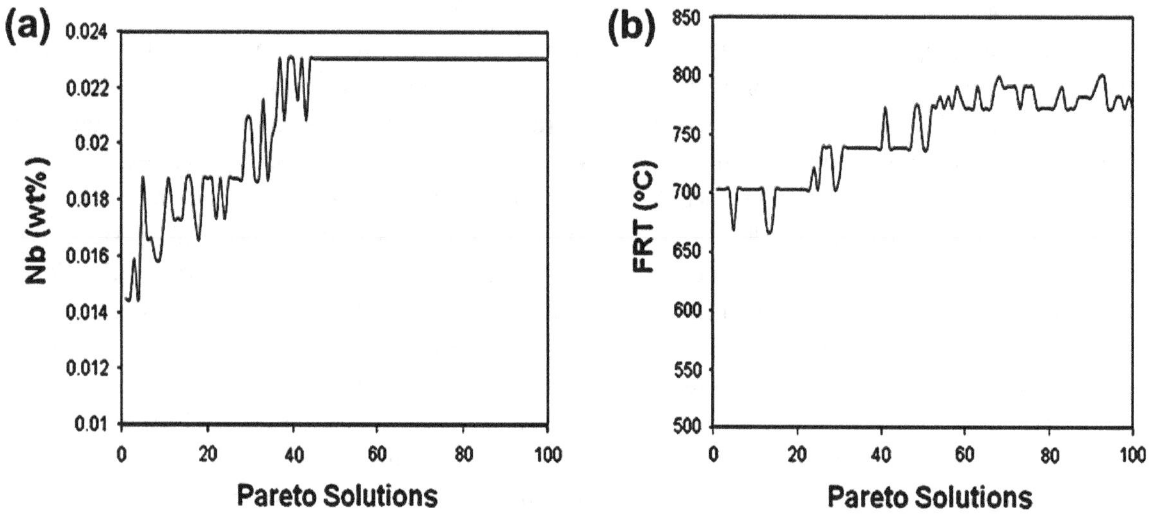

Figure 7. Variation in (a) Nb and (b) FRT with increase in UTS and decrease in r-bar as found in Pareto solutions (I. Mohanty, 2011)

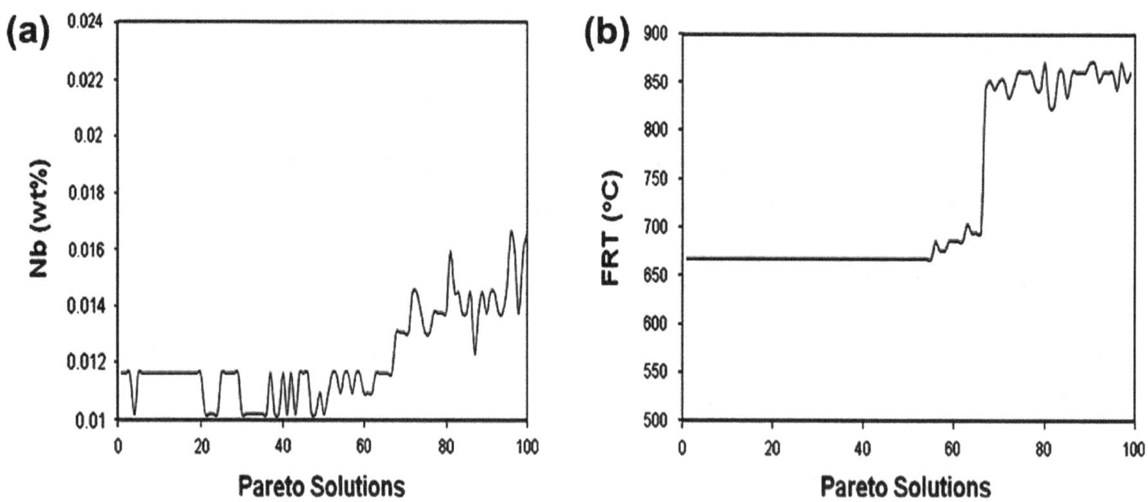

regarding the role of the variables which are most important for designing the chemistry and processing of the steel. The inputs, which have constant or near constant values for all non-dominated solutions, can easily be used as it is for developing steel with optimum strength and formability balance. But the value of inputs, which vary with varying strength/formability, can be selected from the figures given according to the property requirements.

Multi-objective genetic algorithm could easily be used for complicated optimization of mechanical properties of IF steel. The pareto fronts give a detail information about the different of combination input parameters and the corresponding combination of properties which helps in designing the alloy by taking into consideration the process parameters as well as chemistry The detail study of the pareto solutions has helped in understanding the system better which further can be verified computationally as well as by conducting experiments.

Case 4: Improvement of Yield Strength Accuracy in Skin Pass Mill Using Sequence of Artificial Neural Nets for Elongation Set-Point Specification

Ferritic steels show a localized, heterogeneous type of transition from elastic to plastic deformation which produces a yield point in stress-strain curve (Dieter, 1928). During tensile testing of a steel sheet, there is a zone observed in the stress-strain diagram just after yielding where elongation occurs without any increase in stress. The elongation which occurs at about constant load is called yield-point elongation which is shown in Figure 8(a). Yield point elongation creates stretcher strain marks on the formed components made from cold rolled and annealed steel sheets. The normal practice to eliminate the yield-point phenomena is skin passing (X. Fang, 2002) which is a rolling operation with a very small (usually 0.5–2.5%) thickness reduction performed at the final stage of a sheet rolling process. During this process the sheet is strained just above the strain associated with the yield point which effectively increases the yield strength and prevents the yield point elongation phenomena from occurring during forming operation (as shown in Figure 8(b)). Skin passing is a process to produce good surface flatness and low surface roughness in cold rolled sheets. It also helps in improvement of formability by decreasing the yield point and improvement of mechanical properties (Hudd, 1998).

Based on thickness, grade of steel and the shelf life of the coil a certain amount of strain is given during skin pass in cold rolling process to heal the effect created during annealing. But to arrive at the optimised value many trials are to be taken and this method is unable to consider the impact of the variation of different upstream compositional and process parameters. As thickness reduction helps in improving the yield strength of the coil and skin pass being the final step in the entire steel making process chain, there is a justification in fine-tuning this thickness reduction. A new methodology has been proposed in this case study for achieving a narrower tolerance band in yield strength through increased accuracy in elongation settings by appropriate consideration of the variation of nearly all of the afore

Figure 8. (a) Yield point elongation (b) No yield point elongation (Dieter, 1988)

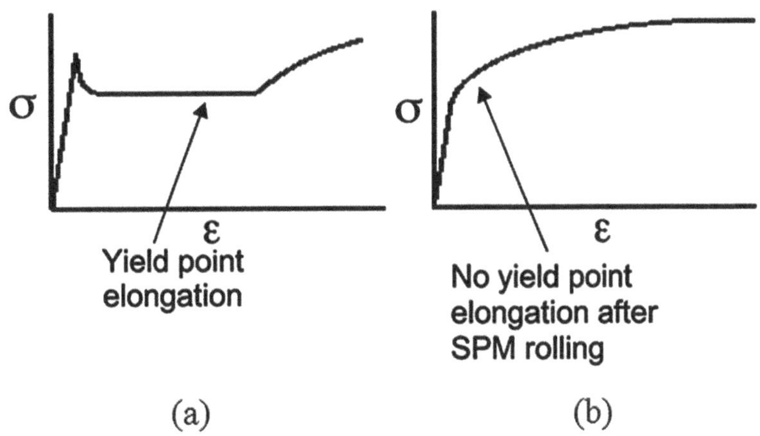

Figure 9. The normal, or "process forward" ANN: FANN

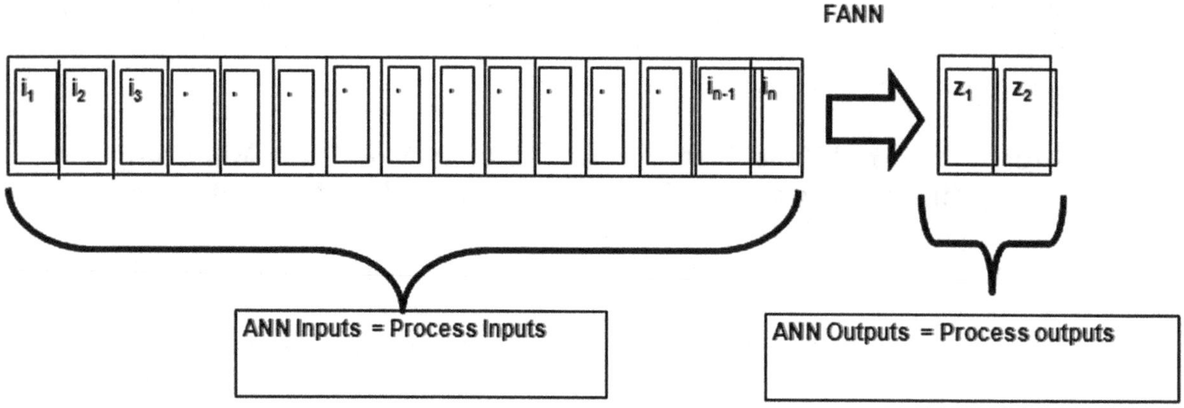

Figure 10. The "Reverse process" ANN: RANN

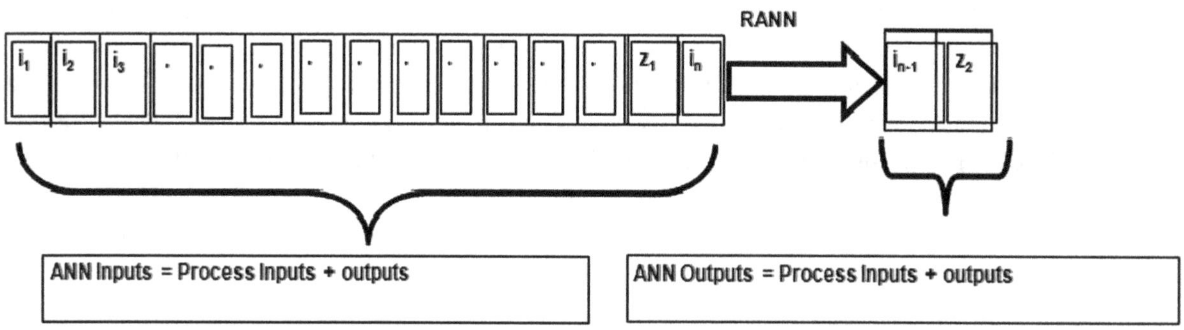

mentioned upstream parameters. The new approach is based on a sequence of artificial neural networks (ANNs). One type of ANN is used to predict the SPM-elongation taking all other coil upstream process parameters and the precise targeted yield strength as inputs. The second type of ANN uses the elongation output generated by the first ANN as one of the inputs to predict the yield strength. If the predicted yield strength is well within limits specified by the customer, the ANN-predicted SPM-elongation is accepted; otherwise the SPM-elongation value set by the ANN is bypassed and the value as suggested by existing method is taken.

Conceptually an ANN is blind to the process features of the data it is fed with as long as input vectors are related to output vectors in a one-to-one manner. The input vector can be actual process inputs and the output can be actual process output. This type of ANN is used to capture the process-directional functionality, i.e. yield strength of a coil can be predicted when the chemistry, SPM-elongation ($\underline{e}$) & other upstream process parameters are given. This is called forward process-ANN or FANN which is illustrated in Figure 9. But equally well the input vector can be a mixture of actual process-inputs and process-outputs while the ANN-output is an input of the process. That is the advantage of using ANNs. This type of ANN is used to capture the process-reverse-directional functionality, i.e., given chemistry and other upstream process parameters, and one or more of outputs (YS), $\underline{e}$ can be predicted. This is called reverse-process ANN or RANN which is shown in Figure 12.

Figure 11. Test Results from Trial 1. New Method reduces violations by 60%

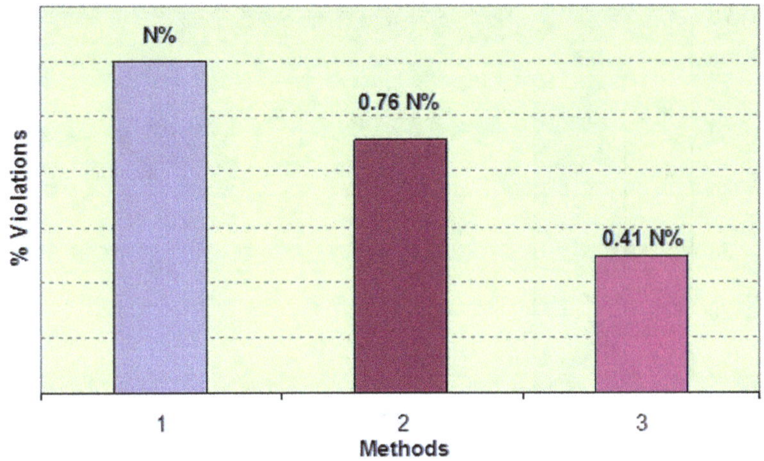

In Figure 11, $i_1, i_2, i_3, \ldots, i_n$ are the ANN inputs that are composed of process inputs like chemistry, other upstream process parameters and <u>**e**</u>. and z_1, z_2 etc are ANN outputs which correspond exactly with process outputs are like yield strength (YS).

In Figure 10, $i_1, i_2, i_3, \ldots, z_1, i_n$ are the ANN inputs which include process inputs like chemistry, other upstream process parameters *and process output* such as yield strength (YS) and $i_{n-1}, z_2, \ldots$etc are ANN outputs which includes process inputs like <u>**e**</u> and other possible outputs.

For this work, production data of five different grades of steel were collected. The data includes the chemistry (% Carbon(C), % Manganese (Mn), % Phosphorus (P), % Sulphur (S), % Silicon (Si), Nitrogen (N), % Niobium (Nb), % Titanium (Ti), % Aluminum (Al)) of the coils, hot rolling process parameters (Finish Rolling Temperature (FRT), Coiling temperature (CT)), cold rolling parameters (thickness, width, % thickness reduction), batch annealing furnace parameters (control heating time, soaking time, hot spot temperature, cold spot temperature), SPM elongation (<u>*e*</u>) and yield strength.

In this work first the principle of using oppositely-oriented pair of ANNs to cross-relate process parameters and mechanical properties of coiled product is done. The ANNs used in the present case are supervised multilayered feed forward networks which are trained and validated using standard gradient descent back-propagation algorithms. The inputs and outputs are connected through two hidden layers. Different network topologies have been investigated to find the appropriate network to simulate the problem. Finally, the best network was chosen as the one with 13-40-40-1 topology; 13 neurons in the input layer, 40 in the first hidden layer, 40 in the second hidden layer and 1 in the output layer. Then the proposed methodology that uses this pair of ANNs to predict yield strength is developed and the results of application of this method is as follows. Two data cases are considered, one for a larger data base of coils involving different grades for which exact figures cannot be expressed in open literature, but comparative values will serve to demonstrate validation. The other case is from coils of a specific grade in a specific month of production, for which numerical data is expressed.

Figure 12. Test Results from Trial 2. New Method brings down violation from above 3% to zero

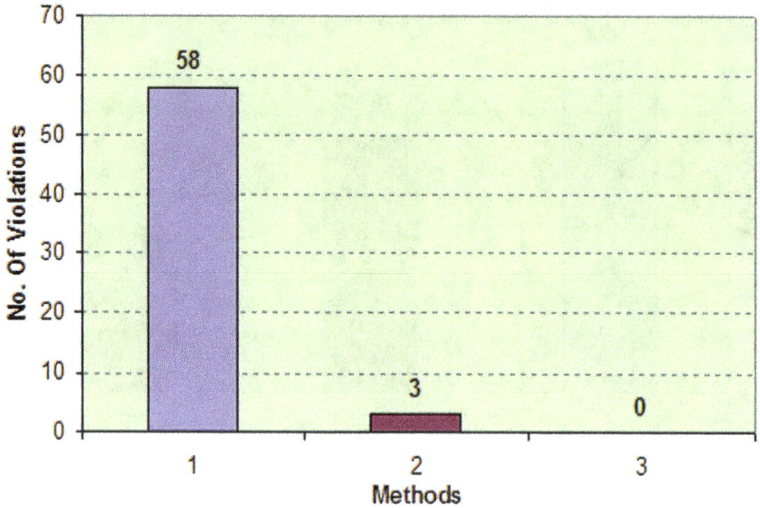

Trial 1

This test has been carried out by taking all kinds of grades for a period of four years and the total no of data set is 7515. Figure 11 shows the efficiency of the current as well as new method through bar charts which is described below.

Method 1

Percentage failure of cases if current plant practice is followed = N% (say).
Here "failure" implies that the coil needs to be reprocessed in the SPM.

Method 2

Percentage Failure of cases when RANN alone is used = 0.76 N%
Here "N" is exactly the same quantity expressed in Method 1 above.

Method 3

Percentage Failure of cases from new method as described in algorithm = 0.41 N%

Trial 2

The same type of comparison as in Trial 1 was carried out for a particular grade of coils in a certain production month in 2009. The no of data sets in this case is 1852. Figure 12 shows the efficacy of the new method, as described in this case, in reducing violations. In percentage terms, violations have been brought down from little above 3% to zero.

Method 1

No of failure cases of present method = 58 (3.131%).

Method 2

No of cases when new method fails = 3 (0.161%)

Method 3

No of Cases when new method, based on algorithm, leads to violation = 0

Application of these two ANNs in sequence, and juxtaposition of the results with the existing practise, has been found to reduce the variations of attained YS from targeted YS by about 60% on the average.

6. CONCLUSION

The computing world has a lot to gain from neural networks. Their ability to learn by example makes them very flexible and powerful. Furthermore, there is no need to devise an algorithm in order to perform a specific task; i.e. there is no need to understand the internal mechanisms of that task. They are also very well suited for real time systems because of their fast response and computational times which are due to their parallel architecture. Data-driven models would be useful in solving a practical problem or modelling a particular system or process if (1) a considerable amount of data describing this problem is available; (2) there are no considerable changes to the modelled system during the period covered by the model. Such models are especially effective if it is difficult to build knowledge-driven simulation models (e.g. due to lack of understanding of the underlying processes), or the available models are not adequate enough. It is of course always useful to have modelling alternatives and to validate the simulation results of physically based models with data-driven ones, or vice versa.

In the case studies discussed here, it has been seen that, neural network models can be used as a good tool for studying the role of various processing and composition variables on the final properties of steel. Findings from this study can be successfully utilized to design the composition and processing of IF steel with desirable properties. On-line availability of this property prediction system helps in reducing the expenses incurred by sample taking, yield loss due to sample cutting and by the delay in delivery to the downstream processes. It also helps in predicting variation of properties over the length of the strip. It has been explicitly shown that an ANN model, trained by comprehensive datasets, can even track and simulate the complicated metallurgical phenomena like the effect slab drop-out temperature, coiling temperature and also the impact of micro-alloying elements in hot rolled steel. By coupling the explanatory insights of neural network with its powerful predictive capabilities, ANN can be efficiently applied to evaluate, understand and predict the complex metallurgical phenomena.

The results in case 2 clearly indicate that multi-objective genetic algorithm could easily be used for complicated optimization of properties steel. The pareto fronts give a wide range of combination of strength and ductility and can be quite helpful for designing the alloy along with its processing parameters. The analysis of the Pareto solutions has generated some useful information regarding the system, which gives further clue for experimental as well as computational verification. ANN is also found to be a good tool for developing the objective functions for highly non-linear cases like properties of steel.

The self organising map is found to be quite capable of clustering the steel data according to its alloying additions. The process of self organising mapping has certain limitation to map huge data concerning all the ten compositional and six process parameters. The data can be successfully mapped according to their dominant strengthening mechanism with certain degree of supervision using Learning Vector Quantisation.

In skin pass rolling process for setting the elongation two neural networks, a Reverse Process ANN or RANN, and a Forward-Process ANN or FANN, have been developed independently to predict the SPM elongation on the basis of targeted Yield Strength (YS) and other process parameters and to predict the YS from process inputs, respectively, with good accuracy. Application of these two ANNs in sequence and juxtaposition of the results with the existing practice has been found to reduce the variations of attained YS from targeted YS.

Even though neural networks have a huge potential the best can be achieved when they are integrated with models of different types and which follow different modelling paradigms (thus constituting hybrid models), including the combination with physically based models in an optimal way. Process knowledge is very much required for the pre- and post- processing of neural network input-output data. In order to make a model successful the selection and application of methods for e.g. selection of relevant inputs, data scaling, elimination of invariances, etc is necessary. One of the challenges for hydro informatitians in this respect is to ensure that data-driven models are properly incorporated into the existing modelling and decision support frameworks.

The long term objective of further application development is the intelligent steel production plant. This plant should one day operate using neural networks and other methods from artificial intelligence at all higher automation levels, including scheduling and management systems. Besides control, automatic diagnosis of automation and technological faults has become a broad field of application of intelligent algorithms. The large potential of these approaches can be exploited to create integrated production techniques that improve the efficiency and competitiveness of entire industrial facilities while reducing demand for resources.

REFERENCES

Abbott, L. F., Varela, K., Sen, K., & Nelson, S. B. (1997). Synaptic depression and cortical gain control. *Science*, *275*(5297), 220–223. doi:10.1126/science.275.5297.221 PMID:8985017

Agam, Y., Bullock, D., & Sekuler, R. (2005). Imitating unfamiliar sequences of connected linear motions. *Journal of Neurophysiology*, *94*(4), 2832–2843. doi:10.1152/jn.00366.2005 PMID:16014793

Andrews, R., Diederich, J., & Tickle, A. B. (1995). Survey and critique of techniques for extracting rules form trained artificial neural networks. *Knowledge-Based Systems*, *8*(6), 373–389. doi:10.1016/0950-7051(96)81920-4

Bhadeshia, H. K. D. H. (1999). Neural Networks in Materials Science. *ISIJ International*, *39*(10), 966–979. doi:10.2355/isijinternational.39.966

Bhadeshia, H. K. D. H. (2008). Mathematical models in materials science. *Materials Science and Technology, 24*(2), 128–136. doi:10.1179/174328407X213107

Bhadeshia, H. K. D. H., MacKay, D. J. C., & Svensson, L.-E. (1995). Impact toughness of C–Mn steel arc welds – Bayesian neural network analysis. *Materials Science and Technology, 11*(10), 1046–1051. doi:10.1179/mst.1995.11.10.1046

Bose, N. K., & Liang, P. (1996). *Neural Network Fundamentals with Graphs, Algorithms, and Applications*. McGraw-Hill, Inc.

Brun, F., Yoshida, T., Robson, J. D., Narayan, V., Bhadeshia, H. K. D. H., & MacKay, D. J. C. (1999). Theoretical design of ferritic creep resistant steels using neural network, kinetic, and thermodynamic models. *Materials Science and Technology, 15*(5), 547–554. doi:10.1179/026708399101506085

Cai, J., Cottis, R. A., & Lyon, S. B. (1999). Phenomenological modelling of atmospheric corrosion using an artificial neural network. *Corrosion Science, 41*(10), 2001–2030. doi:10.1016/S0010-938X(99)00024-4

Cole, D., Martin-Moran, C., Sheard, A., Bhadeshia, H. K. D. H., & MacKay, D. J. C. (2000). Modelling creep rupture strength of ferritic steel welds. *Sci. Technol. Weld. Join., 5*(2), 81–89. doi:10.1179/136217100101538065

Cool, T., Bhadeshia, H. K. D. H., & MacKay, D. J. C. (1997). The yield and ultimate tensile strength of steel welds. *Materials Science and Engineering A, 223*(1-2), 186–200. doi:10.1016/S0921-5093(96)10513-X

Das, P., & Datta, S. (2007). Exploring the non-linearity in empirical modelling of a steel system using statistical and neural network models. *International Journal of Production Research, 45*(3), 699–717. doi:10.1080/00207540600792465

Das, P., Mukherjee, S., Ganguly, S., Bhattacharyay, B. K., & Datta, S. (2009). Genetic algorithm based optimization for multi-physical properties of HSLA steel through hybridization of neural network and desirability function. *Computational Materials Science, 45*(1), 104–110. doi:10.1016/j.commatsci.2008.03.050

Datta (2009). Extraction of knowledge from high strength steel data using soft computing techniques - An overview. *Statistical Analysis and Data Mining, 1*, 329-337.

Datta, S., & Banerjee, M. K. (2004). Kohonen Network Modelling for the Strength of Thermomechanically Processed HSLA Steel. *ISIJ International, 44*(5), 846–851. doi:10.2355/isijinternational.44.846

Datta, S., & Banerjee, M. K. (2004). Optimizing parameters of supervised learning techniques (ANN) for precise mapping of the input-output relationship in TMCP steels. *Scand. J. Met., 33*(6), 310–315. doi:10.1111/j.1600-0692.2004.00699.x

Datta, S., & Banerjee, M. K. (2006). Mapping the input–output relationship in HSLA steels through expert neural network. *Materials Science and Engineering A, 420*(1-2), 254–264. doi:10.1016/j.msea.2006.01.037

Datta, S., & Banerjee, M. K. (2006). Mapping the input–output relationship in HSLA steels through expert neural network. *Materials Science and Engineering A, 420*(1-2), 254–264. doi:10.1016/j.msea.2006.01.037

Datta, S., & Chattopadhyay, P. P. (2013). Soft computing techniques in advancement of structural metals. *International Materials Reviews, 58*(8), 475–504. doi:10.1179/1743280413Y.0000000021

Datta, S., Pettersson, F., Ganguly, S., Saxe'n, H., & Chakraborti, N. (2007). Designing High Strength Multi-phase Steel for Improved Strength–Ductility Balance Using Neural Networks and Multi-objective Genetic Algorithms. *ISIJ International*, *47*(8), 1193–2001. doi:10.2355/isijinternational.47.1195

Datta, S., Pettersson, F., Ganguly, S., Saxén, H., & Chakraborti, N. (2007). Designing high strength multi-phase steel for improved strength-ductility balance using neural networks and multi-objective genetic algorithms. *ISIJ International*, *47*(8), 1193–2001. doi:10.2355/isijinternational.47.1195

Deb. (2001). *Multi-objective optimization using evolutionary algorithms*. John Wiley & Sons Ltd.

Dey, S., Datta, S., Chattopadhyay, P. P., & Sil, J. (2008). Modeling the properties of TRIP steel using AFIS: A distributed approach. *Computational Materials Science*, *43*(3), 501–511. doi:10.1016/j.commatsci.2007.12.009

Dieter, G. E. (1928). *Mechanical Metallurgy*. McGraw Hill Publication.

Dieter, G. E. (1988). *Mechanical Metallurgy, SI Metric Edition*. London: McGraw-Hill Book Company.

Dobrza'nski, L. A., Kowalski, M., & Madejski, J. (2005). Methodology of the mechanical properties prediction for the metallurgical products from the engineering steels using the Artificial Intelligence methods. *Journal of Materials Processing Technology*, *164–165*, 1500–1509. doi:10.1016/j.jmatprotec.2005.02.194

Döll, R., Sörgel, G., Daum, M., & Zouhar, G. (1999). Control of mechanical properties by monitoring microstructure. *Proc. ASIA STEEL*.

Draeger, A., Engell, S., & Ranke, H. (1995). *Model Predictive Control Using Neural Networks*. IEEE Control Systems.

Dumortier, C., Lehert, P., Krupa, P., & Charlier, A. (1998). Statistical modelling of mechanical properties of microalloyed steels by application of artificial neural network. *Materials Science Forum*, *284-286*, 393–400. doi:10.4028/www.scientific.net/MSF.284-286.393

Dunne, D., Tsuei, H., & Sterjovski, Z. (2004). Artificial Neural Networks for Modelling of the Impact Toughness of Steel. *ISIJ International*, *44*(9), 1599–1607. doi:10.2355/isijinternational.44.1599

Evans, M. (1999). Method for improving parametric creep rupture life of 2·25Cr–1Mo steel using artificial neural networks. *Materials Science and Technology*, *15*(6), 647–658. doi:10.1179/026708399101506391

Fang, X., Fan, Z., Ralph, B., Evans, P., & Underhill, R. (2002). Effect of temper rolling on tensile properties of C–Mn steels. *Materials Science and Technology*, 18.

Fujii, H., MacKay, D. J. C., & Bhadeshia, H. K. D. H. (1996). Bayesian Neural Network Analysis of Fatigue Crack Growth Rate in Nickel Base Superalloys. *ISIJ International*, *36*(11), 1373–1382. doi:10.2355/isijinternational.36.1373

Ganguly, S., Datta, S., & Chakraborti, N. (2007). Genetic Algorithms in Optimization of Strength and Ductility of Low-Carbon Steels. *Materials and Manufacturing Processes*, *22*(5), 650–658. doi:10.1080/10426910701323607

Ganguly, S., Datta, S., & Chakraborti, N. (2009). Genetic algorithm-based search on the role of variables in the work hardening process of multiphase steels. *Computational Materials Science*, *45*(1), 158–166. doi:10.1016/j.commatsci.2008.01.074

Ghosh, S. K., Ganguly, S., Chattopadhyay, P. P., & Datta, S. (2005). Modeling the Effect of Copper on Hardness of Microalloyed Dual Phase Steel through Neural Network and Neuro-fuzzy Systems. *ISIJ International*, *45*(9), 1345–1351. doi:10.2355/isijinternational.45.1345

Ghosh, S. K., Ganguly, S., Chattopadhyay, P. P., & Datta, S. (2008). Effect of copper and micro-alloying (Ti, B) on tensile properties of HSLA steels predicted by ANN. *Ironmaking & Steelmaking*.

Ghosh, S. K., Ganguly, S., Chattopadhyay, P. P., & Datta, S. (2009). Effect of copper and microalloying (Ti, B) addition on tensile properties of HSLA steels predicted by ANN technique. *Ironmaking & Steelmaking*, *36*(2), 125–132. doi:10.1179/174328107X254880

Gladman, T., Dulieu, D., & McIvor. (1975). Structure-property relationships in high strength micro alloyed steels. In *Proceedings of an international symposium on HSLA steels, Microalloying '75.*

Goldberg, D. E. (2002). *Genetic algorithms in search, optimization and machine learning*. New Delhi: Pearson-Education.

Goswami, T. (1996). Prediction of Low Cycle Fatigue Lives of Low Alloy Steels. *ISIJ International*, *36*(3), 354–360. doi:10.2355/isijinternational.36.354

Guo, Z., & Sha, W. (2004). Modelling the correlation between processing parameters and properties of maraging steels using artificial neural network. *Computational Materials Science*, *29*(1), 12–28. doi:10.1016/S0927-0256(03)00092-2

Haykin, S. (1994). *Neural Networks: A Comprehensive Foundation*. McMillan.

Hosseini, S. M. K., Zarei-Hanzaki, A., Yazdan Panah, M. J., & Yue, S. (2004). ANN model for prediction of the effects of composition and process parameters on tensile strength and percent elongation of Si–Mn TRIP steels. *Materials Science and Engineering A*, *374*(1-2), 122–128. doi:10.1016/j.msea.2004.01.007

Hunter, A., Kennedy, L., Henry, J., & Ferguson, R. I. (2000). Application of neural networks and sensitivity analysis to improved prediction of trauma survival. *Computer Methods and Algorithms in Biomedicine*, *62*(1), 11–19. doi:10.1016/S0169-2607(99)00046-2 PMID:10699681

Integrated Computational Materials Engineering (ICME). (2010). *Integrating Computational Materials Science and Engineering John Allison July 26, DOE – BES Workshop on CMS*. Author.

Kim, Y. I., Moon, K. C., Kang, B. S., Han, C., & Chang, K. S. (1998). Application of neural network to the supervisory control of a reheating furnace in the steel industry. *Control Engineering Practice*, *6*(8), 1009–1014. doi:10.1016/S0967-0661(98)00098-7

Kohonen, T. (1989). *Self-Organization and Associative Memory* (3rd ed.). Berlin: Springer-Verlag. doi:10.1007/978-3-642-88163-3

Kohonen, T. (1989). *Self-Organizing and Associative Memory*. Berlin, Germany: Springer-Verlag. doi:10.1007/978-3-642-88163-3

Korczak, P., Dyja, H., & Labuda, E. (1998). Using neural network models for predicting mechanical properties after hot plate rolling processes. *Journal of Materials Processing Technology*, *80–81*, 481–486. doi:10.1016/S0924-0136(98)00151-4

Kudin, D. P., Tseloval'nikov, V. M., Ksenzuk, F. A., & Kokin, N. M. (1971). Influence of the rate of skin pass rolling and tension on the amount of deformation. *Metallurgist*, *15*(12), 817–818. doi:10.1007/BF00735606

Kusiak, J., & Kuziak, R. (2002). Modelling of microstructure and mechanical properties of steel using the artificial neural network. *Journal of Materials Processing Technology*, *127*(1), 115–121. doi:10.1016/S0924-0136(02)00278-9

Lalam, S. H., Bhadeshia, H. K. D. H., & MacKay, D. J. C. (2000). Estimation of mechanical properties of ferritic steel welds. Part 1: Yield and tensile strength. *Sci. Technol. Weld. Join.*, *5*(3), 135–147. doi:10.1179/136217100101538137

Larkiola, J., Myllykoski, P., Korhonen, A. S., & Cser, L. (1998). The role of neural networks in the optimization of rolling processes. *Journal of Materials Processing Technology*, *80–81*, 16–23. doi:10.1016/S0924-0136(98)00206-4

Lein, J. K. (1997). Environment Decision Making: An information technology approach. Malden, MA: Blackwell Science.

Llewellyn & Hudd. (1998). *Steels: metallurgy and applications* (3rd ed.). Oxford, UK: Butterworth-Heinemann.

Mahfouf, M., Jamei, M., & Linkens, D. A. (2005, May). Optimal Design of Alloy Steels Using Multi-objective Genetic Algorithms. *Materials and Manufacturing Processes*, *20*(3), 553–567. doi:10.1081/AMP-200053580

Majumdar, G., Oraon, B., Laha, A., Ghosh, S. K., Mohanty, I., & Datta, S. (2010). Developing rules bases on the influence of welding parameters in FCAW process from ANN model. *International Journal of Mechatronics and Manufacturing Systems*, *3*(1), 155–164. doi:10.1504/IJMMS.2010.029887

Martinetz, T., Protzel, P., Gramckow, O., & Sörgel, G. (1995). Neural network control for steel rolling mills. In Proc. Neural networks: artificial intelligence and industrial applications, (pp. 280-286). Springer Verlag. doi:10.1007/978-1-4471-3087-1_55

Miettinen, K. (1999). *Nonlinear Multiobjective Optimization*. Boston: Kluwer Academic Publishers.

Mohanty, I., Bhattacharjee, D., & Datta, S. (2011). Designing cold rolled IF steel sheets with optimized tensile properties using ANN and GA. *Computational Materials Science*, *50*(8), 2331–2337. doi:10.1016/j.commatsci.2011.03.007

Mohanty, I., Datta, S., & Bhattacharjee, D. (2008). Composition-processing-property correlation of cold rolled IF steel sheets using neural network. *Materials and Manufacturing Processes*, *24*(1), 100–105. doi:10.1080/10426910802543947

Mohanty, I., Sarkar, S., Jha, B., Das, S., & Kumar, R. (2014). Online mechanical property prediction system for hot rolled IF steel. *Ironmaking & Steelmaking*, *41*(Sep), 618–627. doi:10.1179/1743281214Y.0000000178

Monajati, H., Asefi, D., Parsapour, A., & Abbasi, Sh. (2010). Analysis of the effects of processing parameters on mechanical properties and formability of cold rolled low carbon steel sheets using neural networks. *Computational Materials Science*, *49*(4), 876–881. doi:10.1016/j.commatsci.2010.06.040

Mukhopadhyay, A., & Iqbal, A. (2005). Prediction of mechanical properties of hot rolled, low-carbon steel strips using artificial neural network. *Materials and Manufacturing Processes*, *20*(5), 793–812. doi:10.1081/AMP-200055140

Narayan, V., Abad, R., Lopez, B., Bhadeshia, H. K. D. H., & Mackay, D. J. C. (1999). Estimation of hot torsion stress strain curves in iron alloys using a neural network analysis. *ISIJ International*, *39*(10), 999–1005. doi:10.2355/isijinternational.39.999

Negm. (2005). Validation Testing for Temporal Neural Networks for RBF Recognition. *Proceedings of World Academy of Science, Engineering and Technology, 5.*

Odejobi & Umoru. (2009). Applications of soft computing techniques in materials engineering: A review. *African Journal of Mathematics and Computer Science Research, 2*(7), 104-131.

Ramuhalli, P., Udpa, L., & Udpa, S. S. (2002). Electromagnetic NDE signal inversion by function-approximation neural networks. *IEEE Transactions on Magnetics*, *38*(6), 3633–3642. doi:10.1109/TMAG.2002.804817

Robinson, A. J. (1994). An Application of Recurrent Nets to Phone Probability Estimation. *IEEE Transactions on Neural Networks*, *5*(2), 298–305. doi:10.1109/72.279192 PMID:18267798

Rumelhart, D., Hinton, G. E., & Williams, R. J. (1986). Learning representations by back-propagating errors. *Nature*, *323*(6088), 533–536. doi:10.1038/323533a0

Schooling, J. M., Brown, M., & Reed, P. A. S. (1999). An example of the use of neural computing techniques in materials science—the modelling of fatigue thresholds in Ni-base superalloys. *Materials Science and Engineering A*, *260*(1-2), 222–239. doi:10.1016/S0921-5093(98)00957-5

Singh, S. B., Bhadeshia, H. K. D. H., MacKay, D. J. C., Carey, H., & Martin, I. (1998). Article. *Ironmaking & Steelmaking*, *25*, 355.

Sinha, A., & Sikdar, S. (2013). Article. *Mater. Des.*, *46*, 227–234.

Sumpter, B. G., & Noid, D. W. (1996). On the Design, Analysis, and Characterization of Materials Using Computational Neural Networks. *Annual Review of Materials Science*, *26*(1), 223–277. doi:10.1146/annurev.ms.26.080196.001255

Svensson, T. (2004). Fatigue Fract. *Eng. Mater. Struct.*, *27*, 981–990. doi:10.1111/j.1460-2695.2004.00819.x

Tamehiro, H., Murata, M., Habu, R., & Nagumo, M. (1987). Optimum Microalloying of Niobium and Boron in HSLA Steel for Mechanical Processing Trans. *ISIJ*, *27*(2), 120–129. doi:10.2355/isijinternational1966.27.120

Wang, Y., & Xiang, B. (2007). Radial basis function network calibration model for near-infrared spectra in wavelet domain using a genetic algorithm. *Analytica Chimica Acta*, *602*(1), 55–65. doi:10.1016/j.aca.2007.09.016 PMID:17936107

Willis, M. J., Montague, G. A., Di Massimo, C., Tham, M. T., & Morris, A. J. (1992). Artificial neural networks in process estimation and control. *Automatica*, *28*(6), 1181–1187. doi:10.1016/0005-1098(92)90059-O

Yoo, Y. S., Jo, C. Y., & Jones, C. N. (2002). Compositional prediction of creep rupture life of single crystal Ni base superalloy by Bayesian neural network. *Materials Science and Engineering A*, *A336*(1-2), 22–29. doi:10.1016/S0921-5093(01)01965-7

Chapter 11
Multi-Objective Evolutionary Algorithms:
Application in Designing Particle Reinforced Mould Materials

A. K. Nandi
CSIR – Central Mechanical Engineering Research Institute, India

K. Deb
Michigan State University, USA

ABSTRACT

The primary objective in designing appropriate particle reinforced polyurethane composite which will be used as a mould material in soft tooling process is to minimize the cycle time of soft tooling process by providing faster cooling rate during solidification of wax/plastic component. This chapter exemplifies an effective approach to design particle reinforced mould materials by solving the inherent multi-objective optimization problem associated with soft tooling process using evolutionary algorithms. In this chapter, first a brief introduction of multi-objective optimization problem with the key issues is presented. Then, after a short overview on the working procedure of genetic algorithm, a well- established multi-objective evolutionary algorithm, namely NSGA-II along with various performance metrics are described. The inherent multi-objective problem in soft tooling process is demonstrated and subsequently solved using an elitist non-dominated sorting genetic algorithm, NSGA-II. Multi-objective optimization results obtained using NSGA-II are analyzed statistically and validated with real industrial application. Finally the fundamental results of this approach are summarized and various perspectives to the industries along with scopes for future research work are pointed out.

DOI: 10.4018/978-1-5225-0290-6.ch011

Figure 1. Development of metal components through RP-ST-wax pattern -investment casting

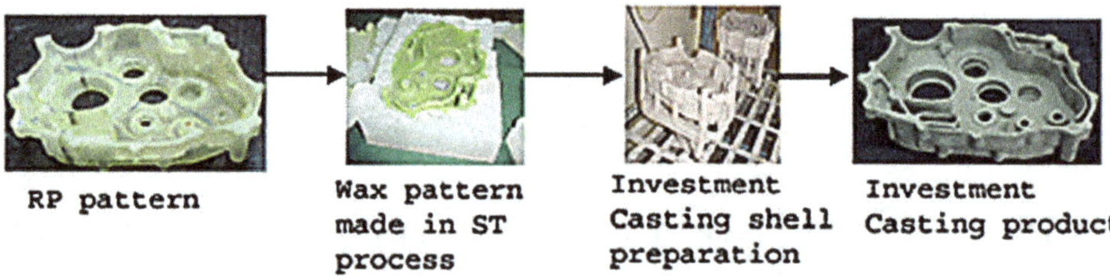

INTRODUCTION

Soft tooling (ST) is a rapid tooling process where polymeric flexible materials are used for making mould (Rosochowski, 2000). ST process is particularly suitable to produce wax patterns in small batch by vacuum assisted or gravity casting method based on RP (Rapid Prototype)/others pattern. This wax component is used in investment casting process as pattern to make ceramic slurry and finally metallic components are produced by fusing the wax pattern. The method of making metal component through different processes (rapid prototyping (RP)-soft tooling-investment casting) is illustrated in Figure 1.

The common polymeric flexible mould materials used in soft tooling process are polyurethane (PU), silicone rubber, etc. However, these mould materials in general possess poor thermal conductivity (TC). For this reason, solidification time of wax/plastic pattern produced in ST process is significantly long because of poor heat flow rate through the mould wall. As a result the rapidity of soft tooling process is reduced to a great extent.

Previous experimental studies (Nandi, Datta, & Orkus, 2012) reveal that similar to other composites (Agrawal et al., 2000; Boudenne et al., 2004; Iqbal, Frormann, Saleem, & Ishaq, 2007; Ishida & Rimdusit, 1998; Keith, King, Lenhart, & Zimny, 2007; Kim, Choi, Lee, & Kang, 2008; Lopes & Felisberti, 2004; Mamunya, Davydenko, Pissis, & Lebedev, 2002; Miller et al., 2006; Mu, Feng, & Diao, 2007; Ng, Lu, & Lau, 2005; Oliveira et al., 2007, Razzaq & Frormann, 2007; Subodh, Manjusha, Philip, & Sebastian, 2008; Sundstrom & Chen, 1970; Tsukuda, Sumimoto, & Ozawa, 1997; Vinod, Varghese, & Kuriakose, 2004; Wang, Li, & Su, 2008; Wang, Gao, Wang, & Hua, 2001; Weidenfeller, Hofer, & Schilling, 2004; Wong & Bollampally, 1999; Xu, Chung, & Mroz, 2001; Yu, Hing, & Hu, 2002;) increase of conductive filler loading raises the equivalent thermal conductivity (ETC) of flexible mould materials. Moreover, with enhancing the filler content, the equivalent modulus of elasticity (EME) of particle reinforced SR also increases, which yields to escalating the stiffness of mould box (Nandi et al., 2012).But high stiffness value of the mould box is not suitable for STP because it produces numerous complications especially when the pattern is to be taken out from the mould box. Moreover presence of filler material augments the equivalent viscosity of melt SR composite (Nandi, Vesterinen, Cingi, Seppala, & Orkus, 2010). In STP, elevated viscosity of liquid FMM is not permitted as it creates the problem of filling up the space between the mould box and pattern. The reason is that high viscosity condenses the flow ability of melt mould material. Thus there is a need to appropriately design composite flexible mould materials to improve the soft tooling process by enhancing its solidification time and keeping other inherent advantages. Since, in the present problem the motivation is to increase the thermal conductivity of mould material and simultaneously to restrict the increases of modulus of elasticity and viscosity, it seems that these objectives are conflicting to each other. Therefore, this problem is tackled using a suitable multi-objective optimization tool. This chapter presents an effective approach to design a particle reinforced polyurethane mould material by solving the intrinsic multi-objective optimization problem associated that is involved in soft tooling process using multi-objective evolutionary algorithms.

MULTI-OBJECTIVE OPTIMIZATION PROBLEM

Basic Concept

Optimization is a technique to arrive at one or more solutions, which correspond to either minimum or maximum values of one or more objectives (in the form of objective or subjective function) satisfying certain conditions. When the optimization problem involves only one objective function, the task of finding the optimal solution is called single objective optimization and if the problem involves two or more objectives, the task of finding one or more optimum solution is known as multi-objective solution. The latter search is also known as *multiple criterion decision making* (MCDM). K. Deb (2001) mentioned that single objective optimization is a degenerate case of multiple optimization problems. The real world problems commonly involve more than one objective. The extreme value principal is not applicable in such situation where all the objectives are equally important. In this case a number of solutions may be produced to trade-off among different objectives. A solution that is extreme with respect to one objective requires a compromise with other objective(s). This restricts the choice of a solution which is optimal with respect to only one objective. Therefore, a number of sets of solutions are obtained and then the designer has to select a set from these sets of solutions, which will serve the purpose originally intended.

Definition of the Multi-Objective Problem

A general Multi-objective problem (MOP) includes a set of *n* parameters (called decision variables), a set of *k* objective functions, a set of *m* inequality constraints and a set of *j* equality constraint. Objective functions and constraints are functions of the decision variables. The optimization goal is as follows:

Maximize/minimize $y=f(x)=\left(f_1(x), f_2(x), \ldots\ldots\ldots\ldots, f_k(x)\right)$ (1)

subject to $e(x)=(e_1(x), e_1(x), \ldots\ldots\ldots e_m(x)) \leq 0$,

$h(x)=(h_1(x), h_1(x), \ldots\ldots\ldots h_j(x))=0$,

where $x=(x_1, x_2, \ldots\ldots\ldots, x_n) \in X$

$y=(y_1, y_2, \ldots\ldots\ldots, x_k) \in Y$

$x_i^L \leq x_i \leq x_i^U, i=1,2,\ldots\ldots\ldots n$

and *x* is the decision vector, *y* is the objective vector, *X* is denoted as the decision space and *Y* is called the objective space. x_i^L and x_i^U are the lower and upper bounds of each decision variable, x_i which constitute the decision variable space, *X*. The constraints and variable bounds determine the set of feasible solutions.

The Feasible Solution Set

The feasible solution set X_f is defined as the set of decision vectors, x that satisfy the constraints $e(x)$ and $h(x)$:

$$X_f = \{x \in X \,|\, e(x) \leq 0, h(x) = 0\} \tag{2}$$

The image of X_f, i.e., the feasible region in the objective space, is denoted as $Y_f = f(X_f) = \bigcup_{x \in X_f} \{f(x)\}$.

In single-objective optimization, the feasible set is completely (totally) ordered according to the objective function f, for two solutions a, $b \in X_f$ either $f(a) \geq f(b)$ or $f(a) \leq f(b)$. The goal is to find the solution(s) that leading to the maximum/minimum value of f (Cohon, 1985). However, when several objectives are involved, the situation changes: X_f is, in general, not totally ordered, but partially ordered (Pareto, 1896).

Concept of Domination and Pareto Dominance

For any two decision vectors a and b of a maximization problem,

$a \succ b$ (a dominates b) iff $f(a) > f(b)$ (3)

$a \succeq b$ (a weakly dominates b) iff $f(a) \geq f(b)$
$a \sim b$ (a is indifferent to b) iff $f(a) \ngeq f(b) \wedge f(b) \ngeq f(a)$

For a minimization problem, the above concepts are analogous. Based on the concept of Pareto dominance, the optimality criterion for MOPs can be introduced.

Multi-objective optimization algorithms use the concept of dominance. Two solutions are compared on the basis of whether one dominates the other solution or not. K. Deb defines the concept of domination in his book (Deb, 2001) in very nice way. His proposed definition of domination is

A solution $x^{(1)}$ is said to dominate the other solution $x^{(2)}$, if under-mentioned both the conditions are true:

- The solution $x^{(1)}$ is no worser than $x^{(2)}$ in all objectives, or $f_j(x^{(1)}) \blacktriangleright f_j(x^{(2)})$ for all $j=1,2,\ldots\ldots..k$
- The solution $x^{(1)}$ is strictly better than at least one objective, or worser than $x^{(2)}$ in all objectives, or $f_j(x^{(1)}) \blacktriangleleft f_j(x^{(2)})$ for at least one $j \in (1,2,\ldots\ldots..k)$

If any of the above conditions fails, the solution $x^{(1)}$ does not dominate the solution $x^{(2)}$. If $x^{(1)}$ dominates the solution $x^{(2)}$ one can make the following statement

- $x^{(2)}$ is dominated by $x^{(1)}$;
- $x^{(1)}$ is non-dominated by $x^{(2)}$, or
- $x^{(1)}$ is non inferior to $x^{(2)}$

Pareto-Optimality and Pareto-Front

Having several objective functions, the idea of "optimum" changes, because in multi-objective problems, we are really searching for good compromises (or "trade-offs") between the objectives rather than a single solution as in the case of global optimization.

The idea of "optimum" in multi-objective optimization problem, that is most commonly adopted is the one originally proposed by Francis Y. Edgeworth (1881). This notion was later generalized by Vilfredo Pareto (1896). Although some authors refer to this idea as Edgeworth-Pareto optimum (Stadler, 1988), but the most authors use most commonly accepted term: Pareto-optimum. Deb et al. (Deb, 2001) define that a vector of decision variables $\vec{x}^{*} \in F$ (F is set of all vectors) is *Pareto-optimal* if there does not exist another $\vec{x} \in F$, such that $f_i\left(\vec{x}\right) \leq f_i\left(\vec{x}^{*}\right)$ for all i=1, ….,k and $f_j\left(\vec{x}\right) < f_j\left(\vec{x}^{*}\right)$ for at least one j. In words, this definition says that $\vec{x}^{*}$ is Pareto-optimal if there exists no feasible vector of decision variables $\vec{x} \in F$, which would decrease some criterion without causing a simultaneous increase in at least one other criterion. Unfortunately, this concept almost always gives not a single solution, but rather a set of solutions called the Pareto-optimal set. The vectors $\vec{x}^{*}$ corresponding to the solutions included in the Pareto-optimal set are called non-dominated. The plot of the objective functions whose non-dominated vectors are in the Pareto-optimal set is called the Pareto-front. Figure 2 shows some of the sample Pareto-front for four combinations of two types of objectives.

The Pareto-optimal set comprises the globally optimal solutions. However, as with SOPs there may also be local optima which constitute a non-dominated set within a certain neighborhood. These correspond to the concepts of global and local Pareto-optimal sets introduced by Deb (1998).

The set A is denoted as a local Pareto-optimal set *iff*

$$\forall a \in A : \nexists x \in X_f : x \succ a \in \|x \succ a\| < \varepsilon \Lambda \|f(x) - f(a)\| < \delta \quad (4)$$

where $\|\cdot\|$ is a corresponding distance metric and $\varepsilon > 0$, $\delta > 0$.

The set A is called a global Pareto-optimal set *iff*

$$\forall a \in A : \nexists x \in X_f : x \succ a \quad (5)$$

The difference between local and global optima is visualized in Figure 3. The dashed line constitutes a global Pareto-optimal front, while the solid line depicts a local Pareto-optimal front. The decision vectors associated with the latter are locally non-dominated though not Pareto-optimal, because the solution related to point Z dominates any of them. Finally, note that a global Pareto-optimal set does not necessarily contain all Pareto-optimal solutions and that every global Pareto-optimal set is also a local Pareto-optimal set.

Figure 2. Continuous curves represent the Pareto-optimal solutions for four combinations of two types of objectives (Deb, 2001)

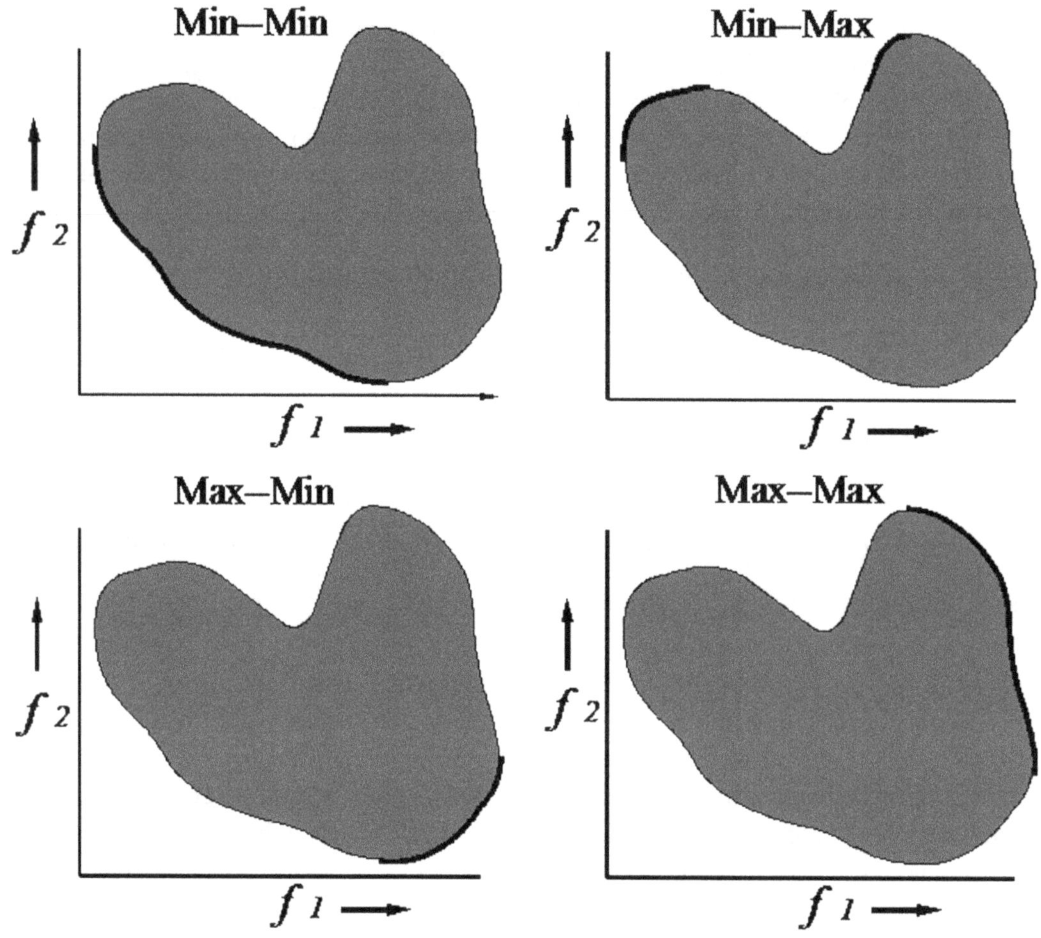

Goal of Multi-Objective Optimization Problem

The search results of multi-objective optimization problem (MOOP) are distinctly of two different classes; one lies in optimal region and others remains in non-optimal regions. In conflicting two objectives case, a number of Pareto-optimal solutions are obtained after solving the problem. In presence of multiple Pareto-optimal solutions, it is difficult to prefer one solution over the other without any further information about the problem. However, in absence of any more information, all Pareto-optimal solutions are equally important. So, in such situation the right approach is to find as many Pareto optimal solutions as possible in a problem. Deb (2001) suggested two different goals in multi-objective optimization:

- To find a set of solutions as close as possible to the Pareto-optimal front
- To find a set of solutions as diverse as possible.

Figure 3. Illustration of locally optimal solution sets and globally optimal solution sets in objective space (Zitzler, 1999)

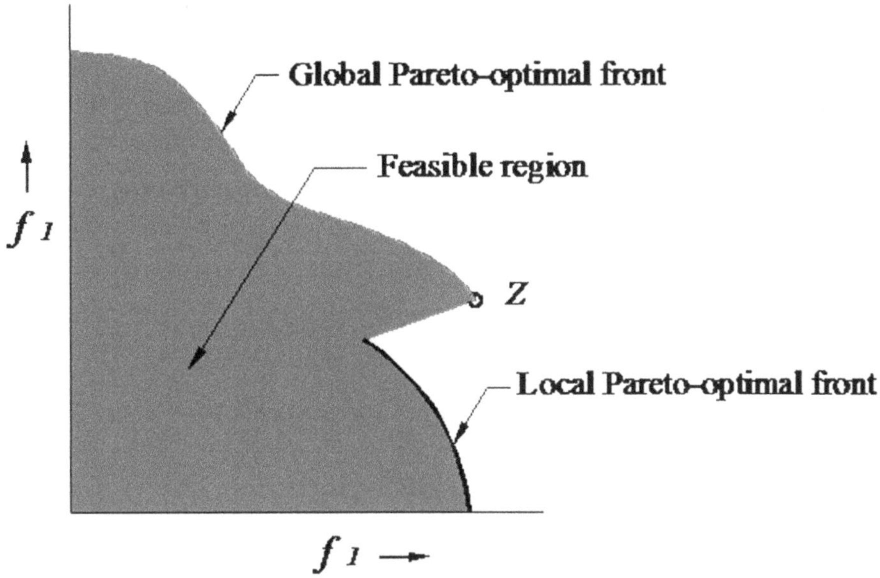

So far discussion on goal of Multi-objective optimization problem restricted with the confliction objectives. However, if the objectives are not conflicting to each other the number of Pareto-optimal set becomes one.

Search and Decision Making

In solving an MOP, two conceptually distinct types of problem difficulty can be identified (Horn, 1997): the search and decision making. The first aspect refers to the optimization process in which the feasible set is sampled for Pareto optimal solutions. As with single-objective optimization, large and complex search spaces can make search difficult and exclude the use of exact optimization methods like linear programming (Steuer, 1986). The second aspect deals with the problem of selecting a suitable compromise solution from the Pareto-optimal set. A human decision maker (DM) is necessary to make the often difficult trade-offs between conflicting objectives. Depending on how optimization and the decision process are combined, multi-objective optimization methods can be broadly classified into three categories (Deb, Sinha, Korhonen, & Wallenius, 2010; Horn, 1997; Hwang & Masud, 1979):

Decision Making Before Search: Apriori Approach

The objectives of the MOP are aggregated into a single objective which implicitly includes preference information given by the DM.

Search Before Decision Making: Aposteriori Approach

Optimization is performed without any preference information given. The result of the search process is a set of (ideally Pareto-optimal) candidate solutions from which the final choice is made by the DM.

Decision Making During Search: Interactive Approach

The DM can articulate preferences during the interactive optimization process. After each optimization step, a number of alternative trade-offs is presented on the basis of which the DM specifies further preference information, respectively guides the search.

Genetic Algorithm

Genetic algorithm is a search technique which mimics the principle of natural selection and natural genetics (Goldberg, 1989) to find the best solution for a specific problem. The genetic algorithm is an approach to solve problems which are not yet fully characterized or too complex to allow full characterization, but for which some analytical evaluation or physical interpretation to evolutes the performance of a solution, is available. GA is a stochastic global search method that mimics the metaphor of natural biological evolution. Genetic algorithms operate on a population of feasible solutions by applying the principle of survival of the fittest to produce better approximations to a solution. At each generation, a new set of approximations is created by the process of selecting individuals according to their level of fitness in the problem domain and breeding them together using operators borrowed from natural genetics. This process leads to the evolution of populations of individuals that are better suited to their environment than the individuals that they were created from, just as in natural adaptation.

The basic concept of a genetic algorithm is to encode a potential solution to a problem as a series of parameter strings, chromosomes, composed over some alphabet, so that chromosome values (genotypic) are uniquely mapped onto the decision variable (*phenotypic*). A single set of parameter string or chromosome is treated as the genetic material of an individual solution. Initially a large population of candidate solutions is created with random parameter values. These solutions are essentially bred with each other for several simulated generations under the principle of survival of the fittest, meaning that the probability that an individual solution will pass on some of its parameter values to subsequent children is directly related to the fitness of individual i.e. how good that solution is relative to the others in the population.

Breeding takes place through the use of recombination operators such as crossover, which simulates basic biological cross-fertilization, and mutation, essentially the introduction of noise. The simple application of these operators with a reasonable selection mechanism has produced surprisingly good results over a wide range of problems. After recombination and mutation, the individual strings are then, if necessary, decoded, the objective function evaluated, a fitness value assigned to each individual and individuals selected for mating according to their fitness, and so the process continues through subsequent generations. In this way, the average performance of individuals in a population is expected to increase, as good individuals are preserved and bred with one another and the less fit individuals die out. The GA is terminated when some criteria are satisfied, e.g. a certain number of generations, a mean deviation in the population, or when a particular point in the search space is encountered.

Genetic Algorithms and the Traditional Methods

The working principle of GA clearly indicates that the GA significantly differs in some very fundamental way from other traditional search and optimization methods. The four major significant differences are:

- GAs search a population of solutions instead of a single point solution as in traditional search methods.
- GAs do not derivative based algorithm. It does not use any derivative information or other auxiliary knowledge; only the objective function and corresponding fitness levels influence the directions of search.
- GAs use probabilistic transition rules instead of deterministic principle.

It is important to note that the GA provides a number of potential solutions to a given problem and the choice of final solution is left to the user. In cases where a particular problem does not have one individual solution, for example a family of Pareto-optimal solutions, as is the case in multi-objective optimization and scheduling problems, then the GA is potentially useful for identifying these alternative solutions simultaneously.

Simple Genetic Algorithm

The schematic shown in Figure 4 illustrates the structure of a simple genetic algorithm (SGA) as described by Goldberg (Goldberg, 1989). GA starts with initial random population consisted of potential solution points called individuals. The decision is made whether the individual is good or bad for the

Figure 4. A schematic representation of simple genetic algorithm

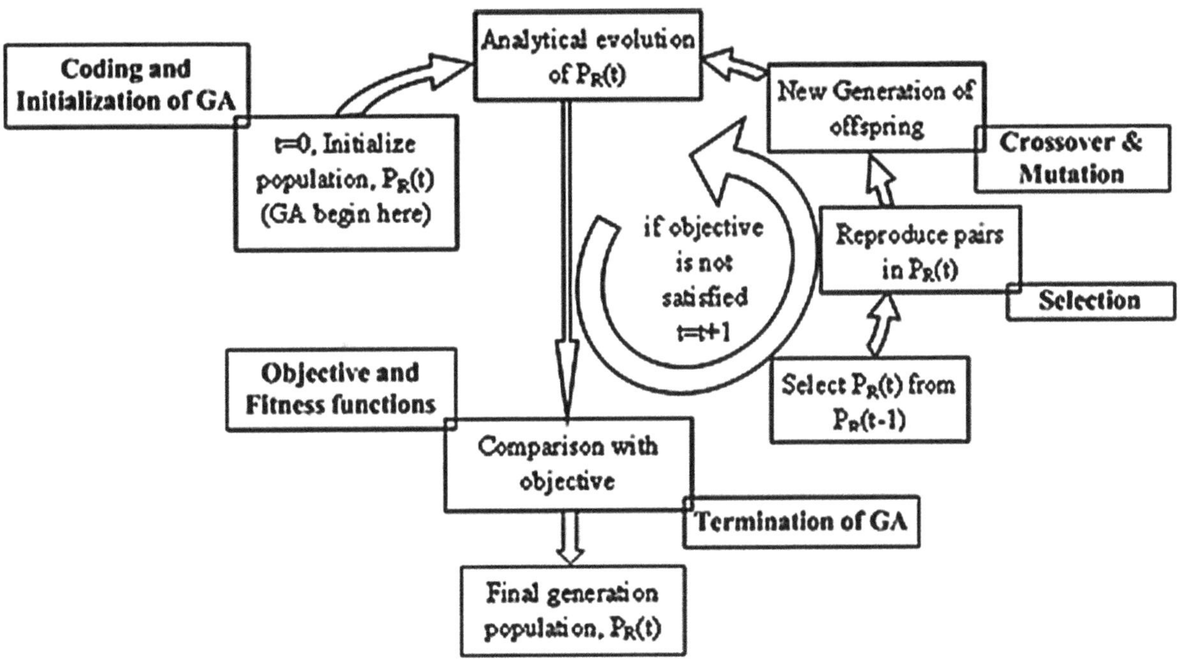

given problem based on the fitness obtained from the evaluation of objective function. Once the fitness value is evaluated and assigned to each individual, then initial population meets the first genetic operator, selection process. This operator provides more chances of survival for the strong individuals and to decay the weakest ones according to their fitness.

Next, crossover operator is performed on selected individuals to build the new individuals by combining the existing ones. Crossover follows reproduction and allows two individuals to swap their structures depending on the probability factor. This result in the creation of a pair of offspring solution contains characteristics of their parents. Then the mutation operator is applied to supply diversity in the population. As the fitness of a population may remain static for a number of generations before a superior individual is found, the application of conventional termination criteria becomes problematic. A common practice is to terminate the GA after a pre-defined number of generations and then test the quality of the best members of the population against the problem definition. If no acceptable solutions are found, the GA may be restarted or a fresh search initiated with more number of generations.

Coding and Initialization of GA

Binary Coding

The most commonly used representation of chromosomes in the GA is that of the single-level binary string. Here, each decision variable (say, *d* and *h*) in the parameter set is encoded as a binary string (s_d and s_h, respectively) and these are concatenated to form a chromosome as shown in Figure 5. Binary-coded GAs are not restricted to use only integer and for the given lower bound (d_{min}) and upper bound (d_{max}) of a variable (say, *d*), the value of the variable (*d*) is calculated from the GA-string using the decoding scheme represented by Equation (6).

$$d = d^{\min} + \frac{d^{\max} - d^{\min}}{2^{l_d} - 1} DV\left(s_d\right) \tag{6}$$

Where l_d is the string length used to code the *d* variable and $DV(s_d)$ is the decoded value of the string s_d. This mapping function allows

Figure 5. A schematic representation of a chromosome with 5 bits for S_d and 5 bits for S_h

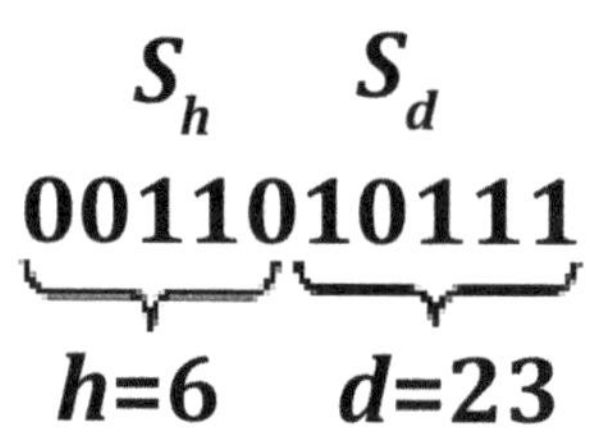

Real Coding

For a continuous search space, binary-coded GA faces many problems such as

- Hamming cliffs associated with certain strings (e.g., 01111 and 10000) from which a transition to a neighboring solution requires the alteration of many bits.
- Inability to achieve any arbitrary precision in the optimal solution. The more the required precision, the larger is the string length, results more computational complexity

In real coding, the variables are directly represented in real type as $\underbrace{25}_{h}\underbrace{8}_{d}$.

Initialization of GA

Initial population of a GA is normally determined at random. With a binary population of N_{ind} individuals whose chromosomes are L_{ind} bits long, $N_{ind} \times L_{ind}$ random numbers uniformly distributed from the set {0, 1} would be produced.

Objective Function and Fitness Function

The objective function is used to provide an analytical measure of how individuals have performed in the problem domain. The objective function, f is a function of decision variable(s). The fitness value of an individual/solution in the population is determined based on the fitness function which consists of the objective function value and the penalty value for constraint violation which is determined by a penalty function, $f_{\text{costraint}}$. Thus, the fitness value is calculated as follows:

$$\text{Fitness value } = f + f_{\text{constraint}} \tag{7}$$

Since GAs mimic the survival-of-the-fittest of the nature to make a search process, GAs are suitable for maximization problem where objective function is directly used as fitness function $F(x)$. In the case of a minimization problem, the fit individuals will have the lowest numerical value of the associated objective function. This situation is handled using a conversion of the objective function into an equivalent maximization problem and used as fitness function so that the optimum point remain unchanged.

Selection

Selection (Reproduction) guides the tool to find the optimized solution by preferring individuals/members of the population with higher fitness over one with lower fitness. It is the operator which generates the mating pool. This operator determines that the number of times a particular individual will be used for reproduction and the number of offspring that an individual will produce. Some of the popularly used selection methods are as follows:

Tournament Selection

In tournament selection a small group of individuals is taken from the population and the individual with best fitness is chosen for reproduction. The size of the group chosen is called the tournament size. A tournament size of two is called binary tournament.

Roulette Wheel Selection Methods

Roulette wheel (proportionate) selection scheme chooses a certain individual with a probability proportional to its fitness.

$$p\left[I_{j,t}\right] = \frac{f\left(I_{j,t}\right)}{\sum_{k=1}^{n} f\left(I_{k,t}\right)} \tag{8}$$

Where $p\left[I_{j,t}\right]$ is the probability of getting selected of any j^{th} individual at a generation t, $f\left(I_{j,t}\right)$ and $\sum_{k=1}^{n} f\left(I_{k,t}\right)$ are corresponding individual fitness and the sum of the fitness of the population with size n, respectively.

The property as represented by Equation (8) is satisfied by applying a random experiment that has some similarity with a generalized roulette game. In the roulette game, the slots are not equally wide, that is why different outcomes occur with different probabilities. Figure 6 gives a graphical representation of how this roulette wheel game works.

Rank Selection

This selection method is similar to the Roulette wheel selection mechanism. However, it chooses a certain individual with a probability proportional to its rank value. The worst solution will have rank 1, the next 2, so on and the best will have a rank equal to N (N is the number of chromosomes in the population).

Figure 6. Graphical representation of the roulette wheel selection mechanism

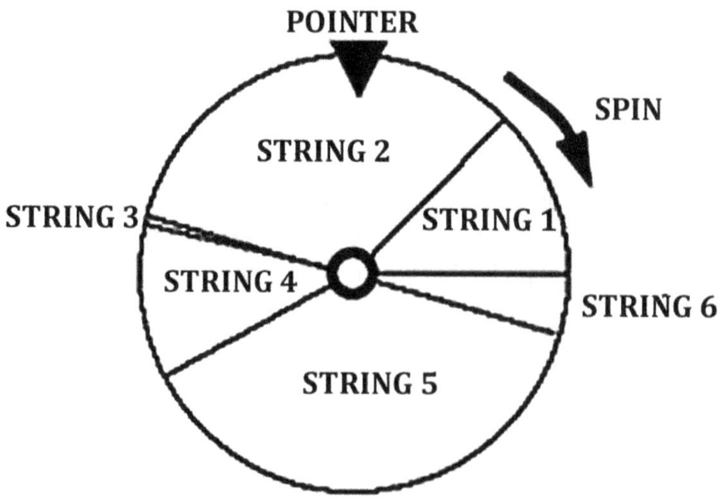

In addition another scheme for selection is applied along with all three selection schemes discussed above which is called 'elitism'. The idea of elitism is to avoid the observed best-fitted individual dies out just by selecting it for the next generation without any random experiment. Elitism significantly influences the speed of the convergence of a GA. But it can lead to premature convergence also.

Crossover

The basic operator for producing new chromosomes in the GA is that of crossover. Like natural reproduction, crossover produces new individuals so that some genes of a new child come from one individual while others comes from the other individual. In essence, crossover is the exchange of genes between the chromosomes of the two parents. The process may be described as cutting two strings at a randomly chosen position and swapping the two tails. It is known as the single-point crossover, and the mechanism is visualized in Figure 7(a). An integer position, i is selected at random with a uniform probability between one and the string length, l, minus one (i.e., $i \in [1, l-1]$). When the genetic information is exchanged among the parent individuals (represented by the strings, P_1 and P_2) about this point, two new offspring (represented by the strings, O_1 and O_2) are produced. The two offspring in Figure 7(a) are produced when the crossover point, $i=4$ is selected.

For multi-point crossover, multiple crossover positions (m) are chosen at random with no duplicates and sorted into ascending order. Then the bits between two successive crossover points are exchanged between the two parents to produce two new offspring. The process of multi-point crossover is illustrated in Figure 7(b)with shaded color. The idea behind multi-point crossover is that the parts of the chromosome representation that contribute to the most to the performance of a particular individual may not necessarily be contained in adjacent substrings. Further, multi-point crossover appears to encourage the exploration of the search space, thus making the search more robust.

Mutation

Mutation is nothing but deformation of the genetic information of an individual (solution) by means of some external influences. The bit-wise mutation operator changes a bit, 1 to 0, and vice versa, with a prescribed probability (called, mutation probability) as shown in Figure 8. In real reproduction, the probability that a certain gene is mutated is almost equal for all genes. So, it is near at hand to use the mutation technique for a given binary string, where there is a given probability that a single gene is modified. The probability should be rather low in order to avoid chaotic behavior of the GA.

Figure 7. A typical example of (a) single point crossover (b) multi-point crossover (with m=5)

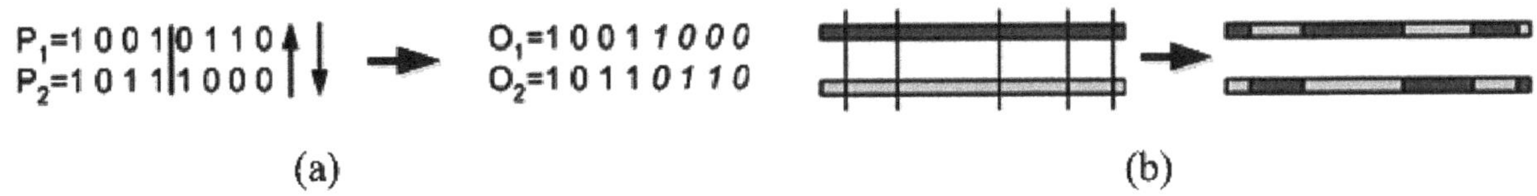

Figure 8. Mutation effect on offspring's strings

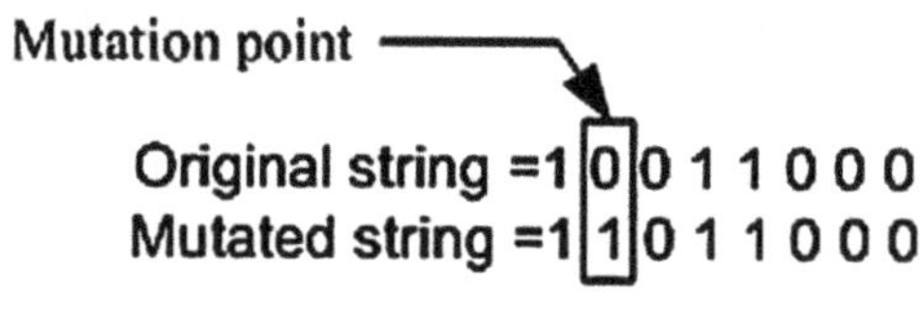

Termination of the GA

Because the GA is a stochastic search method, it is difficult to formally specify convergence criteria. As the fitness of a population may remain static for a number of generations before a superior individual is found, the application of conventional termination criteria becomes problematic. A common practice is to terminate the GA after a pre-specified number of generations and then test the quality of the best members of the population against the problem definition. If no acceptable solutions are found, the GA may be restarted or a fresh search can be initiated. Attaining a pre-specified fitness function value or when best fitness of the population does not change appreciably over successive iterations may be also considered as a termination criteria.

Description of the Working Principle of GA

The working principle of a (binary-coded) GA is described here with an engineering optimization problem, namely designing a can (Deb, 2001) as shown in Figure 9. The objective of the problem is to determine the optimum values of the diameter (*d*, in cm) and height (*h*, in cm) of the can in order to minimize its cost, *f*(*d*, *h*) subject to some constraints and the problem is defined as follows:

Minimize $$f\left(d,h\right)=c\left(\frac{\pi d^{2}}{2}+\pi dh\right), \tag{9}$$

Subject to

$$g\left(d,h\right)\equiv\frac{\pi d^{2}h}{4}\geq 300,$$

$$d_{\min}\leq d\leq d_{\max},\quad h_{\min}\leq h\leq h_{\max}$$

where c is the cost of the can material per square cm, which is taken as 0.005 and the minimum and maximum values of d and h are taken as, $d_{min}=h_{min}=0$ and $d_{max}=h_{max}=31$.

Therefore, in this problem the number of decision variables is two (*d* and *h*). The population size of GA is considered as six and it is kept constant throughout the GA-operation. GA operates in a number

Figure 9. Working principle of a genetic algorithm

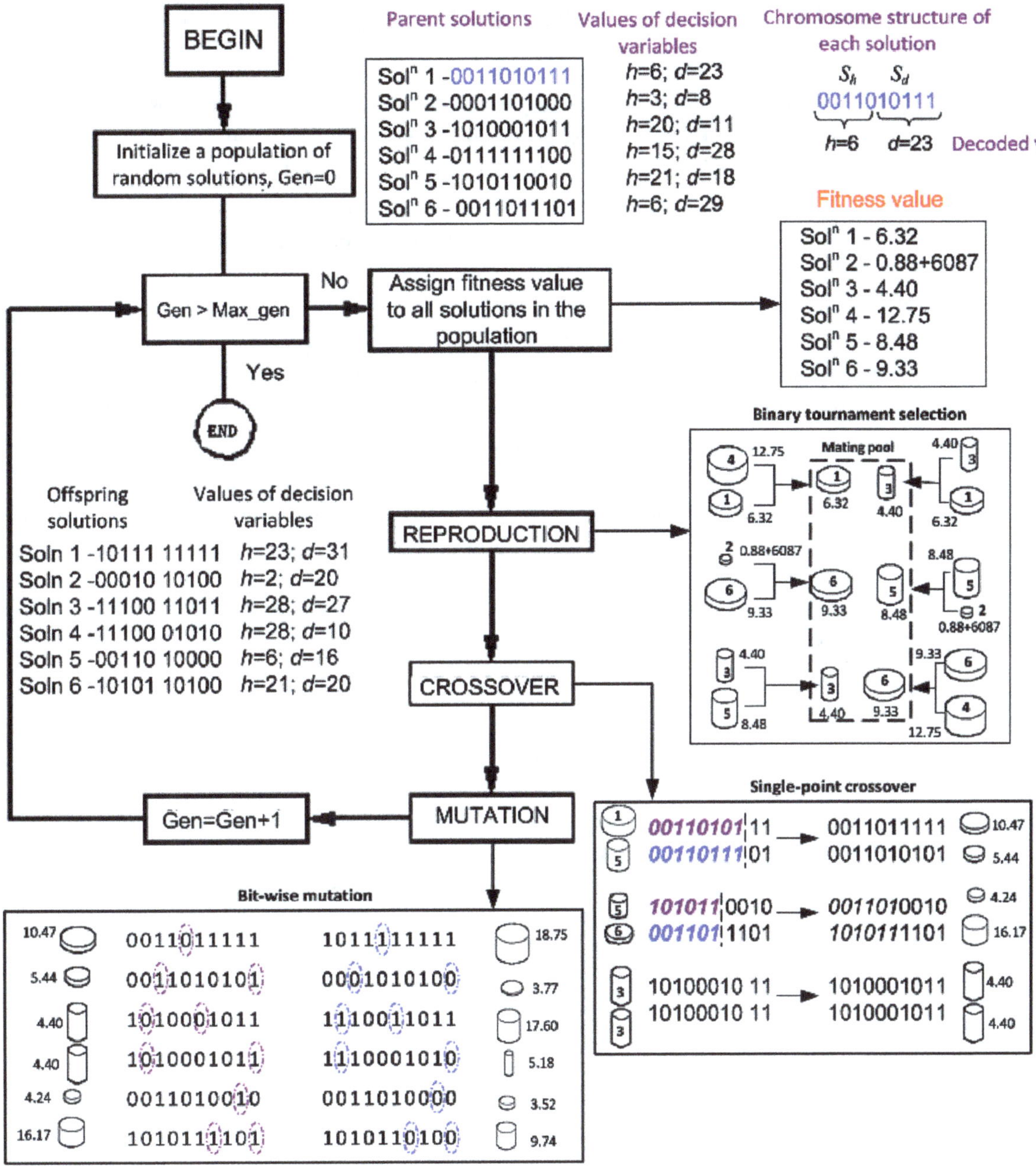

of iterations until a specified termination criterion is satisfied and in the Figure 9, the maximum number of iteration/generation (Max_gen) is treated as the termination criteria.

GA iteration starts with the creation of six random solutions which are treated as the parent solutions. The chromosome structure of each solution is the same as presented in Figure 5. Then, the fitness values of all the parent solutions in the population are calculated using following fitness function.

$$\text{Fitness value } = \begin{cases} f(d,h) + P_p \times [300 - g(d,h)] & \text{if } g(d,h) \leq 300 \\ f(d,h) & \text{if } g(d,h) > 300 \end{cases} \tag{10}$$

where P_p is the penalty parameter and the value of P_p is taken as 25. After that selection/reproduction operation is performed on the parent solutions based on the fitness values of solutions (as depicted in the binary tournament selection table in Figure 9) to form the mating pool. The number of solutions contained in the mating pool is to be equal to the number of parent solution in order to maintain a constant population size throughout the GA-iteration. After that, crossover operator is applied among the two randomly chosen solutions from the mating pool based on a given probability (crossover-probability, say 0.9). Then, bit-wise mutation is carried out on each of the six solutions obtained after employing the crossover operator using a given mutation probability (say 0.02) and produces six new (offspring) solutions. The objective function values corresponding to each solution are depicted in the single-point crossover and bit-wise mutation tables in Figure 9. It completes one iteration/generation of GA. Now, the checking of GA-termination criterion is performed and if it satisfies the termination criteria, stop the GA-iteration process, otherwise starts another generation/iteration with treating the six offspring solutions obtained in the previous iteration as the parent solutions and then assigning fitness value to all the solutions in the population and continue the same iteration procedure as described above.

MULTI-OBJECTIVE EVOLUTIONARY ALGORITHM

From the previous section, it has been observed that evolutionary algorithms (EAs) seem to be especially suited to multi-objective optimization. Because they are able to capture multiple Pareto-optimal solutions in a single simulation run and may exploit similarities of solutions by recombination. Some researchers suggest that multi-objective search and optimization might be a problem area where EAs do better than other blind search strategies (Fonseca & Fleming, 1995; Valenzuela-Rend´on & Uresti-Charre, 1997).

After the first pioneering studies on evolutionary multi-objective optimization appeared in the mid-1980s (Fourman, 1985; Schaffer, 1984; Schaffer, 1985), a few different MOEA (multi-objective evolutionary algorithm) implementations were proposed in the years 1991–1994 (Fonseca & Fleming, 1993; Hajela & Lin, 1992; Horn, Nafpliotis, & Goldberg, 1994; Kursawe, 1991; Srinivas & Deb, 1994). Later, these approaches (and variations of them) were successfully applied to various multi-objective optimization problems (Cunha, Oliviera, & Covas, 1997; Fonseca & Fleming, 1998a; Ishibuchi & Murata, 1996; Parks & Miller, 1998; Valenzuela-Rend´on & Uresti-Charre, 1997). Some researchers have investigated particular topics of evolutionary multi-objective search, such as convergence to the Pareto-optimal front (Veldhuizen & Lamont, 1998; Rudolph, 1998), niching (Obayashi, Takahashi, & Takeguchi, 1998), and elitism (Obayashi, Takahashi, & Takeguchi, 1998; Parks & Miller, 1998), while others have concentrated on developing new evolutionary techniques (Laumanns, Rudolph, & Schwefel, 1998; Lis & Eiben, 1997].

Since the optimization goal in MOP (multi-objective optimization problem) consists of multiple objectives, the basic EAs, such as GA as described in the previous section cannot be used straightforward and requires some modifications. In order to apply EA to multi-objective optimization, two major problems raised those must be addressed: (1) How to accomplish fitness assignment and selection, respectively, in order to guide the search towards the Pareto-optimal set and (2) How to maintain a diverse population in order to prevent premature convergence and achieve a well distributed and well spread non-dominated set.

There are various techniques suggested by many researchers for fitness assignment and selection in MOEA such as Selection by Switching Objectives (Fourman, 1985), Aggregation Selection with Parameter Variation (Ishibuchi & Murata, 1996) and Pareto-based Selection (Goldberg, 1989). Various mechanisms are proposed in order to make Population Diversity in MOEA namely Fitness Sharing (Goldberg & Richardson, 1987), Restricted Mating, Isolation by Distance (Ryan, 1995), Over-specification (Kursawe, 1991), Re-initialization (Fonseca & Fleming, 1998b) and Crowding (De Jong, 1975). De Jong (De Jong, 1975) suggested a policy to always include the best individual of P_t into P_{t+1} in order to prevent losing it due to sampling effects or operator disruption. This strategy, which can be extended to copy the *b* best solutions to the next generation, is denoted as *elitism*. Based on the use of elitism, EAs are classified into two groups such as non-elitist MOEA and elitist MOEA. In this chapter, a well-established and the most popular widely applicable elitist MOEAs such as NSGA II (non-dominated sorting genetic algorithm) (Deb, Pratap, Agarwal, & Meyarivan, 2002) is presented in the following.

NSGA-II

Though the basic working procedure of SGA as described in Figure 9 is followed by NSGAII, there are few modifications including selection operator. In NSGAII as presented in Figure 10, the offspring population Q_t (of size, N) is first created using the parent population P_t (of size, N). Then, both the parent and offspring populations are mixed up to form a combined population, R_t of size 2N. After that, a non-dominated sorting is performed to classify the enter population R_t. Once the non-dominated sorting is over, solution(s) of different non-dominated fronts, one at a time fill the new population of size, N. The fillings start with the best non-dominated front, and so on. Since the overall population size of R_t is 2N, not all fronts may be accommodated in N slots available in the new population. All fronts, which could not be accommodated, are simple deleted. When the last allowed front is being considered there may presence more solutions in the last front than the remaining slots in the new population. Instead of arbitrarily discarding some members from the last front, it would be wise to use a niche-preserving strategy to choose the members of the last front, which decide in the least crowded region in that front. Based on this new population another offspring population Q_t is created using genetic operators like crowded tournament selection, crossover and mutation in the next generation, and this cycle is continued till a specified number of generations is reached or other specified termination criteria is fulfilled.

Two popular niche-preserving strategies for determining the solution(s) to be present in the last front are crowding distance sorting and clustering approach.

In crowding distance sorting approach, the crowding distance for each individual is measure by calculating the cuboid length which is the sum of distances between an individual's two closest neighbors in each objective (as presented in Figure 11(a)). The individuals with higher crowding distance are preferred which is followed in the original NSGA II.

In clustering technique (as shown in Figure 11(b)) first each of the solutions in the non-dominated front is assumed to belong to a separate cluster. Thereafter the distance between all pairs of clusters is calculated by first finding the centroid of each cluster and then calculating the (Euclidean) distance between the centroids. Two clusters having a minimum distance between them are merged together into a bigger cluster, thereby reducing one cluster from the previous number of clusters. Then again the distance between all pairs of present clusters are calculated in the similar way as done before. This procedure is continued until the desired number of clusters (equivalent to the desired number of solu-

Figure 10. A schematic presentation of NSGA II (Deb, 2001)

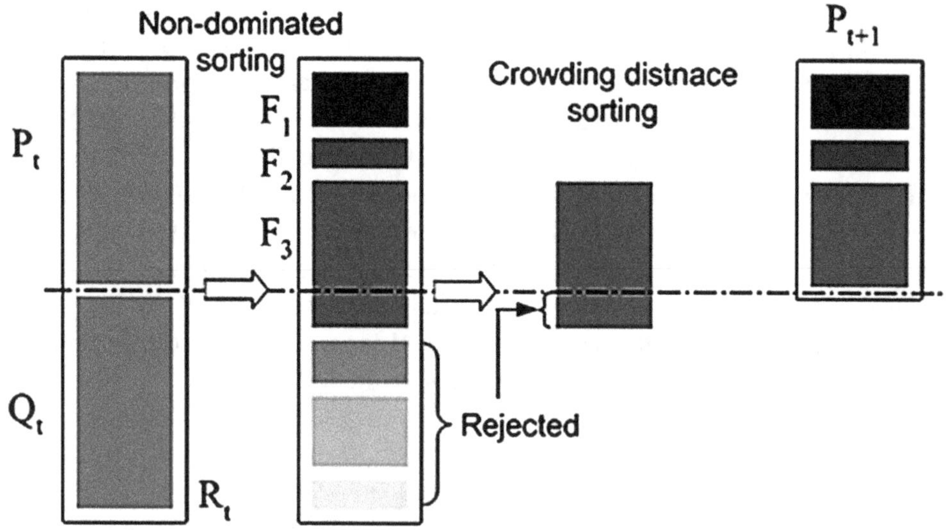

Figure 11. A schematic representation of (a) crowding distance sorting approach (Deb, 2001) (b)clustering approach (Zitzler, 1999)

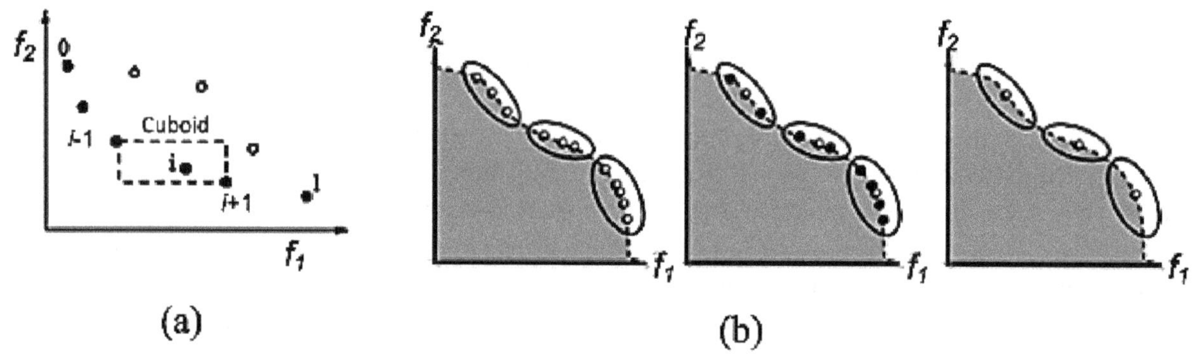

tions) is identified. Finally, the solutions closest to the centroid of the cluster is retained and all other solutions from each cluster are deleted.

Unlike conventional tournament selection operator, in crowded tournament selection, a solution, *i* wins a tournament with other solution(s), *j* if either solution *i* has a better rank compared to solution(s), j or if they have the same rank but solution, i has a crowding distance more than that of solution(s), j. The rank (r_i) of a solution, *i* is defined by r_i=1+n_i, where n_i is the number of solutions dominate solution i. The rank of each solution in a front is same.

In this chapter optimization results obtained by NSGA-II with crowding distance and clustering approaches in designing particle reinforced polyurethane mould material are with the help of certain performance measures. Furthermore, some statistical test, namely Wilcoxon signed-rank test (a non-parametric statistical test for testing hypothesis on median) (Siegel, 1956) is performed on the performance matrices in order to decide whether the performances of NSGA-II with clustering approach and with crowding distance approach are equivalent or not.

Performance Metrics of Multi-Objective Optimization Algorithms

There are two main criteria to be taken into account to assess the performance of a multi-objective optimization algorithm, namely the distance of the obtained solutions to the Pareto-front and the distribution of obtained solutions on the Pareto-front. In most of the practical problems there is a limitation of the knowledge of true Pareto-optimal solutions. Thus, the following three metrics are normally considered in order to make a comparative assessment of the results obtained by multi-objective optimization algorithms.

1. Size of the dominated space
2. Coverage of two Pareto-fronts
3. Non-uniformity of the Pareto-front

The size of the dominated space $S_d(A)$ indicates the measure of how much of the objective space is weakly dominated by the Pareto-front, A. The higher value of $S_d(A)$ indicates the better performance. The measurement, coverage of two Pareto-fronts, A and B ($C(A, B)$) provides the fraction of B weakly dominated by A and is calculated as follows:

$$C(A,B) = \frac{\left|\left\{b \in B | \exists a \in A{:}a \succ b\right\}\right|}{\left|B\right|} \tag{11}$$

Where $\left|B\right|$ is the cardinality of the Pareto-optimal set B and $a \succ b$ means that solution a dominates the solution b. If, $C(A, B) > C(B, A)$, this means that the Pareto-front, A has better solutions than the Pareto-front, B. The measurements, size of the dominated space and coverage of two Pareto-fronts suggest the degree of convergence of a Pareto-front. The measurement of non-uniformity of a Pareto-front, A is calculated as follows:

$$D(A) = \sqrt{\frac{\sum\left(\frac{d_i}{\overline{d}} - 1\right)^2}{\left|A\right| - 1}} \tag{12}$$

Where d_i is the distance between two consecutive solutions along the Pareto-front, A and $\overline{d}$ is the average distance. A lower value of $D(A)$ indicates the better distribution of obtained solutions on the Pareto-front A.

MOOP IN DESIGINING OF COMPOSITE MOULD MATERIAL

To improve the soft tooling process, the primary motivation is to increase the thermal conductivity of mould material by adopting particle reinforced composite flexible polymer. As enhancing the thermal conductivity of mould material by highly conductive particle reinforcement also increases its effective modulus of elasticity and equivalent viscosity, a MO optimization procedure is adopted to decide the appropriate values of decision factors involved in the manufacturing of particle reinforced flexible polymeric composites. The main controlling factors involved towards affecting the composite's effective thermal conductivity, modulus of elasticity and viscosity for a given composite system are the volume fraction of filler particles (V_f), particle size (P_s) and shape factor (S_f) of filler particles and the temperature (T). The proposed procedure of designing the particle reinforced mould material is carried out through two approaches. The formulations of multi-objective problems corresponding to these approaches are presented in the following. In the problem I, we have considered the equivalent thermal conductivity and modulus of elasticity where the equivalent viscosity is assumed simply as a constraint. On the other hand, all the three properties of the mould material are treated as objectives to carry out a deeper investigation by treating them with equal importance in designing the polyurethane composite mould material. In the present work, we attempt to solve the problem step by step by first solving the problem formulated in a simple way (Problem I) and then in a more complex manner (Problem II).

Definition of Problem I

The optimization problem evolved in the ST process in order to make it cost effective by reducing cooling time as well as keeping more ease in the casting process. The objectives are described below:

$$f_1\text{=Maximize } (k_c)\, f_{\text{Thermal Conductivity}_{\text{Equi}}}(k_f, k_p, P_s, S_f, T, V_f) \tag{13}$$

$$f_2\text{=Minimize } (E_c)\, f_{\text{Modulus of Elasticity}_{\text{Equi}}}(G_f, G_p, K_f, K_p, V_f) \tag{14}$$

Following constraints are considered (for a given composite system of flexible mould material and conductive filler material):

$$0 \le P_s \le 500, \tag{15}$$

$$0 \le S_f \le 4.0, \tag{16}$$

$$0 \le V_f \le 100, \tag{17}$$

$$23°\text{C} \le T \le 30°\text{C}, \tag{18}$$

$$\frac{\varphi_m(P_s,S_f) - V_f}{\varphi_m(P_s,S_f)} >= 0, \tag{19}$$

$$\eta_c^{\text{Limiting}} - (\eta_c)\, f_{\text{Viscosity}_{\text{Equi}}}(P_s,V_f,\eta_p) >= 0, \tag{20}$$

where k_f, k_p, G_f, G_p, K_f, K_p and η_p are constant and η_p and η_c are the viscosity of liquid polymer and the effective viscosity of particle reinforced polymer, respectively. η_c^{Limiting} is the limiting value of the effective viscosity of particle reinforced polymer. The function $\varphi_m(P_s, S_f)$ is called the maximum loading level (maximum packing fraction) and $f_{\text{Viscosity}_{\text{Equi}}}(P_s,V_f,\eta_p)$ is the equivalent viscosity function. These are used in the subsequent sections.

Definition of Problem II

In this problem, viscosity is added as an objective function:

$$\text{Maximize } (k_c)\, f_{\text{Thermal Conductivity}_{\text{Equi}}}(k_f,k_p,P_s,S_f,T,V_f) \tag{13}$$

$$\text{Minimize } (E_c)\, f_{\text{Modulus of Elasticity}_{\text{Equi}}}(G_f,G_p,K_f,K_p,V_f) \tag{14}$$

$$\text{Minimize } (\eta_c)\, f_{\text{Viscosity}_{\text{Equi}}}(P_s,S_f,V_f,\eta_p) \tag{21}$$

subject to satisfaction of following constraints (for a given composite system of flexible mould material and conductive filler material):

$$0 \le P_s \le 500, \tag{15}$$

$$0 \le S_f \le 4.0, \tag{16}$$

$$0 \le V_f \le 100, \tag{17}$$

$$23°\text{C} \le T \le 30°\text{C}, \tag{18}$$

$$\frac{\varphi_m(P_s,S_f)-V_f}{\varphi_m(P_s,S_f)}>=0, \tag{19}$$

where $k_f, k_p, G_f, G_p, K_f, K_p$ and η_p are constant.

The upper and lower limits of P_s and S_f are fixed based on the fact that the particles having these sizes and shape factors are most suitable for handling in practical applications. Furthermore, the size and shape factors of common filler materials that are commercially available in particulate form fall in those ranges. In soft tooling process, the preparation of mould is generally carried out at room temperature. Therefore, in this work, the range of temperature is set from 23°Cto 30°C.

VARIOUS MODELS/ OBJECTIVE FUNCTIONS

Equivalent Thermal Conductivity Model

Various models of equivalent thermal conductivity (ETC) of composite materials were proposed in the literatures. However, it was found that all of those models are not appropriate for different composite systems. Previous investigations (Nandi et al., 2012) suggest that for particle reinforced polymeric flexible mould material composite systems, the empirical model, Lewis-Nielsen relation (Lewis & Nielsen, 1970) gives a closer estimate over other models. But this model does not possess a generic nature for particle reinforced polymer composites, because the controlling parameter of this model varies with the shape and orientation of the dispersed particles in the composite. On the other hand, the semi-empirical model proposed by Agari and Uno (Agari & Uno, 1986) shows comparatively better result only in a particular type of filler-matrix combination. Besides, the thermal conductivity (TC) of matrix material (k_p), TC of filler material (k_f) and volume/weight fraction (V_f) of filler in the matrix, there are some other physical aspects (like formation of filler conductive chain, formation of crystallinity of polymer, etc) evolved during manufacturing and curing of (reactive) PCs. But, the mechanism and (numerical) formulation of these physical aspects are highly complex and non-linear, because they depend on such kind of parameters which have certain amount of uncertainty, ambiguity and vagueness in practical situations. By realizing it, in the author's previous work (Nandi et al., 2013) a generic model of equivalent thermal conductivity for particle reinforced polymer composites was suggested which was developed based on a genetic fuzzy system. The proposed genetic fuzzy model of ETC employs TSK-type fuzzy logic rule which is expressed by the following expression.

If x_1 is A_1 and x_2 is A_2 and...... x_4 is A_4, then $y = f(k_f, k_p, x_1, x_2, x_3, x_4)$,

where $A_1, \ldots, A_4$ are the fuzzy subsets of the input variables, $x_1, \ldots, x_4$, respectively and the rule consequent function is obtained based on an analytical approach (Nandi et al., 2013) given as follows:

$$y(\text{ETC}) = k_p\left(\frac{k_f+2k_p+2V_f(k_f-k_p)}{k_f+2k_p-V_f(k_f-k_p)}\right)+c^r\, V_f\left(\frac{P_sS_fk_f}{T}\right)^{p^r}, \tag{22}$$

where c^r and p^r are the coefficient and power term of rth rule, respectively, P_s is the filler particle size, S_f is the shape factor of filler particle, and T is the temperature.

The genetic-fuzzy model of ETC is trained using the experimental data obtained from various literatures. The effectiveness of the model is extensively tested in three stages to establish its practical applicability: based on experimental data (not used in model development), newly measured thermal conductivities of different polymeric flexible mould composites and finally by assessing the feasibility of values of missing data in the reported incomplete data set based on the developed model. It has been found that the developed model affords better prediction in comparison to the existing ones, and also exhibits a generic character applicable to a wide range of two-phase particle-reinforced polymer composites.

Using a genetic linear regression approach (Nandi, 2006), the membership function distributions (MFDs) of controlling (fuzzy) parameters (V_f, P_s, S_f and T) and the corresponding rule base of the TSK-type fuzzy model (whose rule consequent function is defined in Equation (22)) are obtained (Nandi et al., 2013). Based on the rule consequent function, expressed in the form of Equation (22), the optimized MFDs and rule base, the effective thermal conductivity (k_c) of a two-phase polymer composite can be calculated using the following empirical expression.

$$k_c = \frac{\sum_{r=1}^{R^l} \left(\prod_{v=1}^{n} \mu_v(x_v) \right) \times y}{\sum_{r=1}^{R^l} \left(\prod_{v=1}^{n} \mu_v^r(x_{1,\ldots,n}) \right)} \tag{23}$$

where R^l is the number of rule(s) fired out of a total of R rules for a set of given values of inputs (such as k_f, k_p, V_f, P_s, S_f and T).The genetic fuzzy model of equivalent thermal conductivity is adopted here in order to investigate the role of various fillers so as to reduce cooling time in soft tooling process.

Equivalent Modulus of Elasticity Model

In (Nandi et al., 2012), an investigation of validating different existing models of equivalent modulus of elasticity using the experimental results of equivalent modulus of elasticity of different particle reinforced polymeric composites was carried out. It has been observed that the model proposed by Lielens (Lielens et al., 1998) which is formulated based on a normalization of the upper and lower bounds suggested by Hashim-Srikman (Hashin & Shtrikman, 1963) provides close agreements with the experimental data compared to other models, particularly up to a volume fraction of 0.3 of the particulates. In the present investigation, we have considered the Lielen`s model of equivalent modulus of elasticity. Here the equivalent modulus of elasticity may be calculated by Equation (24) by assuming the composite to be (quasi) isotropic and (quasi) homogeneous

$$E_c = \frac{9K_cG_c}{3K_c + G_c}, \tag{24}$$

where E_c(N/mm^2), K_c (N/mm^2) and G_c (N/mm^2) are the equivalent modulus of elasticity, bulk modulus and shear modulus of polymer composite, respectively.

Equivalent Viscosity Model

The equivalent viscosity of particle reinforced melt polymer suspension is described as follows:

$$\eta_c = \eta_p \times \eta_r, \tag{25}$$

where η_r is the relative viscosity of suspension.

In the present work, the model presented in (Krieger & Dougherty, 1959) for concentrated suspensions is considered:

$$\eta_r = \left(1 - \frac{V_f}{\varphi_m}\right)^{-A\varphi_m}, \tag{26}$$

where *A* is the controllable parameters and the value of this parameter is related to the suspension system. The optimum values of *A* for Al and graphite-filled flexible mould material composite systems determined using a genetic algorithm are as follows (Nandi et al., 2010):

1. A=2.8697 for suspension of flexible mould material with Al particles.
2. A=10.3459 for suspension of flexible mould material with graphite particles provided $V_f > \varnothing_m$ and $\eta_r = 100$.

FLEXIBLE MOULD MATERIALS AND PARTICULATE FILLER

Flexible Polymeric Mould Materis used in Soft Tooling Process

The materials used for making mould in soft tooling process are soft and flexible in nature. Since soft tooling process is particularly used for making components made of wax or low melting point plastics, soft reactive thermosetting polymeric materials those are capable to withstand heat are applicable as mould material in this process. In this chapter, the flexible mould materials those are extensively used in soft tooling process such as Polyurethane (which is assumed to behave quasi-isotropic and quasi-homogeneous) is adopted.

Polyurethanes constitute one of the most complete and versatile classes of polymeric materials known nowadays. Polyurethane is a kind of reactive polymer and is characterized by the presence of urethane linkage. A urethane linkage is produced through the reaction of an isocyanate group, ($-N=C=O$) with a hydroxyl (alcohol) group $-OH$. Polyurethane polymers are formed through step-growth polymerization by reacting a monomer containing at least two isocyanate functional groups with another monomer containing at least two hydroxyl (alcohol) groups in the presence of a catalyst. A unique feature of polyurethane is that a wide variety of structural changes could be brought about with different hydroxyl compounds and isocyanates, leading to a wide spectrum of properties in polyurethanes. The polyurethane used in the present study is of type PMC-121/50 Wet, manufactured by Smooth-ON, Inc. USA. It has two parts, Part A and Part B, also called hardener. Both the parts (part A and part B) are mixed in the ratio of 1:1 by weight or volume, respectively. Once the Part A and Part B are mixed, curing process starts even in room temperature. After a certain time (called pot life), the mixture turn into gel form, and after that it gradually becomes in solid rubber type. The polyurethane of the above mentioned type possesses the following properties those are measured experimentally:

Density (of cured polyurethane) =1.04 gm/cc
Modulus of elasticity (of cured polyurethane) = 1631.2 N/mm^2
Poisson's ratio (of cured polyurethane) = 0.496
Shear modulus (of cured polyurethane) = 545.187 N/mm^2
Bulk modulus (of cured polyurethane) = 67966.667 N/mm^2
Viscosity (of the mixture of melt PU, during pot life) at 23 °C = 3.5195 Pas
Thermal conductivity =0.1615 W/m-k

Filler Materials

Through an investigation on the role of various metallic and non-metallic filler particles in reinforced composites of flexible mould materials (Nandi, Datta, & Deb, 2011; Nandi, Deb, Ganguly, & Datta, 2012) it is observed that aluminum (Al) and graphite (Gr) are possessed higher thermal conductivity in conjunction with lower elasticity modulus for the composites, along with low cost. Based on this realization, design of composite polyurethane is carried out using aluminum (Al) fine powder and simple (uncoated) bulk graphite (Gr) granules. The physical properties of Al and graphite particles are shown in Table 1.

MULTI-OBJECTIVE OPTIMIZATION TOWARDS DESIGINING PARTICLE REINFORCED MOULD MATERIALS

Since the decision variables involved in optimization of equivalent thermal conductivity, modulus of elasticity and modulus of viscosity are continuous type, real-coded GA is being used for solving both the problems, Problem I and Problem II as stated in sub-sections 'Definition of Problem I' and 'Definition of Problem II', respectively. The values of GA-parameters adopted for real coded GA are listed in Table 2.

Multi-Objective Optimization for Solving Problem I

In this study, three levels of viscosity of melt mould material composite are considered such as Low=10 Pas; Medium=50 Pas; and High=200 Pas.

Since the results of an evolutionary algorithm depend on random values, twenty independent runs of the algorithm are performed with different values of random seed to make the results obtained by NSGA-II statistically reliable. The mean and standard deviation of the number of Pareto-optimal solutions, $S_d(A)$, $C(A,B)$ and $D(A)$ obtained by NSGA-II with clustering and crowding distance approaches in twenty different trials are listed in Table 3. The size of the dominated space, $S(A)$ is measured by normalizing the objectives taking minimum and maximum values of equivalent thermal conductivity as 0.1 W/m-k and 20 W/m-k, equivalent modulus of elasticity as 1000N/mm^2 and 6000 N/mm^2, respectively in both the composite systems.

Table 1. Properties of filler materials

Properties	Filler Type	
	Aluminum	Graphite
Thermal conductivity, (W/m-k)	237	209.34
modulus of elasticity, (GPa)	68.95	25
Poisson ratio	0.35	0.31
Average aspect ratio	1.82	2.025
Average size (μm)	12.3	296

Table 2. Values of different GA-parameters

GA-Parameter	Value
Population size	500
Number of generations	200
Mutation probability	1/number of decision variables
Crossover probability	0.98
Distribution Index for real-coded crossover	20
Distribution Index for real-coded mutation	100

From Table 3, it has been found that in both composite systems, the difference in mean values of *S*(*A*) of the Pareto-fronts obtained by NSGA-II using clustering and crowding distance approaches are not significant. However, we have observed a considerable difference in the mean values of *C*(*A*, *B*) and *D*(*A*) of the Pareto-fronts obtained by NSGA-II using clustering and crowding distance approaches. It is seen that the spread of solutions is more uniform in the NSGA-II with clustering approach as opposed to the crowding distance approach. However, as described in Table 3, the high value of coverage of two Pareto-fronts indicates that more solutions obtained by NSGA-II with clustering approach are (weakly) dominated by that obtained by NSGA-II with crowding distance approach. In the Table3, a very low value of standard deviation of $S_d(A)$, *C*(*A*, *B*) and *D*(*A*) is observed which indicates that the performance of NSGA-II is statistically more stable.

Furthermore, a non-parametric statistical test, specifically Wilcoxon signed-rank testis carried out on the values of $S_d(A)$, *D*(*A*) and *C*(*A*, *B*) that are calculated based on the results obtained by NSGA-II with crowding distance and clustering approaches. Wilcoxon signed-rank sum test is used to test the null hypothesis (H_0) that both crowding distance and clustering approaches are equivalent corresponding to a performance metric. The test uses the sign and the magnitude of the rank of the difference between pairs of measurements. The result of the test applied on the results obtained by twenty independent runs with different random seeds in problem I is illustrated in Table 4. From the Table 4, the following observations are noticed for both the composite systems with different viscosity levels. The clustering approach and crowding distance approach are found equivalent in respect of the size of the dominated space of Pareto-front. On the other hand, clustering approach is found better in the uniformity of spread of solutions

Table 3. Statistical information of the solutions obtained by twenty independent runs with different random seeds in problem I

Composite	Viscosity label (Pas)	NSGA-II with niching approaches	Number of Pareto-optimal solutions obtained		Size of the dominated space, $S_d(A)$		Coverage of two Pareto-fronts, $C(A, B)$		Non-uniformity of the Pareto-front, $D(A)$	
			Mean	Standard Deviation	Mean	Standard Deviation	Mean	Standard Deviation	Mean	Standard Deviation
Al-PU	10	Crowding distance	500	0	0.3446	0.0032	0.4047	0.1674	0.5982	0.0411
		Clustering	500	0	0.3442	0.0039	0.2695	0.1668	0.2323	0.0098
	50	Crowding distance	500	0	0.4004	0.0052	0.4899	0.1323	0.6035	0.0361
		Clustering	500	0	0.3996	0.0056	0.3035	0.1342	0.2352	0.0056
	200	Crowding distance	500	0	0.4028	0.0034	0.4798	0.1557	0.6175	0.0380
		Clustering	500	0	0.4012	0.0043	0.2688	0.1299	0.2368	0.0106
Gr-PU	10	Crowding distance	500	0	0.1276	0.0025	0.4848	0.1598	0.5747	0.0557
		Clustering	500	0	0.1278	0.0016	0.3115	0.0949	0.2763	0.1114
	50	Crowding distance	500	0	0.2685	0.0037	0.4103	0.1432	0.5977	0.0541
		Clustering	500	0	0.2683	0.0043	0.2782	0.1535	0.2347	0.0185
	200	Crowding distance	500	0	0.3481	0.0036	0.4344	0.1863	0.6081	0.0426
		Clustering	500	0	0.3495	0.0040	0.2808	0.1303	0.2336	0.0110

throughout the entire Pareto-front. However, the crowding distance approach performs better compared to clustering approach in finding solutions towards convergence to the true Pareto-front in problem I.

Figure 12 describes the Pareto-optimal fronts of optimum values of k_c-E_c combinations for different composite systems when the viscosity is considered as a constraint in the multi-objective optimization process carried out using NSGA-II with crowding distance approach. The Pareto-fronts shown in Figure 12 are obtained in a particular run, and for these fronts the number of Pareto-optimal solutions, size of the dominated space and non-uniformity of the Pareto-fronts are listed in Table 5. For low viscosity level, it is found that the Pareto-optimal front of the Al-PU composite overlaps with that of the Gr-PU system. Thus, given a choice between the two combinations of composite systems, one cannot give preferences on any of them when the equivalent viscosity is below 10 Pas. But, for higher viscosity levels (more than 10 Pas), it has been found that the Pareto-front of the Gr-PU composite dominates the solutions of Al-PU composite systems towards the maximization of k_c and minimization E_c. Thus, given a choice between the two combinations of composite systems Gr-filled PU and Al-filled PU, the former is better particularly when the viscosity level is 50Pas and higher. Because one can achieve same thermal conductivity value with lower stiffness for a given values of other parameter in graphite-PU system compared to Al-PU system.

Table 4.Wilcoxon signed-rank test on the results obtained by twenty independent runs with different random seeds in problem I

Composite	Viscosity label (Pas)	Performance metrics	Number of non-zero difference of paired data ($A_{Crowding}$-$A_{Clustering}$)	Sum of the positive ranks (S_{PR})	Sum of the negative ranks (S_{NR})	Value of Test statistics, T (minimum of S_{PR} and S_{NR})	Critical value of Wilcoxon Test	Remark (based on two tailed significance levels of 0.05)
		S_d	19	106	84	84	46	T>46, Cannot reject H_0
	10	C	20	161	49	49	52	T<52, Reject H_0 (i.e., crowding distance approach is better than clustering approach)
		D	20	210	0	0	52	T<52, Reject H_0 (i.e., clustering approach is better than crowding distance approach)
		S_d	19	101.5	88.5	88.5	46	T>46, Cannot reject H_0
Al_PU	50	C	20	173	37	37		T<52, Reject H_0 (i.e., crowding distance approach is better)
		D	20	210	0	0	52	T<52, Reject H_0 (i.e., clustering approach is better)
		S_d	20	134	76	76	52	T>52, Cannot reject H_0
	200	C	20	174	36	36	52	T<52, Reject H_0 (i.e., crowding distance approach is better)
		D	20	210	0	0	52	T<52, Reject H_0 (i.e., clustering approach is better)
		S_d	20	118	92	92	52	T>52, Cannot reject H_0
	10	C	20	174	36	36	52	T<52, Reject H_0 (i.e., crowding distance approach is better)
		D	20	207	3	3	52	T<52, Reject H_0 (i.e., clustering approach is better)
		S_d	19	190	85.5	85.5	46	T>46, Cannot reject H_0
Gr_PU	50	C	20	166	44	44	52	T<52, Reject H_0 (i.e., crowding distance approach is better)
		D	20	210	0	0	52	T<52, Reject H_0 (i.e., clustering approach is better)
		S_d	20	68	142	68	52	T>52, Cannot reject H_0
	200	C	20	176	34	34	52	T<52, Reject H_0 (i.e., crowding distance approach is better)
		D	20	210	0	0	52	T<52, Reject H_0 (i.e., clustering approach is better)

Form the above studies; it is found that all the decision variables (P_s, S_f and T) are constant (P_s =68.884 micron; S_f=1.626 and T=24 °C) except V_f. In Figure 13, the variations of V_f with k_c of Al filled polyurethane composites for different viscosity levels are demonstrated. From Figure 13, V_f seems to be varying linearly with k_c with the following relationship:

Figure 12. Pareto-optimal fronts obtained using NSGA-II with crowding distance approach in optimization of k_c and E_c of polyurethane composites for different viscosity labels: (a) 10Pas (b) 50 Pas and (c) 200 Pas

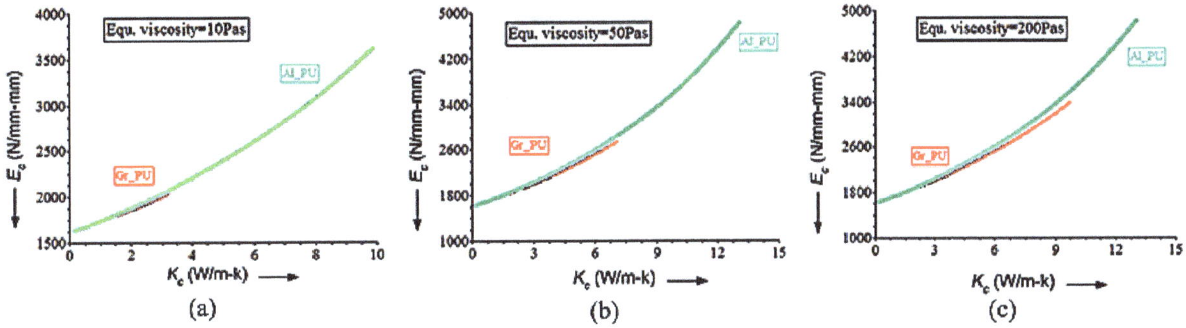

Table 5. The number of Pareto-optimal solutions, size of the dominated space and non-uniformity of the Pareto-optimal fronts shown in Figure 12

Composite	Viscosity label (Pas)	Number of Pareto-optimal solutions obtained	Size of the dominated space, $S_d(A)$	Non-uniformity of the Pareto-front, $D(A)$
Al_PU	10	500	0.3468	0.5958
	50	500	0.4057	0.6296
	200	500	0.4043	0.6271
Gr_PU	10	500	0.1297	0.5723
	50	500	0.2710	0.5276
	200	500	0.3512	0.6120

$$V_f = 2.892k_c - 0.4995 \tag{27}$$

Relationships between V_f and k_c are valid only for the Pareto-solutions, i.e. for some fixed values of other variables.

Similar observations in the variation of decision variables (as found in Al-filled PU composites) are noticed in the parametric study of Gr filler PU composites, which means all the decision variables (P_s, S_f and T) except V_f seem constant (P_s =68.8 micron; S_f=1.6 and T=24 °C). In Figure 14, the variations of V_f with k_c in Gr filled polyurethane composites for different viscosity levels are demonstrated and V_f seems to be varying linearly with k_c with the following relationships (Equation (28)):

$$V_f = 3.0994k_c - 0.4979 \tag{28}$$

In both the PU composites filled with Al and Gr particles, it has been found that the upper limit of V_f reaches lesser value for the lower viscosity level. Increase of upper limit of V_f is found more pronounced in the initial stage of increasing the viscosity level, which gradually diminishes. Later, it does not change for further increase of viscosity level. The increasing rates of V_f with viscosity are different for various filler materials. For smaller viscosity values, the upper limit of V_f in Gr-PU composite system is much

Figure 13. Variation of V_f with k_c of Al filled polyurethane composites for different viscosity labels: (a) 10 Pas (b) 50 Pas and (c) 200 Pas

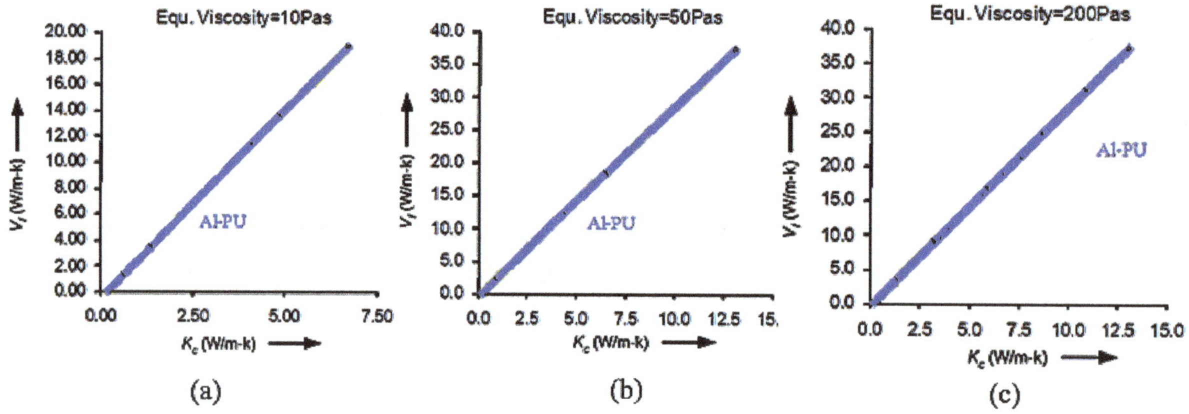

Figure 14. Variation of V_f with k_c of Gr filled polyurethane composites for different viscosity labels: (a) 10 Pas (b) 50 Pas and (c) 200 Pas

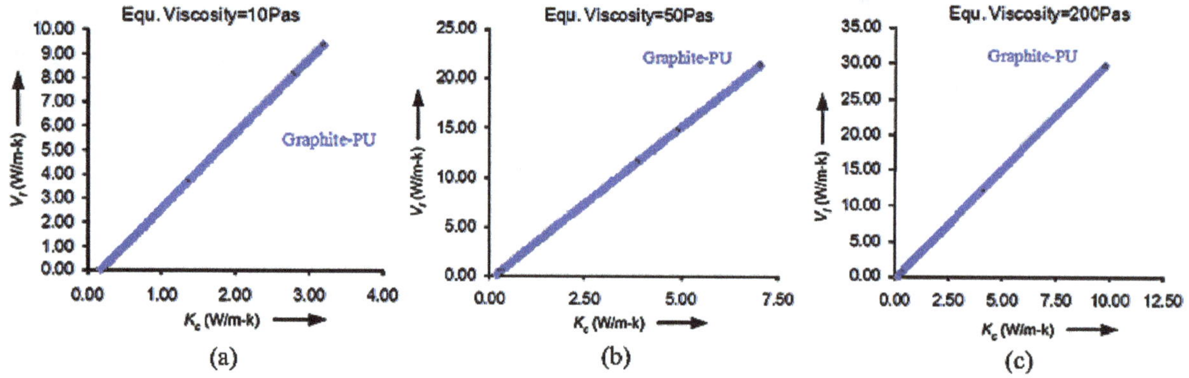

smaller than that in Al-PU composite. But for the higher viscosity (above 50 Pas), the upper limits of V_f in Gr-PU is marginally lower than that in Al-PU composite. Such analysis of the obtained trade-off solutions by a multi-objective optimization study is termed as the task of *innovization* (obtaining innovative principles through optimization) (Deb & Srinivasan, 2006) and is found to reveal salient problem knowledge in many engineering design problems that are difficult to achieve otherwise.

Multi-Objective Optimization for Solving Problem II

Solving the problem I endows one with a notion about the variation of different decision variables with the effective thermal conductivity of mould material. But it has been found that no particular Pareto-solution could be preferred from the optimization results since the Pareto-front changes with different constraint values of effective viscosity. For this reason, Problem II is considered, which is more complicated than Problem I since the effective viscosity is also assumed as an objective.

To analyze the stability of the algorithm, here also we have executed twenty trials of NGSA-II with crowding distance approach and clustering approach for different values of random seed. Table 6 includes the mean and standard deviation of the number of Pareto-optimal solutions, size of the dominated space, coverage of two Pareto-fronts and non-uniformity of the Pareto-front obtained by twenty independent runs with different random seeds for Al-PU and Gr-PU composites. It has been found that for all the trials, NSGA-II with both clustering and crowding distance approaches find 500 Pareto-optimal solutions. The quality of the Pareto-optimal solutions obtained with the two niching approaches into NSGA-II is measured by three qualitative metrics size of the dominated space, coverage of two Pareto-fronts and non-uniformity of the Pareto-front. The size of the dominated space, $S(A)$ is measured by normalizing the objectives taking minimum and maximum values of equivalent thermal conductivity, equivalent modulus of elasticity and equivalent viscosity in two composite systems which are shown in Table 7. From Table 6, it has been found that in both composite systems, the difference in mean values of $S(A)$ and $C(A, B)$ of the Pareto-fronts obtained by NSGA-II using clustering and crowding distance approaches are not significant, which indicates that performance of both the approaches are same in towards convergence to the Pareto-front. But, after comparing the values of $D(A)$ in Table 6 it has been found that the spread of the solutions found with NSGA-II with clustering approach is more uniform than with NSGA-II with crowding distance approach. Moreover, both the niching approaches into NSGA-II are found statistically stable as lower value of standard deviation of different performance metrics is observed.

Table 6. Statistical information of the solutions obtained by twenty independent runs with different random seeds in problem II

Composite	NSGA-II with	Number of Pareto-optimal solutions obtained		Size of the dominated space		Coverage of two Pareto-fronts		Non-uniformity of the Pareto-front	
		Mean	SD	Mean	SD	Mean	SD	Mean	SD
Al_PU	Crowding distance	500	0	0.3175	0.0023	0.1863	0.2355	0.6274	0.0564
	Clustering	500	0	0.3146	0.0044	0.1695	0.2559	0.2517	0.0475
Gr_PU	Crowding distance	500	0	0.3375	0.0045	0.1654	0.2107	0.6664	0.0639
	Clustering	500	0	0.3357	0.0064	0.1447	0.2306	0.2867	0.1517

SD=Standard deviation

Table 7. Minimum and maximum values of equivalent thermal conductivity, equivalent modulus of elasticity and equivalent viscosity considered in calculating size of the dominated space in Table 5

Composite	Equivalent thermal conductivity, W/m-k		Equivalent modulus of elasticity, N/mm^2		Equivalent viscosity, Pas	
	Minimum	Maximum	Minimum	Maximum	Minimum	Maximum
Al_PU	0.1	20	1000	6000	3	20
Gr_PU	0.1	20	1000	6000	3	100

Wilcoxon signed-rank test is carried out on the values of $S_d(A)$, $D(A)$ and $C(A, B)$ that are calculated based on the results obtained by NSGA-II with crowding distance and clustering approaches by twenty independent runs with different random seeds in problem II. The results of Wilcoxon signed-rank test are illustrated in Table 8. From the Table8, it observed that both crowding distance and clustering approaches are found equivalent towards obtaining convergence to the Pareto-front, but clustering approach is better than crowding distance approach to obtain a more uniform spread of solutions in Pareto-front.

The Pareto-optimal fronts for maximization of k_c, minimization of E_c and minimization η_c in PU composite systems reinforced with aluminum (Al) and graphite obtained by NSGA-II with clustering and crowding distance approach in a particular run are shown in Figure 15. For analyzing the multi-objective optimization results in designing of particle reinforced polyurethane mould materials for soft tooling process, we have considered the Pareto-fronts obtained by NSGA-II using clustering approach, and the size of the dominated space and non-uniformity of the Pareto-front are 0.317792 and 0.238658 in Al-PU composite and 0.335859 and 0.220445 in Gr-PU composite, respectively.

From Figure 15, it is observed that in both composite systems, E_c and η_c are correlated to each other, because it has been found that minimization of E_c also minimizes η_c. However, in k_c-E_c combination,

Table 8. Wilcoxon signed-rank test on the results obtained by twenty independent runs with different random seeds in problem II

Composite	Performance metrics	Number of non-zero difference of paired data ($A_{Crowding}$-$A_{Clustering}$)	Sum of the positive ranks (S_{PR})	Sum of the negative ranks (S_{NR})	Value of Test statistics, T (minimum of S_{PR} and S_{NR})	Critical value of Wilcoxon Test	Remark (based on two tailed significance levels of 0.05)
Al_PU	S_d	20	146	64	64	52	T>52, Cannot reject H_0 (i.e. both crowding distance and clustering approaches are equivalent)
	C	20	112	98	98	52	T>52, Cannot reject H_0 (i.e. both crowding distance and clustering approaches are equivalent)
	D	20	210	0	0	52	T<52, Reject H_0 (i.e., clustering approach is better than crowding distance approach)
Gr_PU	S_d	20	150	60	60	52	T>52, Cannot reject H_0 (i.e. both crowding distance and clustering approaches are equivalent)
	C	20	112	98	98	52	T>52, Cannot reject H_0 (i.e. both crowding distance and clustering approaches are equivalent)
	D	20	206	4	4	52	T<52, Reject H_0 (i.e., clustering approach is better than crowding distance approach)

Figure 15. Pareto-optimal fronts in optimization of k_c, E_c and η_c of PU composite system filled with (a) Al and (b) graphite

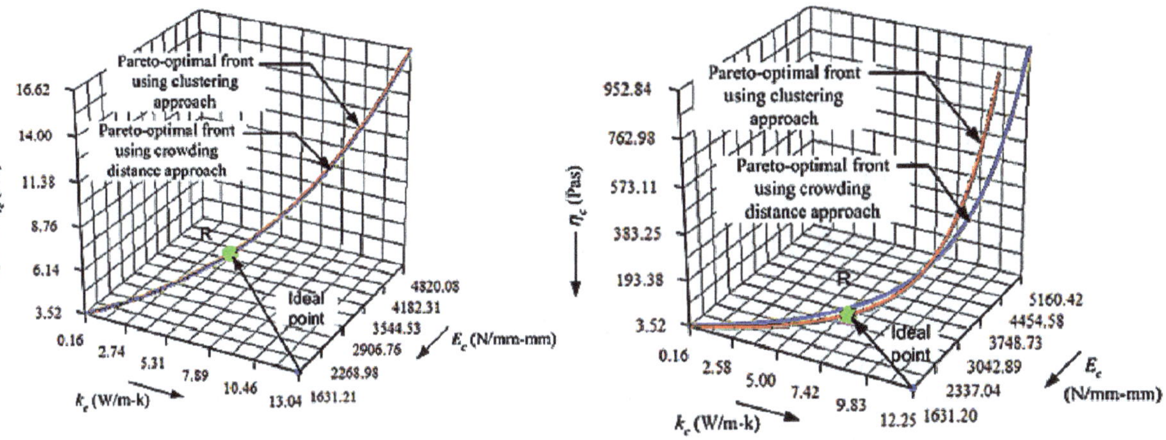

(a) (b)

E_c is not correlated with k_c. Similarly in k_c-η_c combination η_c is also not correlated to k_c. Thus, to find the optimum values of decision variables (P_s, S_f, V_f and T), any of the k_c-E_c or k_c-η_c can be considered.

Form the above study, it is also observed that all the decision variables (P_s, S_f and T) are constant (in case of Al filled PU system: P_s =68.9 micron, S_f=1.626 and T=24.26 °C and in case of graphite filled PU system: P_s =68.8 micron; S_f=1.6 and T=24.1 °C) except V_f. In Figures 16(a) and (b), the variations of V_f with k_c in both PU composites reinforced with Al and graphite, respectively are demonstrated. In Al-PU composite systems V_f seems to be varying linearly with k_c with the relationship (expressed in Equation (29)) as shown in Figure 16(a).

$$V_f = 2.8908k_c - 0.3935 \tag{29}$$

The variation of k_c with respect to V_f looks identical to that found in Equation(27), which suggests the existence of correlation among E_c and η_c in Al filled PU mould material composite systems.

Figure 16. Variation of V_f with k_c of (a) Al filled PU composites (b) graphite filled PU composites

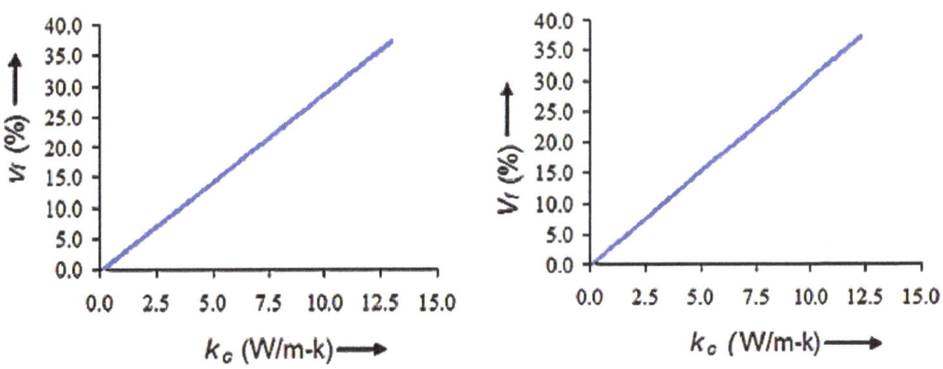

(a) (b)

The relationship of V_f with k_c as shown in Figure 16(b) for the graphite-PU composite systems is expressed by the Equation (30).

$$V_f = 3.0737k_c - 0.414 \tag{30}$$

The existence of correlation among E_c and η_c in graphite filled PU mould material composite systems is proved by observing that the Equation (30) resembles with Equation (28).

Selection of Preferred Solution(s) Based on MCDM

Although there are advantages of knowing the range of each objective for Pareto-optimality and the shape of the Pareto-optimal frontier itself in a problem for an adequate decision-making, the task of choosing a single preferred Pareto-optimal solution is an important task because the practitioners finally adopt the chosen preferred solution for implementation in industrial applications. Now-a-days various MCDM techniques are available and following a classification by Veldhuizen and Lamont (2000) the articulation of preferences may be done either before search (a priori) or after search (a posteriori) or even during search (progressive) of the optimization process. In the present problem, we have adopted two such MCDM techniques after the optimization process: reference point based technique (Miettinen, 1999) to identify a single preferred solution, and selection of a set of (few) preferred solutions by identifying knee-zone (Branke, Deb, Dierolf, & Osswald, 2004) on the Pareto-front based on trade-off and then DM chooses a solution from the knee zone based on higher level information. Typical other methods of selecting a preferred solution from the set of Pareto-optimal solutions are the compromise programming approach (Yu, 1973), the marginal rate of substitution approach (Miettinen, 1999), the pseudo-weight vector approach (Deb, 2001), etc.

DM can decide a reference point based on various aspects concerning to the problem domain. A most well-accepted reference point suitable to all kinds of multi-objective problem is the ideal point. An ideal point is referred to the vector of individual optimal objective values. In the present application, the ideal point is considered as a reference point and a single preferred solution is determined by finding the solution on the Pareto-front having minimum (Euclidian) distance from the reference point.

After a thorough study on the shape of Pareto-optimal front, sometimes it is found that there may be solutions where a small improvement in one objective would lead to a large deterioration in any of the other objectives. These solutions are called 'knees'. The area of the Pareto-front surrounded by such solutions is termed as knee. Without any knowledge about the user's preferences, it may be argued that the region around that knee is most likely to be interesting for the DM.

Ideal point of Al filled PU composite system is identified from the obtained Pareto-optimal front shown in Figure 15(a), as follows: k_c=13.009; E_c=1632.96; η_c =3.52361. The preferred Pareto-optimal solution obtained based on this reference point is designated by 'R' as shown in Figure 15(a) and the values of corresponding objective functions and decision variables are cited in sl. no. 1 in Table 8. The preferred solution, R is determined by considering each of the three properties of Al filled PU composite k_c, E_c and η_c play as an objective in the decision-making procedure. In previous Section, it is seen that there exists a correlation between the objectives, E_c and η_c. Therefore, the preferred solution can also be decided by taking the projection of the Pareto-optimal front obtained in 3D into either k_c-E_c or k_c-η_c plane. The preferred Pareto-optimal solutions which are the solutions with minimum distance from the

Figure 17. Projection of Pareto-optimal front of Al filled PU composite system as shown in Figure 15(a) in (a) k_c-η_c plane (b) E_c-k_c plane and (c) E_c-η_c plane

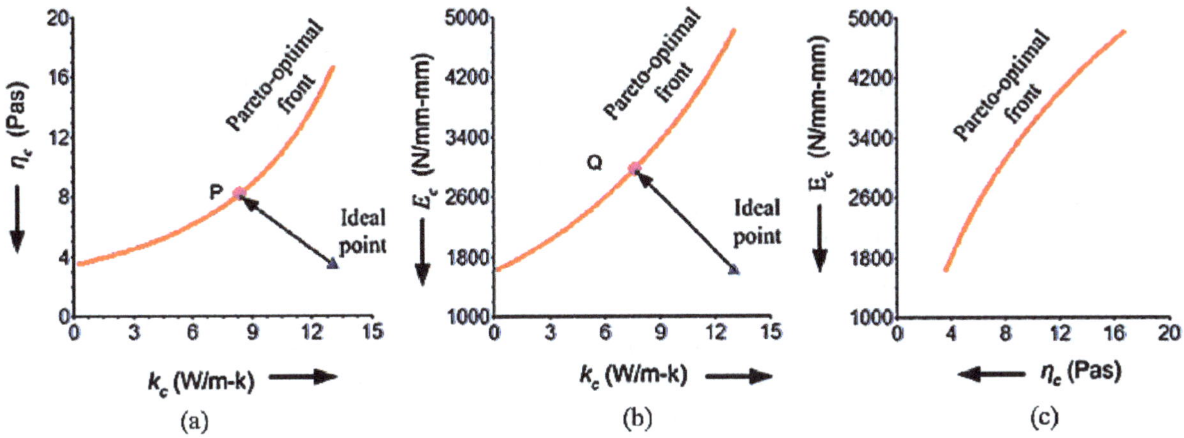

Table 9. List of optimal solutions based on minimum distance from ideal point: Al-PU composite

Sl No.	Ideal point	Optimal values of objectives			Optimal values of decision variables				Preferred Pareto-optimal solution
		k_c	E_c	η_c	P_s	S_f	V_f	T (ºC)	
1	k_c=13.009; E_c=1632.96; η_c =3.52361	6.64504	2742.22	6.6380	68.8854	1.6264	18.844	24.265	Denoted as 'R' in Figure 15(a)
2	k_c=13.009; η_c =3.52361	8.34036	3167.60	8.1567	68.8854	1.6262	23.744	24.259	Denoted as 'P' in Figure 17
3	k_c=13.009; E_c=1632.96	7.59173	2970.56	7.4267	68.8853	1.6266	21.578	24.258	Denoted as 'Q' in Figure 17

ideal (reference) point obtained in the k_c-η_c and k_c-E_c planes are referred to 'P' and 'Q', respectively as shown in Figure 17(a) and Figure 17(b), respectively. The related objective function values and decision variable values corresponding to the preferred points 'P' and 'Q'are enlisted in Table 9. From the Figures 17(a) and 17(b), it is observed that knee-zone is not clearly visible in the entire Pareto-optimal front (except at both the extreme ends) in both the k_c-η_c and k_c-E_c planes. The correlation between E_c and η_c is characterized by the curve as shown in Figure 17(c).

Ideal point (k_c=12.2413; E_c=1633.37; η_c =3.54156) of Gr filled PU composite system considering all the three properties k_c, E_c and η_c as objectives is identified from the obtained Pareto-optimal front shown in Figure 15(b). The preferred Pareto-optimal solution, 'R' is obtained based on this reference point as shown in Figure 15(b).

Like Al filled PU composite system, the preferred Pareto-optimal solutions ('R') obtained based on reference point (Ideal point: k_c=12.2413; E_c=1633.37; η_c=3.54156) in Gr filled PU composite systems are shown in Figure 15(b). The values of corresponding objective functions and decision variables are cited in Table 10. There exist a correlation between the objectives, E_c and η_c as found in previous Section. Thus, the preferred solution is decided by taking the projection of the Pareto-optimal front obtained in 3D into either k_c-E_c or k_c-η_c plane. The preferred Pareto-optimal solutions which are the solutions with minimum distance from the ideal (reference) point obtained in the k_c-η_c and k_c-E_c planes are referred to 'P' and 'Q', respectively as shown in Figure 18(a) and Figure 18(b), respectively. The related objective function values and decision variable values corresponding to the preferred points 'P' and 'Q' are enlisted in Table 10. From the Figure 18(b) no such knee-zone is found in the Pareto-optimal front in k_c-E_c plane. However, in Figure 18(a), it is observed knee-zone (which is marked by dotted ellipse) is clearly visible in the middle of Pareto-optimal front in the k_c-η_c combination. The solutions in this zone are characterized by the fact that a small improvement in either objective will cause a large deterioration in the other objective, which makes moving in either direction not very attractive. If we assume linear preference functions, and (due to the lack of any other information) also assume that each preference function (k_c and η_c) is likely in same fashion, the solutions at the knee are most likely to be the optimal choice of the practitioners. Here the preferred point 'P' is obtained by calculating the minimum distance from the ideal point, which is positioned on the right end of the dotted ellipse, which suggests that the extent of knee-zone on the right side is limited to the preferred point 'P'. The correlation between E_c and η_c is found by the curve as shown in Figure 18(c).

Table 10. List of optimal solutions based on minimum distance from ideal point: Graphite-PU composite

Sl No.	Ideal point	Optimal values of objectives			Optimal values of decision variables				Preferred Pareto-optimal solution
		k_c	E_c	η_c	P_s	S_f	V_f	T (°C)	
1	k_c=12.2413; E_c=1633.37; η_c =3.54156	6.9136	2695.78	46.7194	68.8845	1.6256	20.874	24.021	Denoted as 'R' in Figure 15(b)
2	k_c=12.2413; η_c =3.54156	9.5310	3318.99	175.4140	68.8845	1.6260	28.901	24.023	Denoted as 'P' in Figure 18
3	k_c=12.2413; E_c=1633.37	7.0321	2720.74	49.3424	68.8853	1.6258	21.236	24.019	Denoted as 'Q' in Figure 18

Figure 18. Projection of Pareto-optimal front of graphite filled PU composite system as shown in Figure 15(b) in (a) k_c-η_c plane (b) E_c-k_c plane and (c) E_c- η_c plane

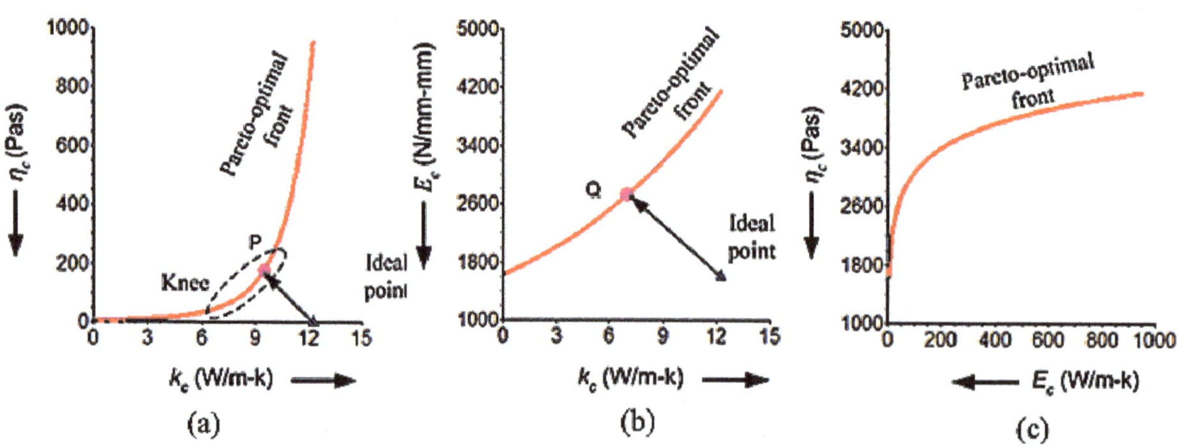

VALIDATION OF RESULTS WITH REAL INDUSTRIAL APPLICATION

The effect of cooling time in soft tooling process is experimentally investigated in order to validate the results of optimum values of design parameters in developing particulate reinforced PU mould materials as obtained through multi-objective evolutionary algorithms. The present experimental study of cooling time is carried out with the PU mould material reinforced with Al particle. Validation of results through the optimization process is made by comparing the cooling time required using Al filled composite mould material with that of using mould material without reinforcement for producing a cubical shaped (size: 50 mm x 50 mm x 50 mm) wax component.

In the present study, investment casting wax (A7-11) supplied by Blayson Olefines Ltd, UK is used. The wax consists of natural wax, synthetic wax and natural resins. The melting point of wax is 55°C – 70°C and the boiling point/flash point is greater than 200°C. The melt wax possesses viscosity of 0.6 Pas at around 100 °C.

Preparation of Mould and Manufacturing of Wax Component

A cubical shaped RP component made using Stereo Lithography Apparatus (SLA) is used as a pattern for making the mould. The pattern is finished to a desired quality as required for the final wax component. The RP pattern is suspended by teflon rod in an enclosed box (made of perspex sheet) and necessary arrangement is made to provide passage in the mould box for pouring the molten wax into the mould box. The weights of part-A and part-B of liquid polymer (PU) and Al filler are calculated based on the respective densities of these components for a certain volume fraction of filler in the composite. The part A (liquid polymer) is first mixed with filler particles in a container. To assure a uniform mixing, stirring of liquid polymer and gradual pouring of filler particles is done simultaneously. Once, the uniform mixing of liquid polymer and filler particles is achieved, the hardener (Part B) is poured into the mixture of liquid polymer and filler, and stirred it properly further for sometimes to attain the particle reinforced mould material. Before pouring the mould material into mould box, release agent is sprayed thoroughly inside the mould box. In the box, the (liquid) mixture of Al particle reinforced flexible mould material is poured all around the pattern. After that, the box with composite mould material is placed inside vacuum casting machine to eliminate any traces of entrapped air in the mould, later it is allowed to cure in room temperature. Gradually, it sets with time and becomes a solid rubbery mass with the pattern inside. The mould box is then cut along a suitable parting line and the RP pattern is removed from within the mould box. Two thermocouple wires are placed on the upper part of the mould box for

Figure 19. Mould box prepared with Al filled polyurethane

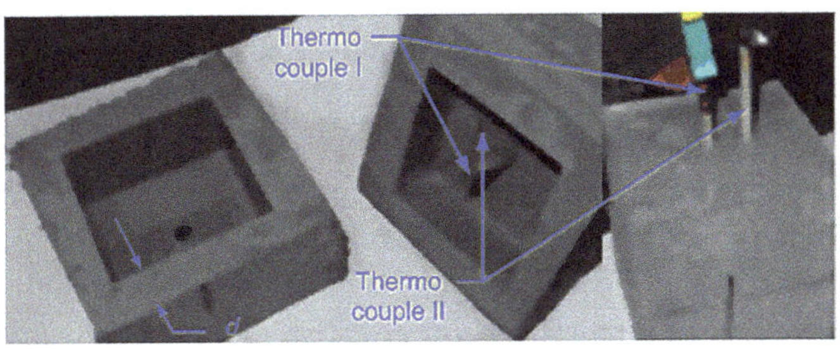

measuring the temperatures. Then, both parts of the mould box are assembled in proper position with the help of adhesive tape (as shown in Figure 19). Once the mould is prepared, liquid wax is poured into the mould and cooling/solidification of liquid wax is done in room temperature.

Experimental Setup

Once the liquid wax is solidified, both parts of the mould box are disassembled by removing the adhesive tape and the wax component is taken out. After cleaning the mould box, it becomes ready for another casting. It is noticed that the surface of wax component solidifies faster than the core (inside) of the component. Depending on the size/shape of component, a certain depth of solidification from the surface may be sufficient so that the wax component can be taken out from the mould box without any damage. Industrial people follow such technique frequently for shortening the processing time to increase the productivity. Thus, in this experimental study, the measurements of temperature are taken at two distinct positions of the wax component, at the center and at a depth of 5 mm from the surface of component. Measurement of temperatures is taken through thermocouple system (as shown in Figure 19). The readings of temperatures are collected in an interval of 1 sec through data acquisition system.

Results and Discussions

The experimentation is carried out with the Al particle reinforced PU mould material composites. A very close to the design value of volume fraction of filler as found in Table 9, i.e., 25% is used to develop the particle reinforced mould material. The behaviors of cooling rate of the cubical shaped wax component, where the mould wall thickness is kept as (d=) 30 mm and the outside temperature of mould box is the ambient condition (25°C), are demonstrated here. The variation of temperatures with time at the location of 5 mm deep from the surface and the core of wax component using PU mould material are plotted in Figure 20(a). In Figure 20(b), variations of temperatures at the same positions of wax component using Al particle reinforced PU mould material are shown. In the experimental study of cooling time in Soft Tooling process, the temperature measurements were carried out through data acquisition system using thermocouple wires. The data acquisition system starts to take reading of temperature before pouring the molten wax into the mould cavity. For this reason, a sudden increase in temperature is noticed in both the Figures 20(a)-(b). Moreover, during experimentation, measurement of temperature and pouring of molten wax into the mould cavity were carried out simultaneously. As a result, more or less in all cases, disturbances of recording of temperature through thermo-couple have occurred. In both the Figures 20(a)-(b), it is noticed that initially cooling rate near to surface of component is much faster than that at the core. This takes place as a result of the heat near the wax pattern being dissipated only through the mould wall. Whereas heat at the core of wax pattern flows through the mould wall as well as (solidified) wax and the thermal conductivity of wax is quite lower than the mould material. After a long time, both the temperatures gradually come to the atmospheric one. The temperature of liquid wax during pouring in both the moulds made of PU and Al particle reinforced PU is kept almost same as 85°C. Figure 20(a) demonstrates that to reach the temperature for instance 40°C at a depth of 5mm from the surface of wax component, it takes around 80 min in case of PU mould, whereas only 25 min is required in case of Al particle reinforced PU as found in Figure 20(b). Thus, it is revealed that the cooling time is reduced significantly in soft tooling process using particle-reinforcement with mould materials. This has happened due to the increase of effective thermal conductivity of mould material.

Figure 20. Cooling rate of wax pattern using PU mould of (d=) 30 mm wall thickness at room temperature (a) without any reinforcement (b) Al particle reinforced

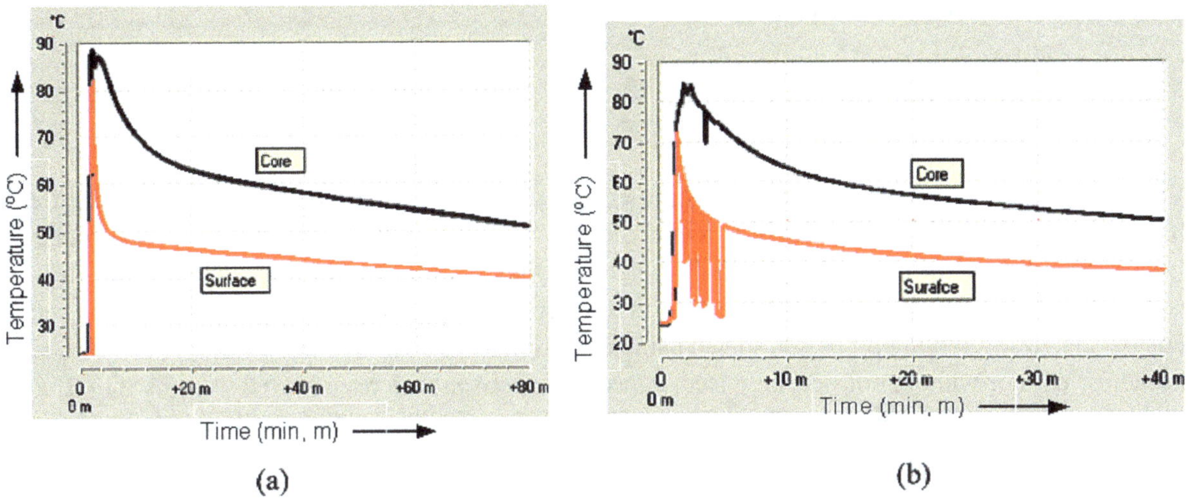

However, besides the effective thermal conductivity of mould material, the cooling time also depends on the temperature difference between the inside and outside of mould (i.e., the ambient temperature and the temperature of melt wax inside the mould), and thickness of mould wall.

The minimum thickness of mould is decided based on the shape and size of the wax component to be manufactured. On the other hand, cooling/solidification of wax in ambient temperature are the most desirable technique in industry due to no additional expenditure. Therefore, the only controllable parameter in ST process is the thermal conductivity of mould material which practitioners can improve in order to solidify the wax component with a minimum time.

FUTURE RESEARCH DIRECTIONS

In this chapter, the geometrical configuration of the pattern which will be produced through soft tooling process is not considered. But, in practical the stiffness of the mould is to be minimum in order to have an ease of taking out the pattern from the mould box. The stiffness of mould depends on the modulus of elasticity of mould material as well as the size of the mould which is related to the pattern's geometrical configuration. Therefore better results may be obtained instead of considering only modulus of elasticity of mould material, the configuration of the pattern as well specifically for bigger size pattern. In the present study, cooling time in soft tooling process is significantly reduced by particle reinforcement with mould material. Moreover, it may further be reduced using forced-cooling system during solidification of wax/plastic pattern. Use of multi-phase polymeric composite mould materials may provide better results in soft-tooling process. The results of present research work may be utilized for further research activities, namely in estimating the cooling time, the Fuzzy Rule Based Modelling of effective thermal conductivity may useful. It should be noted that the model of equivalent thermal conductivity is applicable only to the 2-phase particle reinforced polymer composite. However, the same approach and technique can be adopted for constructing generic model of equivalent thermal conductivity of multi-phase polymer composites though some extra parameters may be involved in that case. But before constructing the model, one should assure that sufficient experimental data of equivalent thermal conductivity of multi-phase polymer composites should be available in hand.

CONCLUSION

A useful approach is presented in this chapter to design and develop a suitable particle reinforced flexible mould material for soft tooling process. In order to reduce the cycle time of soft tooling process, the primary motivation is to solve the multi-objective optimization problem associated consisting of various conflicting mould material properties. In this approach, the multi-objective optimization problem associated soft tooling process is solved using evolutionary algorithms. In this regard, two different problems, with or without considering viscosity as an objective, have been solved using NSGA-II and it has been observed that modulus of elasticity and viscosity of particle reinforced mould material are correlated. Pareto-optimal fronts obtained by solving the multi-objective problems are presented and the roles of various decision variables are analysed using a recently proposed *innovization* technique. The preferred Pareto-optimal solutions are selected by using the reference point MCDM technique as well as by locating `knees' on the Pareto-optimal fronts, if exists. Experimental studies of cooling time with an industrial application revealed that the cooling time is reduced significantly in the soft tooling process using particle-reinforcement with mould materials.

RFERENCES

Agari, Y., & Uno, T. (1986). Estimation on thermal conductivities of filled polymers. *Journal of Applied Polymer Science*, *32*(7), 5705–5712. doi:10.1002/app.1986.070320702

Agrawal, R., Saxena, N. S., Mathew, G., Thomas, S., & Sharma, K. B. (2000). Effective thermal conductivity of three-phase styrene butadiene composites. *Journal of Applied Polymer Science*, *76*(12), 1799–1803. doi:10.1002/(SICI)1097-4628(20000620)76:12<1799::AID-APP10>3.0.CO;2-D

Boudenne, A., Ibos, L., Fois, M., Gehin, E., & Majeste, J. C. (2004). Thermophysical properties of polypropylene/aluminium composites. *Journal of Polymer Science. Part B, Polymer Physics*, *42*(4), 722–732. doi:10.1002/polb.10713

Branke, J., Deb, K., Dierolf, H., & Osswald, M. (2004). Finding knees in multi-objective optimization. LNCS, 3242, 722–731. doi:10.1007/978-3-540-30217-9_73

Cohon, J. L. (1985). Multicriteria programming: brief review and application. In J. S. Gero (Ed.), *Design Optimization* (pp. 163–191). Orlando, FL: Academic Press.

Cunha, A. G., Oliviera, P., & Covas, J. (1997). Use of genetic algorithms in multicriteria optimization to solve industrial problems. In T. B¨ack (Ed.), *Proceedings of the Seventh International Conference on Genetic Algorithms* (pp. 682-688).

De Jong, K. A. (1975). An analysis of the bevavior of a class of genetic adaptive systems (Ph. D. Thesis, Ann Arbor, MI: University of Michigan). Dissertation Abstracts International, 36(10), 5140B (University Microfilms No. 76-9381)

Deb, K. (1998). *Multi-objective genetic algorithms: Problem difficulties and construction of test functions.* Technical Report No. CI-49/98. Department of Computer Science/XI, University of Dortmund, Germany.

Deb, K. (2001). *Multi-Objective Optimization using Evolutionary Algorithms*. John Wiley & Sons Ltd.

Deb, K., Pratap, A., Agarwal, S., & Meyarivan, T. (2002). A fast elitist non-dominated sorting genetic algorithm for multi-objective optimisation: NSGA-II. *IEEE Transactions on Evolutionary Computation*, *6*(2), 182–197. doi:10.1109/4235.996017

Deb, K., Sinha, A., Korhonen, P., & Wallenius, J. (2010). An Interactive Evolutionary Multi-Objective Optimization Method Based on Progressively Approximated Value Functions. *IEEE Transactions on Evolutionary Computation*, *14*(5), 723–739. doi:10.1109/TEVC.2010.2064323

Deb, K., & Srinivasan, A. (2006). Innovization: Innovating design principles through optimization. In *Proceedings of the Genetic and Evolutionary Computation Conference (GECCO-2006)* (pp. 1629-1636), New York: The Association of Computing Machinery (ACM).

Edgeworth, F. Y. (1881). *Mathematical Psychics*. London: Keagan.

Fonseca, C. M., & Fleming, P. J. (1993). Genetic algorithms for multiobjective optimization: Formulation, discussion and generalization. In S. Forrest (Ed.), *Proceedings of the Fifth International Conference on Genetic Algorithms*(pp. 416-423). San Mateo, CA:Morgan Kaufmann.

Fonseca, C. M., & Fleming, P. J. (1995). An overview of evolutionary algorithms in multiobjective optimization. *Evolutionary Computation*, *3*(1), 1–16. doi:10.1162/evco.1995.3.1.1

Fonseca, C. M., & Fleming, P. J. (1998a). Multiobjective optimization and multiple constraint handling with evolutionary algorithms—part ii: Application example. *IEEE Transactions on Systems, Man, and Cybernetics*, *28*(1), 38–47. doi:10.1109/3468.650320

Fonseca, C. M., & Fleming, P. J. (1998b). Multiobjective optimization and multiple constraint handling with evolutionary algorithms—part i: A unified formulation. *IEEE Transactions on Systems, Man, and Cybernetics*, *28*(1), 26–37. doi:10.1109/3468.650319

Fourman, M. P. (1985). Compaction of symbolic layout using genetic algorithms, In J. J. Grefenstette (Ed.), *Proceedings of an International Conference on Genetic Algorithms and Their Applications* (141-153).

Goldberg, D. E. (1989). *Genetic algorithms for search, Optimization, and Machine Learning*. Reading, MA: Addision-Wesley.

Goldberg, D. E., & Richardson, J. (1987). Genetic algorithms with sharing for multimodal function optimization. In J. J. Grefenstette (Ed.), *Genetic Algorithms and their Applications:Proceedings of the Second International Conference on Genetic Algorithms* (pp. 41-49).

Hajela, P., & Lin, C.-Y. (1992). Genetic search strategies in multicriterion optimal design. *Structural Optimization*, *4*(2), 99–107. doi:10.1007/BF01759923

Hashin, Z., & Shtrikman, S. (1963). A variational approach to the theory of the elastic behaviour of multiphase materials. *Journal of the Mechanics and Physics of Solids*, *11*(2), 127–140. doi:10.1016/0022-5096(63)90060-7

Horn, J. (1997). F1.9 multicriteria decision making. In T. B¨ack, D. B. Fogel, & Z. Michalewicz (Eds.), *Handbook of Evolutionary Computation*. Bristol, UK: Institute of Physics Publishing. doi:10.1887/0750308958/b386c85

Horn, J., Nafpliotis, N., & Goldberg, D. E. (1994). A niched pareto genetic algorithm for multiobjective optimization. In *Proceedings of the First IEEE Conference on Evolutionary Computation,IEEE World Congress on Computational Computation* (pp. 82-87).

Hwang, C. L., & Masud, A. S. M. (1979). *Multiple Objectives Decision Making-Methods and Applications: A State-of-the-art Survey*. Berlin: Springer-Verlag. doi:10.1007/978-3-642-45511-7

Iqbal, A., Frormann, L., Saleem, A., & Ishaq, M. (2007). The effect of filler concentration on the electrical, thermal, and mechanical properties of carbon fiber-reinforced poly (styrene-co-acrylonitrile) composites. *Polymer Composites*, *28*(2), 186–197. doi:10.1002/pc.20253

Ishibuchi, H., & Murata, T. (1996). Multi-objective genetic local search algorithm. In *Proceedings of IEEE International Conference on Evolutionary Computation (ICEC'96)* (pp. 119-124). doi:10.1109/ICEC.1996.542345

Ishida, H., & Rimdusit, S. (1998). Very high thermal conductivity obtained by boron nitride-filled polybenzoxazine. *Thermochimica Acta*, *320*(1-2), 177–186. doi:10.1016/S0040-6031(98)00463-8

Keith, J. M., King, J. A., Lenhart, K. M., & Zimny, B. (2007). Thermal conductivity models for carbon/liquid crystal polymer composites. *Journal of Applied Polymer Science*, *105*(6), 3309–3316. doi:10.1002/app.26607

Kim, S. W., Choi, B., Lee, S. H., & Kang, K. H. (2008). Measurement of thermophysical properties of particulate-filled polymer composites. *High Temperatures. High Pressures*, *37*(1), 21–30.

Krieger, I. M., & Dougherty, T. J. (1959). A mechanism for non-newtonian flow in suspensions of rigid spheres. *Transactions of the Society of Rheology*, *3*(1), 137–152. doi:10.1122/1.548848

Kursawe, F. (1991). A variant of evolution strategies for vector optimization. In H.-P. Schwefel & R. M¨anner (Eds.), *Parallel Problem Solving from Nature —Proc. 1st Workshop PPSN* (pp. 193-197). Berlin: Springer. doi:10.1007/BFb0029752

Laumanns, M., Rudolph, G., & Schwefel, H.-P. (1998). A spatial predatorprey approach to multi-objective optimization: A preliminary study. In A. E. Eiben, T. B¨ack, M. Schoenauer, & H.-P. Schwefel (Eds.), *Fifth International Conference on Parallel Problem Solving from Nature (PPSNV)* (pp. 241-249). doi:10.1007/BFb0056867

Lewis, T. B., & Nielsen, L. E. (1970). Dynamic mechanical properties of particulate-filled composites. *Journal of Applied Polymer Science*, *14*(6), 1449–1471. doi:10.1002/app.1970.070140604

Lielens, G., Pirotte, P., Couniot, A., Dupret, F., & Keunings, R. (1998). Prediction of thermo-mechanical properties for compression moulded composites. *Composites. Part A, Applied Science and Manufacturing*, *29*(1-2), 63–70. doi:10.1016/S1359-835X(97)00039-0

Lis, J., & Eiben, A. E. (1997). A multi-sexual genetic algorithm for multiobjective optimization. In *Proceedings of 1997IEEE International Conference on Evolutionary Computation (ICEC'97)* (pp. 59-64). doi:10.1109/ICEC.1997.592269

Lopes, C. M. A., & Felisberti, M. I. (2004). Thermal conductivity of PET/(LDPE/Al) composites determined by MDSC. *Polymer Testing*, *23*(6), 637–643. doi:10.1016/j.polymertesting.2004.01.013

Mamunya, Y. P., Davydenko, V. V., Pissis, P., & Lebedev, E. V. (2002). Electrical and thermal conductivity of polymers filled with metal powders. *European Polymer Journal*, *38*(9), 1887–1897. doi:10.1016/S0014-3057(02)00064-2

Miettinen, K. (1999). *Nonlinear multi-objective optimization*. Boston: Kluwer.

Miller, M. G., Keith, J. M., King, J. A., Edwards, B. J., Klinkenberg, N., & Schiraldi, D. A. (2006). Measuring thermal conductivities of anisotropic synthetic graphite-liquid crystal polymer composites. *Polymer Composites*, *27*(4), 388–394. doi:10.1002/pc.20231

Mu, Q., Feng, S., & Diao, G. (2007). Thermal conductivity of silicone rubber filled with ZnO. *Polymer Composites*, *28*(2), 125–130. doi:10.1002/pc.20276

Nandi, A. K. (2006). TSK-Type FLC using a combined LR and GA: Surface roughness prediction in ultraprecision turning. *Journal of Materials Processing Technology*, *178*(1-3), 200–210. doi:10.1016/j.jmatprotec.2006.03.164

Nandi, A. K., Datta, S., & Deb, K. (2011). Investigating the role of non-metallic fillers in particulate reinforced flexible mould material composites using evolutionary algorithms. *Materials and Manufacturing Processes*, *26*(3), 541–549. doi:10.1080/10426914.2011.560221

Nandi, A.K., Datta, S., & Deb, K., & Orkas, j. (2013). A Genetic Fuzzy Based Modeling of Effective Thermal Conductivity for Polymer Composites. *Journal of Intelligent and Fuzzy Systems*, *25*, 259–270.

Nandi, A. K., Datta, S., & Orkus, J. (2012). Effective properties of particle reinforced flexible mould material towards reducing cooling time in soft tooling process. *Journal of Applied Polymer Science*, *124*(3), 2567–2581. doi:10.1002/app.34654

Nandi, A. K., Deb, K., Ganguly, S., & Datta, S. (2012). Investigating the role of metallic fillers in particulate reinforced flexible mould material composites using evolutionary algorithms. *Applied Soft Computing*, *12*(1), 28–39. doi:10.1016/j.asoc.2011.08.059

Nandi, A. K., Vesterinen, A., Cingi, C., Seppala, J., & Orkas, J. (2010). Studies on equivalent viscosity of particle-reinforced flexible mold materials used in soft tooling process. *Journal of Reinforced Plastics and Composites*, *29*(14), 2081–2098. doi:10.1177/0731684409347808

Ng, H. Y., Lu, X., & Lau, S. K. (2005). Thermal conductivity of boron nitride-filled thermoplastics: Effect of filler characteristics and composites processing conditions. *Polymer Composites*, *26*(6), 778–790. doi:10.1002/pc.20151

Obayashi, S., Takahashi, S., & Takeguchi, Y. (1998). Niching and elitist models for MOGAs. In A. E. Eiben, T. B¨ack, M. Schoenauer, & H.-P. Schwefel (Eds.), *Fifth International Conference on Parallel Problem Solving from Nature (PPSN-V)* (pp. 260-269). doi:10.1007/BFb0056869

Oliveira, F. A., Alves, N., Giacometti, J. A., Constantino, C. J. L., Mattoso, L. H. C., Balan, A. M. O. A., & Job, A. E. (2007). Study of the thermomechanical and electrical properties of conducting composites containing natural rubber and carbon black. *Journal of Applied Polymer Science*, *106*(2), 1001–1006. doi:10.1002/app.26689

Pareto, V. (1896). *CoursD'Economie Politique* (2 vols.). Lausanne: F. Rouge.

Parks, G. T., & Miller, I. (1998). Selective breeding in a multiobjective genetic algorithm. In A. E. Eiben, T. B¨ack, M. Schoenauer, & H.-P. Schwefel (Eds.), *Fifth International Conference on Parallel Problem Solving from Nature (PPSN-V)* (pp. 250-259). doi:10.1007/BFb0056868

Razzaq, M. Y., & Frormann, L. (2007). Thermomechanical studies of aluminium nitride filled shape memory polymer composites. *Polymer Composites*, *28*(3), 287–293. doi:10.1002/pc.20283

Rosochowski, A., & Matuszak, A. (2000). Rapid tooling: The state of the art. *Journal of Materials Processing Technology*, *106*(1-3), 191–198. doi:10.1016/S0924-0136(00)00613-0

Rudolph, G. (1998). On a multi-objective evolutionary algorithm and its convergence to the pareto set. In *IEEE International Conference on Evolutionary Computation (ICEC'98)* (pp. 511-516). doi:10.1109/ICEC.1998.700081

Ryan, C. (1995). Niche and species formation in genetic algorithms. In L. Chambers (Ed.), Practical Handbook of Genetic Algorithms (pp. 57-74). Academic Press.

Schaffer, J. D. (1984). *Multiple Objective Optimization with Vector Evaluated Genetic Algorithms*. (Ph. D. thesis). Vanderbilt University.

Schaffer, J. D. (1985). Multiple objective optimization with vector evaluated genetic algorithms, In J. J. Grefenstette (Ed.), *Proceedings of an International Conference on Genetic Algorithms and Their Applications* (pp. 93-100).

Siegel, S. (1956). *Non-parametric statistics for the behavioral sciences*. New York: McGraw-Hill.

Srinivas, N., & Deb, K. (1994). Multiobjective optimization using nondominated sorting in genetic algorithms. *Evolutionary Computation*, *2*(3), 221–248. doi:10.1162/evco.1994.2.3.221

Stadler, W. (1988). Fundamentals of multicriteria optimization. In W. Stadler (Ed.), *Multicriteria Optimization in Engineering and the Sciences* (pp. 1–25). New York, NY: Plenum Press. doi:10.1007/978-1-4899-3734-6_1

Steuer, R. E. (1986). *Multiple Criteria Optimization: Theory, Computation, and Application*. New York: Wiley.

Subodh, G., Manjusha, M. V., Philip, J., & Sebastian, M. T. (2008). Thermal properties of polytetrafluoroethylene/Sr2Ce2Ti5O16 polymer/ceramic composites. *Journal of Applied Polymer Science*, *108*(3), 1716–1721. doi:10.1002/app.27606

Sundstrom, D.W., & Chen, S.Y. (1970). Thermal Conductivity of Reinforced Plastics. *Journal of Composite Materials, 4*(1), 113-117.

Tsukuda, R., Sumimoto, S., & Ozawa, T. (1997). Thermal conductivity and heat capacity of ABS resin composites. *Journal of Applied Polymer Science*, *63*(10), 1279–1286. doi:10.1002/(SICI)1097-4628(19970307)63:10<1279::AID-APP6>3.0.CO;2-H

Valenzuela-Rend'on, M., & Uresti-Charre, E. (1997). A non-generational genetic algorithm for multiobjective optimization, In T. B¨ack (Ed.), *Proceedings of the Seventh International Conference on Genetic Algorithms* (pp. 658-665), San Francisco, CA:Morgan Kaufmann.

Veldhuizen, D. A. V., & Lamont, G. B. (1998). Evolutionary computation and convergence to a pareto front. In J. R. Koza, W. Banzhaf, K. Chellapilla, K. Deb, M. Dorigo, D. B. Fogel, M. H. Garzon, D. E. Goldberg, H. Iba, & R. Riolo (Eds.), *Genetic Programming 1998:Proceedings of the Third Annual Conference* (pp. 22-25).

Veldhuizen, D. V., & Lamont, G. B. (2000). Multiobjective evolutionary algorithms: Analyzing the state-of-the-art. *Evolutionary Computation Journal*, *8*(2), 125–128. doi:10.1162/106365600568158 PMID:10843518

Vinod, V. S., Varghese, S., & Kuriakose, B. (2004). Aluminium powder filled nitrile rubber composites. *Journal of Applied Polymer Science*, *91*(5), 3156–3161. doi:10.1002/app.13472

Wang, L., Li, F., & Su, Z. (2008). Effective thermal conductivity behaviour of filled vulcanized perfluoromrthyl vinyl ether rubber. *Journal of Applied Polymer Science*, *108*(5), 2968–2974. doi:10.1002/app.27612

Wang, Q., Gao, J., Wang, R., & Hua, Z. (2001). Mechanical and rheological properties of HDPE/Graphite composite with enhanced thermal conductivity. *Polymer Composites*, *22*(1), 97–103. doi:10.1002/pc.10521

Weidenfeller, B., Hofer, M., & Schilling, F. R. (2004). Thermal conductivity, thermal diffusivity, and specific heat capacity of particle filled polypropylene. *Composites. Part A, Applied Science and Manufacturing*, *35*(4), 423–429. doi:10.1016/j.compositesa.2003.11.005

Wong, C. P., & Bollampally, R. S. (1999). Thermal conductivity, elastic modulus, and coefficient of thermal expansion of polymer composites filled with ceramic particles for electronic packaging. *Journal of Applied Polymer Science*, *74*(14), 3396–3403. doi:10.1002/(SICI)1097-4628(19991227)74:14<3396::AID-APP13>3.0.CO;2-3

Xu, Y., Chung, D. D. L., & Mroz, C. (2001). Thermally conducting aluminium nitride polymer-matrix composites. *Composites. Part A, Applied Science and Manufacturing*, *32*(12), 1749–1757. doi:10.1016/S1359-835X(01)00023-9

Yu, P. L. (1973). A class of solutions for group decision problems. *Management Science*, *19*(8), 936–946. doi:10.1287/mnsc.19.8.936

Yu, S., Hing, P., & Hu, X. (2002). Thermal conductivity of polystyrene-aluminium nitride composite. *Composites. Part A, Applied Science and Manufacturing*, *33*(2), 289–292. doi:10.1016/S1359-835X(01)00107-5

Zitzler, E. (1999). *Evolutionary Algorithms for Multiobjective Optimization: Methods and Applications*. (Ph. D. thesis). Swiss Federal Institute of Technology (ETH), Zurich, Switzerland.

Chapter 12
Data-Driven Bi-Objective Genetic Algorithms EvoNN and BioGP and Their Applications in Metallurgical and Materials Domain

Nirupam Chakraborti
Indian Institute of Technology Kharagpur, India

ABSTRACT

Data-driven modeling and optimization are now of utmost importance in computational materials research. This chapter presents the operational details of two recent algorithms EvoNN (Evolutionary Neural net) and BioGP (Bi-objective Genetic Programming) which are particularly suitable for modeling and optimization tasks pertinent to noisy data. In both the approaches a tradeoff between the accuracy and complexity of the candidate models are sought, ultimately leading to some optimum tradeoffs. These novel strategies are tailor-made for constructing models of right complexity, excluding the non-essential inputs. They are constructed to implement the notion of Pareto-optimality using a predator-prey type genetic algorithm, providing the user with a set of optimum models, out of which an appropriate one can be easily picked up by applying some external criteria, if necessary. Several materials related problems have been solved using these algorithms in recent times and a couple of typical examples are briefly presented in this chapter.

INTRODUCTION

This chapter deals with two recent algorithms Evolutionary Neural Net (EvoNN) and Bi-objective Genetic Programming (BioGP) and their applications in metallurgical and materials problems. The EvoNN and BioGP algorithms were developed and proposed by the present author and his collaborators a few years ago (Pettersson et al. 2007; Mondal et al. 2011; Giri et al. 2013a, 2013b). Since then, these algorithms

DOI: 10.4018/978-1-5225-0290-6.ch012

have been applied successfully to numerous materials related problems (Bhattacharya et al. 2009; Rajak et al. 2009, 2011; Bansal et al. 2013, Jha et al. 2014). Both the algorithms share certain common features, which in general terms, could be listed as:

- They are suitable for creating data-driven meta-models out of highly non-linear and/or noisy data.
- They use multi-objective genetic algorithm over other paradigms, namely neural network and genetic programming, to generate a set of optimum models; while using a particular model remains the Decision Maker's (DM) prerogative.
- They are geared to work out an optimum tradeoff between the accuracy and complexity of the model and tend to eliminate incorporation of any superfluous noise in the process.

In recent times the present author has written two other book chapters (Chakraborti 2013, Chakraborti 2014) detailing these algorithms beside a number of journal articles involving their applications. Therefore only the basic features of these algorithms will be presented here and the emphasis will be more on their mode of usage. The source codes of both the algorithms are open and coded in Matlab™ and are available free of cost from the author of this chapter for any non-commercial academic usage. The details of their various features from a user's point of view will be presented in this chapter, along with some pertinent applications.

EvoNN AND BioGP IN A NUTSHELL

The essential features of both the algorithms are presented below:

The Essentials of EvoNN

For any given system, EvoNN tends to create an optimum set of neural network-based models that would tend to eliminate any random noise present in the data, prune any superfluous input avoiding any tendencies to either overfit or underfit (Collet 2007). To achieve this, the EvoNN algorithm deals with a population of neural nets with flexible architecture, allows custom made crossover and mutation processes in the lower part of the network, and being a hybrid algorithm, applies a Linear least Square (LLSQ)

Figure 1. The training strategy in EvoNN

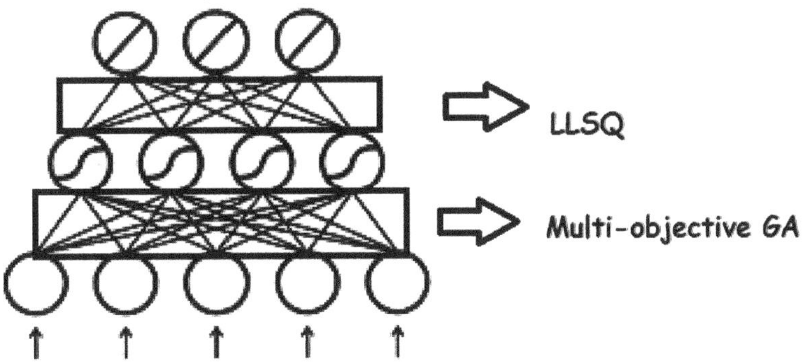

Algorithm in the upper part (Mondal et al. 2011), as shown in Figure 1. In EvoNN it is not mandatory to use all the input nodes, or all the hidden nodes. This flexibility allows using only the essential inputs in the optimized models along with the essential architecture formed by an optimum topology of the neural connections and their associated weights.

In its current version the optimized models are created in EvoNN by applying a predator-prey genetic algorithm (Li 2003; Pettersson et al. 2007;) on a randomly created population of neural nets with flexible architecture. This bi-objective genetic algorithm searches for a tradeoff between the training error of the network and its corresponding complexity expressed through the number of connections used. The approximation of the Pareto frontier (Coello Coello et al. 2002) obtained through this strategy, contains the optimized models; out of which a preferred network is usually picked up by applying a suitable information criterion (Pettersson et al. 2009). This is shown in Figure 2a, where each diamond is a distinct neural net belonging to the family of Pareto-optimum solutions shown in the figure, and in principle, an optimum model in its own level of error and complexity. Applying the external consideration of the Corrected Akaike Information Criterion (AICc) (Pettersson et al. 2009) that penalizes a model for an increase in parameterization and at the same time rewards for an increase in accuracy, a particular model is selected, as shown in Figure 2 (top) and its prediction is shown against the original experimental data in Figure 2 (bottom). Since the original data contained a significant amount of noise, the selected model, as expected from an intelligent modeling strategy, took quite a conservative approach towards modeling the large fluctuations in the original data set and many of the large fluctuations were left uncaptured. Further details of EvoNN are available elsewhere (Chakraborti 2013, Chakraborti 2014).

The Essentials of BioGP

In recent times, Genetic Programming (GP) (Collet 2007; Poli et al. 2008) is increasingly becoming a powerful alternate of neural nets, allowing the users to construct a metal-model with mathematical or logical functions of their own choice, which would be difficult to obtain through the paradigm of neural nets with rather restricted environment of weights, layers and transfer functions. In GP, the encoding is done as a tree structure and the typical evolutionary operators like crossover, selection and mutation are done on a population of trees generated on the basis of some user defined function and terminal sets. In general sense, GP uses the major genetic algorithm (GA) operators like *crossover*, *mutation* and *selection*. However, owing to the unique features of tree encoding and the requirements of learning features, these operators are tailor-made a bit differently in GP paradigm. Some of the essential features of a typical GP algorithm are listed below with some cursory explanations. Further details are available elsewhere (Collet 2007; Poli et al. 2008). All these features are utilized in BioGP as well, though it is substantially different from the conventional GP.

- **Automatically Defined Function (ADF):** Essentially an utilization of the subprogram concept quite common in efficient computing, it refers to a separately evolved sub-tree that can be used by the main tree at times and perhaps in many of its branches, instead of growing its features afresh in those instances and locations.
- **Function Set:** It contains the essential mathematical/logical operators like +, –, ÷, $\sqrt{}$, ∧. Each member of the function set also needs to be associated with an *arity*. For example, $\sqrt{}$ is a function of arity one, an operation like + is usually of arity two and so on.

Figure 2. Top: optimized models in EvoNN and the model selected trough AICc; bottom: predictions of the AICc selected optimized model against the original experimental data set

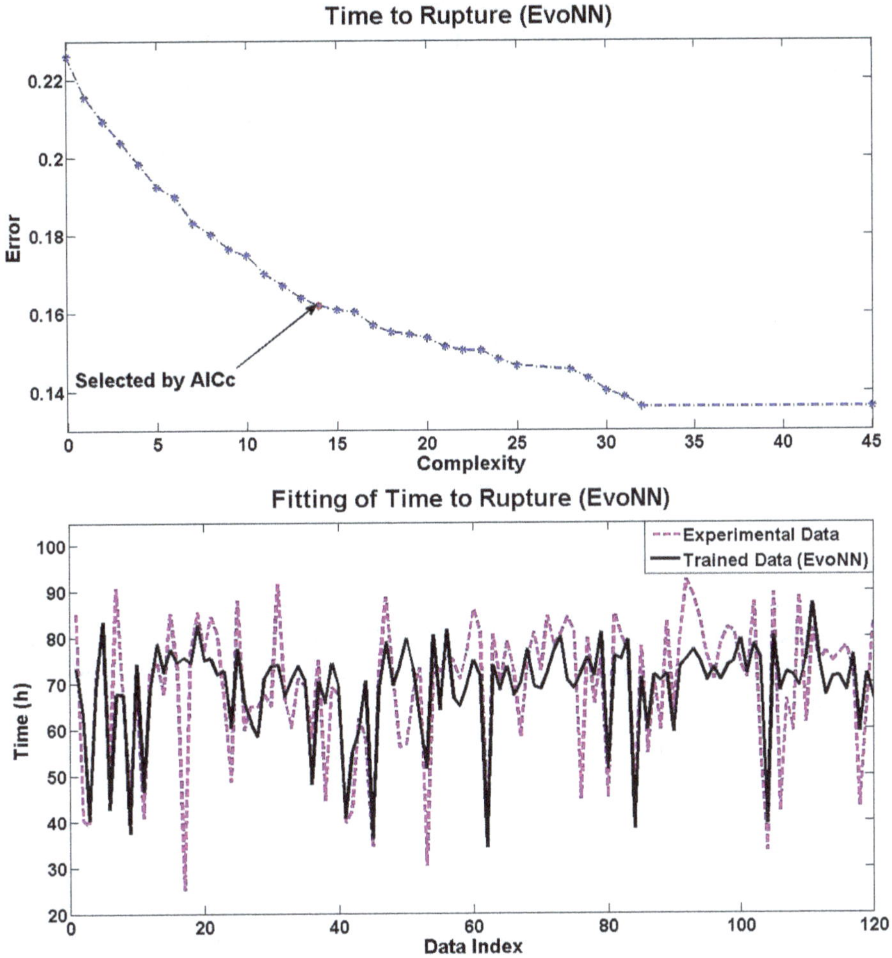

- **Terminal Set:** The terminal set consists of the input variables and constants. The input variables are usually read from a data file and the terminal set accepts any *Ephemeral random floating-point constant* (ERC), as it is known in GP parlance (Collet 2007). By default, any ADF with 0 arity becomes a member of the terminal set.
- **Fitness:** Normally, for training a function, the fitness is taken as a root mean square error (MSE) or a relative root mean square error (RMSE), computed by running the training data set through the GP tree and recording the outputs.
- **Selection:** Like GA *Tournament Selection* is a common selection operator in GP. In GA the tournament size is quite commonly kept as 2. In GP however the trend is to use a population size larger than the conventional GA and the evolution usually takes place for a relatively less number of generations. This necessitates a larger tournament size, e.g. 5.
- **Crossover:** Various forms of crossover operators are available in GP: among them the *Standard,* and *Height fair* crossovers are quite widely used. In Standard crossover two sub-trees are ran-

domly exchanged between the participating parents with some probability, while for height fair crossover the similar exchange takes place at a selected depth.

- **Mutation:** Three types of mutations, *Standard*, *Small* and *Mono-parental exchange,* are commonly used in GP. Standard mutation entails deletion of a sub-tree, which is then randomly grown. In case of Small mutation a terminal is replaced by another, a numerical constant in the terminal is often slightly perturbed. Alternately, an operator is replaced by another of same arity (e.g., a '$\sqrt{}$' by a 'log'). In case of Mono-parental exchange two sub-trees belonging to a same tree are exchanged.

The conventional GP creates a data-driven model by lowering its learning error; expressed either through MSE and RMSE values computed using the training data set. In the process, the trees often tend to grow unmanageably with marginal or no improvement in fitness leading to a problem known as *bloat* (Collet 2007;Poli et al. 2008) in the GP literature. Often, some rogue trees develop with certain branches, returning division by zero or other execution problems of similar nature, rendering the implementation of GP quite complicated. To circumvent such problems and also to work out the correct *fit*, the BioGP algorithm has been proposed recently by the author of his chapter and his collaborators (Giri et al. 2013a, 2013b) and used along with EvoNN. The BioGP involves the tree encoding and the other features of conventional GP. However, each tree evolves in parts as a number of distinct sub-trees, which are ultimately aggregated with some weights and a bias that are optimized with a linear least square algorithm, more in the fashion of a neural network. A parameter named *Error reduction ratio* is implemented in BioGP that quantifies the performance of any individual sub-tree. If any of them demonstrates bloat or any other undesirable features, the error reduction ratio fails to reach a prescribed level. The sub-tree in question can then be easily removed and re-grown without hampering the rest of them. Also, up to a certain number of generations BioGP acts in a single objective fashion, minimizing just the training error. After that, like EvoNN, a predator-prey genetic algorithm kicks in, working out a tradeoff between the training error and complexity, where the measure of complexity is taken as a weighted sum of the depth of the tree and the total number of function nodes. From the members of the Pareto set obtained through this procedure, usually the model with the minimum error is selected (Giri et al. 2013a, 2013b). Further details are provided elsewhere (Giri et al. 2013a, 2013b). Performance of the BioGP algorithm on the data set same as in Figure 2 is shown in Figure 3. The AICc criterion cannot be effectively implemented in a tree structure and therefore the optimum model with lowest error is selected from Figure 3a. Its prediction are compared with the original noisy data and conservative trends, quite similar to EvoNN (Figure 2 (bottom)) are observed.

Combined Training and Testing in EvoNN and BioGP

Expectedly, the quality of a Data-driven model depends on the training data set. However, one also needs a reliable testing data set in order to check the goodness of the model. This is particularly important considering the fact that an overfitted model may lead to an impressive performance in its training set, but it may not work at all in a testing set. A practical problem in this regard is that in a real life situation, often there is a paucity of data and assigning a separate data set, exclusively for the testing purpose, is often difficult; if not impossible in some cases. To circumvent this problem both EvoNN and BioGP have adopted an ingenious strategy of simultaneous training and testing. Here, if deemed necessary, the user can split the available data set into a number of overlapping subsets. In that case separate training

Figure 3. Top: optimized models in BioGP; bottom: predictions of the optimized model with lowest error against the original experimental data set

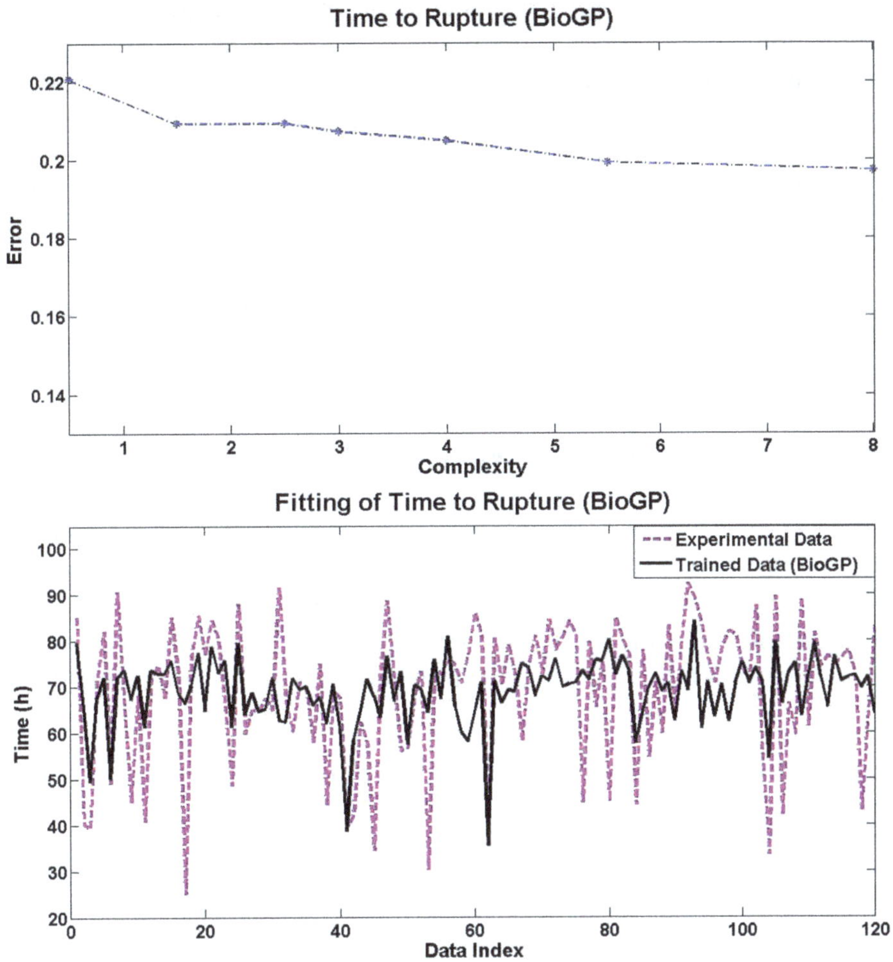

will be conducted using each of the subsets as well as the complete data set: thus leading to N+1 system models, if N partitions are made in the original data set. Each of these models will then be tested N times, using every data set other than the one that has been used for its training, and the performance, in terms of the prediction error, will be systematically tabulated for each of these evaluations. This allows the user to thoroughly evaluate and examine all the models, and enables to pick up the most suitable one out of a total of N+1 alternates. Further details are provided elsewhere (Mondal et al. 2011).

Single Variable Response Analysis in EvoNN and BioGP

Once an optimum Data-driven model is constructed after proper pruning of the redundant variables and selected from the Pareto-frontier it would be necessary to analyze the impact of every single variable on the objective associated with it. This is often a difficult task as the models are often implicitly non-linier and the variables are quite commonly intricately related to each other.

To analyze the response of a single variable towards any particular property, isolating it from the rest, a simple intuitive procedure is adopted in both EvoNN and BioGP (Helle et al. 2006, Chakraborti 2013, Chakraborti 2014). In this approach the variable being examined is perturbed in a systematic and meaningful fashion, maintaining the rest of the input variables constant at the base level. The output generated from this configuration is recorded. If the pattern of variation for any output variable follows exactly the same nature of the perturbation provided to the input variable, they are assumed to be *directly* correlated. Alternately, if increasing the input variable leads to a decrease in the output signal and vice versa, they are considered to be *inversely* correlated. Often, this procedure would lead to a mixed type of response, where a part of the response is direct and the remaining part is of inverse nature. It is also possible to detect a *no dependence* response occasionally, where this simple procedure perhaps is not able to detect any correlation or perhaps there exists none. This procedure is schematically presented in Figure 4. The dashed line shows the planned perturbation in one particular variable and the solid line depicts the corresponding output variable response.

The Bi-Objective Optimization Modules in EvoNN and BioGP

Once the meta-models are created they can be used by an optimization module for computing the Pareto frontiers for two objectives at a time. The predator-prey algorithm that is used in the meta-model construction takes up the optimization task too. Here the prey denotes a population of the evolving solutions, as in any genetic and evolutionary algorithm, and the predators are some artificial entities to cull the weaker among them, following a number of hunting rules in a specifically defined neighbor-

Figure 4. Input variable response shown schematically: Left: Direct response; right: Inverse response; bottom: No dependence

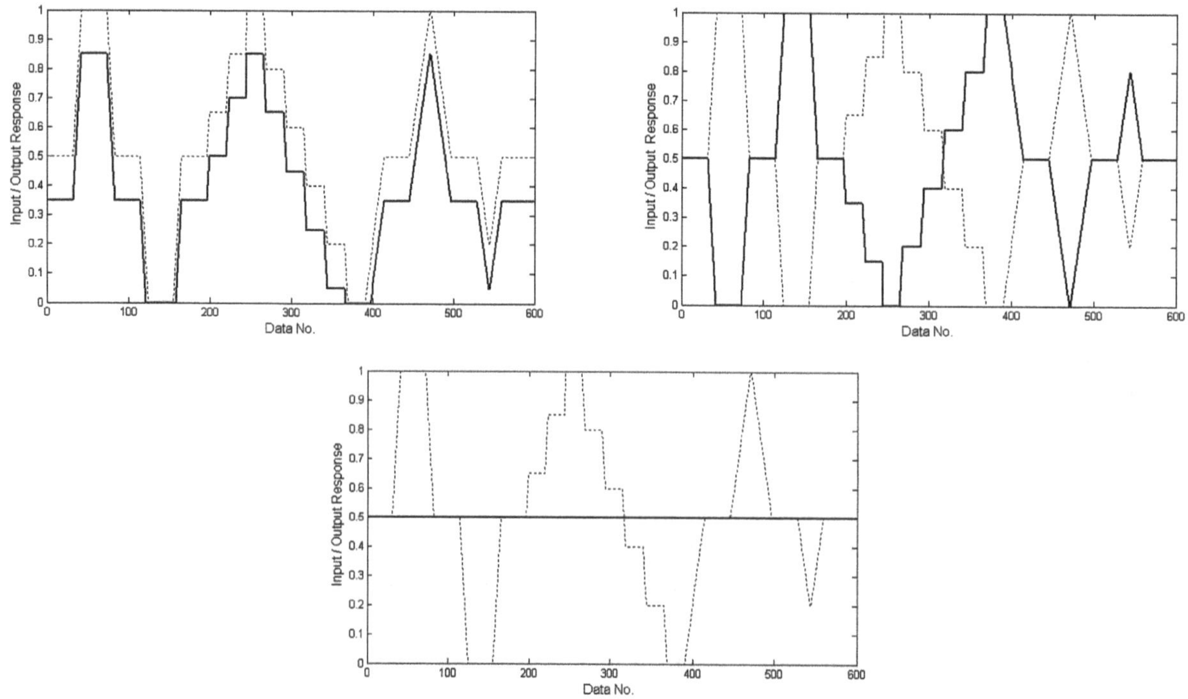

Figure 5. Computational workspace in predator prey algorithm. The predators are denoted by larger and lighter circles and the smaller and darker symbols in the lattice are the prey

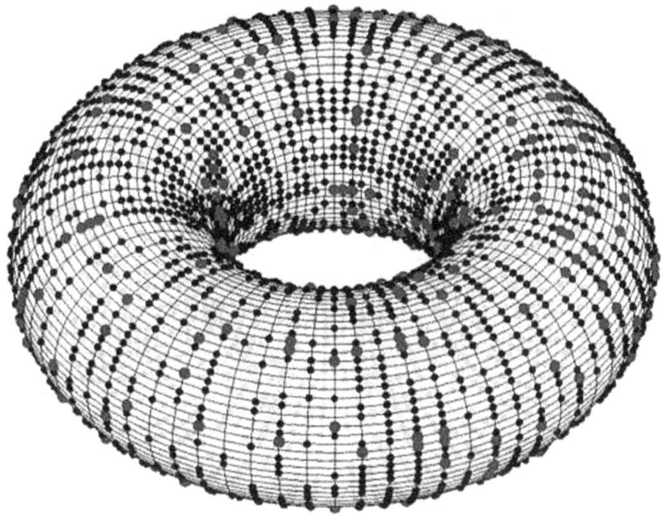

hood. After several generation of evolution through continuous predator attacks, a progeny of the dour prey emerges; this could not be hunted and thus leads to the Pareto frontier. Both the predators and the preys are introduced on the uniformly placed grids at the outer surface of a toroidal construct, as shown in Figure 5. The details of the genetic operators used, and other relevant details are well documented elsewhere (Chakraborti 2013, Chakraborti 2014).

THE EvoNN AND BioGP CODES

Both EvoNN and BioGP are open source software, encoded in MATLAB™ and available free of cost from the author of this chapter for any non-commercial academic usage. Both the algorithms are structured in a similar fashion, keeping the users' convenience in mind. The codes are generic and already tested in a wide variety of problems (Chakraborti 2013, Chakraborti 2014). Although, any part of them can be altered, if need be; for most problems it is simply sufficient to adjust a few user defined parameters. Some of the essential details are briefly provided below.

The Data Training Module in EvoNN

In order to run EvoNN the Files listed in Table 1 need to be present in the current MATLAB™ directory. As and when needed, the relevant files will be used during both data training and optimization.

To run the program the datafile, 'Testdata.xls' needs to be formatted in a particular fashion, as shown in Figure 6 for the training of two functions f1(x1,x2) and f2(x1,x2). The column number where a particular variable or an objective is placed will be used in the code, and in most cases there will be more rows of data than what is shown in Figure 6.

Table 1. The essential files in EvoNN

Folder name/file extension	Filename	Remark
NN_util	NNevalnet	
PP_util	CROW_SORT, FONSECA_RANK, KillBadPrey, MovePredator, MovePrey, NONDOM_SORT, PlotLattice	
util	MUTE, PlotPareto, XOVER	
*.m	f_EvoNN_net, get_trained_data, output, PP_NNGA, PP_NNGA_subsets, PPopt, Prediction_EvoNN, svr	
*.mat	Setslog#-##_*filename*, Testdata, Test_data, trend	The values of '#', '##' and '*filename*' in the Setslog files, which are created during execution of the codes, would depend on the model configuration and the data filename used. The names other than 'Testdata' and 'Test_data' can also be used.
*.xls	Testdata, Test_data	Other names could be used.

Figure 6. Typical data arrangement in the input Excel file

x1	x2	f1	f2
0.1	2.22E-16	0.1	11.04984
0.100311	2.22E-16	0.100311	11.01173
0.100761	2.22E-16	0.100761	10.9567
0.101175	2.22E-16	0.101175	10.90645
0.1019	2.22E-16	0.1019	10.81902
0.102489	2.22E-16	0.102489	10.74864
0.102564	2.22E-16	0.102564	10.7397
0.102989	2.22E-16	0.102989	10.68926
0.103397	2.22E-16	0.103397	10.64112

The file 'Testdata.xls' needs to be opened in the MATLAB™ environment. For the newer versions of MATLAB™, for example the 2013 release of the software, the numerical portion of the data needs to be brought into the MATLB™ workspace as 'Matrix', using the 'Import Data' feature. After that one needs to run the following command in the command window:

```
>> data=Testdata
```

Next, again through the 'Import Data' feature, transfer the text portion of the file to the workspace as 'Cell Array' and then run the following in the command window.

```
>> textdata=Testdata
```

Followed by this the two files containing the text and the numerical information needs to be merged and saved as a file named 'Testdata.mat' using the flowing commands:

```
>> data = [textdata; num2cell(data)];
>> save Testdata.mat data
```

In the older versions of MATLAB™ one needs to put .xls file in the file current folder and a double click on it would generate two 'data' and 'textdata' that need to be concatenated and saved in a .mat form through the commands:

```
>> data = [textdata; num2cell(data)];
>> save Testdata.mat data
```

Once the data file is saved in a .mat format, one needs to run the program PP_NNGA_subsets.m. This program open source and the major user controls are located in the 'Define parameters' region of the code as shown in Figure 7.

Figure 7. User defined parameters in the learning module of EvoNN

```
%==================Define parameters========================================
Data = 'Testdata.mat';          %Data file
in_index = 1:2;                 %independent variables column no.
out_index = 3;                  %dependent variable column no.

Xmin = eps; Setslog.Xmin = Xmin;    %normalization range for variables
Xmax = 1; Setslog.Xmax = Xmax;

Setslog.in_index = in_index;
Setslog.out_index = out_index;

DataSet = importdata(Data);
Setslog.DATA = DataSet;
Setslog.paraname = DataSet(1:2,:);
DataSet(1:2,:) = [];
DataSet = cell2mat(DataSet);
Setslog.DataSet = DataSet;

subsets = 1; overlap = 0;           %number of partitions of datafile and overlap b/w them
set_size = length(DataSet(:,1)); Setslog.set_size = set_size;
Setslog.subsets = subsets;
Setslog.overlap = overlap;
subset_size = round((set_size + (subsets-1)*overlap) / subsets);
Setslog.subset_size = subset_size;

Setslog.generations = 100;  %max generations for evolution
Setslog.nonodes = 5;        %maximum number of nodes
Setslog.noinnodes = length(Setslog.in_index);
Setslog.nooutnodes = length(Setslog.out_index);
Setslog.maxrank = 20;       %maxrank retained at KillInterval
Setslog.KillInterval = 20;  %Interval at which bad preys are eliminated
Setslog.Prey_popsize = 500;        %Initial popsize
Setslog.no_Prey_preferred = 500;   %Desired popsize
Setslog.no_new_Prey = 500;          %new prey introduced every KillInterval
Setslog.Predator_popsize = 200;     %Number of Predators
Setslog.no_x = 50;                  %lattice size (no of rows)
Setslog.no_y = 50;                  %lattice size (no of cols)
Setslog.ploton = 1;                 %set 0 for no plots or 1 for plots at every generation
%======================================================================
```

Although the comments incorporated 'Define parameters' section of PP_NNGA_subsets.m are self explanatory, a few points are highlighted below, providing some further clarification:

- The user gets to set a number of parameters, which are self-explanatory in Figure 7. The computation stops after the number of generations prescribed in the variable 'Setslog.generations' is reached.
- During the execution of the program the system will create a file of the format 'Setslog#-##_*filename*.mat', where the parameter'#' will indicate the number of hidden nodes in the selected model, '##' would indicate the ouput column number in the used dataset, and '*filename*' would indicate the source of the input data.

Once executed with proper parameter settings, 'PP_NNGA_subsets.m' will initiate with a message of following type in the command window:

```
Training Whole Dataset
Generation 1: Predatorpop 200 PredMoves 4
Preypop before: 1334; Preypop after: 711
```

This will be updated after every generation and after training the whole data set, the training of the subsets, if prescribed, will begin. In Figure 7 the corresponding variables are set as follows:

```
subsets = 1; overlap = 0;          %number of partitions of datafile and
overlap b/w them
```

This will require just the training of the whole data set as no subsets or partitions are created. In this case, the program exits with the following message:

```
Trainining Subsets
 1
```

Figure 8. Predator and prey distribution in the computational lattice at a specific point of execution

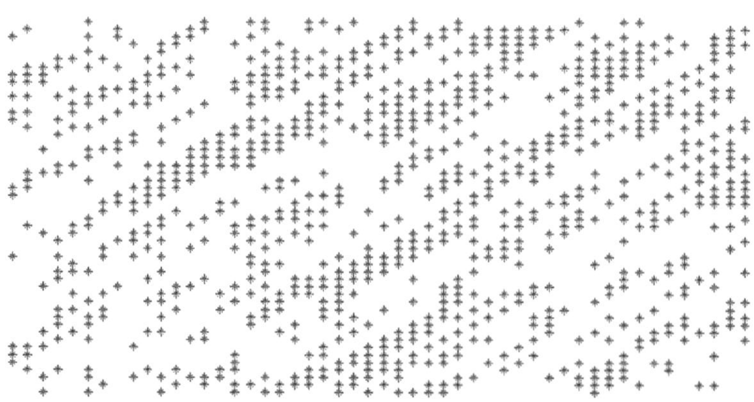

Periodically, during the execution the distribution of the predator and prey population will be shown on screen, as presented in Figure 8. The predators will appear as red crosses on screen, while the prey will be denoted as blue crosses.

The evolving tradeoff between the complexity (F2) and training error (F2) will continuously updated and plotted. The final tradeoff curve will have the AICc selected model is marked in red, as shown in Figure 9. Performance of this model can be evaluated by running the 'output' command on the 'Setslog' file created during the execution of the code. For the Setslof file named ''Setslog5-3_Testdata.mat' the task would running the following in the command window:

```
>> output ('Setslog5-3_Testdata.mat')
```

This will execute with the response:

```
Training data 1
ERROR TABLE for training of f1 by EvoNN:
 1.5321e-13
ans =
 1.5321e-13
```

Figure 9. A typical final tradeoff curve between the complexity (F2) and training error (F1). Each point denotes a distinct optimum model. The model corresponding to the F2 value of 5 is selected through AICc and is specially marked

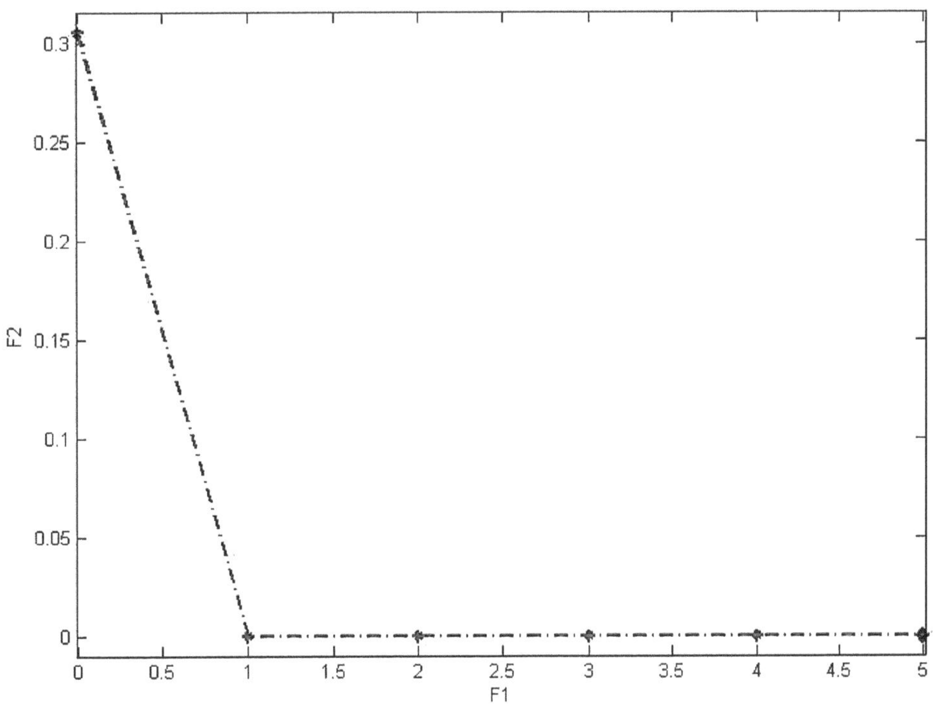

The variable 'ans' provides the training error. Since in this case only one set of data was trained the 'ERROR TABLE' contains just one entry. In case some overlapping data partitions are used the AICc selected models form them will be tested on each other and the corresponding error values will show up in the 'ERROR TABLE', and it will contain the original training errors of each individual data set as diagonal entries. Some plots showing the quality of the models will also be generated, as shown in Figures 10 and 11. In Figure 10 the model predictions are plotted against the original data, and a perfect match between them would lead to a slope of unity. In Figure 11 the original data and model predictions are superimposed on each other by plotting them on a same scale.

Similarly, the single variable responses could be obtained by running the svr command using the same Setslog file. The file 'trend.mat' contains a default insignal profile, which can be readily used, unless the user requires a customized profile. Executing the following command would provide the single variable responses both graphically and also in a tabular form.

```
>> svr('Setslog5-3_Testdata.mat','figname','trend.mat')
```

The selected model can be readily used for predicting the output from a fresh set of data belonging to the same system. For that one needs to execute the program 'Prediction_EvoNN.m' in the command window. The name new data file (e.g. Test_data.mat) and the original Setslog files need to be mentioned in 'Prediction_EvoNN.m', as shown below:

Figure 10. The original data (abscissa) is plotted against the model predictions (ordinate)

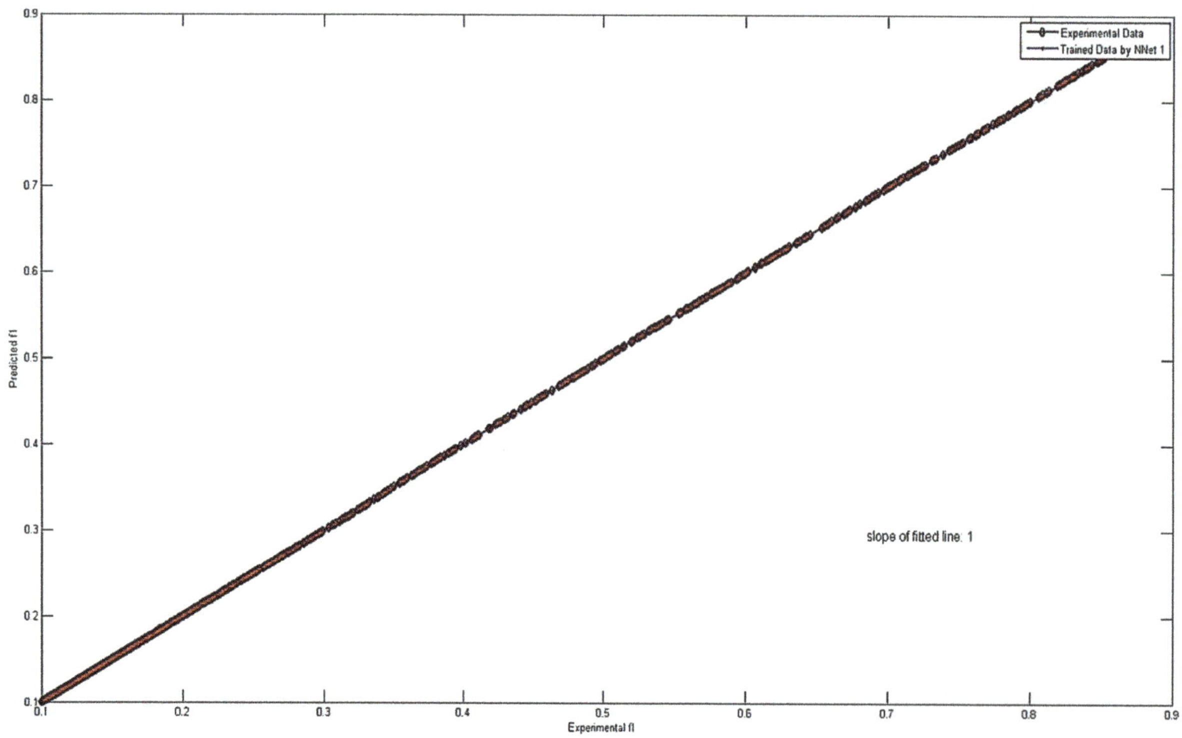

Figure 11. The original data and the model predictions superimposed on each other

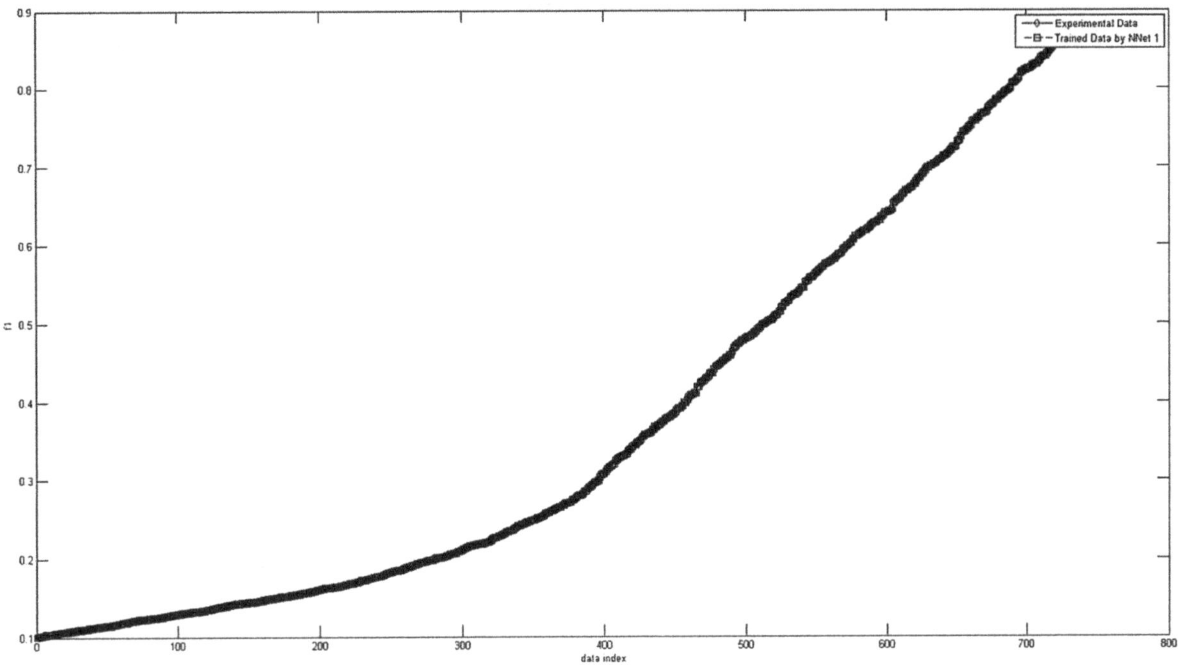


```
load Test_data.mat                    % load the new dataset in .mat and it
must not be in cell form
load Setslog5-3_data_demo.mat    % load the Setslog file for the concerned
objective
Data=Test_data(:,:);
```

The Data Training Module in BioGP

The format of the BioGP source code is quite similar to EvoNN, so that the user of one module can almost readily pickup the other. The essential files for the data training module in BioGP are listed in Table 2. It should be noted that currently the folder 'adf' contains subprograms for the following mathematical operations: +, –, *, ^2, ÷, and log. The users should be able to include any other functions, if needed. The format of the data files are same as in EvoNN Here it is assumed that 'data_demo.xls' and 'Test_data.xls' are data files used respectively for training and testing. Here he format of the 'Setslog' file is is *Setslogm-n-o_p*; where *m* denotes the type of evolution (1 for only bi-objective evolution and 2 for a single objective error minimization, followed by a bi-objective evaluation), *n* is the maximum number of subtrees, *o* is the output column number in the given data set and *p* denotes the corresponding data file name.

In BioGP training module the main file that the user needs to execute is ' PP_GP_subsets.m'. The data file needs to be converted to the '.mat' format exactly in the same way as EvoNN. Like EvoNN the major user control parameters are located in the 'Define parameters' region of ' PP_GP_subsets.m'. Most parameters are similar to EvoNN. Few that are specific for BioGP are listed below.

Table 2. The essential files in BioGP

Folder name/file extension	filename
adf	add, mul, powsq, rdiv, rlog, rsqrt, sub
GP_util	ari2dec, dec2ari, full_tree, GPevaltree, grow_tree, initialize, mutate, recat, rename_ind, snatch_tree, snatch_tree
PP_util	CROW_SORT, FONSECA_RANK, KillBadPrey, MovePredator, MovePrey, NONDOM_SORT, PlotLattice
util	MUTE, PlotPareto, tournament_select, XOVER
*.m	f_BioGP, get_func, get_trained_data, output, plottree, PP_GP, PP_GP_subsets, PPopt, Prediction_BioGP, svr
*.mat	data_demo, Setslog2-3-3_data_demo, Setslog2-3-4_data_demo, Test_data, trend
*.xls	data_demo, Test_data

The type of evolution is specified as:

```
Setslog.evo_type = 2;  %set 1 for only Biobj evolution and 2 for first single
obj followed by Biobj evolution
```

The useful mathematical functions are specified as:

```
F2_set = {'add' 'sub' 'mul' 'rdiv'};     %double arity functions
```

The tree architecture is specified as:

```
Setslog.max_depth = 3;              %max depth to which a tree grows
Setslog.max_roots = 3;              %max subtrees that a tree grows
```

The number of generations for single and bi-objective evolutions, in case of a type 2 evolution, are defined as:

```
Setslog.generation1 = 50;           %generations for single obj evolution set
Setslog.evo_type = 2
Setslog.generations = 50;          %max generations for bi -obj evolution
```

The tournament size for the single objective GP runs are specified as:

```
Setslog.tour_size = 4;              %tournament size for single objective GP
```

The complexity parameter λ is specified as:

```
Setslog.lambda = 0.5;               %parameter to evaluate complexity of tree
= lambda*depth + (1-lambda)*nodes
```

The condition for pruning a rogue subtree by checking the error reduction ratio is specified as:

```
Setslog.err_red_lim = 1e-3;          %any subtree contibuting less than this
value for error reduction is eliminated
Once ' PP_GP_subsets.m' is executed, the program initiates the single objec-
tive training (if used) with the following message:
Training Whole Dataset
Gen 1; Best Ind fit: 1.666e-16
```

Subsequently the bi-objective training sets in with the following message (in this example it was initiated after ten generations of single objective evaluations):

```
Generation 11: Predatorpop 100 PredMoves 6
Preypop before: 1190; Preypop after: 862
```

After the prescribed number of generations the training of the "Whole Dataset' ends and that of the subsets (if prescribed) begins. If no subsets are used the program exits with the message:

```
Trainining Subsets
     1
```

The graphics support in BioGP is quite similar to that in EvoNN. By default, from the final tradeoff plot between the complexity (F1) and training error (F2) the model with the least error is selected as shown in Figure 12.

Figure 12. A typical final tradeoff curve between the complexity (F2) and training error (F1). Each point denotes a distinct optimum model. The model corresponding to the F1 value of 3.5 is selected, as it ensures lowest training error

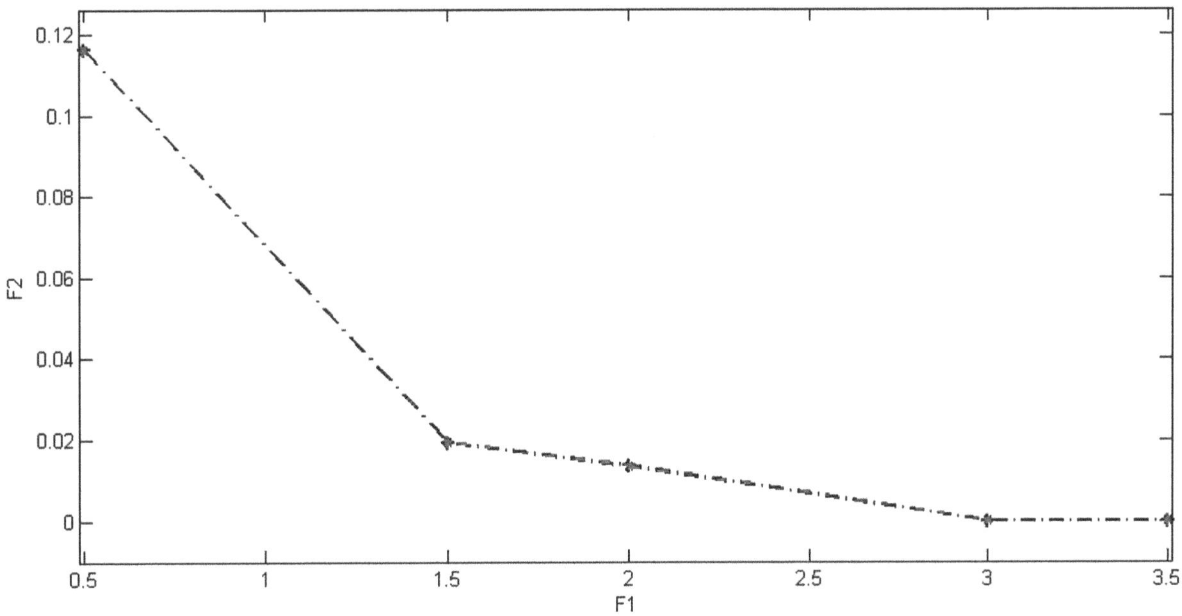

To visualize and extract the exact mathematical formula of any of the optimum trees in the optimum tradeoff curve like Figure 11, one needs to execute the 'plottree' function in the command window. For example:

```
>> plottree('Setslog2-3-6_GA_Data.mat', 1, 4)


Bias: 14.4304
Root1 Weight: 6117.6741; Error Reduction Ratio: 0.49031
Root2 Weight: -0.0045729; Error Reduction Ratio: 0.36523
Root3 Weight: 1799.0772; Error Reduction Ratio: 0.13713
```

The procedure for checking the single variable responses and running the test data remains same as what has been already discussed for EvoNN.

THE PARETO OPTIMIZATION MODULE

The optimization module used by EvoNN and BioGP is 'PPopt.m'. It is based upon Predator-prey Genetic Algorithm (Li 2003; Pettersson et al. 2007). It requires the Setslog files of the two objectives trained through the data training modules, and both could be simultaneously optimized. If any constraints are used, the corresponding Setslog files are needed as well. The user defined parameters in 'PPopt.m' are shown in Figure 13. Since the same Predator-prey Genetic Algorithm is used in both the training and optimization modules, several parameters are actually similar. During the optimization runs any objective can either be maximized or minimized by setting a proper flag (either 1 or -1) for the corresponding variable named 'obj'.

The code 'PPopt.m' executes with the following type of message in every generation:

```
Generation 1: Predatorpop 100 PredMoves 8
Preypop before: 1324; Preypop after: 741
```

It also provides periodic graphical information on the predator-prey distributions in the computational grid, as well as the evolution of Pareto front. The final tradeoff curve is plotted as shown in Figure 14.

SOME MATERIALS RELATED APPLICATIONS

The EvoNN and BioGP algorithms are increasingly being used in various real life applications in the materials domain (Chakraborti 2013). Two representative cases are briefly highlighted below.

In a recent study (Kumar et al. 2012) meta-models were constructed for the various mechanical properties of some micro alloyed steel, using compiled literature information on their experimental data from the work of numerous unrelated groups scattered throughout the globe. The chances of scatter and noise in this data are actually huge. Despite that the meta-models performed remarkably well and were instrumental in discovering a number of hitherto unknown compositions steel of this category

Figure 13. User defined parameters in the Pareto-optimization module of EvoNN/BioGP

```
%===============User Input=========================================
obj(1) = 1;                %set 1 for min and -1 for max
obj(2) = 1;                %set 1 for min and -1 for max
Setslog = {};
Setslog(1) = {'Setslog5-3_data_demo.mat'};
Setslog(2) = {'Setslog5-4_data_demo.mat'};
% if constraint is applicable
%for constraint function introduce Setslog(3) and so on
% Setslog(3) = {'Setslog10-8_data_mod_smbl.mat'};
% Setslog(4) = {'Setslog10-9_data_mod_smbl.mat'};
%loadwlog loads the tree from Setslog file, the prototype:
%loadwlog(Setslog, setno, index);
%setno is the training subset used and index is tree index obtained from
%the pareto front of training, 1 being the tree with lowest error seeting
%setno =0 takes the last training subset and index = 0 sets the default
%tree chosen by BioGP training
wlog(1) = loadwlog(importdata(Setslog{1}), 0, 0);
wlog(2) = loadwlog(importdata(Setslog{2}), 0, 0);
% wlog(3) = loadwlog(importdata(Setslog{3}), 0, 0);
% wlog(4) = loadwlog(importdata(Setslog{4}), 0, 0);
% wlog(3) = loadwlog(importdata(Setslog{3}), 0, 0);
%define extra wlog for checking the constraint and edit the contra function
%as desired
novar = length(wlog(1).in_index);
LB = []; UB = []; svstr = 'pareto_';
for i = 1:2
    LB = [LB; wlog(i).xmin];
    UB = [UB; wlog(i).xmax];
    svstr = [svstr num2str(obj(i)*wlog(i).out_index) '_'];
end
LB = max(LB); UB = min(UB);

Prey_popsize = 500;          %Initial popsize
no_Prey_preferred = 500;     %Desired popsize
no_new_Prey = 200;           %new prey introduced every KillInterval
Predator_popsize = 100;      %Number of Predators

no_generations = 200;        %max generations
P_move_prey = 0.5;           %Prob with which a Prey moves
P_mut = 0.1;                 %prob of choosing a prey for mutation
%Prob of Xover is 1 for every Prey
F_bad = 1e5;                 %fitness assigned to preys performing badly
%2D-lattice
no_x = 50;                   %lattice size (no of rows)
no_y = 50;                   %lattice size (no of cols)
KillInterval = 4;           %Interval at which bad preys are eliminated
maxrank = 4;                %maxrank retained at KillInterval
ploton = 1;                  %set 0 for no plots or 1 for plots at every generation
%============================DND=========================================
```

with superior properties. A typical result is shown in Figure 15 where a tradeoff is sought between two conflicting properties, yield strength and ductility, in an attempt to maximize both. The Pareto solution shown in this figure lists a series of novel alloys with significantly better property combinations than those in the original data set. A typical predicted alloy was actually cast during study and three samples

Figure 14. A Pareto-frontier computed through 'PPopt.m'

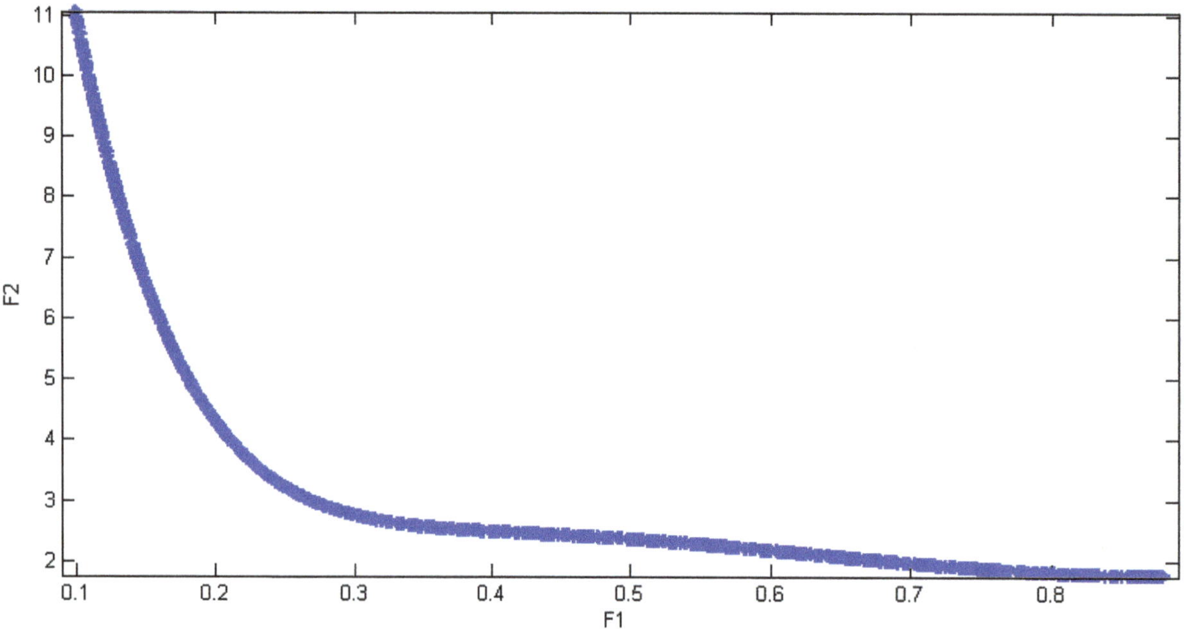

Figure 15. Novel steels of superior properties discovered through data-driven meta-modeling and optimization

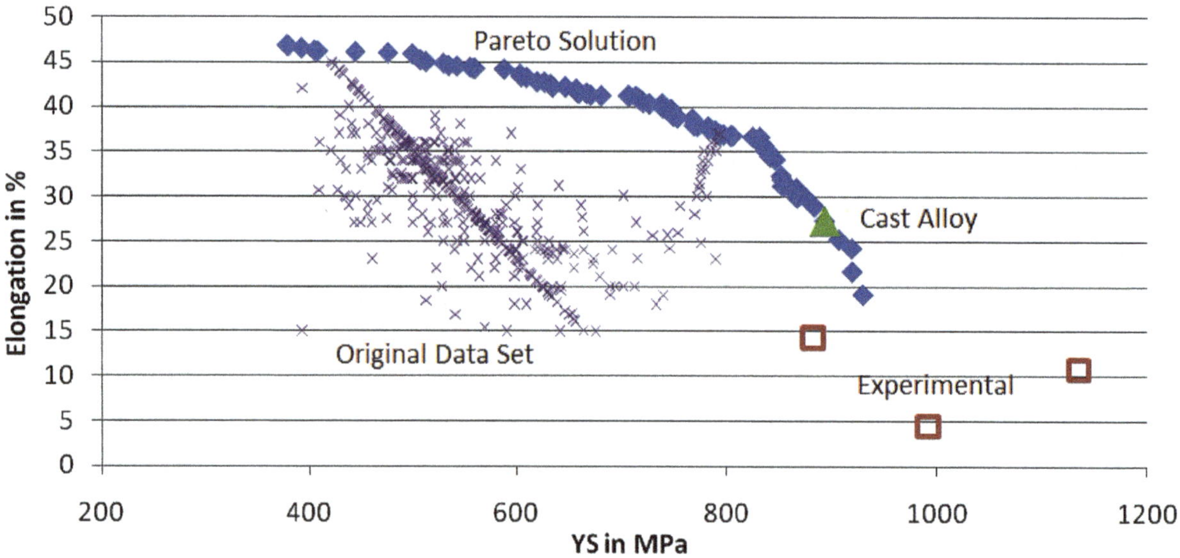

of it were actually heat treated and tested for the mechanical properties. The measured property values were well within the predicted range, corroborating the efficacy of this data driven method as a powerful strategy for alloy design.

Figure 16. Identification of AB_2 compounds based upon their chemical properties

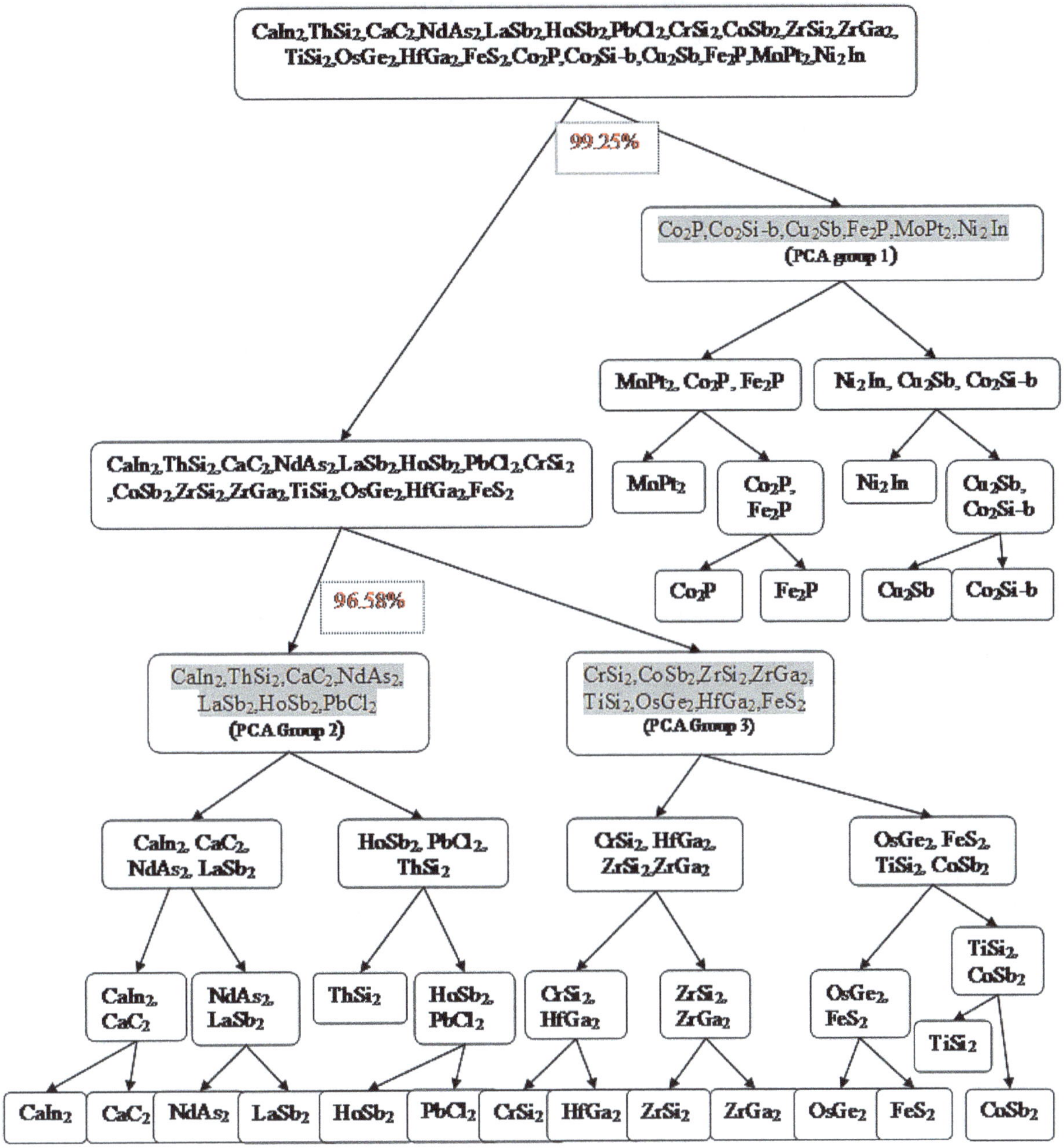

Apart from modeling and optimization resulting in newer materials discovery, the procedure outlined in this chapter, with minor modification can be easily used for classification and identification of materials based upon their common properties. In a recent study (Agarwal et al. 2009) the chemical properties (atomic radius, electron density, chemical potential etc.) of a large number of AB_2 type compounds were analyzed through the EvoNN approach. Using a statistical Principal component analysis (PCA) technique some of the subgroups in these compounds were predetermined. Next, following a decision tree approach, EvoNN in its learning mode could initially identify those subgroups and from there could successfully proceed towards identifying each and every constituent successfully, as shown in Figure 16.

FUTURE RESEARCH DIRECTIONS

At present the optimization module associated with EvoNN and BioGP is suitable for two conflicting objectives and is based upon Pareto-optimality (Coello Coello et al. 2002). In the presence of a large number of objectives it is however extremely difficult to compute the Pareto-optimality in an evolutionary way, as too many sub-optimal non-dominated solutions are usually produced and it becomes almost impossible to maintain the required population diversity. This can be circumvented by introducing some less rigid optimality conditions that have been proposed in recent times and replacing the Pareto condition by any one of them (Farina 2002). Future research to upgrade the procedure for multiple objectives should be directed in this direction.

CONCLUSION

Data-driven modeling and optimization based upon the evolutionary approaches are still emerging in materials research and expected to have a huge potential. The two recent algorithms EvoNN and BioGP discussed in this chapter are now quite ready to take up any challenging problem in this area tackled by the researchers at large.

REFERENCES

Agarwal, A., Pettersson, F., Singh, A., Kong, C. S., Saxén, H., Rajan, K., & Chakraborti, N. et al. (2009). Identification and Optimization of AB2 Phases Using Principal Component Analysis, Evolutionary Neural Nets, and Multiobjective Genetic Algorithms. *Materials and Manufacturing Processes*, *24*(3), 274–281. doi:10.1080/10426910802678321

Bansal, A., Barman, A., Ghosh, S., & Chakraborti, N. (2013). Designing Cu-Zr Glass Using Multiobjective Genetic Algorithm and Evolutionary Neural Network Metamodels-Based Classical Molecular Dynamics Simulation. *Materials and Manufacturing Processes*, *28*(7), 733–740. doi:10.1080/10426914.2013.763961

Bhattacharya, B., Kumar, G. R. D., Agarwal, A., Erkoç, S., Singh, A., & Chakraborti, N. (2009). Analyzing Fe-Zn system using molecular dynamics, evolutionary neural nets and multi-objective genetic algorithms. *Computational Materials Science*, *46*(4), 821–827. doi:10.1016/j.commatsci.2009.04.023

Chakraborti, N. (2014). Strategies for Evolutionary Data Driven Modeling in Chemical and Metallurgical Systems. *Springer*. doi:10.1007/978-3-319-06508-3_4

Chakraborti, N. (2013). Evolutionary Data-Driven Modeling. In Informatics for Materials Science and Engineering. Oxford, UK: Butterworth-Heinemann.

Coello Coello, C. A., Lamont, G. B., & Van Veldhuizen, D. A. (2002). *Evolutionary Algorithms for Solving Multi-Objective Problems*. Norwell, MA: Kluwer. doi:10.1007/978-1-4757-5184-0

Collet, P. (2007). Genetic programming. In Handbook of Research on Nature Inspired Computing for Economics and Management. Hershey, PA: Idea. doi:10.4018/978-1-59140-984-7.ch005

Farina, M., & Amato, P. (2002). On the Optimal Solution Definition for Many-criteria Optimization Problems. In *Proceedings of the NAFIPS-FLINT International Conference* (pp. 233-238). Piscataway, NJ: IEEE Service Center. doi:10.1109/NAFIPS.2002.1018061

Giri, B. K., Hakanen, J., Miettinen, K., & Chakraborti, N. (2013a). Genetic programming through bi-objective genetic algorithms with a study of a simulated moving bed process involving multiple objectives. *Applied Soft Computing*, *13*(5), 2613–2623. doi:10.1016/j.asoc.2012.11.025

Giri, B. K., Pettersson, F., Saxén, H., & Chakraborti, N. (2013b). Genetic Programming Evolved through Bi-Objective Genetic Algorithms Applied to a Blast Furnace. *Materials and Manufacturing Processes*, *28*(7), 776–782. doi:10.1080/10426914.2013.763953

Helle, M., Pettersson, F., Chakraborti, N., & Saxén, H. (2006). Modeling noisy blast furnace data using genetic algorithms and neural networks. *Steel Research International*, *77*(2), 75–81.

Jha, R., Sen, P. K., & Chakraborti, N. (2014). Multi-Objective Genetic Algorithms and Genetic Programming Models for Minimizing Input Carbon Rates in a Blast Furnace Compared with a Conventional Analytic Approach. *Steel Research International*, *85*(2), 219–232. doi:10.1002/srin.201300074

Kumar, A., Chakrabarti, D., & Chakraborti, N. (2012). Data-Driven Pareto Optimization for Microalloyed Steels Using Genetic Algorithms. *Steel Research International*, *83*(2), 69–74. doi:10.1002/srin.201100189

Li, X. (2003). Chapter. In C. Fonseca, P. Fleming, E. Zitzler, L. Thiele, & K. Deb (Eds.), Evolutionary Multi-Criterion Optimization (pp. 207–221). Berlin: Springer. doi:10.1007/3-540-36970-8_15

Mondal, D. N., Sarangi, K., Pettersson, F., Sen, P. K., Saxén, H., & Chakraborti, N. (2011). Cu-Zn separation by supported liquid membrane analyzed through Multi-objective Genetic Algorithms. *Hydrometallurgy*, *107*(3-4), 112–123. doi:10.1016/j.hydromet.2011.02.008

Pettersson, F., Biswas, A., Sen, P. K., & Chakraborti, N. (2009). Analyzing Leaching Data for Low-Grade Manganese Ore Using Neural Nets and Multiobjective Genetic Algorithms. *Materials and Manufacturing Processes*, *24*(3), 320–330. doi:10.1080/10426910802679386

Pettersson, F., Chakraborti, N., & Saxén, H. (2007). A genetic algorithms based multi-objective neural net applied to noisy blast furnace data. *Applied Soft Computing*, *7*(1), 387–397. doi:10.1016/j.asoc.2005.09.001

Poli, R., Langdon, W. B., & McPhee, N. F. (2008). *A field guide to genetic programming*. Retrieved from http://www.gp-field-guide.org.uk

Rajak, P., Tewary, U., Das, S., Bhattacharya, B., & Chakraborti, N. (2011). Phases in Zn-coated Fe analyzed through an evolutionary meta-model and multi-objective Genetic Algorithms. *Computational Materials Science*, *50*(8), 2502–2516. doi:10.1016/j.commatsci.2011.03.034

ADDITIONAL READING

Chakraborti, N. (2004). Genetic algorithms in materials design and processing. *International Materials Reviews*, *49*(3-4), 246–260. doi:10.1179/095066004225021909

Coello Coello, C. A., & Landa Becerra, R. (2009). Evolutionary Multiobjective Optimization in Materials Science and Engineering. *Materials and Manufacturing Processes*, *24*(2), 119–129. doi:10.1080/10426910802609110

Datta, S., & Chattopadhyay, P. P. (2013). Soft computing techniques in advancement of structural metals. *International Materials Reviews*, *58*(8), 475–504. doi:10.1179/1743280413Y.0000000021

Mitra, K. (2008). Genetic algorithms in polymeric material production, design, processing and other applications: A review. *International Materials Reviews*, *53*(5), 275–297. doi:10.1179/174328008X348174

Paszkowicz, W. (2009). Genetic Algorithms, a Nature-Inspired Tool: Survey of Applications in Materials Science and Related Fields. *Materials and Manufacturing Processes*, *24*(2), 174–197. doi:10.1080/10426910802612270

Paszkowicz, W. (2013). Genetic Algorithms, a Nature-Inspired Tool: A Survey of Applications in Materials Science and Related Fields: Part II. *Materials and Manufacturing Processes*, *28*(7), 708–725. doi:10.1080/10426914.2012.746707

KEY TERMS AND DEFINITIONS

Data-Driven Modeling: Modeling strategy based upon experimental, industrial or simulation data, which are usually associated with a significant amount of noise.

Genetic Algorithms: Non-calculus, nature inspired optimization and learning technique

Genetic Programming: Non-calculus, nature inspired learning technique emplying a tree structure.

Materials Design: Seeking novel materials based upon physical principles and existing data, usually employing advanced computing techniques.

Meta-Modeling: Modeling a model. Often amounts to constructing a secondary data-driven model, without directly involving the physics of the process.

Multi-Objective Optimization: To seek the optimality in the presence of at least two mutually conflicting objectives.

Neural Net: Data-driven modeling strategy, roughly mimicking human brain activities.

Pareto-Optimality: Optimal condition in the presence of conflicting multiple objectives prescribed by Vilfredo Pareto.

Chapter 13
Modeling of Steelmaking Processes

Seppo Louhenkilpi
Aalto University, Finland

Subhas Ganguly
National Institute of Technology Raipur, India

ABSTRACT

In the field of experiment, theory, modeling and simulation, the most noteworthy progressions applicable to steelmaking technology have been closely linked with the emergence of more powerful computing tools, advances in needful software's and algorithms design, and to a lesser degree, with the development of emerging computing theory. These have enabled the integration of several different types of computational techniques (for example, quantum chemical, and molecular dynamics, DFT, FEM, Soft computing, statistical learning etc., to name a few) to provide high-performance simulations of steelmaking processes based on emerging computational models and theories. This chapter overviews the general steps and concepts for developing a computational process model including few exercises in the area of steel making. The various sections of the chapter aim to describe how to developed models for various issues related to steelmaking processes and to simulate a physical process starts with the process fundaments. The examples include steel converter, tank vacuum degassing, and continuous casting, etc.

1. INTRODUCTION

Computational models are nowadays increasingly accepted in steel industry to simulate metallurgical processes, such as converters, ladle treatments, continuous casting etc. These models are typically used for process designing, process optimization, process control, forecasting etc. All these day to day largely perfumed industrial metallurgical processes are obviously involved many complex physicochemical reactions. A lot of specific single models for this field have been evolved for specific purposes (Yu & Louhenkilpi 2013; Yu, Miettinen & Louhenkilpi 2013; Yu, Miettinen, & Louhenkilpi n.d.; Yu, Miettinen, & Louhenkilpi 2014; Miettinen, Louhenkilpi & Holappa 1996; Braun & Pfeifer 2007; Louhenkilpi et. al 2005; Clark, Wagner & Trousset 2003; Sahai & Emi 1996; Mazumdar, Yamanoglu & Shankarnarayanan,

DOI: 10.4018/978-1-5225-0290-6.ch013

& Guthrie 1995; Kuzmin & Turek 2004; Charkraborty & Sahai 1991), and stochastic and/or hybrid nature of models based on physicochemical laws but including these non-physical features.

In the very recent times, more attention has been attracted on developing real-time models which can be available for on-line use in industry for development of reliable process control system(Louhenkilpi, Laitinen & Nieminen 1993; Louhenkilpi et. al 2005a; Louhenkilpi et al 2005b. Real-time calculation of an industrial process offers many new possibilities in on-line process control (Allendorf et al 1998). In real-time calculation, many practical requirements will be set for the model. The computing time must, for instance, be short enough and special process conditions must be included in the model. Due to the increasing power of modern computers available, the requirements concerning the computing time can today, however, be met more and more easily.

For developing a comprehensive process model, the modelers and group must have multiple skills: good knowledge about the process to be modeled, must know the general modeling principles including the numerical methods and they must have wide knowledge about thermodynamics, metallurgical kinetics, and transport phenomena. Moreover, they should have needful expertise in the area of modern computational science. One challenge is the coupling of the thermodynamic and kinetic models with transport modeling as with CFD (Väyrynen & Wang 2010; Vapalathi et al 2006; Thomas & Najjar 1991). The other bottleneck is the determination of time when a process phenomena changes from one length scale to another length scale such as atomic regime to molecular or molecular regime to microstructural regime and modeling the interface of the two process over different length scale (Introduction 2005; Kruskopf et al 2012).

This chapter targets to present the overviews of the fundamentals of how to develop a comprehensive metallurgical process model pertinent to the complex industrial situation. The specific focus of the chapter is on the steel making processes only. The fundamental concepts of modeling, modeling techniques and the basic steps for developing a computational process model followed by heat transfer and mass transfer simulation examples are described in the first few sections. Nevertheless, the latter part of the chapter focuses on the mathematical modeling of few important processes in steel making technology. The mathematical model of BOF converter steel making process is discussed. A comprehensive discussion on BOF simulator CONSIM (Kruskopf et al 2012) is also highlighted. The methods of mathematical modeling of degassing process, continuous casting of steel, and solidification and microsegregation of steel along with few more issues in steel making process modeling has been exemplified.

2. FUNDAMENTALS OF MODELLING TECHNIQUES IN STEELMAKING PROCESSES

2.1 General Classification of Modeling Techniques Used in Steel Making Process

There are many kinds of models and modeling concepts implemented in steel industries for modeling various metallurgical processes in steelmaking. They can be grouped broadly into three major classes:

1. **Empirical/Statistical Models**
    a. Based on observations and/or measurements
    b. A lot of experimental data are usually needed (for data analysis)

   c. Methods: Statistical methods such as regression, neural networks, principle component analysis etc, Taguchi concept etc.
   d. Generalization of empirical models is not usually possible, so they are valid only for specific situations
2. **Physical Models**
   a. Physical models with water are called "water models"
   b. Based on four similarity criteria between the prototype and the real process (dimensionless numbers): Geometrical, mechanical, thermal, chemical similarity
   c. A lot of limitations: difficult to meet all the criteria in a prototype
3. **Fundamental (or Mechanistic) Models**
   a. Based on physical and chemical laws such us thermodynamic equilibrium, heat transfer, fluid flow, mass transfer, etc.
   b. Fundamental models are very general in nature and so quite universal compared to empirical models
   c. With rapid advancement of computers and numerical mathematical methods, fundamental models are today widely applied in industry to more and more complex phenomena and these models are step by step replacing the physical models
   d. Examples of mathematical methods applied in fundamental models: Numerical methods as FDM, FVM, FEM and analytical methods.
   e. A typical results of fundamental modeling is represented in Figure 1

Figure 1. A typical results of a fundamental, numerical model

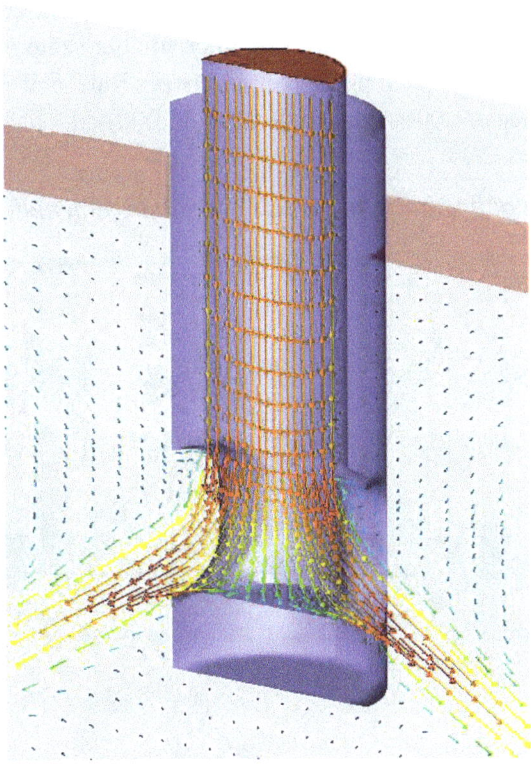

Fundamental models can also be grouped in general into two classes. One is Numerical models, in which physico-chemical model equations are solved using numerical methods such as finite difference method (FDM), finite element method (FEM), finite volume method (FVM) etc (Ueshima et al 1986; Entwistle 2001; Eymard, Gallouët, & Herbin 2000). Numerical models are good for complex cases and are today very widely used and successful to gain significant popularity in industry. On the other hand, analytical models, in which physicochemical model equations are solved using analytical methods (Dynamic 2001; Mieti'inen & Louhenkilpi 1994; Williams 1996; Qi, et al 2015). However, only simple cases can be solved using analytical methods. Analytical methods have been used a lot in the past, but, today numerical methods have more or less replaced the analytical models. Very few partial differential equations have analytical solutions. Fundamental models can also be grouped according to the scale they are developed for. Larger scale models (= process scale models) are based on conservation of mass, momentum, energy, etc. and they are called as continuum models (Kobayashi 1988; Mackenbrock 1996; Miettinen 1997; Bower et al 2015). This means that the equations used are based on balance calculations Such as mass conservation laws for instance, that matter may neither be created nor destroyed. Figure 2 described the wide application of these fundamental models over different length scales. For smaller scale there are several modeling techniques developed, such as:

- Molecular Dynamics Models (MD)
- Monte Carlo models (MC)
- Phase Field models
- Cellular automata

The Phase Field (PF) is quite a new modeling technique for mesoscale phenomena (e.g. dendritic scale). It is also a continuum model and is very popular today for modeling of microstructures like dendrites etc. Anyhow phase field concept as well as other smaller scale models need typically a lot of calculation time and they are not so much popular in industry (Karma & Rappel n.d.; Steinbach 2009; Lee et al 1999; Nestler & Wheeler 2000; Rappaz & Scheid 2000).

Figure 2. Diverse application of fundamental models over different length scales

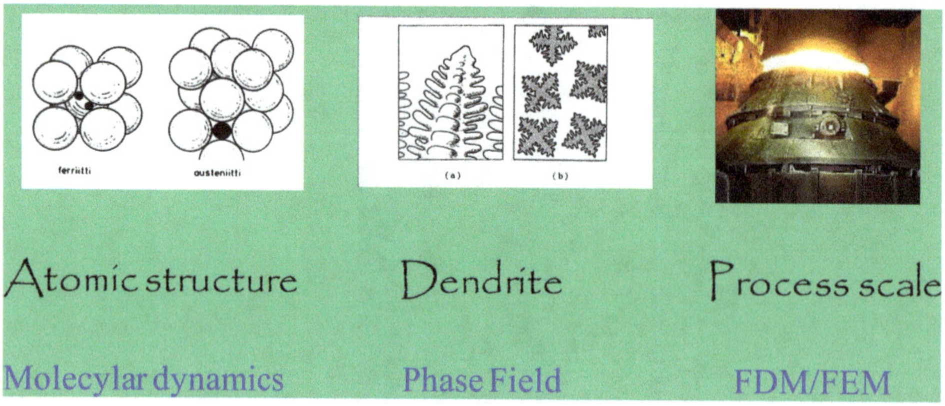

2.2 Steps in Development of Numerical Models Applicable to Process Modeling

Fundamental models are based on physical and chemical laws/principles (thermodynamic, heat transfer, mass transfer, fluid flow, etc.). In nutshell, the following steps are typically needed to develop a fundamental, numerical model.

- Understand the process fundamentally
- Make the conceptual model
- Make the governing equations (PDEs)
- Discretization of the PDEs
- Make the grid
- Write the approximation equations for each node in the grid
- Prepare and Solve the matrix equations
- Validation.

These steps are further described in more details with a simple case example of heat transfer through a medium.

Step 1: Understand the Process at the Fundamental Level

Understand the physical and chemical laws, equations and the fundamental phenomena taking place in the process.

Step 2: Conceptual Modeling

Usually processes involve many complex phenomena and it is not always possible or even needed to model all the phenomena in one model. In this step, one must consider that what are the most important phenomena and which of the model must include. Conceptual modeling refers the visualization of the phenomena as well as the process, making simplifications for the model, problem identification etc. It is also very good to make a literature search for similar types of studies. It is advisable to remember: "the model must be as simply as possible but not simpler".

Step 3: Model Formulation

The fundamental model is a mathematical description of the process and phenomena to be modeled. So, all the phenomena including boundary conditions must be described using mathematical equations. These model equations are called as the governing equations of the model. These governing equations are conservation equations for mass, heat, and so on and they are partial differential equations (PDE). So in this step:

- Define and write the partial differential equations (PDEs) and the associated boundary conditions describing the conceptual process
- Define and identify the material data needed in these equations

In numerical modeling, it is very important to understand PD equations. PDEs can be expressed using many different notations. The example of formulation PDE and subsequent calculations for heat flow problem is explained in the latter section.

Step 4: Discretization of the PDEs and the Boundary Condition Equations

To solve the governing equations, these must be firstly discretized. This means that the PDEs (and boundary conditions) must be changed to a system of discrete algebraic equations. This step is called as the discretization stage. There are many different discretization approaches as FDM, FEM, FVM etc. FDM is the oldest of the methods for the numerical solution of PDEs. In FDM, Taylor series expansions are used to generate FDM approximations to the partial derivates of the governing equations. FDM method is more commonly applied to orthogonal, structural grid with constant grid spacing. The spacing need not be uniform and more complex grid can also be used, but it is easier to make the model if the grid is very uniform. Cylindrical and spherical grid can anyhow also be used in FDM. When the partial derivates are replaced by the FDM approximations, the result is algebraic equation. The partial derivate can be written using Taylor series. An example of a discretization is given below.

$$\frac{\partial T}{\partial x} \approx \frac{T(x+\Delta x)-T(x)}{\Delta x} + \frac{1}{2}\Delta x \frac{\partial^2 T}{\partial x^2} + ... = \frac{T(x+\Delta x)-T(x)}{\Delta x} + O(\Delta x) \quad (1)$$

In FDM, the last terms in the Taylor series are usually neglected and the FDM approximation is the:

$$\frac{\partial T_i}{\partial x} \cong \frac{T_{i+1}-T_i}{\Delta x} \text{ or } \frac{\partial T}{\partial x} \approx \frac{T(x+\Delta x)-T(x)}{\Delta x} \quad (2)$$

Δx is the spacing of this grid points in the x direction, and i, i+1 are neighboring nodal points in the grid. In this case the approximation error is $O(\Delta x)$. This is called the truncation error and it is the bigger the bigger the nodal spacing Δx is in the grid. This kind of approximation is called forward difference. We can also use other ways such as backward difference, central difference, and the error term depends on the way used. In an analogous way one can obtain finite difference approximations to higher order derivatives and differential operators.

$$\text{PDE: } k\frac{\partial^2 T}{\partial x^2} = 0 \quad (3)$$

FDM approximation:

$$k\frac{T_{i+1} - 2T_i + T_{i-1}}{\Delta x^2} = 0 \quad (4)$$

Time derivatives can be obtained similar to the 1D forward differences for space.

$$\frac{\partial T_i}{\partial t} = \frac{T_i^{t+\Delta t} - T_i^t}{\Delta t} + O(\Delta t) \tag{5}$$

The accuracy of this approximation depends on the time step, Δt. If the step is high, the error is bigger. More accurate approximations can be derived.

Step 5: Make the Geometry and the Mesh (Grid) for Your System

This is the computational domain of your process. Mesh can be one-dimensional, two-, or three-dimensional as the case will be. The coordinate system can be Cartesian (x-, y-, z-), cylindrical, spherical or so on. The computational domain is constructed by using different kind of 1-, 2-, or 3-dimensional elements. In FDM the elements are usually, 1D: line, 2D: square or rectangular and 3D: cube or orthogonal 3D volume element. A typical 3D volume element is described in Figure 3.

Step 6: Write the Nodal Approximation Equation for Each Node in the Mesh from the Discrete Algebraic Equations from Step 4

The mesh has mesh nodes $j_x j_y j_z$. Each node has its identification indice(s) and the geometrical location. For instance for 1D -mesh: (i), 2D -mesh: (i, j) and 3D -mesh: (i,j,k). The distance between the neighboring nodes are Δx, Δy andΔz respectively. In this step the algebraic equations from step 4 are written for each node using its indice(s). This leads to a set or system of algebraic equations. The large 3D computational domains may have many million nodes and so many million algebraic equations. These kinds of systems must be solved by computers. If the case is small and we have just a few equations, it might be possible to solve these equations by hand calculation or for instance by general spreadsheet tools and so it is not needed to do the next steps.

Figure 3. An example of a cube 3D volume element, Δx = Δy = Δz. This is a typical volume element in 3D FDM models

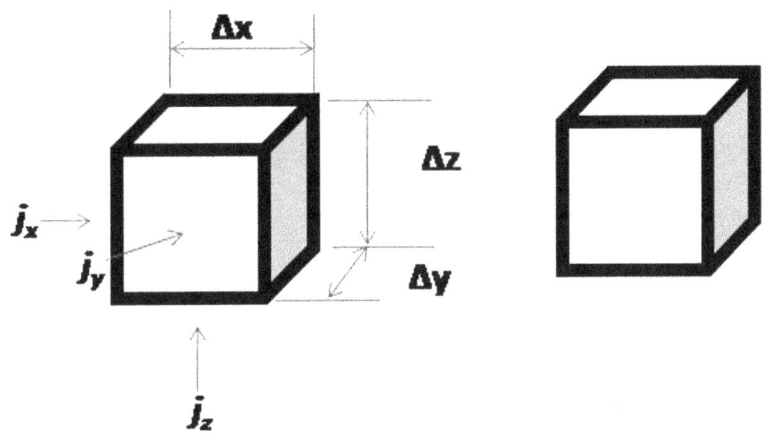

Step 7: Solve the Matrix Equations Numerically

The complexity and size of this system depends on the dimensionality, number of nodes and the governing equations. For linear and non-linear systems there are also different solution methods.

Typically there are two families of numerical solution methods:

- Direct methods (Gaussian elimination,…)
- Iterative methods (Jacobi, Gauss-Siedel,…)

Direct methods can typically be used only for simple, 1D cases.

Step 8: Validation

Validation means that the model results are compared with the measurements from the actual process. This step is important to confirm that all the simplifications, governing equations, discretizations, mesh, material properties, and solving methods etc. are correctly carried out and the model gives results which are accurate enough for the problem it is aimed to. If the result is not accurate enough, the model must be improved. Note that unless model is validated over a wide range of operating conditions, the model cannot be used for process simulation.

2.3 Fundamental of Heat Transfer Calculation

Let us consider the one dimensional heat flow through a solid medium. The corresponding PDE that describes the steady state heat transfer by conduction in X-direction with constant thermal conductivity, k can be written as:

$$k\frac{\partial^2 T}{\partial x^2} = 0 \tag{6}$$

This equation is actually a balance equation for a single, 1D calculation element (line) in the model. It says that heat transfer by conduction into the element is the same as heat transfer out from the element by conduction. The fundamental law for heat conduction is the Fourier's law.

$$q = -k_x \frac{dT}{dx} \tag{7}$$

If the case is two-dimensional and transient, the balance equation can be visualized as:

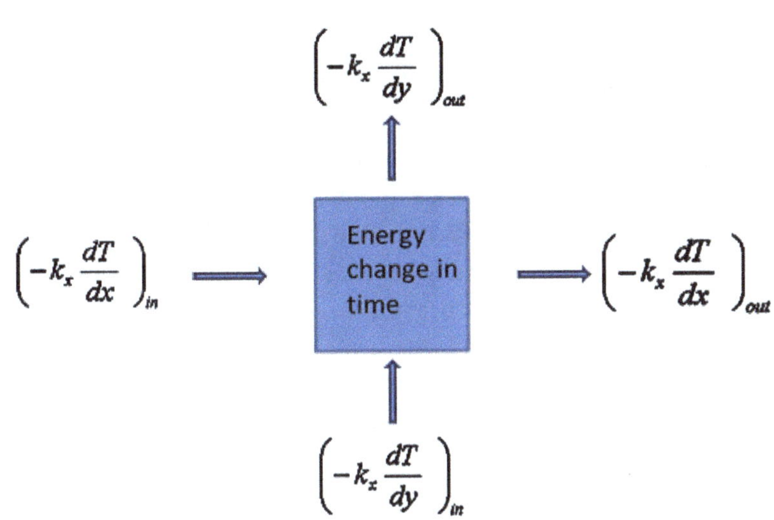

and the relevant PDE is:

$$k\frac{\partial^2 T}{\partial x^2}+k\frac{\partial^2 T}{\partial y^2}=\rho c\frac{\partial T}{\partial t} \tag{8}$$

The last term describes the change of energy (enthalpy) inside the 2D element during a time step. Extension to three dimensions means just that we must add a similar conductivity term for z-direction on the left side as for x- and y-direction. If heat release phenomena inside the element is present, it is taken into account by adding a sink/source term, S, into the PDE. It must be defined separately depending on the type of the case. For instance, in the case of solidification, it takes into account the latent heats releasing during solidification and phase transformations. So three dimensional transient heat transfers by conduction including sink/source term, is described as:

$$k\frac{\partial^2 T}{\partial x^2}+k\frac{\partial^2 T}{\partial y^2}+k\frac{\partial^2 T}{\partial z^2}=\rho c\frac{\partial T}{\partial t}+S \tag{9}$$

For many cases, we also have other phenomenon than just heat conduction. In the case of heat transfer, typically also heat convection. One-dimensional convection (or correctly said advection) heat transfer by the bulk flow can be expressed by the following formula:

$$q_x = u_x(\rho c T) \tag{10}$$

Where q_x is the heat flux, u_x is the x-velocity of the bulk flow, T is the temperature, ρ density and c specific heat, ρcT can be also expressed as H because H = ρ cT. So if we have also advection heat transfer, the advection term must be added to the PDE. In that case, for one-dimensional transient case the PDE will be:

$$k\frac{\partial^2 T}{\partial x^2}+\frac{\partial\left(v_x\rho c T\right)}{\partial x}=\frac{\partial \rho c T}{\partial t} \tag{11}$$

The radiation means that heat energy is carried by electromagnetic waves. All bodies emit electromagnetic radiation or thermal radiation to their surroundings. At the same time they absorb such radiation from the surroundings. The higher the temperature is, the more is the energy emitted. Perfect black surface or black body absorbs all the radiation from the surrounding and perfect white surface emit all the radiation from the surface. So, for instance, houses are typically painted shiny white in hot countries, because white reflects heat radiation and keeps the house cooler and the shinier the surface, the higher is the reflection also. Good emitters are also good absorbers, so black surface emits and absorbs the radiation very well. But a good absorber reflects very little energy. Radiation travels in straight lines, can travel through a vacuum, and travels at the speed of light. If there is gas (O_2, He, H_2, N_2, etc.) in the medium, radiation is not heating it but just goes through. The radiation heats up just the surface it is absorbing. But, if the gas has molecules (CO_2, H_2O, CO, ...) or solid particles, then these are emitting and absorbing the radiation. The modelling of these kinds of phenomena is difficult. The heat emitted from a perfect blackbody surface is described by the Stefan-Boltzmann law.

$$q = \sigma \cdot T_s^4 \tag{12}$$

The law says, that the heat, q, emitted as thermal radiation from a blackbody depends on the surface temperature (T_s) and on Boltzmann's constant, σ. Note that T is expressed in Kelvin. Real surfaces usually emit less than the blackbody, typically between 10 and 90 percent. This is taken into account by introducing a dimensionless factor ε (<1), called emissivity into the equation. So for the perfect black body, the emissivity is 1. Real surfaces usually emit less than the blackbody power, typically between 10 and 90 percent (ε≅ 0.1 - 0.9).

$$q = \varepsilon \cdot \sigma \cdot T_s^4 \tag{13}$$

Usually a body or surface (s_1) also absorbs radiation/heat from the surrounding surface (s_2). Let's assume that the absorbing heat is coming from a surface at the temperature T_o. Then the radiated net heat from a surface (s_1) will be:

$$q = \varepsilon_1 \cdot \sigma \cdot T_{s_1}^4 - \varepsilon_2 \cdot \sigma \cdot T_{s_2}^4 \tag{14}$$

If we assume, that $\varepsilon_1 = \varepsilon_2$, so we get the following formula for the net heat from a surface.

$$q = \varepsilon \cdot \sigma (T_s^4 - T_{ext}^4) \tag{15}$$

The modeling of radiation inside the process is challenging if there are if the gas has molecules (CO_2, H_2O, CO, ...) or solid particles, then these are emitting and absorbing the radiation. The modeling of this radiation will be more complex, if we have surfaces which are not viewing themselves perfectly. In these cases, we must use view factors. The view factor, $F_{A \to B}$, is the proportion of all that radiation which leaves the surface A and strikes the surface B.

Boundary conditions: These are needed to couple the PDEs to the surroundings. In the heat transfer models, many kind of boundary equations can be used. It is quite typical that the boundary conditions affect strongly on the final results of the model calculations. If the boundary conditions are inaccurately determined, the model gives poor results. If the temperature at the surface is directly known, this is called as Dirichlet boundary condition(Mo"es, B'echet & Tourbier 2006). If the heat flux is known, this is called as a Neumann boundary condition (Grote & Kirsch 2004) and is expressed as:

$$-k\frac{\partial T}{\partial n} = q \tag{16}$$

Where q is the known heat flux from or to the surface $\partial \boldsymbol{T} / \partial \boldsymbol{n}$ is the temperature gradient normal to the surface (in the model) and k is the thermal conductivity of the material in the model near the surface. Convective boundary condition describes heat flux from the model surface to the external flow: heat transfer coefficient, h, and the temperature of the external flow must be given. The equation in that case may be written as:

$$-k\frac{\partial T}{\partial n} = h(T_{surface} - T_{ext}) \tag{17}$$

Where, h is the heat transfer coefficient describing the heat flux efficiency from the model surface to the external flow. It typically takes into account the heat transfer by convection and conduction. The value for h might sometimes be difficult to know. It depends, for instance:

- On the conditions in the boundary layer, which are influenced by surface geometry, etc.
- Nature of the external fluid flow (velocity, turbulence,...)
- Fluid properties

The fluid can be gas, forced gas, liquid pool, forced liquid as spray water, some general values can be found in the literature. For many practical applications, more accurate and specific information are needed and so own experimental, specific studies should be done. For radiation boundary condition the above mentioned expression, Stefan-Boltzmann law, where $\varepsilon_1 = \varepsilon_{2,}$ is typically used.

2.3.1 Simulation of Heat Flow through a Wall

Let's assume that the temperature of the inner wall is known and it is 20 °C and the temperature of the outer surface is known, -10 °C. Thermal conductivity coefficient, k, is constant. Heat transfers through the wall takes place by conduction and it is one-dimensional. The situation is under steady state. Let's take the one-dimensional uniform grid with 5 nodal points as shown in Figure 4. The nodes 1 and 5 are boundary nodes, let's say inner and outer surface of the wall.

The governing equations and boundary conditions are:

Figure 4. One-dimensional uniform grid with 5 nodal points

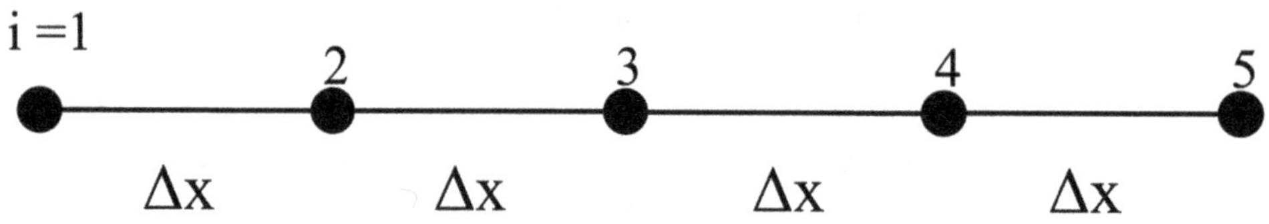

$$k\frac{\partial^2 T}{\partial x^2}=0 \tag{18}$$

$$T_1=20$$

$$T_5=-10$$

Discretization using FDM gives:

$$k\frac{T_{i+1}-2T_i+T_{i-1}}{\Delta x^2}=0 \tag{19}$$

The approximation equations for each node are now:

$$\begin{cases} T_1: & T_1=20 \\ T_2: & k\frac{T_3-2T_2+T_1}{\Delta x^2}=0 \\ T_3: & k\frac{T_4-2T_3+T_2}{\Delta x^2}=0 \\ T_4: & k\frac{T_5-2T_4+T_3}{\Delta x^2}=0 \\ T_5: & T_5=-10 \end{cases} \tag{20}$$

If we have only a few nodes, we can solve the unknown nodal temperatures already without making matrix equations and solving them. But for higher amount of nodes and complex systems, matrix equations are usually applied. To make matrix equations, we write the equations to the following form:

$$\begin{cases} T_1: & T_1=20 \\ T_2: & k\frac{T_3-2T_2+T_1}{\Delta x^2}=0\rightarrow \\ T_3: & k\frac{T_4-2T_3+T_2}{\Delta x^2}=0\rightarrow \\ T_4: & k\frac{T_5-2T_4+T_3}{\Delta x^2}=0\rightarrow \\ T_5: & T_5=-10 \end{cases} \begin{cases} -2T_2+T_3=-20 \\ T_2-2T_3+T_4=0 \\ T_3-2T_4=10 \end{cases} \tag{21}$$

$$\downarrow$$

$$[A][T]=[B]$$

And finally the matrix equation will be:

$$\begin{array}{l}\to \\ \to \\ \to\end{array}\begin{cases}-2T_2 + T_3 = -20 \\ T_2 - 2T_3 + T_4 = 0 \\ T_3 - 2T_4 = 10\end{cases} \tag{22}$$

$$\downarrow$$

$$[A][T] = [B]$$

$$\downarrow$$

$$\begin{bmatrix}-2 & 1 & 0 \\ 1 & -2 & 1 \\ 0 & 1 & -2\end{bmatrix}\begin{bmatrix}T_2 \\ T_3 \\ T_4\end{bmatrix} = \begin{bmatrix}-20 \\ 0 \\ 10\end{bmatrix}$$

Note that in this case, k and Δx have no influence on the results. Finally the matrix equation must be solved. There are many numerical methods to solve matrix equations. Simpler equations can be solved using direct methods, but more complex need iterative methods.

2.3.2 One-Dimensional, Transient Heat Transfer through a Wall by Conduction

In this example the PDE also have the material data such as density, ρ, specific heat, c and thermal conductivity, k. We assume that they are constant in the present context. Therefore the PDE is:

$$\frac{\partial H}{\partial t} = k\frac{\partial^2 T}{\partial x^2} \tag{23}$$

where the heat energy, H=$\rho c T$ and replacing H in the PDE, we have

$$\rho c\frac{\partial T}{\partial t} = k\frac{\partial^2 T}{\partial x^2} \tag{24}$$

We consider Δx to be constant between the nodes, t is time at the present time step and t+ Δt at the next time step. Discretization for the node, i, is:

$$\rho c\frac{T_i^{t+\Delta t} - T_i^t}{\Delta t} = k\frac{T_{i+1}^t - 2T_i^t + T_{i-1}^t}{\Delta x^2} + O(\Delta t, \Delta x^2) \tag{25}$$

When all the temperatures are already calculated and known at the present time step, we have for the next time step:

$$T_i^{t+\Delta t} = \frac{k \cdot \Delta t}{\rho c \cdot \Delta x^2}(T_{i+1}^t - 2T_i^t + T_{i-1}^t) + T_i^t \qquad (26)$$

This discretization scheme is called explicit. The coding of the explicit algorithm is trivial and can be performed easily using a spreadsheet as MS Excel. Explicit means that the unknown nodal temperatures for the new time (t+ Δt) are determined exclusively by known nodal temperatures at the previous time (t). At the time = 0, the temperatures are obtained from prescribed initial conditions. So the calculation begin at t =Δt and continues to the end of the time using intervals of Δt.

The term $k \cdot \Delta t / \rho c \cdot \Delta x^2$ in the equation is referred as grid Fourier number (Fo). The accuracy of the FDM solution may be improved by decreasing the values of Δx and Δt. Of course the number of nodal points increases with decreasing Δx and the number of time intervals also increases with decreasing Δt. Hence, the computing time increases with decreasing Δx and Δt. The choice of Δx is typically based on a compromise between the accuracy of the results and the computing time. Once Δx is chosen, the value of Δt cannot be chosen independently. It is determined by stability requirements. For the present case, the stability criterion is:

$$Fo = \frac{k \cdot \Delta t}{\rho c \cdot \Delta x^2} \leq \frac{1}{2} \longrightarrow \Delta t \leq \frac{\rho c \cdot \Delta x^2}{k \cdot 2} \qquad (27)$$

The explicit algorithm can be illustrated by the following stencil, see Figure 5. It says that the temperature of the node i at the new time, $T_i^{t+\Delta t}$, depends exclusively only by the temperatures of the nodes i, i-1 and i+1 at previous time and by the Fo number.

The discretization in time dependent cases (transient) can be made using explicit or implicit scheme. So far we have considered only explicit cases. As mentioned explicit scheme is trivial and can be easily coded. The implicit discretization of the PDE in this example 2 is:

$$\rho c \frac{T_i^{t+\Delta t} - T_i^t}{\Delta t} \cong k \frac{T_{i+1}^{t+\Delta t} - 2T_i^{t+\Delta t} + T_{i-1}^{t+\Delta t}}{\Delta x^2} \qquad (28)$$

Figure 5. The stencil explaining the explicit algorithm

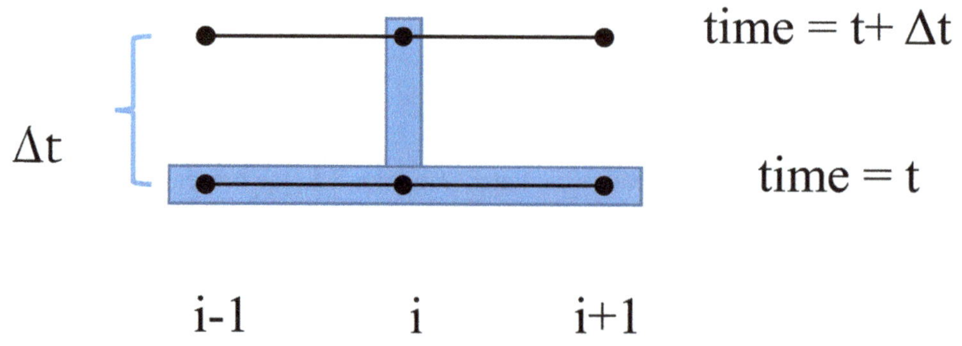

Figure 6. The stencils describing the scheme between explicit, implicit and Crank-Nicolson algorithms

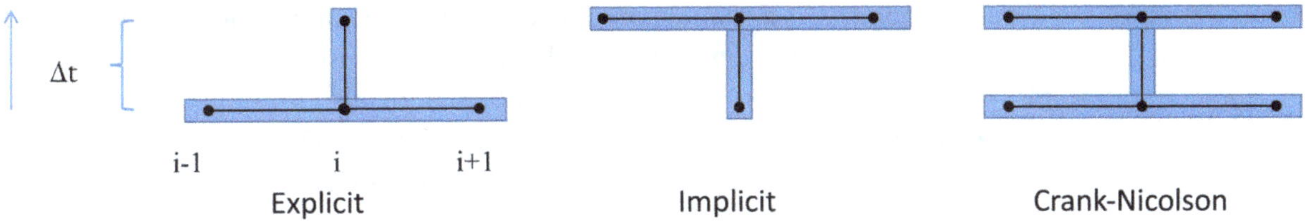

The difference is that in the implicit scheme, also in the right hand side of the equation, time for nodal temperatures is t+ Δt. Because t+ Δt is used also for the right hand side, these nodal temperatures are not known in advance as they are for the time t. This makes the implicit scheme mathematically more complicated and more difficult to solve than the explicit scheme. Anyhow, implicit scheme is more stable and larger time intervals (Δt) can be used than in explicit methods. So a reduction of computation time may often be realized by employing an implicit scheme, rather than explicit scheme. We are not studying here more accurately the implicit scheme, but the required formulas can be found in the general textbooks for FDM. One common scheme between explicit and implicit is also the so called Crank-Nicolson scheme. These schemes are illustrated by the stencils below in Figure 6.

2.4 Fundamentals of Mass Transfer Calculation

When heat transfer takes place by conduction, convection and radiation, mass transfer takes place only by diffusion and convection. Both are very important phenomena in metallurgical processes. In the bulk phase the convection typically dominates but in the boundary layers, the diffusion is important. For calculating convection mass transfer in the bulk phase, the fluid flow must be known. Typically it is calculated by solving the Navier-Stokes equations (laminar or turbulent) coupled with mass balance equation. The mass transfer equations by diffusion and convection are very analogical with the heat transfer equations by conduction and convection and so is the numerical treatment. On-dimensional (x-direction) mass transfer by convection can be described as:

$$J_x = v_x \cdot C \tag{29}$$

The origin of diffusion is the random motion of atoms, ions or molecules in a medium. The mechanisms of diffusion in solid differ strongly from diffusion in gases and liquids. In the gas and liquid the molecules are freer to move to any direction at random and the forces between them are small and can be neglected. In a solid the atoms are bound to their positions in the crystal lattice and the forces are strong between them. Diffusion is possible, but needs more energy. Increasing temperature means more energy for diffusion. If there are composition gradients in the alloy, this causes diffusion and results in mass transfer. The basic law for diffusion relates the net flux of diffusing atoms to the concentration gradient of the atoms in the medium, which can be gas, solid or liquid. This is called as the Fick's first law.

$$J_x = -D_{AB} \frac{dC_A}{dx} \tag{30}$$

Where Jx is the flux or net flow of atoms in x-direction, D is diffusion coefficient or diffusivity, C is concentration of the diffusing atoms, x is a spatial coordinate, the minus sign indicates that the atoms diffuse from higher concentration to lower concentration. Fick's first law is valid in strictly only for diffusion in pure solids (= self diffusion), but it is a good approximation also for dilute binary alloys, where only the solute atoms diffuse. More generally, the diffusion occurs to minimize the free energy. It should therefore be driven by a gradient of free energy and so the thermodynamic force for diffusion is *chemical potential gradient*. For many cases Fick's law is anyhow a good approximation.

The general 1D convection-diffusion mass transfer equation can be described:

$$\frac{\partial C}{\partial t} = D\frac{\partial^2 C}{\partial x^2} + v_x\frac{\partial c}{\partial x} + S \tag{31}$$

Where C is species concentration, D diffusion coefficient, v_x velocity in x direction and S source or sink term. The boundary conditions are also very analogous to boundary conditions in heat transfer. They are typically a know concentration or convection boundary conditions. The later can be described as:

$$-D\frac{\partial C}{\partial n} = \alpha \cdot (C_{surface} - C_{bulk}) \tag{32}$$

α is the mass transfer coefficient and it is equivalent to D_i/δ, where Di is the diffusion coefficient of the species, *i,* and δ is the thickness of the boundary layer.

2.4.1Surface Hardening of Steel by Carburization

Let's study the carburization of the steel surface. The steel carbon content is 0.2%. At time $t \geq 0$, the carbon content at the steel surface will be increased to 1%. How long does it take that the desired concentration 0.6% would be at the distance 1 mm from the surface? The temperature is 1000 °C.

We can assume that the diffusion is one-dimensional. So the PDE is:

$$\frac{\partial C}{\partial t} = D\frac{\partial^2 C}{\partial x^2} \tag{33}$$

The discretizated equation using explicit method for time and forward discretization for space (x):

$$C_i^{n+1} = \frac{D \cdot \Delta t}{\Delta x^2}(C_{i+1}^n - 2C_i^n + C_{i-1}^n) + C_i^n \tag{34}$$

The stability criterion is:

$$\Delta t \leq \frac{(\Delta x)^2}{2D} \tag{35}$$

Figure 7. Seven nodes one dimensional elements representing the thickness of the carburizing specimen

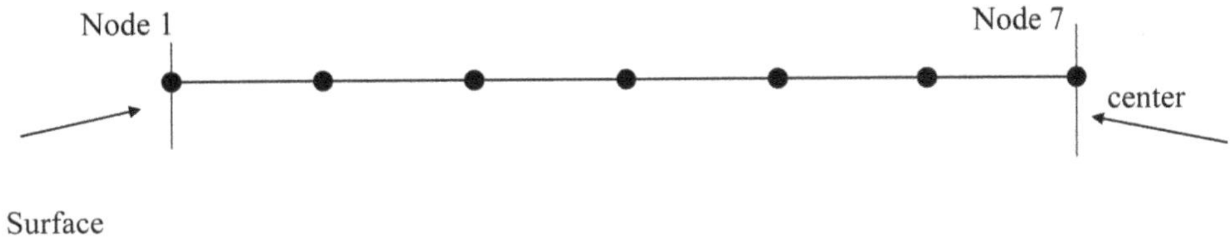

where D is the diffusion coefficient. Let's then make the 1D mesh. So, using this model one can calculate the case by following way (Figure 7).

The thickness is not given, so let's just assume thick enough, say 7 nodes with $\Delta x = 1$ mm. Consider the proper boundary conditions: say,C_1=1% at the surface and at the center we can use symmetry boundary condition, i.e. $C_7 = C_6$. Initial condition: at time = 0, C_1=1%, other nodes $C_{2,3,4,5,6,7} = 0.2\%$. Also assume, diffusion coefficient as: 2.98^{-11} m^2/s. first estimate the Δt, so that the stability criterion is met, and then calculate the case by coding the discretizated equation.

2.4.2 Hydrogen Removal from Slab

If slabs or blooms have too high hydrogen level after continuous casting, they typically are moved to hydrogen removal furnace. In this furnace the hydrogen moves out from the slab/bloom by diffusion. One problem is to know, when the H level is low enough in a slab/bloom, so that it can be moved forward and what is the ideal furnace temperature. This problem can be studied and solved by applying Fick's diffusion law and knowing the hydrogen diffusion coefficient in different steels. Let's consider the situation that hydrogen level is 10 ppm in slab and it is moved to hydrogen removal furnace. What is the ideal furnace temperature and how long the slab should be in the furnace?

Let's make 1D mesh. For slab, 1 D is enough but for bloom, 2D model is needed. The calculation domain is from the slab surface to its center. The equations are the same as in the case of earlier. If we use the same mesh as in the previous example, the boundary conditions are: C_1=0 ppm and at the center we can use symmetry boundary condition, i.e. $C_6 = C_7$. Initial condition: at time = 0, C_1=0 ppm, all other nodes C=10 ppm. Rest of the calculation is same as before.

As mentioned earlier, the Fick's laws are valid for homogeneous phase, where only one element is diffusing because of its concentration gradient. In this case steel is the homogenous bulk phase and hydrogen is diffusing in it because of the gradient (C_1 = 0 ppm). But because there are so many different steels grades, D_H must be determined for each of the steel separately. What is the ideal temperature for the furnace? If we look at the diffusion coefficient picture, we can see that D_H is higher in ferrite than in austenite. So the optimum temperature is near the A_1 temperature of the steel to be treated. At lower temperatures, H gets trapped due to various reasons, which decreases diffusion. During diffusion, H atoms wander in a random manner through the lattice (interstitial diffusion) but tend to get trapped or delayed at sites acting as traps. These sites are potential wells, whose depth is greater than in the undisturbed, defect-free crystal lattice. In steels, there are different kinds of traps caused by several reasons, like grain boundaries, dislocations, strain fields, precipitates and deformation. So the trapping effect should be taken into account especially at the furnace temperature of lower than ≈ 400 °C, because below that the trapping is strongly increasing. Figure 8 shows a typical calculated hydrogen diffusion coefficient in FM35 (C 0.36, Si 0.2, Mn 0.8) steel.

Figure 8. Calculated hydrogen diffusion coefficient in steel FM35 (C 0.36, Si 0.2, Mn 0.8), together with experimental data points. Solid line refers to the effective diffusion with alloying induced trapping and dotted line refers to the lattice diffusion with no trapping

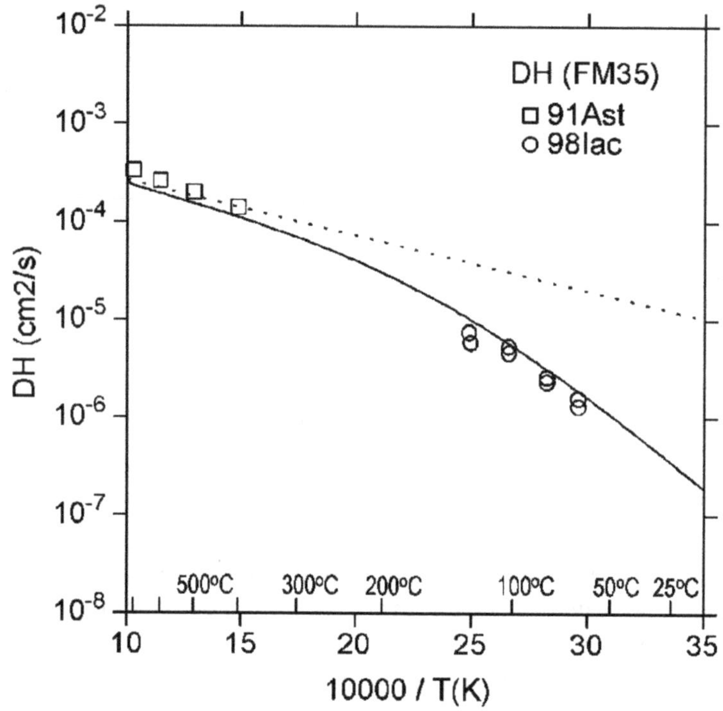

3.0 MATHEMATICAL MODELING OF STEEL CONVERTER PROCESS, BOF

3.1 Fundamental Phenomena in BOF Steel Making Process

Basic oxygen steelmaking converter (BOF) is the most widely used process for producing crude steel (low carbon steel) from molten pig iron (hot metal from blast furnace with high carbon content) under oxidizing and basic environment (basic converter slag).The slag phase plays a crucial role in all metallurgical processes and so also in BOF. An early formation of the liquid slag phase during the beginning of the BOF process is for several reasons of great importance. There exist many different types of BOF converters. Generally they are divided into top-blowing, bottom-blowing and mixed blowing converters. The first, modern converter was developed in 1952-53 at Linz and Donawitz in Austria. It was called as LD converter. It was a top-blowing converter. Pure oxygen is injected at supersonic speed into molten iron through a water cooled, multi-hole lance. Later the bottom-blowing and mixed blowing converters were developed. One big advantage of mixed blowing converters is a very efficient metal stirring during blowing. This leads to faster reactions and it is possible to reach lower carbon content in steel. Almost all top-blown converters are today equipped with bottom blowing facilities. Usually argon or nitrogen (even oxygen) is blown through the vessel bottom for improved stirring. The oxygen converter process is essentially an oxidizing refining process in which high carbon hot metal (3.7% to 4.5% C) is converted to medium or low carbon crude steel with low impurity content. The desired carbon level depends on the

Figure 9. A simple schematic of a BOF type furnace

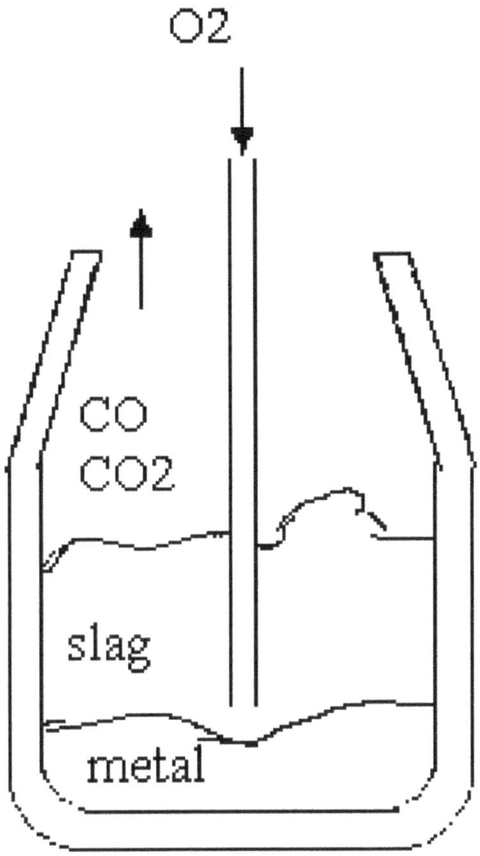

manufactured steel type. The oxidation reactions in BOF are very exothermic and energy is released so much that recycled scrap and/or iron ore additions are used to compensate the too high heating. There are many names for different types of converters but here we use BOF as a general name for all converters. A simple schematic of a BOF type furnace is shown in Figure 9.

Typical inputs:

- Pig iron from blast furnace (S- or P- treatment can be carried out before charging)
- Scrap and ore to adjust the thermal balance to obtain the target steel temperature
- Addition to form basic slag: CaO, CaO-MgO,…
- Oxygen through top lance (or bottom nozzles)
- Argon/nitrogen through bottom nozzles

Outputs:

- Crude steel → to secondary steelmaking
- Waste gas (rich in CO) → recovered → for use
- Slag poured out after the steel → recovered → for use

Table 1. Typical melt compositions and temperature before and after the blowing

	%C	%Mn	%Si	%P	%S	%O	T/°C
Pig Iron (in)	4.6	0.3	0.2	0.08	0.020	0.0	1370
Crude Steel (out)	0.03	0.1	0.0	0.01	0.017	0.06	1700

Fundamental phenomena in BOF are highly complex. The oxygen from lance is dissolving into the molten iron and the oxidation of the C and other dissolving elements is starting. C is reacting with dissolved O to gaseous CO (g) and this is going out of the converter. CO is partially oxidized into CO_2 above the melt. This is called post-combustion. Si, Mn, P, S and Fe are also oxidizing and they form oxides, which are going to the slag. These oxides combine with the other oxides charged and are foaming slag. Under certain conditions these reactions can go in the reverse and the elements go back to the melt. Slopping is the general term used when the foam (liquid slag with gas bubbles) comes out of the converter during blowing due to the excessive foam growth. This is greatly depending on the oxidization state of the slag. Because of the strong mechanical effect of the oxygen jet impact on the melt, melt droplets are formed at the impact area. The droplets go to the foaming slag and the reactions take place also there.

There are in general three typical reaction periods during the blowing. The first period is called as the silicon oxidation period. During that period, most of the Si is oxidizing and some Fe and other elements also which have strong thermodynamic affinity to the oxygen (Ti,...). Most P is also oxidizing. The remainder O reacts with C. The next period is the full decarburization period, and then the final is the carbon diffusion period. During the silicon oxidation period, the basic reactions are (Gordon et al 1975; Turkenitch 1971; Deo & Boom 1993; Bardenheuer et al 1970; Kusnetsov, Sham & Bol'shakov 1972).

- $O_2 \rightarrow 2\,[O] \rightarrow$ to the steel melt
- $[Si] + 2\,[O] \rightarrow (SiO_2) \rightarrow$ reaction with slag components
- $[Mn] + 2\,[O] \rightarrow (MnO_2) \rightarrow$ reaction with slag components
- $[Fe] + [O] \rightarrow (FeO) \rightarrow$ reaction with slag components
- $2\,[P] + 5\,[O] \rightarrow (P2O_5) \rightarrow$ reaction with slag components
- $[C] + [O] \rightarrow CO\,(g) \rightarrow$ post combustion $\rightarrow$ off gas

The second period is the full decarburization period and then mainly all the oxygen reacts with carbon. If the lance position is (too) high, some oxygen reacts with Fe also. On the other hand, if the lance position is too low, FeO from slag is reducing and released O reacts with C again. The basic reactions are:

- $[C] + [O] \rightarrow CO\,(g) \rightarrow$ post combustion $\rightarrow$ off gas
- $[Fe] + [O] \Leftrightarrow (FeO) \rightarrow$ reaction with slag components

The final period can be called as the carbon diffusion period. At the end of the blow, when carbon content is low, the carburization rate decreases and is mainly limited by the diffusion in the metal bath. Bottom blowing (Ar/N_2) is effective to increase the melt stirring and so diffusion/convection. Some other elements may again be oxidizing, as Fe, Mn, P,...

So, in addition to primary reactions taking place in iron bath, secondary oxidation of carbon and minor constituents takes place on the surface of iron droplets circulating in slag (or at the surface of primary oxides penetrated in iron bath) iron oxides as the main oxidant:

$$(Fe_tO)_{slag} + [C]_{Fe} \rightarrow CO(g) + t[Fe]_{Fe}$$

$$\beta(Fe_tO)_{slag} + \alpha[M]_{Fe} \rightarrow (M_\alpha O_\beta)_{slag} + \beta t[Fe]_{Fe}$$

($M = Si, Mn, P, Ti, V$ *and* Fe)

These reactions are as well reduction reactions for iron oxide and other slagged oxides acting as oxidants. Oxidation of carbon monoxide formed in primary oxidation takes place either by direct reaction with oxygen:

$$O_2(g) + 2CO(g) \rightarrow 2CO_2(g)$$

or by oxidizing slag constituents like iron oxides:

$$(Fe_tO)_{slag} + CO(g) \rightarrow CO_2(g) + t[Fe]_{Fe}$$

These two last reactions are called as the post combustion reactions because CO2 is forming from CO.

When starting to develop a numerical model for any system or purpose, the step one is to understand the process at the fundamental level. It is also good to think and define clearly the objective(s) of the model. The objectives define the nature of model and the phenomena that has to be considered. For BOF, the objectives could be:

Objective 1: Find out minimum Carbon and Phosphorous
Objective 2: Predict the final metal temperature
Objective 3: Predict Carbon and temperature at any time
Objective 4: Predict slopping
Objective 5: Study mixing characteristics to improve it
Objective 6: A full holistic process model including all the most important phenomena

The step 2 is the conceptual modeling, i.e. visualization of the modeled phenomena as well as the process, simplifications for the model. It is also very good to make a literature search for similar types of studies. A good rule is that "the model must be as simple as possible but not simpler".

For BOF a lot of specific models for specific purposes have been developed and also many simplified process models are available. However, there is still lack of a general, full process model including all the most important phenomena in one model, as the phenomena in BOF are very complex, so a fundamental model including accurately all the phenomena in one model is challenging task. Typically we need the following main modules to develop a fundamental BOF process model:

- Thermodynamic module,
- Heat transfer module, and
- Mass transfer module.

Especially the mass transfer module is difficult to develop. It should include the fluid flow calculations of the bulk metal flow, effect of top and bottom blowing on it, supersonic oxygen flow calculations and effects on the liquid steel surface on which they impinge, model for the slag formation and its foaming, mass transfer of the dissolved elements in the bulk melt, melting and dissolution of scrap, etc. Many specific CFD models for mass transfer have been developed as supersonic oxygen flow calculations and the effects on the liquid steel surface on which they impinge as well as the effect and configuration of the bottom tuyres for metal flow in the BOF. The latest process models include the combination of the CFD modeling tool with the thermodynamic software or module but with a lot of simplifications. Even with them the simulation can take from several hours to several weeks and cannot be applied for the prediction of all parameters of the converter steelmaking process.

3.2 Thermodynamic Module

Let's consider one single reaction taking place in BOF as an example:

$[Si] + 2\,[O] \rightarrow (SiO_2)$

To calculate this reaction, from thermodynamic point of view, we need to know the reaction Gibbs energy and the activities of the a_{Si}, a_O in the melt and a_{SiO2} in the slag phase. The reaction Gibbs energy and K_e (equilibrium constant for the equation above), can be formulated as:

$$\Delta G_r^o = -RT(\ln K_e) \tag{36}$$

$$K_e = \frac{a_{SiO_2}}{a_{Si} \cdot a_O^2} \tag{37}$$

R is the gas constant and T the absolute temperature. For activity calculations of the elements in the steel melt, we need a thermodynamic solution model for liquid steel and the corresponding thermodynamic databank. There are many different thermodynamic models for liquid steel, one quite general is the Wagner-Lupis-Elliot dilute solution formalism. For calculation activities in the slag phase, we need a thermodynamic solution model for slag and the corresponding databank. The slag model should include the liquid slag phase as well as the solid slag phase, because the slag can be solid in the beginning and for instance solid CaO will be added during the blowing. A lot of different slag models are also available but it is important that the used model includes all the important slag elements. It is quite a big work to code comprehensive solution models for steel and slag including the databanks, so one alternative is to use commercial library tool as ChemApp (Petersen & Hack 2007). We have many reactions and the question is how we know which reaction or reactions, during a calculation time step, take the oxygen and how much, i.e. what is the reaction order. In general we know, that minor elements tend to oxidize in thermodynamic order (Ti, Si, Mn, C, Fe, P). Oxygen tends to be used predominantly in the reactions leading to greatest decrease in the Gibbs energy of the system. There are typically two thermodynamic solutions for that. The first one is the minimization of the Gibbs energies and the second one is the use the thermodynamic reaction affinity equations. The general condition of a chemical system is that the system's Gibbs energy is decreased and the equilibrium state is reached where the Gibbs energy has the minimum value. So searching the minimum Gibbs energy for the system where all the elements of each reaction are available, the equilibrium state is reached. This concept is computationally quite heavy and needs relatively much calculation time and also a good numerical minimization algorithm must be available. Again, one alternative is to use a commercial library tool as ChemApp for this. The second concept is computationally much faster and easier. For each reaction, the activity values and reaction Gibbs energies are determined and the reaction affinity value for oxygen (see case example CONSIM later). The reaction which has the highest affinity value during a time step, takes all the oxygen during that step. If the time step is small, the concept is reasonable. After each time step, the affinity values are calculated again. This concept also leads to thermodynamic equilibrium state but it is possible to take the kinetic effects into account using reaction rate formulas coupled with the affinity equations. In BOF the temperatures are high, so it can be assumed that the reaction rates are fast and chemical reaction kinetics plays no significant role. So, thermodynamic equilibrium condition can be used.

Now we can calculate the reactions to their equilibrium states. We still need to know the reaction temperatures, pressures for the gas elements O_2, CO, CO_2, and we need to resolve, what proportion of the oxygen penetrates the bath to remove the carbon and other melt elements, and what proportion remains in the top space for post-combustion. One more question is, where the reactions take place. There are three or even more environments at which iron bath constituents are oxidized and slagged:

1. The primary oxidation zone at oxygen jet impingement surface,
2. Iron with high oxygen content and even oxides formed in primary oxidation zone penetrate into iron bath and react with the fresh melt from the bath interior,
3. Iron droplets circulating the slag, and
4. In the end of a blow even direct interaction with slag and iron baths may play some role.

The first two reaction environments are in the superficial part of the iron bath and the third one is formed by droplets thrown from the superficial layers of iron melt to slag or in furnace atmosphere. Mass flows in the iron bath are transporting melt with oxidizing elements from the bath interior to the superficial layers. As there is not enough knowledge available to establish a quantitative model to compute the progress of oxidation reactions in all previously mentioned reaction environments separately, simplifications must be done. One concept is that the actual reaction zones (2 to 3) are treated as a one generalized reaction zone, or hot spot zone, which has a homogeneous composition. In this zone to which bath mixing flows transport alloy elements from the bulk of iron bath, the primary oxygen from top lance and secondary oxygen from reduction of slagged oxides (predominantly iron oxide) is distributed between the oxidation reactions of impurity elements and iron under thermodynamic control. The composition of the generalized reaction zone is controlled by mass transfer rate from the bulk and rate of oxidation reactions. This is a bit simplified concept and assumes that the reactions take place in the hot spot area. Some reactions take place in the dispersed slag foam, but this is so difficult to model that simplifications must be done as this one generalized reaction zone. Weak mixing (high lance blowing

Figure 10. The droplets are in an oxygen rich environment and due to their small size increase the effective reaction area strongly

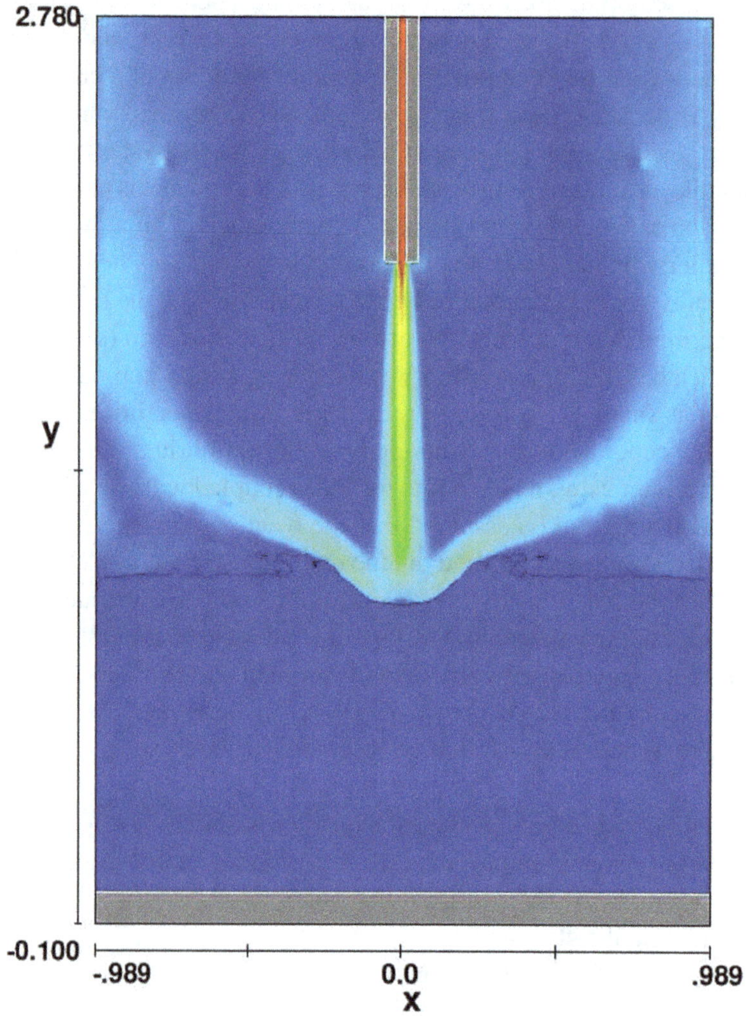

with no bottom stirring) in iron bath leads to a decrease in impurity element concentrations (chemical potentials) in the superficial layers of metal bath and, accordingly, in rates of their oxidation – iron tends to be predominantly oxidized. Intensive mixing tends to enrich the chemical reaction zone with impurity elements and leads to acceleration of their oxidation. The Figure 10 describes a typical simulation results that the droplets are in close vicinity of an oxygen rich environment.

3.3 Mass Transfer Module

Fluid flow in BOFs is highly complex. A fundamentally based multiphase reacting flow model capable of predicting the dynamics of flow and mass transfer in BOF has not been possible till date. For accurate BOF modeling, mass transfer is anyhow also very important. The "hot spot" area where the reactions are fast is quite small. So mass must be transferred from the bulk zone to the hot spot area and/or dissolved oxygen from the hot spot area to the bulk zone! Melt droplets from bath surface to the slag is also important (= strong metal-slag mixing = reactions are faster). CFD models for bulk metal flow can be done, but mass transfer in slag bath is typically not taken into account as it is assumed to be very intensively mixed (during slag foaming period) and as it would be extremely difficult to create a model for mass transfer of slag constituents and its role as rate controller for oxidizing-reduction processes taking place in slag. Slag zone is typically assumed to be homogeneous. Important parameters affecting the mass transfer are the top lance parameters (level position, design, oxygen flow rate,…), bottom blowing parameters (locations, gas amount,…), and converter design. A fundamentally based multiphase mass transfer module should include sub-models for many things as for bulk flow, supersonic oxygen flow and its effects on the baths, turbulence models, energy models, scrap model and its effects on the baths flow, etc. Typically this is too complex still today and many simplifications must be done.

3.4 Heat Transfer Module

A large amount of heat is released due to the oxidation reactions, making the system essentially autogenously. For calculating the heat balance in the system, we need to do an energy or heat balance model for the whole system, including the additions, scrap and heat losses. As for mass transfer, fundamental model for heat transfer also needs many PDEs and these should be coupled with mass transfer and thermodynamic modules. This is also quite a complex task and not been possible till date either. So many simplifications must be done. Temperature of the melt at primary oxidation zone (I) is very high. It is observed that surface temperatures can be up to 2500 °C which is far over the mean bath temperature. Also the slag temperature has been reported to exceed clearly the mean iron melt temperature. On the other hand the temperature of non-melted scrap is lower than the mean temperature of the bath. All these temperatures should be modeled accurately to get good results from the BOF model.

Post-Combustion

The dissolved oxygen and carbon react to carbon monoxide and later above the melt some of CO will oxidize to CO_2. This is called post-combustion. The ratio $CO_2/(CO_2+CO)$ is called the post-combustion ratio. The post-combustion releases about three (3) times more energy that the C+O → CO reaction. So it is important to control and know the post-combustion ratio because post-combustion affects much on the heat balance of the whole furnace. Some converters are equipped with special post-combustion technology to increase the post-combustion ratio and better utilize this energy. Some post combustion takes place also indirectly by the reduction of iron oxide in slag, as presented earlier in reactions.

$$CO(g)+('FeO')_{slag}=CO_2(g)+Fe(l) \tag{38}$$

It is difficult to estimate how substantial part of the carbon dioxide is produced by direct oxidation of carbon monoxide and how substantial part by the reduction of iron oxide. On the other hand, iron oxide doesn't be reduced only by this reaction, but also other impurities in hot metal can in their own oxidation reactions reduce iron oxide.

3.5 Application of CONSIM in BOF Converter Simulation

CONSIM is an in-house BOF converter simulation tool developed in Aalto University in Finland (Jalkanen &Andersin 1989; Andersin & Jalkanen 1990; Suomi et al 1995; Jalkanen, Suomi & Wallgren 1995; Jalkanen, Suomi & Wallren 1997; Jalkanen, Suomi & Wallgren 1998; Jalkanen, irrankoski, & Oksanen 2002; Jalkanen & Kostamo 2004) led by professor Heikki Jalkanen. It can be used off-line but also on-line. CONSIM is developed for the combined blown converter processes with inert gas bottom blow. In this process in addition to top oxygen blow with a lance also bottom stirring of inert gas, argon or cheaper nitrogen, is applied. Bottom stirring improves the mass transfer in hot metal, accelerates the melting of charged scrap and reduces the oxidation and slagging of iron. Oxygen converter simulator CONSIM describes the chemical and thermal evolution of the process predicting the end analysis, mass and temperature of steel and slag. Program consists of a series of subprograms comprising input data files for charged materials and blowing program, mathematical models for various chemical and thermal phenomena with several parameters to be optimised against data from actual blow in an individual converter.CONSIM includes the three main modules:

1. Thermodynamic module,
2. Heat balance module, and
3. Mass transfer module.

The mass transfer is modeled using a phenomenological equation and the fundamental laws are not used. This is a simplification, but makes the model fast and it can be used also on-line. Phenomenological concept includes many parameters, which must be fitted to each particular converter. That is the disadvantage of the simplification. In CONSIM, converter is divided into three generalized calculation areas: reaction zone (or hot spot zone), the bulk metal area and the slag area. The scheme of simulation adapted in CONSIM along with the reaction zone concept is highlighted in Figure 11 and the program flowchart is shown in Figure 12. The composition in these areas is homogenous over the zone. Reactions, as well as mass and heat transfer take place between the areas according to the thermodynamic, mass transfer and heat transfer modules. Dust, fluxes, and oxides from the reactions go to the slag phase, iron melt and scrap to the bulk metal area and O_2 from lance and from slag reductions to the reaction zone.

Thermodynamic affinity of oxidation reactions control the oxygen distribution between the bath constituents including carbon, minor hot metal constituents and iron and all oxygen is consumed in primary oxidation reactions (*oxidation of hot metal minor constituents and iron*) and post combustion (oxidation of primary carbon monoxide to carbon dioxide).For the components to be slagged (M = Si, Mn, P, Ti, V and Fe), the reaction can be written as:

$$\left[O\right]_{Fe} + \frac{\alpha}{\beta}\left[M\right]_{Fe} \rightarrow M_{\alpha}O_{\beta}(s,l) /// M_{\alpha}O_{\beta}(s,l) \rightarrow (M_{\alpha}O_{\beta})_{slag}$$

Affinity $A = \Delta_R G = -\partial G / \partial \xi_j$ controls oxygen distribution between the oxidation reactions (*oxidizing melt constituents, C, Si, Ti, V, Mn, P, Fe etc.*). Affinity equation for the above reaction is:

Figure 11. Scheme of simulation and the principle of the reaction zone concept. The reaction zone area is represented by the dashed line area in the figure

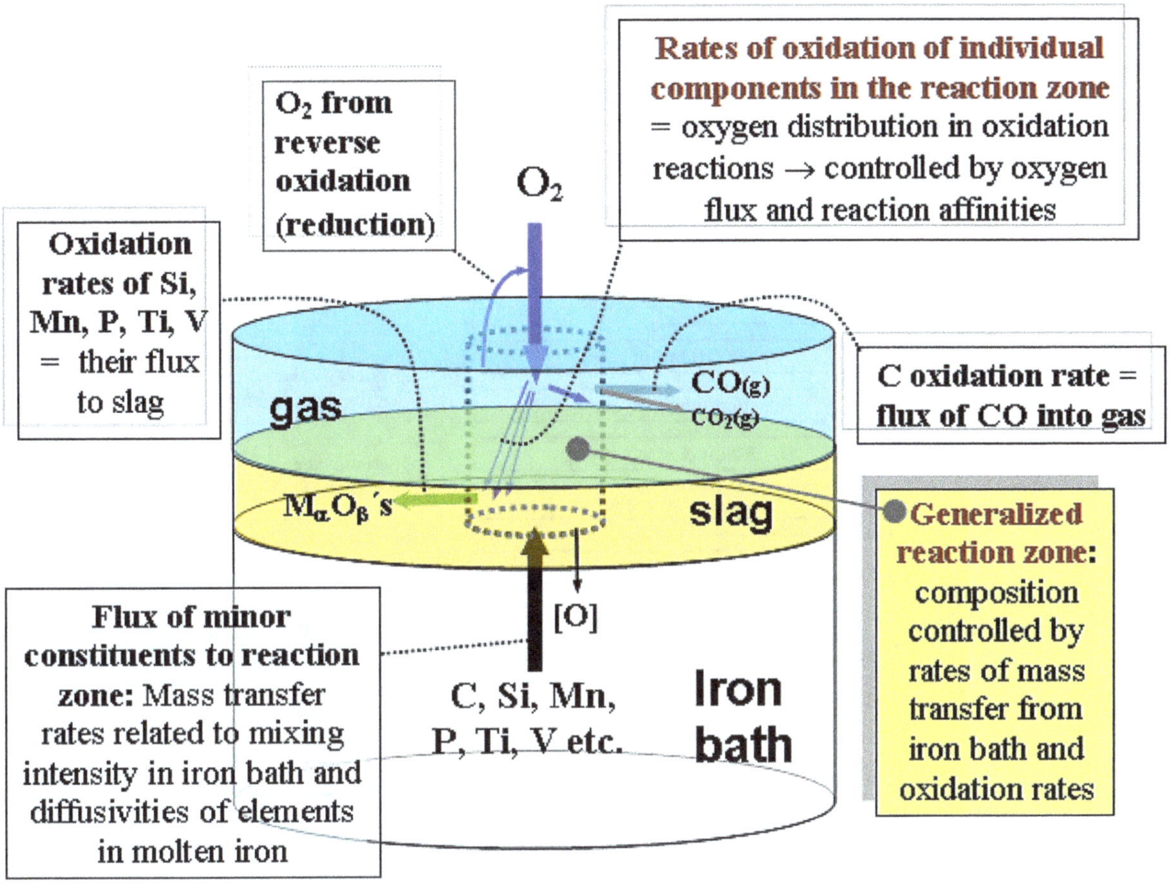

$$A_{(3)} = -\left(\frac{\partial G}{\partial \xi_{(3)}}\right) = -\Delta_R G^o{}_{(3)} - RT\ln\left(\frac{a_{(M_\alpha O_\beta)}^{2/\beta}}{a_{[M]}^{2\alpha/\beta}\cdot(p_{O_2}/p^o)}\right)$$

$$= -\Delta_R G^o{}_{(3)} - RT\ln\left(\frac{a_{(M_\alpha O_\beta)}^{2/\beta}}{\gamma_{[M]}^{2\alpha/\beta}\cdot x_{[M]}^{2\alpha/\beta}\cdot(p_{O_2}/p^o)}\right) \tag{39}$$

Affinity is defined relative to one mole of gaseous oxygen and not relative to dissolved oxygen. This mode has the advantages that oxygen dissolution in iron is of the same form as the oxidation reactions and the program can be more easily modified for simulation of converter processes using inert gas-oxygen mixtures instead of pure oxygen.

The mass balance in CONSIM comprises:

Figure 12. Program flowchart in CONSIM, converter simulation tool developed in Aalto University in Finland

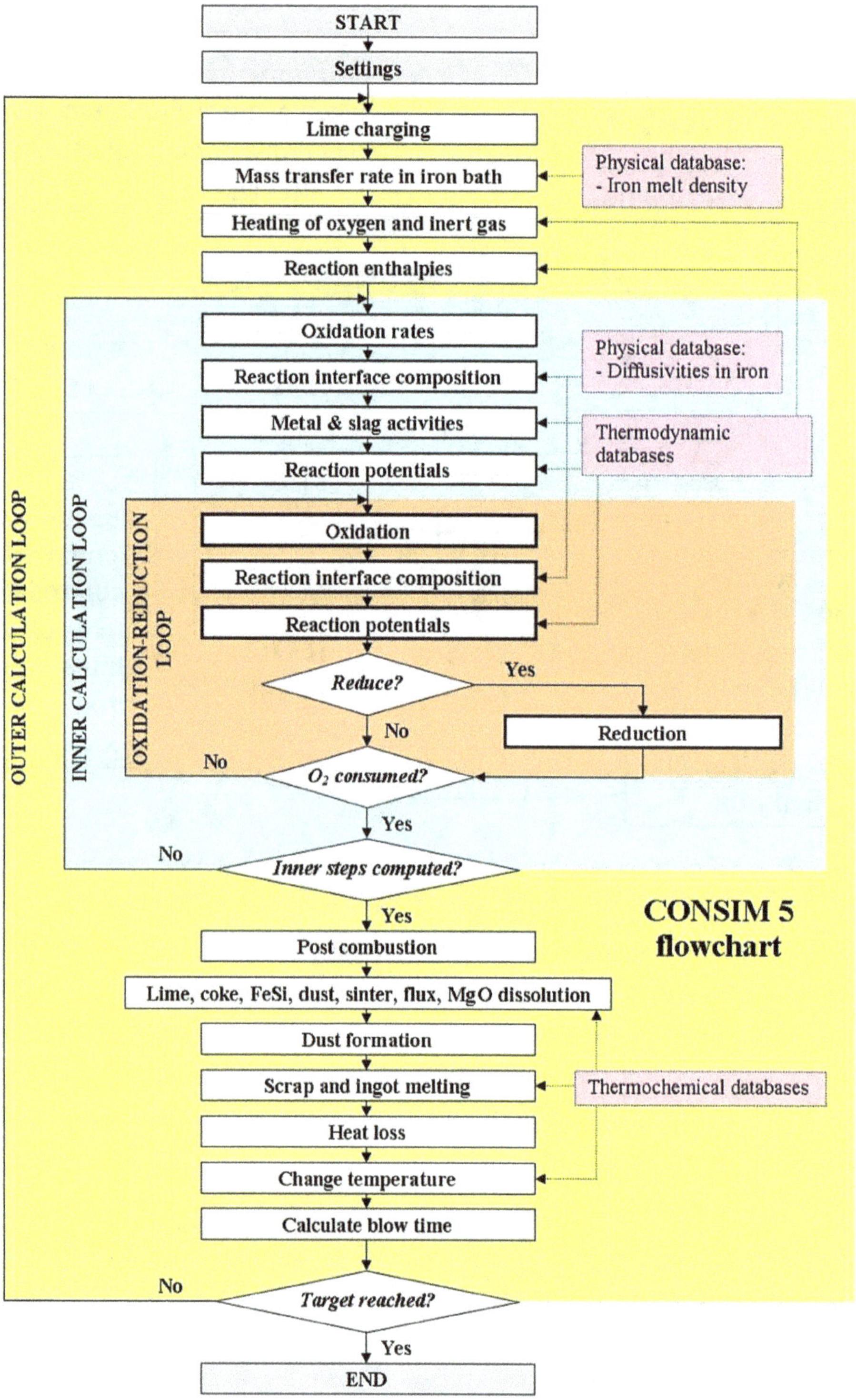

- The materials charged into the converter: hot metal, scrap and other coolants like ingots, and slag formers, lime (or doloma), fluorspar, iron ore, converter dust (the two latter materials are as well coolants).
- Metal bath (steel in the end of the blow) with its constituents.
- Molten slag and share of solid lime (not necessarily completely dissolved even in the end of the blow).
- Off gases and dust brought with it out of the converter.
- Dissolved furnace lining.
- Oxygen blown into the converter is consumed in oxidation reactions including post combustion; minor part of oxidation is due to higher iron oxides and other oxides acting as oxidants brought into the reactor by iron ore or converter dust.

Mass transfer from bulk melt to the reaction zone is calculated by the following phenomenological equation:

$$^{s}x_{i} = {}^{b}x_{i} - \frac{M_{Fe} \cdot Pa}{\varepsilon^{y} \cdot D_{i}^{1/2} \cdot \rho} \cdot {}^{mt}\dot{n}_{i} \tag{40}$$

The heat balance in CONSIM comprises:

1. Heat brought to the converter by hot metal
2. Heat evolved in chemical reactions, in oxidation of main hot metal constituents, carbon, silicon (*could be added also as ferrosilicon*) and iron as well minor elements phosphorus, titanium, vanadium etc., reduction of oxides in ore (*sinter or pellets*) and dust if recycled and post composition heat (*burning the primary carbon monoxide to carbon dioxide*).
3. Heat absorbed in heating and possible melting of charge materials, scrap, ingots, lime, fluorspar, ferrosilicon, coke, iron ore (*sinter or pellets*) and flue dust as well as gases (*oxygen and inert gases*) blown into the converter.
4. Heat loss during the blow by heat flow to the reactor lining and through it to the reactor environments as well as heat lost (*mainly*) by radiation through the furnace opening between the blows.
5. The balance of heat is consumed in heating of iron melt, slag, converter gas and dust escaping with it from the furnace.

4. MATHEMATICAL MODELING OF DEGASSING PROCESS

After primary steelmaking in BOF, the steel is treated in different kind of secondary steelmaking processes and process combinations. The treatments are most commonly performed in ladles. Treatments applied depend among other things on the steel grade and quality requirements. One typical secondary steelmaking process is the vacuum treatment. There are several types of vacuum treatment processes developed. One typical is the tank degasser. In the tank degasser, the ladle is placed in an open-top vacuum tank, which is connected to vacuum pumps. Argon is injected through bottom plugs to decrease the pressure even further but also to cause melt mixing. Mixing is important because mixing brings the reactants

Figure 13. Schematic structure of a vacuum tank degasser

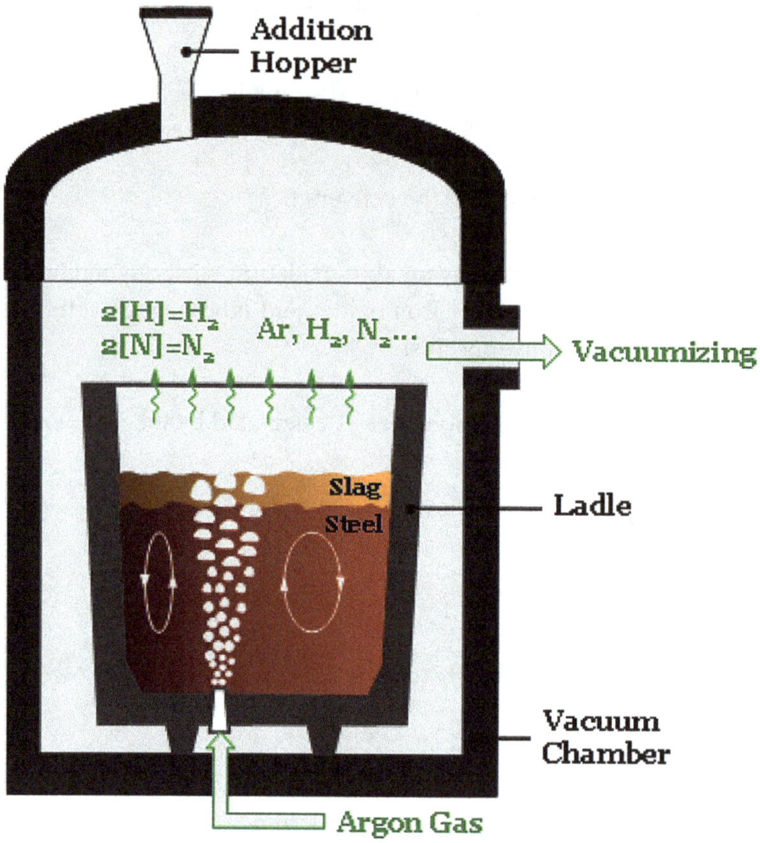

together and removes the products away from the reaction zones. Reactions take place in many zones in the ladle, but primarily at the metal/gas and metal/slag and metal/bath surface interface but liquid-solid interactions also take place at the liquid-refractory interface. The melt surface is covered with a slag layer and various impurities are also removed to the slag layer. Too high gas stirring may lead to problems, because the risk for melt splashing is increasing. Practical treatment time is 20 to 30 minutes. Schematic view of a vacuum tank degasser is shown in Figure 13. Vacuum degassing is applied in the steel industry for several purposes, such as:

- To removal of dissolved gases e.g. oxygen, hydrogen, nitrogen,
- To removal of undesired trace elements with high vapor pressures,
- To produce steel of low carbon content (< 0.03%),
- To produce steels to close chemical composition ranges (including deoxidizers), and
- To remove inclusions from steel.

In general the vacuum-degassing equipments will not probably change drastically in the near future, but the efficiency can be improved by more automation and computer modeling. The main parameters that can be controlled in the tank degassing unit are the pressure, time, steel composition, flow rates of stirring gas injected, and in general the circulation of the melt. The circulation, on the other hand, depends strongly on the ladle design and on the location of the gas injection nozzle(s) in the ladle bottom. Degassing has been modeled quite successfully using both fundamental and statistical concepts. The main reactions taking place in degassing are:

$C+O \rightarrow CO\ (g)$
$H+H \rightarrow H2\ (g)$
$N+N \rightarrow N2\ (g)$

Low vacuum is used to lower the partial pressure of the dissolved elements. When starting to develop a numerical model, the first step is to understand the process at the fundamental level and to think clearly the objective(s) of the model. For degassing unit, the objectives could be:

Objective 1: Predict and optimize the hydrogen or nitrogen or oxygen removal
Objective 2: Predict the inclusion removal
Objective 3: Calculate and study the fluid flow pattern induced by the gas injection
Objective 4: Calculate carbon removal

The step 2 is again the conceptual modeling, i.e. visualization of the modeled phenomena as well as the process, simplifications for the model. For a general degassing model, we need to model the following phenomena:

1. Transfer of the "elements" from bulk phase to the reaction interfaces as to the metal/gas interfaces.
2. Reactions at the interfaces.
3. Transfer of the reaction results from the interfaces to the other bulk phases as the gas.
4. Calculation of the amount of the interface areas.

This means the coupling of the two-phase fluid flow model for steel-gas system together with the thermodynamic and kinetic models. If the aim is to study the heat transfer also, we also need a heat transfer module. So again, we need:

- Thermodynamic module,
- Heat transfer module, and
- Mass transfer module.

4.1 Thermodynamic Module

Thermodynamic module is needed to describe the reactions at the interfaces, for instance at the gas/metal interface. Let's consider the dehydrogenization reaction:

$H+H \rightarrow H2\ (g)$

To calculate this reaction, from thermodynamic point of view, we need to know the reaction Gibbs energy and the activity of the a_H in the steel melt and pressure p_{H2} in the gas bubble. The reaction Gibbs energy and K_e (equilibrium constant for the equation above), can be formulated as:

$$\Delta G_r^o = -RT(\ln K_e) \tag{41}$$

$$K_e = \frac{P_{H_2}}{a_H^2} = \frac{P_{H_2}}{(\gamma \cdot x_H)^2} \tag{42}$$

R is the gas constant and T the absolute temperature. For activity calculations of the elements in the steel melt, we need a thermodynamic solution model for liquid steel and the corresponding thermodynamic databank. If we know a model for activity coefficient (γ), we can use it and the hydrogen composition (x_H). If we model the slag/steel or steel/refractory reactions, we need corresponding models for these reactions.

One more thing needed to model is the interfacial reaction area in each calculation cell. Interfacial area is a very important parameter, because it directly correlates with the amount of mass transfer in the cell. The area can be estimated by using empirical equations but also more fundamental models have been developed as a one-group interfacial area concentration (IAC) model, which takes into account the terms of bubble coalescence induced by random collision, bubble coalescence induced by wake-entrainment, and bubble breakup induced by turbulent impact.

4.2 Heat Transfer Module

During tank degassing, the melt temperature is decreasing a bit mainly because of the gas injection effect and normal heat loss through the ladle walls and top surface. Anyhow this temperature decrease or heat transfer in general does not play so important role in tank degassing, especially in the case of H or N removal. So the model can be isothermal and heat transfer is not needed to model. Empirical models for temperature decrease as a function of time can be applied quite easily if necessary. For more accurate modeling of heat transfer, the energy conservation equations can be coupled with the flow equations. Also accurate boundary conditions are needed. It might be good also to model the heat transfer through the ladle walls so that boundary conditions can be given for the outer surface of the ladle walls, instead of the inner surface.

4.3 Mass Transfer Module

Accurate mass transfer calculation is not possible without detailed calculation of fluid flow. Detailed fluid flow calculation means, in the case of tank degassing, numerical solving of the turbulent two-phase fluid flow model for steel-gas system, i.e. ladle flow phenomena induced by gas stirring. All most commonly used commercial fluid flow software calculates these kinds of systems. Both Euler-Euler and Euler-Lagrangian two-phase flow model approaches have been applied for degassing. The former one treats each phase as interpenetrating continuum. In contrast, the latter one treats dispersed phase as a large number of particles, bubbles, or droplets. To be noticed, the effect of volume fraction of the dispersed phase on the bulk phase flow is typically neglected in this technique, which is questioned if the share of the dispersed phase becomes significant in the flow. One important question is, what is the correct turbulence model for steel-gas two-phase flow. Standard k-ε model is used generally for many purposes, but it seems that it is not accurate enough for gas-steel plume flows. Accordingly, more appropriate models are needed to account correctly the transport of heat, mass, momentum due to turbulence. But in general modeling of turbulence in two-phase/multiphase systems is somewhat uncertain. This is also the case for modeling of the bubble formations and the associate dynamics. One important modeling aspect in gas stirred ladle is the question what happens when the cold gas is injected to the hot steel. How fast the cold gas is heating up? During heating the gas is expanding. The ideal gas law can be roughly applied to calculate the amount of hot gas when the inlet flow rates of the cold gas are known.

When the fluid flow phenomena are correctly modeled, we need individual mass conservation equations for the studied chemical elements as for instance H, N, O, C, and in the gas phase or bubbles H_2, N_2, CO. They are commonly normal convection-turbulent diffusion equations with sink/source terms. Modeling of multiphase flows is very complicated and requires a large computing powder. Many kinds of simplifications can be made. In the case of degassing, under the constant gas flow rate and vacuum conditions, the flow phenomena can be assumed to be constant and so the flow phenomena can be calculated in advance only once. Once a steady state fluid flow conditions are reached, the time-dependent mass conservation or transport equations are then solved using the steady state flow velocities. This reduces significantly the computing times.

At the top free surface, the gas is escaping to the ambient atmosphere. This also leads typically to the formation of a "plume or slag eye". To model correctly the phenomena in degassing, the size of the slag eye should be known. The size can be modeled, but for many purposes it is reasonable just to give it as inlet value because the size can be measured in industry using cameras. Depending of the target and simplifications of the degassing model, the slag layer can be included into the model or it can be excluded. Gas removal reactions take place in the open eye area but some specific reactions as desulfurization at the slag/steel interface. So for correct modeling of desulfurization, the slag layer and the mass transfer also in it should be taken into account. There are also many ways to model the top interface boundary and gas escaping through it. One relatively simple way is to assume the interface fixed and flat and use in the uppermost calculation elements sink terms in the appropriate equations to take the gas away from the calculations.

In the mass conservation equations, we need to model the transfer of the "elements" from bulk phase to the reaction interfaces as to the metal/gas interfaces, model reactions at the interfaces, and model the transfer of the reaction results from the interfaces to the other bulk phases as the gas. Typically the mass transfer in the gas phase can be assumed to be fast. The rate controlling step is either the mass transfer from bulk phase to the interface or interface reactions. The surface active elements (O, S,...) typically prevent especially nitrogen reactions at the interface and so if a lot of O or S is available in the steel, the rate controlling step for nitrogen removal is typical the interface reaction. For hydrogen removal, it is usually the mass transfer form bulk phase to the interface. To model the mass transfer from bulk phase to the interface, we need to know the convective mass transfer coefficient in liquid steel. In practice, Higbie and Danckwerts (eddy-cell) expressions are frequently adopted to compute the mass transfer coefficient in such system. Higbie (1935) assumed that mass transfer in a gas-liquid system is related to the surface renewal time, which is a function of bubble diameter and slip velocity. On the other hand, Danckwerts proposed that the mass transfer in gas-liquid system mainly depends on the motion of small-scale eddies and thus the mass transfer coefficient could be a function of the Laboratory of Metallurgy at Helsinki University of Technology (TKE) dissipation Higbie 1935; Lamont & Scott 1070; Prasher & Wills 1973).

Example Case: In-House Tank Degassing Model for H and N Removal

Hydrogen and nitrogen in steel typically reduce the steel quality. For controlling the N and H content in steel, understanding the behavior of hydrogen and nitrogen during degassing process becomes more important. In order to study degassing process in vacuum tank degasser, a numerical model has been developed. The model is presented here in brief; however more details are available in (Yu, Miettinen, & Louhenkilpi 2013). In the model, both dehydrogenation and denitrogenation are considered. The degassing process principally consists of three steps: (hydrogen or nitrogen) mass transfer from the bulk liquid to the gas-steel interface; degassing reaction at the interface; and mass transfer from the reaction sites to the gas phase (bubble), which is believed to be fast. The process involves many physical phenomena which must be accurately modeled, such as multiphase flow, species transport and chemical reaction. In order to develop a numerical model to describe this complicated process and to reduce the computational demand, some simplifications and assumptions are:

1. Only gas and liquid steel are considered when calculating the multiphase flow phenomena.
2. The model is isothermal. The effect of temperature is neglected.
3. Hydrogen and nitrogen are removed by the stirring gas (bubbles) and through the open-eye that locates at the bath surface. The bath surface is assumed to be flat and the area of the open-eye is adjusted in the corresponding computational cells based on plant observation.
4. The overall dehydrogenation rate is limited by the mass transfer of hydrogen in liquid steel since the interfacial reaction is assumed to be quick. However, the overall denitrogenation rate is determined by the mass transfer of nitrogen in liquid steel and interfacial reaction.
5. The fully hydrated lime, $Ca(OH)_2$, is taken as the main hydrogen source in the bath surface of non-open-eye region which represents the slag covered area. In this study, 250 kg lime is added in ladle furnace before vacuum degassing and 2 wt % hydrogen of the total hydrogen amount in Ca(OH)2 is assumed as the hydrogen source entering the steel in the simulations.
6. The steel is deoxidized before degassing and oxygen content in liquid steel is constant and is set as 6 ppm. No $C+O\rightarrow CO$ reaction takes place. The variations of activity coefficients due to the change in steel composition during vacuum treatment are neglected. The activity coefficients are calculated based on the initial composition with the in-house activity coefficient sub-model.
7. Sulfur and oxygen are considered as surface-active elements. The effect of other elements on dehydrogenation and denitrogenation is taken into account by using the in-house activity coefficient sub-model.

The gas-liquid multiphase flow is solved using the Euler-Euler approach. An interfacial area concentration model is introduced to compute the interfacial area density between liquid steel and the gas bubbles and also, an extended k-ε model, instead of the standard k-ε mixture model, is used to include the influence of gas injection on turbulent properties. The surface of the liquid bath is treated as flat and only gas can escape through it, which is realized by adding a set of sink terms into the conservation equations of gas phase in each control volume adjacent to the surface. The species transport equations are solved for N, H in steel melt and N_2 and H_2 in the gas phase. These species transport equations are solved as a function of time in the steady state multiphase flow conditions calculated firstly. The effect of surface active elements (e.g. sulphur and oxygen) in liquid steel is considered on N removal but not on H removal. It is well known that surface active elements have significant effect on nitrogen removal but it is typically assumed that their effect on hydrogen removal is negligible. The eddy-cell model is applied for mass transfer coefficient. The partial pressure of component (Ar, H_2 and N_2) above the surface is defined instantly and calculated by tracing the amount of each gas escaping from bath. The model structure is presented in Figure 14. In principle, hydrogen removal could take place through the bubble surface and bath surface. These two routes are compared in Figure 15, where the evolutions of dehydrogenation rate by gas bubbles and by bath surface are plotted. It is shown that the bubble surface is the main area for hydrogen removal especially at the beginning of the process as the gas-liquid interfacial area is much higher than that at the bath surface (Figures 14 and 15).

Figure 14. Schematic of model structure /xxx/

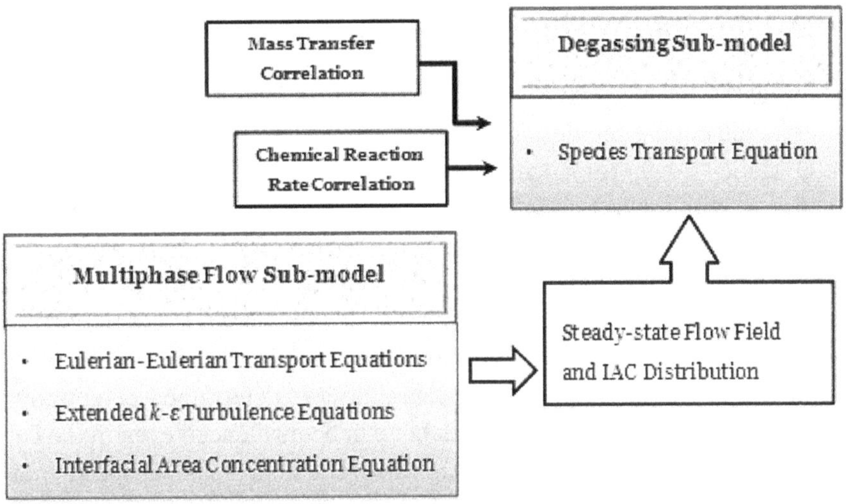

Figure 15. Dehydrogenation rates by bubble surface and bath surface for one calculated case /xxx/

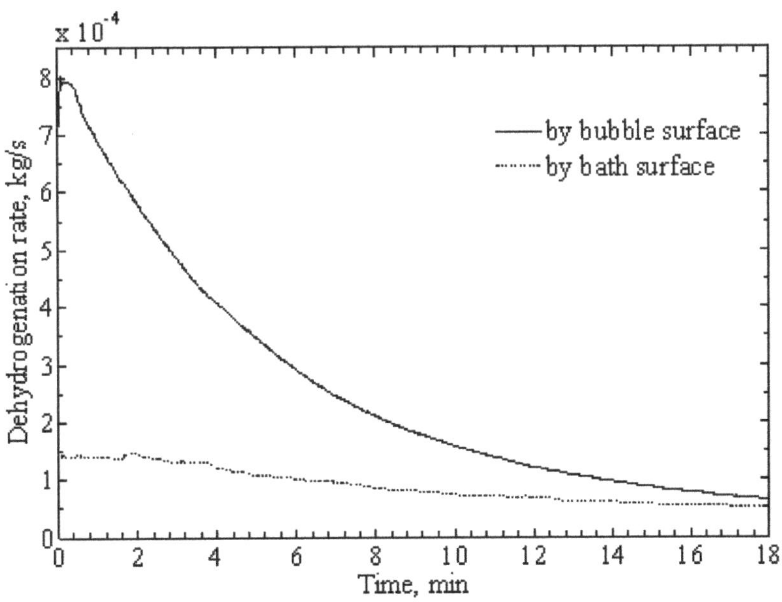

5. MATHEMATICAL MODELING OF CONTINUOUS CASTING

5.1 General and the Main Fundamental Phenomena in Continuous Casting

Continuous casting is the important linking process between steelmaking and rolling. As early as 1856 Henry Bessemer suggested a continuous casting method but just during the 1930s and 1940s continuous casting became a common production method for nonferrous metals and later from the 1960s for steels. The relatively low thermal conductivity of steel and the high casting temperatures meant that many problems had to be solved compared to non-ferrous casting. In the mid 1980's continuous casting grew into the biggest casting method, exceeding the conventional ingot steel casting route.

The principle of the continuous casting method is simple (see Figure 16). The liquid steel is transferred in a ladle to the casting machine. When the casting operation starts, the nozzle at the bottom of the ladle is opened and the steel flows at a controlled rate into the tundish and from the tundish through a submerged entry nozzle (SEN) into one or more moulds (Najjar, Thomas & Hershey 1995). The moulds are generally water-cooled copper moulds. The first solidification takes place at the metal-mould interface (Mills 1994; O'Malley 1999; Yamauchi et al 1993). The thickness of the solidified shell increases progressively when it is withdrawn through the machine. At the mould exit, the shell must be thick enough to support the liquid pool. Below the mould the shell is cooled by spraying water. The mould cooling is called the primary cooling and the spray cooling the secondary cooling. At the machine end the strand is cut off and transferred to a rolling mill. The big challenge in continuous casting is to cast steel continuously without interruptions and without many kinds of defects. Solidification control is important for surface and internal quality. Steel cleanliness is determined essentially already by the preceding operations in the ladle and in the tundish but can be influenced even in the casting operation. Important control parameters in solidification are e.g.: steel chemistry, casting speed, mould level, mould powder, mould oscillation, liquid steel temperature, secondary cooling conditions as well as parameters affecting the flow phenomena in the mould. There are many kinds of continuous casting machines. For steels, typically different kind of vertical casters are used, but for non-ferrous alloys (Al, Cu,…), also horizontal and up-casting machines are available. Today the vertical casters are mostly curved with straight or curved mould. These kinds of machines are lower and casting speed can be higher. Curved machines have many benefits compared to the original vertical machines (Figure 16).

The modeling of solidification systems is a problem of a great mathematical and industrial significance. Like most commercial processes, continuous casting involves many complex physical phenomena and no model can today include all of the phenomena at once. To develop a model, it is essential to select the most important phenomena and make the reasonable assumptions depending what the model is meant for. To obtain reliable results, accurate data on the thermophysical material properties are also needed along with accurate boundary conditions. The main phenomena are fluid flow, free surface, gas bubbling, electromagnetic forces (stirring/EMS, braking/EMBr), heat transfer, solidification, microstructure formation, phase transformations, inclusions, precipitations, stresses, etc.

A lot of fluid flow models for tundish have been published and successfully applied (Thomas 1991; Thomas & Zhang 2011; Thomas 2002; Upton et al 1988; Davies, Blake & Campbell 1988). The models solve the turbulent Navier-Stokes equation and the mass balance equation. Some models have heat transfer included and also the calculations of the particle paths for estimation of the particle removal in tundish. Some open questions today are: what is the correct turbulence model and in the case of heat transfer how to know exactly the heat transfer boundary conditions. It is also good to keep in mind that the

Figure 16. A schematic view of Continuous Casting (CC) process. CCoffers a lot of benefits compared to the conventional ingot casting methods

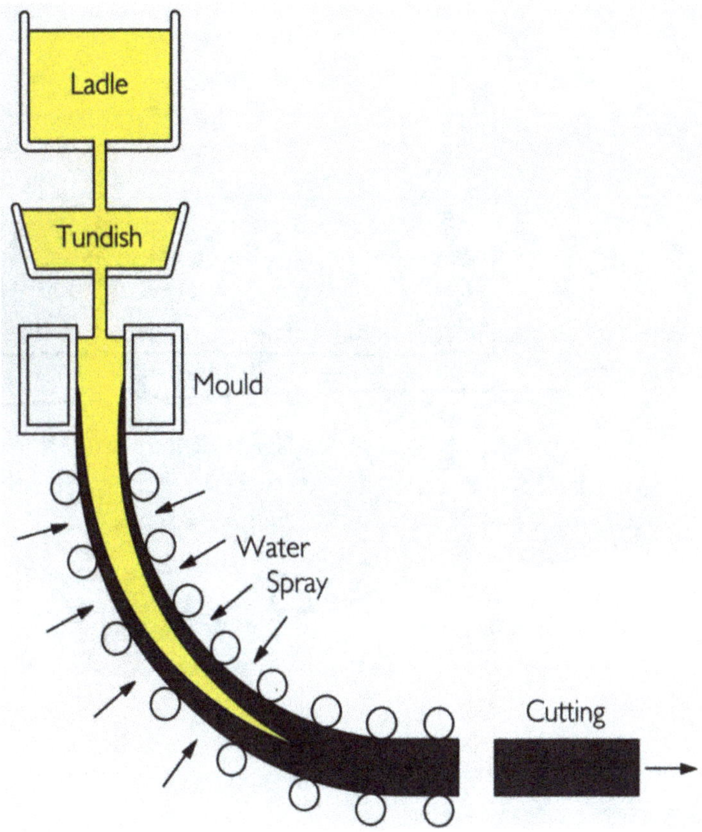

temperature of the inlet steel is not constant during the casting operation and can change a lot especially during grade change operations. High temperature differences in tundish cause natural convection. In the case of particle tracking, one open question still is how to correctly model the removal of particles from melt to the walls or tundish slag.

The mould involves also many physical phenomena. The main phenomena are fluid flow, heat transfer, solidification, movement of the strand, gas bubbling, EMS, particles, free surface, Flow phenomena in mould are typically studied by ignoring the heat transfer and solidification. Only fluid flow models are used. This assumption is believed to be quite reasonable when studying only flow phenomena. The equations are the same as for tundish, but the calculation of the shape of the free surface should be included. For these kinds of models one important question is also how long the calculation domain should be and what is the boundary condition at the exit. These kinds of fluid flow models also may include particle tracking, gas injection, EMS, EMBr, etc. A typical flow of molten steel in tundish is shown in Figure 17.

Anyhow fluid flow, heat transfer and solidification in continuous casting are interlinked with one other. Coupled models calculate the fluid flow, heat transfer and solidification simultaneously. The coupled models include three main equations: the energy equation for temperature including solidification, the Navier-Stokes equation for velocities (momentum) and the mass equation for mass. In the coupled models, it is often assumed that the mushy zone has the flow characteristics of a porous medium, and

Figure 17. Represent the flow of liquid steel in tundish

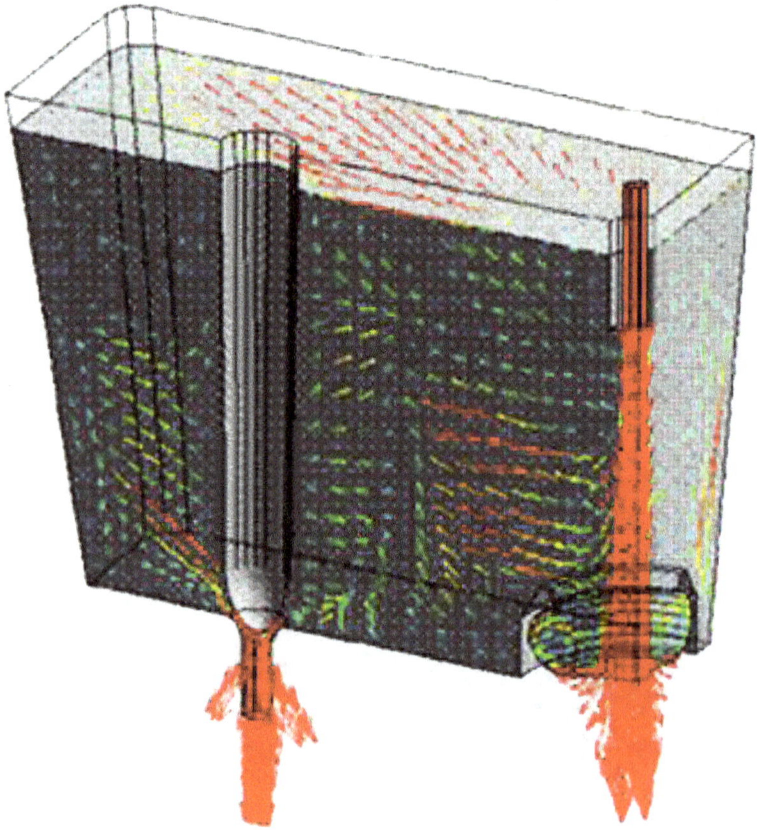

so a Darcy law can be applied for the mushy zone. If the flow is turbulent, a turbulent model must also be added. Normally, a so-called k-ε turbulence model is applied. This increases again the complexity of the system, because two additional equations must be solved. If electromagnetic stirring, gas injection or bulging, for instance, is added to the model, the corresponding model equations must be known and solved and the complexity is again increasing. Simplified approach also exits. Many heat transfer models for continuous casting take the fluid flow effect into account using a simplified approach called as the effective thermal conductivity approach. The convective heat transfer generated by the fluid flow is taken into account by using an effective thermal heat conductivity formula. This simplification is often done due to the fact that the fluid flow does not affect very much the results in the solid shell, but more on the results within the liquid pool. So, if the aim is to study the shell thickness profile, the location of the liquid pool end or the temperatures in the solid part of the strand, this simplification can be done and the models are simpler.

As a conclusion, the continuous casting models can be divided into three categories: (1) heat transfer models based on effective thermal conductivity concept, (2) coupled heat transfer-fluid flow models, and (3) fluid flow models. The effective thermal conductivity models can be used to study the temperatures at the solid part and related data as the shell thickness. Coupled models are quite complicated and a lot of computing time is needed. Fluid flow models are used to study the flow phenomena in the tundish and mould.

5.2 Modeling of Continuous Casting with Effective Thermal Conductivity Approach

In these models, it is assumed that the whole strand (solid and liquid part) is withdrawn through the machine with a velocity, which is the casting speed (Figure 18). The effective thermal conductivity can be defined as presented in following equation (other formulas are also available). It is a function of solid phase fraction (and liquid, because the liquid fraction is 1-fs). In the liquid phase, the thermal conductivity is higher (4-10 times) than the normal conductivity of the melt because of the melt convection and this amount is defined by the constant A. It is difficult to define the value accurately. The value is also different in the mould than deeper in the pool due to the higher flow velocities close to the inlet nozzle. This value should be defined as a function of distance from the meniscus. Electromagnetic stirring also affects this value and the stirring can be taken into account by applying a higher value in the stirred area. The constant A should be fitted by experimental studies. The effective heat conductivity method assumes that fluid flow increases liquid pool heat transport isotropically, which is not the case, because fluid flow only contributes to heat transport in the flow direction.

$$k_{eff} = kf_s + Ak \cdot (1 - f_s) \tag{43}$$

The equation to be solved now is the basic partial differential equation of heat conduction including the removal of latent heat of solidification. In the case of steel continuous casting, the heat conduction in the casting direction is small and thus it can be ignored. This is due to the relatively high casting speed and low thermal conductivity. But in the case of copper or aluminium continuous casting, the heat conduction in the casting direction cannot be ignored. The thermal conductivity of these metals is much higher than that of steels and the casting speed is usually lower.

Figure 18. When applying the effective thermal conductivity modeling concept for continuous casting, it is assumed that the whole strand (solid and liquid part) is withdrawn through the machine with a velocity, which is the casting speed

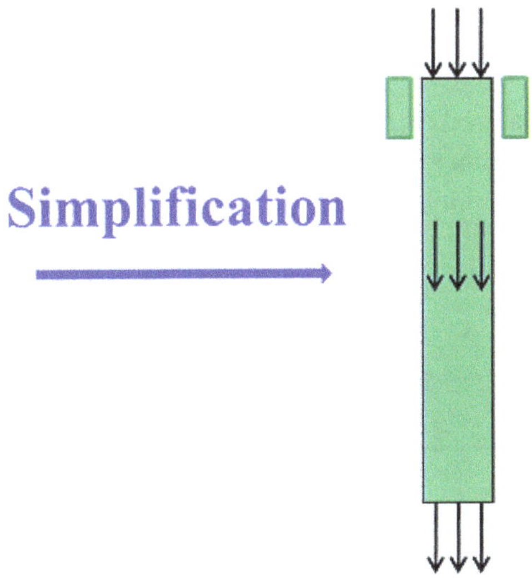

Some years ago, most of the heat transfer models for steels were two-dimensional. The simulated geometry was a two-dimensional strand slice and the models simulated its movement at a constant casting speed through the machine. In these models, the casting direction is treated as time. The equation in these cases can be described by:

$$\rho c \frac{\partial T}{\partial t} = \frac{\partial (k_{eff} \frac{\partial T}{\partial x})}{\partial x} + \frac{\partial (k_{eff} \frac{\partial T}{\partial y})}{\partial y} + Q \tag{44}$$

Actually the slice is not moving but the boundary conditions are changed as a function of time. The time is zero at the meniscus and below the meniscus it is calculated by equation t=z/v, where v is the casting speed and z the distance from the meniscus. In above equation, ρ is the density, c is the specific heat, k is the thermal conductivity, and Q is a term describing the rate of energy released by phase transformations. For solidification phase change, it is usually defined as:

$$Q = \rho L \frac{\partial f_s}{\partial t} \tag{45}$$

Here, L is the latent heat of solidification, and *fs=fs(T)* is called the solidified fraction in the mushy zone. The later term describes the way in which the latent heat is released during solidification. This way depends strongly on the chemical composition of the material to be cast. The solid-state transformations can be treated in a similar way. Therefore, the material data needed are the density, the specific heat, the thermal conductivity, the latent heats, and the fractions during phase transformations. Equation 45 can also be expressed in another form using a so-called enthalpy formulation. This method is recommended, because in specific method some enthalpy may be missed. The enthalpy, H, is defined as the sum of sensible ($\int$c dT) and the latent heats (Q).The equation now becomes:

$$\rho \frac{\partial H}{\partial t} = \frac{\partial (k_{eff} \frac{\partial T}{\partial x})}{\partial x} + \frac{\partial (k_{eff} \frac{\partial T}{\partial y})}{\partial y} \tag{46}$$

In this case, the necessary material data are the density, the enthalpy, and the thermal, conductivity. The enthalpy now includes all the other data, except the density and the thermalconductivity. This equation can be solved more easily by applying Kirchhoff's transformation (Figures 18 and 19):

$$K(T) = \int_0^T k_{eff} dt \tag{47}$$

The equation becomes now:

Figure 19. Some years ago, most of the heat transfer models for steels were two-dimensional slice models. The simulated geometry was a two-dimensional strand slice and the models simulated its movement at a constant casting speed through the machine. In these models, the casting direction is treated as time

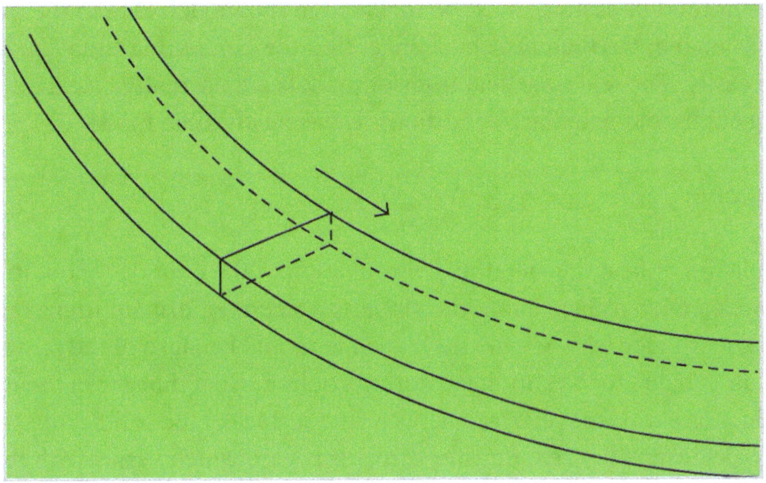

$$\rho\frac{\partial H}{\partial t}=\frac{\partial^2(K)}{\partial x^2}+\frac{\partial^2(K)}{\partial y^2} \tag{48}$$

When the computers have developed, nowadays tree-dimensional steady state models as well as three dimensional transient, on-line models have been developed. They are no more slice models. Applying the Kirchhoff's transformation and the enthalpy formulation the three-dimensional models can be described by:

$$\frac{\partial^2(K)}{\partial x^2}+\frac{\partial^2(K)}{\partial y^2}+\frac{\partial^2(K)}{\partial^2 z}+\frac{\partial u_z\rho H}{\partial z}=0 \tag{49}$$

The three first terms describe the heat conduction in the three spatial directions, the fourth term the convective heat transfer in the casting direction, caused by the movement of the strand. In the case of steel, the heat conduction in the casting direction is small and thus it can be ignored. This is due to the relatively high casting speed and low thermal conductivity. But in the case of copper or aluminium, the heat conduction in the casting direction cannot be ignored. The thermal conductivity of these metals is much higher than that of steels and the casting speed is usually lower. In the case of transient models, the transient term must be added to the equation, and it becomes:

$$\rho\frac{\partial H}{\partial t}+\frac{\partial u_z\rho H}{\partial z}=\frac{\partial^2(K)}{\partial x^2}+\frac{\partial^2(K)}{\partial y^2}+\frac{\partial^2(K)}{\partial^2 z} \tag{50}$$

Real-time calculation of the actual shell thickness profile and strand temperatures offers new possibilities in on-line process control. In real-time calculation, many practical requirements will be set for the model. The computing time must, for instance, be short enough and the special process conditions, such as the start and the end of casting, must be included in the model. Due to the increasing power of modern computers available, the requirements concerning the computing time can today, however, be met more and more easily. The real time heat transfer models are normally based on the same basic assumptions, same equations and boundary conditions as the steady state models.

Boundary Conditions

For the correct simulation of heat transfer, the determination of the boundary conditions describing the heat transfer phenomena taking place along the strand surface is of crucial importance. In the mould, there exists a thermal resistance between the shell and the mould because of the powder lubrication and the formation of an air gap due to the shrinkage during cooling. Heat transfer in the mould is controlled mainly by heat conduction across the interface between the surface of the solidifying shell and the mould. It is quite difficult to determine the heat transfer across this gap, which varies with time and position of the mould. The gap is a function of casting variables (casting speed, superheat, casting powder, etc.), steel composition, mould taper, etc. Advanced models simulate not only the strand but also the mould around the strand and the heat transfer across the gap is determined by the gap heat transfer coefficient. These advanced models actually consist of two models: the mould model and the strand model. The boundary equation for the strand model can be expresses as:

$$-k_{shell}\frac{\partial T}{\partial n} = h_{gap}(T_s - T_m) \tag{51}$$

and for the mould model:

$$-k_{mould}\frac{\partial T}{\partial n} = h_{gap}(T_m - T_s) \tag{52}$$

Here, Ts and Tm are the surface temperatures of the shell and the mould; h_{gap} is the gap heat transfer coefficient. It is usually defined as a function of strand surface temperature $h_{gap} = h_{gap}(Ts)$. The models are running iteratively. The corresponding temperatures in the models must become the same. At the cold face of the casting mould, a convective heat transfer coefficient is used. It can be determined from the flow rate of the water and an empirical formula is often used. If the mould model is not used, it is difficult to know, the external temperature, T_m, in the boundary equation for the strand. In the mould the average heat flux, Q_{ave}, extracted from the mould can be calculated from the heat balance of the mould cooling water. The gap heat transfer coefficient can be determined or validated using experimental measurements (thermocouple measurements in the mould and heat flux measurements of the cooling water). The mould heat flux (measured from the mould cooling water) can also be used directly as the boundary condition in the mould, but the accurate heat flux distribution is quite difficult to know

Below the mould, water sprays and rolls extract heat from the surface and the boundary conditions are usually expressed as heat transfer coefficients. This, on the other hand, must be determined as a function of actual cooling parameters. This determination can be done in a laboratory set-up or directly in the casting machine using strand surface temperature measurements. A lot of empirical formulas have been derived in this way. However, the formulas are valid only for the particular type of casting machine and steel grade and cooling conditions, for which they are derived. They cannot simply be transferred directly to other casting machines. Below the mould, the heat flux from the strand surface to the environment is usually defined by the following equation:

$$-k\frac{\partial T}{\partial n} = h(T - T_{ext}) + \varepsilon\sigma(T^4 - T_{ext2}{}^4) \tag{53}$$

Here $h=h_{eff}$ is the effective heat transfer coefficient, T is the strand surface temperature, T_{ext} and T_{ext2} the external temperature for cooling water or air, ε is emissivity and σ is the Stefan-Bolzmann constant. The effective heat transfer coefficient takes into account the other heat transfer mechanisms except radiation. The last term represents the heat flux by radiation. The input parameters in this equation are: h_{eff}, T_{ext}, $T_{ext2}$and ε. Before this boundary condition can be applied in actual computations, the relationship between the effective heat transfer coefficient and the cooling parameters must be determined. Many models define the cooling zones to smaller areas to accurately define the boundary conditions. The subareas (Figure13) can be 1) roller contact area, 2) pre-nozzle area, 3) spraying area, and after spray and pool water area. For each area the boundary data must be defined: h, T_{ext}, ε, T_{ext2}. They also depend on the roll types, nozzle types, etc (Figure 20).

Figure 20. Many models define the cooling zones to smaller areas to accurately define the boundary conditions. The subareas can be 1) roller contact area, 2) pre-nozzle area, 3) spraying area, and after spray and pool water area

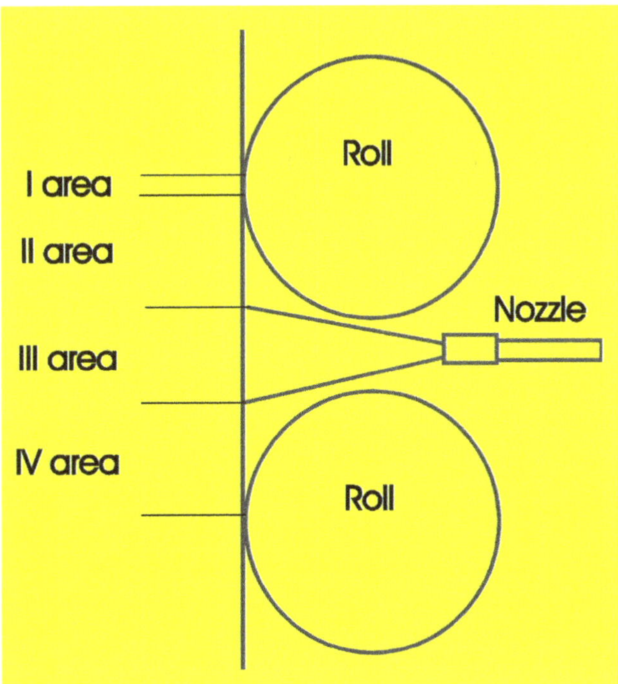

It is well known that with decreasing surface temperature a critical value, the Leidenfrost temperature, is reached whereupon the mechanism of heat transfer changes from film boiling to transition boiling. At higher temperatures the heat transfer coefficient is relatively insensitive to surface temperature (under conditions of film boiling), but increases sharply below the Leidenfrost temperature. Above the Leidenfrost temperature the surface temperature is sufficiently hot to create a steam layer through which the water droplets do not fully penetrate. As the temperature decreases, a change in the heat transfer mechanism occurs with boiling occurring on the surface. The Leidenfrost temperature depends on the surface quality, amount of water, etc. In continuous casting of steel it is approximately in the range of 700-900 °C. But at higher flows, it can be much higher and at very low flows much smaller, respectively. The Leidenfrost effect is very important, and it should be taken into account when defining the relationship between the effective heat transfer coefficient and the cooling parameters.

Material Data

To obtain reliable results from the heat transfer simulations, accurate data on the thermo-physical material properties are also needed. Typical data needed are the density, the thermal conductivity, and the specific heat. Other important data are the phase transformation temperatures and the corresponding latent heats, and also the way in which the latent heats are released during the phase transformations. If the enthalpy formulation is used, the enthalpy values can be used directly if they are known. The enthalpy then includes all the other data except the thermal conductivity and the density. The material data are not only functions of the temperature and the chemical composition but also of the cooling rate. This is because the kinetics of phase transformations depends on the cooling rate and the thermophysical properties are related to the phases formed. Thus, for accurate simulation of solidification and cooling processes, one should know the material data as a function of temperature, composition, and cooling rate. Although material data have been measured for a great number of steel grades, most of these data are valid for special steel grades and/or for the low temperature region only and there are only a little data for higher temperatures up to the liquid phase. So it is seldom possible to find all the data needed. This is especially the case for carbon and low alloyed steel grades, because in these steels even smaller variations in the composition might have a significant effect on the phase transformations and so on, the thermophysical material properties.

In the heat transfer simulations, the equations are usually solved using fixed grids, i.e. the strand dimensions are the same everywhere in the machine. This means that the contraction of the steel is not calculated and taken into account. In these kinds of models, it is important to take care of the correct mass and heat balance. One method to take care of this is that the density is not varied either. In such cases the density should be that of the initial liquid and constant. However, during solidification the fluid in the interdendritic space is free to move and it more or less compensates the solidification contractions. To take this feeding into account, the density should be constant and that of the solidus temperature. Two calculations for a slab caster were made to test the effect of the density. The first calculation with the changing density gave 17.50 m for pool length and the second with the constant density 20.85 m. So, the difference in the pool length was over 3 meters. So, it is important to take care of the correct mass and heat balance.

6.0 MODELING OF SOLIDIFICATION AND MICROSEGREGATION

Models Using a Fixed Calculation Domain for Dendrites/Grains

In classical solidification models, the solidification phenomena are typically modeled in a given volume element. One example is given in the Figure 20, where the solidification phenomena are modeled in one-volume-element set on the side of a dendrite arm having a hexagonal arm arrangement. Perpendicular to the growth direction, the volume element is described with a triangle. This is a typical volume element for most solute redistribution models. For simulation of solidification and phase transformations, we typically need:

- Mass transfer in solid.
- Mass transfer in liquid.
- Distribution of alloying elements at the interface.
- Mass balance equations for the interface movements.
- Phase stability model.

At the interface, the distribution of the alloying elements can be based on the distribution coefficients or as in the more advance models on the chemical potentials, i.e chemical-potential-equality equations at the interface. So for each element i:

$$\mu_i^S = \mu_i^L \tag{54}$$

where μ_i^S means the chemical potential of the element i at the interface in the solid side and μ_i^L in the liquid side. This condition means that thermodynamic equilibrium takes place at the interface. For rapid solidification processes, more energy should be taken into account as the solid-liquid interfacial energy and the kinetics energies dealing with movement of the atoms at the interface. Chemical potentials are calculated from Gibbs energies. This means that the model must include a thermodynamic solution model and the corresponding thermodynamic data bank for calculating the chemical potentials for different steel compositions. In addition to the chemical-potential-equality equations, material balance equations are also needed for the volume element for the determination of the compositional changes in the volume element, resulting from the phase interface movements. By applying together the chemical-potential-equality equations and the material balance equations, phase stabilities of different phases as for example ferrite, austenite and liquid are determined. Mass transfer equations are balance equations describing how the solute elements of the liquid during the fractional moving of the solid/liquid interface during a time step will be divided between the liquid phase and solid phase. In the equations, also the amount of material diffusing from the phase interface should be taken into account. See ref (Réger et al 2002; Louhenkilpi et al 2004) for more detailed description of this. Finally, one needs so simulate the solute diffusion in the solid and liquid phases. This can be made by applying Fick's 2nd law. In many models the diffusion in the liquid phase is anyhow assumed to be infinite. As a result of these calculations, one gets the fractions and the compositions of each phase and the amount of the microsegregations as a function of the decreasing temperature.

Typical assumptions in these kinds of models are for instance:

- The dendritic structure is regular and has a hexagonal arm arrangement.
- There is no solute change between the volume element and its surroundings.
- There is complete solute mixing (no local solute gradients) in the liquid phase.
- Thermodynamic equilibrium is reached at the phase interfaces.
- The diffusion of solutes is independent of the presence of other solutes.
- The differences in the molar volumes of the phases are negligible.
- During solidification, a new phase begins to form as soon as it is thermodynamically possible

These assumptions are quite reasonable. Anyhow, if the solidification speed is high, as in rapid solidification processes, the diffusion in the melt can also be considered to be finite. In these cases also the assumption about thermodynamic equilibrium at the phase interface is then not exactly correct because, as mentioned above, more additional phenomena should also be taken into account. The assumption that the diffusion of solutes is independent of the presence of other solutes means, that diffusion can be calculated by Fick's law and diffusion coefficients for binary systems (element-steel) are used. More advanced models are using multi-diffusion equations and mobilities (Figure 21).

Figure 21. A volume element on the side of a dendrite arm

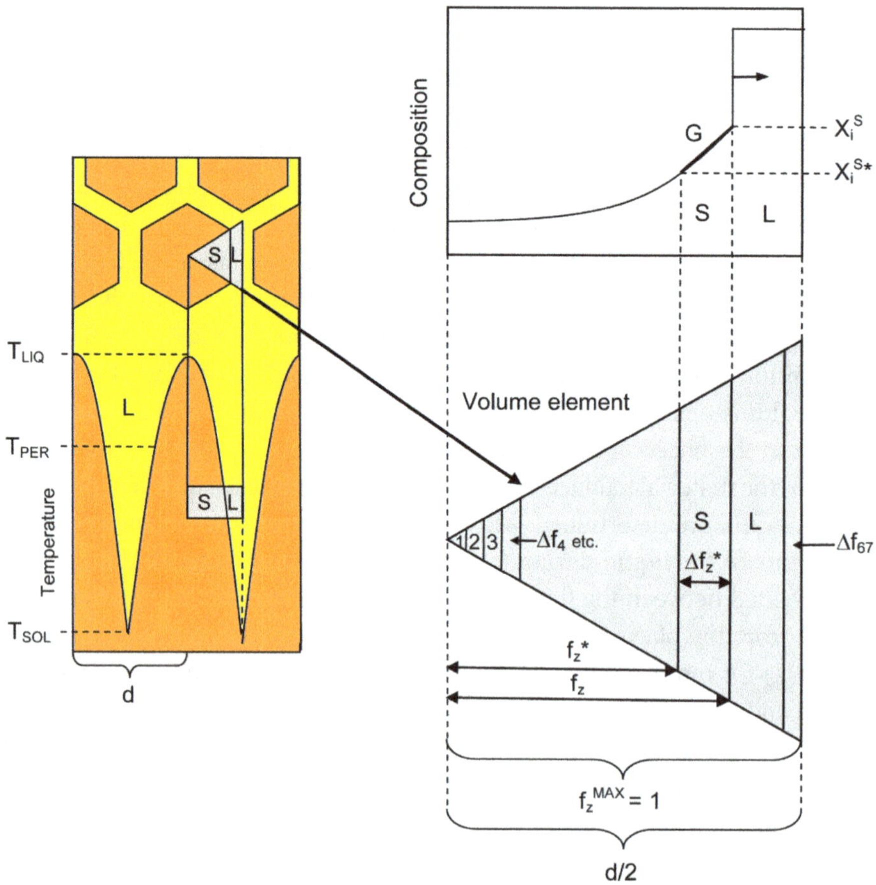

Simpler Microsegregation Models

Microsegregation is a fundamental phenomenon taking place during solidification in the scale of dendrites. The degree of microsegregation of an element in an alloy depends on the following main factors: 1) Its equilibrium partition coefficient, which is defined as a ratio between the C_s/C_l calculated from the solid and liquid lines. Typically it is smaller than one. If the liquidus and solidus lines are straight, the coefficient is constant during the whole solidification. 2) The diffusion coefficient of the element in the solid phase of the alloy. Equilibrium solidification for binary alloys follows phase diagrams: mass transfers are infinite (D_s, D_l) and distribution coefficient and liquid fractions are taken from the phase diagram. This is not the fact in reality and so the actual solidification in casting machines differs a lot from the equilibrium solidification. Typically the microsegregation is much higher in real casting conditions than calculated from the phase diagrams. The advanced solidification models calculate microsegregation more fundamentally and accurately. Anyhow some simpler model has also been developed to simulate the non-equilibrium solidification. One of the first models presented was the Scheil's analytical segregation equation (Eq. 55). It has many assumptions and simplifications e.g. that the diffusion is infinite in the melt but zero at the solid and a constant partition coefficient (k) is assumed. For multi-component alloys, it is also difficult to know the value of k for each element. F_s is the solid fraction.

$$C_l = C_o\,(1-F_s)^{k-1} \tag{55}$$

Many improvements have been presented like the model of Brody and Flemings (Eq. 56). It takes into account the diffusion in the solid, called back-diffusion in this model.

$$C_S = k^*C_0^*(1-F_S/(1+\alpha^*k))^{k-1} \text{ (where } \alpha = 4D_S t_f/d^2) \tag{56}$$

D_S is the diffusion coefficient of the solute element in solidifying steel, t_f is the local solidification time and d the half of the secondary dendrite arm spacing.

Dendrite Arm Spacing

Steel solidification in castings and ingots is dendritic. Fineness of dendritic structure is usually measured as dendrite arm spacing. A lot of different kind of empirical equations for estimation of secondary (λ_2) or primary (λ_1) arm spacing were published(Louhenkilpi et al 2004; Räisänen et al 2004; Louhenkilpi et al n.d.; Louhenkilpi et al 2006; Louhenkilpi et al 2003; Louhenkilpi 1995; Miettinen et al 1999; Kytönen et al 1998). The equations are mainly based on experimental measurements and statistical modeling. Secondary arm spacing is determined as the mean value measured between secondary arms whose attachments to primary arms are clearly visible. In general, experimental measurements show good agreement with the time in which the arms are in contact with the liquid. This time is represented by the local solidification time (t_f), defined as the time spent between the liquidus and solidus temperatures. Instead of the local solidification time, it can also be used cooling rate (C_R), temperature gradient (G) or growth rate (R).

The equations of arm spacing are valid only for the conditions and composition range used in the study. It is generally found that the secondary arm spacing is related to the local solidification time or cooling rate by power laws as Eq. (58) shows. The primary arm spacing is usually described with G and R according to the equation 1. The typical equations used for λ_1 andλ_2 have the forms:

$$\lambda_1 = N \cdot G^{-l} \cdot R^{-m} \tag{57}$$

$$\lambda_2 = B \cdot C_R^{-n} = B(G \cdot R)^{-n} = M \cdot t_f^{\ d} \tag{58}$$

where *N, l, m, B, n, M* and *d* are parameters which depend on mainly the alloy composition. The solidification time is usually determined as:

$$t_f = \frac{T_L - T_S}{C_R} \tag{59}$$

where the T_L and T_S are the liquidus and solidus temperature, respectively. Most measurements of secondary arms have been performed on a small directionally solidified ingots equipped with thermocouples. The value of t_f can be read directly from the temperature/time recording. One problem is that the solidus temperature is difficult to determine accurately from the experiments because it is difficult to evaluate at which temperature the solidification is complete. In actual solidification process, it is also difficult to define the cooling rate, because it is not constant. So it is necessary to be careful when extrapolating experimental results to an industrial scale, since the solidification conditions are not the same.

CONCLUSION

The numerical models are successful in solving the problems in steel making processes such as modeling the continuous casting, vacuum degassing of steel, steel converter melting, solidification and grain development kind of problems, to a certain extent. Transport theories i.e., mass, momentum and heat transfers are the basic fundamental inputs to address these problems. One of bottle neck in generating the reliable results in simulation of heat and mass transfer is to obtain reliable data on the thermo-physical material properties. On the other hand, as in the case with all other simulations approaches, iterative runs are required for simulation results to be more reliable and to describe the real time practice of the process.

Finally, to make more practical use of these simulations, developing interface for real time applications is the utmost need of present circumstances. The future research effort may be given to develop more real time application models including emerging computational technologies compatible with the latest development of the steel melting technologies.

REFERENCES

Allan, F. (2015). A continuum model of deformation, transport and irreversible changes in atomic structure in amorphous lithium–silicon electrodes. *Acta Materialia*, *98*, 229–241. doi:10.1016/j.actamat.2015.07.036

Allendorf, S., Ottesen, D., Hardesty, D., Goldstein, D., Smith, C., & Malcolmson, A. (1998). Laser based sensor for real-time measurement ofoff-gas composition and temperature in BOF steelmaking. *AISE Steel Technology*, *4*, 31–35.

Andersin, E., & Jalkanen, H. (1990). Physicochemical Modeling of Converter Steelmaking. In *Proceedings of The 6th International Iiron and Steel Congress.*

Bardenheuer, F. (1970). Contribution to the metallurgy of the LD process. *Blast Furnace and Steel Plant., 58*(39), 401–406.

Braun, A., & Pfeifer, H. (2007). *Investigations of Non-isothermal Flow Conditions in a Two Strand Tundish Water Modelusing DPIV and PLIF-technique.* SteelSim 2007 - 2nd International Conference of Simulation and Modeling of Metallurgical Processes in Steelmaking, Graz, Austria.

Chakraborty, S., & Sahai, Y. (1991). Effect of Varying Ladle Stream Temperature on the Melt Flow and Heat Transfer in Continuous Casting Tundishes. *ISIJ International, 31*(9), 960–967. doi:10.2355/isijinternational.31.960

Clark, Wagner, & Trousset. (2003, January). Modelling for Tundish Applications. *Refractories Engineer*, 9-12.

Davies, R., Blake, N., & Campbell, P. (1988). Solidification Modelling--an Aid to Continuous Casting. *4th International Conference Continuous Casting 2.*

Deo, B., & Boom, R. (1993). Fundamentals of Steelmaking Metallurgy. New York: Prentice Hall International.

Drugan, W. J. (2001). Dynamic fragmentation of brittle materials: Analytical mechanics-based models. *Journal of the Mechanics and Physics of Solids, 49*(6), 1181–1208. doi:10.1016/S0022-5096(01)00002-3

Entwistle, K. M. (2001). Basic Principles of the Finite Element Method. Maney Publishing.

Eymard, R., Gallouët, T., & Herbin, R. (2000). Finite volume methods. In Handbook of Numerical Analysis, (vol. 7, pp. 713-1018). Academic Press.

Gordon, H. (1975) Theory of BOF Reaction Rates. In BOF Steelmaking (vol. 2). New York: The Iron and Steel Society.

Higbie, R. (1935). The Rate of Absorption of a Pure Gas into a Still Liquid during Short Periods of Exposure. *Transactions of the American Institute of Chemical Engineers, 31*, 365–389.

Introduction to Materials Modelling. (2005). London: Institute of Materials, Minerals and Mining.

Jalkanen, H., & Andersin, E. (1989). Thermodynamic modeling of converter steelmaking. Paper presented in the Conference Modern Technologies for Making and Casting of Steel.

Jalkanen, H., & Kostamo, T. (2004). Simulation of Oxygen Converter Process. In P. Palfy, M. Solc, & E. Vircikova (Eds.), *Metallurgy Refractories and Environment* (pp. 87–99). Kosice: Harlequin.

Jalkanen, H., Suomi, M.-L., & Wallgren, M. (1997). Simulation of complex chemical high temperature processes - experiences and problems in BOP modelling. In *Proceedings of IPMM´97, Australasia-Pacific Forum on Intelligent Processing and Manufacturing of Materials.*

Jalkanen, H., Suomi, M.-L., & Wallgren, M. (1998). *Simulation of Oxygen Converter Process (BOF).* Paper presented in Joint Finnish-South African Symposium on Metallurgical Research, Johannesburg, South Africa.

Jalkanen, H., Virrankoski, R., & Oksanen, T. (2002). Simulator of Oxygen Converter Process (BOF). Paper presented in Pyrometallurgy ´02, Cape Town, South Africa.

Karma & Rappel. (n.d.). Quantitative phase-field modeling of dendritic growth in two and three dimensions. *Physical Review D, 57*, 4323-4349.

Kobayashi, S. (1988). Solute redistribution during solidification with diffusion in solid phase; A theoretical analysis. *Journal of Crystal Growth, 88*(1), 87–89. doi:10.1016/S0022-0248(98)90010-0

Kruskopf, A., Väyrynen, P., Vallejo, V., Jalkanen, H., & Louhenkilpi, S. (2012). *Modeling of BOF Converter: CONSIM Simulator and Its Further Development*. Academic Press.

Kusnetsov, Sham, & Bol´shakov. (1972). *Steel in the USSR*. Academic Press.

Kuzmin, D., & Turek, S. (2004). Numerical Simulation of Turbulent Bubbly Flow. In *Proceedings of 3rd International Symposium on Two-Phase Flow Modelling and Experimentation*.

Kytönen, H., Tolvanen, M., Louhenkilpi, S., & Holappa, L. (1998). Modeling of Casting, Welding and Advanced Solidification Processes VIII. The Minerals, Metals & Materials Society.

Lamont, J., & Scott, D. (1970). An Eddy Cell Model of Mass Transfer into the Surface of a Turbulent Liquid. *AIChE Journal. American Institute of Chemical Engineers, 16*(4), 513–519. doi:10.1002/aic.690160403

Lee, J. S., Kim, S. G., Kim, W. T., & Suzuki, T. (1999). Numerical simulation of peritectic reaction using a multi-phase-field model. *ISIJ International, 39*(7), 730–736. doi:10.2355/isijinternational.39.730

Louhenkilpi, S., Laine, J., Räisänen, T., & Hätönen, T. (2004). *2nd Int. Conf.and Exhibition on New Developments in Metallurgical Process Technology*.

Louhenkilpi, S., Laitinen, E., & Nieminen, R. (1993). Real-Time Simulation of Heat Transfer in Continuous Casting. *Metallurgical Transactions. B, Process Metallurgy, 24*(4), 685–693. doi:10.1007/BF02673184

Louhenkilpi, S., Mäkinen, M., Vapalahti, S., Räisänen, T., & Laine, J. (2005). 3D steady state and transient simulation tools for heat transfer and solidification in continuous casting. *Materials Science and Engineering A, 413-414*, 135–138. doi:10.1016/j.msea.2005.08.153

Louhenkilpi, S., Mäkinen, M., Vapalahti, S., Räisänen, T., & Laine, J. (2005). 3D steady state and transient simulation tools for heat transfer and solidification in continuous casting. *Materials Science and Engineering A, 413-414*, 135–138. doi:10.1016/j.msea.2005.08.153

Louhenkilpi, S., Mäkinen, M., Vapalahti, S., Räisänen, T., & Laine, J. (n.d.). ICASP Advances in Solidification Processes. *Proc. Int. Conf.*

Louhenkilpi, S., Räisänen, T., Vapalahti, S., Laine, J., & Mäkinen, M. (2006). *Modeling of Casting, Welding and Advanced Solidification Processes XI*. Warrendale, PA: TMS.

Louhenkilpi, S., Uoti, M., Kytönen, H., & Vapalahti, S. (2003). *Modeling of Casting, Welding and Advanced Solidification Processes X*. Warrendale, PA: TMS.

Louhenkilpi, S., Vapalahti, S., Laine, J., Mäkinen, M., & Räisänen, T. (2005). 3D steady state and transient simulation tools for heat transfer and solidification in continuous casting. *Materials Science and Engineering A*, *414*, 135–138. doi:10.1016/j.msea.2005.08.153

Mackenbrock, A., & Tacke, K.-H. (1996). Generalized enthalpy method for multicomponent phase change. *Metallurgical and Materials Transactions. B, Process Metallurgy and Materials Processing Science*, *27*(5), 871–881. doi:10.1007/BF02915617

Marcus, J. (2004). Grote, Christoph Kirsch, Dirichlet-to-Neumann boundary conditions for multiple scattering problems. *Journal of Computational Physics*, *10*, 630–650.

Mazumdar, D., Yamanoglu, G., Shankarnarayanan, R., & Guthrie, R. I. L. (1995). Similarity Consideration in the Physical Modelling of Steel Tundish Systems. *Steel Research*, *66*, 14–19.

Mieti'inen, J., & Louhenkilpi, S. (1994). Calculation of Thermophysical Properties of Carbon and Low Alloyed Steels for Modeling of Solidification Processes. *Metallurgical and Materials Transactions. B, Process Metallurgy and Materials Processing Science*, *25*(6), 909–911. doi:10.1007/BF02662773

Miettinen, J. (1997). Calculation of solidification-related thermophysical properties for steels. *Metallurgical Transactions. B, Process Metallurgy*, *28*(2), 281–297. doi:10.1007/s11663-997-0095-2

Miettinen, J., Kytönen, H., Louhenkilpi, S., & Laine, J. (1999). *Proc. of 12th IAS Steelmaking Seminar*.

Miettinen, J., Kytönen, H., Louhenkilpi, S., & Laine, J. (1999). *12th IAS Steelmaking Seminar*. Instituto Argentino de Siderurgia.

Miettinen, J., Louhenkilpi, S., & Holappa, L. (1996). Coupled Simulation of Heat Transfer and Phase Transformation in Continuous Casting of Steel. *ISIJ International*, *36*(Suppl), S183–S186. doi:10.2355/isijinternational.36.Suppl_S183

Miettinen, J., Louhenkilpi, S., Kytönen, H., & Laine, J. (2010). IDS: Thermodynamic-kinetic-empirical tool for modelling of solidification, microstructure and material properties. *Mathematics and Computers in Simulation*, *80*(7), 1536–1550. doi:10.1016/j.matcom.2009.11.002

Mills, K. C. (1994). *The Performance of Casting Powders*. Steel Technol. Int.

Mo¨es, N., B'echet, E., & Tourbier, M. (2006). Imposing Dirichlet boundary conditions in the extended finite element method. *International Journal for Numerical Methods in Engineering*, *67*(12), 1641–1669. doi:10.1002/nme.1675

Najjar, F. M., Thomas, B. G., & Hershey, D. E. (1995). Numerical study of steady turbulent flow through bifurcated nozzles in continuous casting. *Metallurgical and Materials Transactions. B, Process Metallurgy and Materials Processing Science*, *26*(4), 749–765. doi:10.1007/BF02651721

Nestler, B., & Wheeler, A. A. (2000). A multi-phase-field model of eutectic and peritectic alloys: Numerical simulation of growth structures. *Physica D. Nonlinear Phenomena*, *138*(1-2), 114–133. doi:10.1016/S0167-2789(99)00184-0

O' Malley, R. J. (1999). Observations of various steady state and dynamic thermal behaviors in a continuous casting mold. *82nd Steelmaking Conference*.

Petersen, S., & Hack, K. (2007). The Thermochemistry Library ChemApp and its applications Int. J. Mat. Res. (formerly. *Zeitschrift fur Metallkunde*, *98*, 935–945.

Prasher, B. D., & Wills, G. B. (1973). Mass Transfer in an Agitated Vessel, Industrial & Engineering. *Chemistry Process Design and Development*, *12*(3), 351–354. doi:10.1021/i260047a023

Qi, X. B., Chen, Y., Kang, X. H., Li, D. Z., & Du, Q. (2015). An analytical approach for predicting as-cast grain size of inoculated aluminum alloys. *Acta Materialia*, *99*, 337–346. doi:10.1016/j.actamat.2015.08.006

Räisänen, T., Louhenkilpi, S., Hätönen, T., Toivanen, J., Laine, J., & Kekäläinen, M. (2004). *European Cong. on Computational Methods in Applied Sciences and Engineering ECCOMAS 2004*. University of Jyväskylä.

Rappaz, J., & Scheid, J. F. (2000). Existence of solutions to a phase-field model for the isothermal solidification process of a binary alloy. *Mathematical Methods in the Applied Sciences*, *23*(6), 491–513. doi:10.1002/(SICI)1099-1476(200004)23:6<491::AID-MMA120>3.0.CO;2-4

Réger, M., Verő, B., Kytönen, H., & Louhenkilpi, S. (2002). *Proc. of 4th European Continuous Casting Conference.*

Sahai, Y., & Emi, T. (1996). Melt Flow Characterization in Continuous Casting Tundishes. *ISIJ International*, *36*(6), 667–672. doi:10.2355/isijinternational.36.667

Steinbach, I. (2009). Phase-field models in materials science Ingo Steinbach. *Modelling and Simulation in Materials Science and Engineering*, *17*(7), 31. doi:10.1088/0965-0393/17/7/073001

Suomi, M.-L., Jalkanen, H., Andersin, E., & Wallgren, M. (1995). CONSIM-2 Program for Oxygen Converter Simulation. In *Proceedings of the 2nd Colloquium on Process Simulation.*

Suomi, M.-L., Jalkanen, H., & Wallgren, M. (1995). Simulation of Oxygen Converter Process. In *Proceedings of the 2nd Colloquium on Process Simulation.*

Thomas, B. G. (2002). Modeling of the continuous casting of steel - past, present and future. *Metallurgical and Materials Transactions. B, Process Metallurgy and Materials Processing Science*, *33*(6), 795–812. doi:10.1007/s11663-002-0063-9

Thomas, B. G. (1991). Mathematical Modeling of the Continuous Slab Casting Mold: a State of the Art Review. *1991 Steelmaking Conference.*

Thomas, B. G., & Najjar, F. M. (1991). Finite element modelling of turbulent fluid flow and heat transfer in continuous casting. *Applied Mathematical Modelling*. http://doi.org/10.1016/0307-904X(91)90001-6

Thomas, B. G., & Zhang, L. (2001). Mathematical modeling of fluid flow in continuous casting. *ISIJ International*, *41*(10), 1181–1193. doi:10.2355/isijinternational.41.1181

Turkenitch, D. I. (1971). *Upravlenie plavkoi stali v konvertere*. Moskva: Metallurgiya.

Ueshima, Y., Mizoguchi, S., Matsumiya, T., & Kajioka, H. (1986). Analysis of solute distribution in dendrites of carbon steel with δ/γ transformation during solidification. *Metallurgical Transactions. B, Process Metallurgy*, *17*(4), 845–859. doi:10.1007/BF02657148

Upton, E. A., Rao, T. R. S., Dauby, P. H., & Knechtges, R. C. (1988). Physical Metallurgy and Mathematical Modeling as Tools for Continuous Casting Optimization at LTV Steel. *Ironmaking & Steelmaking*, *15*, 51–57.

Vapalahti, S., Louhenkilpi, S., & Räisänen, T. (2006). The Effect of Fluid Flow on Heat Transfer and Shell Growth in Continuous Casting of Copper. *Materials Science Forum*, *508*, 503–508. doi:10.4028/www.scientific.net/MSF.508.503

Väyrynen, P., & Wang, S. (2010). Control of Fluid Flow, Heat Transfer and Inclusions in Continuous Casting: CFD and Neural Network Studies. The Minerals, Metals & Ma. Retrieved from http://66.221.229.18/pdfs/tp/cast_tp/FloSci-Bib02-10.pdf

Williams, J. A. (1996). Analytical models of scratch hardness. *Tribology International*, *29*(8), 675–694. doi:10.1016/0301-679X(96)00014-X

Yamauchi, A., Sorimachi, K., Sakuraya, T., & Fujii, T. (1993). Heat Transfer Between Mold and Strand Through Mold Flux Film in Continuous Casting of Steel. *ISIJ International*, *33*(1), 140–147. doi:10.2355/isijinternational.33.140

Yu, S., & Louhenkilpi, S. (2013). Numerical Simulation of Dehydrogenation of Liquid Steel in the Vacuum Tank Degasser. *Metallurgical and Materials Transactions. B, Process Metallurgy and Materials Processing Science*, *44*(2), 459–468. doi:10.1007/s11663-012-9782-8

Yu, Miettinen, & Louhenkilpi. (2013). Numerical Study on the Removal of Hydrogen and Nitrogen from the Melt of Medium Carbon Steel in Vacuum Tank Degasser. *Materials Science Forum, 762*.

Yu, Miettinen, & Louhenkilpi. (2014a). Modeling Study of Nitrogen Removal from the Vacuum Tank Degasser. *Steel Research International, 85*, 1393–1402.

Yu, Miettinen, Shao, & Louhenkilpi. (2014b). Mathematical Modeling of Nitrogen Removal from the Vacuum Tank Degasser. *Steel Research International, 85*.

Compilation of References

Abbott, L. F., Varela, K., Sen, K., & Nelson, S. B. (1997). Synaptic depression and cortical gain control. *Science*, *275*(5297), 220–223. doi:10.1126/science.275.5297.221 PMID:8985017

Agam, Y., Bullock, D., & Sekuler, R. (2005). Imitating unfamiliar sequences of connected linear motions. *Journal of Neurophysiology*, *94*(4), 2832–2843. doi:10.1152/jn.00366.2005 PMID:16014793

Agari, Y., & Uno, T. (1986). Estimation on thermal conductivities of filled polymers. *Journal of Applied Polymer Science*, *32*(7), 5705–5712. doi:10.1002/app.1986.070320702

Agarwal, K., & Shivpuri, R. (2013). A Computational Framework for Integrated Process Design of High Performance Parts. 2nd *World Congress on Integrated Computational Materials Engineering.*

Agarwal, A., Pettersson, F., Singh, A., Kong, C. S., Saxén, H., Rajan, K., & Chakraborti, N. et al. (2009). Identification and Optimization of AB2 Phases Using Principal Component Analysis, Evolutionary Neural Nets, and Multiobjective Genetic Algorithms. *Materials and Manufacturing Processes*, *24*(3), 274–281. doi:10.1080/10426910802678321

Agrawal, R., Saxena, N. S., Mathew, G., Thomas, S., & Sharma, K. B. (2000). Effective thermal conductivity of three-phase styrene butadiene composites. *Journal of Applied Polymer Science*, *76*(12), 1799–1803. doi:10.1002/(SICI)1097-4628(20000620)76:12<1799::AID-APP10>3.0.CO;2-D

Ahrens, L. H. (1952). Anion Affinity and Polarizing Power of Cations. *Nature*, *169*(4298), 463–463. doi:10.1038/169463a0

Al-Abbasi, F. M., & Nemes, J. A. (2003). Micromechanical modeling of dual phase steels. *International Journal of Mechanical Sciences*, *45*(9), 1449–1465. doi:10.1016/j.ijmecsci.2003.10.007

Allan, F. (2015). A continuum model of deformation, transport and irreversible changes in atomic structure in amorphous lithium–silicon electrodes. *Acta Materialia*, *98*, 229–241. doi:10.1016/j.actamat.2015.07.036

Allen, J., Mistree, F., Panchal, J., Gautham, B., Singh, A., Reddy, S., . . . Kumar, P. (2013). Integrated Realization of Engineered Materials and Products: A Foundational Problem. 2nd *World Congress on Integrated Computational Materials Engineering.*

Allendorf, S., Ottesen, D., Hardesty, D., Goldstein, D., Smith, C., & Malcolmson, A. (1998). Laser based sensor for real-time measurement ofoff-gas composition and temperature in BOF steelmaking. *AISE Steel Technology*, *4*, 31–35.

Allison, J., Li, M., Wolverton, C., & Su, X. M. (2006). Virtual Aluminium Castings: An Industrial Application of ICME. JOM, 28-35.

Allred, A. L., & Rochow, E. G. (1958). A scale of electronegativity based on electrostatic force. *Journal of Inorganic and Nuclear Chemistry*, *5*(4), 264–268. doi:10.1016/0022-1902(58)80003-2

Anand, G., Datta, S., & Chattopadhyay, P. P. (2013). Deterministic Approach for Microstructurally Engineered Formable Steels. *International Journal of Metallurgical Engineering*, *2*(1), 69–78.

Anand, G., Dey, P., Kok, P. J. J., Chakraborty, D., & Chattopadhyay, P. P. (2014). Architectured microstructures in steel. *Materials Science and Technology*, *30*(9), 1086–1093. doi:10.1179/1743284714Y.0000000535

Anapagaddi, R., Shukla, R., Goyal, S., Singh, A. K., Allen, J. K., Panchal, J. H., & Mistree, F. (2014). Exploring the Design Space to Determine the Operating Set Points for a Tundish. *International Design Engineering Technical Conferences & Computers and Information in Engineering Conference IDETC/CIE 2014*.

Andersin, E., & Jalkanen, H. (1990). Physicochemical Modeling of Converter Steelmaking. In *Proceedings of The 6th International Iiron and Steel Congress*.

Andrews, R., Diederich, J., & Tickle, A. B. (1995). Survey and critique of techniques for extracting rules form trained artificial neural networks. *Knowledge-Based Systems*, *8*(6), 373–389. doi:10.1016/0950-7051(96)81920-4

Andriotis, A. N., Lisenkov, S., & Menon, M. (2011). Ferromagnetic interactions in hosted bipartite materialsgeneralized-double-exchange and generalized superexchange interactions. *Journal of Physics Condensed Matter*, 23.

Andriotis, A. N., & Menon, M. (2011). Magnetic coupling in dilute magnetic semiconductors: A new perspective.[b]. *Phys. St. Sol.*, *248*, 2032.

Andriotis, A. N., & Menon, M. (2012). The synergistic character of the defect-induced magnetism in diluted magnetic semiconductors and related magnetic materials. *Journal of Physics Condensed Matter*, *24*, 455801. PMID:23086142

Andriotis, A. N., Menon, M., Sheetz, R. M., & Chernozatonskii, L. (2003). Magnetic properties of c60 polymers. *Physical Review Letters*, *90*, 026801. PMID:12570566

Andriotis, A. N., Mpourmpakis, G., Lisenkov, S., Sheetz, R. M., & Menon, M. (2013). U-calculation of the lsda+u functional using the hybrid b3lyp and hse functionals. *Phys. Status Solidi B*, *250*, 356.

Andriotis, A. N., Sheetz, R. M., Richter, E., & Menon, M. (2005). Defect-originated magnetism in carbon-based and non-traditional inorganic compounds: A new class of magnetic materials. *Europhysics Letters*, *72*, 658.

Assadi, M. H. N., Zhang, Y. B., & Li, S. (2009). N codoping induced ferromagnetism in zno:co (1010) thin films. *Journal of Applied Physics*, *106*, 093911.

Azzi, V. D., & Tsai, S. W. (1965). Anisotropic strength of composites. *Experimental Mechanics*, *5*(9), 283–288. doi:10.1007/BF02326292

Babuska, I., & Melenk, J. M. (1997). The partition of unity method. *International Journal for Numerical Methods in Engineering*, *40*(4), 727–758. doi:10.1002/(SICI)1097-0207(19970228)40:4<727::AID-NME86>3.0.CO;2-N

Baker, C. H., & Norris, P. M. (2015). Effect of long- and short-range order on SiGe alloy thermal conductivity: Molecular dynamics simulation. *Physical Review B: Condensed Matter and Materials Physics*, *91*(18), 180302. doi:10.1103/PhysRevB.91.180302

Balachandran, P. V. (2011). *Statistical learning for chemical crystallography.* (Doctoral dissertation). Iowa State University, Ames, IA.

Balachandran, P. V., Broderick, S. R., & Rajan, K. (2011). Identifying the "inorganic gene" for high-temperature piezoelectric perovskites through statistical learning. *Proceedings of the Royal Society A: Mathematical, Physical, and Engineering Science, 465*, 2155-2175.

Balachandran, P. V., Puggioni, D., & Rondinelli, J. M. (2014). Crystal-chemistry guidelines for noncentrosymmetric A_2BO_4 Ruddlesden-Popper oxides. *Inorganic Chemistry*, *53*(1), 336–348. doi:10.1021/ic402283c PMID:24320755

Balachandran, P. V., & Rajan, K. (2012). Structure maps for $A^I_6A^{II}_4(BO_4)_6X_2$ apatites via data mining. *Acta Crystallographica. Section B, Structural Science*, *68*(1), 24–33. doi:10.1107/S0108768111054061 PMID:22267555

Balachandran, P. V., Rajan, K., & Rondinelli, J. M. (2014). Electronically driven structural transitions in $A_{10}(PO_4)_6F_2$apatites (A=Ca, Sr, Pb, Cd, and Hg). *Acta Crystallographica. Section B, Structural Science*, *70*(3), 612–615. doi:10.1107/S2052520614003710 PMID:24892609

Balachandran, P. V., & Rondinelli, J. M. (2013). Interplay of octahedral rotations and breathing distortions in charge ordering perovskite oxides. *Physical Review B: Condensed Matter and Materials Physics*, *88*(5), 054101. doi:10.1103/PhysRevB.88.054101

Bandyopadhyay, N. R., & Datta, S. (2004). Effect of manganese partitioning on transformation induced plasticity characteristics in microalloyed dual phase steels. *ISIJ International*, *44*(5), 927–934. doi:10.2355/isijinternational.44.927

Bansal, A., Barman, A., Ghosh, S., & Chakraborti, N. (2013). Designing Cu-Zr Glass Using Multiobjective Genetic Algorithm and Evolutionary Neural Network Metamodels-Based Classical Molecular Dynamics Simulation. *Materials and Manufacturing Processes*, *28*(7), 733–740. doi:10.1080/10426914.2013.763961

Barbero, E. J., & de Vivo, L. (2001). A constitutive model for elastic damage in fiber-reinforced PMC laminae. *International Journal of Damage Mechanics*, *10*(1), 73–93. doi:10.1106/6PQ6-31JW-F69K-74LU

Bardenheuer, F. (1970). Contribution to the metallurgy of the LD process. *Blast Furnace and Steel Plant.*, *58*(39), 401–406.

Bazant, Z. P., & Planas, J. (1998). *Fracture and size effect in concrete and other quasibrittle materials*. Boca Raton: CRC Press.

Behan, A. J., Mokhtari, A., Blythe, H. J., Score, D., Xu, X. H., Neal, J. R., & Gehring, G. A. et al. (2008). Two magnetic regimes in doped zno corresponding to a dilute magnetic semiconductor and a dilute magnetic insulator. *Physical Review Letters*, *100*, 047206. PMID:18352328

Belsky, A., Hellenbrandt, M., Karen, V. L., & Luksch, P. (2002). New developments in the inorganic crystal structure database (ICSD): Accessibility in support of materials research and design. *Acta Crystallographica. Section B, Structural Science*, *58*(3), 364–369. doi:10.1107/S0108768102006948 PMID:12037357

Belytschko, T., Gracie, R., & Ventura, G. (2009). A review of extended/generalized finite element methods for material modeling. Modelling and Simulation in Materials Science and Engineering, 17(4).

Benveniste, Y. (1987). A new approach to the application of Mori-Tanaka's theory in composite materials. *Mechanics of Materials*, *6*(2), 147–157. doi:10.1016/0167-6636(87)90005-6

Berdichevskii, V. (1979). Variational-asymptotic method of constructing a theory of shells. *Journal of Applied Mathematics and Mechanics*, *43*(4), 711–736. doi:10.1016/0021-8928(79)90157-6

Bermejo, R., Baudín, C., Moreno, R., Llanes, L., & Sánchez-Herencia, A. J. (2007). Processing optimisation and fracture behaviour of layered ceramic composites with highly compressive layers. *Composites Science and Technology*, *67*(9), 1930–1938. doi:10.1016/j.compscitech.2006.10.010

Berryman, J. G., & Berge, P. A. (1993). Rock elastic properties: Dependence on microstructure. In C. S. Chang & J. W. Ju (Eds.), Homogenization and constitutive modeling for heterogeneous materials. AMD.

Bhadeshia, H. K. D. H. (1999). Neural Networks in Materials Science. *ISIJ International*, *39*(10), 966–979. doi:10.2355/isijinternational.39.966

Bhadeshia, H. K. D. H. (2008). Mathematical models in materials science. *Materials Science and Technology*, *24*(2), 128–136. doi:10.1179/174328407X213107

Bhadeshia, H. K. D. H., MacKay, D. J. C., & Svensson, L.-E. (1995). Impact toughness of C–Mn steel arc welds – Bayesian neural network analysis. *Materials Science and Technology*, *11*(10), 1046–1051. doi:10.1179/mst.1995.11.10.1046

Bhalla, A. S., Guo, R., & Roy, R. (2000). The perovskite structure - a review of its role in ceramic science and technology. *Materials Research Innovations*, *4*(1), 3–26. doi:10.1007/s100190000062

Bhatia, D., McCabe, R. W., Harold, M. P., & Balakotaiah, V. (2009). Experimental and kinetic study of NO oxidation on model Pt catalysts. *Journal of Catalysis*, *266*(1), 106–119. doi:10.1016/j.jcat.2009.05.020

Bhattacharya, B., Kumar, G. R. D., Agarwal, A., Erkoç, S., Singh, A., & Chakraborti, N. (2009). Analyzing Fe-Zn system using molecular dynamics, evolutionary neural nets and multi-objective genetic algorithms. *Computational Materials Science*, *46*(4), 821–827. doi:10.1016/j.commatsci.2009.04.023

Born, M., & Oppenheimer, R. (1927). Zur quantentheorie der molekeln. *Annalen der Physik*, *389*(20), 457–484. doi:10.1002/andp.19273892002

Bortz, A. B., Kalos, M. H., & Lebowitz, J. L. (1975). A new algorithm for Monte Carlo simulation of Ising spin systems. *Journal of Computational Physics*, *17*(1), 10–18. doi:10.1016/0021-9991(75)90060-1

Bose, N. K., & Liang, P. (1996). *Neural Network Fundamentals with Graphs, Algorithms, and Applications*. McGraw-Hill, Inc.

Bouchard, P. O., Bay, F., Chastel, Y., & Tovena, I. (2000). Crack propagation modelling using an advanced remeshing technique. *Computer Methods in Applied Mechanics and Engineering*, *189*(3), 723–742. doi:10.1016/S0045-7825(99)00324-2

Boudenne, A., Ibos, L., Fois, M., Gehin, E., & Majeste, J. C. (2004). Thermophysical properties of polypropylene/aluminium composites. *Journal of Polymer Science. Part B, Polymer Physics*, *42*(4), 722–732. doi:10.1002/polb.10713

Bouquerel, J., Verbeken, K., & DeCooman, B. C. (2006). Microstructurebased model for the static mechanical behavior of multiphase steels. *Acta Materialia*, *54*(6), 1443–1456. doi:10.1016/j.actamat.2005.10.059

Bragard, J., Karma, A., Lee, Y. H., & Plapp, M. (2002). Linking phase-field and atomistic simulations to model dendritic solidification in highly undercooled melts. *Interface Science*, *10*(2-3), 121–136. doi:10.1023/A:1015815928191

Brandon, D. G. (1966). The structure of high-angle grain boundaries. *Acta Metallurgica*, *14*(11), 1479–1484. doi:10.1016/0001-6160(66)90168-4

Branke, J., Deb, K., Dierolf, H., & Osswald, M. (2004). Finding knees in multi-objective optimization. LNCS, 3242, 722–731. doi:10.1007/978-3-540-30217-9_73

Braun, A., & Pfeifer, H. (2007). *Investigations of Non-isothermal Flow Conditions in a Two Strand Tundish Water Modelusing DPIV and PLIF-technique*. SteelSim 2007 - 2nd International Conference of Simulation and Modeling of Metallurgical Processes in Steelmaking, Graz, Austria.

Brese, N. E., & O'Keeffe, M. (1991). Bond-valence parameters for solids. *Acta Crystallographica. Section B, Structural Crystallography and Crystal Chemistry*, *47*(2), 192–197. doi:10.1107/S0108768190011041

Breslow, R., Jaun, B., Kluttz, R. Q., & Xia, C. Z. (1982). Ground state pi-electron triplet molecules of potential use in the synthesis of organic ferromagnets. *Tetrahedron*, *38*, 863.

Broughton, J. Q., & Gilmer, G. H. (1986). Molecular dynamics investigation of the crystal–fluid interface. VI. Excess surface free energies of crystal–liquid systems. *The Journal of Chemical Physics*, *84*(10), 5759–5768. doi:10.1063/1.449884

Brun, F., Yoshida, T., Robson, J. D., Narayan, V., Bhadeshia, H. K. D. H., & MacKay, D. J. C. (1999). Theoretical design of ferritic creep resistant steels using neural network, kinetic, and thermodynamic models. *Materials Science and Technology*, *15*(5), 547–554. doi:10.1179/026708399101506085

Bucholz, E. W., Kong, C. S., Marchman, K. R., Sawyer, W. G., Phillpot, S. R., Sinnott, S. B., & Rajan, K. (2012). Data-driven model for estimation of friction coefficient via informatics methods. *Tribology Letters*, *47*(2), 211–221. doi:10.1007/s11249-012-9975-y

Bulatov, V., Abraham, F. F., Kubin, L., Devincre, B., & Yip, S. (1998). Connecting atomistic and mesoscale simulations of crystal plasticity. *Nature*, *391*(6668), 669–672. doi:10.1038/35577

Burgess, D. (2011). Thermochemical Data. In *NIST Chemistry WebBook*. NIST Standard Reference Database Number 69. National Institute of Standards and Technology. Retrieved February 10, 2014, from http://webbook.nist.gov

Cahn, J. W., & Allen, S. M. (1977). A microscopic theory for domain wall motion and its experimental verification in Fe-Al alloy domain growth kinetics. *Journal de Physique. Colloque*, *38*(C7), C7–C51. doi:10.1051/jphyscol:1977709

Cahn, J. W., & Hilliard, J. E. (1958). Free energy of a nonuniform system. I. Interfacial free energy. *The Journal of Chemical Physics*, *28*(2), 258–267. doi:10.1063/1.1744102

Cai, J., Cottis, R. A., & Lyon, S. B. (1999). Phenomenological modelling of atmospheric corrosion using an artificial neural network. *Corrosion Science*, *41*(10), 2001–2030. doi:10.1016/S0010-938X(99)00024-4

Cai, P. Z., Green, D. J., & Messing, G. L. (1998). Mechanical characterization of Al2O3/ZrO2 hybrid laminates. *Journal of the European Ceramic Society*, *18*(14), 2025–2034. doi:10.1016/S0955-2219(98)00151-4

Campbell, B. J., Stokes, H. T., Tanner, D. E., & Hatch, D. M. (2006). ISODISPLACE: A web-based tool for exploring structural distortions. *Journal of Applied Crystallography*, *39*(4), 607–614. doi:10.1107/S0021889806014075

Catalan, G. (2008). Progress in perovskite nickelate research. *Phase Transitions*, *81*(7-8), 729–749. doi:10.1080/01411590801992463

Ceder, G. (1998). Computational materials science: Predicting properties from scratch. *Science*, *277*, 1237–1242.

Ceperley, D. M., & Alder, B. J. (1980). Ground state of the electron gas by a stochastic method. *Physical Review Letters*, *45*(7), 566–569. doi:10.1103/PhysRevLett.45.566

Chakhalian, J., Rondinelli, J. M., Liu, J., Gray, B. A., Kareev, M., Moon, E. J., & Freeland, J. W. et al. (2011). Asymmetric orbital-lattice interactions in ultrathin correlated oxide films. *Physical Review Letters*, *107*(11), 116805. doi:10.1103/PhysRevLett.107.116805 PMID:22026694

Chakraborti, N. (2013). Evolutionary Data-Driven Modeling. In Informatics for Materials Science and Engineering. Oxford, UK: Butterworth-Heinemann.

Chakraborti, N. (2014). Strategies for Evolutionary Data Driven Modeling in Chemical and Metallurgical Systems. *Springer*. doi:10.1007/978-3-319-06508-3_4

Chakraborty, S., & Sahai, Y. (1991). Effect of Varying Ladle Stream Temperature on the Melt Flow and Heat Transfer in Continuous Casting Tundishes. *ISIJ International*, *31*(9), 960–967. doi:10.2355/isijinternational.31.960

Chang, Y. A., Chen, S., Zhang, F., Yan, X., Xie, F., Schmid-Fetzer, R., & Oates, W. A. (2004). Phase diagram calculation: Past, present and future. *Progress in Materials Science*, *49*(3), 313–345. doi:10.1016/S0079-6425(03)00025-2

Chattopadhyay, P. P. (2014). Microstructure based design of formable steels. *The Banaras Metallurgist*, 43-52.

Chattopadhyay, S., & Natha, T. K. (2010). Room temperature enhanced positive magnetoresistance in pt and carrier induced zn(fe)o and zn(fe,al)o dilute magnetic semiconductors junction. *Journal of Applied Physics*, *108*, 083904.

Cheetham, A. K. (1994). Advanced inorganic materials: An open horizon. *Science*, *264*(5160), 794–795. doi:10.1126/science.264.5160.794 PMID:17794719

Chen, C. R., Pascual, J., Fischer, F. D., Kolednik, O., & Danzer, R. (2007). Prediction of the fracture toughness of a ceramic multilayer composite – Modeling and experiments. *Acta Materialia*, *55*(2), 409–421. doi:10.1016/j.actamat.2006.07.046

Chen, H. C., Era, H., & Shimizu, M. (1989). Effect of phosphorus on the formation of retained austenite and mechanical properties in Si low-carbon steel sheet. *Metallurgical Transactions. A, Physical Metallurgy and Materials Science*, *20*(3), 437–445. doi:10.1007/BF02653923

Chen, L. Q. (2002). Phase-field models for microstructure evolution. *Annual Review of Materials Research*, *32*(1), 113–140. doi:10.1146/annurev.matsci.32.112001.132041

Chen, L. Q., & Wang, Y. (1996). The continuum field approach to modeling microstructural evolution. *JOM*, *48*(12), 13–18. doi:10.1007/BF03223259

Chen, L. Q., Wolverton, C., Vaithyanathan, V., & Liu, Z. K. (2001). Modeling solid-state phase transformations and microstructure evolution. *MRS Bulletin*, *26*(03), 197–202. doi:10.1557/mrs2001.42

Chen, L. Q., & Yang, W. (1994). Computer simulation of the domain dynamics of a quenched system with a large number of nonconserved order parameters: The grain-growth kinetics. *Physical Review B: Condensed Matter and Materials Physics*, *50*(21), 15752–15756. doi:10.1103/PhysRevB.50.15752 PMID:9975940

Chen, Z. Z., Kioussis, N., Tu, K., Ghoniem, N., & Yang, J. (2010). Inhibiting adatom diffusion through surface alloying. *Physical Review Letters*, *105*(1), 015703. doi:10.1103/PhysRevLett.105.015703 PMID:20867465

Choi, H., McDowell, D. L., Allen, J. K., Rosen, D., & Mistree, F. (2008). An Inductive Design Exploration Method for Robust Multiscale Materials Design. *Journal of Mechanical Design*, *130*(031402), 1–13.

Chokshi, A. H., Rosen, A., Karch, J., & Gleiter, H. (1989). On the validity of the Hall-Petch relationship in nanocrystalline materials. *Scripta Metallurgica*, *23*(10), 1679–1683. doi:10.1016/0036-9748(89)90342-6

Chorkendorff, I., & Niemantsverdriet, J. W. (2007). *Concepts of Modern Catalysis and Kinetics* (2nd ed.). Wiley-VCH.

Chow, C. L., & Wang, J. (1987). An anisotropic theory of elasticity for continuum damage mechanics. *International Journal of Fracture*, *33*(1), 3–16. doi:10.1007/BF00034895

Christensen, R. M. (1979). *Mechanics of composite materials*. New York: Wiley.

Chung, D., Chun, H., & Keleş, S. (2013). *spls: Sparse Partial Least Squares (SPLS) Regression and Classification*. R package version 2.2-1. Retrieved from http://CRAN.R-project.org/package=spls

Chun, H., & Keleş, S. (2010). Sparse partial least squares regression for simultaneous dimension reduction and variable selection. *Journal of the Royal Statistical Society. Series B, Statistical Methodology*, *72*, 3–25. doi:10.1111/j.1467-9868.2009.00723.x PMID:20107611

Clark, Wagner, & Trousset. (2003, January). Modelling for Tundish Applications. *Refractories Engineer*, 9-12.

Clarke, R., Ressom, H. W., Wang, A., Xuan, J., Liu, M. C., Gehan, E. A., & Wang, Y. (2008). The properties of high-dimensional data spaces: Implications for exploring gene and protein expression data. *Nature Reviews. Cancer*, *8*(1), 37–49. doi:10.1038/nrc2294 PMID:18097463

Clegg, W. J., Kendall, K., Alford, N. M., Button, T. W., & Birchall, J. D. (1990). A simple way to make tough ceramics. *Nature*, *347*(6292), 455–457. doi:10.1038/347455a0

Clementi, E., & Raimondi, D. L. (1963). Atomic Screening Constants from SCF Functions. *The Journal of Chemical Physics*, *38*(11), 2686. doi:10.1063/1.1733573

Cococcioni, M., & de Gironcoli, S. (2005). Linear response approach to the calculation of the effective interaction parameters in the lda+u method. *Physical Review B: Condensed Matter and Materials Physics*, *71*, 035105.

Coello Coello, C. A., Lamont, G. B., & Van Veldhuizen, D. A. (2002). *Evolutionary Algorithms for Solving Multi-Objective Problems*. Norwell, MA: Kluwer. doi:10.1007/978-1-4757-5184-0

Cohen, M. (1976). Unknowables in the essence of materials science and engineering. *Materials Science and Engineering*, *25*, 3–4. doi:10.1016/0025-5416(76)90043-4

Cohon, J. L. (1985). Multicriteria programming: brief review and application. In J. S. Gero (Ed.), *Design Optimization* (pp. 163–191). Orlando, FL: Academic Press.

Cole, D., Martin-Moran, C., Sheard, A., Bhadeshia, H. K. D. H., & MacKay, D. J. C. (2000). Modelling creep rupture strength of ferritic steel welds. *Sci. Technol. Weld. Join.*, *5*(2), 81–89. doi:10.1179/136217100101538065

Collet, P. (2007). Genetic programming. In Handbook of Research on Nature Inspired Computing for Economics and Management. Hershey, PA: Idea. doi:10.4018/978-1-59140-984-7.ch005

Collins, J. B., & Levine, H. (1985). Diffuse interface model of diffusion-limited crystal growth. *Physical Review B: Condensed Matter and Materials Physics*, *31*(9), 6119–6122. doi:10.1103/PhysRevB.31.6119 PMID:9936623

Cook, J., Gordon, J. E., Evans, C. C., & Marsh, D. M. (1964). A mechanism for the control of crack propagation in all-brittle systems. *Proceedings of the Royal Society of London. Series A, Mathematical and Physical Sciences*, *282*(1391), 508–520. doi:10.1098/rspa.1964.0248

Cool, T., Bhadeshia, H. K. D. H., & MacKay, D. J. C. (1997). The yield and ultimate tensile strength of steel welds. *Materials Science and Engineering A*, *223*(1-2), 186–200. doi:10.1016/S0921-5093(96)10513-X

Council, N. S. T. (2011). *Materials Genome Initiative for global competitiveness*. Washington, D.: Office of Science and Technology Policy.

Cunha, A. G., Oliviera, P., & Covas, J. (1997). Use of genetic algorithms in multicriteria optimization to solve industrial problems. In T. B¨ack (Ed.), *Proceedings of the Seventh International Conference on Genetic Algorithms* (pp. 682-688).

Curtarolo, S., Setyawan, W., Wang, S., Xue, J., Yang, K., Taylor, R. H., & Levy, O. et al. (2012). AFLOWLIB.ORG: A distributed materials properties repository from high-throughput *ab initio* calculations. *Computational Materials Science*, *58*, 227–235. doi:10.1016/j.commatsci.2012.02.002

Curtze, S., & Kuokkala, V. T. (2010). Dependence of tensile deformation behavior of TWIP steels on stacking fault energy, temperature and strain rate. *Acta Materialia*, *58*(15), 5129–5141. doi:10.1016/j.actamat.2010.05.049

Cutler, W. A., Zok, F. W., & Lange, F. F. (1996). Mechanical behavior of several hybrid ceramic-matrix-composite laminates. *Journal of the American Ceramic Society*, *79*(7), 1825–1833. doi:10.1111/j.1151-2916.1996.tb08002.x

Cwudzinski, A., & Jowsa, J. (2012). Mathematical Modeling of Behaviour a Liquid Steel Flow and a Non-metallic Inclusions in the One Strand Wedge Type Tundish with a Subflux Turbulence Controller and a Low Dam. *Materials Science Forum, 706-709*, 1379–1384. doi:10.4028/www.scientific.net/MSF.706-709.1379

Dalton, L. A., & Dougherty, E. R. (2011). Bayesian minimum mean-square error estimation for classification error – Part 1: Definition and the Bayesian MMSE error estimator for discrete classification. *IEEE Transactions on Signal Processing*, *59*(1), 115–129. doi:10.1109/TSP.2010.2084572

Damle, C., & Sahai, Y. (1995). The Effect of Tracer Density on Melt Flow Characterization in Continuous Casting Tundishes – A Modeling Study. *ISIJ International*, *35*(2), 163–169. doi:10.2355/isijinternational.35.163

Daniel, I. M., & Ishai, O. (2006). *Engineering mechanics of composite materials* (2nd ed.). New York: Oxford University Press.

Dash, M., & Liu, H. (1997). Feature Selection for Classification. *Intelligent Data Analysis*, *1*(1-4), 131–156. doi:10.1016/S1088-467X(97)00008-5

Das, P., & Datta, S. (2007). Exploring the non-linearity in empirical modelling of a steel system using statistical and neural network models. *International Journal of Production Research*, *45*(3), 699–717. doi:10.1080/00207540600792465

Das, P., Mukherjee, S., Ganguly, S., Bhattacharyay, B. K., & Datta, S. (2009). Genetic algorithm based optimization for multi-physical properties of HSLA steel through hybridization of neural network and desirability function. *Computational Materials Science*, *45*(1), 104–110. doi:10.1016/j.commatsci.2008.03.050

Datta (2009). Extraction of knowledge from high strength steel data using soft computing techniques - An overview. *Statistical Analysis and Data Mining, 1*, 329-337.

Datta, S., & Banerjee, M. K. (2004). Kohonen Network Modelling for the Strength of Thermomechanically Processed HSLA Steel. *ISIJ International*, *44*(5), 846–851. doi:10.2355/isijinternational.44.846

Datta, S., & Banerjee, M. K. (2004). Optimizing parameters of supervised learning techniques (ANN) for precise mapping of the input-output relationship in TMCP steels. *Scand. J. Met.*, *33*(6), 310–315. doi:10.1111/j.1600-0692.2004.00699.x

Datta, S., & Banerjee, M. K. (2005a). Fuzzy modeling of strength–composition–process parameters relationships of HSLA steels. *Materials and Manufacturing Processes*, *20*(5), 761–776. doi:10.1081/AMP-200055136

Datta, S., & Banerjee, M. K. (2005b). Neuro-fuzzy modeling of the strength of thermomechanically processed HSLA steels. *Indian Journal of Physics*, *79*, 473–483.

Datta, S., & Banerjee, M. K. (2006). Mapping the input–output relationship in HSLA steels through expert neural network. *Materials Science and Engineering A*, *420*(1-2), 254–264. doi:10.1016/j.msea.2006.01.037

Datta, S., & Chattopadhyay, P. P. (2013). Soft Computing Techniques in Advancement of Structural Metals. *International Materials Reviews*, *58*(8), 475–504. doi:10.1179/1743280413Y.0000000021

Datta, S., Mahfouf, M., Zhang, Q., Chattopadhyay, P. P., & Sultana, N. (2016). Imprecise knowledge based design and development of titanium alloys for prosthetic applications. *Journal of the Mechanical Behavior of Biomedical Materials*, *53*, 350–365. doi:10.1016/j.jmbbm.2015.08.039

Datta, S., Pettersson, F., Ganguly, S., Saxe'n, H., & Chakraborti, N. (2007). Designing High Strength Multi-phase Steel for Improved Strength–Ductility Balance Using Neural Networks and Multi-objective Genetic Algorithms. *ISIJ International*, *47*(8), 1193–2001. doi:10.2355/isijinternational.47.1195

Davies, R., Blake, N., & Campbell, P. (1988). Solidification Modelling--an Aid to Continuous Casting. *4th International Conference Continuous Casting 2.*

Davies, C. H. J. (1995). The effect of neighbourhood on the kinetics of a cellular automaton recrystallisation model. *Scripta Metallurgica et Materialia, 33*(7), 1139–1143. doi:10.1016/0956-716X(95)00335-S

Davila, C. G., & Camanho, P. P. (2003). *Failure criteria for FRP laminates in plane stress.* Technical Report NASA/TM-2003-212663. NASA Langley Research Center.

Davila, C. G., Camanho, P. P., & Rose, C. A. (2005). Failure criteria for FRP laminates. *Journal of Composite Materials, 39*(4), 323–345. doi:10.1177/0021998305046452

Davila, C., Jaunky, N., & Goswami, S. (2003). *Failure criteria for FRP laminates in plane stress.* Norfolk, VA: Structural Dynamics, and Materials Conference.

De Jong, K. A. (1975). An analysis of the bevavior of a class of genetic adaptive systems (Ph. D. Thesis, Ann Arbor, MI: University of Michigan). Dissertation Abstracts International, 36(10), 5140B (University Microfilms No. 76-9381)

Deb, K. (1998). *Multi-objective genetic algorithms: Problem difficulties and construction of test functions.* Technical Report No. CI-49/98. Department of Computer Science/XI, University of Dortmund, Germany.

Deb. (2001). *Multi-objective optimization using evolutionary algorithms.* John Wiley & Sons Ltd.

Deb, K. (2001). *Multi-Objective Optimization using Evolutionary Algorithms.* John Wiley & Sons Ltd.

Deb, K., Pratap, A., Agarwal, S., & Meyarivan, T. (2002). A fast elitist non-dominated sorting genetic algorithm for multi-objective optimisation: NSGA-II. *IEEE Transactions on Evolutionary Computation, 6*(2), 182–197. doi:10.1109/4235.996017

Deb, K., Sinha, A., Korhonen, P., & Wallenius, J. (2010). An Interactive Evolutionary Multi-Objective Optimization Method Based on Progressively Approximated Value Functions. *IEEE Transactions on Evolutionary Computation, 14*(5), 723–739. doi:10.1109/TEVC.2010.2064323

Deb, K., & Srinivasan, A. (2006). Innovization: Innovating design principles through optimization. In *Proceedings of the Genetic and Evolutionary Computation Conference (GECCO-2006)* (pp. 1629-1636), New York: The Association of Computing Machinery (ACM).

Deo, B., & Boom, R. (1993). Fundamentals of Steelmaking Metallurgy. New York: Prentice Hall International.

Derlet, P. M., & Van Swygenhoven, H. (2004). High-frequency vibrational properties of metallic nanocrystalline grain boundaries. *Physical Review Letters, 92*(3), 035505. doi:10.1103/PhysRevLett.92.035505 PMID:14753887

Devarajan, S. P., Hinojosa, J. A. Jr, & Weaver, J. F. (2008). STM study of high-coverage structures of atomic oxygen on Pt(1 1 1): *P*(2×1) and Pt oxide chain structures. *Surface Science, 602*(19), 3116–3124. doi:10.1016/j.susc.2008.08.008

Dey, A., Dey, P., Datta, S., Bandyopadhyay, N. R., Mukhopadhyay, A. K., & Maiti, H. S. (2007). A new theoretical model for development of damage tolerant composite. In *International Seminar on Ceramics, "CeraTec 2007"*. Indian Ceramic Society.

Dey, A., Dey, P., Datta, S., Bandyopadhyay, N. R., Mukhopadhyay, A. K., & Maiti, H. S. (2008b). A new theoretical model for development of damage tolerant composites. *Transaction of the Indian Ceramic Society, 67*(2), 63–74. doi:10.1080/0371750X.2008.11078643

Dey, A., Dey, P., Datta, S., & Mukhopadhyay, A. K. (2008a). A new model for multilayer ceramic composites. *Materials and Manufacturing Processes, 23*(5), 513–527. doi:10.1080/10426910802104278

Dey, P., Bible, J., Datta, S., Broderick, S., Jasinski, J., Sunkara, M., & Rajan, K. et al. (2014). Informatics-aided bandgap engineering for solar materials. *Computational Materials Science*, *83*, 185–195. doi:10.1016/j.commatsci.2013.10.016

Dey, P., Dey, S., Datta, S., & Sil, J. (2011). Dynamic discreduction using Rough Sets. *Applied Soft Computing*, *11*, 3887–3897. doi:10.1016/j.asoc.2011.01.015

Dey, S., Datta, S., Chattopadhyay, P. P., & Sil, J. (2008). Modeling the properties of TRIP steel using AFIS: A distributed approach. *Computational Materials Science*, *43*(3), 501–511. doi:10.1016/j.commatsci.2007.12.009

Dharan, C. K. H. (1978). Fracture mechanics of composite materials. *Journal of Engineering Materials and Technology*, *100*(3), 233–247. doi:10.1115/1.3443485

Dieter, G. E. (1928). *Mechanical Metallurgy*. McGraw Hill Publication.

Dieter, G. E. (1988). *Mechanical Metallurgy, SI Metric Edition*. London: McGraw-Hill Book Company.

Ding, N., Bao, Y., Sun, Q., & Wang, L. (2011). Optimization of Flow Control Devices in a Single-Strand Slab Continuous Casting Tundish. *International Journal of Minerals, Metals and Materials*, *18*(3), 292–297. doi:10.1007/s12613-011-0436-7

Dobrza'nski, L. A., Kowalski, M., & Madejski, J. (2005). Methodology of the mechanical properties prediction for the metallurgical products from the engineering steels using the Artificial Intelligence methods. *Journal of Materials Processing Technology*, *164–165*, 1500–1509. doi:10.1016/j.jmatprotec.2005.02.194

Döll, R., Sörgel, G., Daum, M., & Zouhar, G. (1999). Control of mechanical properties by monitoring microstructure. *Proc. ASIA STEEL.*

Donoho, D. L., & Grimes, C. (2003). Hessian eigenmaps: Locally linear embedding techniques for high-dimensional data. *Proceedings of the National Academy of Sciences of the United States of America*, *100*(10), 5591–5596. doi:10.1073/pnas.1031596100 PMID:16576753

Draeger, A., Engell, S., & Ranke, H. (1995). *Model Predictive Control Using Neural Networks.* IEEE Control Systems.

Drugan, W. J. (2001). Dynamic fragmentation of brittle materials: Analytical mechanics-based models. *Journal of the Mechanics and Physics of Solids*, *49*(6), 1181–1208. doi:10.1016/S0022-5096(01)00002-3

Duan, L. B., Rao, G. H., Wang, Y. C., Yu, J., & Wang, T. (2008). Magnetization and raman scattering studies of (co,mn) codoped zno nanoparticles. *Journal of Applied Physics*, *104*, 013909.

Dudarev, S. L., Botton, G. A., Savrasov, S. Y., Humphreys, C. J., & Sutton, A. P. (1998). Electron-energy-loss spectra and the structural stability of nickel oxide:an lsda+u study. *Physical Review B: Condensed Matter and Materials Physics*, *57*, 1505.

Dumay, A., Chateau, J. P., Allain, S., Migot, S., & Bouaziz, O. (2008). Influence of addition elements on the stacking-fault energy and mechanical properties of an austenitic Fe–Mn–C steel. *Materials Science and Engineering A*, *483*, 184–187. doi:10.1016/j.msea.2006.12.170

Dumortier, C., Lehert, P., Krupa, P., & Charlier, A. (1998). Statistical modelling of mechanical properties of microalloyed steels by application of artificial neural network. *Materials Science Forum*, *284-286*, 393–400. doi:10.4028/www.scientific.net/MSF.284-286.393

Dunne, D., Tsuei, H., & Sterjovski, Z. (2004). Artificial Neural Networks for Modelling of the Impact Toughness of Steel. *ISIJ International*, *44*(9), 1599–1607. doi:10.2355/isijinternational.44.1599

Dvorak, G. J., & Laws, N. (1987). Analysis of progressive matrix cracking in composite laminates II. first ply failure. *Journal of Composite Materials*, *21*(4), 309–329. doi:10.1177/002199838702100402

Echaabi, J., Trochu, F., & Gauvin, R. (1996). Review of failure criteria of fibrous composite materials. *Polymer Composites, 17*(6), 786–798. doi:10.1002/pc.10671

Edgeworth, F. Y. (1881). *Mathematical Psychics.* London: Keagan.

Efron, B. (2013). Bayes' theorem in the 21st Century. *Science, 340*(6137), 1177–1178. doi:10.1126/science.1236536 PMID:23744934

Efron, B., & Gong, G. (1983). A leisurely look at the Bootstrap, the Jackknife, and Cross-Validation. *The American Statistician, 37*(1), 36–48.

Efron, B., Hastie, T., Johnstone, I., & Tibshirani, R. (2004). Least angle regression. *Annals of Statistics, 32*, 407. doi:10.1214/009053604000000067

Entwistle, K. M. (2001). Basic Principles of the Finite Element Method. Maney Publishing.

Ercolessi, F., & Adams, J. B. (1994). Interatomic potentials from first-principles calculations: The force-matching method. *EPL, 26*(8), 583–588. doi:10.1209/0295-5075/26/8/005

Erdogan, F. (1972). Fracture problems in composite materials. *Engineering Fracture Mechanics, 4*(4), 811–840. doi:10.1016/0013-7944(72)90018-5

Eschrig, H. (2003). *The Fundamentals of Density Functional Theory.* Leipzig, Germany: Edition am Gutenbergplatz Leipzig.

Eshelby, J. D. (1957). The determination of the elastic field of an ellipsoidal inclusion, and related problems. *Proceedings of the Royal Society of London. Series A, Mathematical and Physical Sciences, 241*(1226), 376–396. doi:10.1098/rspa.1957.0133

Evans, M. (1999). Method for improving parametric creep rupture life of 2·25Cr–1Mo steel using artificial neural networks. *Materials Science and Technology, 15*(6), 647–658. doi:10.1179/026708399101506391

Eymard, R., Gallouët, T., & Herbin, R. (2000). Finite volume methods. In Handbook of Numerical Analysis, (vol. 7, pp. 713-1018). Academic Press.

Fang, X., Fan, Z., Ralph, B., Evans, P., & Underhill, R. (2002). Effect of temper rolling on tensile properties of C–Mn steels. *Materials Science and Technology*, 18.

Farina, M., & Amato, P. (2002). On the Optimal Solution Definition for Many-criteria Optimization Problems. In *Proceedings of the NAFIPS-FLINT International Conference* (pp. 233-238). Piscataway, NJ: IEEE Service Center. doi:10.1109/NAFIPS.2002.1018061

Feuerbacher, M., Heidelmann, M., & Thomas, C. (2015). Hexagonal high-entropy alloys. *Materials Research Letters, 3*(1), 1–6. doi:10.1080/21663831.2014.951493

Fischbacher, T., Franchin, M., Bordignon, G., & Fangohr, H. (2007). A systematic approach to multiphysics extensions of finite-element-based micromagnetic simulations: Nmag. *Magnetics. IEEE Transactions on, 43*(6), 2896–2898.

Fitzsimmons, M. R., Eastman, J. A., Müller-Stach, M., & Wallner, G. (1991). Structural characterization of nanometer-sized crystalline Pd by x-ray-diffraction techniques. *Physical Review B: Condensed Matter and Materials Physics, 44*(6), 2452–2460. doi:10.1103/PhysRevB.44.2452 PMID:9999814

Foiles, S. M., & Hoyt, J. J. (2006). Computation of grain boundary stiffness and mobility from boundary fluctuations. *Acta Materialia, 54*(12), 3351–3357. doi:10.1016/j.actamat.2006.03.037

Folsom, C. A., Zok, F. W., & Lange, F. F. (1994). Flexural properties of brittle multilayer materials: I, Modeling. *Journal of the American Ceramic Society*, *77*(3), 689–696. doi:10.1111/j.1151-2916.1994.tb05350.x

Folsom, C. A., Zok, F. W., Lange, F. F., & Marshall, D. B. (1992). Mechanical behavior of a laminar ceramic/fiber-reinforced epoxy composite. *Journal of the American Ceramic Society*, *75*(11), 2969–2975. doi:10.1111/j.1151-2916.1992.tb04373.x

Fong, D. D., Stephenson, G. B., Streiffer, S. K., Eastman, J. A., Auciello, O., Fuoss, P. H., & Thompson, C. (2004). Ferroelectricity in ultrathin perovskite films. *Science*, *304*(5677), 1650–1653. doi:10.1126/science.1098252 PMID:15192223

Fonseca, C. M., & Fleming, P. J. (1993). Genetic algorithms for multiobjective optimization: Formulation, discussion and generalization. In S. Forrest (Ed.), *Proceedings of the Fifth International Conference on Genetic Algorithms*(pp. 416-423). San Mateo, CA:Morgan Kaufmann.

Fonseca, C. M., & Fleming, P. J. (1995). An overview of evolutionary algorithms in multiobjective optimization. *Evolutionary Computation*, *3*(1), 1–16. doi:10.1162/evco.1995.3.1.1

Fonseca, C. M., & Fleming, P. J. (1998a). Multiobjective optimization and multiple constraint handling with evolutionary algorithms—part ii: Application example. *IEEE Transactions on Systems, Man, and Cybernetics*, *28*(1), 38–47. doi:10.1109/3468.650320

Fonseca, C. M., & Fleming, P. J. (1998b). Multiobjective optimization and multiple constraint handling with evolutionary algorithms—part i: A unified formulation. *IEEE Transactions on Systems, Man, and Cybernetics*, *28*(1), 26–37. doi:10.1109/3468.650319

Fourman, M. P. (1985). Compaction of symbolic layout using genetic algorithms, In J. J. Grefenstette (Ed.), *Proceedings of an International Conference on Genetic Algorithms and Their Applications* (141-153).

Freeman, A. J., & Wimmer, E. (1995). Density functional theory as a major tool in computational materials science. *Annual Review of Materials Science*, *25*(1), 7–36. doi:10.1146/annurev.ms.25.080195.000255

Frenkel, D., & Smit, B. (2002). *Understanding molecular simulation from algorithms to applications*. San Diego, CA: Academic Press.

Froese, F. C. (1972). Average-energy-of-configuration Hartree-Fock results for the atoms helium to radon. *Atomic Data and Nuclear Data Tables*, *4*, 301–399. doi:10.1016/S0092-640X(72)80008-1

Fu, C. C., Torre, J. D., Willaime, F., Bocquet, J.-L., & Barbu, A. (2004, December19). Multiscale modelling of defect kinetics in irradiated iron. *Nature Materials*, *4*(1), 68–74. doi:10.1038/nmat1286

Fujii, H., MacKay, D. J. C., & Bhadeshia, H. K. D. H. (1996). Bayesian Neural Network Analysis of Fatigue Crack Growth Rate in Nickel Base Superalloys. *ISIJ International*, *36*(11), 1373–1382. doi:10.2355/isijinternational.36.1373

Gai, Y., Li, J., Li, S. S., Xia, J. B., & Wei, S. H. (2009). Design of narrow-gap tio2: A passivated codoping approach for enhanced photoelectrochemical activity. *Physical Review Letters*, *102*, 036402. PMID:19257373

Ganguly, S., Datta, S., & Chakraborti, N. (2007). Genetic Algorithms in Optimization of Strength and Ductility of Low-Carbon Steels. *Materials and Manufacturing Processes*, *22*(5), 650–658. doi:10.1080/10426910701323607

Ganguly, S., Datta, S., & Chakraborti, N. (2009). Genetic algorithm-based search on the role of variables in the work hardening process of multiphase steels. *Computational Materials Science*, *45*(1), 158–166. doi:10.1016/j.commatsci.2008.01.074

Ganguly, S., Datta, S., Chattopadhyay, P. P., & Chakraborti, N. (2009). Designing the Multiphase Microstructure of Steel for Optimal TRIP Effect: A Multiobjective Genetic Algorithm Based Approach. *Materials and Manufacturing Processes*, *24*(1), 31–37. doi:10.1080/10426910802540398

Gao, J., & Thompson, R. G. (1996). Real time-temperature models for Monte Carlo simulations of normal grain growth. *Acta Materialia*, *44*(11), 4565–4570. doi:10.1016/1359-6454(96)00079-1

Gardner, J. P. (1994). *Micromechanical modeling of composite materials in finite element analysis using an embedded cell approach* (Masters thesis, MIT, USA). Retrieved from http://dspace.mit.edu/bitstream/handle/1721.1/43275/31369665.pdf?...1

Gautham, B., Kulkarni, N., Khan, D., Zagade, P., Reddy, S., & Uppaluri, R. (2013). Knowledge Assisted Integrated Design of a Component and Its Manufacturing Process. *2nd World Congress on Integrated Computational Materials Engineering*.

Gens, A., & Potts, D. M. (1988). Critical state models in computational geomechanics. *Engineering Computations*, *5*(3), 178–197. doi:10.1108/eb023736

Georges, A., Kotliar, G., Krauth, W., & Rozenberg, M. J. (1996). Dynamical mean-field theory of strongly correlated fermion systems and the limit of infinite dimensions. *Reviews of Modern Physics*, *68*(1), 13–125. doi:10.1103/RevModPhys.68.13

Germain, N., Besson, J., & Feyel, F. (2007). Composite layered materials: Anisotropic nonlocal damage models. *Computer Methods in Applied Mechanics and Engineering*, *196*(41–44), 4272–4282. doi:10.1016/j.cma.2007.04.009

Getman, R.B., Schneider, W.F., Smeltz, A.D., Delgass, W.N., Ribeiro, F.H. (2009). Article. *Physical Review Letters*, *78*(102), 076101.

Getman, R. B., & Schneider, W. F. (2010). DFT-Based Coverage-Dependent Model of Pt-Catalyzed NO Oxidation. *ChemCatChem*, *2*(11), 1450–1460. doi:10.1002/cctc.201000146

Getman, R. B., Xu, Y., & Schneider, W. F. (2008). Thermodynamics of Environment-Dependent Oxygen Chemisorption on Pt(111). *The Journal of Physical Chemistry C*, *112*(26), 9559–9572. doi:10.1021/jp800905a

Ghosh, A. (2008). Hygrothermal effects on the initiation and propagation of damage in composite shells. *Aircraft Engineering and Aerospace Technology*, *80*(4), 386–399. doi:10.1108/00022660810882755

Ghosh, A., & Sinha, P. K. (2004). Dynamic and impact response of damaged laminated composite plates. *Aircraft Engineering and Aerospace Technology*, *76*(1), 29–37. doi:10.1108/00022660410514982

Ghosh, S. K., Ganguly, S., Chattopadhyay, P. P., & Datta, S. (2005). Modeling the Effect of Copper on Hardness of Microalloyed Dual Phase Steel through Neural Network and Neuro-fuzzy Systems. *ISIJ International*, *45*(9), 1345–1351. doi:10.2355/isijinternational.45.1345

Ghosh, S. K., Ganguly, S., Chattopadhyay, P. P., & Datta, S. (2008). Effect of copper and micro-alloying (Ti, B) on tensile properties of HSLA steels predicted by ANN. *Ironmaking & Steelmaking*.

Ghosh, S. K., Ganguly, S., Chattopadhyay, P. P., & Datta, S. (2009). Effect of copper and microalloying (Ti, B) addition on tensile properties of HSLA steels predicted by ANN technique. *Ironmaking & Steelmaking*, *36*(2), 125–132. doi:10.1179/174328107X254880

Ghosh, S., Guha, A., Krishna, K. M., Mukhopadhyay, A. K., & Maiti, H. S. (2006). Tape cast multilayer composite of nano zirconia with high toughness. *Materials and Manufacturing Processes*, *21*(7), 662–668. doi:10.1080/10426910600611680

Ghosh, S., Guha, A., Mukhopadhyay, A. K., & Maiti, H. S. (2005). Tape cast porous nano alumina multilayer composites. *Transactions of the Indian Ceramic Society*, *64*(2), 101–108. doi:10.1080/0371750X.2005.11012195

Ghosh, S., & Moorthy, S. (2004). Three dimensional Voronoi cell finite element model for microstructures with ellipsoidal heterogeneties. *Computational Mechanics*, *34*(6), 510–531. doi:10.1007/s00466-004-0598-5

Giannozzi, P. (2009). Quantum espresso: A modular and open-source software project for quantum simulation of materials. *Journal of Physics Condensed Matter*, *21*, 395502. PMID:21832390

Gillespie, D. T. (2001). Approximate accelerated stochastic simulation of chemically reacting systems. *The Journal of Chemical Physics*, *115*(4), 1716–1733. doi:10.1063/1.1378322

Giri, B. K., Hakanen, J., Miettinen, K., & Chakraborti, N. (2013a). Genetic programming through bi-objective genetic algorithms with a study of a simulated moving bed process involving multiple objectives. *Applied Soft Computing*, *13*(5), 2613–2623. doi:10.1016/j.asoc.2012.11.025

Giri, B. K., Pettersson, F., Saxén, H., & Chakraborti, N. (2013b). Genetic Programming Evolved through Bi-Objective Genetic Algorithms Applied to a Blast Furnace. *Materials and Manufacturing Processes*, *28*(7), 776–782. doi:10.1080/10426914.2013.763953

Gladman, T., Dulieu, D., & McIvor. (1975). Structure-property relationships in high strength micro alloyed steels. In *Proceedings of an international symposium on HSLA steels, Microalloying '75*.

Gleiter, H. (2000). Nanostructured materials: Basic concepts and microstructure. *Acta Materialia*, *48*(1), 1–29. doi:10.1016/S1359-6454(99)00285-2

Gokhale, A. A., Kandoi, S., Greeley, J. P., Mavrikakis, M., & Dumesic, J. A. (2004). Molecular-level descriptions of surface chemistry in kinetic models using density functional theory. *Chemical Engineering Science*, *59*(22-23), 4679–4691. doi:10.1016/j.ces.2004.09.038

Golanski, D., Grzes, J., Kikuchi, N., & Kozłowski, M. (2000). Modeling of residual stresses in metal matrix composites using digital image based technique and homogenization method.*Proceedings of the International Conference on Mathematical Modeling and Simulation of Metal Technologies*.

Goldberg, D. E., & Richardson, J. (1987). Genetic algorithms with sharing for multimodal function optimization. In J. J. Grefenstette (Ed.), *Genetic Algorithms and their Applications:Proceedings of the Second International Conference on Genetic Algorithms* (pp. 41-49).

Goldberg, D. E. (1989). *Genetic algorithms for search, Optimization, and Machine Learning*. Reading, MA: Addision-Wesley.

Goldberg, D. E. (2002). *Genetic algorithms in search, optimization and machine learning*. New Delhi: Pearson-Education.

Gol'denblat, I. I., & Kopnov, V. A. (1965). Strength of glass-reinforced plastics in the complex stress state. *Polymer Mechanics*, *1*(2), 54–59. doi:10.1007/BF00860685

Gonze, X. (1997). First-principles responses of solids to atomic displacements and homogeneous electric fields: Implementation of a conjugate-gradient algorithm. *Physical Review B: Condensed Matter and Materials Physics*, *55*(16), 10337–10354. doi:10.1103/PhysRevB.55.10337

Gordon, H. (1975) Theory of BOF Reaction Rates. In BOF Steelmaking (vol. 2). New York: The Iron and Steel Society.

Gordy, W. (1946). A New Method of Determining Electronegativity from Other Atomic Properties. *Physical Review*, *69*(11-12), 604–607. doi:10.1103/PhysRev.69.604

Goswami, T. (1996). Prediction of Low Cycle Fatigue Lives of Low Alloy Steels. *ISIJ International*, *36*(3), 354–360. doi:10.2355/isijinternational.36.354

Grafe, U., Böttger, B., Tiaden, J., & Fries, S. G. (2000a). Coupling of multicomponent thermodynamic databases to a phase field model: Application to solidification and solid state transformations of superalloys. *Scripta Materialia, 42*(12), 1179–1186. doi:10.1016/S1359-6462(00)00355-9

Grafe, U., Böttger, B., Tiaden, J., & Fries, S. G. (2000b). Simulations of the initial transient during directional solidification of multicomponent alloys using the phase field method. *Modelling and Simulation in Materials Science and Engineering, 8*(6), 871–879. doi:10.1088/0965-0393/8/6/308

Grassel, O., Frommeyer, G., Derder, C., & Hofmann, H. (1997). Phase transformations and mechanical properties of Fe-Mn-Si-Al TRIP-steels. *Journal de Physique IV, 7*(C5), 383-388.

Grinberg, I., West, D. V., Torres, M., Gou, G., Stein, D. M., Wu, L., & Rappe, A. M. et al. (2013). Perovskite oxides for visible-light-absorbing ferroelectric and photovoltaic materials. *Nature, 503*(7477), 509–512. doi:10.1038/nature12622 PMID:24213630

Guo, S., Ng, C., Lu, J., & Liu, C. T. (2011). Effect of valence electron concentration on stability of fcc or bcc phase in high entropy alloys. *Journal of Applied Physics, 109*(10), 103505. doi:10.1063/1.3587228

Guo, Z., & Sha, W. (2004). Modelling the correlation between processing parameters and properties of maraging steels using artificial neural network. *Computational Materials Science, 29*(1), 12–28. doi:10.1016/S0927-0256(03)00092-2

Gupta, A. K., Patel, B. P., & Nath, Y. (2012). Continuum damage mechanics approach to composite laminated shallow cylindrical/conical panels under static loading. *Composite Structures, 94*(5), 1703–1713. doi:10.1016/j.compstruct.2011.12.002

Gupta, A. K., Patel, B. P., & Nath, Y. (2013a). Nonlinear static analysis of composite laminated plates with evolving damage. *Acta Mechanica, 224*(6), 1285–1298. doi:10.1007/s00707-013-0875-5

Gupta, A. K., Patel, B. P., & Nath, Y. (2013b). Continuum damage modeling of composite laminated plates using higher order theory. *Composite Structures, 99*, 141–151. doi:10.1016/j.compstruct.2012.11.038

Gupta, A. K., Patel, B. P., & Nath, Y. (2014). Postbuckling response of composite laminated plates with evolving damage. *International Journal of Damage Mechanics, 23*(2), 222–244. doi:10.1177/1056789513489322

Gupta, P. K. (1996). Non-crystalline solids: Glasses and amorphous solids. *Journal of Non-Crystalline Solids, 195*(1), 158–164. doi:10.1016/0022-3093(95)00502-1

Gupta, V., Yuan, J., & Martinez, D. (1993). Calculation, measurement, and control of interface strength in composites. *Journal of the American Ceramic Society, 76*(2), 305–315. doi:10.1111/j.1151-2916.1993.tb03784.x

Haile, J. M. (1997). *Molecular Dynamics Simulation: Elementary Methods*. New York, NY: Wiley.

Hairie, A., Hairie, F., Lebouvier, B., Nouet, G., Paumier, E., Ralantoson, N., & Sutton, A. P. (1994). Free energies of interfaces from quasi-harmonic theory: A critical appraisal. *Interface Science, 2*(1), 17–29. doi:10.1007/BF00188816

Hajela, P., & Lin, C.-Y. (1992). Genetic search strategies in multicriterion optimal design. *Structural Optimization, 4*(2), 99–107. doi:10.1007/BF01759923

Hall, E. O. (1951). The deformation and ageing of mild steel: III discussion of results. *Proceedings of the Physical Society. Section B, 64*(9), 747–753. doi:10.1088/0370-1301/64/9/303

Hamilton, J. C., Siegel, D. J., Daruka, I., & Léonard, F. (2003). Why do grain boundaries exhibit finite facet lengths? *Physical Review Letters, 90*(24), 246102. doi:10.1103/PhysRevLett.90.246102 PMID:12857204

Han, J., & Kamber, M. (2006). *Data Mining: Concepts and Techniques. The Morgan Kaufmann Series in Data Management Systems* (2nd ed.). San Francisco: Morgan Kaufmann.

Han, S. J., Song, J. W., Yang, C. H., Park, S. H., Park, J., Jeong, Y. H., & Rhie, K. W. (2002). A key to room-temperature ferromagnetism in fe-doped zno. *Cu. Appl. Phys. Lett.*, *81*, 4212.

Hashin, Z. (1980). Failure criteria for unidirectional fiber composites. *Journal of Applied Mechanics*, *47*(2), 329–334. doi:10.1115/1.3153664

Hashin, Z. (1983). Analysis of Composite Materials-A Survey. *Journal of Applied Mechanics*, *50*(3), 481–505. doi:10.1115/1.3167081

Hashin, Z., & Shtrikman, S. (1963). A variational approach to the theory of the elastic behaviour of multiphase materials. *Journal of the Mechanics and Physics of Solids*, *11*(2), 127–140. doi:10.1016/0022-5096(63)90060-7

Hastie, T., Tibshirani, R., & Friedman, J. (2001). *The Elements of Statistical Learning*. New York, NY, USA: Springer New York Inc.

Hawkins, D. M. (2004). The problem of overfitting. *Journal of Chemical Information and Computer Sciences*, *44*(1), 1–12. doi:10.1021/ci0342472 PMID:14741005

Hawkins, J. M., Weaver, J. F., & Asthagiri, A. (2009). Density functional theory study of the initial oxidation of the Pt(111) surface. *Physical Review B: Condensed Matter and Materials Physics*, *79*(12), 125434. doi:10.1103/PhysRevB.79.125434

Hayakawa, K., Murakami, S., & Liu, Y. (1998). An irreversible thermodynamics theory for elastic-plastic-damage materials. *European Journal of Mechanics. A, Solids*, *17*(1), 13–32. doi:10.1016/S0997-7538(98)80061-2

Haykin, S. (1994). *Neural Networks: A Comprehensive Foundation*. McMillan.

Hedin, L. (1965). New method for calculating the one-particle Green's function with application to the electron-gas problem. *Physical Review*, *139*(3A), A796–A823. doi:10.1103/PhysRev.139.A796

Helle, M., Pettersson, F., Chakraborti, N., & Saxén, H. (2006). Modeling noisy blast furnace data using genetic algorithms and neural networks. *Steel Research International*, *77*(2), 75–81.

Henkelman, G., & Jónsson, H. (1999). A dimer method for finding saddle points on high dimensional potential surfaces using only first derivatives. *The Journal of Chemical Physics*, *111*(15), 7010. doi:10.1063/1.480097

Henkelman, G., & Jónsson, H. (2000). Improved tangent estimate in the nudged elastic band method for finding minimum energy paths and saddle points. *The Journal of Chemical Physics*, *113*(22), 9978. doi:10.1063/1.1323224

Henkelman, G., & Jónsson, H. (2001). Long time scale kinetic Monte Carlo simulations without lattice approximation and predefined event table. *The Journal of Chemical Physics*, *115*(21), 9657. doi:10.1063/1.1415500

Hernandez, G., Allen, J. K., & Mistree, F. (2001). The Compromise Decision Support Problem: Modelling the Deviation Function as in Physical Programming. *Engineering Optimization*, *33*(4), 445–471. doi:10.1080/03052150108940928

Herring, C. (1949). *The physics of powder metallurgy* (R. Gomer & C. S. Smith, Eds.). New York: McGraw-Hill.

Hesselbarth, H. W., & Göbel, I. R. (1991). Simulation of recrystallization by cellular automata. *Acta Metallurgica et Materialia*, *39*(9), 2135–2143. doi:10.1016/0956-7151(91)90183-2

Heyden, A., Bell, A. T., & Keil, F. J. (2005). Efficient methods for finding transition states in chemical reactions: Comparison of improved dimer method and partitioned rational function optimization method. *The Journal of Chemical Physics*, *123*(22), 224101. doi:10.1063/1.2104507 PMID:16375464

Higbie, R. (1935). The Rate of Absorption of a Pure Gas into a Still Liquid during Short Periods of Exposure. *Transactions of the American Institute of Chemical Engineers*, *31*, 365–389.

Hill, R. (1948). A Theory of the yielding and plastic flow of anisotropic metals. *Proceedings - Royal Society. Mathematical, Physical and Engineering Sciences, 193*(1033), 281–297. doi:10.1098/rspa.1948.0045

Hill, R. (1963). Elastic properties of reinforced solids: Some theoretical principles. *Journal of the Mechanics and Physics of Solids, 11*(5), 357–372. doi:10.1016/0022-5096(63)90036-X

Hinton, M. J., Kaddour, A. S., & Soden, P. D. (2004). A further assessment of the predictive capabilities of current failure theories for composite laminates: Comparison with experimental evidence. *Composites Science and Technology, 64*(3–4), 549–588. doi:10.1016/S0266-3538(03)00227-6

Hirvonen, J. P., Torri, P., Lappalainen, R., Likonen, J., Kung, H., Jervis, J. R., & Nastasi, M. (1998). Oxidation of MoSi2/SiC nanolayered composite. *Journal of Materials Research, 13*(04), 965–973. doi:10.1557/JMR.1998.0135

Hohenberg, P., & Kohn, W. (1964). Inhomogeneous Electron Gas. *Physical Review, 136*(3B), B864–B871. doi:10.1103/PhysRev.136.B864

Holm, E. A., & Battaile, C. C. (2001). The computer simulation of microstructural evolution. *JOM, 53*(9), 20–23. doi:10.1007/s11837-001-0063-2

Honkala, K., Hellman, A., Remidiakis, I., Logadottir, A., Carlsson, A., Dahl, S., & Nørskov, J. et al. (2005). Ammonia synthesis from First-Principles calculations. *Science, 307*(5709), 555–558. doi:10.1126/science.1106435 PMID:15681379

Hore, S., Das, S. K., Banerjee, S., & Mukherjee, S. (2013, February). Monte Carlo simulation of microstructure evolution during thermo-mechanical rolling of steel using grid computing technology. In *Parallel Computing Technologies (PARCOMPTECH), 2013 National Conference on* (pp. 1-7). IEEE. doi:10.1109/ParCompTech.2013.6621406

Horn, J., Nafpliotis, N., & Goldberg, D. E. (1994). A niched pareto genetic algorithm for multiobjective optimization. In *Proceedings of the First IEEE Conference on Evolutionary Computation,IEEE World Congress on Computational Computation* (pp. 82-87).

Hornbogen, E. (1984). Overview no. 37: On the microstructure of alloys. *Acta Metallurgica, 32*(5), 615–627. doi:10.1016/0001-6160(84)90135-4

Hornbogen, V. I. (1968). Über die Begriffe “Gefiige” und “Phase”, Prakt. *Metallographie, 5*, 51.

Horn, J. (1997). F1.9 multicriteria decision making. In T. B¨ack, D. B. Fogel, & Z. Michalewicz (Eds.), *Handbook of Evolutionary Computation*. Bristol, UK: Institute of Physics Publishing. doi:10.1887/0750308958/b386c85

Hosseini, S. M. K., Zarei-Hanzaki, A., Yazdan Panah, M. J., & Yue, S. (2004). ANN model for prediction of the effects of composition and process parameters on tensile strength and percent elongation of Si–Mn TRIP steels. *Materials Science and Engineering A, 374*(1-2), 122–128. doi:10.1016/j.msea.2004.01.007

Hotop, H., & Lineberger, W. C. (1985). Binding Energies in Atomic Negative Ions: II. *Journal of Physical and Chemical Reference Data, 14*(3), 731. doi:10.1063/1.555735

Howard, C. J., & Stokes, H. T. (1998). Group-theoretical analysis of octahedral tilting in perovskites. *Acta Crystallographica. Section B, Structural Crystallography and Crystal Chemistry, 54*(6), 782–789. doi:10.1107/S0108768198004200

Huang, H., Gilmer, G. H., & Díaz de la Rubia, T. (1998). An atomistic simulator for thin film deposition in three dimensions. *Journal of Applied Physics, 84*(7), 3636–3649. doi:10.1063/1.368539

Huang, K. (1987). *Statistical Mechanics* (2nd ed.). Hoboken, NJ: John Wiley & Sons, Inc.

Hunter, A., Kennedy, L., Henry, J., & Ferguson, R. I. (2000). Application of neural networks and sensitivity analysis to improved prediction of trauma survival. *Computer Methods and Algorithms in Biomedicine*, *62*(1), 11–19. doi:10.1016/S0169-2607(99)00046-2 PMID:10699681

Hwang, C. L., & Masud, A. S. M. (1979). *Multiple Objectives Decision Making-Methods and Applications: A State-of-the-art Survey*. Berlin: Springer-Verlag. doi:10.1007/978-3-642-45511-7

Integrated Computational Materials Engineering (ICME). (2010). *Integrating Computational Materials Science and Engineering John Allison July 26, DOE – BES Workshop on CMS*. Author.

Introduction to Materials Modelling. (2005). London: Institute of Materials, Minerals and Mining.

Iqbal, A., Frormann, L., Saleem, A., & Ishaq, M. (2007). The effect of filler concentration on the electrical, thermal, and mechanical properties of carbon fiber-reinforced poly (styrene-co-acrylonitrile) composites. *Polymer Composites*, *28*(2), 186–197. doi:10.1002/pc.20253

Ishibuchi, H., & Murata, T. (1996). Multi-objective genetic local search algorithm. In *Proceedings of IEEE International Conference on Evolutionary Computation (ICEC'96)* (pp. 119-124). doi:10.1109/ICEC.1996.542345

Ishida, H., & Rimdusit, S. (1998). Very high thermal conductivity obtained by boron nitride-filled polybenzoxazine. *Thermochimica Acta*, *320*(1-2), 177–186. doi:10.1016/S0040-6031(98)00463-8

Jain, A. K. (2010). Data clustering: 50 years beyond K-means. *Pattern Recognition Letters*, *31*(8), 651–666. doi:10.1016/j.patrec.2009.09.011

Jain, A., Ong, S. P., Hautier, G., Chen, W., Richards, W. D., Dacek, S., & Persson, K. A. et al. (2013). Commentary: The Materials Project: A materials genome approach to accelerating materials innovation. *APL Materials*, *1*(1), 011002. doi:10.1063/1.4812323

Jalkanen, H., & Andersin, E. (1989). Thermodynamic modeling of converter steelmaking. Paper presented in the Conference Modern Technologies for Making and Casting of Steel.

Jalkanen, H., Suomi, M.-L., & Wallgren, M. (1997). Simulation of complex chemical high temperature processes - experiences and problems in BOP modelling. In *Proceedings of IPMM'97, Australasia-Pacific Forum on Intelligent Processing and Manufacturing of Materials*.

Jalkanen, H., Suomi, M.-L., & Wallgren, M. (1998). *Simulation of Oxygen Converter Process (BOF)*. Paper presented in Joint Finnish-South African Symposium on Metallurgical Research, Johannesburg, South Africa.

Jalkanen, H., Virrankoski, R., & Oksanen, T. (2002). Simulator of Oxygen Converter Process (BOF). Paper presented in Pyrometallurgy ´02, Cape Town, South Africa.

Jalkanen, H., & Kostamo, T. (2004). Simulation of Oxygen Converter Process. In P. Palfy, M. Solc, & E. Vircikova (Eds.), *Metallurgy Refractories and Environment* (pp. 87–99). Kosice: Harlequin.

Jang, J.-S. R., Sun, C.-T., & Mizutani, E. (1997). *Neuro-Fuzzy and Soft Computing*. Englewood Cliffs, NJ: Prentice Hall.

Janot, C. (1992). *Quasicrystals*. Oxford, UK: Clarendon Press.

Janssens, K. G. (2003). Random grid, three-dimensional, space-time coupled cellular automata for the simulation of recrystallization and grain growth. *Modelling and Simulation in Materials Science and Engineering*, *11*(2), 157–171. doi:10.1088/0965-0393/11/2/304

Janssens, K. G. F., Raabe, D., Kozeschnik, E., Miodownik, M. A., & Nestler, B. (2007). *Computational Materials Engineering: An Introduction To Microstructure Evolution*. Amsterdam, Netherlands: Elsevier.

Janssens, K. G., Olmsted, D., Holm, E. A., Foiles, S. M., Plimpton, S. J., & Derlet, P. M. (2006). Computing the mobility of grain boundaries. *Nature Materials*, *5*(2), 124–127. doi:10.1038/nmat1559 PMID:16400330

Jayakumar, O. D., Gopalakrishnan, I. K., & Kulshreshtha, S. K. (2006). Surfactant-assisted synthesis of co-and li-doped zno nanocrystalline samples showing room-temperature ferromagnetism. *Advanced Materials*, *18*, 1857.

Jenkins, G. A. (1972). 4 The systems approach. *Systems Behaviour*, 56.

Jha, P. K., Dash, S. K., & Kumar, S. (2001). Fluid Flow and Mixing in a Six Strand Billet Caster Tundish: A Parametric Study. *ISIJ International*, *41*(12), 1437–1446. doi:10.2355/isijinternational.41.1437

Jha, R., Sen, P. K., & Chakraborti, N. (2014). Multi-Objective Genetic Algorithms and Genetic Programming Models for Minimizing Input Carbon Rates in a Blast Furnace Compared with a Conventional Analytic Approach. *Steel Research International*, *85*(2), 219–232. doi:10.1002/srin.201300074

Jiménez-Piqué, E., Ceseracciu, L., Chalvet, F., Anglada, M., & de Portu, G. (2005). Hertzian contact fatigue on alumina/alumina-zirconia laminated composites. *Journal of the European Ceramic Society*, *25*(15), 3393–3401. doi:10.1016/j.jeurceramsoc.2004.09.011

John, G. H., Kohavi, R., & Pfleger, K. (1994). Irrelevant features and the subset selection problem. In *Proceedings of the Eleventh International Conference on Machine Learning*. doi:10.1016/B978-1-55860-335-6.50023-4

Johnson, R. A. (1989). Alloy models with the embedded-atom method. *Physical Review B: Condensed Matter and Materials Physics*, *39*(17), 12554–12559. doi:10.1103/PhysRevB.39.12554 PMID:9948120

Jones, R. O. (2015). Density functional theory: Its origins, rise to prominence, and future. *Reviews of Modern Physics*, *87*(3), 897–924. doi:10.1103/RevModPhys.87.897

Kachanov, L. (1999). Rupture time under creep conditions. *International Journal of Fracture*, *97*(1-4), 11-18.

Kane, M. H., Strassburg, M., Fenwick, W. E., Asghar, A., Payne, A. M., Gupta, S., & Ferguson, I. T. et al. (2006). Correlation of the structural and ferromagnetic properties of ga1-xmnxn grown by metalorganic chemical vapor deposition. *Journal of Crystal Growth*, *287*, 591.

Karma & Rappel. (n.d.). Quantitative phase-field modeling of dendritic growth in two and three dimensions. *Physical Review D*, *57*, 4323-4349.

Kattner, U. R. (1997). The thermodynamic modeling of multicomponent phase equilibria. *JOM*, *49*(12), 14–19. doi:10.1007/s11837-997-0024-5

Kawamoto, T. (1997). A theoretical model for ferromagnetism of tdae-c60. *Sol. St. Commun.*, *101*, 231.

Kawamoto, T., & Suzuki, N. (1997). Pressure effects in ferromagnetism of orbital ordering ferromagnet. *Phys. Soc. Japan*, *66*, 2487.

Keith, J. M., King, J. A., Lenhart, K. M., & Zimny, B. (2007). Thermal conductivity models for carbon/liquid crystal polymer composites. *Journal of Applied Polymer Science*, *105*(6), 3309–3316. doi:10.1002/app.26607

Kendall, K. (1975). Transition between cohesive and interfacial failure in a laminate. *Proceedings of the Royal Society of London. Series A, Mathematical and Physical Sciences*, *344*(1637), 287–302. doi:10.1098/rspa.1975.0102

Khachaturyan, A. G. (1983). *Theory of Structural Transformations in Solids*. New York: Wiley and Sons.

Khachaturyan, A. G., & Shatalov, G. A. (1969). Elastic-interaction potential of defects in a crystal. *Sov Phys Solid State*, *11*(1), 118–123.

Khatri, P., & Huda, M. N. (2014). Application of attractive potential by dft + u to predict the electronic properties of materials without highly localized bands. *Computational Materials Science*, *81*, 290.

Kim, D. K., & Kriven, W. M. (2003). Mullite–Aluminum phosphate laminated composite fabricated by tape casting. *Journal of the American Ceramic Society*, *86*(11), 1962–1964. doi:10.1111/j.1151-2916.2003.tb03590.x

Kim, E. Y., Yang, H. S., Han, S. H., Kwak, J. H., & Choi, S. H. (2012). Effect of initial microstructure on strain-stress partitioning and void formation in DP980 steel during uniaxial tension. *Metals and Materials International*, *18*(4), 573–582. doi:10.1007/s12540-012-4030-y

Kim, S. W., Choi, B., Lee, S. H., & Kang, K. H. (2008). Measurement of thermophysical properties of particulate-filled polymer composites. *High Temperatures. High Pressures*, *37*(1), 21–30.

Kim, Y. I., Moon, K. C., Kang, B. S., Han, C., & Chang, K. S. (1998). Application of neural network to the supervisory control of a reheating furnace in the steel industry. *Control Engineering Practice*, *6*(8), 1009–1014. doi:10.1016/S0967-0661(98)00098-7

Kittilstved, K. R., Norberg, N. S., & Gamelin, D. R. (2005). Chemical manipulation of high-tc ferromagnetism in zno diluted magnetic semiconductors. *Physical Review Letters*, *94*, 147209. PMID:15904107

Kobayashi, R. (1993). Modeling and numerical simulations of dendritic crystal growth. *Physica D. Nonlinear Phenomena*, *63*(3), 410–423. doi:10.1016/0167-2789(93)90120-P

Kobayashi, S. (1988). Solute redistribution during solidification with diffusion in solid phase; A theoretical analysis. *Journal of Crystal Growth*, *88*(1), 87–89. doi:10.1016/S0022-0248(98)90010-0

Kohn, W., & Sham, L. (1965). Self-consistent equation including exchange and correlation effects. *Physical Review*, *140*(4A), A1133–A1138. doi:10.1103/PhysRev.140.A1133

Kohonen, T. (1989). *Self-Organization and Associative Memory* (3rd ed.). Berlin: Springer-Verlag. doi:10.1007/978-3-642-88163-3

Koinuma, H., & Takeuchi, I. (2004). Combinatorial solid-state chemistry of inorganic materials. *Nature Materials*, *3*(7), 429–438. doi:10.1038/nmat1157 PMID:15229491

Komorowski, J., Pawlak, Z., Polkowski, L., & Skowron, A. (2002). *Rough Sets: A Tutorial.* Retrieved from http://alfa.mimuw.edu.pl/prace/1999/D5/Tutor06_09.ps

Korczak, P., Dyja, H., & Labuda, E. (1998). Using neural network models for predicting mechanical properties after hot plate rolling processes. *Journal of Materials Processing Technology*, *80–81*, 481–486. doi:10.1016/S0924-0136(98)00151-4

Kotliar, G., Savrasov, S. Y., Haule, K., Oudovenko, V. S., Parcollet, O., & Marianetti, C. (2006). Electronic structure calculations with dynamical mean-field theory. *Reviews of Modern Physics*, *78*(3), 865–951. doi:10.1103/RevModPhys.78.865

Koza, J. R. (1992). *Genetic programming: on the programming of computers by means of natural selection* (Vol. 1). MIT press.

Kreidler, E. R., & Hummel, F. A. (1970). The crystal chemistry of apatite: Structure fields of fluor and chlorapatite. *The American Mineralogist*, *55*, 170–184.

Kresse, G., & Furthmuller, J. (1996). Efficient iterative schemes for *ab initio* total-energy calculations using a plane-wave basis set. *Physical Review B: Condensed Matter and Materials Physics*, *54*(16), 11169–11186. doi:10.1103/PhysRevB.54.11169 PMID:9984901

Kresse, G., & Hafner, J. (1993). Ab initio molecular dynamics for liquid metals. *Physical Review B: Condensed Matter and Materials Physics*, *47*, 558. PMID:10004490

Kresse, G., & Joubert, D. (1999). From ultrasoft pseudopotentials to the projector augmented-wave method. *Physical Review B: Condensed Matter and Materials Physics*, *59*, 1758.

Krieger, I. M., & Dougherty, T. J. (1959). A mechanism for non-newtonian flow in suspensions of rigid spheres. *Transactions of the Society of Rheology*, *3*(1), 137–152. doi:10.1122/1.548848

Krushke, J. K. (2011). *Doing Bayesian Data Analysis*. Oxford, UK: Academic Press, Elsevier.

Kruskopf, A., Väyrynen, P., Vallejo, V., Jalkanen, H., & Louhenkilpi, S. (2012). *Modeling of BOF Converter: CONSIM Simulator and Its Further Development*. Academic Press.

Kudin, D. P., Tseloval'nikov, V. M., Ksenzuk, F. A., & Kokin, N. M. (1971). Influence of the rate of skin pass rolling and tension on the amount of deformation. *Metallurgist*, *15*(12), 817–818. doi:10.1007/BF00735606

Kumar, A., Chakrabarti, D., & Chakraborti, N. (2012). Data-Driven Pareto Optimization for Microalloyed Steels Using Genetic Algorithms. *Steel Research International*, *83*(2), 69–74. doi:10.1002/srin.201100189

Kumar, A., Mazumdar, D., & Koria, S. C. (2008). Modeling of Fluid Flow and Residence Time Distribution in a Four-Strand Tundish for Enhancing Inclusion Removal. *ISIJ International*, *48*(1), 38–47. doi:10.2355/isijinternational.48.38

Kuroda, S., Nishizawa, N., Takita, K., Mitime, M., Bando, Y., Osuch, K., & Dietl, T. (2007). Origin and control of high-temperature ferromagnetism in semiconductors. *Nature Materials*, *6*, 440. PMID:17515918

Kursawe, F. (1991). A variant of evolution strategies for vector optimization. In H.-P. Schwefel & R. M¨anner (Eds.), *Parallel Problem Solving from Nature —Proc. 1st Workshop PPSN* (pp. 193-197). Berlin: Springer. doi:10.1007/BFb0029752

Kusiak, J., & Kuziak, R. (2002). Modelling of microstructure and mechanical properties of steel using the artificial neural network. *Journal of Materials Processing Technology*, *127*(1), 115–121. doi:10.1016/S0924-0136(02)00278-9

Kusnetsov, Sham, & Bol´shakov. (1972). *Steel in the USSR*. Academic Press.

Kuzmin, D., & Turek, S. (2004). Numerical Simulation of Turbulent Bubbly Flow. In *Proceedings of 3rd International Symposium on Two-Phase Flow Modelling and Experimentation.*

Kytönen, H., Tolvanen, M., Louhenkilpi, S., & Holappa, L. (1998). Modeling of Casting, Welding and Advanced Solidification Processes VIII. The Minerals, Metals & Materials Society.

Lai, Z. W. (1990). Theory of ordering dynamics for Cu 3 Au. *Physical Review B: Condensed Matter and Materials Physics*, *41*(13), 9239–9256. doi:10.1103/PhysRevB.41.9239 PMID:9993266

Lalam, S. H., Bhadeshia, H. K. D. H., & MacKay, D. J. C. (2000). Estimation of mechanical properties of ferritic steel welds. Part 1: Yield and tensile strength. *Sci. Technol. Weld. Join.*, *5*(3), 135–147. doi:10.1179/136217100101538137

Lamont, J., & Scott, D. (1970). An Eddy Cell Model of Mass Transfer into the Surface of a Turbulent Liquid. *AIChE Journal. American Institute of Chemical Engineers*, *16*(4), 513–519. doi:10.1002/aic.690160403

Langer, J. S. (1986). Models of pattern formation in first-order phase transitions. In Directions in Condensed Matter Physics. World Scientific.

Lan, Y. J., Xiao, N. M., Li, D. Z., & Li, Y. Y. (2005). Mesoscale simulation of deformed austenite decomposition into ferrite by coupling a cellular automaton method with a crystal plasticity finite element model. *Acta Materialia*, *53*(4), 991–1003. doi:10.1016/j.actamat.2004.10.045

Larkiola, J., Myllykoski, P., Korhonen, A. S., & Cser, L. (1998). The role of neural networks in the optimization of rolling processes. *Journal of Materials Processing Technology*, *80–81*, 16–23. doi:10.1016/S0924-0136(98)00206-4

Lathiotakis, N. N., Andriotis, A. N., & Menon, M. (2008a). Codoping: A possible pathway for inducing ferromagnetism in zno. *Physical Review B: Condensed Matter and Materials Physics*, *78*, 193311.

Laumanns, M., Rudolph, G., & Schwefel, H.-P. (1998). A spatial predatorprey approach to multi-objective optimization: A preliminary study. In A. E. Eiben, T. B¨ack, M. Schoenauer, & H.-P. Schwefel (Eds.), *Fifth International Conference on Parallel Problem Solving from Nature (PPSNV)* (pp. 241-249). doi:10.1007/BFb0056867

Lee, C.-H., Orloff, N. D., Birol, T., Zhu, Y., Goian, V., Rocas, E., & Schlom, D. G. et al. (2013). Exploiting dimensionality and defect mitigation to create tunable microwave dielectrics. *Nature*, *502*(7472), 532–536. doi:10.1038/nature12582 PMID:24132232

Lee, C., Yang, W. T., & Parr, R. (1988). Development of the Colle-Salvetti correlation-energy formula into a functional of the electron density. *Physical Review B: Condensed Matter and Materials Physics*, *37*(2), 785–789. doi:10.1103/PhysRevB.37.785 PMID:9944570

Lee, H., Peng, K., & Wang, J. (1985). An anisotropic damage criterion for deformation instability and its application to forming limit analysis of metal plates. *Engineering Fracture Mechanics*, *21*(5), 1031–1054. doi:10.1016/0013-7944(85)90008-6

Lee, J. S., Kim, S. G., Kim, W. T., & Suzuki, T. (1999). Numerical simulation of peritectic reaction using a multi-phase-field model. *ISIJ International*, *39*(7), 730–736. doi:10.2355/isijinternational.39.730

Lee, W., Howard, S. J., & Clegg, W. J. (1996). Growth of interface defects and its effect on crack deflection and toughening criteria. *Acta Materialia*, *44*(10), 3905–3922. doi:10.1016/S1359-6454(96)00068-7

Lein, J. K. (1997). Environment Decision Making: An information technology approach. Malden, MA: Blackwell Science.

Lemaitre, J. (1996). A course on damage mechanics (2nd ed.). Berlin: Springer.

Lemaitre, J. (1971). Evaluation of dissipation and damage in metals submitted to dynamic loading.*Proceedings of the ICM-1.*

Le, T., Epa, C., Burden, F. R., & Winkler, D. A. (2012). Quantitative structure-property relationship modeling of diverse materials properties. *Chemical Reviews*, *112*(5), 2889–2919. doi:10.1021/cr200066h PMID:22251444

Lewis, T. B., & Nielsen, L. E. (1970). Dynamic mechanical properties of particulate-filled composites. *Journal of Applied Polymer Science*, *14*(6), 1449–1471. doi:10.1002/app.1970.070140604

Li, X. (2003). Chapter. In C. Fonseca, P. Fleming, E. Zitzler, L. Thiele, & K. Deb (Eds.), Evolutionary Multi-Criterion Optimization (pp. 207–221). Berlin: Springer. doi:10.1007/3-540-36970-8_15

Lieb, E. H. (1989). Two theorems on the hubbard model. *Physical Review Letters*, *62*, 1201. PMID:10039602

Lielens, G., Pirotte, P., Couniot, A., Dupret, F., & Keunings, R. (1998). Prediction of thermo-mechanical properties for compression moulded composites. *Composites. Part A, Applied Science and Manufacturing*, *29*(1-2), 63–70. doi:10.1016/S1359-835X(97)00039-0

Linero Segrera, D. L., Oliver, J., & Huespe, A. E. (2010). Numerical modeling of the fracture process in reinforced concrete by means of the continuum strong discontinuity approach. Part I: Formulation. *Ingeniería e Investigación*, *30*(2), 5–15.

Lin, H. T., Chin, T. S., Shih, J. C., Lin, S. H., Hong, T. M., Huang, R. T., & Kai, J. J. et al. (2004). Enhancement of ferromagnetic properties in zn1-xcoxo by additional cu doping. *Applied Physics Letters*, *85*, 621.

Lin, Y. H., Ying, M., Li, M., Wang, X., & Nan, C. W. (2007). Room-temperature ferromagnetic and ferroelectric behavior in polycrystalline zno-based thin films. *Applied Physics Letters*, *90*, 222110.

Lis, J., & Eiben, A. E. (1997). A multi-sexual genetic algorithm for multiobjective optimization. In *Proceedings of 1997IEEE International Conference on Evolutionary Computation (ICEC'97)* (pp. 59-64). doi:10.1109/ICEC.1997.592269

Li, S., & Ghosh, S. (2006). Extended Voronoi cell finite element model for multiple cohesive crack propagation in brittle materials. *International Journal for Numerical Methods in Engineering*, *65*(7), 1028–1067. doi:10.1002/nme.1472

Lisenkov, S., Andriotis, A. N., & Menon, M. (2012). Magnetic anisotropy and engineering of magnetic behavior of the edges in co embedded graphene nanoribbons. *Physical Review Letters*, *108*, 187208. PMID:22681116

Lisenkov, S., Andriotis, A. N., Sheetz, R. M., & Menon, M. (2011). Effects of codoping on the ferromagnetic enhancement in zno. *Physical Review B: Condensed Matter and Materials Physics*, *83*, 235203.

Liu, G. R., & Quek, S. S. (2013). *The Finite Element Method: A Practical Course*. Oxford, UK: Butterworth-Heinemann.

Liu, H., & Hsu, S. M. (1996). Fracture behavior of multilayer silicon nitride/boron nitride ceramics. *Journal of the American Ceramic Society*, *79*(9), 2452–2457. doi:10.1111/j.1151-2916.1996.tb08996.x

Liu, H., & Yu, L. (2005). Towards integrating feature selection algorithms for classification and clustering. *IEEE Transactions on Knowledge and Data Engineering*, *17*(4), 491–502. doi:10.1109/TKDE.2005.66

Liu, Z. K. (2009). First-principles calculations and CALPHAD modeling of thermodynamics. *Journal of Phase Equilibria and Diffusion*, *30*(5), 517–534. doi:10.1007/s11669-009-9570-6

Liu, Z. K., Chen, L. Q., Raghavan, P., Du, Q., Sofo, J. O., Langer, S. A., & Wolverton, C. (2004). An integrated framework for multi-scale materials simulation and design. *Journal of Computer-Aided Materials Design*, *11*(2-3), 183–199. doi:10.1007/s10820-005-3173-2

Llegbusi, O. J., & Szekely, J. (1989). Effect of Magnetic Field on Flow, Temperature and Inclusion Removal in Shallow Tundishes. *ISIJ International*, *29*(12), 1031–1039. doi:10.2355/isijinternational.29.1031

Llewellyn & Hudd. (1998). *Steels: metallurgy and applications* (3rd ed.). Oxford, UK: Butterworth-Heinemann.

Lopes, C. M. A., & Felisberti, M. I. (2004). Thermal conductivity of PET/(LDPE/Al) composites determined by MDSC. *Polymer Testing*, *23*(6), 637–643. doi:10.1016/j.polymertesting.2004.01.013

Louhenkilpi, S., Laine, J., Räisänen, T., & Hätönen, T. (2004). *2nd Int. Conf.and Exhibition on New Developments in Metallurgical Process Technology.*

Louhenkilpi, S., Laitinen, E., & Nieminen, R. (1993). Real-Time Simulation of Heat Transfer in Continuous Casting. *Metallurgical Transactions. B, Process Metallurgy*, *24*(4), 685–693. doi:10.1007/BF02673184

Louhenkilpi, S., Mäkinen, M., Vapalahti, S., Räisänen, T., & Laine, J. (2005). 3D steady state and transient simulation tools for heat transfer and solidification in continuous casting. *Materials Science and Engineering A*, *413-414*, 135–138. doi:10.1016/j.msea.2005.08.153

Louhenkilpi, S., Mäkinen, M., Vapalahti, S., Räisänen, T., & Laine, J. (n.d.). ICASP Advances in Solidification Processes. *Proc. Int. Conf.*

Louhenkilpi, S., Räisänen, T., Vapalahti, S., Laine, J., & Mäkinen, M. (2006). *Modeling of Casting, Welding and Advanced Solidification Processes XI*. Warrendale, PA: TMS.

Louhenkilpi, S., Uoti, M., Kytönen, H., & Vapalahti, S. (2003). *Modeling of Casting, Welding and Advanced Solidification Processes X*. Warrendale, PA: TMS.

Luo, A. (2013). Application of Computational Thermodynamics and CALPHAD in Magnesium Alloy Development. *2nd World Congress on Integrated Computational Materials Engineering*.

Mackay, D. (1998). Introduction to Monte Carlo Methods. In M. Jordan (Ed.), Learning in Graphical Models (pp.175-204). Kluwer Academic Press. doi:10.1007/978-94-011-5014-9_7

Mackenbrock, A., & Tacke, K.-H. (1996). Generalized enthalpy method for multicomponent phase change. *Metallurgical and Materials Transactions. B, Process Metallurgy and Materials Processing Science*, *27*(5), 871–881. doi:10.1007/BF02915617

Mahfouf, M., Jamei, M., & Linkens, D. A. (2005, May). Optimal Design of Alloy Steels Using Multiobjective Genetic Algorithms. *Materials and Manufacturing Processes*, *20*(3), 553–567. doi:10.1081/AMP-200053580

Maiwald-Immer, T., Gohler, T., Fischersworring-Bunk, A., Korner, C., Osmanlic, F., & Bauerei, A. (2013). Application of ICME Methods for the Development of Rapid Manufacturing Technologies. *2nd World Congress on Integrated Computational Materials Engineering*.

Majumdar, G., Oraon, B., Laha, A., Ghosh, S. K., Mohanty, I., & Datta, S. (2010). Developing rules bases on the influence of welding parameters in FCAW process from ANN model. *International Journal of Mechatronics and Manufacturing Systems*, *3*(1), 155–164. doi:10.1504/IJMMS.2010.029887

Mammoli, A. A., Graham, A. L., Reimanis, I. E., & Tullock, D. L. (1995). The effect of flaws on the propagation of cracks at bi-materials inferfaces. *Acta Metallurgica et Materialia*, *43*(3), 1149–1156. doi:10.1016/0956-7151(94)00308-5

Mamunya, Y. P., Davydenko, V. V., Pissis, P., & Lebedev, E. V. (2002). Electrical and thermal conductivity of polymers filled with metal powders. *European Polymer Journal*, *38*(9), 1887–1897. doi:10.1016/S0014-3057(02)00064-2

Mannhart, J., & Schlom, D. (2010). Oxide interfaces – an opportunity for electronics. *Science*, *327*(5973), 1607–1611. doi:10.1126/science.1181862 PMID:20339065

Marat-Mendes, R. M., & Freitas, M. M. (2010). Failure criteria for mixed mode delamination in glass fibre epoxy composites. *Composite Structures*, *92*(9), 2292–2298. doi:10.1016/j.compstruct.2009.07.017

Marcus, J. (2004). Grote, Christoph Kirsch, Dirichlet-to-Neumann boundary conditions for multiple scattering problems. *Journal of Computational Physics*, *10*, 630–650.

Marquis, E. A., Seidman, D. N., Asta, M., & Woodward, C. (2006). Composition evolution of nanoscale Al 3 Sc precipitates in an Al–Mg–Sc alloy: Experiments and computations. *Acta Materialia*, *54*(1), 119–130. doi:10.1016/j.actamat.2005.08.035

Marquis, E. A., Seidman, D. N., Asta, M., Woodward, C., & Ozoliņš, V. (2003). Mg Segregation at A l/A l 3 S c Heterophase Interfaces on an Atomic Scale: Experiments and Computations. *Physical Review Letters*, *91*(3), 036101. doi:10.1103/PhysRevLett.91.036101 PMID:12906429

Martinetz, T., Protzel, P., Gramckow, O., & Sörgel, G. (1995). Neural network control for steel rolling mills. In Proc. Neural networks: artificial intelligence and industrial applications, (pp. 280-286). Springer Verlag. doi:10.1007/978-1-4471-3087-1_55

Martínez, D., & Gupta, V. (1994). Energy criterion for crack deflection at an interface between two orthotropic media. *Journal of the Mechanics and Physics of Solids*, *42*(8), 1247–1271. doi:10.1016/0022-5096(94)90034-5

Martínez, E., Monasterio, P. R., & Marian, J. (2011). Billion-atom synchronous parallel kinetic Monte Carlo simulations of critical 3D Ising systems. *Journal of Computational Physics*, *230*(4), 1359–1369. doi:10.1016/j.jcp.2010.11.006

Martin, R. M. (2004). *Electronic structure: Basic Theory and Practical Methods*. Cambridge, UK: Cambridge University Press. doi:10.1017/CBO9780511805769

MATLAB and Statistics Toolbox Release. (2012b). TheMathWorks, Inc.

Matzenmiller, A., Lubliner, J., & Taylor, R. L. (1995). A constitutive model for anisotropic damage in fiber-composites. *Mechanics of Materials*, *20*(2), 125–152. doi:10.1016/0167-6636(94)00053-0

Mawdsley, J. R., Kovar, D., & Halloran, J. W. (2000). Fracture behavior of alumina/monazite multilayer laminates. *Journal of the American Ceramic Society*, *83*(4), 802–808. doi:10.1111/j.1151-2916.2000.tb01278.x

Mazumdar, D., & Guthrie, R. I. L. (1999). The Physical and Mathematical Modelling of Continuous Casting Tundish System. *ISIJ International*, *39*(6), 524–547. doi:10.2355/isijinternational.39.524

Mazumdar, D., Yamanoglu, G., Shankarnarayanan, R., & Guthrie, R. I. L. (1995). Similarity Consideration in the Physical Modelling of Steel Tundish Systems. *Steel Research*, *66*, 14–19.

McConnell, H. M. (1967). Informal remarks. *Proc. R. A. Welch Found. Chem. Res.*, *11*, 144.

McConnell, H. M. (1963). Ferromagnetism in solid free radicals. *The Journal of Chemical Physics*, *39*, 1910.

McDonald, I. R. (1972). NPT-ensemble Monte Carlo calculations for binary liquid mixtures. *Molecular Physics*, *23*(1), 41–58. doi:10.1080/00268977200100031

McDowell, D. L., Panchal, J. H., Choi, H. J., Seepersad, C. C., Allen, J. K., & Mistree, F. (2010). *Integrated Design of Multiscale Materials and Products*. New York: Elsevier.

Melenk, J. M., & Babuska, I. (1996). The partition of unity finite element method: Basic theory and applications. *Computer Methods in Applied Mechanics and Engineering*, *139*(1–4), 289–314. doi:10.1016/S0045-7825(96)01087-0

Mendelev, M. I., & Srolovitz, D. J. (2002). Impurity effects on grain boundary migration. *Modelling and Simulation in Materials Science and Engineering*, *10*(6), R79–R109. doi:10.1088/0965-0393/10/6/201

Mercier, P. H. J., Dong, Z., Baikie, T., Le Page, Y., White, T. J., Whitfield, P. S., & Mitchel, L. D. (2007). *Ab initio* constrained crystal-chemical Rietveld refinement of $Ca_{10}(V_xP_{1-x}O_4)_6F_2$ apatites. *Acta Crystallographica. Section B, Structural Crystallography and Crystal Chemistry*, *63*(1), 37–48. doi:10.1107/S0108768106045538 PMID:17235192

Mercier, P. H. J., Le Page, Y., Whitfield, P. S., Mitchell, L. D., Davidson, I. J., & White, T. J. (2005). Geometrical parameterization of the crystal chemistry of *P6₃/m* apatites: Comparison with experimental data and *ab initio* results. *Acta Crystallographica. Section B, Structural Crystallography and Crystal Chemistry*, *61*(6), 635–655. doi:10.1107/S0108768105031125 PMID:16306671

Mevik, Wehrens, & Liland (2013). *PLS: Partial Least Squares and Principal Component regression*. R package version 2.4-3. Retrieved from http://CRAN.R-project.org/package=pls

Meyer, M. D., Vanderschueren, D., & Cooman, B. D. (1999). The influence of the substitutionof Si by Al on the properties of cold rolled C-Mn-Si TRIP steels. *ISIJ International*, *39*(8), 813–822. doi:10.2355/isijinternational.39.813

Mieti'inen, J., & Louhenkilpi, S. (1994). Calculation of Thermophysical Properties of Carbon and Low Alloyed Steels for Modeling of Solidification Processes. *Metallurgical and Materials Transactions. B, Process Metallurgy and Materials Processing Science*, *25*(6), 909–911. doi:10.1007/BF02662773

Miettinen, J., Kytönen, H., Louhenkilpi, S., & Laine, J. (1999). *12th IAS Steelmaking Seminar.* Instituto Argentino de Siderurgia.

Miettinen, J. (1997). Calculation of solidification-related thermophysical properties for steels. *Metallurgical Transactions. B, Process Metallurgy*, *28*(2), 281–297. doi:10.1007/s11663-997-0095-2

Miettinen, J., Kytönen, H., Louhenkilpi, S., & Laine, J. (1999). *Proc. of 12th IAS Steelmaking Seminar.*

Miettinen, J., Louhenkilpi, S., & Holappa, L. (1996). Coupled Simulation of Heat Transfer and Phase Transformation in Continuous Casting of Steel. *ISIJ International*, *36*(Suppl), S183–S186. doi:10.2355/isijinternational.36.Suppl_S183

Miettinen, J., Louhenkilpi, S., Kytönen, H., & Laine, J. (2010). IDS: Thermodynamic-kinetic-empirical tool for modelling of solidification, microstructure and material properties. *Mathematics and Computers in Simulation*, *80*(7), 1536–1550. doi:10.1016/j.matcom.2009.11.002

Miettinen, K. (1999). *Nonlinear Multiobjective Optimization.* Boston: Kluwer Academic Publishers.

Miettinen, K. (1999). *Nonlinear multi-objective optimization.* Boston: Kluwer.

Miki, Y., & Thomas, B. G. (1999). Modeling of Inclusion Removal in a Tundish. *Metallurgical Transaction B*, *30B*(4), 639-654.

Miller, M. G., Keith, J. M., King, J. A., Edwards, B. J., Klinkenberg, N., & Schiraldi, D. A. (2006). Measuring thermal conductivities of anisotropic synthetic graphite-liquid crystal polymer composites. *Polymer Composites*, *27*(4), 388–394. doi:10.1002/pc.20231

Millett, P. C., Selvam, R. P., & Saxena, A. (2007). Stabilizing nanocrystalline materials with dopants. *Acta Materialia*, *55*(7), 2329–2336. doi:10.1016/j.actamat.2006.11.028

Mills, G., & Jónsson, H. (1994). Quantum and Thermal Effects in H_2 Dissociative Adsorption: Evaluation of Free Energy Barriers in Multidimensional quantum systems. *Physical Review Letters*, *72*(7), 1124–1127. doi:10.1103/PhysRevLett.72.1124 PMID:10056623

Mills, G., Jónsson, H., & Schenter, G. K. (1995). Reversible work transition state theory: Application to dissociative adsorption of hydrogen. *Surface Science*, *324*(2-3), 305–337. doi:10.1016/0039-6028(94)00731-4

Mills, K. C. (1994). *The Performance of Casting Powders.* Steel Technol. Int.

Ming Yuan, H., Evans, A. G., & Hutchinson, J. W. (1994). Crack deflection at an interface between dissimilar elastic materials: Role of residual stresses. *International Journal of Solids and Structures*, *31*(24), 3443–3455. doi:10.1016/0020-7683(94)90025-6

Ming-Yuan, H., & Hutchinson, J. W. (1989). Crack deflection at an interface between dissimilar elastic materials. *International Journal of Solids and Structures*, *25*(9), 1053–1067. doi:10.1016/0020-7683(89)90021-8

Mishin, Y. (2004). Atomistic modeling of the γ and γ'-phases of the Ni–Al system. *Acta Materialia*, *52*(6), 1451–1467. doi:10.1016/j.actamat.2003.11.026

Mishin, Y., Asta, M., & Li, J. (2010). Atomistic modeling of interfaces and their impact on microstructure and properties. *Acta Materialia*, *58*(4), 1117–1151. doi:10.1016/j.actamat.2009.10.049

Mistree, F., Hughes, O. F., & Bras, B. (1992). Compromise Decision Support Problem and the Adaptive Linear Programming Algorithm. In M. P. Kamat (Ed.), *Structural Optimization: Status and Promise* (pp. 251–290). Washington, DC: AIAA.

Mi, W. B., Bai, H. L., Liu, H., & Sun, C. Q. (2007). Microstructure, magnetic, and optical properties of sputtered mn-doped zno films with high-temperature ferromagnetism. *Journal of Applied Physics*, *101*, 023904.

Mo¨es, N., B'echet, E., & Tourbier, M. (2006). Imposing Dirichlet boundary conditions in the extended finite element method. *International Journal for Numerical Methods in Engineering*, *67*(12), 1641–1669. doi:10.1002/nme.1675

Moës, N., Dolbow, J., & Belytschko, T. (1999). A finite element method for crack growth without remeshing. *International Journal for Numerical Methods in Engineering*, *46*(1), 131–150. doi:10.1002/(SICI)1097-0207(19990910)46:1<131::AID-NME726>3.0.CO;2-J

Mohanty, I., Bhattacharjee, D., & Datta, S. (2011). Designing cold rolled IF steel sheets with optimized tensile properties using ANN and GA. *Computational Materials Science*, *50*(8), 2331–2337. doi:10.1016/j.commatsci.2011.03.007

Mohanty, I., Datta, S., & Bhattacharjee, D. (2008). Composition-processing-property correlation of cold rolled IF steel sheets using neural network. *Materials and Manufacturing Processes*, *24*(1), 100–105. doi:10.1080/10426910802543947

Mohanty, I., Sarkar, S., Jha, B., Das, S., & Kumar, R. (2014). Online mechanical property prediction system for hot rolled IF steel. *Ironmaking & Steelmaking*, *41*(Sep), 618–627. doi:10.1179/1743281214Y.0000000178

Monajati, H., Asefi, D., Parsapour, A., & Abbasi, Sh. (2010). Analysis of the effects of processing parameters on mechanical properties and formability of cold rolled low carbon steel sheets using neural networks. *Computational Materials Science*, *49*(4), 876–881. doi:10.1016/j.commatsci.2010.06.040

Mondal, D. N., Sarangi, K., Pettersson, F., Sen, P. K., Saxén, H., & Chakraborti, N. (2011). Cu-Zn separation by supported liquid membrane analyzed through Multi-objective Genetic Algorithms. *Hydrometallurgy*, *107*(3-4), 112–123. doi:10.1016/j.hydromet.2011.02.008

Moorthy, S., & Ghosh, S. (1996). A model for analysis of arbitrary composite and porous microstructures with voronoi cell finite elements. *International Journal for Numerical Methods in Engineering*, *39*(14), 2363–2398. doi:10.1002/(SICI)1097-0207(19960730)39:14<2363::AID-NME958>3.0.CO;2-D

Moorthy, S., & Ghosh, S. (2000). Adaptivity and convergence in the Voronoi cell finite element model for analyzing heterogeneous materials. *Computer Methods in Applied Mechanics and Engineering*, *185*(1), 37–74. doi:10.1016/S0045-7825(99)00349-7

Mooser, E., & Pearson, W. B. (1959). On the crystal chemistry of normal valence compounds. *Acta Crystallographica*, *12*(12), 1015–1022. doi:10.1107/S0365110X59002857

Mori, T., & Tanaka, K. (1973). Average stress in matrix and average elastic energy of materials with misfitting inclusions. *Acta Metallurgica*, *21*(5), 571–574. doi:10.1016/0001-6160(73)90064-3

Mo, Y., Varga, K., Kaxiras, E., & Zhang, Z. (2005). Kinetic Pathway for the Formation of Fe Nanowires on Stepped Cu(111) Surfaces. *Physical Review Letters*, *94*(15), 155503. doi:10.1103/PhysRevLett.94.155503 PMID:15904159

Mukherjee, M., Singh, S. B., & Mohanty, O. N. (2006). Neural network analysis of straininduced transformation behaviour of retained austenite in TRIP-aided steels. *Materials Science and Engineering A*, *434*(1-2), 237–245. doi:10.1016/j.msea.2006.06.076

Mukhopadhyay, A., & Iqbal, A. (2005). Prediction of mechanical properties of hot rolled, low-carbon steel strips using artificial neural network. *Materials and Manufacturing Processes*, *20*(5), 793–812. doi:10.1081/AMP-200055140

Mulla, S., Chen, N., Cumaranatunge, L., Blau, G., Zemlyanov, D., Delgass, W., & Ribeiro, F. et al. (2006). Reaction of NO and O_2 to NO_2 on Pt: Kinetics and catalyst deactivation. *Journal of Catalysis*, *241*(2), 389–399. doi:10.1016/j.jcat.2006.05.016

Mulla, S., Chen, N., Delgass, W., Epling, W., & Ribeiro, F. (2005). NO_2 inhibits the catalytic reaction of NO and O_2 over Pt. *Catalysis Letters*, *100*(3-4), 267–270. doi:10.1007/s10562-004-3466-1

Muller, O., & Roy, R. (1974). *The major ternary structural families.Crystal chemistry of non- metallic materials*. New York: Springer-Verlag. doi:10.1007/978-3-642-65706-1

Mulliken, R. S. (1934). A New Electroaffinity Scale: Together with Data on Valence States and on Valence Ionization Potentials and Electron Affinities. *The Journal of Chemical Physics*, *2*(11), 782. doi:10.1063/1.1749394

Mu, Q., Feng, S., & Diao, G. (2007). Thermal conductivity of silicone rubber filled with ZnO. *Polymer Composites*, *28*(2), 125–130. doi:10.1002/pc.20276

Murari, V., & Upadhyay, C. S. (2012). Micromechanics based ply level material degradation model for unidirectional composites. *Composite Structures*, *94*(2), 671–680. doi:10.1016/j.compstruct.2011.09.001

Murari, V., & Upadhyay, C. S. (2013). Micromechanics based diffuse damage model for unidirectional composites. *Composite Structures*, *96*, 419–432. doi:10.1016/j.compstruct.2012.09.008

Myers, R. H., Montgomery, D. C., & Anderson-Cook, C. M. (2009). *Response Surface Methodology: Process and Product Optimization using Designed Experiments*. New York: John Wiley & Sons.

Naeem, M., Hasanain, S. K., Afgan, S. S., & Rumaiz, A. (2008). Competing effects of cu ionic charge and oxygen vacancies on the ferromagnetism of (zn, co)o nanoparticles. *Journal of Physics Condensed Matter*, *20*, 255223.

Najjar, F. M., Thomas, B. G., & Hershey, D. E. (1995). Numerical study of steady turbulent flow through bifurcated nozzles in continuous casting. *Metallurgical and Materials Transactions. B, Process Metallurgy and Materials Processing Science*, *26*(4), 749–765. doi:10.1007/BF02651721

Nandi, A. K. (2006). TSK-Type FLC using a combined LR and GA: Surface roughness prediction in ultraprecision turning. *Journal of Materials Processing Technology*, *178*(1-3), 200–210. doi:10.1016/j.jmatprotec.2006.03.164

Nandi, A. K., Datta, S., & Deb, K. (2011). Investigating the role of non-metallic fillers in particulate reinforced flexible mould material composites using evolutionary algorithms. *Materials and Manufacturing Processes*, *26*(3), 541–549. doi:10.1080/10426914.2011.560221

Nandi, A. K., Datta, S., & Orkus, J. (2012). Effective properties of particle reinforced flexible mould material towards reducing cooling time in soft tooling process. *Journal of Applied Polymer Science*, *124*(3), 2567–2581. doi:10.1002/app.34654

Nandi, A. K., Deb, K., Ganguly, S., & Datta, S. (2012). Investigating the role of metallic fillers in particulate reinforced flexible mould material composites using evolutionary algorithms. *Applied Soft Computing*, *12*(1), 28–39. doi:10.1016/j.asoc.2011.08.059

Nandi, A. K., Vesterinen, A., Cingi, C., Seppala, J., & Orkas, J. (2010). Studies on equivalent viscosity of particle-reinforced flexible mold materials used in soft tooling process. *Journal of Reinforced Plastics and Composites*, *29*(14), 2081–2098. doi:10.1177/0731684409347808

Nandi, A.K., Datta, S., & Deb, K., & Orkas, j. (2013). A Genetic Fuzzy Based Modeling of Effective Thermal Conductivity for Polymer Composites. *Journal of Intelligent and Fuzzy Systems*, *25*, 259–270.

Narayan, V., Abad, R., Lopez, B., Bhadeshia, H. K. D. H., & Mackay, D. J. C. (1999). Estimation of hot torsion stress strain curves in iron alloys using a neural network analysis. *ISIJ International*, *39*(10), 999–1005. doi:10.2355/isijinternational.39.999

Narymbetov, B., Omerzu, A., Kabanov, V. V., Tokumoto, M., Kobayashi, H., & Michailovic, D. (2000). Origin of ferromagnetic exchange interactions in a fullereneorganic compound. *Nature*, *407*, 883. PMID:11057661

Nasser, S., & Hori, M. (1993). *Micromechanics: overall properties of heterogeneous materials*. Amsterdam: North-Holland.

Negm. (2005). Validation Testing for Temporal Neural Networks for RBF Recognition. *Proceedings of World Academy of Science, Engineering and Technology, 5*.

Negulyaev, N. N., Stepanyuk, V. S., Hergert, W., Bruno, P., & Kirschner, J. (2008). Atomic-scale self-organization of Fe nanostripes on stepped Cu(111) surfaces: Molecular dynamics and kinetic Monte Carlo simulations. *Physical Review B: Condensed Matter and Materials Physics*, *77*(8), 085430. doi:10.1103/PhysRevB.77.085430

Nestler, B., & Wheeler, A. A. (2000). A multi-phase-field model of eutectic and peritectic alloys: Numerical simulation of growth structures. *Physica D. Nonlinear Phenomena*, *138*(1-2), 114–133. doi:10.1016/S0167-2789(99)00184-0

Neugebauer, J., & Hickel, T. (2013). Density functional theory in materials science. *Wiley Interdisciplinary Reviews: Computational Molecular Science*, *3*(5), 438–448. PMID:24563665

Ng, H. Y., Lu, X., & Lau, S. K. (2005). Thermal conductivity of boron nitride-filled thermoplastics: Effect of filler characteristics and composites processing conditions. *Polymer Composites*, *26*(6), 778–790. doi:10.1002/pc.20151

Nørskov, J., Bligaard, T., Logadottir, A., Bahn, S., Hansen, L., Bollinger, M., & Jacobsen, C. et al. (2002). Universality in heterogeneous catalysis. *Journal of Catalysis*, *209*(2), 275–278. doi:10.1006/jcat.2002.3615

Norton, D. P., Pearton, S. J., Hebard, A. F., Theodoropoulou, N., Boatner, L. A., & Wilson, R. G. (2003). Ferromagnetism in mn-implanted zno:sn single crystals. *Applied Physics Letters*, *82*, 239.

Nyagards, M. (2003). *Microstructural Finite Element Modeling of Metals*. (Unpublished dissertation). Royal Institute of Technology, Sweden.

O' Malley, R. J. (1999). Observations of various steady state and dynamic thermal behaviors in a continuous casting mold. *82nd Steelmaking Conference*.

Obayashi, S., Takahashi, S., & Takeguchi, Y. (1998). Niching and elitist models for MOGAs. In A. E. Eiben, T. B¨ack, M. Schoenauer, & H.-P. Schwefel (Eds.), *Fifth International Conference on Parallel Problem Solving from Nature (PPSN-V)* (pp. 260-269). doi:10.1007/BFb0056869

Odejobi & Umoru. (2009). Applications of soft computing techniques in materials engineering: A review. *African Journal of Mathematics and Computer Science Research, 2*(7), 104-131.

Olden, J. D., Joy, M. K., & Death, R. G. (2004). An accurate comparison of methods for quantifying variableimportance in artificial neural networks using simulated data. *Ecological Modelling*, *178*(3-4), 389–397. doi:10.1016/j.ecolmodel.2004.03.013

Oliveira, F. A., Alves, N., Giacometti, J. A., Constantino, C. J. L., Mattoso, L. H. C., Balan, A. M. O. A., & Job, A. E. (2007). Study of the thermomechanical and electrical properties of conducting composites containing natural rubber and carbon black. *Journal of Applied Polymer Science*, *106*(2), 1001–1006. doi:10.1002/app.26689

Oliver, J. (2000). On the discrete constitutive models induced by strong discontinuity kinematics and continuum constitutive equations. *International Journal of Solids and Structures*, *37*(48–50), 7207–7229. doi:10.1016/S0020-7683(00)00196-7

Oliver, J., Huespe, A. E., Pulido, M. D. G., & Chaves, E. (2002). From continuum mechanics to fracture mechanics: The strong discontinuity approach. *Engineering Fracture Mechanics*, *69*(2), 113–136. doi:10.1016/S0013-7944(01)00060-1

Olmsted, D. L., Foiles, S. M., & Holm, E. A. (2007). Grain boundary interface roughening transition and its effect on grain boundary mobility for non-faceting boundaries. *Scripta Materialia*, *57*(12), 1161–1164. doi:10.1016/j.scriptamat.2007.07.045

Olson, G. B. (1997). Computational design of hierarchically structured materials. *Science*, *277*(5330), 1237–1242. doi:10.1126/science.277.5330.1237

Olson, G. B. (2013). Genomic materials design: The ferrous frontier. *Acta Materialia*, *61*(3), 771–781. doi:10.1016/j.actamat.2012.10.045

Orogengoa, D., Capillas, C., Aroyo, M. I., & Perez-Mato, J. M. (2009). AMPLIMODES: Symmetry-mode analysis on the Bilbao Crystallographic Server. *Journal of Applied Crystallography*, *42*(5), 820–833. doi:10.1107/S0021889809028064

Ovchinnikov, A. A. (1978). Multiplicity of the ground state of large alterrnant organic molecules with conjugated bonds. *Theoret. Chim. Acta (Berl.)*, *47*, 297.

Ozaki, N., Nishizawa, N., Marcet, S., Kuroda, S., Eryu, O., & Takita, K. (2006). Significant enhancement of ferromagnetism in zn1xcrxte doped with iodine as an n-type dopant. *Physical Review Letters*, *97*, 037201. PMID:16907538

Ozaki, N., Okabayashi, I., Kumekawa, T., Nishizawa, N., Kuroda, S., & Takita, K. (2005). Suppression of ferromagnetism due to hole doping in zn1xcrxte grown by molecular-beam epitaxy. *Applied Physics Letters*, *87*, 192116.

Papaefthymiou, S., Bleck, W., Kruijver, S., Sietsma, J., Zhao, L., & van der Zwaag, S. (2004). Influence of intercritical deformation on microstructure of TRIP steels containingAl. *Materials Science and Technology*, *20*, 201–206.

Pareto, V. (1896). *CoursD'Economie Politique* (2 vols.). Lausanne: F. Rouge.

Park, D. S., Roh, T. W., Han, B. D., Kim, H. D., & Park, C. (2002). Microstructure and mechanical properties of silicon nitride with tri-laminate structures. *Journal of the European Ceramic Society*, *22*(4), 535–543. doi:10.1016/S0955-2219(01)00310-7

Park, H. C., Yang, T. Y., & Stevens, R. (2003). Fracture strength and damage resistance of slip cast Y-TZP/Ce-TZP layered composites. *Journal of Materials Science*, *38*(3), 561–564. doi:10.1023/A:1021849909830

Park, M. S., & Min, B. I. (2003). Ferromagnetism in zno codoped with transition metals: Zn1-x(feco)xo and zn1-x(fecu)xo. *Physical Review B: Condensed Matter and Materials Physics*, *68*, 224436.

Parks, G. T., & Miller, I. (1998). Selective breeding in a multiobjective genetic algorithm. In A. E. Eiben, T. B¨ack, M. Schoenauer, & H.-P. Schwefel (Eds.), *Fifth International Conference on Parallel Problem Solving from Nature (PPSN-V)* (pp. 250-259). doi:10.1007/BFb0056868

Parr, R. G., & Yang, W. (1995). Density-functional theory of the electronic structure of molecules. *Annual Review of Physical Chemistry*, *46*(1), 701–728. doi:10.1146/annurev.pc.46.100195.003413 PMID:24341393

Patel, B. P., & Gupta, A. K. (2014). An investigation on nonlocal continuum damage models for composite laminated panels. *Composites. Part B, Engineering*, *60*, 485–494. doi:10.1016/j.compositesb.2013.12.073

Patel, B. P., Shukla, K. K., & Nath, Y. (2005). Thermal postbuckling characteristics of laminated conical shells with temperature-dependent material properties. *AIAA Journal*, *43*(6), 1380–1388. doi:10.2514/1.13259

Pauling, L. (1932). The nature of the chemical bond. IV: The energy of single bonds and the relative electronegativity of atoms. *Journal of the American Chemical Society*, *54*(9), 3570–3582. doi:10.1021/ja01348a011

Paxton, A. T. (2014). From quantum mechanics to physical metallurgy of steels. *Materials Science and Technology*, *30*(9), 1063–1070. doi:10.1179/1743284714Y.0000000521

Pearson, R. G. (1986). Absolute electronegativity and hardness correlated with molecular orbital theory. *Proceedings of the National Academy of Sciences of the United States of America*, *83*(22), 8440–8441. doi:10.1073/pnas.83.22.8440 PMID:16578791

Pelatt, B. D., Ravichandran, R., Wager, J. F., & Keszler, D. A. (2011). Atomic Solid State Energy Scale. *Journal of the American Chemical Society*, *133*(42), 16852–16860. doi:10.1021/ja204670s PMID:21861503

Pemmaraju, C. D., Archer, T., Sanchez-Portal, D., & Sanvito, S. (2007). Atomic-orbital-based approximate self-interaction correction scheme for molecules and solids. *Physical Review B: Condensed Matter and Materials Physics*, *75*, 045101.

Pemmaraju, C. D., Hanafin, R., Archer, T., Braun, H. B., & Sanvito, S. (2008). Impurity-ion pair induced high-temperature ferromagnetism in co-doped zno. *Physical Review B: Condensed Matter and Materials Physics*, *78*, 054428.

Peng, H., Long, F., & Ding, C. (2005). Feature selection based on mutual information: Criteria of max-dependency, max-relevance, and min-redundancy. *IEEE Transactions on Pattern Analysis and Machine Intelligence*, *27*, 1226–1238. doi:10.1109/TPAMI.2005.159

Peng, P., Soh, A. K., Yang, R., & Hu, Z. Q. (2006). First-principles study of alloying effect of Re on properties of Ni/Ni 3 Al interface. *Computational Materials Science*, *38*(2), 354–361. doi:10.1016/j.commatsci.2005.10.010

Perdew, J. P., Burke, K., & Ernzerhof, M. (1996). Generalized Gradient Approximation Made Simple. *Physical Review Letters*, *77*(18), 3865–3868. doi:10.1103/PhysRevLett.77.3865 PMID:10062328

Perdew, J. P., & Wang, Y. (1992). Accurate and simple analytic representation of the electron-gas correlation energy. *Physical Review B: Condensed Matter and Materials Physics*, *45*(23), 13244–13249. doi:10.1103/PhysRevB.45.13244 PMID:10001404

Perdew, J. P., & Zunger, A. (1981). Self-interaction correlation to density-functional approximations for many-electron systems. *Physical Review B: Condensed Matter and Materials Physics*, *23*(10), 5048–5079. doi:10.1103/PhysRevB.23.5048

Perez-Mato, J. M., Orobengoa, D., & Aroyo, M. I. (2010). Mode crystallography of distorted structures. *Acta Crystallographica. Section A, Foundations of Crystallography*, *66*(5), 558–590. doi:10.1107/S0108767310016247 PMID:20720321

Petch, N. J. (1953). The cleavage strength of polycrystals. *Journal of the Iron and Steel Institute*, *174*, 25–28.

Petersen, S., & Hack, K. (2007). The Thermochemistry Library ChemApp and its applications Int. J. Mat. Res. (formerly. *Zeitschrift fur Metallkunde*, *98*, 935–945.

Pettersson, F., Biswas, A., Sen, P. K., & Chakraborti, N. (2009). Analyzing Leaching Data for Low-Grade Manganese Ore Using Neural Nets and Multiobjective Genetic Algorithms. *Materials and Manufacturing Processes*, *24*(3), 320–330. doi:10.1080/10426910802679386

Pettersson, F., Chakraborti, N., & Saxén, H. (2007). A genetic algorithms based multi-objective neural net applied to noisy blast furnace data. *Applied Soft Computing*, *7*(1), 387–397. doi:10.1016/j.asoc.2005.09.001

Pettifor, D. G. (1995). *Bonding and Structure of Molecules and Solids*. Oxford, UK: Oxford University Press.

Phillipps, A. J., Clegg, W. J., & Clyne, T. W. (1993a). Fracture behaviour of ceramic laminates in bending—I. Modelling of crack propagation. *Acta Metallurgica et Materialia*, *41*(3), 805–817. doi:10.1016/0956-7151(93)90014-J

Phillipps, A. J., Clegg, W. J., & Clyne, T. W. (1993b). Fracture behaviour of ceramic laminates in bending—II. Comparison of model predictions with experimental data. *Acta Metallurgica et Materialia*, *41*(3), 819–827. doi:10.1016/0956-7151(93)90015-K

Phillipps, A. J., Clegg, W. J., & Clyne, T. W. (1993c). The correlation of interfacial and macroscopic toughness in SiC laminates. *Composites*, *24*(2), 166–176. doi:10.1016/0010-4361(93)90013-X

Phillpot, S. R., & Sinnott, S. B. (2009). Simulating multifunctional structures. *Science*, *325*(5948), 1634–1635. doi:10.1126/science.1177794 PMID:19779181

Piana, S., Reyhani, M., & Gale, J. (2005). Simulating micrometer-scale crystal growth from solution. *Nature*, *438*(7064), 70–73. doi:10.1038/nature04173 PMID:16267550

Pinho, S. T., Davila, C. G., Camanho, P. P., Iannucci, L., & Robinson, P. (2005). *Failure models and criteria for FRP under in-plane or three dimensional stress states including shear nonlinearity.* Technical Report NASA/TM-2005-213530. NASA Langley Research Center.

Poletti, M. G., & Battezzati, L. (2014). Electronic and thermodynamic criteria for the occurrence of high entropy alloys in metallic systems. *Acta Materialia*, *75*, 297–306. doi:10.1016/j.actamat.2014.04.033

Poli, R., Langdon, W. B., & McPhee, N. F. (2008). *A field guide to genetic programming*. Retrieved from http://www.gp-field-guide.org.uk

Pompidou, S., & Lamon, J. (2007). Analysis of crack deviation in ceramic matrix composites and multilayers based on the Cook and Gordon mechanism. *Composites Science and Technology*, *67*(10), 2052–2060. doi:10.1016/j.compscitech.2006.11.013

Popa, A. I., Vavilova, E., Arango, Y. C., Kataev, V., Taschner, C., Klauss, H. H., & Klingeler, R. et al. (2009). High-temperature ferromagnetism of li-doped vanadium oxide nanotubes. *Europhysics Letters*, *88*, 57002.

Pornprasertsuk, R., Cheng, J., Huang, H., & Prinz, F. B. (2007). Electrochemical impedance analysis of solid oxide fuel cell electrolyte using kinetic Monte Carlo technique. *Solid State Ionics*, *178*(3-4), 195–205. doi:10.1016/j.ssi.2006.12.016

Potyrailo, R., Rajan, K., Stoewe, K., Takeuchi, I., Chislom, B., & Lam, H. (2011). Combinatorial and high-throughput screening of materials libraries: Review of state of the art. *ACS Combinatorial Science*, *13*(6), 579–633. doi:10.1021/co200007w PMID:21644562

Prasher, B. D., & Wills, G. B. (1973). Mass Transfer in an Agitated Vessel, Industrial & Engineering. *Chemistry Process Design and Development*, *12*(3), 351–354. doi:10.1021/i260047a023

Presland, A. E. B. (1960). *Mechanical Properties of Intermetallic Compounds*. New York: Wiley.

Provatas, N., & Elder, K. (2010). *Phase-Field Methods in Materials Science and Engineering*. Weinheim, Germany: Wiley-VCH. doi:10.1002/9783527631520

Puck, A., Kopp, J., & Knops, M. (2002). Guidelines for the determination of the parameters in Puck's action plane strength criterion. *Composites Science and Technology*, *62*(3), 371–378. doi:10.1016/S0266-3538(01)00202-0

Puck, A., & Schürmann, H. (1998). Failure analysis of FRP laminates by means of physically based phenomenological models. *Composites Science and Technology*, *58*(7), 1045–1067. doi:10.1016/S0266-3538(96)00140-6

Qi, L., & Li, J. (2012). Adsorbate interactions on surface lead to a flattened Sabatier volcano plot in reduction of oxygen. *Journal of Catalysis*, *295*, 59–69. doi:10.1016/j.jcat.2012.07.019

Qin, R. S., & Bhadeshia, H. K. (2010). Phase field method. *Materials Science and Technology*, *26*(7), 803–811. doi:10.1179/174328409X453190

Qi, X. B., Chen, Y., Kang, X. H., Li, D. Z., & Du, Q. (2015). An analytical approach for predicting as-cast grain size of inoculated aluminum alloys. *Acta Materialia*, *99*, 337–346. doi:10.1016/j.actamat.2015.08.006

R Core Team. (2013). *R: A language and environment for statistical computing*. R Foundation for Statistical Computing. Retrieved from http://www.R-project.org/

R Development Core Team. (2012). *R: A language and environment for statistical computing*. Vienna, Austria: R Foundation for Statistical Computing.

Raabe, D. (2001). Mesoscale simulation of recrystallization textures and microstructures. *Advanced Engineering Materials*, *3*(10), 745. doi:10.1002/1527-2648(200110)3:10<745::AID-ADEM745>3.0.CO;2-C

Raabe, D. (2002). Cellular automata in materials science with particular reference to recrystallization simulation. *Annual Review of Materials Research*, *32*(1), 53–76. doi:10.1146/annurev.matsci.32.090601.152855

Raabe, D. (2007). Multiscale recrystallization models for the prediction of crystallographic textures with respect to process simulation. *The Journal of Strain Analysis for Engineering Design*, *42*(4), 253–268. doi:10.1243/03093247JSA219

Raabe, D., & Becker, R. C. (2000). Coupling of a crystal plasticity finite-element model with a probabilistic cellular automaton for simulating primary static recrystallization in aluminium. *Modelling and Simulation in Materials Science and Engineering*, *8*(4), 445–462. doi:10.1088/0965-0393/8/4/304

Rabe, K. M., Phillips, J. C., Villars, P., & Brown, I. D. (1992). Global multinary structural chemistry of stable quasicrystals, high-Tc ferroelectrics, and high-Tc superconductors. *Physical Review B: Condensed Matter and Materials Physics*, *45*(14), 7650–7676. doi:10.1103/PhysRevB.45.7650 PMID:10000573

Rabotnov, Y. N. (1969). *Creep problems in structural members*. Amsterdam: North-Holland Press.

Radhakrishnan, B., Sarma, G. B., & Zacharia, T. (1998). Modeling the kinetics and microstructural evolution during static recrystallization—Monte Carlo simulation of recrystallization. *Acta Materialia*, *46*(12), 4415–4433. doi:10.1016/S1359-6454(98)00077-9

Räisänen, T., Louhenkilpi, S., Hätönen, T., Toivanen, J., Laine, J., & Kekäläinen, M. (2004). *European Cong. on Computational Methods in Applied Sciences and Engineering ECCOMAS 2004*. University of Jyväskylä.

Rajak, P., Tewary, U., Das, S., Bhattacharya, B., & Chakraborti, N. (2011). Phases in Zn-coated Fe analyzed through an evolutionary meta-model and multi-objective Genetic Algorithms. *Computational Materials Science*, *50*(8), 2502–2516. doi:10.1016/j.commatsci.2011.03.034

Rajan, K. (2005). Materials Informatics. *Materials Today*, *8*(10), 38–45. doi:10.1016/S1369-7021(05)71123-8

Rajan, K. (2010). Data Mining and Inorganic Crystallography. In D. W. M. Hoffmann & L. N. Kuleshova (Eds.), *Data Mining in Crystallography134, Structure & Bonding. Springer Berlin/Heidelberg.*

Rajasekaran, S., & Murray, D. W. (1973). Incremental finite element matrices. *Journal of the Structural Division*, *99*(12), 2423–2438.

Rak, M., Izdebski, M., & Brozi, A. (2001). Kinetic Monte Carlo study of crystal growth from solution. *Computer Physics Communications*, *138*(3), 250–263. doi:10.1016/S0010-4655(01)00238-7

Ramaswami & Bhaskaran. (2009). A Study on Feature Selection Techniques in Educational Data Mining. *Journal of Computing*, *1*, 7–11.

Ramuhalli, P., Udpa, L., & Udpa, S. S. (2002). Electromagnetic NDE signal inversion by function-approximation neural networks. *IEEE Transactions on Magnetics*, *38*(6), 3633–3642. doi:10.1109/TMAG.2002.804817

Randle, V. (2004). Twinning-related grain boundary engineering. *Acta Materialia*, *52*(14), 4067–4081. doi:10.1016/j.actamat.2004.05.031

Rappaz, J., & Scheid, J. F. (2000). Existence of solutions to a phase-field model for the isothermal solidification process of a binary alloy. *Mathematical Methods in the Applied Sciences*, *23*(6), 491–513. doi:10.1002/(SICI)1099-1476(200004)23:6<491::AID-MMA120>3.0.CO;2-4

Rappaz, M., & Gandin, C. A. (1993). Probabilistic modelling of microstructure formation in solidification processes. *Acta Metallurgica et Materialia*, *41*(2), 345–360. doi:10.1016/0956-7151(93)90065-Z

Rashid, M. M. (1998). The arbitrary local mesh replacement method: An alternative to remeshing for crack propagation analysis. *Computer Methods in Applied Mechanics and Engineering*, *154*(1–2), 133–150. doi:10.1016/S0045-7825(97)00068-6

Razzaq, M. Y., & Frormann, L. (2007). Thermomechanical studies of aluminium nitride filled shape memory polymer composites. *Polymer Composites*, *28*(3), 287–293. doi:10.1002/pc.20283

Reddy, J. N. (2004). *Mechanics of laminated composite plates and shells: theory and analysis* (2nd ed.). Boca Raton, FL: CRC Press.

Reddy, J. N. (2006). *An introduction to the finite element method* (3rd ed.). New York: McGraw-Hill.

Reddy, J. N., & Gartling, D. K. (2010). *The finite element method in heat transfer and fluid dynamics*. CRC Press.

Reddy, R., Smith, W. F., Mistree, F., Bras, B. A., Chen, W., Malhotra, A., & Patel, P. et al. (1992). *DSIDES. User Manual Systems*. Design Laboratory, Department of Mechanical Engineering, University of Houston.

Reed, M. J., Arkun, F. E., Berkman, E. A., Elmasry, N. A., Zavada, J., Luen, M. O., & Bedair, S. M. et al. (2005). Effect of doping on the magnetic properties of gamnn: Fermi level engineering. *Applied Physics Letters*, *86*, 102504.

Réger, M., Verő, B., Kytönen, H., & Louhenkilpi, S. (2002). *Proc. of 4th European Continuous Casting Conference*.

Requena, J., Moreno, R., & Moya, J. S. (1989). Alumina and alumina/zirconia multilayer composites obtained by slip casting. *Journal of the American Ceramic Society*, *72*(8), 1511–1513. doi:10.1111/j.1151-2916.1989.tb07692.x

Reuss, A. (1929). Berechnung der Fließgrenze von Mischkristallen auf Grund der Plastizitätsbedingung für Einkristalle. *ZAMM - Journal of Applied Mathematics and Mechanics. Zeitschrift für Angewandte Mathematik und Mechanik*, *9*(1), 49–58. doi:10.1002/zamm.19290090104

Reuter, K., & Scheffler, M. (2003). First-Principles atomistic thermodynamics for oxidation catalysis: Surface phase diagrams and catalytically interesting regions. *Physical Review Letters*, *90*(4), 046103. doi:10.1103/PhysRevLett.90.046103 PMID:12570437

Rice, J. R., & Wang, J. S. (1989). Embrittlement of interfaces by solute segregation. *Materials Science and Engineering A*, *107*, 23–40. doi:10.1016/0921-5093(89)90372-9

Rickman, J. M., & LeSar, R. (2002). Free-energy calculations in materials research. *Annual Review of Materials Research*, *32*(1), 195–217. doi:10.1146/annurev.matsci.32.111901.153708

Ringner, M. (2008). What is principal component analysis? *Nature Biotechnology*, *26*(3), 303–304. doi:10.1038/nbt0308-303 PMID:18327243

Robbins, D. H. Jr, Reddy, J. N., & Rostam-Abadi, F. (2005a). An efficient continuum damage model and its application to shear deformable laminated plates. *Mechanics of Advanced Materials and Structures*, *12*(6), 391–412. doi:10.1080/15376490500259202

Robbins, D. H. Jr, Reddy, J. N., & Rostam-Abadi, F. (2005b). Layerwise modeling of progressive damage in fiber-reinforced composite laminates. *International Journal of Mechanics and Materials in Design*, *2*(3-4), 165–182. doi:10.1007/s10999-006-9001-3

Robinson, A. J. (1994). An Application of Recurrent Nets to Phone Probability Estimation. *IEEE Transactions on Neural Networks*, *5*(2), 298–305. doi:10.1109/72.279192 PMID:18267798

Rogal, J., Reuter, K., & Scheffler, M. (2007). First-Principles statistical mechanics study of the stability of a subnanometer thin surface oxide in reactive environments: CO oxidation at Pd(100). *Physical Review Letters*, *98*(4), 046101. doi:10.1103/PhysRevLett.98.046101 PMID:17358787

Rollett, A. D., & Mullins, W. W. (1997). On the growth of abnormal grains. *Scripta Materialia*, *36*(9), 975–980. doi:10.1016/S1359-6462(96)00501-5

Rondinelli, J. M., May, S. J., & Freeland, J. W. (2012). Control of octahedral connectivity in perovskite oxide heterostructures: An emerging route to multifunctional materials discovery. *MRS Bulletin*, *37*(03), 261–270. doi:10.1557/mrs.2012.49

Rondinelli, J. M., Stengel, M., & Spaldin, N. (2008). Carrier-mediated magnetoelectricity in complex oxide heterostructures. *Nature Nanotechnology*, *3*(1), 46–50. doi:10.1038/nnano.2007.412 PMID:18654450

Rosochowski, A., & Matuszak, A. (2000). Rapid tooling: The state of the art. *Journal of Materials Processing Technology*, *106*(1-3), 191–198. doi:10.1016/S0924-0136(00)00613-0

Roters, F., Eisenlohr, P., Bieler, T. R., & Raabe, D. (2011). *Crystal plasticity finite element methods: in materials science and engineering*. John Wiley & Sons.

Roth, L. M. (1966). Simple narrow-band model of ferromagnetism due to intraatomic exchange. *Physical Review*, *149*, 306.

Roweis, S. T. & Saul, L.K. (2000). Nonliner dimensionality reduction by local linear embedding, *Science, 22*, 2323-2326.

Ruddlesden, S. N., & Popper, P. (1957). New compounds of the K_2NiF_4 type. *Acta Crystallographica*, *10*(8), 538–539. doi:10.1107/S0365110X57001929

Rudolph, G. (1998). On a multi-objective evolutionary algorithm and its convergence to the pareto set. In *IEEE International Conference on Evolutionary Computation (ICEC'98)* (pp. 511-516). doi:10.1109/ICEC.1998.700081

Rumelhart, D., Hinton, G. E., & Williams, R. J. (1986). Learning representations by back-propagating errors. *Nature*, *323*(6088), 533–536. doi:10.1038/323533a0

Ryan, C. (1995). Niche and species formation in genetic algorithms. In L. Chambers (Ed.), Practical Handbook of Genetic Algorithms (pp. 57-74). Academic Press.

Saal, J. E., Kirklin, S., Aykol, M., Meredig, B., & Wolverton, C. (2013). Materials Design and Discovery with High-Throughput Density Functional Theory: The Open Quantum Materials Database (OQMD). *Journal of Metals*, *65*, 1501–1509.

Safavian, S. R., & Landgrebe, D. (1991). A survey of decision tree classifier methodology. *IEEE Transactions on Systems, Man, and Cybernetics*, *21*, 660–674. doi:10.1109/21.97458

Sahai, Y., & Emi, T. (1996). Melt Flow Characterization in Continuous Casting Tundishes. *ISIJ International*, *36*(6), 667–672. doi:10.2355/isijinternational.36.667

Sakuma, Y., Matsumura, O., & Akisue, O. (1991). Influence of C content and annealing temperatureon microstructure and mechanical properties of 400°C transformed steel containing retained austenite. *ISIJ International*, *31*(11), 1348–1353. doi:10.2355/isijinternational.31.1348

Sanchez, J. M. (1993). Cluster expansions and the configurational energy of alloys. *Physical Review B: Condensed Matter and Materials Physics*, *48*(18), 14013–14015. doi:10.1103/PhysRevB.48.14013 PMID:10007810

Sanderson, R. T. (1952). Electronegativities in inorganic chemistry. *Journal of Chemical Education*, *29*(11), 539. doi:10.1021/ed029p539

Sansoz, F., & Molinari, J. F. (2005). Mechanical behavior of Σ tilt grain boundaries in nanoscale Cu and Al: A quasi-continuum study. *Acta Materialia*, *53*(7), 1931–1944. doi:10.1016/j.actamat.2005.01.007

Santos, J. P. T., Marques, M., Teles, L. K., & Ferreira, L. G. (2010). Antiferromagnetism with spin polarization of gan-based diluted magnetic semiconductors. *Physical Review B: Condensed Matter and Materials Physics*, *81*, 115209.

Sato, K., Bergqvist, L., Kudrnovsky, J., Dederichs, P. H., Eriksson, O., Turek, I., & Zeller, H. K. R. et al. (2010). First-principles theory of dilute magnetic semiconductors. *Reviews of Modern Physics*, *82*, 1633.

Saxena, S. K. (1993). *Thermodynamic data on oxides and silicates: an assessed dataset based on thermochemistry and high pressure phase equilibrium*. Berlin: Springer-Verlag. doi:10.1007/978-3-642-78332-6

Sbaizero, O., & Lucchini, E. (1996). Influence of residual stresses on the mechanical properties of a layered ceramic composite. *Journal of the European Ceramic Society*, *16*(8), 813–818. doi:10.1016/0955-2219(96)00005-2

Schaffer, J. D. (1984). *Multiple Objective Optimization with Vector Evaluated Genetic Algorithms*. (Ph. D. thesis). Vanderbilt University.

Schaffer, J. D. (1985). Multiple objective optimization with vector evaluated genetic algorithms, In J. J. Grefenstette (Ed.), *Proceedings of an International Conference on Genetic Algorithms and Their Applications* (pp. 93-100).

Schiøtz, J., Di Tolla, F. D., & Jacobsen, K. W. (1998). Softening of nanocrystalline metals at very small grain sizes. *Nature*, *391*(6667), 561–563. doi:10.1038/35328

Schmidt, C., Finnis, M. W., Ernst, F., & Vitek, V. (1998). Theoretical and experimental investigations of structures and energies of Σ= 3,[112] tilt grain boundaries in copper. *Philosophical Magazine A*, *77*(5), 1161–1184. doi:10.1080/01418619808214246

Schooling, J. M., Brown, M., & Reed, P. A. S. (1999). An example of the use of neural computing techniques in materials science—the modelling of fatigue thresholds in Ni-base superalloys. *Materials Science and Engineering A*, *260*(1-2), 222–239. doi:10.1016/S0921-5093(98)00957-5

Schultz, A. J., & Kofke, D. A. (2011). Algorithm for constant-pressure Monte-Carlo simulation of crystalline solids. *Physical Review E: Statistical, Nonlinear, and Soft Matter Physics*, *84*(4), 046712. doi:10.1103/PhysRevE.84.046712 PMID:22181312

Setyawan, W., & Curtarolo, S. (2010). High-throughput electronic structure calculations: Challenges and tools. *Computational Materials Science*, *49*(2), 299–312. doi:10.1016/j.commatsci.2010.05.010

Sglavo, V. M., & Bertoldi, M. (2006). Design and production of ceramic laminates with high mechanical resistance and reliability. *Acta Materialia*, *54*(18), 4929–4937. doi:10.1016/j.actamat.2006.06.019

Shahani, A. R., & Amini Fasakhodi, M. R. (2009). Finite element analysis of dynamic crack propagation using remeshing technique. *Materials & Design*, *30*(4), 1032–1041. doi:10.1016/j.matdes.2008.06.049

Shannon, R. (1976). Revised effective ionic radii and systematic studies of interatomic distances in halides and chalcogenides. *Acta Crystallographica. Section A, Foundations of Crystallography*, *32*(5), 751–767. doi:10.1107/S0567739476001551

Sheppard, D., Terrell, R., & Henkelman, G. (2008). Optimization methods for finding minimum energy paths. *The Journal of Chemical Physics*, *128*(13), 134106. doi:10.1063/1.2841941 PMID:18397052

Shigegaki, Y., Brito, M. E., Hirao, K., Toriyama, M., & Kanzaki, S. (1996). Processing of a novel multilayered silicon nitride. *Journal of the American Ceramic Society*, *79*(8), 2197–2200. doi:10.1111/j.1151-2916.1996.tb08959.x

Shim, Y., & Amar, J. G. (2005). Rigorous synchronous relaxation algorithm for parallel kinetic Monte Carlo simulations of thin film growth. *Physical Review B: Condensed Matter and Materials Physics*, *71*(11), 115436. doi:10.1103/PhysRevB.71.115436

Shin, J. M., Lee, H. S., Cha, S. Y., Lee, S., Kim, J. Y., Park, N., & Jeong, S. et al. (2012). Strong ferromagnetism in pt-coated zncoo: The role of interstitial hydrogen. *Applied Physics Letters*, *100*, 172409.

Shobu, K., Watanabe, T., & Tsuji, K. (1989). Effects of Mo2B5 addition to MoSi2 ceramics. *Journal of the Ceramic Society of Japan*, *97*(10), 1311–1314. doi:10.2109/jcersj.97.1311

Sidoroff, F. (1981). Description of anisotropic damage application to elasticity.*Proceedings of the IUTAM Colloquium on Physical Nonlinearities in Structural Analysis*. doi:10.1007/978-3-642-81582-9_35

Siegel, S. (1956). *Non-parametric statistics for the behavioral sciences*. New York: McGraw-Hill.

Simo, J. C. (1997). TJR Hughes Computational inelasticity, Interdisciplinary Applied Mathematics. *Mechanics of Materials*, 7.

Singh, A. K., Paradeshi, R., & Goyal, S. (2011). Integrated Modeling of Tundish and Continuous Caster to meet Quality Requirement of Cast Steels.*1st World Congress on Integrated Computational Materials Engineering*. doi:10.1002/9781118147726.ch11

Singh, S. B., Bhadeshia, H. K. D. H., MacKay, D. J. C., Carey, H., & Martin, I. (1998). Article. *Ironmaking & Steelmaking*, *25*, 355.

Singh, S. B., & Kumar, A. (1999). Postbuckling response and strength of laminates under combined in-plane loads. *Composites Science and Technology*, *59*(5), 727–736. doi:10.1016/S0266-3538(98)00125-0

Sinha, A., Allen, J., Panchal, J., & Mistree, F. (2013). Microstructure Mediated Design of Material and Product.*2nd World Congress on Integrated Computational Materials Engineering*. doi:10.1002/9781118767061.ch7

Sinha, A., & Sikdar, S. (2013). Article. *Mater. Des.*, *46*, 227–234.

Skriver, H. L. (2004). *Materials Science Databases, CAMP DTU*. Retrieved from http://guglen.dk/databases/hlsPT/ptable.php

Slater, J. C. (1930). Atomic Shielding Constants. *Physical Review*, *36*(1), 57–64. doi:10.1103/PhysRev.36.57

Smeltz, A. D., Delgass, W. N., & Ribeiro, F. H. (2010). Oxidation of NO with O_2 on Pt(111) and Pt(321) Large Single Crystals. *Langmuir*, *26*(21), 16578–16588. doi:10.1021/la101653x PMID:20666498

Smith, K., Chen, L., Darling, R., Garosshen, T., Jaworowski, M., Opalka, S., . . . Zafiris, G. (2013). ICME Approach to Corrosion Pit Growth Prediction. 2nd *World Congress on Integrated Computational Materials Engineering.*

Smith, W. F., Milisavljevic, J., Sabeghi, M., Allen, J. K., & Mistree, F. (2014). Accounting for Uncertainty and Complexity in the Realization of Engineered Systems.*International Conference on Complex Systems Design & Management.*

Soden, P., Hinton, M., & Kaddour, A. (1998). A comparison of the predictive capabilities of current failure theories for composite laminates. *Composites Science and Technology*, *58*(7), 1225–1254. doi:10.1016/S0266-3538(98)00077-3

Sola, J., & Sevilla, J. (1997). Importance of input data normalization for the application of neural networks to complex industrial problems. *IEEE Transactions on Nuclear Science*, *44*(3), 1464–1468. doi:10.1109/23.589532

Solovyev, I. V., Dederichs, P. H., & Anisimov, V. I. (1994). Corrected atomic limit in the localdensity approximation and the electronic structure of d impurities in rb. *Physical Review B: Condensed Matter and Materials Physics*, *50*, 16861. PMID:9976082

Srikrishna, A., Reddy, B. E., & Srinivas, V. S. (2013). Automatic Feature subset selection using Genetic Algorithm for clustering. *International Journal on Recent Trends in Engineering and Technology*, *9*, 85–89.

Srinivas, N., & Deb, K. (1994). Multiobjective optimization using nondominated sorting in genetic algorithms. *Evolutionary Computation*, *2*(3), 221–248. doi:10.1162/evco.1994.2.3.221

Srolovitz, D. J., Grest, G. S., & Anderson, M. P. (1986). Computer simulation of recrystallization—I. Homogeneous nucleation and growth. *Acta Metallurgica*, *34*(9), 1833–1845. doi:10.1016/0001-6160(86)90128-8

Srolovitz, D. J., Grest, G. S., Anderson, M. P., & Rollett, A. D. (1988). Computer simulation of recrystallization—II. Heterogeneous nucleation and growth. *Acta Metallurgica*, *36*(8), 2115–2128. doi:10.1016/0001-6160(88)90313-6

Stachurski, Z. H. (2011). On structure and properties of amorphous materials. *Materials (Basel)*, *4*(9), 1564–1598. doi:10.3390/ma4091564

Stadler, W. (1988). Fundamentals of multicriteria optimization. In W. Stadler (Ed.), *Multicriteria Optimization in Engineering and the Sciences* (pp. 1–25). New York, NY: Plenum Press. doi:10.1007/978-1-4899-3734-6_1

Steinbach, I. (2009). Phase-field models in materials science Ingo Steinbach. *Modelling and Simulation in Materials Science and Engineering*, *17*(7), 31. doi:10.1088/0965-0393/17/7/073001

Steuer, R. E. (1986). *Multiple Criteria Optimization: Theory, Computation, and Application*. New York: Wiley.

Stokes, H. T., Hatch, D. M., & Campbell, B. J. (2007). *Isotropy*. Retrieved from stokes.byu.edu/isotropy.html

Stokes, H. T., Campbell, B. J., & Cordes, R. (2013). Tabulation of irreducible representations of the crystallographic space groups and their superspace extensions. *ActaCrystallographica Section A*, *69*, 388–395. PMID:23778095

Subodh, G., Manjusha, M. V., Philip, J., & Sebastian, M. T. (2008). Thermal properties of polytetrafluoroethylene/Sr2Ce2Ti5O16 polymer/ceramic composites. *Journal of Applied Polymer Science*, *108*(3), 1716–1721. doi:10.1002/app.27606

Suh, C., & Rajan, K. (2009). Informatics for chemical crystallography. *JOM Journal of the Minerals. Metals and Materials Society*, *61*(1), 48–53. doi:10.1007/s11837-009-0009-7

Suli, E. (2000). *Lecture Notes on Finite Element Methods for Partial Differential Equations*. Oxford, UK: University of Oxford.

Sumpter, B. G., & Noid, D. W. (1996). On the Design, Analysis, and Characterization of Materials Using Computational Neural Networks. *Annual Review of Materials Science*, *26*(1), 223–277. doi:10.1146/annurev.ms.26.080196.001255

Sundstrom, D.W., & Chen, S.Y. (1970). Thermal Conductivity of Reinforced Plastics. *Journal of Composite Materials*, *4*(1), 113-117.

Sunkara, S., KalyanVendra, V., Jasinski, J., Deutsch, T., Andriotis, A. N., Menon, M., & Sunkara, M. (2014). New visible light absorbing materials for solar fuels, ga(sbx)n1−x alloy. *Advanced Materials*. doi:10.1002/adma.201305083

Suomi, M.-L., Jalkanen, H., & Wallgren, M. (1995). Simulation of Oxygen Converter Process. In *Proceedings of the 2nd Colloquium on Process Simulation*.

Suomi, M.-L., Jalkanen, H., Andersin, E., & Wallgren, M. (1995). CONSIM-2 Program for Oxygen Converter Simulation. In *Proceedings of the 2nd Colloquium on Process Simulation.*

Sutton, A. P., Mulheran, P. A., & Stoneham, A. M. (1992). Direct Free Energy Minimization Methods: Application to Grain Boundaries [and Discussion]. *Philosophical Transactions of the Royal Society of London A: Mathematical, Physical and Engineering Sciences, 341*(1661), 233-245.

Sutton, A. P., & Balluffi, R. W. (1995). *Interfaces in crystalline materials.* Oxford, UK: Clarendon press.

Svensson, T. (2004). Fatigue Fract. *Eng. Mater. Struct., 27*, 981–990. doi:10.1111/j.1460-2695.2004.00819.x

Swiniarski, R. W., & Skowron, A. (2003). Rough set methods in feature selection and recognition. *Pattern Recognition Letters, 24*, 833–849. doi:10.1016/S0167-8655(02)00196-4

Takeguchi, T., Yamanaka, T., Takahashi, H., Watanabe, H., Kuroki, T., Nakanishi, H., & Ueda, W. et al. (2013). Layered perovskite oxide: A reversible air electrode for oxygen evolution/reduction in rechargeable metal-air batteries. *Journal of the American Chemical Society, 135*(30), 11125–11130. doi:10.1021/ja403476v PMID:23802735

Tamehiro, H., Murata, M., Habu, R., & Nagumo, M. (1987). Optimum Microalloying of Niobium and Boron in HSLA Steel for Mechanical Processing Trans. *ISIJ, 27*(2), 120–129. doi:10.2355/isijinternational1966.27.120

Tang, T. (2008). *Variational asymptotic micromechanics modeling of composite materials* (Ph.D. thesis, Utah State University, Logan, Utah). Retrieved from http://digitalcommons.usu.edu/cgi/viewcontent.cgi?article=1071&context=etd

Tang, H., van der Ven, A. F., & Trout, B. L. (2004, July). Phase diagram of oxygen atoms adsorbed on platinum(111) by first principles investigation. *Physical Review B: Condensed Matter and Materials Physics, 70*(4), 045420. doi:10.1103/PhysRevB.70.045420

Tan, P.-N., Steinback, M., & Kumar, V. (2005). *Introduction to Data Mining.* Addison-Wesley.

Temperley, H. N. V. (1981). *Graph Theory and Applications.* New York: Wiley.

Tenenbaum, J. B., Silva, V., & Langford, J. C. (2000). A Global Geometric Framework for Nonlinear Dimensionality Reduction. *Science, 290*(5500), 2319–2323. doi:10.1126/science.290.5500.2319 PMID:11125149

Terada, K., Miura, T., & Kikuchi, N. (1997). Digital image-based modeling applied to the homogenization analysis of composite materials. *Computational Mechanics, 20*(4), 331–346. doi:10.1007/s004660050255

Thomas, B. G., & Najjar, F. M. (1991). Finite element modelling of turbulent fluid flow and heat transfer in continuous casting. *Applied Mathematical Modelling.* http://doi.org/10.1016/0307-904X(91)90001-6

Thomas, B. G. (1991). Mathematical Modeling of the Continuous Slab Casting Mold: a State of the Art Review. *1991 Steelmaking Conference.*

Thomas, B. G. (2002). Modeling of the continuous casting of steel - past, present and future. *Metallurgical and Materials Transactions. B, Process Metallurgy and Materials Processing Science, 33*(6), 795–812. doi:10.1007/s11663-002-0063-9

Thomas, B. G., & Zhang, L. (2001). Mathematical modeling of fluid flow in continuous casting. *ISIJ International, 41*(10), 1181–1193. doi:10.2355/isijinternational.41.1181

Tian, T., & Burrage, K. (2004). Binomial leap methods for simulating stochastic chemical kinetics. *The Journal of Chemical Physics, 121*(21), 10356–10364. doi:10.1063/1.1810475 PMID:15549913

Tian, Y. P., & Fu, Y. M. (2008). Elasto-plastic postbuckling of damaged orthotropic plates. *Applied Mathematics and Mechanics, 29*(7), 841–853. doi:10.1007/s10483-008-0702-y

Tian, Y. P., Fu, Y. M., & Wang, Y. (2009a). Nonlinear dynamic response of piezoelectric elasto-plastic laminated plates with damage. *Composite Structures*, *88*(2), 169–178. doi:10.1016/j.compstruct.2008.02.019

Tian, Y. P., Fu, Y. M., & Wang, Y. (2009b). Nonlinear dynamic response and vibration active control of piezoelectric elasto-plastic laminated plates with damage. *Journal of Vibration and Control*, *15*(10), 1463–1492. doi:10.1177/1077546309103265

Tibshirani, R., Johnstone, I., Hastie, T., & Efron, B. (1996). Regression shrinkage and selection via the lasso. *Journal of the Royal Statistical Society. Series A, (Statistics in Society)*, *58*, 267.

Tibshirani, R., Walther, G., & Hastie, T. (2001). Estimating the number of clusters in a data set via the gap statistic. *Journal of the Royal Statistical Society. Series B, Statistical Methodology*, *63*(2), 411–423. doi:10.1111/1467-9868.00293

Timoshenko, S., & Gere, J. M. (1972). *Mechanics of materials*. New York: Van Nostrand Reinhold Co.

Tomaszewski, H., Weglarz, H., Boniecki, M., & Rećko, W. M. (2000). Effect of barrier layer thickness and composition on fracture toughness of layered zirconia/alumina composites. *Journal of Materials Science*, *35*(16), 4165–4176. doi:10.1023/A:1004823429336

Tomaszewski, H., Węglarz, H., Wajler, A., Boniecki, M., & Kalinski, D. (2007). Multilayer ceramic composites with high failure resistance. *Journal of the European Ceramic Society*, *27*(2–3), 1373–1377. doi:10.1016/j.jeurceramsoc.2006.04.030

Tong, M., Li, D., & Li, Y. (2004). Modeling the austenite–ferrite diffusive transformation during continuous cooling on a mesoscale using Monte Carlo method. *Acta Materialia*, *52*(5), 1155–1162. doi:10.1016/j.actamat.2003.11.006

Trautt, Z. T., & Upmanyu, M. (2005). Direct two-dimensional calculations of grain boundary stiffness. *Scripta Materialia*, *52*(11), 1175–1179. doi:10.1016/j.scriptamat.2004.12.033

Tsai, S. W., & Hahn, H. T. (1980). *Introduction to composite materials*. Westport, CT: Technomic Publishing.

Tsai, S. W., & Wu, E. M. (1971). A general theory of strength for anisotropic materials. *Journal of Composite Materials*, *5*(1), 58–80. doi:10.1177/002199837100500106

Tschopp, M. A., & McDowell, D. L. (2007a). Structures and energies of Σ 3 asymmetric tilt grain boundaries in copper and aluminium. *Philosophical Magazine*, *87*(22), 3147–3173. doi:10.1080/14786430701255895

Tschopp, M. A., & McDowell, D. L. (2007b). Asymmetric tilt grain boundary structure and energy in copper and aluminium. *Philosophical Magazine*, *87*(25), 3871–3892. doi:10.1080/14786430701455321

Tsuchida, Y., Watanabe, T., Kato, T., & Seto, T. (2010). Effect of hydrogen absorption on strain-induced low-cycle fatigue of low carbon steel. *Procedia Engineering*, *2*(1), 555–561. doi:10.1016/j.proeng.2010.03.060

Tsukuda, R., Sumimoto, S., & Ozawa, T. (1997). Thermal conductivity and heat capacity of ABS resin composites. *Journal of Applied Polymer Science*, *63*(10), 1279–1286. doi:10.1002/(SICI)1097-4628(19970307)63:10<1279::AID-APP6>3.0.CO;2-H

Tuffé, S., & Wilkinson, D. S. (1995). MoSi2-based sandwich composite made by tape casting. *Journal of the American Ceramic Society*, *78*(11), 2967–2972. doi:10.1111/j.1151-2916.1995.tb09071.x

Tullock, D. L., Reimanis, I. E., Graham, A. L., & Petrovic, J. J. (1994). Deflection and penetration of cracks at an interface between two dissimilar materials. *Acta Metallurgica et Materialia*, *42*(9), 3245–3252. doi:10.1016/0956-7151(94)90422-7

Turkenitch, D. I. (1971). *Upravlenie plavkoi stali v konvertere*. Moskva: Metallurgiya.

Ueshima, Y., Mizoguchi, S., Matsumiya, T., & Kajioka, H. (1986). Analysis of solute distribution in dendrites of carbon steel with δ/γ transformation during solidification. *Metallurgical Transactions. B, Process Metallurgy*, *17*(4), 845–859. doi:10.1007/BF02657148

Upmanyu, M., Srolovitz, D. J., Shvindlerman, L. S., & Gottstein, G. (1999). Misorientation dependence of intrinsic grain boundary mobility: Simulation and experiment. *Acta Materialia*, *47*(14), 3901–3914. doi:10.1016/S1359-6454(99)00240-2

Upton, E. A., Rao, T. R. S., Dauby, P. H., & Knechtges, R. C. (1988). Physical Metallurgy and Mathematical Modeling as Tools for Continuous Casting Optimization at LTV Steel. *Ironmaking & Steelmaking*, *15*, 51–57.

Vaithyanathan, V., Wolverton, C., & Chen, L. Q. (2002). Multiscale modeling of precipitate microstructure evolution. *Physical Review Letters*, *88*(12), 125503. doi:10.1103/PhysRevLett.88.125503 PMID:11909473

Valenzuela-Rend'on, M., & Uresti-Charre, E. (1997). A non-generational genetic algorithm for multiobjective optimization, In T. B¨ack (Ed.), *Proceedings of the Seventh International Conference on Genetic Algorithms* (pp. 658-665), San Francisco, CA:Morgan Kaufmann.

van de Walle, A. (2013). Methods for First-Principles alloy thermodynamics. *JOM*, *65*(11), 1523–1532. doi:10.1007/s11837-013-0764-3

van de Walle, A., & Asta, M. (2002c). Self-driven lattice-model Monte Carlo simulations of alloy thermodynamic properties and phase diagrams. *Modelling and Simulation in Materials Science and Engineering*, *10*(5), 521–538. doi:10.1088/0965-0393/10/5/304

van de Walle, A., Asta, M., & Ceder, G. (2002a). The alloy theoretic automated toolkit: A User Guide. *CALPHAD Journal*, *26*(4), 539–553. doi:10.1016/S0364-5916(02)80006-2

Van De Walle, A., & Ceder, G. (2002). The effect of lattice vibrations on substitutional alloy thermodynamics. *Reviews of Modern Physics*, *74*(1), 11–45. doi:10.1103/RevModPhys.74.11

van de Walle, A., & Ceder, G. (2002b). Automating First-Principles phase diagram calculations. *Journal of Phase Equilibria*, *23*(4), 348–359. doi:10.1361/105497102770331596

Vandeperre, L. J., Kristofferson, A., Carlström, E., & Clegg, W. J. (2001). Thermal shock of layered ceramic structures with crack-deflecting interfaces. *Journal of the American Ceramic Society*, *84*(1), 104–110. doi:10.1111/j.1151-2916.2001.tb00615.x

Vapalahti, S., Louhenkilpi, S., & Räisänen, T. (2006). The Effect of Fluid Flow on Heat Transfer and Shell Growth in Continuous Casting of Copper. *Materials Science Forum*, *508*, 503–508. doi:10.4028/www.scientific.net/MSF.508.503

Väyrynen, P., & Wang, S. (2010). Control of Fluid Flow, Heat Transfer and Inclusions in Continuous Casting: CFD and Neural Network Studies. The Minerals, Metals & Ma. Retrieved from http://66.221.229.18/pdfs/tp/cast_tp/FloSciBib02-10.pdf

Veldhuizen, D. A. V., & Lamont, G. B. (1998). Evolutionary computation and convergence to a pareto front. In J. R. Koza, W. Banzhaf, K. Chellapilla, K. Deb, M. Dorigo, D. B. Fogel, M. H. Garzon, D. E. Goldberg, H. Iba, & R. Riolo (Eds.), *Genetic Programming 1998:Proceedings of the Third Annual Conference* (pp. 22-25).

Veldhuizen, D. V., & Lamont, G. B. (2000). Multiobjective evolutionary algorithms: Analyzing the state-of-the-art. *Evolutionary Computation Journal*, *8*(2), 125–128. doi:10.1162/106365600568158 PMID:10843518

Villars, P., Brandenburg, K., Berndt, M., LeClair, S., Jackson, A., Pao, Y.-H., & Iwata, S. et al. (2001). Binary, ternary, and quaternary compound former/nonformer prediction via Mendeleev number. *Journal of Alloys and Compounds*, *317-318*, 26–38. doi:10.1016/S0925-8388(00)01410-9

Villars, P., Onodera, N., & Iwata, S. (1998). The Linus Pauling file (LPF) and its application to materials design. *Journal of Alloys and Compounds*, *279*(1), 1–7. doi:10.1016/S0925-8388(98)00605-7

Vinod, V. S., Varghese, S., & Kuriakose, B. (2004). Aluminium powder filled nitrile rubber composites. *Journal of Applied Polymer Science*, *91*(5), 3156–3161. doi:10.1002/app.13472

Voigt, W. (1889). Uber die Beziehung zwischen den beiden Elastizitätskonstanten isotroper Körper. *Wiedmanns Annalen Physik and Chemie*, *38*(12), 573–587. doi:10.1002/andp.18892741206

Voter, A. F. (1997). Hyperdynamics: Accelerated molecular dynamics of infrequent events. *Physical Review Letters*, *78*(20), 3908–3911. doi:10.1103/PhysRevLett.78.3908

Waber, J. T., & Cromer, D. T. (1965). Orbital radii of atoms and ions. *The Journal of Chemical Physics*, *42*(12), 4116. doi:10.1063/1.1695904

Wang, L., Li, F., & Su, Z. (2008). Effective thermal conductivity behaviour of filled vulcanized perfluoromrthyl vinyl ether rubber. *Journal of Applied Polymer Science*, *108*(5), 2968–2974. doi:10.1002/app.27612

Wang, Q., Gao, J., Wang, R., & Hua, Z. (2001). Mechanical and rheological properties of HDPE/Graphite composite with enhanced thermal conductivity. *Polymer Composites*, *22*(1), 97–103. doi:10.1002/pc.10521

Wang, Q., Sun, Q., Jena, P., & Kawazoe, Y. (2004). Carrier-mediated ferromagnetism in n codoped (zn,mn)o (1010) thin films. *Physical Review B: Condensed Matter and Materials Physics*, *70*, 052408.

Wang, X., Lau, K. C., Turner, C. H., & Dunlap, B. I. (2010). Kinetic Monte Carlo simulation of the elementary electrochemistry in a hydrogen-powered solid oxide fuel cell. *Journal of Power Sources*, *195*(13), 4177–4184. doi:10.1016/j.jpowsour.2009.12.130

Wang, Y. M., Chen, M. W., Sheng, H. W., & Ma, E. (2002). Nanocrystalline grain structures developed in commercial purity Cu by low-temperature cold rolling. *Journal of Materials Research*, *17*(12), 3004–3007. doi:10.1557/JMR.2002.0436

Wang, Y. Z., & Chen, L. Q. (1999). *Methods of Materials Research.* New York: Wiley and Sons.

Wang, Y., Chen, D., & Zhang, X. (2000). Calculated equation of state of Al, Cu, Ta, Mo, and W to 1000 GPa. *Physical Review Letters*, *84*(15), 3220–3223. doi:10.1103/PhysRevLett.84.3220 PMID:11019055

Wang, Y., Curtarolo, S., Jiang, C., Arroyave, R., Wang, T., Ceder, G., & Liu, Z. K. (2004). Ab initio lattice stability in comparison with CALPHAD lattice stability. *Calphad*, *28*(1), 79–90. doi:10.1016/j.calphad.2004.05.002

Wang, Y., Liu, Z. K., & Chen, L. Q. (2004). Thermodynamic properties of Al, Ni, NiAl, and Ni 3 Al from first-principles calculations. *Acta Materialia*, *52*(9), 2665–2671. doi:10.1016/j.actamat.2004.02.014

Wang, Y., & Xiang, B. (2007). Radial basis function network calibration model for near-infrared spectra in wavelet domain using a genetic algorithm. *Analytica Chimica Acta*, *602*(1), 55–65. doi:10.1016/j.aca.2007.09.016 PMID:17936107

Weaver, J. F., Chen, J.-J., & Gerrard, A. L. (2005). Oxidation of Pt(1 1 1) by gas-phase oxygen atoms. *Surface Science*, *592*(1-3), 83–103. doi:10.1016/j.susc.2005.07.010

Weidenfeller, B., Hofer, M., & Schilling, F. R. (2004). Thermal conductivity, thermal diffusivity, and specific heat capacity of particle filled polypropylene. *Composites. Part A, Applied Science and Manufacturing*, *35*(4), 423–429. doi:10.1016/j.compositesa.2003.11.005

Wei, S. H., Ferreira, L. G., Bernard, J. E., & Zunger, A. (1990). Electronic properties of random alloys: Special quasirandom structures. *Physical Review B: Condensed Matter and Materials Physics*, *42*(15), 9622–9649. doi:10.1103/PhysRevB.42.9622 PMID:9995203

Weiss, B. M., & Iglesia, E. (2009). NO Oxidation Catalysis on Pt Clusters: Elementary Steps, Structural Requirements, and Synergistic Effects of NO_2 Adsorption Sites. *The Journal of Physical Chemistry C, 113*(30), 13331–13340. doi:10.1021/jp902209f

White, T. J., Ferraris, C., Kim, J., & Madhavi, S. (2005). Apatite - An Adaptive Framework Structure. *Reviews in Mineralogy and Geochemistry, 57*(1), 307–401. doi:10.2138/rmg.2005.57.10

White, T.J. & ZhiLi, D. (2003). Structural derivation and crystal chemistry of apatites. *ActaCrystallographica Section B, 59*(1), 1–16. PMID:12554967

Williams, J. A. (1996). Analytical models of scratch hardness. *Tribology International, 29*(8), 675–694. doi:10.1016/0301-679X(96)00014-X

Williams, P. L., Mishin, Y., & Hamilton, J. C. (2006). An embedded-atom potential for the Cu–Ag system. *Modelling and Simulation in Materials Science and Engineering, 14*(5), 817–833. doi:10.1088/0965-0393/14/5/002

Willis, M. J., Montague, G. A., Di Massimo, C., Tham, M. T., & Morris, A. J. (1992). Artificial neural networks in process estimation and control. *Automatica, 28*(6), 1181–1187. doi:10.1016/0005-1098(92)90059-O

Witten, I. H., & Frank, E. (2000). *Data Mining: Practical Machine Learning Tools and Techniques with Java Implementation. The Morgan Kaufmann Series in Data Management Systems.* San Francisco: Morgan Kaufmann.

Wolf, D., Wang, J., Phillpot, S. R., & Gleiter, H. (1995). Phonon-induced anomalous specific heat of a nanocrystalline model material by computer simulation. *Physical Review Letters, 74*(23), 4686–4689. doi:10.1103/PhysRevLett.74.4686 PMID:10058573

Wong, C. P., & Bollampally, R. S. (1999). Thermal conductivity, elastic modulus, and coefficient of thermal expansion of polymer composites filled with ceramic particles for electronic packaging. *Journal of Applied Polymer Science, 74*(14), 3396–3403. doi:10.1002/(SICI)1097-4628(19991227)74:14<3396::AID-APP13>3.0.CO;2-3

Wu, C., Schmidt, D., Wolverton, C., & Schneider, W. (2012). Accurate coverage-dependence incorporated into First-Principles kinetic models: Catalytic NO oxidation on Pt(111). *Journal of Catalysis, 286*, 88–94. doi:10.1016/j.jcat.2011.10.020

Wu, E. M. (1972). Optimal experimental measurements of anisotropic failure tensors. *Journal of Composite Materials, 6*(4), 472–489. doi:10.1177/002199837200600404

Wu, K., Chen, S., Zhang, F., & Chang, Y. A. (2009). Integrating CALPHAD into phase field simulations for practical applications. *Journal of Phase Equilibria and Diffusion, 30*(5), 571–576. doi:10.1007/s11669-009-9567-1

Wu, Z. X., Zhang, Y. W., & Srolovitz, D. J. (2009). Grain boundary finite length faceting. *Acta Materialia, 57*(14), 4278–4287. doi:10.1016/j.actamat.2009.05.026

Xu, Y., Chung, D. D. L., & Mroz, C. (2001). Thermally conducting aluminium nitride polymer-matrix composites. *Composites. Part A, Applied Science and Manufacturing, 32*(12), 1749–1757. doi:10.1016/S1359-835X(01)00023-9

Yadav, M. K., Sanyal, B., & Mookerjee, A. (2009). Stabilization of ferromagnetism in mn doped zno with c co-doping. *Journal of Magnetism and Magnetic Materials, 321*, 273.

Yamakov, V., Wolf, D., Phillpot, S. R., Mukherjee, A. K., & Gleiter, H. (2002). Dislocation processes in the deformation of nanocrystalline aluminium by molecular-dynamics simulation. *Nature Materials, 1*(1), 45–49. doi:10.1038/nmat700 PMID:12618848

Yamauchi, A., Sorimachi, K., Sakuraya, T., & Fujii, T. (1993). Heat Transfer Between Mold and Strand Through Mold Flux Film in Continuous Casting of Steel. *ISIJ International, 33*(1), 140–147. doi:10.2355/isijinternational.33.140

Yeh, J. W., Chen, S. K., Lin, S. J., Gan, J. Y., Chin, T. S., Shun, T. T., & Chang, S. Y. (2004). Nanostructured High-Entropy Alloys with Multiple Principal Elements: Novel Alloy Design Concepts and Outcomes. *Advanced Engineering Materials*, *6*(5), 299–303. doi:10.1002/adem.200300567

Ye, L. H., & Freeman, A. J. (2006). Defect compensation, clustering, and magnetism in cr-doped anatase tio2. *Physical Review B: Condensed Matter and Materials Physics*, *73*, 081304.

Yoo, Y. S., Jo, C. Y., & Jones, C. N. (2002). Compositional prediction of creep rupture life of single crystal Ni base superalloy by Bayesian neural network. *Materials Science and Engineering A*, *A336*(1-2), 22–29. doi:10.1016/S0921-5093(01)01965-7

Yu, Miettinen, & Louhenkilpi. (2013). Numerical Study on the Removal of Hydrogen and Nitrogen from the Melt of Medium Carbon Steel in Vacuum Tank Degasser. *Materials Science Forum, 762*.

Yu, Miettinen, & Louhenkilpi. (2014a). Modeling Study of Nitrogen Removal from the Vacuum Tank Degasser. *Steel Research International, 85*, 1393–1402.

Yu, Miettinen, Shao, & Louhenkilpi. (2014b). Mathematical Modeling of Nitrogen Removal from the Vacuum Tank Degasser. *Steel Research International, 85*.

Yu, P. L. (1973). A class of solutions for group decision problems. *Management Science*, *19*(8), 936–946. doi:10.1287/mnsc.19.8.936

Yu, S., Hing, P., & Hu, X. (2002). Thermal conductivity of polystyrene-aluminium nitride composite. *Composites. Part A, Applied Science and Manufacturing*, *33*(2), 289–292. doi:10.1016/S1359-835X(01)00107-5

Yu, S., & Louhenkilpi, S. (2013). Numerical Simulation of Dehydrogenation of Liquid Steel in the Vacuum Tank Degasser. *Metallurgical and Materials Transactions. B, Process Metallurgy and Materials Processing Science*, *44*(2), 459–468. doi:10.1007/s11663-012-9782-8

Yu, W., & Tang, T. (2007a). A variational asymptotic micromechanics model for predicting thermoelastic properties of heterogeneous materials. *International Journal of Solids and Structures*, *44*(22–23), 7510–7525. doi:10.1016/j.ijsolstr.2007.04.026

Yu, W., & Tang, T. (2007b). Variational asymptotic method for unit cell homogenization of periodically heterogeneous materials. *International Journal of Solids and Structures*, *44*(11–12), 3738–3755. doi:10.1016/j.ijsolstr.2006.10.020

Zackay, V. F., Parker, E. R., Fahr, D., & Bush, R. (1967). The enhancement of ductility in high strength steels. *Transactions of the American Society for Metals*, *60*, 252–259.

Zackay, V., Parker, E., Fahr, D., & Busch, R. (1967). The enhancement of ductility in high-strength steels.*Transactions of the ASM*, *60*,252–259.

Zadeh, L. A. (1965). Fuzzy Sets. *Information and Control*, *8*(3), 338–353. doi:10.1016/S0019-9958(65)90241-X

Zaefferer, S. (2005, July). Application of orientation microscopy in SEM and TEM for the study of texture formation during recrystallisation processes. *Materials Science Forum*, *495*, 3–12. doi:10.4028/www.scientific.net/MSF.495-497.3

Zhang, L. (2006). Fluid Flow, Heat Transfer and Inclusion Motion in Molten Steel Continuous Casting Tundishes. *Fifth International Conference on CFD in the Process Industries*.

Zhang, L., Taniguchi, S., & Cai, K. (2000). Fluid Flow and Inclusion Removal in Continuous Casting Tundish. *Metallurgical and Material Transaction B*, *31B*, 253-266.

Zhang, G. J., Yue, X. M., & Watanabe, T. (1999). Al2O3/TiC/(MoSi2+Mo2B5) multilayer composites prepared by tape casting. *Journal of the European Ceramic Society*, *19*(12), 2111–2116. doi:10.1016/S0955-2219(99)00031-X

Zhang, W., & Cai, Y. (2010). *Continuum damage mechanics and numerical applications*. Berlin: Springer. doi:10.1007/978-3-642-04708-4

Zhou, J.-S., Goodenough, J. B., & Dabrowski, B. (2005). Exchange interaction in the insulating phase of $RNiO_3$. *Physical Review Letters*, *95*(12), 127204. doi:10.1103/PhysRevLett.95.127204 PMID:16197104

Zhou, X. W., Wadley, H. N. G., Johnson, R. A., Larson, D. J., Tabat, N., Cerezo, A., & Kelly, T. F. (2001). Atomic scale structure of sputtered metal multilayers. *Acta Materialia*, *49*(19), 4005–4015. doi:10.1016/S1359-6454(01)00287-7

Zhu, R., Pan, E., & Chung, W. (2007). Fast multiscale kinetic Monte Carlo simulations of three-dimensional self-assembled quantum dot islands. *Physical Review B: Condensed Matter and Materials Physics*, *75*(20), 205339. doi:10.1103/PhysRevB.75.205339

Zhu, W., Qiu, X., Iancu, V., Chen, X. Q., Pan, H., Wang, W., & Zhang, Z. et al. (2009). Band gap narrowing of titanium oxide semiconductors by noncompensated anion-cation codoping for enhanced visible-light photoactivity. *Physical Review Letters*, *103*, 226401. PMID:20366114

Zhu, X., Birringer, R., Herr, U., & Gleiter, H. (1987). X-ray diffraction studies of the structure of nanometer-sized crystalline materials. *Physical Review B: Condensed Matter and Materials Physics*, *35*(17), 9085–9090. doi:10.1103/PhysRevB.35.9085 PMID:9941303

Ziegler, M., Kröger, J., Berndt, R., Filinov, A., & Bonitz, M. (2008). Scanning tunneling microscopy and kinetic Monte Carlo investigation of cesium superlattices on Ag(111). *Physical Review B: Condensed Matter and Materials Physics*, *78*(24), 245427. doi:10.1103/PhysRevB.78.245427

Zienkiewicz, O. C., & Taylor, R. L. (1977). *The finite element method* (Vol. 3). London: McGraw-hill.

Zitzler, E. (1999). *Evolutionary Algorithms for Multiobjective Optimization: Methods and Applications*. (Ph. D. thesis). Swiss Federal Institute of Technology (ETH), Zurich, Switzerland.

Zou, H., & Hastie, T. (2012). *elasticnet: Elastic-Net for Sparse Estimation and Sparse PCA*. R package version 1.1. Retrieved from http://CRAN.R-project.org/package=elasticnet

Zou, H., & Hastie, T. (2005). Regularization and variableselection via the elastic net. *Journal of the Royal Statistical Society. Series A, (Statistics in Society)*, *67*(2), 301–320. doi:10.1111/j.1467-9868.2005.00503.x

Zunger, A. (1980). Systematization of the stable crystal structure of all AB-type binary compounds: A pseudopotential orbital-radii approach. *Physical Review B: Condensed Matter and Materials Physics*, *22*(12), 5839–5872. doi:10.1103/PhysRevB.22.5839

About the Contributors

Shubhabrata Datta received his PhD from the Bengal Engineering and Science University, Shibpur, presently known as IIEST, Shibpur, India in 2003. His research area includes modeling and optimization in materials systems, particularly application of computational intelligence technique for design of materials. Dr. Datta has worked in different laboratories all over the world as visiting scientist. He has more than hundred publications, has acted as guest editor in many international journal special issues and is member of the editorial board of several international journals.

J. Paulo Davim received his PhD in Mechanical Engineering from the University of Porto in 1997, the Aggregate title from the University of Coimbra in 2005 and a DSc from London Metropolitan University in 2013. Currently, he is Professor at the Department of Mechanical Engineering of the University of Aveiro. He has about 30 years of teaching and research experience in manufacturing, materials and mechanical engineering with special emphasis in Machining & Tribology. Recently, he has also interest in Management/Industrial Engineering and Higher Education for Sustainability. He is the Editor in Chief of eight international journals, Guest Editor of journals, books Editor, book Series Editor and Scientific Advisory for many international journals and conferences. Presently, he is an Editorial Board member of 30 international journals and acts as reviewer for more than 80 prestigious Web of Science journals. In addition, he has also published as author and co-author more than 10 books, 60 book chapters and 350 articles in journals and conferences (more 200 articles in journals indexed in Web of Science, h-index 34+).

***

Janet K. Allen holds the John and Mary Moore Chair and is Professor of Industrial and Systems Engineering at the University of Oklahoma. She came to the University of Oklahoma in August 2009 where she and Professor Farrokh Mistree have established the Systems Realization Laboratory @ The University of Oklahoma which is home to researchers from Mechanical Engineering and Industrial and Systems Engineering. The focus of Dr. Allen's research is engineering design and especially the management of uncertainty and computational methods to facilitate engineering design of complex systems. This has led to her interest in the design of networks and supply chains. Dr. Allen and her students study/have studied aerospace systems, materials, energy systems, mechanical systems, and design methods. Dr. Allen is a Fellow of ASME and a Senior Member of AIAA.

Gautam Anand did his Bachelors in Engineering in Metallurgy and Materials Engineering from Bengal Engineering and Science University, Shibpur (now IIEST, Shibpur) in 2011 and Masters of Science in Materials Science and Engineering from The University of Sheffield on Commonwealth scholarship in 2014. He is presently carrying out his doctoral research at The University of Sheffield. He has written six papers in SCI journals so far. His area of interest is Computational Materials Science.

Ravikiran Anapagaddi received his Master's of Technology degree in Chemical Engineering in 2011 from the Indian Institute of Technology, Madras. Post his graduation, he has worked with TCS Research, India in the field of Integrated Computational Materials Engineering till October 2014. Presently, he is working with Grace Davison Chemical India Private Limited, Chennai. Ravikiran is particularly interested in developing the physico-chemical and metallurgical process modeling and simulations, mathematical modeling of multiphase systems, thermodynamic and reaction engineering modeling and process optimization. Ravikiran has also worked in the field of simplified modeling and design exploration of steel making.

Antonis N. Andriotis was born in September 23, 1947 in Mytilene, Greece. He is a Greek citizen, married with two children. After completing his B.Sc (1970) at the University of Athens, Greece, he received his BSc (Honours) and the Ph.D degrees from the University of Cape Town, S. Africa in 1971 and 1976, respectively. On his return to Greece, he fulfilled a 28 month military service and since 1979 has worked in the National Research Centers of Greece (in National Hellenic Research Foundation, Athens, Greece from 1979-1989 and in the Institute of Electronic Structure and Laser (IESL) - FORTH, Heraklion, Crete, Greece from 1989-present day). Andriotis' field of Interest is Theoretical and Computational Condensed Matter Physics and Materials Research and since 2002 has the title of Research Director at IESL/FORTH. He has published more than one hundred and thirty papers and is the author of six books (http://esperia.iesl.forth.gr/~andriot).

Prasanna V. Balachandran is currently a postdoctoral research associate in the Theoretical Division at Los Alamos National Laboratory, USA. He earned his Ph.D. in Materials Science and Engineering from Iowa State University. His research interests include application of first-principles based electronic structure methods and informatics-based methods for accelerated design and discovery of new materials with targeted properties.

Debashish Bhattacharjee is currently Group Director for Research and Development at Tata Steel. He is responsible for developing research strategy, collaboration network and executing a portfolio of research that is balanced between aligning to current business needs and developing technology options for the future. Dr Bhattacharjee is a Fellow of the Indian National Academy of Engineering, and a Fellow of the Indian Institute of Metals. He is in active contact with academia by being Industrial Professor at Warwick University, UK, Visiting Professor at Imperial College London, and Guest Professor at University of Science and Technology, Beijing.

Joe Bible is currently a postdoctoral fellow at the Eunice Kennedy Shriver National Institute of Child Health and Human Development. He received his PhD in Biostatistics from the University of Louisville in 2015. His research interests include longitudinal data analysis, adolescent substance abuse, periodontal disease and materials science.

Professor Partha Protim Chattopadhyay Obtained Bachelor of Engineering and Master of Engineering degrees from University of Calcutta in 1984 and 1989, respectively and obtained the PhD degree from Indian Institute of Technology, Kharagpur in 2000. Professor Chattopadhyay served in Hindustan Motors Limited, Uttarpara, West Bengal as Foundry Engineer from 1985-90 and as Promotion Officer in Ministry of Industries (SISI), GoI during 1990-91. Subsequently, he joined erstwhile Bengal Engineering College and served in different capacities including Professor and Head of Metallurgy and Materials and Engineering and Dean (Faculty of Engineering and Technology) in Bengal Engineering and Science University, Shibpur. Professor Chattopadhyay has supervised 11 doctoral scholars and published about 127 papers in international Journal of repute and 36 peer reviewed papers in conference proceedings. He has received several awards out of which worth mentioning are guests scientist under the funding of Gottfried Wilhelm Leibniz program at University of Ulm, Germany (2001-02), Commonwealth Academic Fellowship award at University of Cambridge in 2004, Metallurgist of the year award by Ministry of steel (GoI), Eminent Engineer award by The Institution of Engineers (India) in 2014 and MRSI Medal in 2014. He has acted as Principal and Co-Principal Investigator of different research projects amount in about Rs. 9 crores. He is member of several professional bodies including Council Member of Indian Institute of Metals and Fellow of Institution of Engineers (India). His research interest includes phase transformation in metal and alloys and microstructure property correlation of materials.

Nirupam Chakraborti obtained his Ph.D from University of Washington in 1983, and he is a Professor at Indian Institute of Technology, Docent at Åbo Akademi University, Finland and Adjunct Professor at POSTECH, Korea. He also taught and conducted research at several other universities worldwide. Considered to be a leading international authority of evolutionary computation applied to materials research, he has written numerous papers and book chapters in the area. Since 2003 he has been regularly guest editing the genetic algorithms special issues of Materials and Manufacturing Processes, a major journal published by Taylor and Francis. EvoNN (Evolutionary Neural Net) and BioGP (Bi-objective Genetic Programming) are two algorithms developed by him and his collaborators that are described in detail in this chapter.

Somnath Datta is Professor of Biostatistics at the University of Florida. Dr. Datta received his PhD in Statistics and Probability from the Statistics Department at Michigan State University in 1988. He is an elected member of the International Statistical Institute, a fellow of the American Statistical Association, and a fellow of the Institute of Mathematical Statistics. He has served on editorial boards of numerous peer reviewed journals in statistics and bioinformatics. Currently, he is founding Editor-in-Chief of a Springer book series. Dr. Datta's research concentrations include theory and methods of statistics, biostatistics and bioinformatics, in addition to collaborative research in spinal cord injury rehabilitation, dentistry, cancer informatics, plat science and materials science.

Kalyanmoy Deb is the Herman E. & Ruth J. Koenig Endowed Chair in the Department of Electrical and Computing Engineering at Michigan State University. Before joining Michigan State University in 2013, he was Deva Raj Endowed Chair and Gurmukh and Veena Mehta Endowed Chair in the Department of Mechanical Engineering at the Indian Institute of Technology, Kanpur. Deb established the Kanpur Genetic Algorithms Laboratory in 1997 and the Computational Optimization and Innovation (COIN) Laboratory at Michigan State University in 2013. In 2001, Wiley published a book titled *Multi-Objective Optimization using Evolutionary Algorithms* written by Deb.

Arjun Dey is presently working as 'Scientist' at ISRO Satellite Centre, Bangalore. Working as Senior Research Fellow (SRF) at CSIR-Central Glass and Ceramic Research Institute, Kolkata, he earned Doctoral Degree in 'Materials Science and Engineering' from Indian Institute of Engineering Science and Technology (IIEST), Shibpur, Howrah. He published 165 publications (includes 75 SCI Journal Publications). He has co-authored 2 books "Nanoindentation of Brittle Solids" and "Microplasma Sprayed Hydroxyapatite Coatings" from CRC Press. He already obtained several prestigious awards e.g., "Dr. R. L. Thakur Memorial Award" from The Indian Ceramic Society, "DST-Fast Track Young Scientist Scheme Project Grant Award" from Department of Science and Technology (DST), India, "Young Engineers Award" and "Metallurgical and Materials Engineering Division Prize" from The Institution of Engineers (India) and "Young Scientist Award" from Materials Research Society of India, Kolkata Chapter and CSIR-SRFship Award, from Council of Scientific and Industrial Research (CSIR) to accomplish his PhD.

Partha Dey is presently Associate Professor at the Department of Mechanical Engineering of Academy of Technology, Adisaptagram. Dr. Dey received his PhD in Engineering from Bengal Engineering and Science University (now IIEST), Shibpur in 2007. He was the recipient of National Talent Search Fellowship from NCERT, and the Fast Track Fellowship (Young Scientist Award) from DST, Government of India. Having done a Postdoctoral Fellowship at Iowa State University with Prof. Krishna Rajan as advisor, Dr. Dey is currently in full-time teaching. His research interests include Finite Element Analysis, and applications of Soft Computing and Computational Intelligence in different domains including Materials Engineering.

Swati Dey has completed her Ph.D. (Engineering) in 2015 from Indian Institute of Engineering Science and Technology, Shibpur, Howrah. Her M. Tech. (Materials Engineering) was completed from Bengal Engineering and Science University, Shibpur, Howrah in 2007 and B.E. (Electrical Engineering) was from Nagpur University in 2001. She has worked as a project fellow in N.M.L. Jamshedpur (August 2002-July2005), worked in IBM India Pvt. Ltd. (December2007-December 2008), was a CSIR-SRF for 4 years (January 2009-December 2012). She has 11 publications in International and 1 publication in National Journal. Her research interest is Artificial/Computational Intelligence and its application in materials domain. She has worked using various soft computing tools like Artificial Neural Network, Fuzzy Logic, Rough Set, Genetic Programming and Genetic Algorithm.

Subhas Ganguly is presently Assistant Professor in National Institute of Technology, Raipur, India. He received his PhD from the Bengal Engineering and Science University, Shibpur, presently IIEST, Shibpur, India. His research interests include informatics in alloy design and development, soft computing methods in materials application, data mining, computational optimization and evolutionary computing. His research contribution has been recognized by the Institution of Engineers (India) Young Engineers award and Best Paper award in 2008 and 2009, respectively.

Ankur Kumar Gupta is working as 'Scientist' at Structures Group, ISRO Satellite Centre, Indian Space Research Organisation (ISRO), Bangalore. Dr. Gupta earned the Masters Degree in 'Materials Science and Engineering at Motilal Nehru National Institute of Technology, Allahabad and received his Ph.D. in 2013 from Applied Mechanics Dept., Indian Institute of Technology (I.I.T.) Delhi, India. His thesis work involved continuum damage modeling of composite laminated panels. Dr. Gupta has published more than 24 peer reviewed research articles in scientific journals and conferences. He has also invited at Nanjing University of Aeronautics and Astronautics, China to present his research work. His main research interests are finite element analysis, computational mechanics, damage mechanics, mechanics of composite, plates and shells. His recent research activity includes the finite element simulation of nanoindentation process of coatings, thin films and bulk materials.

Seppo Louhenkilpi is Professor and Head of the Group Metallurgy in the Department of Materials Science and Engineering, Aalto University, Finland. He received his doctor degree from Aalto University (formerly Helsinki University of Technology) in 1995. His main research subjects are modeling of metallurgical processes and phenomena including continuous casting, solidification, materials microstructures and properties. His group is extensively networked with the international research community and Prof.Louhenkilpi has been coordinator for several national and international research projects since 1984. He has 165 research papers and received Professor Hasse Fredrikssons award (Royal Institute of Technology, KTH, Sweden) for 2016 for his work in solidification and continuous casting.

Madhu Menon is the Associate Director of Center for Computationals Sciences at the University of Kentucky. He received his BSc (Honors) from University of Bombay, India and PhD from the University of Notre Dame, USA. He has been involved in the development of formalisms and computational methods in theoretical condensed matter physics. His area of research include computational nanotechnology including fullerenes, nanotubes, graphene and nanowires; magnetism of diluted magnetic semiconductors and solar energy conversion materials. He has published over 180 articles in peer reviewed journals.

Farrokh Mistree holds the L. A. Comp Chair in the School of Aerospace and Mechanical Engineering at the University of Oklahoma in Norman, Oklahoma. Farrokh's current research focus is *model-based realization of complex systems by managing uncertainty and complexity.* The key question he is investigating is *what are the principles underlying rapid and robust concept exploration when the analysis models are incomplete and possibly inaccurate?* He has co-authored two textbooks, one monograph and more than 400 technical papers dealing with the design of material, mechanical, thermal and structural systems; ships and aircraft. Farrokh is a Fellow of ASME and an Associate Fellow of AIAA.

Itishree Mohanty is currently working as a principal researcher in R&D, Tata Steel, India. She works in materials modelling and product design group and her functional expertise lies in the area of process modelling where control, prediction and simulation of plant processes is done using soft computing technologies. She has also done a significant amount of work in the area of process automation like design, development and implementation of systems in plants.

Anoop Kumar Mukhopadhyay earned a Ph D in 1988 from Jadavpur University, Kolkata, India Dr. Anoop Kumar Mukhopadhyay, presently a Chief Scientist and Head, Advanced Mechanical and Materials Characterization Division at CSIR-Central Glass and Ceramic Research Institute, Kolkata, India has more than 115 SCI Journal Publications, 125 National/International Conference Publications, 3 Book Chapters, 2 Books co-authored and published in 2014 and 2015 from CRC Press, USA plus 7 Indian Patents to his credit. He has guided more than 7 PhD students with 3 PhDs awarded, more than 50 B.Tech. and 10 M.Tech. Dissertation Theses. A recipient of many Awards and Fellowships, he was awarded the Materials Research Society of India Medal in 2014 for his contributions in understanding deformation and fracture of ceramics. He has been Invited Speaker and Chair/Co-Chair of Technical Sessions in numerous National/International Conferences.

Arup Kumar Nandi is currently a principle scientist at CSIR-Central Mechanical Engineering Research Institute, Durgapur, India. His main research focus is in the area of artificial intelligence (Fuzzy Logic, Neural Network, Evolutionary Algorithms) and its application in manufacturing and engineering design. He has published more than 30 articles refereed international journal paper and presented more than 12 papers in national and international conferences.

Jitesh Panchal is an Associate Professor in the School of Mechanical Engineering at Purdue University. He received his B. Tech (2000) from Indian Institute of Technology (IIT) Guwahati, and MS (2003) and PhD (2005) in Mechanical Engineering from Georgia Institute of Technology. Dr. Panchal's research interests are in computational design of complex engineering systems with focus on three areas a) concurrent products and materials design, b) collective systems innovation, and c) cyber-physical systems for design and manufacturing. He is a co-author of the book titled Integrated Design of Multiscale, Multifunctional Materials and Products. He is a recipient of CAREER award from the National Science Foundation (NSF), Young Engineer Award and two best paper awards from ASME CIE division, and a university silver medal from IIT Guwahati.

James Rondinelli is an assistant professor at Northwestern University in the Materials Science and Engineering (MSE) Department. His interests are in the first-principles design and understanding of functional complex inorganic materials. In 2014 he was recognized by the American Chemical Society as an Emerging Leader in Solid-State Chemistry, and also received the Ross Coffin Purdy Award from the American Ceramic Society in 2012. Additional honors include an NSF-CAREER Award (2015), DARPA Young Faculty Award (2012), and ARO Young Investigator Program (YIP) award (2012). Dr. Rondinelli has (co)-authored more than 70 peer-reviewed publications and holds 1 patent. He received a B.S. in MSE from Northwestern (2006) and a Ph.D. in Materials from the University of California, Santa Barbara (2010). From 2010-2011, he was the Joseph Katz Named Fellow in the X-ray Science Division at Argonne National Laboratory. Prior to joining Northwestern, he was an assistant professor at Drexel University (2011-14).

Bishnupada Roy is presently working as Assistant Professor in the Department of Chemistry, Bankura Unnayani Institute of Engineering, India. Beside his teaching assignment, he is pursuing his research at Visva-Bharati University, Santiniketan, India in the field of biological application of metal/polymer nanocomposites. He already has several journal and conference publications in his credit. He was awarded for the best oral presentation in International Conference on Nano Science & Engineering Application-2014 (ICONSEA 2014), organized by Institute of Science & Technology, Jawaharlal Nehru Technological University, Hyderabad. He has organized National Workshop on Nano Science & Technology (NWNST 2015) at Bankura Unnayani Institute of Engineering campus as convener in January, 2015.

Rishabh Shukla received his B. Tech degree in Metallurgical and Materials Engineering in 2012 from the Indian Institute of Technology, Roorkee. After his graduation, he is working with TCS Research, India in the field of Integrated Computational Materials Engineering. Rishabh is particularly interested in integrating models at various length scales and bridging the gap between design and manufacturing. Rishabh has also worked in the field of simplified modeling and design exploration of steel making and continuous casting process. His areas of interest are steel making, reduced order modelling, corrosion science and process optimization.

Amarendra Singh is currently with the Department of Materials Science and Engineering at IIT Kanpur and his research interests are in the areas of Mathematical Modeling of Metallurgical Operations, Steelmaking, Solidification Processing and Integrated Computational Materials Engineering. Prior to joining IIT Kanpur in April 2015, he was with TCS Research, India for 23 years where he had successfully demonstrated the power and utility of mathematical modelling in optimization manufacturing operations leading to significant productivity gains, quality improvements, energy saving and/or environment compliance and has delivered several successful innovative solutions to several industries. Prof. Singh has received several honors including Metallurgist of the Year Award from the Ministry of Steel, Govt. of India in 2014.

Index

K

L

M

N

O

P

R

S

T

V

www.ingramcontent.com/pod-product-compliance
Lightning Source LLC
LaVergne TN
LVHW080533300926
849162LV00007B/1592